Oxford Resources for IB

Diploma Programme

2023 EDITION

PHYSICS

COURSE COMPANION

David Homer

Maciej Piętka

William Heathcote

OXFORD
UNIVERSITY PRESS

Acknowledgements

The "In cooperation with IB" logo signifies the content in this textbook has been reviewed by the IB to ensure it fully aligns with current IB curriculum and offers high-quality guidance and support for IB teaching and learning.

The Publisher would like to thank the following members of the DP Science 2023 Research Panel for sharing their insights, expertise, and feedback:

Urška Manners, Dr. Vaughn C. Vick, Edite Briosa, Adrian Slack, Kurt Amundson, Ronald Lim, Aleksandra Radić, Meena Rajesh, Alejandro Casariego, and Joana Domingues.

The Publisher wishes to thank the International Baccalaureate Organization for permission to reproduce their intellectual property.

The publisher would like to thank the following for permissions to use copyright material:

Cover: Andrew John Wells/Getty Images.
Photos: p6: Vladimir Ivanovic/Alamy Stock Photo; **p7(t):** William Farquhar/Shutterstock; **p7(m):** shihina/Shutterstock; **p7(b):** MarcelClemens/Shutterstock; **p9:** Rawpixel.com/Shutterstock; **p13:** Andrei Nekrassov/Shutterstock; **p23:** University of New South Wales; **p41(t):** Juanan Barros Moreno/Shutterstock; **p41(b):** MidoSemsem/Shutterstock; **p43:** Prachaya Roekdeethaweesab/Shutterstock; **p49:** Ljupco Smokovski/Shutterstock; **p56:** Pictorial Press Ltd / Alamy Stock Photo; **p59(t):** James Steidl/Shutterstock; **p59(b):** Yulia Koltyrina/123RF; **p65:** William Heathcote; **p68:** GRANGER - Historical Picture Archvie / Alamy Stock Photo; **p69:** Dr Gary Settles / Science Photo Library; **p73:** Liviu Ionut Pantelimon/Shutterstock; **p74:** Ukrinform / Alamy Stock Photo; **p78:** Vasily Smirnov/Shutterstock; **p83:** Visionsi/Shutterstock; **p87:** sciencephotos / Alamy Stock Photo; **p89:** Helen H. Richardson/The Denver Post/Getty Images; **p91(t):** Image Source/Getty Images; **p91(b):** Hieronymus Ukkel/Shutterstock; **p94:** Henri Silberman/Getty Images; **p100:** TopFoto; **p101:** Findlay / Alamy Stock Photo; **p104:** Patty Lagera/Getty Images; **p106:** Iakov Filimonov/Shutterstock; **p110(t):** Photo 12 / Alamy Stock Photo; **p110(b):** Cyrille Arroyo/Shutterstock; **p114:** Public Domain; **p127:** Lawrence Livermore National Laboratory (CC BY-NC-SA 4.0); **p130:** Brand X/Getty Images; **p135:** Copyright 2022 Ricardo; **p136:** Naeblys/Shutterstock; **p138:** Elizabethsalleebauer/RooM/Getty Images; **p150:** Newscom / Alamy Stock Photo; **p151:** Picture Partners / Alamy Stock Photo; **p160:** World History Archive / Alamy Stock Photo; **p163:** ThamKC/Shutterstock; **p166:** Sueddeutsche Zeitung Photo / Alamy Stock Photo; **p170:** VladKK/Shutterstock; **p196:** Swen Stroop/Shutterstock; **p197(t):** J. Palys/Shutterstock; **p197(b):** beboy/Shutterstock; **p198:** djgis/Shutterstock; **p200:** Jan Wlodarczyk / Alamy Stock Photo; **p205:** Aivolie/Shutterstock; **p214:** Potential Filmmaker/Shutterstock; **p223:** Geopix / Alamy Stock Photo; **p224:** Irina Fischer/Shutterstock; **p225:** TTstudio/Shutterstock; **p230:** GL Archive / Alamy Stock Photo; **p231:** Giovanni Benintende/Shutterstock; **p232:** Everett Historical/Shutterstock; **p237:** Alexey Bykov/Shutterstock; **p240:** JP Chretien/Shutterstock; **p241:** Oligo/Shutterstock; **p242(l):** MarcelClemens/Shutterstock; **p242(r):** Photodisc/Getty Images; **p243:** pockygallery/Shutterstock; **p246:** Public Domain; **p252:** Photodisc/Getty Images; **p254:** liverbird/123RF; **p262:** Andrew Mcclenaghan/Science Photo Library; **p272:** Dutourdumonde Photography/Shutterstock; **p273:** Richard Semik/Shutterstock; **p274(l):** Yuri Korchmar/Shutterstock; **p274(r):** Alice Nerr/Shutterstock; **p285:** yotily/Shutterstock; **p296:** Daderot at English Wikipedia, CC BY-SA 3.0/Wikimedia Commons; **p297:** Photo Image/Shutterstock; **p298:** PHOTOCREO Michal Bednarek/Shutterstock; **p299:** humbak/Shutterstock; **p300:** Anna Omelchenko/Shutterstock; **p305(tl):** yiargo/Shutterstock; **p305(tr):** Melica/Shutterstock; **p305(b):** studioVin/Shutterstock; **p307:** Soumitra Giri/Shutterstock; **p308(t):** Yes058 Montree Nanta/Shutterstock; **p308(b):** Volodymyr Krasyuk/Shutterstock; **p310:** Science & Society Picture Library/Getty Images; **p322:** Trevor Clifford Photography/Science Photo Library; **p325:** Hadrian/Shutterstock; **p343:** Alexa Mat/Shutterstock; **p347:** Trevor Clifford Photography/Science Photo Library; **p366:** Ahmed Abid/500px/Getty Images; **p367(t):** Ocean/Corbis; **p367(b):** V_E/Shutterstock; **p369(t):** William Heathcote/Shutterstock; **p369(b):** imagefactory/Shutterstock; **p374:** Oskari Porkka/Shutterstock; **p375:** John_T/Shutterstock; **p388:** GRANGER - Historical Picture Archvie / Alamy Stock Photo; **p390:** hikrcn/Shutterstock; **p401:** Mopic/Shutterstock; **p403:** Photodisc/Getty Images; **p406:** Science History Images / Alamy Stock Photo; **p407:** INSADCO Photography / Alamy Stock Photo; **p409:** Harvard Natural Sciences Lecture Demonstrations; **p411(t):** Wikimedia Commons; **p411(b):** Volgi archive/Alamy Stock Photo; **p417:** sciencephotos / Alamy Stock Photo; **p418:** Verbcatcher, CC BY-SA 4.0/Wikimedia Commons; **p425:** GIPHOTOSTOCK / Science Photo Library; **p429:** GIPPHOTOSTOCK/Science Photo Library; **p432:** GIPHOTOSTOCK / Science Photo Library; **p435(t):** Menno van der Haven/Shutterstock; **p435(m):** sknol/Shutterstock; **p435(b):** Calin Tatu/Shutterstock; **p436:** Parkpoom Doungkaew/Shutterstock; **p439:** Andrew Lambert Photography / Science Photo Library; **p442:** Adil Celebiyev StokPhoto/Shutterstock; **p446:** Lebrecht Music & Arts / Alamy Stock Photo; **p448:** PictureLux / The Hollywood Archive / Alamy Stock Photo; **p449:** chanonnat srisura/Shutterstock; **p451:** William Heathcote; **p456:** faustasyan/Shutterstock; **p457:** Cleanfotos/Shutterstock; **p472:** Brian Maudsley/Alamy Stock Photo; **p473(t):** Gorodenkoff/Shutterstock; **p473(b):** clearviewstock/Shutterstock; **p474:** John A Davis/Shutterstock; **p475(t):** American Photo Archive/Alamy Stock Photo; **p475(b):** Artokoloro/Alamy Stock Photo; **p476:** Christian Offenberg/Alamy Stock Photo; **p479:** Arthurmarris/Wikimedia; **p482:** Vintage_Space/Alamy Stock Photo; **p488:** Brian Donovan/Shutterstock; **p492:** © Crown Copyright and database rights 2022. Ordnance Survey License Number 100048957. Supplied by: www.ukmapcentre.com; **p504:** MikhailSh/Shutterstock; **p506:** Charistoone-images/Alamy Stock Photo; **p508:** PRISMA ARCHIVO/Alamy Stock Photo; **p522:** Pavel L Photo and Video/Shutterstock; **p539:** Peter Hermes Furian/Shutterstock; **p540:** Biletskiyevgeniy.com/Shutterstock; **p547:** Georgios Kollidas/Shutterstock; **p555:** Sciencephotos/Alamy Stock Photo; **p558:** H.S. Photos/Alamy Stock Photo; **p559:** Carl D. Anderson (1905–1991)/Wikimedia; **p560:** J. Lekavicius/Shutterstock; **p567:** Oliver Hoffmann/Shutterstock; **p580(t):** Regina Erofeeva/Shutterstock; **p580(b):** Philip Steury Photography/Shutterstock; **p583:** Ziga Cetric/Shutterstock; **p586:** Science History Images/Alamy Stock Photo; **p587(t):** Zenobillis/Shutterstock; **p587(b):** Sergey Nivens/Shutterstock; **p588:** IBM Research / Science Photo Library; **p589:** Pavel L Photo and Video/Shutterstock; **p592(t):** Albert Lozano/Shutterstock; **p592(b):** Oscity/Shutterstock; **p596:** Len Collection/Alamy Stock Photo; **p598(t):** Phil Degginger / Alamy Stock Photo; **p598(m):** Phil Degginger / Alamy Stock Photo; **p598(b):** Science Photo Library; **p599:** Morphart Creation/Shutterstock; **p602(t):** BearFotos/Shutterstock; **p602(b):** World History Archive/Alamy Stock Photo; **p607:** Taylon/Shutterstock; **p612:** Mark Mason; **p626:** Andrew Lambert Photography / Science Photo Library; **p629:** CBW/Alamy Stock Photo; **p631(t):** The History Collection/Alamy Stock Photo; **p631(b):** Science History Images/Alamy Stock Photo; **p632:** Everett Historical/Shutterstock; **p633:** N. Feather / Science Photo Library; **p634:** Pictorial Press Ltd/Alamy Stock Photo; **p658:** Yok_onepiece/Shutterstock; **p667:** Everett Collection Historical/Alamy Stock Photo; **p670:** Hornyak/Shutterstock; **p673:** Parilov/Shutterstock; **p675:** Nordroden/Shutterstock; **p676:** imageBROKER/Alamy Stock Photo; **p677:** Everett Collection/Shutterstock; **p678:** Oxford University Press; **p679:** Triff/Shutterstock; **p691:** EHT Collaboration; **p692:** Science Photo Library.

Artwork by Q2A Media, Six Red Marbles, Aptara, Greengate Publishing, Thomson, Stewart Miller, Wearset Ltd, HL Studios, James Stayte, and Oxford University Press.

Although we have made every effort to trace and contact all copyright holders before publication this has not been possible in all cases. If notified, the publisher will rectify any errors or omissions at the earliest opportunity. Links to third party websites are provided by Oxford in good faith and for information only. Oxford disclaims any responsibility for the materials contained in any third party website referenced in this work.

Contents

A Space, time and motion		6
A.1	Kinematics	8
A.2	Forces and momentum	40
A.3	Work, energy and power	106
A.4	Rigid body mechanics	130
A.5	Galilean and special relativity	160
	End-of-theme questions	194

B The particulate nature of matter		196
B.1	Thermal energy transfers	198
B.2	Greenhouse effect	232
B.3	Gas laws	252
B.4	Thermodynamics	272
B.5	Current and circuits	298
	End-of-theme questions	330

Tools for physics		332
	Mathematical tools for physics	333
	Experimental tools for physics	343
	Data analysis and modelling physics	348

C Wave behaviour		366
C.1	Simple harmonic motion	368
C.2	Wave model	388
C.3	Wave phenomena	406
C.4	Standing waves and resonance	436
C.5	Doppler effect	456
	End-of-theme questions	470

D Fields		472
D.1	Gravitational fields	474
D.2	Electric and magnetic fields	504
D.3	Motion in electromagnetic fields	540
D.4	Induction	560
	End-of-theme questions	584

E Nuclear and quantum physics		586
E.1	Structure of the atom	588
E.2	Quantum physics	611
E.3	Radioactive decay	630
E.4	Fission	666
E.5	Fusion and stars	678
	End-of-theme questions	698

Extended-response questions	700
The inquiry process and internal assessment (IA)	702
Index	708

Answers: https://www.oxfordsecondary.com/ib-science-support

Introduction

The Diploma Programme (DP) Physics course is for 16-19 year old students. This course aims to develop a conceptual understanding of physics and the nature of science, enabling students to apply their knowledge in familiar and unfamiliar contexts. The course also encourages students to further apply and strengthen the IB Learner Profile attributes.

The course is split into 5 'themes' which are labeled A to E. These cover broad areas of knowledge, such as 'Wave behaviour' or 'Fields'. Each theme is then broken down into four or five 'topics' which are labeled with numbers, for example Topic B.3. Each topic focuses on a more specific area of knowledge, such as 'Wave phenomena' or 'Gravitational fields'.

Physics deals with the structure and interactions of the matter that makes up the observable Universe, and in studying physics, we aim to formulate universal principles that explain the many different phenomena around them.

Three over-riding scientific concepts used to explain physics are: particles, forces, and energy, and the 2023 DP Physics course uses these broad concepts to bind together all of the topics it covers.

Physics uses the idea of a **particle** – a small piece of matter – to describe nature at both the macroscopic (large scale) and microscopic (small scale) levels. Both terms recur throughout the course as you reconcile practical observations of the macroscopic world with explanations that attempt to model it at the microscopic level. While the nature of the 'particle' used in the modelling varies from theme to theme, the concept of a small piece of matter and the idea that we can model the average behaviour of many such particles is repeated throughout.

Particles interact through the **forces** that act between them. A 'force' is often described as a push or a pull. A more sophisticated description is that force is the concept linking a particle property to the acceleration that the particle experiences. Physics still has difficulty in describing the origins of some forces even though it can describe the effects of these forces well. For example, in Topic D.1 we have a convincing link between gravitational force/acceleration and mass, but only a poor understanding of what causes gravity.

One of the reasons for the technological progress of our species is our ability to control the transfer of **energy**. Cooking, heating, transport and communication all rely on civilization maintaining control of this transfer. Again, we find it difficult to describe what energy is other than to say that it is the ability to do work. Topic A.3 tells us that energy comes in many forms and that we recognise its presence best when it is moving from one state to another. The way we treat particles, forces, and energy are linked by the five themes of this book. The introduction to each theme describes the links between these concepts and the individual theme. The topics that make up that theme then examine how the three concepts are developed through the ideas, theories and laws discussed in that individual topic.

Course book definition

The IB Diploma Programme course books are resource materials designed to support students throughout their two-year Diploma Programme course of study in a particular subject. They will help students gain an understanding of what is expected from the study of an IB Diploma Programme subject while presenting content in a way that illustrates the purpose and aims of the IB. They reflect the philosophy and approach of the IB and encourage a deep understanding of each subject by making connections to wider issues and providing opportunities for critical thinking.

The books mirror the IB philosophy of viewing the curriculum in terms of a whole-course approach; the use of a wide range of resources, international mindedness, the IB learner profile and the IB Diploma Programme core requirements, theory of knowledge, the extended essay, and creativity, activity, service (CAS).

IB mission statement

The International Baccalaureate aims to develop inquiring, knowledgeable and caring young people who help to create a better and more peaceful world through intercultural understanding and respect.

To this end, the organization works with schools, governments and international organizations to develop challenging programmes of international education and rigorous assessment.

These programmes encourage students across the world to become active, compassionate and lifelong learners who understand that other people, with their differences, can also be right.

Nature of Science

Science has features that make it different from other pursuits such as the arts, social sciences, mathematics, or the study of language. Science has particular methodologies and purposes.

We return many times in this course to the Nature of Science (NOS). We illustrate the work you need to understand with the methodology and philosophy of scientists' work. We examine how they measure its impact on science and on society. The effective pursuit of modern scientific work and its theories depends on the **Nature of Science**, which can be summarized in the following eleven aspects:

- **Observations and experiments**
  Sometimes the observations in experiments are unexpected and lead to serendipitous results.

- **Measurements**
  Measurements can be qualitative or quantitative, but all data are prone to error. It is important to know the limitations of your data.

- **Evidence**
  Scientists learn to be sceptical about their observations and they require their knowledge to be fully supported by evidence.

- **Patterns and trends**
  Recognition of a pattern or trend forms an important part of the scientist's work whatever the science.

- **Hypotheses**
  Patterns lead to a possible explanation. The hypothesis is this provisional view and it requires further verification.

- **Falsification**
  Hypotheses can be proved false using other evidence, but they cannot be proved to be definitely true. This has led to paradigm shifts in science throughout history.

- **Models**
  Scientists construct models as simplified explanations of their observations. Models often contain assumptions or unrealistic simplifications, but the aim of science is to increase the complexity of the model, and to reduce its limitations.

- **Theories**
  A theory is a broad explanation that takes observed patterns and hypotheses and uses them to generate predictions. These predictions may confirm a theory (within observable limitations) or may falsify it.

- **Science as a shared activity**
  Scientific activities are often carried out in collaboration, such as peer review of work before publication or agreement on a convention for clear communication.

- **Global impact of science**
  Scientists are responsible to society for the consequences of their work, whether ethical, environmental, economic or social. Scientific knowledge must be shared with the public clearly and fairly.

How to use this book

The aim of this book is to develop conceptual understanding, aid in skills development and provide opportunities to cement knowledge and understanding through practice.

Feature boxes and sections throughout the book are designed to support these aims, by signposting content relating to particular ideas and concepts, as well as opportunities for practice. This is an overview of these features:

Developing conceptual understanding

Guiding questions

Each topic begins with two or more guiding questions to get you thinking. When you start studying a topic, you might not be able to answer these questions confidently or fully, but by studying that topic, you will be able to answer them with increasing depth. Hence, you should consider these as you work through the topic and come back to them when you revise your understanding.

These boxes in the margin will direct you to other parts of the book where a concept is explored further or in a different context. They may also direct you to prior knowledge or a skill you'll need, or give a different way to think about something.

Linking questions

Linking questions within each topic highlight the connections between content discussed there and other parts of the course. Physicists often connect dissimilar phenomena using similar approaches, both conceptual and mathematical.

Nature of Science

These illustrate NOS using issues from both modern science and science history, and show how the ways of doing science have evolved over the centuries. There is a detailed description of what is meant by NOS and the different aspects of NOS on the previous page. The headings of NOS feature boxes show which of the eleven aspects they highlight.

Theory of knowledge

This is an important part of the IB Diploma course. It focuses on critical thinking and understanding how we arrive at our knowledge of the world. The TOK features in this book pose questions for you that highlight these issues.

Parts of the book have a coloured bar on the edge of the page or next to a question. This indicates that the material is for students studying at DP Physics Higher Level. AHL means "additional higher level".

AHL

Developing skills

ATL Approaches to learning

These ATL features give examples of how famous scientists have demonstrated the ATL skills of communication, self-management, research, thinking and social skills, and prompt you to think about how to develop your own strategies.

Physics skills

These contain ways to develop your mathematical, experimental or inquiry skills, especially through experiments and practical work. Some of these can be used as springboards for your Internal Assessment: don't be afraid to modify these to suit the experimental setup available to you.

- The bullet points at the top of these boxes link the content to the skills it helps you develop.

Tools for physics, the inquiry process and internal assessment

These three section of the book are full of reference material for all the essential mathematical and experimental tools required for DP Physics, details on data analysis and modelling physics, as well as guidance on how to use the inquiry process in the study of the subject and to work through your internal assessment (IA). Flick to this section as your working through the rest of the book for more information. Links in the margin throughout the book will direct you towards it too.

Practicing

Worked examples

These are step-by-step examples of how to answer questions or how to complete calculations. You should review these examples carefully, preferably after attempting the question yourself.

Practice questions

These are designed to give you further practice at using your physics and to allow you to check your own understanding and progress.

Data-based questions

Part of your final assessment requires you to answer questions that are based on the interpretation of data. Use these questions to prepare for this. They are also designed to make you aware of the possibilities for data acquisition and analysis for day-to-day experiments and for your IA.

End-of-theme questions

Use these questions at the end of each theme to draw together concepts from that theme and other parts of the book, and to practise answering exam-style questions. Many of these are past IB physics exam questions. You will also find some practice extended-response questions near the back of the book.

The IB Learner Profile

The aim of all IB programmes to develop internationally minded people who work to create a better and more peaceful world. The aim of the programme is to develop this person through ten learner attributes, as described below.

Inquirers: They develop their natural curiosity. They acquire the skills necessary to conduct inquiry and research and snow independence in learning. They actively enjoy learning and this love of learning will be sustained throughout their lives.

Knowledgeable: They explore concepts, ideas and issues that have local and global significance. In so doing, they acquire in-depth knowledge and develop understanding across a broad and balanced range of disciplines.

Thinkers: They exercise initiative in applying thinking skills critically and creatively to recognize and approach complex problems, and to make reasoned, ethical decisions.

Communicators: They understand and express ideas and information confidently and creatively in more than one language and in a variety of modes of communication. They work effectively and willingly in collaboration with others.

Principled: They act with integrity and honesty, with a strong sense of fairness, justice and respect for the dignity of the individual, groups and communities. They take responsibility for their own action and the consequences that accompany them.

Open-minded: They understand and appreciate their own cultures and personal histories, and are open to the perspectives, values and traditions of other individuals and communities. They are accustomed to seeking and evaluating a range of points of view, and are willing to grow from the experience.

Caring: They show empathy, compassion and respect towards the needs and feelings of others. They have a personal commitment to service, and to act to make a positive difference to the lives of others and to the environment.

Risk-takers: They approach unfamiliar situations and uncertainty with courage and forethought, and have the independence of spirit to explore new roles, ideas and strategies. They are brave and articulate in defending their beliefs.

Balanced: They understand the importance of intellectual, physical and emotional ballance to achieve personal well-being for themselves and others.

Reflective: They give thoughtful consideration to their own learning and experience. They are able to assess and understand their strengths and limitations in order to support their learning and personal development.

A note on academic integrity

It is of vital importance to acknowledge and appropriately credit the owners of information when that information is used in your work. After all, owners of ideas (intellectual property) have property rights. To have an authentic piece of work, it must be based on your individual and original ideas with the work of others fully acknowledged. Therefore, all assignments, written or oral, completed for assessment must use your own language and expression. Where sources are used or referred to, whether in the form of direct quotation or paraphrase, such sources must be appropriately acknowledged.

How do I acknowledge the work of others?

The way that you acknowledge that you have used the ideas of other people is through the use of footnotes and bibliographies.

Footnotes (placed at the bottom of a page) or endnotes (placed at the end of a document) are to be provided when you quote or paraphrase from another document or closely summarize the information provided in another document. You do not need to provide a footnote for information that is part of a 'body of knowledge'. That is, definitions do not need to be footnoted as they are part of the assumed knowledge.

Bibliographies should include a formal list of the resources that you used in your work. 'Formal' means that you should use one of the several accepted forms of presentation. This usually involves separating the resources that you use into different categories (e.g. books, magazines, newspaper articles, internet-based resources, and works of art) and providing full information as to how a reader or viewer of your work can find the same information. A bibliography is compulsory in the Extended Essay.

What constitutes malpractice?

Malpractice is behaviour that results in, or may result in, you or any student gaining an unfair advantage in one or more assessment component. Malpractice includes plagiarism and collusion.

Plagiarism is defined as the representation of the ideas or work of another person as your own. The following are some of the ways to avoid plagiarism:

- words and ideas of another person to support one's arguments must be acknowledged

- passages that are quoted verbatim must be enclosed within quotation marks and acknowledged

- email messages, websites on the internet and any other electronic media must be treated in the same way as books and journals

- the sources of all photographs, maps, illustrations, computer programs, data, graphs, audio-visual and similar material must be acknowledged if they are not your own work

- when referring to works of art, whether music, film dance, theatre arts or visual arts and where the creative use of a part of a work takes place, the original artist must be acknowledged.

Collusion is defined as supporting malpractice by another student. This includes:

- allowing your work to be copied or submitted for assessment by another student

- duplicating work for different assessment components and/or diploma requirements.

Other forms of malpractice include any action that gives you an unfair advantage or affects the results of another student. Examples include, taking unauthorized material into an examination room, misconduct during an examination and falsifying a CAS record.

Experience the future of education technology with Oxford's digital offer for DP Science

You're already using our print resources, but have you tried our digital course on Kerboodle?

Developed in cooperation with the IB and designed for the next generation of students and teachers, Oxford's DP Science offer brings together the IB curriculum and future-facing functionality, enabling success in DP and beyond. Use both print and digital components for the best blended teaching and learning experience.

Learn anywhere with mobile-optimized onscreen access to student resources and offline access to the digital Course Book

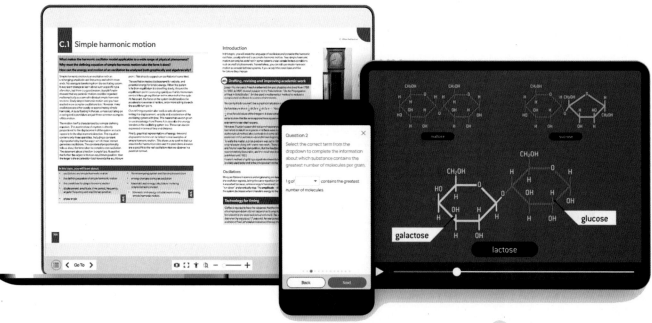

Encourage motivation with a variety of engaging content including interactive activities, vocabulary exercises, animations, and videos

Embrace independent learning and progression with adaptive technology that provides a personalized journey so students can self-assign auto-marked assessments, get real-time results and are offered next steps

↓

Deepen understanding with intervention and extension support, and spaced repetition, where students are asked follow-up questions on completed topics at regular intervals to encourage knowledge retention

↓

↑

Enhance reporting with rich data collected to support responsive teaching at an individual and class level

kerboodle

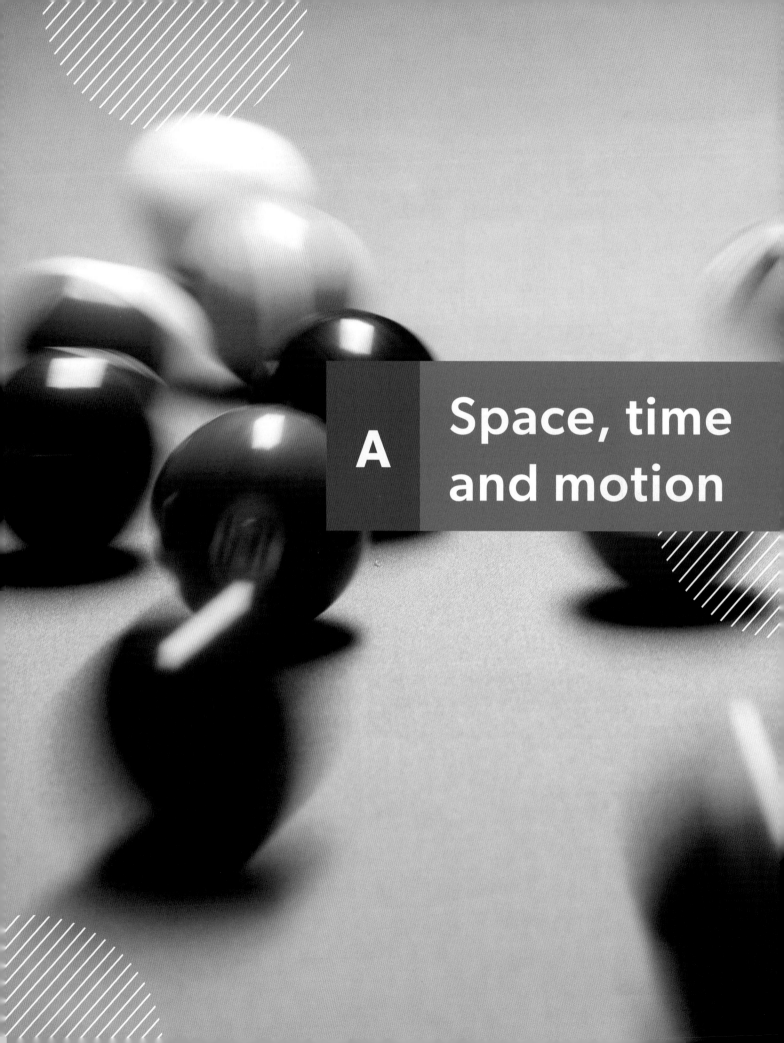

A Space, time and motion

Introduction

A ball is thrown into the air. The Earth revolves on its axis in its journey around the Sun. The Sun moves within the Milky Way. Galaxies separate.

Understanding motion lies at the heart of physics. Descriptions of motion have formed the bases for scientific hypotheses since the science of the Ancient Greeks.

Theme A links the concepts of **particle**, **force** and **energy transfer**. First, we consider motion in the context of a single particle with mass but without size or shape. On this basis we can take first steps to define the parameters of motion in Topic A.1: velocity, displacement, acceleration. These are linked by the important idea of rate of change for the first – but certainly not the last – time in this course. Force links to acceleration in Topic A.2, and an important conserved quantity known as momentum is introduced. Energy transfers that arise from the changes of Topics A.1 and A.2 are discussed in Topic A.3.

Scientists link subject areas together to help extend their insights. In Topic A.4 the collective motion of particles formed into a solid object is discussed in the contexts of momentum, force and energy. The theory underpinning this relies heavily on the first three topics. We see how scientists use concepts in linear motion to inform the description of rotational motion.

Up until the beginning of the 20th century, physics decoupled space and time treating them as distinct properties. The work of Einstein and others shows that space and time are not separate concepts but are linked through the speed of electromagnetic radiation in a vacuum. This is one of the most profound paradigm shifts in physics there has ever been. Topic A.5 introduces special relativity and explains how it changes our views of Galilean space and time.

Throughout the long development of scientific thought about motion and its effects, scientists have relied on the careful accumulation of evidence and observation. This is a crucial stage in the Inquiry Cycle and one that allows a reflection of the truth or otherwise of a scientific **hypothesis**.

This theme is a foundation for your understanding of IB Diploma Programme physics. The important ideas and concepts of this theme allow us to **predict** the behaviour of the average particle in a gas. We will be able to analyse the behaviour of a wave and an oscillating system. Knowing how forces behave permits an understanding of electric and gravitational field theory. The study of Theme A leads to theories of the behaviour of the very largest and the very smallest objects in the universe.

A.1 Kinematics

How can the motion of a body be described quantitatively and qualitatively?

How can the position of a body in space and time be predicted?

How can the analysis of motion in one and two dimensions be used to solve real-life problems?

The word "kinematics" comes from the Greek word *kinēsis*, meaning "movement" or "motion". So "kinematics" means the study of motion and the quantities used to describe it. To describe the motion of a particle, you need quantitative ways to describe its position in the world. You also need to describe the rate at which its position changes (its speed). Then you will need the rate at which the speed itself can also change (its acceleration). You need to know the precise meanings of these terms and be able to distinguish between the size (magnitude) of the quantities and their direction.

With a language to describe distance, speed and acceleration, you can formulate rules to predict future changes. These rules are based on both observation and deduction.

However, to arrive at these rules, you will also need assumptions. As your understanding changes, you can change these assumptions and evolve more complex descriptions of motion. For example, you can change from describing one-dimensional motion (an object moving in a straight line) to two-dimensional motion (an object projected into the air close to Earth's surface). You can understand two-dimensional motion more broadly by separating it into the horizontal and vertical motion. By removing the assumption of negligible friction, you can investigate realistic cases of air resistance acting on an object as it moves through the air.

In this topic, you will learn about:

- describing and analysing motion through space and time, using position, velocity and acceleration vectors

- displacement and the difference between distance and displacement

- velocity and acceleration

- instantaneous and average values of velocity, speed and acceleration, and how they are determined

- the kinematic equations of motion for solving problems with uniform acceleration

- motion with uniform and non-uniform acceleration

- the behaviour of projectiles when there is no fluid resistance

- resolving motion into vertical and horizontal components

- the effects of fluid resistance on projectiles.

Describing motion

Vectors and scalars

It is important to arrive at school or college on time for your class. To do this, you need to determine the distance and deduce a time for your journey using a reasonable estimate for the speed at which you travel.

Figure 2 shows the journey a student takes to travel from home (A) to school (B). Table 1 gives the time at various parts of the journey.

▲ Figure 1 Many devices use GPS to help you navigate. By knowing the distance to your destination and assuming a speed, they calculate an estimated time to arrival.

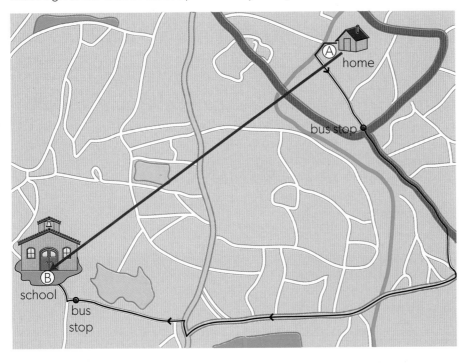

◀ Figure 2 A journey to school. The map shows the direct route (the displacement) and the actual journey (the distance).

Journey leg	Time	Distance for leg / m
leave home	08.10.00	0
arrives at bus stop	08.20.15	800
bus arrives at stop	08.24.30	0
bus arrives near school	08.31.10	2400
walk from bus to school	08.34.00	200

◀ Table 1 Distances and times for different stages of the journey from home to school.

The total **distance** for this route on foot and by bus is 3.4 km, including all the twists and turns. The distance is the same for the return journey too. Distance is a **scalar** quantity. It only has magnitude (size).

The direct line from the student's home to school is also shown on Figure 2. This quantity is known as the **displacement**. It is the change in position of the student between the start and the end of their journey.

The direct line from school to home is the same length as the direct line from home to school, but it is in the reverse direction. The two displacements—to and from school—are not the same. This is because displacement is a **vector** quantity. Vectors have both a magnitude and a direction.

A scalar is a physical quantity that has size but no direction.

A vector is a physical quantity that has size and direction.

You can find out about the properties of vectors and scalars in the *Tools for physics* section on page 339.

You need to know how to work with units as almost every quantity you deal with in IB physics has its own unit. You can find out about the units in the course and how to use and manipulate them on page 334.

The student's journey in Figure 2 is from A to B, which can be written as $\overrightarrow{AB}$. The direction line is 1.7 km long and is in a south-west direction. The displacement must be given as the magnitude together with the direction: 1.7 km on a bearing of about 225° from north.

The displacement of the student's journey home from school is the vector $\overrightarrow{BA}$. This vector is still 1.7 km long but is in the opposite direction at a bearing of about 45° from north.

Measurements—Vectors and scalars

The distinction between vectors and scalars is important. Some physical quantities, such as force, electric field strength and acceleration, have direction built into them; they are the vectors. Other quantities, such as mass and energy, only have magnitude with no direction assigned to them; they are the scalars.

Sometimes a vector quantity might have a minus sign, which indicates its direction. For example, $-2\,\text{m s}^{-1}$ might mean a speed of $2\,\text{m s}^{-1}$ in the reverse or backwards direction. It may be tempting to think that any quantity that has a minus sign must be a vector, but this is not the case. As an example, consider the amount of money in a bank account. A positive amount means that you are in credit, whereas a negative amount means that you are in debt and owe the bank money. Although the balance of the account can have a negative quantity, money itself is still a scalar quantity.

Charge and energy are another two examples of scalar quantities that can have negative values. Can you think of any others?

Measurements—Moving in three dimensions

Displacement is described here in terms of a journey across the flat, two-dimensional landscape of Figure 2. Do changes in height alter things? In fact, only one thing changes, and that is the number of pieces of information required to specify the final position relative to the start. Three pieces are now required:

- the magnitude plus its unit

- the heading (direction)

- the overall change in height during the journey.

Specifying motion in three dimensions thus requires three numbers or coordinates. You will already be familiar with the idea of coordinates from drawing and using graphs. There is flexibility in how the three numbers can be chosen. You may have seen three-dimensional graphs with three axes each at 90° to the others. In this case, coordinate numbers give the distance along each axis. Another option is to use spherical coordinates (Figure 3). Here, a distance r and two angles are required. One angle is the bearing ϕ from north, called the azimuth in astronomy. The other is the angle known as the altitude or elevation θ needed to look directly up or down from the horizontal to the object above (or below) you.

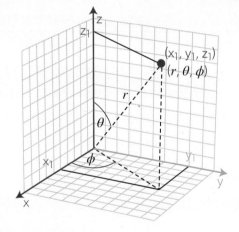

▲ Figure 3   A Cartesian coordinate system with three distances and a spherical coordinate system with two angles and a distance.

In some circumstances, even the distance itself may not be required. Sailors use latitude and longitude when they are navigating. They stay on the surface of the sea and this is effectively a constant distance from the centre of Earth. Astronomers use just azimuth and elevation as the distance to the star is irrelevant for observation.

Science as a shared endeavour—Units of distance

Throughout the world there are many units of length: metres, miles and kilometres are all common ways to specify distance. Different countries and professions use alternative units depending on what is traditional or convenient. For example, surveyors use chains and astronomers use light years (a unit of length, not time)

and the astronomical unit which is the distance from Earth to the Sun. However, in your examination, lengths will be in multiples and sub-multiples of the metre or in a well-recognized scientific unit such as the light year.

There is a list of some of the quantities used in astronomy in the data booklet.

Worked example 1

A ball is dropped from rest from an initial height of 1.2 m and rebounds to 75% of the initial height. Calculate, for the instant when the ball is at its maximum height after the bounce:

a. the displacement of the ball

b. the distance moved by the ball.

Solutions

a. The height after the bounce is $0.75 \times 1.2 = 0.9$ m. The displacement is the change in the ball's position, $0.9 - 1.2 = -0.3$ m. The minus sign indicates that the displacement is directed downward.

b. The distance is the length of the path moved by the ball in both phases of motion, $1.2 + 0.9 = 2.1$ m. The distance is a scalar quantity and does not consider direction.

Worked example 2

A cyclist rides 250 m up a slope that makes an angle of 8.0° to the horizontal. Calculate:

a. the change in height during the ride

b. the horizontal component of the cyclist's displacement.

Solutions

a. The components of the displacement vector can be calculated using trigonometry. The change in height is equal to the vertical displacement, $250 \times \sin 8.0° = 35$ m.

b. Horizontal displacement $= 250 \times \cos 8.0° = 248$ m. Even on a relatively steep road, there is little difference between the horizontal distance and the distance along the slope!

Practice questions

1. A boat travels due east for 2.5 km, then travels due north for a further 3.8 km. Calculate:

   a. the distance travelled by the boat

   b. the magnitude of the displacement of the boat

   c. the bearing from north of the displacement of the boat.

2. The minute hand of a wall clock is 15 cm long. Calculate, to the nearest cm, the distance travelled by the tip of the minute hand and the magnitude of its displacement:

   a. from 12.00 to 12.15

   b. from 12.00 to 12.30.

ATL Using symbols consistently

Distances and displacements can be represented using many different symbols. Sometimes a distance may be labelled as d. When it is a height, then h may be used. In three dimensions, Δx, Δy and Δz may denote the displacements in each direction. Why is s often used to represent distance or displacement in physics? The Latin for distance is *spatium*. It was also useful not to use the letter d since it may be confusing when writing derivatives such as $v = \dfrac{ds}{dt}$. Another Latin abbreviation is the use of c for speed, usually for the speed of light; c stands for *celeritas*—Latin for speed.

Speed and velocity

There are scalar and vector measures of how quickly an object moves.

The scalar quantity is **speed**, which is defined as:

$$speed = \frac{\text{distance travelled on a journey}}{\text{time taken for the journey}}$$

or speed = change of distance per unit time.

You will already be familiar with units of speed such as metre per second ($m\,s^{-1}$) and kilometres per hour ($km\,h^{-1}$), but you can combine any distance unit with any time unit to give speed units.

Velocity is the vector equivalent—it is the speed in a given direction. So to describe the velocity, you need the magnitude and the direction, as for displacement. For example, "$4.2\,m\,s^{-1}$ due north" or "$55\,km\,h^{-1}$ at N 22.5 E".

The definition of velocity is change of displacement per unit time or the rate of change of an object's position.

An object moving at a constant speed covers equal distances in equal times. A passenger train that travels $2400\,m$ in one minute has a speed of $40\,m\,s^{-1}$. In one hour, the train will travel ($3600 \times 40 =$) $144000\,m$. So $40\,m\,s^{-1} \equiv 144\,km\,h^{-1}$.

Using symbols rather than numbers, the magnitude of the velocity of an object (its speed) v is the distance travelled s divided by the time taken t:

$$v = \frac{s}{t} \text{ and } s = vt.$$

Worked example 3

A train moving at a constant speed of $280\,km\,h^{-1}$ takes $2.3\,s$ to pass a signal pole. Calculate the length of the train.

Solution

The speed of the train needs to be converted to $m\,s^{-1}$. Speed $= \dfrac{280 \times 10^3}{60 \times 60} = 77.8\,m\,s^{-1}$.
Length of the train $= 77.8 \times 2.3 = 180\,m$.

Worked example 4

The astronomical unit (AU) is approximately equal to $1.50 \times 10^{11}\,m$. Assuming that Earth moves around the Sun in a circular orbit of a radius 1 AU, estimate the orbital speed of Earth. Give the answer in $km\,s^{-1}$.

Solution

The circumference of Earth's orbit is $2\pi \times 1.50 \times 10^{11} = 9.42 \times 10^{11}\,m$. Earth travels this distance in one year.
Orbital speed $= \dfrac{9.42 \times 10^{11}}{365 \times 24 \times 60 \times 60} = 2.99 \times 10^4\,m\,s^{-1} = 29.9\,km\,s^{-1}$.

Practice questions

3. Two cyclists, Ada and Matt, start from the same point and ride in opposite directions along a straight road, each at a constant speed. After one minute, they are $580\,m$ apart. Ada rides at a speed of $20\,km\,h^{-1}$. Determine the speed at which Matt rides.

4. The speed of light in a vacuum is $3.0 \times 10^8\,m$. Sirius, the brightest star, is approximately 5.5×10^5 AU from Earth (1 AU $= 1.50 \times 10^{11}\,m$). Calculate the distance to Sirius in light years (1 light year, (ly) is the distance travelled by light in one year).

🧪 Measuring speed

- Tool 1: Understand how to accurately measure mass and time to an appropriate level of precision.
- Inquiry 1: Demonstrate independent thinking, initiative, or insight.

To measure speed, you need to know the distance travelled (using a "ruler") and the time taken (using a "clock"). The trick is to choose the best "ruler" and the best "clock" for the speed being measured.

A 30 cm ruler and a digital wristwatch are fine when a biologist measures the speed of an earthworm. But, to measure the speed of a 100 m sprinter, a measured distance on the ground, a stopwatch measuring to 0.1 s and a human observer are barely good enough. Even then, the observer (at the finish) must be careful to watch the smoke from the starting pistol (at the start) and not wait to hear the sound of the gun.

To measure the speed of a soccer ball after a penalty kick, the stopwatch-plus-human method is no longer adequate. You need to use a video camera taking image frames at a known rate (the clock) and a scale, visible on the video, near the path of the ball (the ruler). When measuring the speed of a jet aircraft, the equipment needs to change again.

Choosing the best equipment and method for the task in hand is all part of designing the experiments for an internal assessment.

Graphing motion I—Distance–time

To calculate speeds and velocities, you need two variables: distance and time. Figure 2 showed a map of a student's journey to school. The student does not travel at the same speed throughout the journey as part of it is on foot, part by bus.

A distance–time graph is a good way to display such data visually. The distance travelled by the student is plotted on the y-axis of Figure 5, while the time since the beginning of the journey is plotted on the x-axis.

The gradient of the graph changes for the different parts of the journey: small for the walking sections of the journey, horizontal (zero gradient) for stationary at the bus stop, and steep for the bus journey. What will the graph for the journey home from school look like, assuming that the time for each segment of the journey is the same as in the morning?

Information can easily be extracted from this graph. The gradient of the graph is the speed. Add the overall direction to this speed and you have the velocity too.

For the first walk to the bus stop, the distance was 800 m and the time taken was 615 s. The constant walking speed was therefore $\frac{800}{615}$, which is $1.3\,\mathrm{m\,s^{-1}}$.

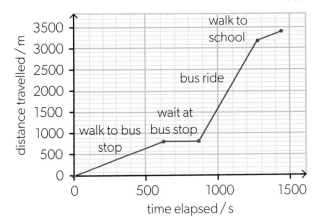

Measurement and experiment are essential tests of scientific knowledge. Galileo was one of the first to appreciate the use of a simple pendulum as a timing instrument. He conducted many early kinematic experiments. Before Galileo, the ability of physicists to study practical kinematics experiments was limited; they made deductions in other ways. Aristotle and other Greek thinkers used logical reasoning to construct persuasive arguments.

A group of scholars in the 14th century started to use mathematical methods to make scientific progress. This approach caused them to be termed the Oxford Calculators. They were sometimes known as the Merton School due to their association with Merton College, Oxford. Their methods influenced many scholars around Europe at that time, and mathematical approaches are still an important way for physicists to test theories.

▲ Figure 4 Scholars at Merton College, Oxford, were among the first to apply mathematical methods to physics and philosophy.

What are the tests of truth or knowledge in other disciplines, such as literature, history or the law?

◀ Figure 5 The Figure 2 journey as a distance–time graph. The graph has the clock time for the journey translated into time (in seconds) since the start.

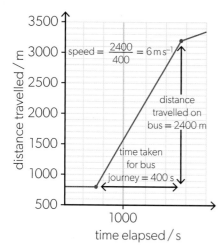

 Figure 6 Using the distance–time graph to calculate average speed.

How does graphical analysis allow for the determination of physical quantities? (NOS)

Throughout the IB Diploma Programme physics course, you use graphs to help you understand the significance of data or the relationship between two quantities. As well as its shape, the graph contains valuable information about other quantities.

The gradient of a graph gives information about the ratio of the change in the *y*-axis quantity to the change in the *x*-axis quantity. The area under a graph indicates the size of the product of the quantities on the axes.

The *Tools for physics* section contains more information about the area and gradient of a graph and how to determine them (pages 360–361).

The gradient of the segment for the bus journey is $\frac{2400}{400} = 6.0$ and so the speed is $6.0\,\text{m s}^{-1}$. The way this is worked out is shown on Figure 6.

Practice questions

5. The graph shows how the distance travelled by an underground train varies with time.

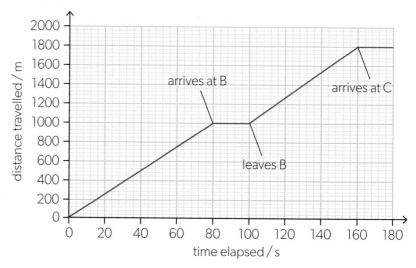

The train departs station A at 0 s, arrives at station B at 80 s, departs from station B at 100 s and arrives at station C at 160 s. Calculate:

a. the speed, in km h^{-1}, of the train between stations A and B

b. the distance between stations B and C

c. the speed, in km h^{-1}, of the train between stations B and C.

6. Louise kicks a ball towards a wall that is 4.0 m away from her. The ball moves at a constant speed of $10\,\text{m s}^{-1}$ when travelling towards the wall. The ball returns to Louise 0.90 s after it was kicked.

a. Calculate:

    i.   the time taken for the ball to reach the wall

    ii. the speed of the ball after it bounces off the wall, assuming that it is constant.

b. Sketch a graph to show how the distance travelled by the ball varies with time.

Instantaneous and average speed

Representing the student's journey by joining data points with straight lines is simplistic. Real journeys rarely have a completely constant speed. You need ways to handle varying speeds and velocities. For real journeys, the distance–time graph will be curved because the speed will be different at different times.

The speedometer in the bus tells the driver the speed. This measure is the **instantaneous speed**: the speed at the instant in time at which it is determined. The instantaneous speed is also the gradient of the distance–time graph at the instant concerned. You can calculate the instantaneous speed from a curved distance–time graph by drawing a tangent at that point and finding the gradient of that tangent.

Figure 7 shows a more realistic distance–time graph for the bus journey to school. The original red line for the bus has been replaced by a green line that is more realistic for the motion of a real bus. The speed varies as the driver gradually speeds up and slows down or negotiates traffic. Figure 7 shows the instantaneous speed calculated at time 1000 s:

Use as large a tangent line as possible.

Gradient = change in values on y-axis ÷ change in values on x-axis

Change in distance = 2500 m – 500 m = 2000 m

Change in time = 1300 s – 900 s = 400 s

Gradient = 2000 ÷ 400 = 5.0 m s^{-1}

From a mathematical point of view, the instantaneous speed is the **rate of change of position with respect to time**.

A mathematician will write this as $\frac{ds}{dt}$, where s is the distance travelled and t is the time. You may also have seen this written as $\frac{\Delta s}{\Delta t}$ where the symbol Δ means "change in". $\frac{\Delta s}{\Delta t}$ is shorthand for $\frac{\text{change in distance}}{\text{change in time}}$.

There is another useful measure of speed. This is the **average speed** and is the speed calculated over the whole the journey without regard to variations in speed. As an equation, this is:

$$\text{average speed} = \frac{\text{distance travelled over the whole journey}}{\text{time taken for the whole journey}}$$

In terms of the distance–time graph, the average speed is equal to the gradient of the straight line that joins the beginning and the end of the time interval concerned (that is, the red line on Figure 7). For the part of the student's journey up to the moment when the bus arrives at the stop near home, the distance travelled is 800 m and the time taken is 870 s (including the wait at the stop), so the average speed is 0.92 m s^{-1}.

Everything written here about average and instantaneous speeds can also refer to average and instantaneous velocities. Remember, of course, to include the directions when quoting these measurements.

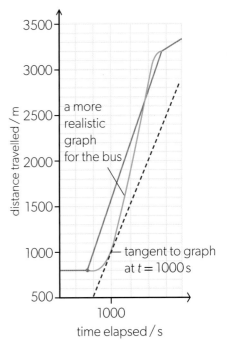

▲ Figure 7 A portion of the distance–time graph for the bus.

Worked example 5

The graph shows how the distance travelled by an object varies with time.

a. Calculate the instantaneous speed of the object at:

   i.   10 s

   ii. 30 s.

b. Calculate the average speed for the whole 80 s of the motion.

c. Outline why the information on the graph is insufficient to determine the average velocity of the object.

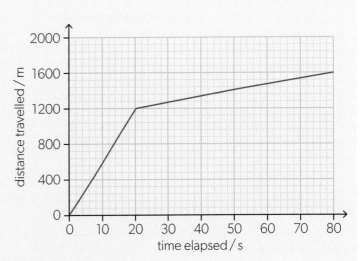

Solutions

a. i.   The speed of the object is constant from 0 to 20 s and the instantaneous speed at 10 s is equal to the gradient of the first straight section of the graph.

         Speed at 10 s $= \dfrac{1200}{20} = 60$ m s^{-1}.

ii. The speed of the object is constant again from 20 to 80 s and equal to the gradient of the second straight section of the graph.

$$\text{Speed at 30 s} = \frac{1600 - 1200}{80 - 20} = 6.7\,\text{m s}^{-1}.$$

b. The average speed is $\dfrac{\text{total distance}}{\text{time taken}} = \dfrac{1600}{80} = 20\,\text{m s}^{-1}.$

c. The graph shows the distance along the path travelled by the object but not whether the object changes direction as it moves, and any change in the direction would affect the displacement and the average velocity.

Worked example 6

Emma runs along a straight track with a constant speed of $2.6\,\text{m s}^{-1}$ for 30 s. She then stops for 5.0 s and runs in the opposite direction with a constant speed of $3.8\,\text{m s}^{-1}$ for a further 15 s. Calculate:

a. the total distance run by Emma

b. her displacement at the end of the run

c. the average speed

d. the average velocity.

Solutions

a. Emma ran $2.6 \times 30 = 78$ m in the first part and $3.8 \times 15 = 57$ m in the second part. The total distance is therefore 135 m.

b. The displacement is $78 - 57 = 21$ m. The run ended 21 m from the starting point.

c. The total time is 50 s so the average speed is $\dfrac{\text{total distance}}{\text{time taken}} = \dfrac{135}{50} = 2.7\,\text{m s}^{-1}.$

d. The average velocity is $\dfrac{\text{displacement}}{\text{time taken}} = \dfrac{21}{50} = 0.42\,\text{m s}^{-1}$ and has the same direction as her original velocity in the first part of the run.

Practice questions

7. An ice hockey puck is hit and slides across ice. The distance–time graph for the puck is shown.

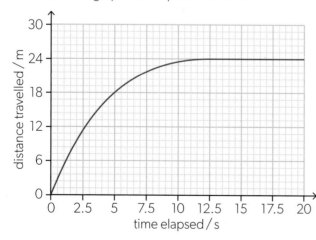

a. Estimate the:
    i. initial speed of the puck
    ii. speed of the puck at 5.0 s.
b. The puck stops after 12.5 s. Calculate the average speed of the puck during the motion.

8. Stephen runs half a lap on a circular track of radius 25 m in 19 s. Calculate his:
    a. average speed
    b. average velocity.

Acceleration

In real journeys, instantaneous speeds and velocities change frequently. Again, you need a mathematical language that helps you to understand the rates of change.

The rate of change of velocity is called **acceleration**. Acceleration is a vector. It is derived from the vector quantity velocity. Sometimes you can write the "magnitude of the acceleration" meaning the size of the acceleration ignoring its direction.

The definition of acceleration is:

$$\text{acceleration} = \frac{\text{change in velocity}}{\text{time taken for the change}}$$

so the units of acceleration are $\frac{m\,s^{-1}}{s}$, which is written as $m\,s^{-2}$. Sometimes you will see this written as m/s^2. However, $m\,s^{-2}$ is preferred in IB Diploma Programme physics, as using the solidus (/) can be ambiguous.

It is important to understand what acceleration means, not just to be able to use it in an equation. When an object has an acceleration of $5\,m\,s^{-2}$, then, for every second it travels, its velocity increases in magnitude by $5\,m\,s^{-1}$ in the direction of the acceleration vector.

For example, the Japanese N700 train has a quoted acceleration of $0.72\,m\,s^{-2}$. Assume that this is a constant value (very unlikely). One second after starting from rest, the speed of the train will be $0.72\,m\,s^{-1}$. One second later (at 2 s from the start) the speed will be $0.72 + 0.72 = 1.44\,m\,s^{-1}$. At 3 s it will be $2.16\,m\,s^{-1}$ and so on. Each second the speed increases by $0.72\,m\,s^{-1}$.

In a similar way to finding average and instantaneous values for speed and velocity, you can find **average acceleration** and **instantaneous acceleration**.

$$\text{average acceleration} = \frac{\text{overall change in velocity}}{\text{time taken for the overall change}}$$

whereas the instantaneous acceleration is the gradient of the tangent to a speed (or velocity)–time graph and is represented symbolically as $\frac{dv}{dt}$ or $\frac{\Delta v}{\Delta t}$.

Worked example 7

How many seconds will it take the N700 to reach its maximum speed of $300\,km\,h^{-1}$ on the Sanyo Shinkansen route?

Solution

$300\,km\,h^{-1} \cong 83.3\,m\,s^{-1}$

Time taken to reach the maximum speed: $\frac{83.3}{0.72} = 116\,s$, just under 2 minutes.

🧪 Spreadsheet models

- Tool 2: Use spreadsheets to manipulate data.
- Tool 2: Represent data in a graphical form.
- Tool 3: Determine the effect of changes to variables on other variables in a relationship.
- Inquiry 2: Interpret diagrams, graphs and charts.

→

One powerful way to help you think about acceleration (and other quantities that change in a predictable way) is to model them using a spreadsheet.

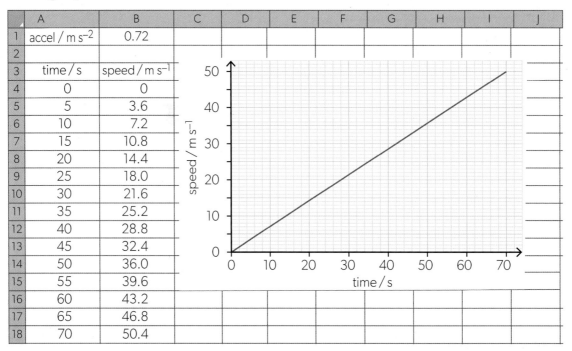

	A	B	C	D	E	F	G	H	I	J
1	accel / m s⁻²	0.72								
2										
3	time / s	speed / m s⁻¹								
4	0	0								
5	5	3.6								
6	10	7.2								
7	15	10.8								
8	20	14.4								
9	25	18.0								
10	30	21.6								
11	35	25.2								
12	40	28.8								
13	45	32.4								
14	50	36.0								
15	55	39.6								
16	60	43.2								
17	65	46.8								
18	70	50.4								

This is a spreadsheet model for the N700 train.

- The value of the acceleration is in cell B1.

- Cells A4 to A18 give the time in increments (increases) of 5.0 s; the computed speed at each of these times is in cells B4 to B18. The speed is calculated by taking the *change* in time between the present cell and the one above it, and then multiplying by the acceleration.

- The formula in cell B5 is "=B4+$B$1*(A5-A4)" and this is copied vertically down cell-by-cell so that cell B6 is "=B5+$B$1*(A6-A5)" and the last cell B18 is "=B17+$B$1*(A18-A17)". (The acceleration is

written as $B$1 so that the spreadsheet uses this cell every time and does not change the cell every time the new speed is calculated. This is achieved by using the dollar sign $ in the cell reference.)

- Finally, a graph is inserted into the spreadsheet to show speed against time and the equation of the straight line. The equation for the line can be computed by the program and added to the graph to confirm that the gradient of the line is 0.72 m s⁻².

- Construct this spreadsheet model yourself. Try changing the value of the acceleration in cell B1 and seeing the effect on the graph.

Graphing motion II—Velocity–time

Distance–time graphs are convenient for displaying speed and velocity changes. Similarly, if you plot speed (or velocity) against time, this displays the changes in acceleration.

Figure 8 shows the journey of a bicycle that is travelling in a straight line. For the first 10 s, the bicycle accelerates at a uniform rate to a velocity of +4 m s⁻¹. A positive sign here means that the velocity (and later the acceleration) is directed to the right. From 10 s to 45 s the bicycle moves at a constant velocity of +4 m s⁻¹. At 45 s the cyclist applies the brakes so that the bicycle stops in 5 s. The bicycle is then stationary for 10 s.

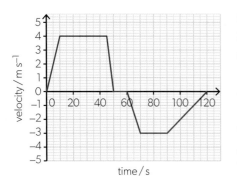

▲ Figure 8 The velocity–time graph for the bicycle ride.

The velocity from 60 s onwards is negative, meaning that the bicycle is travelling in the opposite direction (to the left). The pattern is similar. The bicycle accelerates until the velocity is $-3\,\mathrm{m\,s^{-1}}$, moves at constant velocity for a period and, finally, decelerates to a stop at 120 s.

The magnitude of the **gradient of a velocity–time graph** gives the magnitude of the acceleration, and the sign of the gradient gives the direction in which the acceleration acts.

From 45 s to 50 s the velocity goes from $4\,\mathrm{m\,s^{-1}}$ to 0, and so the acceleration is

$$\frac{\text{final speed} - \text{initial speed}}{\text{time taken for speed change}} = \frac{0 - 4}{5} = -0.80\,\mathrm{m\,s^{-2}}$$

From 90 s to 120 s the magnitude of the acceleration is

$$\frac{\text{speed change}}{\text{time taken}} = \frac{3.0}{30} = 0.10\,\mathrm{m\,s^{-2}}$$

Care is needed with the sign of the acceleration. The gradient of the graph is positive, as is the acceleration, because the bicycle is moving in the negative direction and is slowing down. This simply means that a force acts to the right on the bicycle, in the positive direction, which is slowing the bicycle down. You will see how force leads to acceleration in Topic A.2.

The **area under a velocity–time graph** provides more information. It tells you the total displacement of the moving object. Remember that the product of *velocity × time* is a displacement (and that the product of *speed × time* is a distance). The units tell you this too: when the units of speed and time are multiplied, the seconds cancel to leave only metres:

$$\frac{\text{metre}}{\text{second}} \times \text{second} \rightarrow \text{metre}$$

In the case of a graph with uniform accelerations, the areas, and hence the displacements (distances), are straightforward to calculate. Divide the graph into right-angled triangles and rectangles and then work out the areas for each individual part. This working is shown in Figure 9.

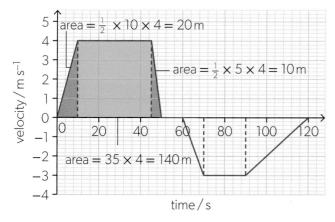

▲ **Figure 9** The velocity–time graph of Figure 8 broken up into its areas.

The individual areas and their calculations are shown on the diagram. Up to 60 s the area is $(20 + 140 + 10 + 0) = 170\,\mathrm{m}$. The area from 60 s until the end is $(-15 - 60 - 45) = -120\,\mathrm{m}$. As usual, the negative sign indicates motion in the opposite direction to the original.

The total distance travelled by the cyclist is $170 + 120 = 290\,\text{m}$ (this is the total ground covered as usual for "distance"). The total displacement after 120 s is $170 - 120 = 50\,\text{m}$. The cyclist (who was travelling along a straight line) is 50 m from the starting point after 120 s.

When a velocity–time graph is non-linear, you:

- estimate the number of squares
- determine the area (distance) for one square
- multiply the number of squares by the area of one square.

This will usually give you an estimate of the overall distance.

Figure 10 gives an example of how this is done.

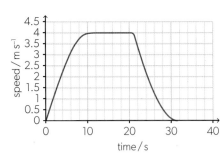

▲ **Figure 10** When a graph line is curved, count the number of squares to estimate the area under the curve.

There are approximately 85 squares between the time axis and the line. Each of the squares is 2 s along the time axis and $0.5\,\text{m s}^{-1}$ along the speed axis. The area of one square is equivalent to $(2 \times 0.5) = 1.0\,\text{m}$ of distance. The total distance travelled is 85 m (or, at least, somewhere between 80 and 90 m).

Worked example 8

The velocity–time graph of an object moving in a straight line is shown. A positive velocity means that the object is moving to the right.

a. Describe the motion of the object from 0 to 50 s.

b. Calculate:

   i.   the total distance travelled to the right

   ii. the acceleration from 20 s to 70 s

   iii. the displacement at 100 s.

c. Determine at what time the object passes the starting position.

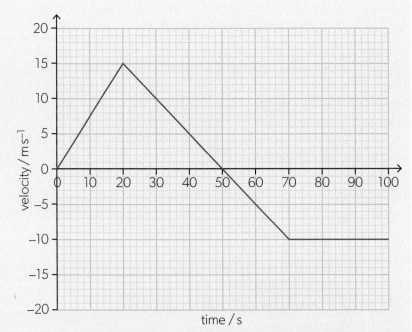

Solutions

a. The object starts at rest and accelerates uniformly to the right for the first 20 s, then decelerates uniformly for the next 30 s, still moving to the right. At 50 s the object is instantaneously at rest.

b. i.   The object is moving to the right whenever its velocity is positive, so the total distance travelled in this direction is equal to the area under the velocity–time graph from 0 to 50 s. Distance $= \dfrac{1}{2} \times 50 \times 15 = 375\,\text{m}$.

   ii. The acceleration is constant from 20 to 70 s and equals the slope of this section of the graph.

   $$\text{Acceleration} = \frac{\text{final velocity} - \text{initial velocity}}{\text{time taken}} = \frac{(-10) - (+15)}{70 - 20} = -0.50\,\text{m s}^{-2}.$$

   The negative sign indicates that the velocity is becoming more negative, but note that the object first decelerates (while moving to the right) from 20 s to 50 s and then accelerates (to the left) from 50 s to 70 s.

iii. From 50 to 100 s the object is moving to the left and you need to subtract the distance moved in this direction from the result of part (i).

The distance is equal to the area between the graph and the time axis from 50 s to 100 s:

$\frac{1}{2} \times 20 \times 10 + 30 \times 10 = 400$ m.

The final displacement = $375 - 400 = -25$ m. The negative sign indicates that the object is to the left from the starting position.

c. At 100 s the object is moving away from the starting position at a constant speed. It takes 2.5 s to travel 25 m at a speed of 10 m s^{-1}, so the object must have passed the starting position at 97.5 s.

Practice questions

9. The graph shows how the speed of a bicycle varies with time.

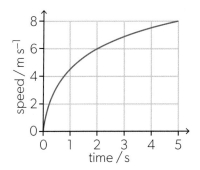

What is the best estimate of the distance travelled by the bicycle during the first 5 s?

A.   15 m   B.   20 m
C.   30 m   D.   40 m

10. For the bicycle in question 9, what is the best estimate of the instantaneous acceleration at 2.0 s?

A.   1.0 m s^{-2} B.   2.0 m s^{-2}
C.   3.0 m s^{-2} D.   6.0 m s^{-2}

11. The speed–time graph of a sprint runner is shown.

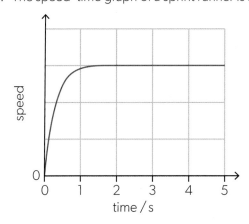

What are the instantaneous speed and instantaneous acceleration of the runner at a time of 4.0 s?

	Instantaneous speed at 4 s	Instantaneous acceleration at 4 s
A.	greater than the average speed	zero
B.	equal to the average speed	non-zero
C.	greater than the average speed	non-zero
D.	equal to the average speed	zero

12. The graph shows the variation with time of the velocity of a cart moving along a straight track. The cart starts from rest.

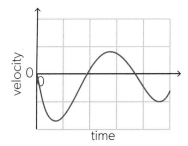

How many times does the cart change its direction of motion?

A.   1   B.   2   C.   3   D.   4

13. The graph shows how the velocity of a cart moving on an inclined straight track varies with time.

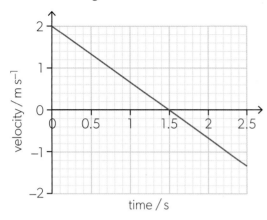

a. Describe the motion of the cart during the first 2.5 s.

b. Calculate:
   i. the acceleration of the cart
   ii. the maximum distance of the cart from its starting position
   iii. the displacement of the cart at 2.5 s.

c. Sketch a graph to show the variation of the displacement of the cart with time.

suvat equations

- Tool 3: Select and manipulate equations.

- Tool 3: Identify and use symbols stated in the guide and the data booklet.

The kinematic equations use a consistent set of symbols for the quantities. The table gives the list.

Symbol	Quantity
s	displacement/distance
u	initial (starting) velocity/speed
v	final velocity/speed
a	acceleration
t	time taken to travel distance s

The list of symbols in the table spell out *suvat*. The equations are sometimes also known by this name.

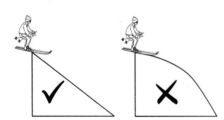

▲ Figure 11 Motion down a slope of constant gradient means you can use the kinematic equations. When the slope changes, they should not be used.

The kinematic (*suvat*) equations of motion

The graphs of distance–time and speed–time give a set of **kinematic equations of motion** that predict the values of the parameters in motion. These also help you to understand the connection between the various quantities introduced so far.

The kinematic equations *only* apply when the acceleration is constant, when the graph of speed against time for the motion is straight.

It is easy to forget the strict rule of constant acceleration for the use of *suvat* equations. Figure 11 shows two examples that look similar but need to be treated in completely different ways. You can use the *suvat* equations in the left-hand diagram because the skier's acceleration is constant.

Deriving the kinematic equations

The derivation begins from a simple graph of speed against time for a constant acceleration from an initial velocity u to a final velocity v during a time t (Figure 12).

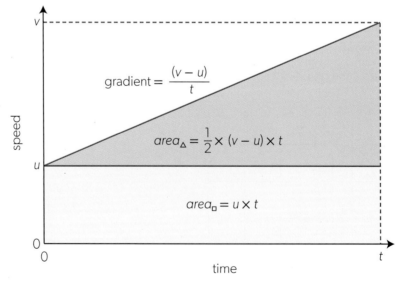

▲ Figure 12 Deriving the first two kinematic equations.

When can problems on projectile motion be solved by applying conservation of energy instead of kinematic equations?

In the left-hand diagram a skier is about to ski down a slope of constant gradient. As you will see in Topic A.2, this means that the force is constant and that the kinematic equations can be used because the acceleration is constant. In the right-hand diagram the slope varies throughout the ski-run. It is poor physics to try to use *suvat* here because the acceleration is not constant. You must, strictly speaking, use the energy transfers and the concept of energy conservation (which you will meet in Topic A.3).

The gradient of the graph gives the acceleration: $a = \dfrac{\text{change in speed}}{\text{time taken for speed change}}$

The change in speed is $v - u$, the time taken is t, so $a = \dfrac{v-u}{t}$. This can be rearranged to:

$$v = u + at \qquad \textbf{first equation of motion}$$

The area under the speed–time graph gives the distance. The graph from 0 to t is made up of two parts: the lower rectangle, $area_\square$ and the upper right-angled triangle, $area_\triangle$.

$$area_\triangle = \frac{1}{2} \times base \times height = \frac{1}{2} \times t \times (v-u) = \frac{1}{2} \times t \times at$$

$$area_\square = \frac{1}{2} \times base \times height = ut$$

distance $s = $ total area $= area_\square + area_\triangle = ut + \dfrac{1}{2} \times (at) \times t$

leading to

$$s = ut + \frac{1}{2}at^2 \qquad \textbf{second equation of motion}$$

The first equation of motion does not contain the distance s. The second equation has no final velocity v. There are three more possible equations, one with a missing t and one with a missing a. The third has a missing u but is not often used.

To eliminate t from the first and second equations, rearrange the first equation in terms of t:

$$t = \frac{v-u}{a}$$

This can then be substituted into the second equation:

$$s = u\left(\frac{v-u}{a}\right) + \frac{1}{2}a\left(\frac{v-u}{a}\right)^2$$

so that

$$as = u(v-u) + \frac{1}{2}(v-u)^2 = uv + \frac{1}{2}v^2 + \frac{1}{2}u^2 - \frac{1}{2} \times 2uv$$

and therefore

$$2as = v^2 - u^2$$

or

$$v^2 = u^2 + 2as \qquad \textbf{third equation of motion}$$

ATL **Research skills**

Modern scientists are careful to reference their sources. However, this has not always been so.

The fourth equation of motion is often called the mean-speed theorem. It is attributed to Galileo (1564–1642), but it has its origins many centuries before. It was proved by Nicole Oresme (c.1320–1382) and was known by the Oxford Calculators (see page 13). The work of these early scholars spread around Europe and influenced many other thinkers, but it was not often attributed to them.

The theorem's origins may lie centuries before that. A translation of a Babylonian tablet suggests that the ancient Babylonians were using a version of this rule to calculate the position of Jupiter. Whether they were the first to use this rule is unknown.

▲ Figure 13 A Babylonian tablet showing calculations of the position of Jupiter.

The derivation of the final equation is left to you as an exercise:

$$s = \left(\frac{v+u}{2}\right)t \qquad \text{fourth equation of motion}$$

There are two ways to approach this fourth proof. One way is to think about the meaning of the speed that corresponds to $\frac{v+u}{2}$ (it is the average speed over the time t) which applies over the whole of the motion. The second way is to take the third equation and amalgamate it with the first.

You will not be expected to remember these proofs or the equations themselves (which appear in the data booklet). They illustrate how useful graphs and equations are for solving kinematic problems.

Worked example 9

A driver of a car travelling at $25\,\text{m s}^{-1}$ along a road applied the brakes. The car comes to a stop in $150\,\text{m}$ with a uniform deceleration. Calculate:

a. the time the car takes to stop

b. the deceleration of the car.

Solutions

a. Start by writing down which values in the *suvat* equations you do and don't know from the question.

$s = 150\,\text{m}, u = 25\,\text{m s}^{-1}, v = 0, a = ?, t = ?$

To work out t, you need the fourth equation: $s = \left(\frac{v+u}{2}\right)t$

which rearranges to $t = \left(\frac{2}{v+u}\right)s$

Substituting the values in the question gives $t = \left(\frac{2}{25}\right)150 = 2 \times 6 = 12\,\text{s}$

b. To find a the equation $v^2 = u^2 + 2as$ is best.

Substituting: $0 = 25^2 + 2 \times a \times 150$

$a = -\frac{25 \times 25}{300} = -\frac{25}{12} = -2.1\,\text{m s}^{-2}$

The minus sign shows that the car is decelerating rather than accelerating.

Worked example 10

A cyclist slows uniformly from a speed of $7.5\,\text{m s}^{-1}$ to a speed of $2.5\,\text{m s}^{-1}$ in a time of $5.0\,\text{s}$. Calculate:

a. the acceleration

b. the distance moved in the $5.0\,\text{s}$.

Solutions

a. $s = ?; u = 7.5\,\text{m s}^{-1}; v = 2.5\,\text{m s}^{-1}; a = ?; t = 5.0\,\text{s}$

Use $v = u + at$ and therefore $2.5 = 7.5 + a \times 5.0$

So, $a = -\frac{5.0}{5.0} = -1.0\,\text{m s}^{-2}$

The negative sign shows this is a deceleration.

b. $s = ut + \frac{1}{2}at^2 = 7.5 \times 5.0 - \frac{1}{2} \times 1.0 \times 5.0^2 = 37.5 - 12.5 = 25\,\text{m}$

Practice questions

14. A cart is launched up a frictionless ramp with an initial speed of $3.0\,\mathrm{m\,s^{-1}}$. The cart moves with a constant acceleration of $1.8\,\mathrm{m\,s^{-2}}$ directed down the ramp. Calculate:
    a. the time the cart takes to return to the starting point
    b. the maximum distance from the starting point.

15. An aircraft starts its takeoff roll from rest and accelerates uniformly to a speed of $100\,\mathrm{km\,h^{-1}}$ in a time of $16\,\mathrm{s}$.
    a. Calculate, in $\mathrm{m\,s^{-2}}$, the acceleration of the aircraft.

    To take off, the aircraft must achieve a speed of $250\,\mathrm{km\,h^{-1}}$.
    b. Calculate the minimum length of the runway required for takeoff, giving the answer to the nearest $100\,\mathrm{m}$. Assume that the acceleration is constant during the entire takeoff.

16. A car needs a distance of $25\,\mathrm{m}$ to slow down to a speed of $12\,\mathrm{m\,s^{-1}}$ with a constant deceleration of $4.3\,\mathrm{m\,s^{-2}}$. Calculate the initial speed of the car.

17. An underground train can accelerate and slow down at a constant rate of $1.3\,\mathrm{m\,s^{-2}}$. The distance between two underground stations is $720\,\mathrm{m}$. Determine:
    a. the maximum speed the train can achieve between the stations
    b. the minimum time of travel between the stations.

18. The graph shows how the displacement of a uniformly accelerated bicycle varies with time.
    a. What is the acceleration of the bicycle and its instantaneous speed at $2.0\,\mathrm{s}$?

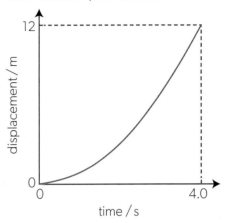

	Acceleration	Speed at 2.0 s
A.	$0.75\,\mathrm{m\,s^{-2}}$	$1.5\,\mathrm{m\,s^{-1}}$
B.	$0.75\,\mathrm{m\,s^{-2}}$	$3.0\,\mathrm{m\,s^{-1}}$
C.	$1.5\,\mathrm{m\,s^{-2}}$	$1.5\,\mathrm{m\,s^{-1}}$
D.	$1.5\,\mathrm{m\,s^{-2}}$	$3.0\,\mathrm{m\,s^{-1}}$

    b. Sketch the velocity–time graph for the bicycle.

 How effectively do the equations of motion model Newton's laws of dynamics?

In Topic A.2, you will meet the laws of motion constructed by Isaac Newton. These laws can be applied whenever the speed of the objects is much less than the speed of light (as explained in Topic A.5). The kinematic equations only apply when acceleration is uniform. Newton's second law of motion suggests that the kinematic equations will therefore only be applicable when the force is constant or the mass of the object does not change.

These two conditions are surprisingly rare in practice. The mass of moving objects changes (an automobile consumes fuel as it moves). The forces acting on an object change for all sorts of reasons: variations in mass, friction or air resistance, and so on. Modelling real motion is more difficult than the kinematic equations suggest.

 How are the equations for rotational motion related to those for linear motion?

In Topic A.4, you will meet the equations that apply to rotational motion—when objects are rotating about an axis. The definitions that set up rotational motion are deliberately chosen to mirror those of linear motion.

The concepts are similar and the use of these definitions leads to a parallel set of equations. This makes them easier to learn and use.

Projectile motion

Watch a dog catching a ball thrown high into the air. It is a remarkable feat of coordination by the animal. What is the physics of the motion? The ball is moving in two dimensions, and it is subject to the vertical acceleration of gravity and the deceleration of air resistance. How do physicists treat this complex situation? The trick they use is to split it up into horizontal and vertical components.

Acceleration due to gravity

Topic D.1 looks in greater detail at gravitational fields, but for the moment you can assume that there is a constant acceleration that acts on all bodies close to the surface of Earth.

When an object is released close to Earth's surface, it accelerates downwards. The force of gravity acts on the object, pulling it towards the centre of Earth. Equally, the object pulls with the same force on Earth in the opposite direction. Not surprisingly, with small objects, the effect of the force on Earth is so small that you do not notice it.

The acceleration due to gravity at Earth's surface is given the symbol g. The accepted value varies from place to place on the surface. For example, in Kuala Lumpur g is $9.776\,\mathrm{m\,s^{-2}}$ whereas in Stockholm it is $9.818\,\mathrm{m\,s^{-2}}$. This is because Earth is not a perfect sphere (it is slightly flattened at the poles) and the densities of the rocks in different locations vary. The different tangential speeds of Earth at different latitudes also have an effect. It is better to buy gold by weight at the equator and sell it at the North Pole rather than the other way round—of course, buying by mass makes no difference!

Data-based questions

In 2012, the Red Bull Stratos project set the record for the highest altitude parachute jump when Felix Baumgartner jumped from an altitude of almost 39 km. This record was subsequently broken by Alan Eustace in 2014.

$h\,/\,\mathrm{m}$	$v\,/\,\mathrm{m\,s^{-1}}$
38 965	6.7
38 960	11.4
38 955	14.4
38 949	18.3
38 945	20.3
38 940	21.7
38 936	23.6

The table shows Baumgartner's speed v at different altitudes h above Earth's surface.

- Plot a graph of v^2 against h.

- Find the gradient of the graph.

- Deduce what the gradient represents. (*Hint*: use $v^2 = u^2 + 2as$.)

- The absolute uncertainty in the given speeds is $\pm 1\,\mathrm{m\,s^{-1}}$. Calculate the uncertainties in the values of v^2 and add error bars to your graph.

- By considering the maximum and minimum gradients of your graph, deduce whether the data are consistent with an acceleration due to gravity of $9.8\,\mathrm{m\,s^{-2}}$.

- When you have studied Topic D.1 you will be able to calculate the acceleration due to gravity g at an altitude of 39 km. Show that it is only about 1% lower than g at the surface of Earth.

Models—The object's shape

You can find out about using error bars in the *Tools for physics* section on page 358.

Scientists use models to represent natural phenomena. The kinematic equations make up a particularly simple model. The requirement of uniform acceleration is already clear, but are there others?

The answer is yes. The equations implicitly ignore the shape of the object. For now, the objects are treated as moving points. The point in question can be the centre of mass. You will meet this idea later in Topic A.4, where you will re-examine the linear kinematic equations and draw parallels with a similar set of equations for rotational motion.

Once forces are acting on an object and the acceleration is changing, then you cannot use *suvat* anymore. The kinematic model has broken down and it needs to be improved. This is where Newton's second law of motion (Topic A.2) comes in because it allows you to link an acting force to an acceleration and, later, to the changing momentum of an object.

Measuring *g*

- Tool 2: Use sensors.

- Tool 3: Interpret features of graphs including gradient and intercepts.

- Inquiry 2: Collect and record sufficient relevant quantitative data.

- Inquiry 3: Compare the outcomes of an investigation to the accepted scientific context.

There are several ways to measure *g*. The first method uses a data logger to collect data. One of the problems with measuring *g* "by hand" is that the experiment happens quickly on Earth. Manual collection of the data is difficult.

Method 1

- An ultrasound sensor should be mounted to sense objects below it.

- The logging system must measure the speed of the object over a time of about 1 s. The time interval between measurements will need to be set. An interval of between 0.1 s and 0.01 s is best. The object, large enough to be detected by the sensor, is dropped vertically.

- The data logger should be set to output a speed–time graph. This is likely to be a straight line. The logger's software may be able to calculate the gradient for you.

- You could extend this experiment by testing objects of different mass but similar size and shape to confirm a suggestion by Galileo that such differences do not affect the drop.

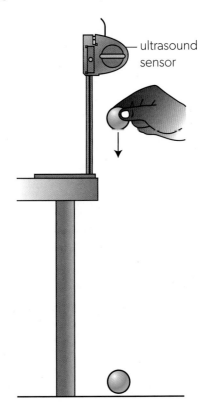

ultrasound sensor

▲ Figure 14 Measuring *g* using an ultrasound sensor.

Method 2

Another option does not involve a sensor and data logger.

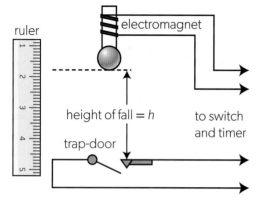

ruler

electromagnet

height of fall $= h$

trap-door

to switch and timer

▲ **Figure 15** When the current to the electromagnet is switched off, the steel ball falls and opens the trapdoor.

- A magnetic field is used to hold a small steel sphere (such as a ball bearing) between two metal contacts. The magnetic field is produced by a coil of wire with an electric current in it. When the current is switched off, the field disappears, and the sphere falls vertically.

- As the sphere leaves the metal contacts, an electronic circuit starts a clock. The clock stops when the sphere opens a small trapdoor and breaks the

connection between the terminals of a timing clock or computer. (The exact details of these connections will depend on your equipment.)

- This system measures the sphere's time of flight t between the contacts and the trapdoor.

- The distance h from the bottom of the sphere to the top of the trapdoor is needed. (You should think about why these are the appropriate measurement points.)

- One possible analysis for the data is to measure t for one value of h—with repeat readings for the same h. Then, to calculate g, use $h = \frac{1}{2}at^2$ as $u = 0$. However, this is a one-off measurement that is prone to error. Think of some reasons why.

- A way to reduce the errors is to change the vertical distance h between the sphere and trapdoor and to plot a graph of h against t^2. The gradient of the graph is $\frac{g}{2}$. What do you think an intercept on the h-axis represents?

Method 3

A further method to estimate g could include making a video of a falling object against a fixed calibrated scale. The image should include a clock. There is an example of such an image later in this topic.

Projectile motion in two-dimensions

So far, you have assumed that things are moving in one dimension—along a straight line and with no air resistance. While this is often the case, there are also important examples of objects that move in a circle (see Topic A.2) or that are projected into the air. The rest of this topic looks at objects moving in two dimensions, with and without air resistance.

A baseball is thrown vertically upwards with an initial speed U. Gravity acts on the baseball from the moment of its release, slowing it down until it stops for an instant at the top of its motion. Gravity continues to act and the baseball now accelerates downwards to reach the ground with the same speed at which it was released. Without air resistance, the displacement–time graph would look like Figure 16.

The ball goes vertically up and then down to land in the same spot from which it was projected. The path in the air is called the **trajectory** and is a vertical line up and down for this case.

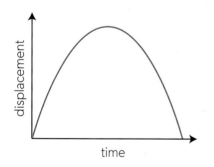

displacement

time

▲ **Figure 16** A displacement–time graph for a baseball thrown vertically upwards in the air. Remember that this is a graph of vertical displacement against time, not the shape of the path the ball makes in the air.

A distance–time graph would look different (Figure 17). It gives similar information but without the direction part of the displacement and velocity vectors. Make sure that you understand the difference between these graphs.

The *suvat* equations introduced earlier can be used to analyse this motion. The initial vertical speed is U, the time to reach the highest point is T, the maximum height is H and the acceleration of the ball is $-g$; g has a negative sign because upwards is the positive direction. As the acceleration due to gravity is downwards,

g must have the opposite, that is, negative, sign. The kinematic equations are printed again but with differences to reflect the vertical motion to the highest point:

$$0 = U - gT \qquad \text{which comes from } v = u + at$$

$$H = UT - \frac{1}{2}gT^2 \qquad \text{which comes from } s = ut + \frac{1}{2}at^2$$

$$0 = U^2 - 2gH \qquad \text{which comes from } v^2 = u^2 + 2as$$

The time for the entire motion (that is, up to the highest point and then back to Earth again) is simply $2T$.

Figure 18(a) shows the speed–time graph for the baseball, while Figure 18(b) shows the velocity–time graph.

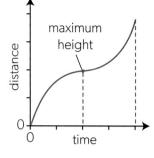

▲ Figure 17 The distance–time graph for the motion of Figure 16.

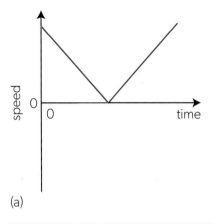

(a)

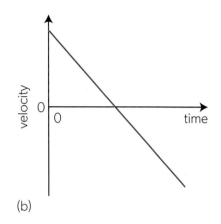

(b)

◄ Figure 18 (a) The speed-time and (b) the velocity-time graphs for an object thrown vertically upwards.

Worked example 11

A student drops a stone from rest at the top of a well. She hears the stone splash into the water at the bottom of the well 2.3 s after releasing the stone. Ignore the time taken for the sound to reach the student from the bottom of the well. The acceleration due to gravity g is $9.8\,\text{m s}^{-2}$.

a. Calculate the depth of the well.

b. Calculate the speed at which the stone hits the water surface.

c. Explain why the time taken for the sound to reach the student can be ignored.

Solutions

a. $u = 0; t = 2.3\,\text{s}$

$$s = ut + \frac{1}{2}at^2$$

$$s = 0 + \frac{1}{2} \times 9.8 \times 2.3^2 = 26\,\text{m}$$

b. $v = u + at = 0 + 9.8 \times 2.3 = 23\,\text{m s}^{-1}$

c. The speed of sound is about $300\,\text{m s}^{-1}$ and so the time to travel about 25 m is about 0.08 s. This is only about 4% of the time taken for the stone to fall.

Worked example 12

A hot-air balloon is rising vertically at a constant speed of $5.0\,\text{m s}^{-1}$. A small object is released from rest relative to the balloon when the balloon is 30 m above the ground. Calculate:

a. the maximum height of the object above the ground

b. the time taken to reach the maximum height

c. the total time taken for the object to reach the ground.

Signs

The signs of the quantities in the kinematic equations are important. For vertical motion, treat upwards as positive and downwards as negative. Something else that is easy to forget is that, at the top of the motion, the vertical speed of the object is zero.

Solutions

a. The object is moving upwards at +5.0 m s⁻¹ when it is released. The acceleration due to gravity is –9.8 m s⁻².

When the object is released, it will continue to travel upwards, but this upwards speed will decrease under the influence of gravity. When it reaches its maximum height, it will stop moving and then begin to fall.

$$v^2 = u^2 + 2as, \text{ so } s = \frac{v^2 - u^2}{2a} = \frac{0 - 5^2}{2 \times (-9.8)} = +1.3\,m$$

This shows that the object rises a further 1.3 m above its release point, and is therefore 31.3 m above ground at the maximum height.

b. $v = u + at$ so $t = \frac{v - u}{a} = \frac{0 - 5}{-9.8} = +0.51\,s$ (The plus sign shows that this is 0.51 s after release.)

c. After reaching the maximum height (at which point the speed is zero) the object falls with the acceleration due to gravity.

$s = -31.3\,m, u = 0, v = ?, a = -9.8\,m s^{-1}, t = ?$

Notice that s is negative because it is in the opposite direction to the upwards + direction.

Using $s = ut + \frac{1}{2}at^2$ gives a value for t of ± 2.53 s. The positive value is the one to use. Think about what the negative value stands for.

So the total time is the 0.51 s to get to the maximum height together with the 2.53 s to fall back to Earth.

This gives a total of 3.04 s, which rounds to 3.0 s.

Notice that, in this example, if you carry the signs through consistently, they give you information about the motion of the object.

Calculating horizontal and vertical motion

You can assume that:

- The surface of Earth is large enough for its surface to be considered locally flat.
- There is no friction or air resistance.

Gravity acts vertically and does not affect motion in the horizontal direction. This will be important when you combine horizontal and vertical motions later. Because the horizontal acceleration is zero, the *suvat* equations are simple.

For horizontal motion:

- The horizontal velocity does not change.
- The horizontal distance travelled is *horizontal speed × time for the motion*.

Topic D.1 goes into more details of what happens if Earth's surface is not considered to be flat.

A student throws a ball horizontally. Figure 19 shows multiple stroboscopic images of the ball every 0.10 s as it moves through the air. The figure also shows, for comparison, the image of a similar ball dropped vertically at the same moment as the ball is thrown.

It is obvious which ball was thrown horizontally. Careful examination of the images of this ball should convince you that the *horizontal* distance between them is constant. When the time interval between images and the distance scale on the picture are known, then you can work out the initial (and unchanging) horizontal speed.

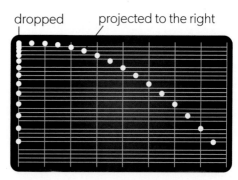

▲ Figure 19 These are multiple images of a ball's motion taken every tenth of a second. A stroboscope is used in a dark room to illuminate the ball at regular time intervals.

The images tell you about the vertical speeds too. Concentrate on the ball that was dropped vertically. The distance between vertical images (strictly, between the same point on the ball in each image) is increasing. The distance s travelled varies with time t from release varies as $s \propto t^2$. This means that when t doubles, then s increases by factor of four. Does it look as though this is what happens? Careful measurements from Figure 21 followed by a plot of s against t^2 could help you to confirm this.

Horizontal motion and vertical motion are completely independent of each other.

The horizontal speed continues unchanged (remember that you are assuming no air resistance) while the vertical speed changes as gravity accelerates the ball. This independence allows a straightforward analysis of the motion. The horizontal and the vertical parts of the motion can be split up and treated separately. Then they are re-combined to determine the velocity and the displacement for the whole of the motion.

The position is summed up in Figure 20 which is a simplified version of Figure 19. At two positions along the trajectory, the separate components of velocity are shown in solid green for the ball projected to the right. From these the resultant vectors (the velocity including direction) are drawn as dashed arrows.

To achieve the best range (best overall horizontal distance travelled), a ball should be projected upwards into the air at an angle to the horizontal. The general principles above still allow you to analyse the situation using the kinematic equations.

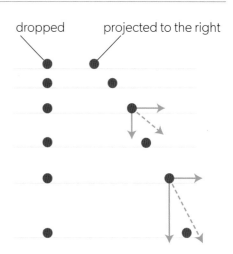

▲ Figure 20 The motion of the ball in Figure 19 with velocity vectors added to scale to two images.

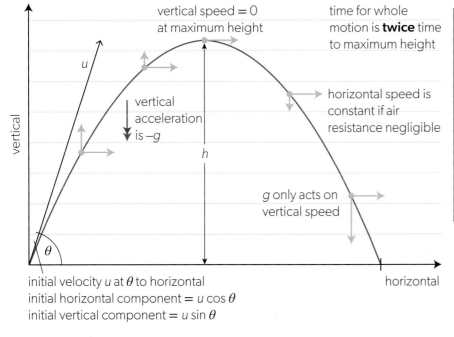

vertical speed = 0 at maximum height

time for whole motion is **twice** time to maximum height

vertical acceleration is $-g$

h

horizontal speed is constant if air resistance negligible

g only acts on vertical speed

initial velocity u at θ to horizontal
initial horizontal component $= u \cos \theta$
initial vertical component $= u \sin \theta$

horizontal

maximum height:

$$h = \frac{u^2 \sin^2 \theta}{2g}$$

range:

$$R = 2Tu \cos \theta$$

T is the time to reach the highest point:

$$T = \frac{u \sin \theta}{g}$$

◀ Figure 21 An analysis of projectile motion.

Figure 21 shows the trajectory of an object (vertical distance against horizontal distance) that has an initial speed u at an angle of θ to the ground. As usual, you assume no air resistance. The first step in the analysis of this motion is to resolve the initial velocity into two components, horizontal and vertical.

The horizontal speed is $u \cos \theta$. The initial vertical speed is $u \sin \theta$.

Horizontal motion

The horizontal speed is constant throughout and remains at its initial value of $u \cos \theta$. When the time that the object takes to reach the highest point is T, then the horizontal distance travelled R (known as the **range**) is $2Tu \cos \theta$.

Analysis of the motion requires the resolution and addition of vectors. Details of how to do this are shown on page 340 of the *Tools for physics* section.

Vertical motion

This is more complicated because gravity acts on the object throughout the motion. The kinematic equations must be used to calculate the vertical position of the object for any time t.

- The vertical speed v_{vert} at t is $(u \sin \theta - gt)$ [from $v = u + at$].
- The vertical position s_{vert} at t is $ut \cos \theta - \dfrac{gt^2}{2}$.
- At the highest point of the trajectory, $v_{vert} = 0$ so $u \sin \theta = gt$, which leads to
  $$T = \frac{u \sin \theta}{g}.$$
- The highest point occurs halfway through the motion which is reached at time T so that $2T$ (the time taken for the object to return to the ground) is given by
  $$2T = \frac{2u \sin \theta}{g}.$$
- The highest point h is given by $h = \dfrac{u^2 \sin^2 \theta}{2g}$ from solving the third kinematic equation $v_{vert}^2 = 0 = (u \sin \theta)^2 - 2gh$.

The **trajectory** is a **parabola** because the vertical position of the object varies with t^2 whereas the horizontal position varies with t.

Study Figure 21 carefully and apply the ideas in it to any projectile problems you need to solve.

In Topic A.3, you will see how problems of this type can be solved by considering energy transfers instead of kinematic equations. The energy-transfer approach is essential when the acceleration is not constant.

Worked example 13

An arrow is fired horizontally from the top of a tower 35 m above the ground. The initial horizontal speed is $30 \, \mathrm{m \, s^{-1}}$. Assume that air resistance is negligible. Calculate:

a. the time for which the arrow is in the air

b. the distance from the foot of the tower at which the arrow strikes the ground

c. the velocity at which the arrow strikes the ground.

Solutions

a. The time taken to reach the ground depends on the vertical motion of the arrow. At the instant when the arrow is fired, the vertical speed is zero. The time to reach the ground can be found using
$$s = ut + \frac{1}{2}at^2$$
$$t^2 = \frac{2 \times 35}{9.8}, \text{ so } t = 2.67 \, \mathrm{s} \text{ or } 2.7 \, \mathrm{s} \text{ to 2 s.f.}$$

b. The distance from the foot of the tower depends only on the horizontal speed.
$s = ut = 30 \times 2.67 = 80.1 \, \mathrm{m} \approx 80 \, \mathrm{m}$.

c. To calculate the velocity, the horizontal and vertical components are required. The horizontal component remains at $30 \, \mathrm{m \, s^{-1}}$. The vertical speed is calculated using $v = u + at = 0 + 9.8 \times 2.67 = 26.2 \, \mathrm{m \, s^{-1}}$.

The speed in the direction of travel can be found using Pythagoras' theorem because the vertical and horizontal components are at $90°$ to each other:
$$\sqrt{30^2 + 26.2^2} = 39.8 \, \mathrm{m \, s^{-1}} \approx 40 \, \mathrm{m \, s^{-1}}$$

The angle at which the arrow strikes the ground is $\tan^{-1}\left(\dfrac{26.2}{30}\right) = 41°$.

Worked example 14

An object is thrown horizontally from a ship and strikes the sea 1.6 s later at a distance of 37 m from the ship. Calculate:

a. the initial horizontal speed of the object

b. the height of the object above the sea when it was thrown.

Solutions

a. The object travelled 37 m in 1.6 s and the horizontal speed was $\frac{37}{1.6} = 23\,\text{m s}^{-1}$.

b. Distance above sea, $s = ut + \frac{1}{2}at^2 = 0 + 0.5 \times 9.8 \times 1.6^2 = 12.5\,\text{m}$.

Worked example 15

A stone is thrown towards a wall with an initial speed of $13\,\text{m s}^{-1}$ at an angle of $35°$ to the horizontal. The stone is initially 2.0 m above the ground and its horizontal distance to the wall is 12 m. The wall is 4.5 m high.

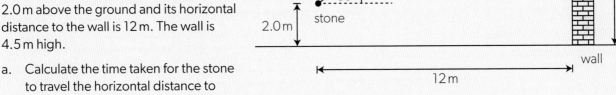

a. Calculate the time taken for the stone to travel the horizontal distance to the wall.

b. Show that the stone will hit the wall.

c. Deduce whether the stone reaches its maximum height before or after hitting the wall.

d. Calculate the speed with which the stone hits the wall.

Solutions

a. The horizontal speed of the stone is $13\cos 35° = 10.6\,\text{m s}^{-1}$ and remains constant during the motion, so the time to travel to the wall is $\frac{12}{13\cos 35°} = 1.1\,\text{s}$.

b. Find the height of the stone above the ground at the time 1.1 s and compare it to the height of the wall. The initial vertical speed of the stone is $13\sin 35° = 7.5\,\text{m s}^{-1}$ and the vertical motion is uniformly accelerated, with the acceleration $-9.8\,\text{m s}^{-2}$. At the wall, the vertical displacement (relative to the initial position) is given by

$13\sin 35° \times 1.1269 - \frac{1}{2} \times 9.8 \times (1.1269)^2 = 2.2\,\text{m}$, so the height above the ground is

$2.2 + 2.0 = 4.2\,\text{m}$. This is just below the top of the wall.

Note that the time value used in the calculation has a higher precision than the answer in part a. This avoids the accumulation of a rounding error.

c. The vertical velocity of the stone when it hits the wall is $13\sin 35° - 9.8 \times 1.1269 = -3.6\,\text{m s}^{-1}$. It is negative, which means that the stone is in the downward phase of motion and therefore it must have reached the maximum height before hitting the wall.

d. The speed is the magnitude of the velocity vector and can be calculated from its horizontal and vertical components. Speed $= \sqrt{10.6^2 + 3.6^2} = 11\,\text{m s}^{-1}$. This is less than the initial speed because the stone is now at a greater height.

Practice questions

19. The diagram shows trajectories of two projectiles launched from the same point. The projectiles reach the same maximum height.

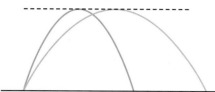

Which quantity has the same value for both projectiles?

A. initial speed

B. time of flight

C. horizontal component of initial velocity

D. speed at maximum height

20. An arrow is fired horizontally from height h and travels a horizontal distance s before striking the ground. Another arrow is fired with the same velocity and travels a horizontal distance $2s$.

What is the height from which the second arrow is fired?

A. $\sqrt{2}h$ B. $2h$

C. $2\sqrt{2}h$ D. $4h$

21. A stone is thrown vertically downwards from the top of a tower with an initial speed of $4.0\,\text{m s}^{-1}$. The stone hits the ground 1.9 s later. What is the height of the tower?

A. 7.6 m   B. 10 m

C. 18 m   D. 25 m

22. A steel ball rolls off a table of height h with a horizontal initial velocity v. What is the horizontal distance moved by the ball, from the edge of the table to the point where the ball hits the floor?

A. $\sqrt{\dfrac{h}{2g}} \times v$ B. $\sqrt{\dfrac{h}{g}} \times v$

C. $\sqrt{\dfrac{2h}{g}} \times v$ D. $\sqrt{\dfrac{h}{g}} \times 2v$

23. A football player kicks a ball into the air. The ball reaches its maximum height 0.90 s after the kick, and its horizontal displacement at that instant is 16 m.

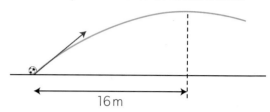

16 m

Calculate, for the ball:

a. its vertical component of the initial velocity

b. its initial speed

c. the angle to the horizontal at which it leaves the ground

d. its maximum height.

24. A tennis player serves a tennis ball towards the net. The ball leaves the racquet horizontally at a height of 2.70 m above the ground. The net is 0.900 m high and a distance of 12.0 m from the player. Determine the minimum initial speed of the ball so that it passes above the net.

25. Olaf throws a dart with a speed of $9.0\,\text{m s}^{-1}$ at an angle of 4.0° above the horizontal. The dart hits the bullseye on a dart board, at a height 0.25 m below the initial height of the dart.

a. Show that the time of flight of the dart is approximately 0.3 s.

b. Calculate the:

   i. distance between Olaf and the dart board

   ii. speed of the dart when it hits the bullseye.

c. Sketch graphs to show the variation with time of the dart's:

   i. horizontal velocity

   ii. vertical velocity.

How does the motion of a mass in a gravitational field compare to the motion of a charged particle in an electric field?

In Topic D.3 you will study the motion of charged objects moving in an electric field. When the electric field is uniform, then the force acting on the charged object is constant. You already know what will happen! The force and therefore the acceleration will be constant, so the motion will be parabolic; the *displacement* will vary with *time*^{2}. One branch of physics helps another!

How does the motion of an object change within a gravitational field?

The parabola is a member of the family of curves known as the conic sections. This is because each member of the family can be cut from part of a cone (Figure 22).

It is no accident that each of these shapes has an important role in motion. The parabola, ellipse and hyperbola all have a part in the orbits and escape trajectories of satellites and rockets. There is more detail about orbits and motion in a gravitational field in Topic D.1.

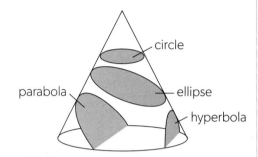

▲ **Figure 22** The various shapes that can be made from a cone. These can all be possible trajectories for an object moving in a gravitational field.

The parabolic shape arises from the nature of the gravitational field near to Earth's surface. This is because the (approximately) constantgravitational force leads to a uniform acceleration and the vertical displacement must be proportional to the (*time of flight*)^2. The horizontal displacement is proportional to time of flight, so (*vertical displacement*) $\propto$ (*horizontal displacement*)^2.

The other conic sections can be seen in other types of motion in a gravitational field and are explored in Topic D.1.

This result helps you to predict the trajectory for other types of field where the force is constant. An example is a charged particle that is moving perpendicular to a uniform electric field. The mathematics of the trajectory is similar and the particle also moves in a parabola. This is treated in Topic D.3.

How does a gravitational force allow for orbital motion?

The gravitational attraction between a satellite and Earth is always towards the centre of mass of the objects. The satellite is subjected to a force directed to Earth's centre. This is precisely the condition for the satellite to orbit Earth in a circular or an elliptical orbit. The nature of the force (called a centripetal force) is explored in Topic A.2.

Moving through fluids

The assumption that air resistance is negligible is usually unrealistic—ask any cyclist. An object that travels through a fluid "stirs" the fluid up. Both gases and liquids are called **fluids**. The object moving in the fluid is subject to a **viscous drag force**. The stirring process is complex so that, even after introducing some simple assumptions about the resistance of a fluid, it is difficult to give a complete analysis.

> The effects of the viscous drag forces themselves are explored in more detail in Topic A.2.

In energy terms, **air resistance** comes from the transfer of some of the kinetic energy of the moving body into the fluid through which it is moving. Some fluids absorb this energy better than others: swimming through water is much more tiring than running through the air.

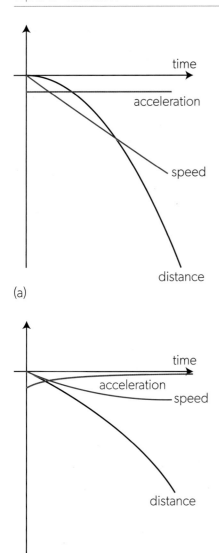

(a)

(b)

▲ **Figure 23** Graphs of distance, speed and acceleration against time for an object falling vertically from rest with (a) no air resistance and (b) air resistance that varies $\propto v^2$. The graphs are drawn to the same scale.

Air resistance in one dimension

Before tackling the more difficult case of projectile motion, it is worth looking at the effects of air resistance on the motion of an object falling vertically after being released from rest.

The effects of air resistance increase with the speed of an object. Eventually, at a high enough speed, the force resisting the movement of the object becomes equal to the gravitational pull downwards. The object cannot accelerate to a greater speed and so has reached a speed maximum. This value is known as the **terminal speed**.

Figure 23 shows graphs of the vertical motion for an object without air resistance (Figure 23(a)) and with air resistance (Figure 23(b)).

Figure 23(a) shows:

- a horizontal line for the acceleration because it is constant at $-9.8\,\mathrm{m\,s^{-2}}$

- a gradient for the speed graph that is constant; the speed is proportional to the time elapsed

- a curved parabolic line for the distance fallen as the displacement is proportional to (time elapsed)^2.

All of these are expected from earlier parts of this topic.

Figure 23(b) shows very different behaviour:

- The acceleration is $-9.8\,\mathrm{m\,s^{-2}}$ when time $t = 0$ but it then falls to zero as the drag force increases (with speed).

- The initial gradient of the speed graph is the same as that for the case without drag but then decreases. The speed becomes constant as the acceleration reaches zero. This constant speed is the terminal speed.

- The distance fallen graph is no longer parabolic and becomes straighter as the time increases. Note that the distance travelled in Figure 23(b) is much less at any given time when drag acts.

The result of all these changes is that:

- The time to fall a particular distance (the time of flight) increases.

- The acceleration varies, falling from g to zero as the speed increases.

- There is a terminal speed that depends on the dimensions of the object (this is explored in Topic A.2).

Worked example 16

The graph shows how the speed of a raindrop falling vertically to the ground varies with time.

a. Outline which feature of the graph indicates that the raindrop falls with a decreasing acceleration.

b. State the terminal speed of the raindrop.

c. Estimate the distance fallen by the raindrop during the first 2.0 s.

d. Calculate the distance fallen during the first 2.0 s by an object dropped from rest in the absence of air resistance. Compare this with your answer to part c.

e. The raindrop is formed in a cloud 240 m above the ground. Calculate the time the raindrop takes to fall to the ground.

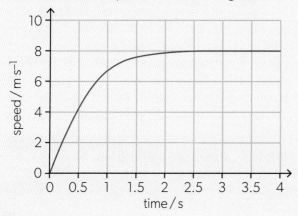

Solutions

a. Acceleration is equal to the gradient of the speed–time graph. The graph has a decreasing gradient, which means that the acceleration of the raindrop quickly decreases from the initial value of $9.8 \, \mathrm{m\,s^{-2}}$ to zero.

b. The terminal speed is $8 \, \mathrm{m\,s^{-1}}$, as indicated by the straight section of the graph after approximately 2 s.

c. The distance can be estimated by counting grid squares between the graph and the time axis. There are approximately 11.5 squares from 0 to 2 s, and each square is equivalent to $2 \, \mathrm{m\,s^{-1}} \times 0.5 \, \mathrm{s} = 1 \, \mathrm{m}$. Estimated distance $= 11.5 \, \mathrm{m}$.

d. Calculate the distance fallen in the absence of air resistance using kinematic (*suvat*) equations. Distance $= \frac{1}{2} \times 9.8 \times 2^2 = 19.6 \, \mathrm{m}$. For the raindrop, air resistance decreases this distance by more than 8 m.

e. Assume that after 2 s the raindrop falls with a constant speed of $8 \, \mathrm{m\,s^{-1}}$. The distance left is $240 - 11.5 = 228.5 \, \mathrm{m}$ and the time to reach the ground is therefore $2 + \frac{228.5}{8} = 31 \, \mathrm{s}$.

Observing the effects of air resistance

- Tool 3: Calculate mean and range.
- Tool 3: Construct and interpret tables and graphs for raw and processed data including scatter graphs and line and curve graphs.
- Inquiry 1: Justify the range and quantity of measurements.
- Inquiry 2: Interpret diagrams, graphs and charts.

This experiment compares the distance–time graphs for two objects that experience different amounts of air resistance. You need two objects to drop:

- One object should be small and dense such as a ball bearing. This object should experience negligible air resistance when dropped through small heights.
- The other object should be larger and lighter: a balloon with a mass of 2 or 3 g attached to it works well.

You are going to drop both objects from a range of heights. First, decide on this range. The maximum height will depend on the room you have available. A maximum height of 2 m will be sufficient, although 3 or 4 m would be better. You need at least eight different height measurements. Divide your maximum height by 8 to give suitable height increments which should ideally be larger than 20 cm.

- Drop the ball bearing from each height and measure the time taken to hit the floor. You should use a suitable method for the time measurement.

- Repeat each measurement three times and record your results.
- Repeat the experiment with the balloon.
- Take averages of your results.
- Plot a distance–time graph for the two objects on the same axes. (Note that time should be on the horizontal axis even though it is the dependent variable.)

Compare the distance–time graphs for the two objects. How do they differ? Is there evidence that the balloon reached a terminal speed?

You can also calculate the average speed over each interval and plot a speed–time graph. To do this, assume that the average speed for each interval occurs at the mid-point of each time interval. An example is shown in the table. The mid-point time for the ball falling between 0 and 0.2 m is 0.23 s (the average of 0 and 0.45 s). The average speed at this time is 0.44 m s^{-1} since the ball travelled 0.2 m in the 0.45 s time interval.

Height / m	Average time / s	Mid-point time / s	Average speed over interval / m s^{-1}
0	0		
0.2	0.45	0.23	0.44
0.4	0.61	0.53	1.25
0.6	0.71	0.66	2.00

Air resistance in two dimensions

The results are similar in two dimensions, as shown for the two cases in Figure 24. Here, the trajectory is shown for an object projected at 45° to the horizontal in a vacuum and with a drag force.

Comparisons of the two trajectories (Figure 24) and the two distance–time graphs (Figure 23) show that when air resistance is taken into account:

- The range is decreased.
- The trajectory is no longer parabolic.
- The maximum height is reduced.
- The vertical acceleration is no longer constant.
- There is now a horizontal deceleration.
- Speeds are reduced.
- The determination of the time of flight becomes complex and depends on the initial conditions of the motion.

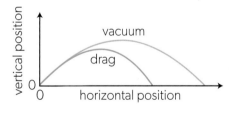

▲ Figure 24 Effect of drag on an object projected at 45° to the horizontal.

Worked example 17

A football is kicked with a speed of 25 m s^{-1} at an angle of 16° to the horizontal.

a. Assuming that air resistance is negligible, calculate for the football:

   i.   the maximum height reached

   ii. the range.

b. State the effect of air resistance on your answers to part a.

c. Discuss the effect of air resistance on the vertical acceleration of the ball.

d. Sketch a possible graph to show how the horizontal speed of the ball varies with time, including the effects due to air resistance.

Solutions

a. Use equations derived on page 31.

   i.   Maximum height $= \dfrac{25^2 \sin^2 16°}{2 \times 9.8} = 2.4\,\text{m}.$

   ii. Time taken to return to the ground $= \dfrac{2 \times 25 \times \sin 16°}{9.8} = 1.4\,\text{s}.$

        Range $= 25 \times 1.4 \times \cos 16° = 34\,\text{m}.$

b. Both the maximum height and the range will be reduced.

c. Air resistance always acts against the motion of the ball, but its effect on the acceleration depends on the phase of motion. When the ball moves up, the vertical component of air resistance acts downwards and *increases* the vertical acceleration (or rather deceleration—the upward speed of the ball decreases *faster* than under the force of gravity alone). When the ball goes down, the opposite happens: the vertical component of air resistance acts upwards and *decreases* the vertical acceleration of the ball.

d. The horizontal speed is initially $25 \times \cos 16° = 24\,\text{m s}^{-1}$ and decreases gradually towards zero due to air resistance. How quickly and by how much depends on factors such as the mass and the diameter of the ball, but a possible graph may look like the one below. Note that the time of flight will be reduced, too, but the graph only shows the first 1.2 s of the motion.

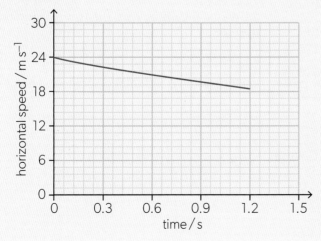

Using simulations and models

The graphs of an object subject to air resistance in Figure 23 were produced using modelling software. This is one way to investigate a system that is complicated or where the situation would involve complex mathematics. You may want to use modelling in your scientific investigation for the internal assessment.

Methods for solving problems such as this are shown in the *Tools for physics* section of this book (on page 364). This can be done using a spreadsheet such as Excel or modelling software such as *Modellus X, Desmos* or *Geogebra*. Modelling techniques are used frequently in scientific research. The powerful computers that enable multi-factor simulations are very much part of the modern nature of science.

How can forces acting on a system be represented both visually and algebraically?

How can Newton's laws be modelled mathematically?

How can knowledge of forces and momentum be used to predict the behaviour of interacting bodies?

For the motion of an object to change, a force must act on it. This is the first and most basic of Newton's laws of motion. At first glance, this law seems obvious. Nevertheless, many questions arise: What does force mean? What is the connection between the force acting on, and the change in movement of, an object?

Scientists need a language to describe forces. They are vector quantities, so they require both magnitude and direction in their description. You can visualize them using scale drawing which works well in situations where more than one force acts on an object.

Sometimes multiple forces combine to give no resultant change in an object's motion. What does our visual approach tell us about this situation? You can extend your understanding beyond purely visual descriptions, to use the mathematics of vectors.

This area of physics is known as classical mechanics. It was the product of the amazing insight of Isaac Newton and others in the 17th century. He formulated a mathematical framework that lasted for 200 years and that is still reliable for many applications today. This framework links force to kinematics for both linear and (as you will see in Topic A.4) rotational motion.

As well as size and direction, the time period over which a force acts on an object is significant. This leads to a new and fundamental concept in physics: momentum. Changes in momentum help you to understand the collisions between objects and particles.

In this topic, you will learn about:

- Newton's laws of motion

- forces as interactions between bodies

- free-body diagrams and how they are used to find the resultant force on a system

- the nature of these contact forces:

  - normal force

  - surface frictional force

  - elastic restoring force

  - viscous drag force

  - buoyancy force

- the nature of these field forces:

  - gravitational force

  - electric force

  - magnetic force

- linear momentum and the conservation of linear momentum

- impulse and change in momentum

- the elastic and inelastic collisions of two bodies and explosions

- angular velocity

- centripetal force and centripetal acceleration.

Describing forces

Introduction

We depend on forces and their effects in all aspects of our life. Forces are sometimes thought of as "pushes or pulls", but the concept of a force goes well beyond this simple description. Forces change the motion of an object and deform the shapes of objects. Forces can act at a distance when there is no contact between a system that produces a force and the object on which it is acting.

▲ Figure 1 Swimmers experience many forces during a race. They exert a force backwards on the water which, by Newton's third law, exerts a forwards force on the swimmer. The swimmer also experiences viscous drag and buoyancy.

This paves the way for your understanding of the nature of a gas in Theme B and provides a key to many of the interactions of nuclear particles in Theme E.

Aristotle and the concept of force

Discussion about the meaning of "force" goes back to the times of earliest scientific thought.

Aristotle, a Greek philosopher who lived about 2300 years ago, had an overarching view of the world (called the Aristotelian cosmology) and he is regarded as important in the development of science. The German philosopher Heidegger wrote that there would have been no Galileo without Aristotle before him. However, despite Aristotle's importance to us, he would not be regarded as a scientist in our modern sense. He is, for example, not known to have performed any experiments to verify his ideas.

Aristotle believed in the "nature" of all objects, including living things. He thought

▲ Figure 2 A statue of the Greek philosopher Aristotle.

that all objects have a natural state which is to be motionless on the surface of Earth and that all objects, when left alone, try to attain this state. Then he distinguished between "natural motion" in which objects fall downwards and "unnatural" or "forced" motion in which objects need to have a force continually applied if they are to remain anywhere other than their natural state. Unfortunately for those learning physics, this is a very persuasive idea because you know intuitively that, when you hold something in your hand, your muscles have to keep "working" in order to do this.

There are many other examples of Aristotelian thought that needed to be overturned before later scientists could move our thinking forward. But it is important to remember the contribution that Aristotle made to science, even if some of his ideas are now disregarded. What do you think students in a century's time or a millennium from now will make of our physics?

Newton's laws of motion

Scientists in the time of Galileo had begun to realize that things were not as simple as the Greek philosophers thought. They were coming to the view that moving objects have **inertia**, meaning a resistance to stopping and that, once in motion, objects continue to move.

Thought experiments

Sometimes in the history of science, scientists have made progress by using experiments to prove or disprove a theory. However, there are also times when they have made progress by pursuing "thought experiments".

A thought experiment is when you take a known situation in which it is easy to imagine the outcome. You then extend the scope of that situation to imagine how the outcome might change. The result may lead to a new theory which can itself be tested using practical experiments.

Galileo himself was reputed to have dropped two cannon balls from the Leaning Tower of Pisa to demonstrate that the ancient Greeks' ideas about falling objects was wrong. There is no contemporary evidence that he did this. It is more likely that it was an idea that he used to explain his ideas of forces. This would encourage others to pursue further thought experiments. There is evidence that a similar experiment was performed in Delft in the Netherlands, possibly earlier than Galileo's version. Galileo demonstrated the link between weight and Newton's second law by persuading people that objects in free-fall accelerate at the same rate.

Einstein is also famous for a thought experiment in which he imagined riding along the crest of a light wave. By pursuing this idea, he conceived his theory of special relativity which is explored in Topic A.5.

Galileo carried out an experiment with inclined planes and spheres. In fact, this may have been a thought experiment—this was often the way forward in those days—but in any event it is easy to see what Galileo was trying to suggest.

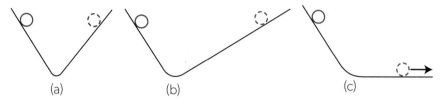

▲ Figure 3 Galileo's thought experiment.

Figure 3 shows a sphere rolled down the left-hand arm of a double inclined plane. In the first experiment (Figure 3(a)), both arms of the inclined plane are at the same angle and the sphere rolls the same distance up the slope as it rolled down (assuming no losses). In the second experiment (Figure 3(b)), the second arm is at a lower angle than before. The sphere rolls up the right-hand plane to the same height as that from which it was released. Galileo suggested that, when the right-hand plane is horizontal (Figure 3(c)), the sphere will continue to roll for ever because it will never be able to climb to the original release height.

Newton's first law

Newton included Galileo's idea in his **first law of motion**, which says:

An object remains stationary or moves at a constant velocity unless an external force acts on it.

Galileo suggested that the sphere on the horizontal plane never stopped. However, Newton realized that there is more to say than this. Unless something from outside applies a force to change it, the velocity of an object (both its speed and its direction) must remain the same. This directly opposed Aristotle's view that a force had to keep pushing constantly at a moving object for the speed to remain the same.

 Theories—Overthrowing Aristotle

Aristotle's methods were based on making observations rather than scientific experiments and creating logical explanations. It is a measure of his success that his ideas dominated science for over a thousand years.

Some of the first scientists to question Aristotle's ideas lived in the Islamic Golden Age. While Europe was in the Dark Ages, Islamic scholarship flourished between the 8th and 14th centuries.

Aristotle's ideas of motion suggested that a moving object would stop once it had found its natural place. The Persian polymath Ibn Sīna (ابن سينا) who lived from c. 980 to 1037 considered the motion of projectiles and concluded that motion without air resistance would continue indefinitely—a precursor to Newton's first law.

Later, the Islamic physicist Abu'l-Barakāt Hibat Allah ibn Malkā al-Baghdādī (أبو البركات هبة الله بن ملكا البغدادي), who lived c. 1080—c. 1165, proposed a relationship between force and acceleration—a precursor to Newton's second law.

→ Al-Baghdādī, building on Ibn Sīna's work, had started to formulate the idea of experimental methods involving repeated observations as a method of investigation. This was an early version of what is now known as the scientific method.

▲ Figure 4 Ibn Sīna as depicted on a banknote from Tajikistan.

Newton's second law

The next step is to ask: when a force acts on an object, in what way does the velocity of the object change?

Newton proposed a fundamental equation that connected force and rate of change of velocity (acceleration). This is contained in his second law. The law can be written in two ways (see later in this topic for the alternative form of the law).

Newton's second law, in its simpler form, says that:

force = mass × acceleration

In symbols, this is written $F = m \times a$, where a is the acceleration of an object of mass m when force F is acting on it.

The appropriate SI units are force in newtons (N), mass in kilograms (kg) and acceleration in metres per second squared ($m\,s^{-2}$). The newton is a derived unit in SI. In terms of fundamental units, it is $kg\,m\,s^{-2}$.

Two things arise from Newton's equation:

- Mass is a scalar quantity, so multiplying a by m will not change the direction of the acceleration. The direction of the force and the direction of the acceleration will be the same. Conversely, applying a force to a mass will change the velocity *in the same direction* as that of the force.

- One way to think about the mass in this equation is as the force required per unit of acceleration for a given object. This provides a way to standardize force units. When an object of mass 1 kg is observed to accelerate with an acceleration of $1\,m\,s^{-2}$, then one unit of force (1 N) must have acted on it.

Definition of a newton

Newton's second law of motion leads directly to the definition of the unit of force—the newton (N):

> 1 N is the applied force needed to accelerate 1 kg of mass at the rate of $1\,m\,s^{-2}$ in the direction of the applied force.

Inertial and gravitational mass

What exactly is *mass*? The mass used in Newton's second law of motion is **inertial mass** ($F = m_i \times a$). This is the property through which an object resists the effects of a force that is trying to change its motion. It also appears later in this topic in the definition of momentum ($p = m_i \times v$).

However, the quantity called mass when considering weight ($W = m_g \times g$) is different. This is the response of matter to the effects of gravity and is called **gravitational mass**.

Galileo is said to have dropped two cannonballs of different mass from the Leaning Tower of Pisa. He observed that they accelerated at the same rate, and thus showed that inertial mass and gravitational mass were the same. This is shown through the application of Newton's second law which gives:

$$m_i a = m_g g$$

When inertial mass and gravitational mass are equal, then this becomes:

$$a = g$$

Objects in free-fall (in the absence of other forces such as air resistance) will experience an acceleration of g.

Even though these two quantities seem to be the same, is there any reason why they should be equivalent?

The equivalence of inertial and gravitational mass has been experimentally verified to better than 1 part in 10^{15}.

Does this mean that what you understand by mass varies with the context? Can you imagine what would happen if inertial and gravitational masses were not equal?

🧪 Force, mass and acceleration

- Tool 2: Use sensors.
- Tool 3: Select and manipulate equations.
- Tool 3: Construct and interpret tables and graphs for raw and processed data including scatter graphs and line and curve graphs.
- Inquiry 2: Collect and record sufficient relevant quantitative data

This series of experiments can help in your understanding of Newton's second law.

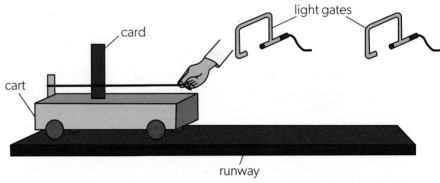

▲ Figure 5 A cart is subject to a constant force provided by an elastic string. The light gates time the cart over a known distance.

Experiment 1

The acceleration of a cart of constant mass is measured when it is pulled by different numbers of identical elastic threads, each one extended by the same amount.

- The motion can be timed using light gates and an electronic timer as shown in Figure 5. Alternatively, the timing can be done using a data logger connected to an ultrasound (or other) sensor. You could even use your mobile phone with a suitable app.
- Accelerate the cart with one elastic thread attached to the rear of the cart. The thread must be extended by the same amount for each run. Devise a way to do this: your hand needs to move at the same speed as the cart so that the thread is the same length throughout the run for the acceleration to be constant.

The card on the cart is of a known length. The time taken by the card to break the light beam can be used to calculate, first, the average speed at each gate and then (using the distance apart of the gates), the acceleration of the cart. The kinematic equations are used for this—so it is essential that the force and acceleration are constant.

- Alternatively, a data logger will give you a direct output of speed against time.
- Repeat the experiment with two, three and possibly four elastic threads, all identical, all extended by the same amount. This means that you will be using one unit of force (with one thread), two units of force (with two threads), and so on.
- Plot a graph of calculated acceleration against number of force units. Is your graph straight? Does it go through the origin? Remember that this experiment has several possible uncertainties when you judge your line of best fit.

→

Experiment 2

The setup is essentially the same as for Experiment 1. However, this time, you use a constant force (possibly two elastic threads).

Measure the acceleration with a constant force and different masses of carts.

This time, Newton's second law predicts that mass $\propto \dfrac{1}{\text{acceleration}}$. Plot a graph of acceleration against $\dfrac{1}{\text{mass}}$. Is it a straight line?

Experiments—Valid or invalid

Experiments are important in determining the validity of hypotheses. But the experiments must be carefully designed so that the results measure what is intended. An experiment that is a good measure of a physical situation is said to be "valid", whereas an experiment that gives misleading results is "invalid".

An experiment may be invalid when a control variable is not properly controlled. Allowing such a variable to change affects the collected data and may lead to a false conclusion.

As an example, consider this experiment. A student wants to test Newton's second law and attaches a trolley to a string. The string passes over a pulley and its other end is attached to some weights on a hanger. The weights produce a tension in the string and accelerate the trolley. The trolley is released from rest and accelerates over a distance d. The acceleration can be calculated when the travel time is known. The student intends to vary the weight on the hanger and measure the effect on the acceleration of the trolley. The independent variable is the weight; the dependent variable is the acceleration a. The control variables are the distance d and the mass of the trolley m.

While this may seem a reasonable experiment, there is a reason why it will not give the results that you might expect. The trolley, the string and the weights are all accelerating. As a result, the mass of this system is changed when different weights are added and the control variable is not constant. To make the experiment valid, the mass of the trolley and the mass of the weights must both be included (assuming the mass of the string is negligible). A typical way of doing this is by transferring weights from the trolley to the hanger.

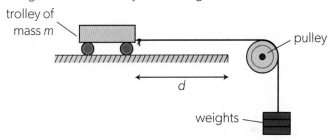

▲ Figure 6 To what extent will this experiment be valid?

Data-based questions

The diagram shows an experiment where an object of mass 0.20 kg is connected via a string and a pulley to another object of mass m. In the experiment, m was varied and the amount of time t that the object took to fall a distance h was measured.

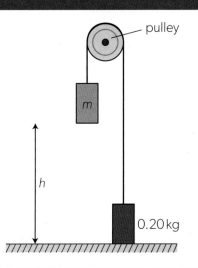

A table of the results is given.

m / kg	t / s		
	1st trial	2nd trial	3rd trial
0.30	0.99	1.14	1.25
0.40	1.00	0.78	0.86
0.50	0.80	0.88	0.66
0.60	0.74	0.62	0.80
0.70	0.60	0.78	0.66
0.80	0.78	0.66	0.54
0.90	0.56	0.74	0.62
1.00	0.64	0.69	0.52

- Copy the table and add a column to find the average time.

- Plot a graph of t (y-axis) against m (x-axis). Include error bars for t.

  It can be shown that: $t^2 = \dfrac{2h}{m}\left(\dfrac{m+0.20}{m-0.20}\right)$

  Prove this relationship.

- Plot a graph of t^2 against $\dfrac{m+0.2}{m-0.2}$.

- Add error bars to your new graph. Remember that the percentage uncertainty in t^2 is double the percentage uncertainty in t.

- Find the gradient of the graph. Using maximum and minimum gradients, find the uncertainty in your graph.

- Use the equation for t^2, to deduce h for this experiment. Quote the uncertainty in h as part of your value.

> You can find explanations for error bars and uncertainties in the *Tools for physics* section on page 358.

Worked example 1

A car with a mass of 1500 kg accelerates uniformly from rest to a speed of 28 m s^{-1} (about 100 km h^{-1}) in a time of 11 s. Calculate the average force that acts on the car to produce this acceleration.

Solution

Acceleration $= a = \dfrac{v-u}{t} = \dfrac{28}{11} = 2.55\,\text{m s}^{-2}$.

Force $= ma = 1500 \times 2.55 = 3.8\,\text{kN}$.

Worked example 2

An aircraft of mass 3.3×10^5 kg takes off from rest in a distance of 1.7 km. The maximum thrust of the engines is 830 kN.

a. Calculate the take-off speed.

b. Discuss the assumptions you have made in part a.

Solutions

a. Acceleration $= \dfrac{8.3 \times 10^5}{3.3 \times 10^5} = 2.5151...\,\text{m s}^{-2}$

   $v^2 = u^2 + 2as$ so $v^2 = 0 + 2 \times 2.5151... \times 1700 = 8552$. Therefore, $v = \sqrt{8552} = 92.5\,\text{m s}^{-1}$.

b. The student has to assume that the thrust of the engines is constant and no other forces act on the aircraft in the direction of its motion. They have to ignore effects of air resistance and assume that the runway is horizontal.

Practice questions

1. An object is projected vertically upwards. Which is correct about the net force acting on the object, at the instant when the object reaches its maximum height? The net force:

   A. is zero

   B. is a maximum

   C. changes direction

   D. is directed downwards

2. An air rifle pellet of mass 2 g is fired at an initial speed of 200 m s^{-1} into a stationary block of clay. The pellet penetrates the block for a distance of 10 cm before coming to rest. What is the average force acting on the pellet from the block of clay?

   A. 4×10^0 N

   B. 4×10^1 N

   C. 4×10^2 N

   D. 4×10^3 N

3. A stationary tennis ball of mass 58 g is hit by a racket. The ball is in contact with the racket for a time of 10 ms and leaves the racket with a speed of 15 m s⁻¹. Calculate the average force exerted by the racket on the ball.

4. A car of mass 1200 kg decelerates from a speed of 80 km h⁻¹ to a speed of 45 km h⁻¹ over a distance of 18 m. Calculate:

   a. the average force acting on the car during deceleration

   b. the time taken to decelerate.

5. A constant force of 4.0×10^2 N acts on an initially stationary railway carriage of mass 1.1×10^4 kg for 8.0 s. Calculate, for a time of 8.0 s:

   a. the velocity of the carriage

   b. the distance travelled.

6. An electron moving horizontally at an initial speed of 8.0×10^6 m s⁻¹ enters a region where a constant vertical force of 6.4×10^{-17} N acts on it.

   a. Outline why the horizontal component of the velocity of the electron remains constant.

   b. The electron travels a horizontal distance of 25 cm before leaving the region of the force. The mass of the electron is 9.11×10^{-31} kg. Determine:

      i. the vertical displacement of the electron as it leaves the region of the force

      ii. the angle that the electron's velocity makes with the horizontal.

Newton's third law

Newton's third law of motion can be expressed in several equivalent ways.

One common way to write Newton's third law of motion is:

Every action has an equal and opposite reaction.

The first point in this formulation of the law is that the words "action" and "reaction" mean "action force" and "reaction force". The law refers directly to the effects of forces.

A second point is that the action–reaction pair must be matching types. A gravitational action force must correspond to a gravitational reaction. A gravitational action cannot link to an electrostatic force.

The third law suggests that forces must appear in pairs, but it is important to identify all the possible force pairs in a situation and then to pair them up correctly. Take, as an example, a rubber ball resting on a table (Figure 7).

The obvious action force here is the weight of the ball, that is, Earth's gravitational pull acting on it. This force acts downwards and—if the table were not there—the ball would accelerate downwards towards the floor according to Newton's second law. What is the reaction force? Given that action force and reaction force must pair up like for like, the reaction is the gravitational force that the ball exerts on Earth. This is the same size as the pull of Earth on the ball, but is in the opposite direction. This gravitational force pair is shown as red vectors in Figure 7(a).

What prevents the ball accelerating downwards? There must be a force exerted by the table on the ball. Because the ball is not accelerated, this upwards force is equal and opposite to the downwards gravitational pull of Earth on the ball.

However, the upwards table force is *not* the reaction force to the ball's weight—that reaction force is the gravitation pull on Earth. The origin of the table force lies in the electrostatic forces between atoms. As the ball rests on the table, it deforms the horizontal surface very slightly. Imagine pushing downwards on the middle of a metre ruler suspended by supports at its end. The ruler bends in

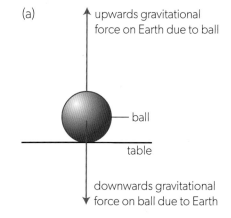

(a) upwards gravitational force on Earth due to ball

ball

table

downwards gravitational force on ball due to Earth

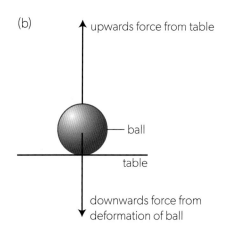

(b) upwards force from table

ball

table

downwards force from deformation of ball

▲ Figure 7 The forces acting on a ball on a table. The forces should all lie on the same vertical line that goes through the centre of the ball.

But are these laws at all?

Can Newton's laws of motion be proved? The answer is that they cannot. Strictly speaking they are assertions, as Newton himself recognized. In his famous *Principia* (written in Latin as was the custom in the 17th century) he writes *Axiomata sive leges motus* [meaning: *the axioms or laws of motion*].

In physics, a law differs from a theory in that a law makes no attempt to explain itself. It is only based on the results of observation. A theory, on the other hand, justifies itself with an explanation.

However, Newton's laws of motion allow us to predict most types of motion. They remained unchallenged for about 200 years until Einstein formulated two theories of relativity at the turn of the 20th century. Einstein showed that the rules Newton proposed were only approximate. However, for human speeds, the laws are reliable to a high degree and are good enough most of the time.

a spring-like way to provide a response to the force acting downwards. The dent in the table surface is the response of the atoms in the table to the weight lying on it. Remove the ball and the table will become flat again. This upwards force, trying to return the table surface to the horizontal, is pushing upwards on the ball. There is a corresponding downwards force from the deformed ball (the ball will become slightly flattened as a response to the gravitational pull). So here is the second action–reaction pair (the black vectors in Figure 7(b)) between two forces that are electrostatic in origin.

To get a feeling for this (literally), take a one-metre laboratory ruler and suspend it between two lab stools. Press down gently with your finger in the centre of the ruler so that it becomes curved. The ruler will bend; you will be able to feel it resisting your efforts to deform it. Remove your finger and the ruler returns to its original shape.

ATL Thinking skills—The language of force

In explaining Newton's third law for a particular example, you must remember to emphasize the nature, the size and the direction of the force you are describing. A common example is that of a rocket in space. Students sometimes write that "…by Newton's third law, the rocket pushes on the atmosphere to accelerate" but this shows a weak understanding of how the propulsion works.

First, of course, the rocket does not "push" on anything. This is proved by the fact that a rocket can accelerate in space where there is no atmosphere.

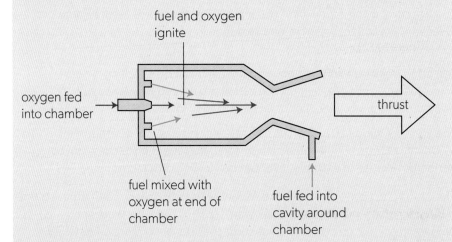

▲ Figure 8   The combustion chamber of a rocket.

Inside the rocket, chemicals react in a combustion chamber to produce a gas with a high temperature and large pressure (Figure 8). The chamber has exhaust nozzles through which this gas escapes. At one end of the chamber, the gas molecules rebound off the end wall and exert a force on it. As a result, they reverse their direction. In principle, the rebounding molecules could then travel down the rocket and leave through the nozzles. There is an action–reaction pair here: the force forwards that the gas molecules exert on the chamber (and therefore the rocket) and the force that the chamber exerts on the gas molecules. It is the first of these two forces that accelerates the rocket. If the chamber were completely sealed and the gas could exert an equal and opposite force at the back of the rocket, then the forward force would be exactly countered by the backwards force and no acceleration would occur.

Later you will interpret this acceleration in a different way. But this explanation will still apply at a microscopic level.

Think about the following situations and discuss them with fellow students. You may wish to return to these when you have a further perspective on mechanics, at the end of Topic A.3.

- A fire fighter must exert considerable force on a fire hose to keep it pointing in the direction that sends water to the correct place.

- There is a suggestion to power space travel to deep space by ejecting ions from a spaceship.

- A sailing dinghy moves forward when the wind blows into the sails.

Worked example 3

Describe action–reaction force pairs according to Newton's third law for the following situations:

a. a helicopter hovering above the ground

b. a dog pulling on its leash (consider forces on the dog).

Solutions

a. The helicopter's weight and the gravitational force exerted by the helicopter on Earth are one such pair. Another pair describes the interaction of the helicopter blades with the surrounding air: the blades exert a downward force on the air and consequently, the air exerts an equal but opposite force on the blades. This is the upward lift force that prevents the helicopter from falling!

b. There are at least three action–reaction pairs that involve forces acting on the dog:

- The dog's weight and the gravitational force from the dog on Earth.

- The tension in the leash acts on the dog in the direction parallel to the leash; the dog exerts an equal but opposite force on the leash.

- The reaction force from the ground acts on the dog. This force has a forward and upward component and forms a force pair with the force exerted by the dog on the ground.

Worked example 4

A brick of mass 4.0 kg lies on a floor. A force of 20 N is applied downwards to the brick. Calculate the magnitude of the force exerted by the brick on the floor.

Solution

The total downward force acting on the brick is the sum of the brick's weight and the externally applied force, $4.0 \times 9.8 + 20 = 59$ N. For the brick to remain at rest, this force must be balanced by an upward reaction force of 59 N exerted by the floor. By Newton's third law, the brick exerts an equal but opposite force on the floor, 59 N downwards.

There is an introduction to the use of vector diagrams on page 340.

Free-body force diagrams

Force is a vector quantity and can be represented by an arrow. The scaled length of the arrow gives the magnitude of the force, and the arrow direction gives the force direction. In simple cases where there are only a few forces acting, this works well. As the situations become more complex, diagrams showing all the forces can become complicated. You can avoid this problem by drawing a **free-body force diagram**.

The rules for a free-body diagram for a body are as follows:

- The diagram is used for one body only. The force vectors are represented by arrows.

- Only the forces acting on the body are drawn.

- The force (vector) arrows are drawn to scale originating at a point that represents the centre of mass of the body

- All forces must be clearly labelled.

🧪 Drawing a free-body diagram

- Tool 3: Draw and interpret free-body diagrams showing forces at point of application or centre of mass as required.

- Tool 3: Add and subtract vectors in the same plane (limited to three vectors).

- Inquiry 2: Interpret diagrams, graphs and charts.

Begin by sketching the general situation with all the bodies that interact in the situation.

Select the body of interest and draw it again removed from the situation. The body may be simply represented as a point (at its centre of mass) or a box or circle.

Draw all the forces acting on this body to scale and label them.

Add the force vectors together (either by drawing or calculation) to give the net force acting on the body. This sum can be used later to draw other conclusions about the motion of the object.

(a)   **situation diagram**

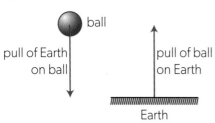

(b) **free-body diagram for the ball**

▲ Figure 9 (a) The situation diagram and (b) the free-body diagram for a ball falling freely under gravity.

Examples of free-body diagrams

1. *A ball falling freely under gravity with no air resistance*

There are two gravitational forces acting: Earth pulling on the ball and the ball pulling on Earth. For the free-body diagram of the ball (Figure 9), we are only interested in the first of these. The free-body diagram is a particularly simple one, showing the object and one force. Notice that the ball is not represented as a real object, but as a point that marks the centre of mass of the object.

2. *The same ball resting on the ground*

   As we saw earlier for the case of the ball on the table, four forces act:

   - the weight of the ball downwards
   - the reaction of Earth to this weight
   - the upwards force from the ground because the ground has been deformed
   - the reaction of the ball to this "spring-like" force.

   The **normal force** (often labelled F_N) is the component of the **contact force** that acts perpendicular to the surface. In this situation this normal force is the force on the ball from the ground.

   A free-body diagram (Figure 10) helps here because, by restricting the diagram to the forces acting only *on* the ball, the four forces reduce to two: the weight of the ball and the upwards force on it due to the deformation of the ground. These two forces are equal and opposite. The net resultant (vector sum) of the forces is zero and there is therefore no acceleration.

(a) **situation diagram**

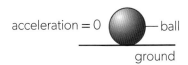

(b) **free-body diagram for the ball**

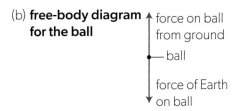

▲ Figure 10 The free-body diagram for a ball at rest on the surface of a table.

3. *An object accelerating upwards in a lift*

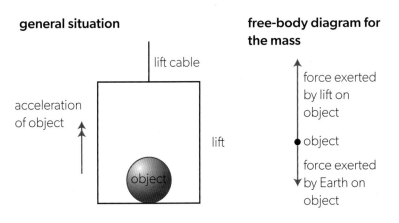

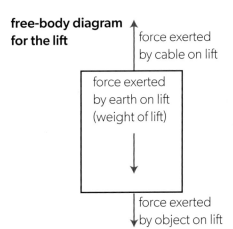

▲ Figure 11 The free-body diagram for an object in a lift.

The weight of the object acts downwards (Figure 11). The magnitude of this force is the same as when the object is on Earth's surface. However, the upwards force of the floor of the lift on the object is now larger than the weight and the resultant force of the two has a net upwards component. The object is accelerated upwards.

For the lift, there is an upward force in the lift cable and a downwards weight of the lift together with the weight of the object. The resultant force is upwards and is equal to the force in the cable less the weight forces of the lift and object.

Worked example 5

A person stands on a weighing scale placed on the floor of a lift. The weighing scale reads the force in newtons.

a. Explain, by reference to the forces acting on the person, why the reading of the weighing scale depends on the vertical acceleration of the lift.

b. The mass of the person is 75 kg and the lift is moving with a downward acceleration of $1.5\,\text{m s}^{-2}$.

   i. Draw a free-body diagram for the person.

   ii. Calculate the reading of the weighing scale.

Solutions

a. By Newton's third law, the reading of the weighing scale is equal to the magnitude of the normal force F_N exerted by the scale on the person. The only other force acting on the person is the person's weight mg, so the net force on the person is $F_N - mg$ (taking forces and accelerations directed upwards as positive, and those directed downwards as negative). The net force is related to the acceleration a of the lift (and the person), $ma = F_N - mg$. Hence, $F_N = m(g + a)$. This shows that the reading of the weighing scale depends on the acceleration of the lift. Upward acceleration ($a > 0$) gives the reading greater than the person's weight; downward acceleration ($a < 0$) gives the reading less than the person's weight.

b. i.   The net force is downwards, so the normal force has a smaller magnitude than the weight.

ii. Downward acceleration has a negative sign in the formula derived in part a.

The reading of the scale is $75(9.8 - 1.5) = 620\,\text{N}$.

Note that 'downward acceleration' means that the lift is either moving upwards with a decreasing velocity or moving downwards with an increasing velocity.

normal force, F_N

person

weight, mg

Practice questions

7. For the person in the worked example above, draw a free-body diagram and calculate the reading of the weighing scale when the lift is:

a. moving with a constant vertical speed

b. accelerating upwards at $2.0\,\text{m s}^{-2}$.

There is also **rotational equilibrium** where something is rotating at a constant angular speed; this is discussed in Topic A.4.

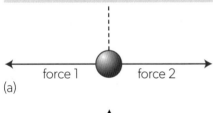

force 1   force 2

(a)

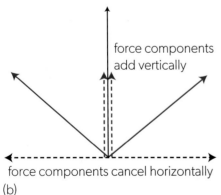

force components add vertically

force components cancel horizontally

(b)

▲ Figure 12 (a) Force 1 and force 2 are equal in size and opposite in direction; the next force is zero. (b) The two forces are no longer in the same line of action and cancel horizontally but not vertically.

Translational equilibrium

When an object is in **translational equilibrium**, it is either at rest or moving at a constant velocity (not just constant *speed* in this case). "Translational" here means moving in a straight line.

Newton's first and second laws remind us that when there is no change of velocity then there must be zero force acting on the object. This zero force can often be the **resultant** (addition) of more than one force. In this section, we will examine what equilibrium implies when there is more than one force.

The simplest case is that of two forces. When they are equal in size and opposite in direction, they will cancel out and be in equilibrium (Figure 12(a)).

When the forces are equal in size, but not acting in the same direction, then equilibrium is not possible. In Figure 12(b), the two horizontal components of the force vectors are equal and opposite, and there is no resultant force component in this direction. Vertically, the two vector components point in the same direction. Overall, there will be an unbalanced vertical force. The object will accelerate upwards in response to this unbalanced force.

This gives a clue as to how we should proceed when there are three or more forces.

situation diagram

free-body diagram

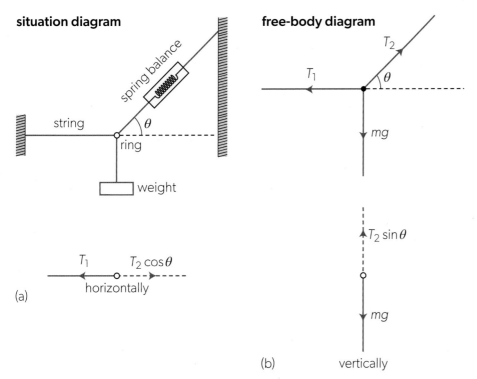

(a)

(b)

▲ Figure 13 The (a) situation diagram and (b) free-body diagram for the forces acting on a small ring.

Figure 13 shows a small ring on which three forces act. The forces are provided by a weight, a spring balance, and the tension in a string. For equilibrium to occur, the three components must add up to zero in any direction in which they are resolved. This is an example of the **resolution of vectors**. The principles behind this are stated on page 342.

Figure 13(b) shows that horizontal and vertical are two good directions for this **resolution of vectors**, because two forces are aligned with these directions and one disappears in each direction chosen:

- The vertical force mg has no component in the horizontal direction.

- The horizontal force T_1 has no vertical component.

Whichever direction is chosen, all the forces must cancel for there to be no resultant force.

There is one more consequence of this idea. Figure 14 shows the forces drawn, as usual, to scale and in the correct direction (in red). The forces can be moved, as shown by the green arrows, into a new arrangement (shown in black). What is special in the rearrangement is that the three forces form a closed triangle where all the arrows meet. This is called the **triangle of forces**. Whenever you can draw the vectors for a system in this way, then the system *must* be in translational equilibrium.

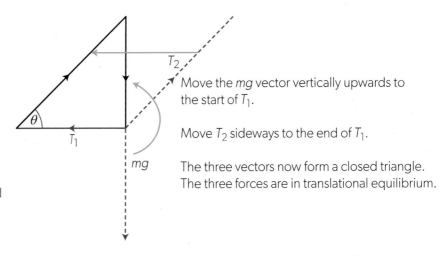

Move the mg vector vertically upwards to the start of T_1.

Move T_2 sideways to the end of T_1.

The three vectors now form a closed triangle. The three forces are in translational equilibrium.

▲ Figure 14 Moving the force vectors so that they act as a triangle of forces.

Worked example 6

A cart of mass 0.80 kg is placed on a ramp that makes an angle of 20°
to the horizontal. The cart is held in equilibrium by a thread.

a. Draw a free-body diagram for the cart.

b. Calculate the magnitude of:

   i.   the normal force from the ramp on the cart

   ii. the tension in the thread.

c. The thread breaks. Calculate the acceleration of the cart.

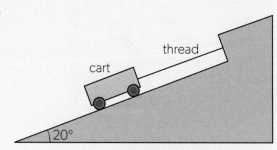

Solutions

a. The forces acting on the cart are the cart's weight, the tension
    in the thread and the normal force from the ramp.

b. It is convenient to resolve the weight of the cart into components parallel and
    perpendicular to the ramp. In equilibrium, the tension is equal and opposite to the
    parallel component of the weight, and the normal reaction is equal and opposite to
    the perpendicular component of the weight.

   i. $F_N = 0.80 \times 9.8 \cos 20° = 7.4\,N$

   ii. $T = 0.80 \times 9.8 \sin 20° = 2.7\,N$

c. When the thread breaks, the net force on the cart is
    equal to the parallel component of the weight.

    Acceleration $= \dfrac{mg \sin 20°}{m} = g \sin 20° = 3.4\,m\,s^{-2}$

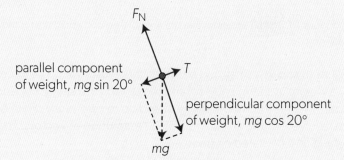

Practice questions

8. An object of mass 2.0 kg is suspended in equilibrium
    by two threads of equal length attached to the ceiling.

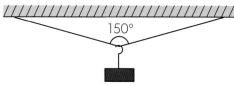

   a. Calculate the tension in each thread.

   b. Explain why the threads are more likely to break
       when the angle between them is increased.

9. The diagram shows an arrangement of two objects
    suspended by threads of negligible mass.

    The mass of object A is M and the
    mass of object B is $2M$.

   a. Draw a free-body diagram for
       object A.

   b. Calculate, in terms of M, the
       tension in each thread.

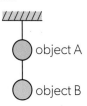

10. A ball of weight 0.50 N is suspended by a thread of
    negligible mass. A horizontal force F acts on the ball.
    In equilibrium, the thread makes an angle of 40° to
    the vertical.

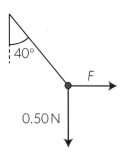

   a. Determine the magnitude of:

      i.   the force F

      ii. the tension in the thread.

   b. The thread breaks and the horizontal force remains
       unchanged. Describe the initial motion of the ball.

Investigating a triangle of forces

- Tool 1: Understand how to accurately measure angles an appropriate level of precision.

- Tool 3: Construct and interpret tables and graphs for raw and processed data including scatter graphs and line and curve graphs.

- Inquiry 2: Interpret diagrams, graphs and charts.

For this experiment you need weights, string and a pulley. Assemble them as shown in the diagram. The pulley should be mounted on a retort stand. It may be easier if the horizontal string, attached to the wall, is also attached to another retort stand so that you can adjust that as well.

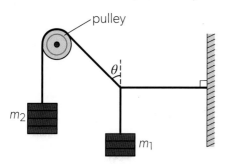

▲ Figure 15 Investigating the triangle of forces.

- Place a known mass m_2 as shown. This mass will remain constant throughout the experiment.

- Place a mass m_1 as shown (m_1 should be less than m_2). Adjust the position of the pulley so that the string attached to the wall is horizontal.

- Measure the angle marked θ.

- Change m_1, repeat the adjustment of the pulley and measure θ.

- Take repeats for each mass.

- Construct a table of your results.

- Add a column for the average angle and for $\cos\theta$.

- Plot a graph of $\cos\theta$ against m_1.

- What does the gradient of your graph represent?

Types of forces

There are many types of force:

- **Non-contact forces** occur when there are forces acting between objects that do not touch. For example, magnetic, electrostatic (electromagnetic) and gravitational forces.

- **Contact forces** are forces that we observe when two objects are touching in some way. We can usually imagine how the forces arise.

The three contact forces that you study in IB Diploma Programme physics are:

- elastic restoring forces

- buoyancy forces

- friction between solids (solid friction) and friction between solids and fluid (viscous drag forces).

 How can knowledge of electrical and magnetic forces allow the prediction of changes to the motion of charged particles?

The study of forces and their effects is not just about spheres bouncing and strings pulling. All forces have a direct impact.

Charged particles from the Sun enter the upper atmosphere and come under the influence of Earth's magnetic field. They interact with the magnetic field and their motions change. The result is the Aurora Borealis or the Aurora Australis—depending on the hemisphere in which you live. Topics D.2 and D.3 will put some of the work from this topic to use in linking the electrical and magnetic forces to the motion of these particles.

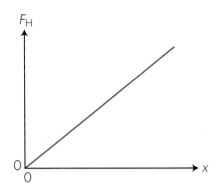

▲ Figure 16 The elastic restoring force–extension graph for a spring that obeys Hooke's law.

Elastic restoring forces

The shape of a solid can be changed by applying a force to it. Different materials respond to a given force in different ways.

A stretched spring exerts an **elastic restoring force** F_H on the objects attached to each end of the extended spring (Figure 16). F_H is equal and opposite to the force that extended the spring initially. The **extension** is the change in length from the spring's initial unextended length; it usually has the symbol x.

Figure 16 shows the variation in the tension produced by the spring against the extension of the spring. When the graph is a straight line through the origin, then $F_H \propto x$ and the tension is directly proportional to the extension. This is known as **Hooke's law** because Robert Hooke is thought to be the first person to formulate it.

Science as a shared endeavour

Robert Hooke (1625–1703) was an English scientist, architect and polymath. He was an assistant to Robert Boyle and constructed the air pump that allowed the discovery of Boyle's law (see page 256).

Hooke argued with Isaac Newton over which of them should take credit for various discoveries, including Newton's theory of gravitation.

He published his discovery of Hooke's law as an anagram, revealing the solution two years later. Scientists would do this to prove that they were the discoverer without disclosing to their rivals what their discoveries were.

Nowadays, scientists work more collaboratively. They publish their works promptly so that other scientists might confirm and improve their findings. Scientists are careful to reference work carried out by other scientists. A system of **peer review**—where other scientists check articles before publication—ensures that full credit is given to other scientists' work.

▶ Figure 17 Hooke's *De Potentia Restitutiva* in which he published the solution to his anagram and described Hooke's law.

LECTURES
De· Potentia Reſtitutiva,·
OR OF
SPRING
Explaining the Power of Springing Bodies.
To which are added ſome
COLLECTIONS
Viz.
A Deſcription of Dr.Pappins Wind-Fountain and Force-Pump. .
Mr. Young's Obſervation concerning natural Fountains.
Some other Conſiderations concerning that Subject.
Captain Sturmy's remarks of a Subterraneous Cave and Ciſtern.
Mr. G. T. Obſervations made on the Pike of Teneriff, 1674.
Some Reflections and Conjectures occaſioned thereupon.
A Relation of a late Eruption in the Iſle of Palma.

By *ROBERT HOOKE.* S.R.S.

·LONDON,
Printed for *John Martyn* Printer to the *Royal Society·,*
at the Bell in St. *Paul's* Church-Yard, 1678.

You will see more about the forces in springs and how this links to elastic potential energy in Topic A.3.

The law can be written as: $F_H = -k \times x$, where k is the constant of proportionality known as the **spring constant**.

The units of k are $N\,m^{-1}$; in fundamental SI units, this is $kg\,s^{-2}$.

The stiffer the spring, the larger the value of k. The spring constant k is known also as the **elastic constant** and the **force constant**.

The negative sign in $F_H = -kx$ reminds you that the force that the spring exerts is in the opposite direction to the extension. Figure 18 shows this for a spring that can be longer or shorter than its unstretched length. When the spring is lengthened and the extension is to the *right* (the middle diagram), the spring is exerting a force to the *left* to try to return to its unextended (equilibrium) length. When the spring is compressed and shorter so that the compression is to the left, the spring exerts a force to the right.

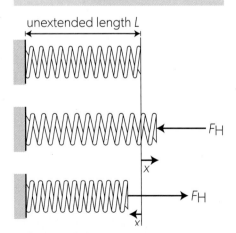

▲ Figure 18 Extending and compressing a spring of unextended length L.

How does a restoring force acting on a particle result in simple harmonic motion?

This emphasis on the direction in which a force acts as predicted by its sign is important. In simple harmonic motion (discussed in Topic C.1), you will see how a force directed to the centre of a motion leads to the prediction of an oscillation.

The spring in Figure 18 is an obvious example of this. Imagine a mass attached to the right-hand end of the spring. The spring is always trying to return the mass to its equilibrium position. The mass accelerates and overshoots the equilibrium position and now the force on it due to the spring acts in the opposite direction. In the absence of frictional losses, the oscillation—compression and extension of the spring—will continue forever without stopping.

Worked example 7

An object of mass 1.5 kg is suspended by a spring attached to the ceiling. The spring extends by 3.0 cm from the unstretched length. The mass of the spring is negligible.
Calculate the elastic constant of the spring.

Solution

The object is in equilibrium so the elastic force from the spring and the weight of the object have equal magnitudes, $k\Delta L = mg$. From here, $k = \dfrac{1.5 \times 9.8}{3.0 \times 10^{-2}} = 490\,\mathrm{N\,m^{-1}}$.

Worked example 8

A student investigates how the spring force F varies with extension ΔL of a spring. The graph shows the experimental results and the line of best fit.

a.   Estimate the elastic constant of the spring.

b.   The student wants to use the graph to predict the extension of the spring when stretched by a force of 50 N. Comment on the assumptions needed.

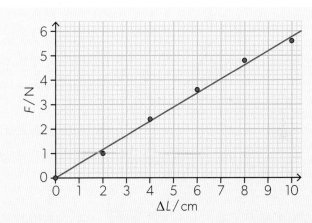

Solutions

a.   The elastic constant is equal to the gradient of the line of best fit. We use the coordinates of two extreme points on the best fit line to calculate the gradient. Elastic constant $= \dfrac{5.8 - 0}{(10.0 - 0) \times 10^{-2}} = 58\,\mathrm{N\,m^{-1}}$.

b.   The linear relationship needs to be extrapolated beyond the range of the experimental data. For a given spring, Hooke's law is only valid in a certain range of extensions (this is called the elastic region of the spring), and the student has to assume that the extension corresponding to the force of 50 N is within this range, which may not be justified for this particular spring!

Investigating Hooke's law

*   Tool 3: Construct and interpret tables and graphs for raw and processed data including scatter graphs and line and curve graphs.

*   Inquiry 1: Demonstrate independent thinking, initiative, or insight.

- Inquiry 2: Collect and record sufficient relevant quantitative data.

- Arrange a spring of known unstretched length with a weight hanging on the end of the spring.

- Devise a way to measure the extension of the string for several weights of increasing size that hang on the end of the spring.

- Repeat the measurements as you remove the weights to check.

- Plot a graph of force (weight) acting on the spring (y-axis) against the extension (x-axis). This is not the obvious way to draw the graph, but it is the way often used. Normally, we would plot the dependent variable (the extension in this case) on the y-axis.

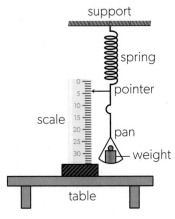

▲ Figure 19 Investigating Hooke's law.

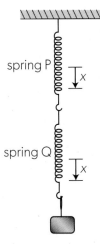

▲ Figure 20 Two identical springs in series give double the extension of one spring for the same force.

> In Topic B.5, you will see two quantities combined using a reciprocal relationship in a similar way.

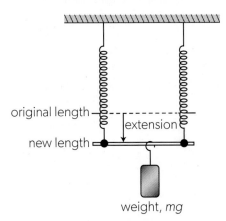

▲ Figure 21 Two identical springs in parallel give half the extension of one spring for the same force.

Combining springs

In series

Two identical springs P and Q, both of spring constant k, are connected in series, and a weight mg is hung on the bottom of Q (Figure 20).

Each spring has the same tension (mg) and therefore each extends by $x = \dfrac{mg}{k}$. The total extension for the pair of springs is $\dfrac{2mg}{k}$. Hooke's law for the springs is written as:

$$mg = k'\frac{2mg}{k}$$

where k' is the new spring constant for both springs acting together. This leads to $k' = \dfrac{k}{2}$, where the new paired spring constant is half of the spring constant for one spring alone.

When springs P and Q have different spring constants k_P and k_Q then the new spring constant k' is given by:

$$\frac{1}{k'} = \frac{1}{k_P} + \frac{1}{k_Q}.$$

In parallel

This time the identical springs are connected in parallel (Figure 21).

The weight is "shared" between them. To see this, remember that the force downwards is mg and, when the springs are in equilibrium, then the total force upwards is mg. Each spring will only contribute one-half of this force, so that is $\dfrac{mg}{2}$ for each spring. With a full load of mg, each spring extends by x. With a load of half of this, each spring extends by $\dfrac{x}{2}$ and the effective spring constant is $2k$, as the arrangement appears to be twice as stiff.

With different spring constants:

$$k' = k_P + k_Q$$

Practice questions

11. When a load is applied to a spring of negligible mass, the spring extends by a distance L. The spring is now cut into two halves of equal length and the halves of the spring are arranged in parallel.
What is the extension of the parallel system under unchanged load?

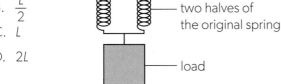

A. $\frac{L}{4}$

B. $\frac{L}{2}$

C. L

D. $2L$

two halves of the original spring

load

12. Spring 1 has the elastic constant k and spring 2 has the elastic constant $2k$. The springs are connected in series and extended by a combined distance L. What is the effective spring constant of the serial system and what is the extension of spring 1?

	Effective spring constant	Extension of spring 1
A.	$\frac{k}{3}$	$\frac{L}{3}$
B.	$\frac{k}{3}$	$\frac{2L}{3}$
C.	$\frac{2k}{3}$	$\frac{L}{3}$
D.	$\frac{2k}{3}$	$\frac{2L}{3}$

Buoyancy forces

How does floating—or buoyancy as it is properly called—work? Buoyancy is related to the density of an object. Many people think that objects float because their weight is small, and that heavy objects sink because they are heavy. Objects that are denser than the fluid sink, objects less dense than the fluid float. The question is: why?

 ## Density and pressure

- Tool 3: Select and manipulate equations.

- Tool 3: Identify and use symbols stated in the guide and the data booklet.

- Tool 3: Express derived units in terms of SI units.

The ideas of pressure and density are important for the work on gas properties in Theme B.

Density is defined as the mass for each unit volume of an object. When the mass is m and the volume is V, then the density ρ is given by $\rho = \frac{m}{V}$

The unit of density is $kg\,m^{-3}$. The density of water is $1000\,kg\,m^{-3}$. The density of steel is about $8000\,kg\,m^{-3}$.

Pressure is the normal (perpendicular) force that acts on a surface for each unit area of the surface (Figure 23). When a normal force F acts on an area A, the pressure P is given by $p = \frac{F}{A}$

The SI unit of pressure is $N\,m^{-2}$ which is usually written as Pa (where Pa stands for pascal). Thus $1\,N\,m^{-2} \equiv 1\,Pa$. The pressure of the atmosphere acting at the surface of Earth is about $100\,000\,N\,m^{-2}$ ($10^5\,Pa$).

▲ Figure 22 Some examples of floating objects.

Pressure is a scalar quantity. In a fluid it acts in all directions. When defined for a solid, pressure acts in the same direction as the weight.

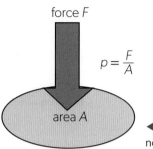

force F

$$p = \frac{F}{A}$$

area A

◄ Figure 23 A force F is acting normal to a surface of area A.

The **buoyancy force** (also called the **upthrust**), F_b, is equal to $\rho V g$ where ρ is the density of the liquid, V is the volume of displaced fluid and g is the acceleration due to gravity. The buoyancy force arises from the pressure difference between the top and bottom of the floating object.

The height of an object floating above the surface of a fluid depends on the relative densities of the fluid and the object.

*   When the densities are equal, the object has **neutral buoyancy**. The object remains where it is in the fluid.

*   When the object density is greater than the fluid density, the net force on the object is downwards and the object sinks.

*   When the object density is less than the fluid density, the net force is upwards and the cube floats at the surface. The volume of the object below the surface has displaced its volume of fluid (displaced means "moved out of the way"). The displaced fluid has a weight equal to the total weight of the object (Figure 24).

This leads to **Archimedes' principle**: The upward buoyancy force on an object, completely or partially submerged in a fluid, is equal to the weight of fluid displaced by the object.

Objects float, so the fraction of their volume that is below the water must be equal to $\dfrac{\text{density of object}}{\text{density of fluid}}$.

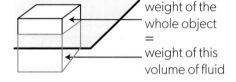

weight of the whole object = weight of this volume of fluid

▲ Figure 24 An object floats on the surface so that its weight is equal to the weight of fluid it displaces.

Global impact of science—The Plimsoll line

The density of seawater is not constant around the world. The salt concentration varies from around 30 g in every litre of seawater to about 40 g l⁻¹. The density also varies with temperature. This has implications for ocean-going ships.

The outside of the ship's hull is marked to show the position where the water level ought to be in various oceans with different salinities and temperatures (Figure 25). This ensures that the ships are not loaded excessively. The ship's master must ensure that the vessel does not have the relevant line under the water surface before leaving harbour.

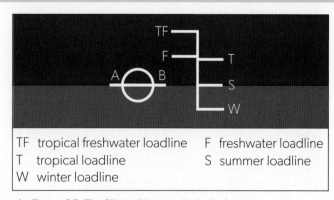

TF   tropical freshwater loadline   F   freshwater loadline
T   tropical loadline   S   summer loadline
W   winter loadline

▲ Figure 25 The Plimsoll line on the hull of a ship.

Submariners and divers need to be aware of the implications of buoyancy:

- A submarine has buoyancy tanks that are filled with air when the ship is on the surface. To submerge, water is allowed to flood the buoyancy tanks to replace the air. This increases the weight of the submarine, and it goes below the surface. To travel back to the surface, compressed air stored inside the submarine pushes the water out of the tanks, so the weight of the submarine decreases.

- Scuba divers need to be able to adjust their position relative to the water surface. They usually wear weights to achieve neutral buoyancy in the water. Otherwise, the diver will have to use too much energy swimming to remain at a particular depth. The weight required to keep an individual diver neutral will, of course, depend on the salinity and temperature of the water.

Worked example 9

A buoy of a uniform cross-sectional area of $0.20\,m^2$ floats in fresh water, with $0.75\,m$ of the height submerged below the water line. The density of fresh water is $1000\,kg\,m^{-3}$.

a. Calculate the mass of the buoy.

b. The buoy is now placed in seawater of greater density than fresh water. Explain the change, if any, in the submerged height of the buoy.

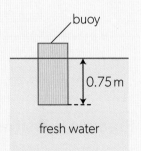

Solutions

a. Newton's first law says that, since the buoy floats without sinking, the buoyancy force must be equal to the weight of the buoy. From Archimedes' principle, this force is equal to the weight of the water displaced by the buoy. By combining the two laws, we can deduce that the mass of the buoy is the same as the mass of the displaced water.

$$m = \rho_{water}V = 1000 \times 0.75 \times 0.20 = 150\,kg$$

b. A smaller volume is needed to displace the mass of seawater equal to the mass of the buoy. Hence, a smaller height of the buoy will be submerged in seawater.

Practice questions

13. Two objects float in water. 25% of the volume of object 1 and 50% of the volume of object 2 remain above the surface of the water.

    What is $\dfrac{\text{density of object 1}}{\text{density of object 2}}$?

    A. 1.5   B. 2.0

    C. 2.5   D. 3.0

    Hot-air balloons rise because their total weight is less than the weight of the air they are displacing. The physics of gas pressures and temperatures is examined in Topic B.3.

14. A container, open at the top, in the shape of a cuboid of dimensions $25\,cm \times 25\,cm \times 10\,cm$ floats in water so that $2.0\,cm$ of the height is submerged below the water line. The density of water is $1.0 \times 10^3\,kg\,m^{-3}$.

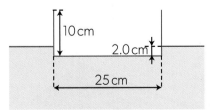

    a. Calculate the mass of the container.

    b. An iron bar of mass $4.0\,kg$ is placed in the container. Predict whether the container will sink.

Worked example 10

A balloon can be modelled as a sphere of radius 8.5 m. The balloon is filled with hot air of average density 0.92 kg m^{-3} and anchored to the ground with a mooring line. The density of the surrounding air is 1.2 kg m^{-3}.

a. Calculate the maximum combined mass of the balloon and the load that the balloon can lift.

b. The actual mass of the balloon and its load is 650 kg. The mooring line is released. Determine the initial vertical acceleration of the balloon.

Solutions

a. The mass of the displaced cold air is $1.2 \times \frac{4}{3}\pi \times 8.5^3 = 3090$ kg. The mass of the hot air inside of the balloon is $0.92 \times \frac{4}{3}\pi \times 8.5^3 = 2370$ kg. The combined mass of the balloon and the load that can be lifted is the difference, $3090 - 2370 = 720$ kg.

b. The upward force on the system is the difference between the force of buoyancy due to the displaced cold air and the combined weight of the hot air and the balloon with its load, $(3090 - 2370 - 650) \times 9.8 = 690$ N. The mass that is being accelerated by this force includes the hot air inside the balloon in addition to the balloon itself.

$$\text{Acceleration} = \frac{\text{force}}{\text{mass}} = \frac{690}{2370 + 650} = 0.23 \, \text{m s}^{-2}$$

Practice questions

Assume that atmospheric air has a density of 1.2 kg m^{-3}.

15. A spherical balloon of radius 0.15 m and mass 0.014 kg is filled with a gas. The balloon floats in air in the state of neutral buoyancy.
    Determine the density of the gas in the balloon.

16. A foam ball of radius 3.0 cm and mass 1.8 g is suspended by a thread of negligible mass.

    a. Calculate the magnitude of the buoyancy force acting on the ball.

    b. Determine the percentage change in the tension force in the thread when the system is placed in a vacuum chamber.

Solid friction

Friction is the force that occurs between two surfaces in contact. If you happen to live in a part of the world where there is snow and ice, then you will know that when the friction between your shoes and the ice disappears it can be a good thing (for skiing), or a bad thing (for falling over).

Figure 26(a) shows an experiment to illustrate the physics of solid friction.

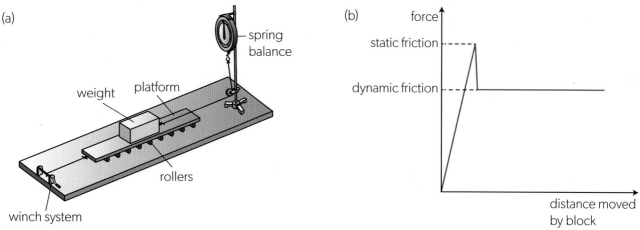

(a)

(b)

▲ Figure 26 As the winch handle is turned, static friction occurs between the weight and the platform. Eventually this is replaced by dynamic friction and the reading on the spring balance decreases.

A metal object rests on a block of wood. One end of the metal is connected to a spring balance (newton-meter). The other end is connected to a winch. Both connections are made using strings that do not stretch. The wood sits on a series of rollers.

When both connecting strings are slack, the spring balance reads zero. The winch is then turned, and the string attached to the wood is wound in. The wood starts to move towards the winch, and initially there is no relative movement between metal and wood.

At first, the spring balance reading (the tension in the strings) increases. The metal and the wood do not move relative to each other as the tension is not enough to overcome the friction between the metal and wood. Eventually, however, the pull is sufficiently great that the wood and the metal begin to slide relative to each other. At this point, the reading on the spring balance decreases, showing that there is now less force required to keep the metal moving relative to the wood.

Figure 26(b) shows how the force registered by the spring balance varies with distance moved by the wooden block. The tension in the string at the instant when the metal begins to slip is greater than the tension when the metal has begun to move.

In summary:

- As the force on the spring balance increases from zero, the platform does not begin to move relative to the weights immediately.

- The platform suddenly begins to move at a particular value of force. At this instant, the force, as shown by the spring balance, drops to a lower value.

- This new value is then maintained as the platform moves steadily.

- "Stick–slip" behaviour occurs where the platform alternately sticks and then jumps to a new sticking position. This behaviour is associated with two values of friction, but this may be too difficult to observe.

- The friction forces depend on the magnitude of the weights on the platform.

The observations lead to a description of frictional forces as:

- **static friction** (in the first part of the experiment, when there is no relative movement between the surfaces), or

- **dynamic friction** (once the metal and wood are moving relative to each other).

These two friction values are observed for most surface pairs.

Any mathematical description of friction needs to take account of this change in behaviour. Such a theory will be empirical.

A **surface frictional force** F_f acts in a direction that is parallel to the plane of contact between a body and the surface on which it rests (Figure 27). The force acts at the plane surface and is given empirically by $F_f \leq \mu F_N$, where F_f is the frictional force exerted by the surface on the block. F_N is the normal reaction of the surface on the block. This is the weight of the block when there is no vertical acceleration.

The symbol μ changes its subscript and its value depending on whether the friction is static or dynamic. It is known as the coefficient of friction. It has no units because it is the ratio of two forces.

Empirical or theoretical?

Empirical means that the hypothesis arises purely from experimental results. **Theoretical** means that a physical model (which may or may not be mathematical) has been constructed to explain results.

The two equations for static and dynamic friction do not arise from a study of the interatomic forces between the surfaces; they are derived purely from experiments with bulk materials. You will see this approach again in Theme B where the gas laws are empirical, but the kinetic model of gases is a theory.

Are there fundamental differences in these two approaches in science? Is one form a "better" type of knowledge than the other? Was Greek science empirical or theoretical?

Static friction

The force expression has the form $F_f \leq \mu_s F_N$; μ_s is the **coefficient of static friction**.

The "less than or equal to" sign ($\leq$) is meant to show that the static friction force can vary from zero up to a maximum value. Between these limits F_f is equal to the pull on the block. Once the pull on the block is equal to F_f, then the block is just about to move.

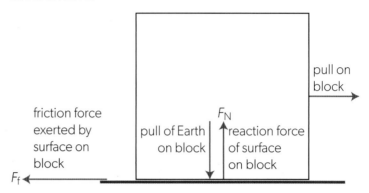

▲ **Figure 27** Forces acting on a block subject to friction at its bottom surface.

Dynamic friction

Once the pull exceeds $\mu_s F_N$, then the block begins to slide. For forces $> F_f$, dynamic friction is at work. The force expression becomes $F_f = \mu_d F_N$, where μ_d is the **coefficient of dynamic friction**.

Dynamic friction only applies when the surfaces move relative to each other. The friction drops from its maximum static value and remains at a constant value. This fixed value depends on the total reaction force acting on the surface but (according to simple theory) is not thought to depend on the relative speed between the two surfaces.

The values of μ_s and μ_d vary greatly depending on the pair of surfaces being used and the condition of the surfaces (for example, whether they are lubricated or not). A few typical values are given in Table 1. If you want to investigate a wider range of surfaces, there are many sources of the coefficient values on the Internet—search for "Coefficients of friction".

Each friction coefficient is a ratio $\left(\dfrac{F_f}{F_N}\right)$ and so has no units.

It is possible for the coefficients to be greater than 1 for some surface pairs. This reflects the fact that for these surfaces the friction is very strong and greater than the weight of the block. Remember that the surfaces are being pulled sideways by a horizontal force, whereas the reaction force is vertical, so we are not really comparing like with like in these empirical rules for friction.

Surface 1	Surface 2	μ_s	μ_d
glass	metal	0.7	0.6
rubber	concrete	1.0	0.8
rubber	wet tarmac	0.6	0.4
rubber	ice	0.3	0.2
metal	metal (lubricated)	0.15	0.06

▲ **Table 1** Typical values of the coefficients of static and dynamic friction for different surfaces.

Friction between a block and a ramp

- Tool 1: Understand how to accurately measure angles to an appropriate level of precision.

- Tool 3: Carry out calculations involving fractions and trigonometric ratios.

- Tool 3: Draw and interpret free-body diagrams showing forces at point of application or centre of mass as required.

- Tool 3: Resolve vectors (limited to two perpendicular components).

One way to measure the static coefficient of friction between two surfaces in a school laboratory is to use the two surfaces as part of a ramp system. Figure 28 shows the arrangement and a free-body diagram for the block.

The upper surface of the ramp is one of the two materials under test. The bottom of the block is the other surface. Resolving at 90° to the plane, $F_N = mg \cos\theta$; resolving along the plane, $F_f = mg \sin\theta$.

Combining these gives $\tan\theta = \dfrac{F_f}{F_N} = \mu_s$

a block on a ramp

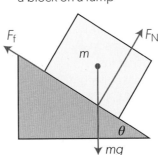

free-body diagram of just the block

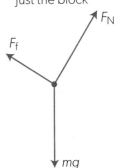

▲ Figure 28 Measuring μ_s using a block on a ramp.

- Begin with the ramp in a horizontal position.

- Gradually raise one end of the ramp until the block just starts to slip.

- The angle between the ramp and the horizontal is measured and the tangent of this angle is equal to the coefficient of static friction.

Data-based questions

The picture shows an experiment to investigate dynamic friction. A block of wood rests on a sanding belt which is moving towards the right. The wood is attached to a newton-meter. Weights are added on top of the wooden block and the frictional force is measured on the newton-meter and recorded. A table of data is given.

- Plot a graph of the data and add error bars.

- How could the graph be used to find:

  ○ the coefficient of friction, μ_d

  ○ the mass of the block?

Weight placed on wooden block / N	Frictional force (±0.2) / N
0.0	1.0
1.0	1.4
2.0	2.0
3.0	2.6
4.0	3.2
5.0	3.8
6.0	4.4
7.0	5.2
8.0	6.0

- The uncertainty in the reading on the newton-meter was ± 0.2 N. Use this to find the uncertainty in your measurement of the coefficient of friction.

You can find explanations for error bars and uncertainties in the *Tools for physics* section on page 358.

Worked example 11

Sophie investigates the frictional force between a metal block and a wooden table. She pulls the block with an inextensible horizontal string connected to a force sensor and records the tension force in the string. The block is initially at rest. Sophie gradually increases the force until the block starts moving. The graph shows the data collected. The block starts moving at about 4.6 s. The mass of the block is 3.5 kg.

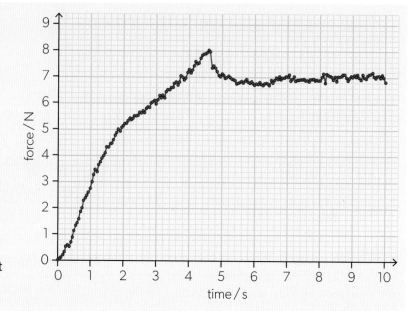

a. Estimate the coefficient of static friction between the block and the table.

b. After 7.0 s the block is moving at a constant speed. Estimate the coefficient of dynamic friction between the block and the table.

Solutions

a. The maximum static frictional force recorded just before the block starts moving is about 8.0 N. The coefficient of static friction is therefore $\mu_s = \dfrac{8.0}{3.5 \times 9.8} = 0.23$

b. When the block is moving at a constant speed, the dynamic frictional force is equal and opposite to the tension in the string. The magnitude of this force is about 7.0 N, so $\mu_d = \dfrac{7.0}{3.5 \times 9.8} = 0.20$

Worked example 12

A box is pushed across a level floor at a constant speed with a force of 280 N at 45° to the floor. The mass of the box is 50 kg. Calculate:

a. the vertical component of the force

b. the weight of the box

c. the horizontal component of the force

d. the coefficient of dynamic friction between the box and the floor.

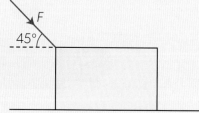

Solutions

a. The vertical component $= 280 \sin 45° = 198$ N.

b. The weight of the box $= mg = 50 \times 9.8 = 490$ N.

c. The horizontal component $= 280 \cos 45° = 198$ N.

d. The vertical component of the force exerted by the floor in the box $= 490 + 198 = 688$ N.
    The friction force = the horizontal component (the box is travelling at a steady speed), so $\mu_d = \dfrac{198}{688} = 0.29$.

Worked example 13

A skier places a pair of skis on a snow slope that is at an angle of 1.7° to the horizontal. The coefficient of static friction between the skis and the snow is 0.025. Determine whether the skis will slide away by themselves.

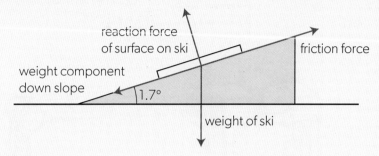

reaction force
of surface on ski

friction force

weight component
down slope

The diagram not to scale

1.7°

weight of ski

Solution

Call the weight of the skis W. The component of weight down the slope $= W\sin 1.7°$ and the reaction force of the surface of the ski $= W\cos 1.7°$.

Therefore, the maximum friction force up the slope $= \mu_s W\cos 1.7°$. The skis will slide if $\mu_s W\cos 1.7° < W\sin 1.7°$. In other words, if $\mu_s < \tan 1.7°$.

The value of $\tan 1.7°$ is 0.0297 and this is greater than the value of μ_s which is 0.025, so the skies will slide away.

Practice questions

17. A box is placed on the horizontal floor of a truck moving at an initial speed of $50\,\text{km h}^{-1}$. The coefficient of static friction between the box and the floor is 0.45. The truck decelerates uniformly and comes to rest in a distance of 25 m.

    a. Calculate the magnitude of the acceleration of the truck.

    b. Deduce whether the box will start sliding against the floor.

18. A hockey puck is sent across an ice rink with an initial speed of $8.0\,\text{m s}^{-1}$ towards a goal that is 16 m away. The coefficient of dynamic friction between the puck and the ice is 0.10. Determine the speed of the puck as it reaches the goal.

19. A box of mass 2.0 kg is pulled up a ramp that makes an angle of 15° to the horizontal with a constant force of 8.0 N. The box starts from rest and travels a distance of 0.50 m in 3.8 s.

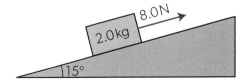

8.0N

2.0kg

15°

Calculate:

a. the acceleration of the box

b. the component of the weight of the box parallel to the ramp

c. the magnitude of the frictional force on the box

d. the coefficient of dynamic friction.

20. A book of weight 12 N is pushed against a vertical wall with a horizontal force F.

    The coefficient of static friction between the book and the wall is 0.75 and the coefficient of dynamic friction is 0.60.

    book

    F

    a. The book is initially at rest. Calculate the minimum magnitude of F so that the book does not slide down the wall.

    b. For the value of F you have calculated in part a, draw a scaled free-body diagram for the book.

    c. The force F is now reduced to 10 N and the book starts sliding. Determine the acceleration of the book.

Theories

Friction originates at the interface between the two materials. The actual causes of friction are still being investigated today and the explanation here is simplified. Leonardo da Vinci mentions friction in his notebooks and some of the next scientific writings about friction appear in works by Guillaume Amontons from around 1700 (Figure 29).

Surfaces that seem very smooth to us are not smooth at all under a microscope. At the atomic level, surfaces consist of peaks and troughs of atoms (Figure 30(a)). When two surfaces are at rest relative to each other, then the atomic peaks rest in the troughs, and it needs a certain level of force to deform or break the peaks sufficiently for sliding to begin. This accounts for static friction.

Once relative motion has started and dynamic friction occurs, then the top surface rises above the deformed peaks. The frictional force is less. The irregularities on the surface are small and of atomic size, which means that even the small forces applied in our lab experiments can cause large stresses to act on the peaks. The peaks then deform like a soft plastic whether the material is hard steel or soft rubber.

Moving surfaces are often coated with a lubricant to reduce wear due to friction (Figure 30(b)). The lubricant fills the space between the two surfaces and either prevents the peaks and troughs of atoms from touching or reduces the amount of contact. In either event, the atoms from the surfaces do not interact as much as before and the friction force and the coefficient are reduced.

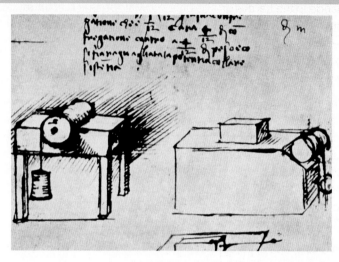

▲ Figure 29 Leonardo da Vinci mentions friction in his notebooks.

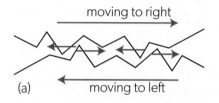

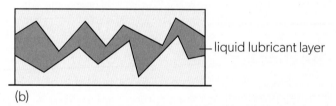

▲ Figure 30 Irregularities at the surfaces of two materials cause the friction.

These friction forces originate in the complex electronic properties of the atoms at the surface. However, this simple theory should give you some understanding of how friction arises. The properties of bulk materials that we perceive on the macroscopic scale arise from microscopic properties that operate at the atomic level.

Air resistance and drag force

Topic A.1 described the effects that air resistance has on the motion of an object falling vertically or being projected at an angle to the horizontal. It will now be clear that the effects are due to a drag force that acts on objects when they move.

In 2014, the skydiver Alan Eustace jumped safely from a record-breaking height of 41.4 km above New Mexico, USA to reach a top speed of about 1300 km h^{-1}, which is faster than the speed of sound. A skydive from more usual heights does not take place at such high speeds. These are usually less than 200 km h^{-1}. The difference between Eustace's speed and the lower value is because air resistance varies with height above Earth.

Figure 31 shows two of the forces that act on a skydiver. The weight of the diver acts vertically downwards and is effectively constant (because there is little change in the Earth's gravitational field strength at the height of the dive, even when 41 km above the surface). The drag force acts in the opposite direction to the motion of the diver and, for a diver falling vertically, this drag will be vertically upwards. (Other forces acting on the diver include the upwards buoyancy caused by the displacement of air by the diver.)

When the skydiver initially jumps from the aircraft, the vertical speed is almost zero and there is almost no air resistance in this direction. Air resistance increases as the speed increases so that, as the diver goes faster and faster, the resistance force becomes larger and larger. The net force therefore decreases and, consequently, the acceleration of the diver downwards also decreases. Eventually the weight force downwards and the resistance force upwards are equal in magnitude and opposite in direction. At this point there is zero acceleration, and the diver has reached **terminal speed**.

A typical graph of vertical speed against time for the skydiver is shown in Figure 32.

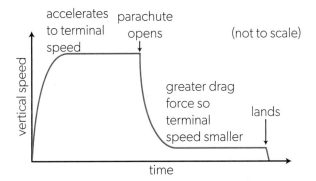

▲ Figure 32 Vertical speed–time graph for a skydiver.

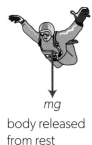
mg
body released
from rest

drag force
mg
forces on body
during acceleration

drag force
mg
forces on body
at terminal speed

▲ Figure 31 Forces acting on a skydiver.

Eventually the skydiver opens the parachute. The parachute envelope has a large surface area. The upwards resistive force is much larger than before and is now greater than the weight. As a result, the directions of the net force and acceleration are also upwards, so the vertical velocity decreases in magnitude. Once again, a balance will be reached where the upward and downward forces are equal and opposite—but at a much lower speed than before (about 12 m s^{-1} for a safe landing).

Stokes' law

We can estimate the terminal speed using Stokes' law. In 1851, the Irish scientist George Stokes derived the theory for the drag force (called more properly the **viscous force**) acting on a small sphere that is moving through a viscous fluid. So Stokes' law can apply to both gases and liquids. A **viscous fluid** with a high viscosity (such as a concentrated sugar solution or an engine oil) has a high resistance to deformation—it will pour slowly. A low-viscosity liquid (such as water) has a low resistance to deformation and pours quickly with little drag.

Stokes had to make some assumptions to derive his theory. These are:

- That the flow is **laminar** so that layers in the fluid flow smoothly past each other without mixing. (The alternative is **turbulent flow** where the layers mix and the whole of the fluid swirls about in a random manner.)

- That the moving particles are spheres with smooth surfaces.

- That the fluid is homogeneous (it is uniform in composition).

- The particles do not interact.

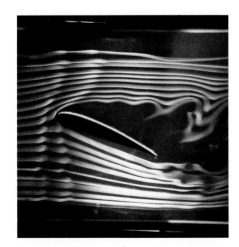

▲ Figure 33 A model of an aircraft wing in a wind tunnel. The wing is at an angle such that the air flow around it becomes turbulent.

Stokes suggested that the drag force F_d is given by $F_d = 6\pi\eta rv$, where r is the radius of the sphere and v is its speed. The constant η is the dynamic **viscosity** of the fluid through which the sphere is moving; it has the units Pa s. For example, the dynamic viscosity of water is 1×10^{-3} Pa s, whereas the dynamic viscosity of honey is up to 20 Pa s. Viscosity is also temperature dependent. For water, η varies from about 1.8 mPa s to 0.3 mPa s as the temperature rises from 0 to 100°C.

In real life, air resistance is dominated by turbulent, rather than laminar, flow as turbulence becomes established at very small speed gradients. The mathematics of turbulence indicates that the turbulent drag force is proportional to $(speed)^2$ rather than the $(speed)$ of the Stokes equation.

Moving through fluids

Both viscous drag and buoyancy effects need to be considered when an object moves through a fluid. A sphere dropping under its own weight through a fluid is subject to three forces as, shown on Figure 34:

- weight W downwards $= mg$, where m is the mass of the sphere

- buoyancy force B upwards $= \rho_f gV$ where ρ_f is the density of the fluid and V is the volume of the sphere

◀ Figure 34 A sphere dropping through a viscous liquid has its weight W, a buoyancy force B, and a drag force D acting on it.

- drag force D upwards $= 6\pi\eta rv$, where r is the radius of the sphere and v is its speed downwards.

Summing the forces gives a net force downwards on the sphere of $W - B - D$ or $mg - \rho_f gV - 6\pi\eta rv$.

The mass m can be replaced by the sphere's density ρ_s and volume V so that the net force is also $\rho_s Vg - \rho_f gV - 6\pi\eta rv$ or $(\rho_s - \rho_f)gV - 6\pi\eta rv$

As the sphere falls, the total upward force increases because Stokes' law means that the drag force is proportional to the speed. Eventually the sum of the buoyancy and drag will equal the weight downwards. This is the condition for terminal speed v_t (see Topic A.1).

Setting the net force to zero and rearranging gives $v_t = \dfrac{(\rho_s - \rho_f)gV}{6\pi\eta r}$.

Measuring the coefficient of viscosity of a fluid

- Tool 3: Understand the significance of uncertainties in raw and processed data.

- Tool 3: Use basic arithmetic and algebraic calculations to solve problems.

- Inquiry 1: Demonstrate independent thinking, initiative, or insight.

- Inquiry 2: Collect and record sufficient relevant quantitative data.

Stokes' equation can be used to determine η for a liquid with the apparatus shown in Figure 35.

- Five or six balls are needed. They should be made from the same material but with different radii. Some trials will be required to match the ball radius to the fluid. It is important to have a small fractional uncertainty in the time measurement—why?

▶ Figure 35
A small sphere drops through a viscous liquid at its terminal speed. The timer measures the time to drop between two light gates to determine the speed of the sphere.

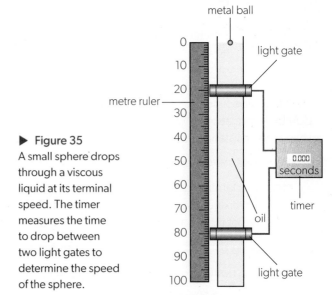

- Measure the density of the liquid and of the material from which the balls are made. Then pour the liquid into a tall, wide, transparent container—for example a large measuring cylinder.

- Devise a way to measure the time for each ball to travel a known vertical distance in the liquid. Each ball needs to be travelling at its terminal speed *before* it reaches the start of the measurement distance. The measurement can be using a stopwatch or a light-gate arrangement (as shown in the diagram).

- Use the equation $v_t = \dfrac{(\rho_s - \rho_f)gV}{6\pi\eta r}$ to calculate η.

- How could this be developed into an internal assessment?

Worked example 14

A ball of radius 8.0 mm and mass 1.3 g is released from rest from the bottom of a long vertical tube filled with oil. The ball rises towards the surface of the oil. The diagram shows how the vertical speed of the ball varies with time.

a. Estimate the initial acceleration of the ball.

b. Hence, calculate the magnitude of the buoyancy force on the ball.

c. Draw a free-body diagram for the ball at a time of 1.5 s.

d. Determine the coefficient of viscosity of the oil.

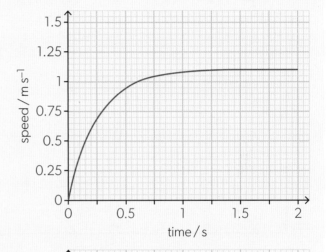

Solutions

a. The initial acceleration is the gradient of the tangent to the speed-time graph at $t = 0$.
   $$\text{Acceleration} = \frac{1.5}{0.35} = 4.3\,\text{m s}^{-2}.$$

b. Initially, the only forces acting on the ball are the buoyancy force F_b and the ball's weight. The net force is the difference between the two and is related to the acceleration by Newton's second law.
   $F_b - mg = ma$, so
   $F_b = 1.3 \times 10^{-3}(9.8 + 4.3) = 1.8 \times 10^{-2}\,\text{N}.$

c. At $t = 1.5$ s the ball is moving upwards with a constant terminal speed and the viscous drag force acts downwards so that the net force on the ball is zero.

d. At terminal speed, the drag force is $F_d = F_b - mg = 1.3 \times 10^{-3} \times 4.3 = 5.6 \times 10^{-3}\,\text{N}.$
   The drag force $F_d = 6\pi\eta r v$, where the terminal speed v is approximately $1.1\,\text{m s}^{-1}$.
   Combining equations gives
   $$\eta = \frac{5.6 \times 10^{-3}}{6\pi \times 8.0 \times 10^{-3} \times 1.1} = 3.4 \times 10^{-2}\,\text{Pa s}.$$

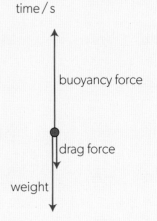

Worked example 15

Calculate the upwards force acting on a skydiver of mass 80 kg who is falling at a constant speed.

Solution

The weight of the skydiver is $80 \times 9.8 = 784$ N. Because the skydiver is falling at a constant speed (that is, terminal speed) the upwards drag force is equal to the downwards weight. Therefore, the upwards force is 780 N to 2 sf.

Practice questions

21. A skydiver is falling vertically towards the ground and opens the parachute a short time after jumping. The graph shows how the speed of the skydiver varies with time.

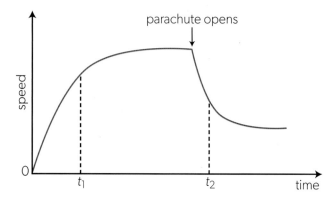

Which of the following correctly compares the directions of the velocity and of the acceleration of the skydiver at times t_1 and t_2?

	Directions of velocity at t_1 and t_2	Directions of acceleration at t_1 and t_2
A.	same	same
B.	same	different
C.	different	same
D.	different	different

22. Two balls of radius R and $2R$, made from the same type of steel, fall through a liquid that exerts a viscous drag force on the balls. The smaller ball reaches the terminal speed v. What is the terminal speed of the larger ball?

    A. v B. $2v$ C. $4v$ D. $8v$

23. A ball of weight 1.2 N falls through a liquid at a constant speed. The density of the ball is 1.5 times greater than the density of the liquid. What is the magnitude of the drag force acting on the ball?

    A. 0.4 N   B. 0.6 N   C. 0.8 N   D. 1.2 N

24. A steel ball falls in a long vertical tube filled with vegetable oil.

    a. Explain how the ball reaches a terminal speed.

    The following data are given.

    > density of the vegetable oil $= 920$ kg m^{-3}
    > viscosity of the vegetable oil $= 8.4 \times 10^{-2}$ Pa s
    > density of steel $= 8000$ kg m^{-3}
    > radius of the ball $= 2.0$ mm

    b. Calculate:

       i.   the weight of the ball

       ii. the buoyancy force acting on the ball in the oil.

    c. Determine the terminal speed of the ball in the oil, assuming that the ball is affected by a viscous drag force.

Force and momentum

Introduction

Many sports involve throwing and catching a ball. Compare catching a table-tennis (ping-pong) ball with catching a baseball travelling at the same speed. One of these is a more painful experience than the other! The velocity may be the same in both cases, but the combination of velocity and mass makes a substantial difference. Equally, comparing the experience of catching a baseball when gently tossed from one person to another with that of catching a firm hit from a strong player tells you that changes in velocity make a difference too.

Momentum

The product of the mass m of an object and its instantaneous velocity v is called the **momentum** p of the object ($p = m \times v$). This quantity has far-reaching consequences in physics.

- Momentum is *mass* × *velocity* **never** *mass* × *speed*.

- Momentum has direction (Figure 36). Mass is a scalar quantity, but velocity is a vector. When mass and velocity are multiplied together, the momentum is also a vector with the same direction as the velocity. Think of the mass as "scaling" the velocity—in other words, just making it bigger or smaller by a factor equal to the mass of the object.

- The unit of momentum is the product of the units of mass and velocity: $kg\,m\,s^{-1}$. You will see a alternative unit later.

- When velocity or mass is changing, then the momentum must also be changing.

- When a net resultant force acts on an object, the object accelerates, and the velocity must change. This means a change in momentum too. A net force leads to a change in momentum.

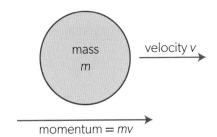

▲ Figure 36 The momentum of a moving object is the product of its mass and its velocity.

Worked example 16

A ball of mass 0.25 kg is moving to the right at a speed of 7.4 m s⁻¹. It strikes a wall at 90° and rebounds from the wall, leaving it with a speed of 5.8 m s⁻¹ moving to the left. Calculate:

a. the momentum of the ball before it strikes the wall

b. the change in momentum after the ball strikes the wall.

Solutions

a. Initial momentum, $p = mv = 0.25 \times 7.4 = 1.85\,kg\,m\,s^{-1}$

b. Final momentum, $p = 0.25 \times 5.8 = 1.45\,kg\,m\,s^{-1}$ to the left. So taking the direction to the right as being positive, the change in momentum $= -1.45 - (+1.85) = -3.3\,kg\,m\,s^{-1}$ to the right (or, alternatively, $+3.3\,kg\,m\,s^{-1}$ to the left).

Collisions and changing momentum

You may have seen a "Newton's cradle". Newton did not invent this device (it was developed in the 20th century as an executive toy), but it helps us to visualize some important rules relating to his laws of motion.

One of the balls (the right-hand one in Figure 37) is moved up away from the remaining four. When released, the ball falls back and hits the second ball. The right-hand ball stops moving, and the left-most ball moves off to the left. It is as though the motion of the original ball transfers through the middle three—which remain stationary—and appears at the left-hand end.

▲ Figure 37 A Newton's cradle.

This is a transfer of momentum. Think about a simpler case where only two spheres are in contact (in the toy, three balls can be lifted out of the way). The right-hand sphere gains momentum as it falls from the top of its swing. When it collides with the other sphere, the momentum appears to be transferred to the second sphere. The first sphere now has zero momentum (it is stationary) and the second has gained momentum. What rules govern this transfer of momentum?

Often in mechanics it is possible to explain an observation in more than one way. The motion of the balls in a Newton's cradle can be explained using energy ideas from Topic A.3.

These interactions between the balls in the Newton's cradle are called **collisions**. This is the term given to any interaction where momentum is transferred or shared between moving objects. Examples of collisions include:

- firing a gun

- hitting a ball with a bat in sport

- two toy cars running into each other

- a pile driver sinking vertical cylinders into the ground on a construction site.

Impulse and momentum

The change in velocity of the spheres in the Newton's cradle can be interpreted either as a change in momentum or as the effect of a force that acts between two colliding spheres for a given time. It is possible to link momentum and force using Newton's second law of motion. This involves a new quantity known as impulse.

Earlier in this topic, Newton's second law of motion was written as $F = ma$, where the symbols have their usual meaning. This equation can be rearranged using one of the kinematic equations: $a = \dfrac{v - u}{t}$. Eliminating a from Newton's second law gives $F = ma = \dfrac{m(v - u)}{t}$ which, in words, means that

$$\text{force} = \frac{\text{change in momentum}}{\text{time taken for change}}.$$

We use the convention that "Δ" means "change in". In symbols, the equation now becomes:

$$F \times \Delta t = \Delta p \text{ or } F = \frac{\Delta p}{\Delta t},$$ where p is the symbol for momentum and t (as usual) means time.

▲ **Figure 38** In 2021, a team of eight strong men pulled the world's heaviest plane 4.3 m in 73 s. The plane has an unloaded mass of 285 000 kg. If we assume that the plane accelerated uniformly from rest, we can calculate the unbalanced force on the aircraft. Frictional forces mean that a much larger force would have been needed.

This equation gives the relationship between force and momentum and provides a further clue to the real meaning of the concept of momentum. The equation shows that we can change the momentum of an object (in other words, accelerate it) by exerting a large force for a short time or by exerting a small force for a long time. A small number of people can get a heavy vehicle moving at a reasonable speed, but they must push for a much longer time than the vehicle itself would take if powered by its own engine (which produces a larger force).

The product of *force* and *time* is called **impulse** and is given the symbol J.

Impulse is the product of the *average resultant force acting on an object F* and the *contact time Δt over which the force acts*. In symbols this is:

$$J = F\Delta t$$

The units of impulse are newton seconds (N s). Impulse is equivalent to change in momentum, and N s gives us an alternative to $kg\,m\,s^{-1}$ as a unit for momentum.

 Resultant force

In most cases where an impulse acts, the force F varies throughout the contact time Δt.

Where the force varies, then F is the average resultant force that is acting throughout the time of contact.

Worked example 17

An impulse of 85 N s acts on a body of mass 5.0 kg that is initially at rest. Calculate the distance moved by the body in 2.0 s after the impulse has been delivered.

Solution

The change in momentum is 85 kg m s⁻¹ so that the final speed is $\dfrac{85}{5} = 17\,m\,s^{-1}$. In 2.0 s the distance travelled is 34 m.

Worked example 18

Jonathan strikes a tennis ball moving with a horizontal initial velocity of $45\,m\,s^{-1}$. The ball leaves the racket in the opposite direction at a speed of $65\,m\,s^{-1}$. The mass of the ball is 58 g.

a. Calculate the impulse that the racket delivered to the ball.

The ball is in contact with the racket for 20 ms.

b. Calculate the average force that the racket exerted on the ball.

Solutions

a. The initial and final velocities of the ball have opposite directions, so the change in velocity is $\Delta v = 65 - (-45) = 110\,m\,s^{-1}$. The impulse is the change in the ball's momentum, $J = m\Delta v = 0.058 \times 110 = 6.4\,N\,s$.

b. $Force = \dfrac{impulse}{time\ taken} = \dfrac{6.4}{0.020} = 320\,N$

Practice questions

25. A ball of mass 0.40 kg moves in a direction at right angles to a wall. The ball hits the wall at a speed of $9.0\,m\,s^{-1}$ and rebounds at a speed of $6.0\,m\,s^{-1}$. The contact time between the ball and the wall is 50 ms. Calculate the average values of:

    a. the acceleration of the ball during the collision with the wall

    b. the force between the wall and the ball.

26. An air rifle pellet of mass 2.0 g is fired at an initial speed of $180\,m\,s^{-1}$ into a stationary block of clay and becomes embedded in the block. The average force acting on the pellet from the block is 750 N.

    a. Calculate:

        i.   the change in the momentum of the pellet

        ii. the stopping time of the pellet in the clay block.

    b. Estimate the distance that the pellet penetrated.

    c. Outline why the answer in part b. is an estimate.

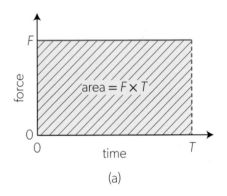

(a)

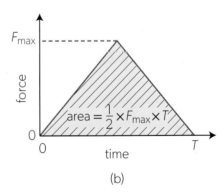

(b)

Force–time graphs

So far we have assumed that forces are constant and do not change with time. This is rarely the case in real life. We need a way to cope with changes in momentum when the force is not constant. The equation $F = \dfrac{\Delta p}{\Delta t}$ helps here. It suggests that a force–time graph can be useful.

When a constant force acts on a mass, then the graph of force against time will look like Figure 39(a). The change in momentum is $F \times T$. This is the hatched area.

The area under a force–time graph is equal to the change in momentum.

Another straightforward case that may more plausible than a constant force is the one in Figure 39(b), where the force rises to a maximum F_{max} and then decreases to zero in a total time T. The area under the graph this time is $\frac{1}{2}F \times T$ and this is the change in momentum in this case.

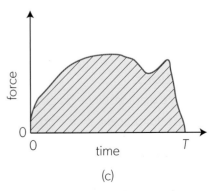

(c)

▲ Figure 39 Force–time graphs.

The final case (Figure 39(c)) is one where there is no obvious mathematical relationship between *F* and *t*, but nevertheless there is a graph of the variation of force with time. You will need to estimate the number of squares under the graph and use the area of one square to evaluate the total change in momentum.

Worked example 19

The sketch graph shows how the force acting on an object varies with time.

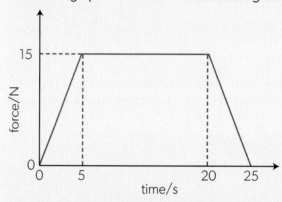

The mass of the object is 50 kg and its initial speed is zero. Calculate the final speed of the object.

Solution

The total area under the force–time graph is

$$2 \times (\frac{1}{2} \times 15 \times 5) + (15 \times 15) = 75 + 225 = 300\,\text{N s}.$$

This is the change in momentum, so the final speed is $\frac{300}{50} = 6.0\,\text{m s}^{-1}$.

Worked example 20

The graph shows how the momentum of an object of mass 40 kg moving along a straight line varies with time.

a.   Explain why the acceleration of the object remains constant.

b.   Calculate the acceleration of the object.

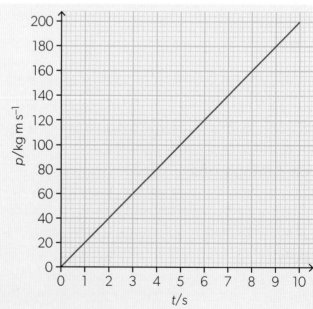

Solutions

a.   The gradient of the graph is $\frac{\Delta p}{\Delta t}$ and this is equal to the net force acting on the object. The gradient is constant. Hence, the net force is constant. Since the mass of the object does not change, the acceleration is also constant.

b.   The net force is $\frac{\Delta p}{\Delta t} = \frac{200}{10} = 20\,\text{N}$. The acceleration is therefore $a = \frac{F}{m} = \frac{20}{40} = 0.50\,\text{m s}^{-2}$.

Practice questions

27. An ice hockey puck of mass 150 g moves at a constant initial speed of 8.0 m s⁻¹ across a horizontal ice surface. A hockey stick hits the puck and it bounces off in the opposite direction. The graph shows how the horizontal force on the puck varies with time.

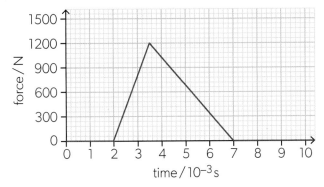

What is the speed of the puck immediately after the hit?

A. 8.0 m s⁻¹

B. 12 m s⁻¹

C. 20 m s⁻¹

D. 28 m s⁻¹

28. A ball rolling on a floor rebounds at right angles from a vertical wall. The graph shows how the contact force between the ball and the wall varies with time.

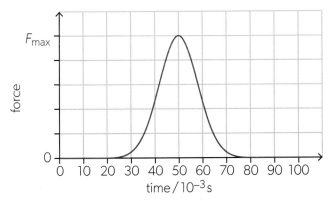

F_{max} is the maximum force acting on the ball. The change in the momentum of the ball is 10 N s.

What is the best estimate of F_{max}?

A. 20 N

B. 50 N

C. 200 N

D. 500 N

29. A ball is dropped vertically onto a floor and rebounds. The graph shows how the momentum of the ball varies with time. The direction upwards is positive.

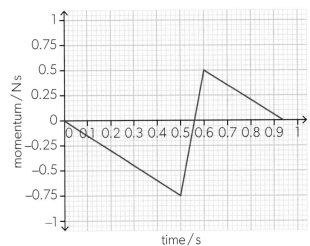

The ball falls freely between 0 and 0.5 s. Air resistance is negligible.

a. Calculate:

 i. the weight of the ball

 ii. the speed of the ball just before impact.

b. The ball is in contact with the floor for a time of 0.10 s.
 Determine the magnitude of the average contact force between the floor and the ball.

Rockets and impulse

Earlier in this topic we discussed the acceleration of a rocket and looked at the situation from the perspective of Newton's second and third laws. A similar analysis is possible in terms of impulse.

Revisiting Newton's second law

- Tool 3: Determine rates of change.

- Tool 3: Derive relationships algebraically.

- Tool 3: Identify and use symbols stated in the guide and the data booklet.

Earlier, we used $F = ma$ to show that F also equals $\dfrac{\Delta p}{\Delta t}$. Using the full expression for momentum p gives $F = \dfrac{\Delta(m \times v)}{\Delta t}$

This can be written as (using the product rule): $F = m\dfrac{\Delta v}{\Delta t} + v\dfrac{\Delta m}{\Delta t}$

You may have to take this algebra on trust, but you can understand the physics that it represents by thinking through what the two terms stand for:

- The first term on the left-hand side is just:

$$mass \times \frac{change\ in\ velocity}{time\ taken\ for\ change}$$

which you will recognize as $mass \times acceleration$—our original form of Newton's second law of motion.

- The second term on the right-hand side is something new. It is:

$$instantaneous\ velocity \times \frac{change\ in\ mass}{time\ taken\ for\ change}$$

Our first version of Newton's second law was a simpler form of the law than this new version. The extra term takes account of what happens when the mass of the accelerating object is also changing.

Our later form of Newton's second law helps with a rocket because the rocket is always losing mass (as the propellant escapes), so m in the equation is not constant.

The second term in the equation contains the ejection speed of the fuel relative to the rocket v, and the rate at which mass is lost from the system $\dfrac{\Delta m}{\Delta t}$. The acceleration of the rocket is therefore:

$$a = \frac{\Delta v}{\Delta t} = -\frac{v\Delta m}{m\Delta t} = -\frac{v}{m} \times \frac{\Delta m}{\Delta t}.$$

The negative sign reminds us that the rocket is *losing* mass while *gaining* speed.

▲ Figure 40 The launch of the Soyuz TMA-10 mission taking astronauts to the International Space Station. The rocket has a mass of 309 000 kg on the launch pad, but burns about 158 000 kg of fuel in the first 2 minutes of its flight. The exhaust gases leave the rocket at 2.6 km s⁻¹. What is the initial acceleration of the rocket? Why will the acceleration increase?

Rockets operate effectively in the absence of an atmosphere. All rockets release a liquid or gas that leaves the rocket at high speed (Figure 40). The fluid can be an extremely hot gas generated in the combustion of a solid chemical (as in a domestic firework rocket) or from the chemical reaction when two gases are mixed and react. It can also be a fluid stored under pressure inside the rocket. In each case, fluid escapes from the combustion or storage chamber through nozzles at the base of the rocket.

As a result, the rocket accelerates in the opposite direction to the direction of fluid ejection. The impulse on the rocket is equal and opposite to the impulse on the fuel as they form a closed system. Therefore, the rate of loss of momentum from the rocket in the form of high-speed fluid must be equal to the rate of gain in momentum of the rocket.

Worked example 21

A small firework rocket has a mass of 65 g. The initial rate at which hot gas is lost from the firework after it has been lit is 3.5 g s⁻¹ and the speed of release of this gas from the rear of the rocket is 130 m s⁻¹.

Calculate the initial acceleration of the rocket.

Solution

$a = \dfrac{\Delta v}{\Delta t} = -\dfrac{v\Delta m}{m\Delta t}$, where v is the release speed of the gas, $\dfrac{\Delta m}{\Delta t}$ is the rate of loss of gas and m is the mass of the rocket.

$$a = \frac{130 \times 3.5}{65} = 7.0\,\text{m s}^{-2}$$

Practice question

30. A spacecraft is initially at rest in outer space. The spacecraft is propelled by a rocket engine that ejects exhaust products at a constant rate of $2.8\,\text{kg s}^{-1}$ with a speed of $3.6 \times 10^3\,\text{m s}^{-1}$ relative to the spacecraft. The initial mass of the spacecraft and its fuel is $4.0 \times 10^4\,\text{kg}$.

   a. Calculate:

      i.   the thrust force of the rocket engine

      ii. the initial acceleration of the spacecraft.

   The rocket engine is fired for 25 minutes. The spacecraft accelerates along a straight line.

   b. Explain how the acceleration of the spacecraft varies with time.

   c. Calculate the final speed of the spacecraft.

   d. The thrust force of the engine must be briefly increased to 65 kN without changing the relative speed of the exhaust products. Calculate the mass that will have to be ejected from the engine per second.

Momentum and Newton's third law

A consequence of Newton's third law of motion is that when two objects A and B interact with object A producing an impulse on object B, then object B must produce an impulse on A. The two impulses will be equal and opposite. Imagine two toy cars colliding in a straight line and then rebounding.

The forces F that the cars exert on each other are equal and opposite and the cars are in contact for identical times Δt. This means that the magnitude of the product $F \times \Delta t$ is the same on both cars, but because the two forces are in opposite directions the signs of $F \times \Delta t$ are different. This product of average resultant external force and the contact time is the change in momentum of each car. It follows that the change of momentum of car A is equal in magnitude and opposite in direction to the change of momentum of car B. When we think of the system as consisting of both cars together, then there has been no change in momentum of the system combined during the collision.

We say that the **momentum has been conserved**.

Experiments to compare the momentum before a collision with the momentum after a collision show, within the experimental uncertainty of the measurements, that the total momentum in a system does not change when no resultant external forces act on it.

Is momentum conserved in a laboratory experiment?

- Inquiry 1: Appreciate when and how to reduce friction.

- Inquiry 1: Demonstrate creativity in the designing, implementation or presentation of the investigation

- Inquiry 3: Compare the outcomes of an investigation to the accepted scientific context.

- Inquiry 3: Identify and discuss sources and impacts of random and systematic errors.

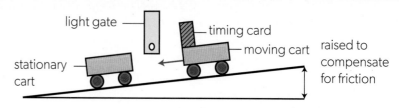

▲ Figure 41 One cart moves down the track at constant speed and collides with another stationary cart. In the collision, the momentum of the system can be shown to be conserved within experimental error.

The exact details of this experiment will depend on the apparatus you have in your school (Figure 41). You may have alternatives to carts on runways, such as air pucks floating on an air table.

- The experiment consists of measuring the speed of a cart of known mass which hits another cart, also of known mass, that is initially stationary. You are likely to have carts of almost identical mass, as this is a particularly easy case to begin with.

- You need a way to measure the velocity of the carts just before and just after the collision. This could be done in various ways:
  - using a data logger with motion sensors
  - using a paper-tape system where a tape attached to the cart is pulled through a device that makes dots at regular time intervals on the tape
  - using a video camera and computer software
  - using a stopwatch to measure the time taken to cover a short, known distance before and after the collision.

- If your carts roll on a track or runway, you can allow for the friction at the cart axles and air resistance. Raise the end of the track so that, when pushed, a cart runs at a constant speed. The friction at the bearings and the air resistance will be exactly compensated by the component of the weight of the cart down the track.

- For the first part of the experiment, arrange the two carts so that they will stick together after colliding.

This can be done using modelling clay, by attaching a pin to one cart that enters a cork on the other cart, or by attaching an attracting magnet to each cart.

- Make the first (moving) cart collide with, and stick to, the second (stationary) cart.

- Measure the speed of the first cart before the collision and the combined speed of the carts after the collision.

- Repeat the experiment several times and think carefully about the likely errors in the results.

- The initial momentum is:

  (*mass of first cart*) × (*velocity of first cart*).

- The final momentum is:

  (*mass of first cart* + *mass of second cart*) × (*combined velocity of both carts*).

- What can you say about the total momentum of the system before the collision compared with the total momentum after the collision? You should consider the experimental errors in the experiment before making your judgement.

- If you have done the experiment carefully, you should find that the momentum before and after the collision are approximately equal.

- Now extend your experiment to different cases:
  - where the carts do not stick together
  - where they are both moving before the collision
  - where the masses are not the same
  - and so on.

- You may need to alter how you measure the velocity to cope with the different cases.

An important point here is that there must be no external force from outside the system that acts on the colliding objects that make up the system. External forces produce accelerations, and these will change the velocity and hence the momentum of the whole system.

In the colliding-carts experiment, gravitational force is acting on the carts. However, because the gravitational force is not being allowed to do any work when the track is horizontal, the force does not contribute to the interaction because the carts are not moving vertically.

Momentum is always constant when no resultant external force acts on the system.

This is known as the **principle of conservation of linear momentum**, the word "linear" is here because the objects concerned move in a straight line. This rule is always confirmed in experiments and is one of the important conservation rules that are true throughout the universe (as far as we know). The history of nuclear physics shows that scientists have needed to propose the existence of new particles in experiments where momentum was apparently not conserved. The proposed particles were subsequently found to exist.

Momentum conservation in practice

Momentum conservation is such an important rule that it is worth us considering a few different situations to see how momentum conservation works. In each of these cases we assume that the centres of the objects lie on a straight line so that the collision happens in one dimension.

1 Two objects with the same mass, one initially stationary, when no energy is lost

This is known as an **elastic collision**. No permanent deformation occurs in the objects that collide, and no energy can be released as internal energy (through friction), as sound or in any other way. The spheres in the Newton's cradle earlier lose only a little energy every time they collide, and this explains why the apparatus is a reasonable demonstration of momentum conservation.

Figure 42 shows the arrangement when the first moving object collides with the second stationary object. The first object stops and remains at rest while the second moves off at the speed that the first object had before the collision. (Try flicking a coin across a smooth table to hit an identical coin head-on to see this happen.) In this case, momentum is conserved because (using an obvious set of symbols for the mass m of objects 1 and 2 and their velocities u (before the collision) and v (after the collision)):

$$m_1 \times u = m_2 \times v$$

Because $m_1 = m_2$, then $u = v$, so the velocity of one mass before the collision is equal to the velocity of the second mass afterwards. The kinetic energy of the moving mass (whichever mass is moving) is $\frac{1}{2}mu^2$ and does not change either. Energy is conserved meaning that the collision is elastic.

The discovery of the electron antineutrino is a case where the energy spectrum of the emitted beta-minus particle (β^-) in beta decay led to the prediction of the neutrino and its subsequent discovery. You will find the details of this discovery in Topic E.3.

 How are concepts of equilibrium and conservation applied to understand matter and motion from the smallest atom to the whole universe?

In this course, there are many references to the conservation laws of charge, energy and momentum. Other conservation rules exist in physics, too. The concept of equilibrium is also universally applicable. The equilibrium of the skin of a balloon and the steady state of stars in the middle of their lives arise from a balance of forces, inwards and outwards. You will find references to conservation in every topic in the course and to equilibria in Themes B and E.

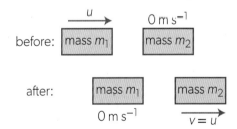

▲ Figure 42 An elastic collision between two identical masses, one of them initially at rest.

 Energy and momentum

Kinetic energy—the energy that a body has when it is moving—is considered in Topic A.3. It is likely, though, that you will have met kinetic energy in an earlier course. The kinetic energy of translational motion is $E_k = \frac{1}{2}mv^2$, where m is the mass of the moving object and v is its linear speed.

The unit of energy is the joule (J). An object of mass 1.0 kg moving at a speed of 1.0 m s^{-1} has a kinetic energy of 0.50 J.

There is a convenient link between kinetic energy and momentum. The equations $E_k = \frac{1}{2}mv^2$ and momentum $p = mv$ can be combined using $p^2 = m^2v^2$ to give $E_k = \frac{p^2}{2m}$. This is often useful in calculations.

If experimental measurements contain uncertainties, how can laws be developed based on experimental evidence? (NOS)

No measurement in science is certain. The Heisenberg uncertainty principle shows this at a fundamental level when it predicts that the product of uncertainty in energy measurement (ΔE) and uncertainty in time determination (Δt) must always be greater than $\dfrac{h}{4\pi}$, where h is the Planck constant. This limitation in experimental measurement is accepted in science and does not inhibit the inquiry cycle of scientific discovery. Uncertainties in measurement are factored into our imprecision in knowledge and do not prevent the generation of theory.

Worked example 22

An astronaut of mass 90 kg (including his gear) is initially at rest outside a spaceship. The astronaut throws a tool of mass 1.5 kg at a speed of 3.0 m s⁻¹ away from him. Calculate

a. the speed of the astronaut immediately after he releases the tool

b. the ratio $\dfrac{\text{kinetic energy of the tool}}{\text{kinetic energy of the astronaut}}$.

Solutions

a. The momentum of the system of the astronaut and the tool is zero; hence $90v - 1.5 \times 3.0 = 0$, where v is the speed of the astronaut. The minus sign indicates that the astronaut and the tool are moving in opposite directions.

$$v = \frac{1.5 \times 3.0}{90} = 0.050 \, \text{m s}^{-1}.$$

b. The ratio of the kinetic energies is equal to $\left(\dfrac{P^2_{\text{tool}}}{2m_{\text{tool}}}\right) \div \left(\dfrac{P^2_{\text{astronaut}}}{2m_{\text{astronaut}}}\right) = \dfrac{m_{\text{astronaut}}}{m_{\text{tool}}}$.

Because the tool and the astronaut have equal magnitude of momentum,

$P_{\text{tool}} = P_{\text{astronaut}}$. Therefore, $\dfrac{\text{kinetic energy of the tool}}{\text{kinetic energy of the astronaut}} = \dfrac{90}{1.5} = 60$.

The tool has a much greater kinetic energy than the astronaut, even if their momenta have equal magnitudes. Note that the answer only depends on the masses involved, not on the speed with which the tool has been thrown.

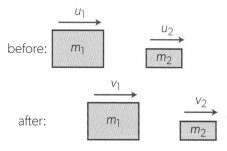

▲ Figure 43 Two moving objects with different mass in an elastic collision.

2 Two objects with different masses when no energy is lost

This time (as you may have seen in an experiment) the situation is more complicated (Figure 43).

Again, $m_1u_1 + m_2u_2 = m_1v_1 + m_2v_2$ but, this time, we cannot eliminate the mass terms from the expression so easily, as m_1 and m_2 are not the same. What we do know is that kinetic energy is conserved. The kinetic energy before the collision must equal the kinetic energy after the collision (because no energy is lost).

This means that (summing the kinetic energies before and after the collision):

$$\frac{1}{2}m_1u_1^2 + \frac{1}{2}m_2u_2^2 = \frac{1}{2}m_1v_1^2 + \frac{1}{2}m_2v_2^2.$$

The momentum and kinetic energy equations can be solved to show that:

$$v_1 = \left(\frac{m_1-m_2}{m_1+m_2}\right)u_1 + \left(\frac{2m_2}{m_1+m_2}\right)u_2 \text{ and } v_2 = \left(\frac{m_2-m_1}{m_1+m_2}\right)u_2 + \left(\frac{2m_1}{m_1+m_2}\right)u_1$$

🧪 Different cases

- Tool 3: Determine the effect of changes to variables on other variables in a relationship.

- Tool 3: Select and manipulate equations.

There are two interesting cases when mass m_2 is initially stationary and m_1 collides with it:

- When m_1 is much smaller than m_2. In the v_1 equation, when m_1 is small, the first term becomes roughly $\left(\dfrac{-m_2}{m_2}\right) u_1$ which is approximately $-u_1$; the second term is zero because $u_2 = 0$;. The small m_1 mass "bounces off" the large mass (this is shown by the minus sign). The large mass gains speed in the forward direction. The magnitude of the speed of the larger mass is roughly $\left(\dfrac{2m_1}{m_2}\right) u_1$. This is a small fraction of the original small-mass speed.

- When m_1 is much greater than m_2, $(m_1 \gg m_2)$. This time the original mass loses hardly any speed. The momentum lost by m_1 is given to m_2 which moves off in the same direction, but at about twice the original speed of m_1. Look at the v_1 and v_2 equations and satisfy yourself that this is true.

Figure 44 shows the effects of the mass ratio for a golf club and ball.

▲ Figure 44 In golf, the head of the driver is heavier than the golf ball. A professional golfer club might strike the ball with the club moving at about 50 m s⁻¹ but the ball might travel off at about 75 m s⁻¹.

3 Two objects colliding when energy is lost

When a moving object collides with a stationary one and the two objects stick together, moving off at the same speed, then some of the initial kinetic energy is lost. After the collision, there is a single object with an increased (combined) mass and a single common velocity. This is an **inelastic collision** (Figure 45).

This is a case you can easily study experimentally.

The momentum equation this time is $m_1 u_1 = (m_1 + m_2)v_1$. A rearrangement shows that $v_1 = \dfrac{m_1}{(m_1 + m_2)} u_1$ and, as we might expect, the final velocity is in the same direction as before, but is always smaller than the initial velocity.

As for energy loss, the incoming kinetic energy is $\frac{1}{2}m_1 u_1^2$ and the final kinetic energy (substituting for v_1) is $\frac{1}{2}(m_1 + m_2) \dfrac{m_1^2}{(m_1 + m_2)^2} u_1^2$.

This is $\dfrac{1}{2}\dfrac{m_1^2}{(m_1 + m_2)} u_1^2$ and the ratio $\dfrac{\text{final kinetic energy}}{\text{initial kinetic energy}} = \dfrac{m_1}{(m_1 + m_2)}$.

How do collisions between charge carriers and the atomic cores of a conductor result in thermal energy transfer?

In Topic B.5, electrical resistance is related to the collisions between electrons moving through a conductor and the transfer of energy from them to the atoms of a metal. Momentum must be conserved in all these interactions. This is the case where an elastic collision occurs between a small fast-moving object (the electron) and a very massive object (the atom). Although the energy transfer is not very efficient, there are very many interactions every second and so an appreciable amount of energy is transferred to the metal resistor.

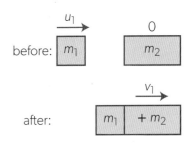

▲ Figure 45 An inelastic collision between two objects.

before: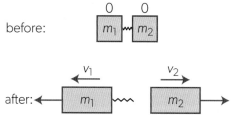

after:

▲ Figure 46 Energy is gained in this collision where a spring between the objects is released.

 Popper and falsifiability

No-one has yet observed a case where the momentum in an isolated system is not conserved, but we should continue to look!

Karl Popper, a philosopher of the 20th century, argued that the test of whether a theory was truly scientific was whether it was capable of being falsified. By this he meant that there must be an experiment that could, in principle, contradict the hypothesis being tested. Popper argued that psychoanalysis was not a science because it could not be falsified by experiment.

Popper also applied his ideas to scientific induction. He said that, although we cannot prove that the Sun will rise tomorrow, because it always has we can use the theory that the Sun rises in the morning until the day when it fails to do so. At that point we must revise our theory.

What should happen when momentum appears not to be conserved in an experiment? Is it more sensible to look for a new theory or to suggest that something has been overlooked?

4 Two objects when energy is gained

There are many occasions when two initially stationary masses gain kinetic energy. Some laboratory dynamics carts have a way to demonstrate this with a plunger and spring inside the cart. Another easy way is to attach two small strong magnets to the front of two carts with the like poles of the magnets facing each other. When the carts are released after being held together, the magnets repel and drive the carts apart.

The analysis is straightforward in this case (Figure 46). The initial momentum is zero as neither object is moving. After the collision, therefore, the momentum is $m_1v_1 + m_2v_2 = 0$. This means that $m_1v_1 = -m_2v_2$.

The objects move apart in opposite directions. When the masses are equal, the speeds will be the same, with one velocity the negative of the other. When the masses are not equal, then $\dfrac{m_1}{m_2} = -\dfrac{v_2}{v_1}$.

Worked example 23

A rail truck of mass 4500 kg moving at a speed of 1.8 m s^{-1} collides with a stationary truck of mass 1500 kg. The two trucks couple together. Calculate the speed of the trucks immediately after the collision.

Solution

Initial momentum = $4500 \times 1.8 = 8100$ kg m s^{-1}.
Final momentum = $(4500 + 1500) \times v$, where v is the final speed.

Momentum is conserved so $v = \dfrac{8100}{4500 + 1500} = 1.4$ m s^{-1}.

Worked example 24

Stone A of mass 0.5 kg travelling at 3.8 m s^{-1} across the surface of a frozen pond collides with a stationary stone B of mass 3.0 kg. Stone B moves off at a speed of 0.65 m s^{-1} in the same original direction as stone A. Calculate the final velocity of stone A.

Solution

Initial momentum = $0.5 \times 3.8 = 1.9$ kg m s^{-1}.
Final momentum = $3.0 \times 0.65 + 0.5v_A$.

Momentum is conserved so $v_A = \dfrac{1.9 - (3.0 \times 0.65)}{0.5} = -0.1$ m s^{-1}.

The minus sign shows that the final velocity of stone A is opposite to its original motion.

Practice questions

31. A ball of mass m rolling on a floor with speed v rebpunds from a wall with an unchanged speed and at the same angle θ to the wall. The diagram shows the top view of the collision.

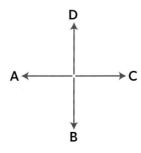

Δp is a vector representing the change in the linear momentum of the ball.

a. What is the magnitude of Δp?

    A. $mv \sin \theta$ B. $mv \cos \theta$

    C. $2mv \sin \theta$ D. $2mv \cos \theta$

b. Which arrow correctly represents the direction of Δp?

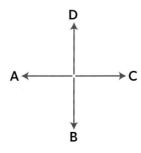

32. Two rail trucks of equal masses move towards each other with speeds $2v$ and v.
The trucks collide and stick together. What is the speed of the trucks immediately after the collision?

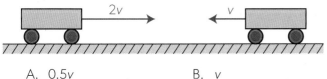

    A. $0.5v$ B. v

    C. $1.5v$ D. $2v$

33. An object of mass 2.0 kg slides without friction on a horizontal ice surface at a speed of 6.0 m s^{-1}. The object explodes into two pieces of masses 0.5 kg and 1.5 kg. Immediately after the explosion, the smaller piece stops relative to the ice.

Calculate:

a. the speed of the larger piece immediately after the explosion

b. the gain in kinetic energy of the system as a result of the explosion.

34. An air rifle pellet of mass 2.0 g is fired horizontally at a block of clay of mass 50 g that rests on a frictionless horizontal surface. The pellet passes through the block with no change in the direction of motion and emerges with a speed of 150 m s^{-1}. Immediately after the pellet emerges from the block, the block is moving at a speed of 2.4 m s^{-1}.

a. Calculate the initial speed of the pellet.

b. It takes 1.5×10^{-4} s for the pellet to travel through the block.

Calculate:

    i. the average acceleration of the pellet

    ii. the average force between the pellet and the block.

35. Two rail trucks of masses 6000 kg and 2000 kg collide head-on at equal speeds v.

Immediately after the collision, the 6000 kg truck stops relative to the ground and the 2000 kg truck moves off with a speed of 6.0 m s^{-1}.

a. Determine the initial speed v of the trucks.

b. Show that the collision is elastic.

Momentum conservation in two dimensions

The momentum and energy considerations above were applied to cases where the motions of the colliding objects were all in the same line. This is motion in one dimension. The physics of two-dimensional collisions is identical to the simpler case of linear motion:

    Momentum is conserved; this is true in every direction.

What assumptions about the forces between molecules of gas allow for ideal gas behaviour? (NOS)

In Topic B.3, there is an important microscopic analysis of the particles in a gas that links the motion of these particles to the macroscopic measurements that we can make of the gas. This analysis relies heavily on the ideas of force and momentum conservation that are introduced in this topic.

To carry out the analysis, assumptions are required. Collisions between the gas particles and the wall are assumed to be elastic, for example. When you have studied the mechanics used in the kinetic theory of Topic B.3, review this topic, and link the ideas there to those here. Cross-linking topics in this way will improve your understanding of both areas of the subject.

Figure 47 shows an air puck travelling towards a stationary puck on an air hockey table.

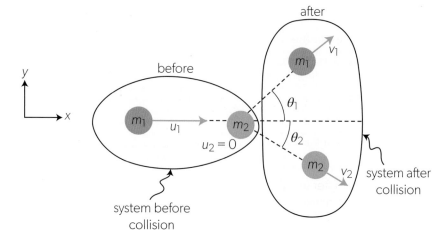

▲ Figure 47 Momentum conservation in two dimensions. A moving object collides off-centre with a stationary object. After the elastic collision the objects move apart with an angle of $\theta_1 + \theta_2$ between them.

The centres of the pucks are not in line with the direction of motion of the first puck and so the pucks move off in different directions, as shown. We assume that the pucks do not rotate either before or after the motion. Otherwise, we would have to allow for the rotational energy and rotational momentum of the pucks; this is left for Topic A.4.

Immediately before the collision, the first puck which has a mass m_1 is moving with speed u_1. The pucks move as shown in Figure 47 immediately after the collision with speeds v_1 and v_2 at angles θ_1 and θ_2.

In problems such as this, a good tip is to choose two axes at right angles that make the subsequent analysis as straightforward as possible. For the example here, one axis is in the same direction as u_1 and the other axis is at right angles to it.

For the axis in the same direction as u_1 (horizontally along the page), applying the conservation of momentum immediately before and after the collision and resolving the velocities gives

$$m_1 \times u_1 = m_1 \times v_1 \cos\theta_1 + m_2 \times v_2 \cos\theta_2$$

For the axis at 90° to u_1 (vertically up the page), momentum conservation gives

$$0 = m_1 \times v_1 \sin\theta_1 - m_2 \times v_2 \sin\theta_2 \text{ so that } m_1 \times v_1 \sin\theta_1 = m_2 \times v_2 \sin\theta_2$$

So far, we have two equations, and we can use these to determine two unknown values. For example, suppose v_2 and θ_2 are unknown. It is possible to show that

$$\theta_2 = \tan^{-1}\left(\frac{v_1 \sin\theta_1}{u_1 - v_1 \cos\theta_1}\right) \text{ and } v_2 = \frac{m_1 v_1 \sin\theta_1}{m_2 \sin\theta_2}$$

which, when θ_2 has been calculated, allows v_2 to be worked out too.

When the collision is elastic (no energy lost from the system), then there is an additional equation from equating kinetic energies before and after the collision:

$$\frac{1}{2}m_1 u_1^2 = \frac{1}{2}m_1 v_1^2 + \frac{1}{2}m_2 v_2^2$$

With the further assumption that the puck masses are the same ($m_1 = m_2$), there is an interesting result because the combination of the momentum equations leads to

$$\frac{1}{2}mu_1^2 = \frac{1}{2}mv_1^2 + \frac{1}{2}mv_2^2 + mv_1v_2\cos(\theta_1 + \theta_2).$$

For the elastic interaction of two objects with identical masses, conservation of momentum and energy indicates that there are three outcomes:

- $v_2 = 0$; in other words, the pucks do not collide and $v_1 = u_1$.

- $v_1 = 0$; the collision was head-on; the first puck stops dead and the second puck continues with the initial velocity of the first puck.

- $\cos(\theta_1 + \theta_2) = 0$; the angle $\theta_1 + \theta_2$, which is the angle of separation, must be equal to 90° after the collision. The pucks move apart with an angle of 90° between them.

Figure 48 shows a time-lapse photograph of the collision between two pucks, one moving and one stationary before the impact.

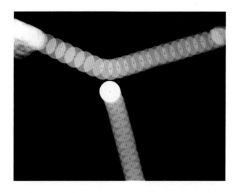

▲ Figure 48 A photograph of an almost elastic collision between two air pucks. The puck that was initially stationary moves off to the bottom of the image. The angle between the two velocity vectors after the collision is a right angle.

Worked example 25

An air puck of mass 0.20 kg collides with a stationary air puck of mass 0.30 kg. After the collision, the second puck moves away with a speed of 0.50 m s⁻¹. The paths of the pucks make angles of 37° and 45° with the original direction of the first puck.

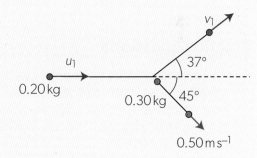

a. Determine the speeds u_1 and v_1 of the first puck before and after the collision.

b. Calculate the percentage change in the kinetic energy of the system.

Solutions

a. The speed of the first puck after the collision can be determined by considering the vertical component of the momentum of the system, which is zero before and after the collision.

$$0.20 \times v_1 \times \sin 37° = 0.30 \times 0.50 \times \sin 45° \Rightarrow v_1 = \frac{0.30 \times 0.50 \times \sin 45°}{0.20 \times \sin 37°} = 0.8812... \simeq 0.88 \text{ m s}^{-1}.$$

The horizontal component of momentum is also conserved, which leads to an equation for the initial speed of the puck: $0.20 \times u_1 = 0.20 \times v_1 \times \cos 37° + 0.30 \times 0.50 \times \cos 45°$.

$$u_1 = v_1 \cos 37° + \frac{0.30}{0.20} \times 0.50 \times \cos 45° = 1.234... \simeq 1.2 \text{ m s}^{-1}.$$

b. The kinetic energy before the collision is $\frac{1}{2} \times 0.20 \times 1.234^2 = 0.152$ J.

After the collision, it is $\frac{1}{2} \times 0.2 \times 0.8812^2 + \frac{1}{2} \times 0.3 \times 0.50^2 = 0.115$ J.

The kinetic energy of the system has decreased by $0.152 - 0.115 = 0.037$ J. This is about 24% of the initial KE of the first puck.

Worked example 26

A snooker ball moving at a speed of $1.20\,\text{m s}^{-1}$ collides elastically with another stationary snooker ball of the same mass. After the collision, the balls move apart in perpendicular directions. The path of the first ball makes an angle of $30°$ to the original direction.

Determine the speeds v_1 and v_2 of the balls after the collision.

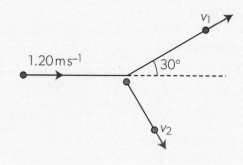

Solution

The second ball recoils at an angle of $60°$ to the original direction of the first ball. The total momentum in the vertical direction is zero:

$$mv_1 \sin 30° = mv_2 \sin 60° \Rightarrow v_2 = \frac{\sin 30°}{\sin 60°} v_1.$$

Applying the conservation of momentum in the horizontal direction:

$$m \times 1.20 = mv_1 \cos 30° + mv_2 \cos 60° = mv_1 \left(\cos 30° + \frac{\sin 30° \cos 60°}{\sin 60°} \right) \approx 1.15 mv_1.$$

From here, $v_1 = \dfrac{1.20}{1.15} = 1.04\,\text{m s}^{-1}$ and $v_2 = \dfrac{\sin 30°}{\sin 60°} \times 1.04 = 0.60\,\text{m s}^{-1}$.

Practice questions

36. A body of mass $1.0\,\text{kg}$ moving at a speed of $1.6\,\text{m s}^{-1}$ collides with an initially stationary body of mass $2.0\,\text{kg}$. After the collision, the first body moves at right angles to the original direction of motion with a speed of $0.80\,\text{m s}^{-1}$.

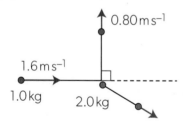

    a. Determine the velocity (magnitude and direction) of the second body after the collision.

    b. Deduce whether the collision is elastic.

37. A particle of mass m moving at a speed of $200\,\text{m s}^{-1}$ collides elastically with a stationary particle of mass $4m$. After the collision, the first particle moves with a speed of $160\,\text{m s}^{-1}$ at an angle of $82.8°$ to the original direction of motion.

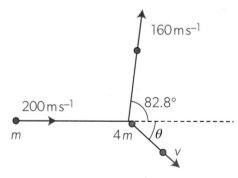

    a. Determine the speed v of the second particle after the collision. Hint: the kinetic energy is conserved.

    b. Calculate the angle θ that the path of the second particle makes with the original direction of motion of the first particle.

Momentum conservation in practice

Recoil of a gun

Figure 49 shows a gun being fired to trigger a snow fall. This prevents a more dangerous avalanche. When the gun fires its shell, the gun moves backwards in the opposite direction to the shell. You should be able to explain this in terms of momentum conservation.

Initially, both gun and shell are stationary; the initial total momentum is zero. When the gun is fired, the shell is propelled in the forwards direction through the gun barrel by the expansion of hot gas. This is the case discussed earlier, in which energy is gained.

The explosion in the barrel is a force internal to the system. The explosion in the chamber behind the shell generates gas at high pressure. The gas exerts a force on the interior of the shell chamber and hence a force on the gun as well. The explosion releases energy and this is transferred into the kinetic energies of both the shell and the gun.

The initial linear momentum was zero and no *external* force has acted on the system. The momentum must continue to be zero and this can only be true when the gun and the shell move in opposite directions with the same magnitude of momentum. The shell will go fast because it has a small mass compared with the gun; the gun moves relatively slowly.

While the gun recoils with equal and opposite momentum to the shell, the kinetic energies are not evenly divided. The kinetic energy of the shell is much larger than the kinetic energy of the gun's recoil since the kinetic energy depends on *speed*[2].

▲ Figure 49 Firing a shell from gun into a snow mass to cause a minor avalanche.

Water hoses

Watch fire fighters extinguish a fire using a high-pressure hose and you will see the effect of water leaving the system. Often two or more fire fighters are needed to keep the hose on target because there is a large force on the hose in the opposite direction to which the water emerges. This can be seen when a garden hose that is free to move starts to shoot backwards in unpredictable directions when the water tap is turned on.

The cross-sectional area of the hose is greater than that of the nozzle through which the water emerges. The mass of water flowing past a point in the hose every second is the same as the mass that emerges from the nozzle every second. The speed of the water emerging from the nozzle must be greater than the water speed along the hose itself. The water gains momentum as it leaves the hose because of this increase in exit speed compared with the flow speed in the hose.

The momentum of the system must be constant and so there must be a force backwards on the end of the hose which needs to be countered by the fire fighters. The water pump (or other source of pressure) that feeds water to the fire hose supplies the kinetic energy and the momentum.

The momentum lost by the system per second is

(*mass of water leaving per second*) × (*speed at which water leaves the nozzle – speed in the hose*)

The mass of water lost per second is $\frac{\Delta m}{\Delta t}$, so the momentum lost per second is

$\dot{p} = \frac{\Delta m}{\Delta t}(v - u)$, where v is the speed of water as it leaves the nozzle and u is the speed of the water in the hose (Figure 50).

When we know the cross-sectional area A of the nozzle of the hose and the density of the water, ρ, then $\frac{\Delta m}{\Delta t}$ can be determined. Figure 50 shows what happens inside the hose during a one-second time interval. Every second you can imagine a cylinder of water leaving the hose. This cylinder is v long and has an area A. The volume leaving per second is therefore Av.

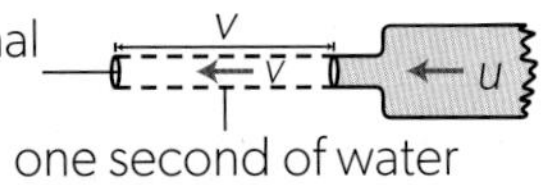

A, cross-sectional area of nozzle

one second of water

▲ Figure 50 Momentum conservation when water leaves a hose.

The mass of the water leaving in one second is

$$\frac{\Delta m}{\Delta t} = (density\ of\ water) \times (volume\ of\ cylinder) = \rho Av.$$

Because the mass entering and leaving the nozzle per second must be the same, this means that

$$the\ change\ of\ momentum\ in\ one\ second = \frac{\Delta m}{\Delta t} \times (v - u) = \rho Av(v - u).$$

When $u \ll v$, then the expression simplifies to ρAv^2.

The hose is an example of where it is important to look at the whole system. Consider what happens when the water is directed at a vertical wall. The water strikes the wall, loses all its horizontal momentum, and trickles vertically down the wall. The momentum must have been absorbed by the wall, its foundations and, therefore, the ground. We might conclude that Earth itself has gained momentum and that we can speed up Earth's rotation by using a garden hose. This is not true, because the water originally gained momentum from a pump. This gain in momentum at the pump must have given some momentum to Earth too. The amount of momentum Earth gained at the pump is equal and opposite to the momentum gained by Earth when the water strikes the wall.

Worked example 27

A mass of 0.48 kg of water leaves a garden hose every second. The nozzle of the hose has a cross-sectional area of $8.4 \times 10^{-5}\,m^2$. The water flows in the hose at a speed of $0.71\,m\,s^{-1}$. The density of water is $1000\,kg\,m^{-3}$. Calculate:

a.   the speed at which water leaves the hose

b.   the force on the hose.

Solutions

a.   Volume of water leaving the hose per second $= \frac{0.48}{1000} = 4.8 \times 10^{-4}\,m^3$.

This leaves through a nozzle of area $8.4 \times 10^{-5}\,m^2$, so the speed must be $\frac{4.8 \times 10^{-4}}{8.4 \times 10^{-5}} = 5.71\,m\,s^{-1}$.

b.   The force on the hose $=$ mass lost per second $\times$ change in speed $= 0.48 \times (5.71 - 0.71) = 2.4\,N$.

Practice question

38.   Water flows in a garden hose with inner diameter 14 mm at a rate of 9.0 litres per minute.

a.   Calculate the speed of water in the hose.

The cross-sectional area of the nozzle of the hose is 12 times smaller than the cross-sectional area of the hose itself.

b.   Determine the change of momentum of the water leaving the nozzle in one second.

c.   Explain why a force is needed to keep the nozzle of the hose stationary.

Helicopters

Helicopters are aircraft that can take off and land vertically and can hover motionless above a point on the ground (Figure 51). There were many attempts to build flying machines on the helicopter principle over the centuries, but the first commercial aircraft flew in the 1930s.

▲ Figure 51 The rotors of a helicopter allow it to take off and land vertically as well as hover above a point on the ground. This uses the principle of conservation of momentum.

A helicopter uses the principle of conservation of linear momentum to hover. The rotating blades exert a force on air that was originally stationary, causing it to move towards the ground, gaining momentum in the process. As a result of Newton's third law, there is an upward force on the helicopter through the rotors.

Global impact of science—Momentum and safety

Both seat belts and airbags restrain the occupants of a car, preventing them from striking the windscreen or the hard areas around it when there is an accident. But there is more to the physics of the air bag and the seat belt than this.

On the face of it, someone in a car loses the same amount of kinetic energy and momentum whether they are stopped abruptly by the windscreen or restrained by the seat belt. The difference

▲ Figure 52 In the case of a car crash, the front of the car is designed to crumple rather than remain rigid. This enables the car to slow down over a longer time.

between the two cases is the time during which the kinetic energy and momentum are lost. Without the seat belt or air bag, the time taken by the passenger to stop will be extremely short and the deceleration will therefore be large. A large deceleration implies a large force acting on the passenger and it is the magnitude of the stopping force that determines the amount of damage they sustain in an accident.

Seat belts and air bags increase the time taken by the occupants of the car to stop and as *force × time = momentum change*, for a constant change in momentum, a long stopping time will imply a smaller, and less damaging, force.

@ **How is conservation of momentum relevant to the workings of a nuclear power station?**

A nuclear power station demonstrates many aspects of the conservation of momentum. The simple example of a hose here can be extended to the more complex situation of a jet of steam at high pressure striking the blades of the rotating turbine with the transfer of kinetic energy from steam to turbine.

The process of moderation in the containment vessel of the reactor also relies on momentum transfer as the neutrons interact (collide) with the moderator atoms. This is again the case where a moving object collides with a stationary object. For maximum energy transfer the masses of the objects should be the same. In practice, in reactors the moderator atoms are often more massive than the neutrons.

Nuclear engineers use conservation of momentum to predict the number of collisions required before a neutron has lost sufficient energy (speed) to be effective in promoting further fissions. This number informs the engineers about the shape and size of the moderator in the reactor.

What other examples of momentum transfer and conservation can you think of for the nuclear power station?

Worked example 28

A toy helicopter of mass 0.80 kg hovers motionless above the ground. The rotating blades of the helicopter force the initially stationary air to move downwards with a speed of 9.0 m s^{-1}.

a.   Calculate the mass of air pushed downwards in one second so that the helicopter can remain stationary.

b.   The density of air is 1.2 kg m^{-3}. The helicopter has a single rotor. Estimate:

   i.   the surface area spanned by the blades of the rotor

   ii.   the radius of a blade.

Solutions

a.   The lift force exerted by the air on the helicopter is equal to the rate of change of the momentum of the air, $v\frac{\Delta m}{\Delta t}$.

   When the helicopter hovers, the magnitude of this force is equal to the weight of the helicopter.

   $$9.0\frac{\Delta m}{\Delta t} = 0.80 \times 9.8; \text{ therefore } \frac{\Delta m}{\Delta t} = \frac{0.80 \times 9.8}{9.0} = 0.87 \text{ kg}.$$

b.   i. $\text{Area} = \dfrac{\text{volume of air passing the blades per second}}{\text{speed of air}} = \dfrac{0.87 \div 1.2}{9.0} = 8.1 \times 10^{-2} \text{ m}^2.$

   ii.   The area spanned by the blades is a circle with radius equal to the length of a blade.

   $$\pi r^2 = 8.1 \times 10^{-2} \Rightarrow r = \sqrt{\frac{8.1 \times 10^{-2}}{\pi}} = 16 \text{ cm}.$$

Practice questions

39. A helicopter of mass 3.0×10^3 kg hovers motionless above the ground.

   a.   Calculate the magnitude of the lift force acting on the helicopter.

   The rotor of the helicopter pushes initially stationary air with a downward speed v. The surface area of the rotor is $A = 95$ m^2. The density of air is $\rho = 1.2$ kg m^{-3}.

   b.   Show that the mass of air pushed downwards per second is $\rho A v$.

   c.   Calculate the speed v.

   The air is now forced to move with a higher speed u so that the helicopter ascends with an initial upward acceleration of 1.2 m s^{-2}.

   d.   Estimate the new speed u.

40. A skier of mass 64 kg crashes at a speed of 45 km h^{-1} into a safety net installed at an edge of a ski slope. The skier will avoid injury if the force on her from the safety net is less than 6.0 kN.

   a.   Calculate the minimum time needed to stop the skier safely.

   b.   Estimate the distance by which the safety net deflects if the skier stops in the minimum safe time. Assume that the force on the skier is constant during deceleration.

Momentum and sport

This topic began with a suggestion that it was less painful to catch a table-tennis ball than a baseball. You should now be able to understand the reason for the difference. You should also realize why good technique in many sports hinges on the application of momentum change. Think about a sport you play or watch, and how effective use of momentum change helps the player.

Many sports in which an object—usually a ball—is struck by hand, foot or bat rely on the efficient transfer of momentum. This transfer is often enhanced by a "follow through", which increases the contact time between bat and ball. The player can maintain the same force but for a longer time interval, so the impulse on the ball increases, increasing the momentum change as well.

 ## Force acting on a soccer ball

- Tool 2: Use sensors.

- Tool 3: Select and manipulate equations.

- Inquiry 3: Explain realistic and relevant improvements to an investigation.

You can estimate the force used to kick a soccer ball in a laboratory. It uses many of the ideas contained in this topic and is a good place to conclude our study of momentum. The basis of the method is a measurement of the contact time between the foot and the ball and the subsequent change in momentum of the ball (Figure 53).

The use of

$$force \times contact\ time = change\ in\ momentum$$

allows the force to be calculated.

- **To measure the contact time**: Stick some metal foil to the shoe of the person who is to kick the ball and to the soccer ball itself. Set up a data logger or fast timer so that it only times while the two pieces of foil are in contact.

- **To measure the change in momentum**: The ball starts from rest so all you need to estimate is the magnitude of the final momentum. The ball should be kicked horizontally from a lab bench.

  - Measure the distance s from where the ball is kicked to where it lands.

  - Measure the distance h from the bottom of the ball on the bench to the floor.

  - Use projectile motion and the kinematic equations to calculate the time t taken for the ball to reach the floor: $h = \frac{1}{2}gt^2$ and therefore $t = \sqrt{\frac{2h}{g}}$.

  - Use this value of t to estimate the initial speed u of the ball as $u = \frac{s}{t}$.

  - Measure the mass of the ball M; therefore the change in momentum is Mu which is equal to the *force on the ball* × T.

This method can be modified to estimate the impact forces involved in many sports including hockey, baseball and golf. Consider this as a possible preliminary experiment leading to your IA. Think about how you might develop the experiment within an inquiry cycle (see the *Tools for physics* section).

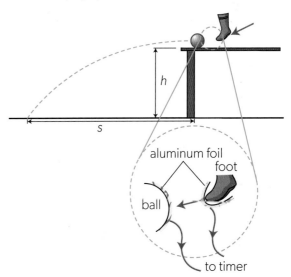

▲ Figure 53 An estimation of the force that acts on a soccer ball when it is kicked requires a measurement of the contact time and the change in momentum.

▲ Figure 54 A fairground carousel—the physics of circular motion in action.

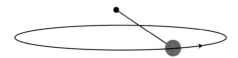

▲ Figure 55 What keeps the object moving in the horizontal circle? Why is the string not horizontal?

Motion in a circle

Theme-park rides are popular with many people. Such rides often include movement around a circle, sometimes moving horizontally, sometimes vertically (Figure 54). What is the physics of circular motion (Figure 55)?

Imagine a small object whirling around in a horizontal circle with a constant speed held at the end of a piece of string.

The choice of the words "constant speed" is deliberate. When motion is circular, you can say that the "speed is constant" but not that "velocity is constant". Motion around a circle at a constant speed is known as **uniform circular motion**.

Velocity—a vector quantity—has both magnitude *and* direction. The object on the string has a constant speed but the direction in which the object is moving is changing all the time. The velocity has a constant *magnitude* but a changing *direction*. When either of the two elements that make up a vector change, then the vector can no longer be regarded as constant.

As the velocity changes (even if only the direction is changed) then the object has been accelerated. Understanding the physics of this acceleration is the key to understanding circular motion. But before looking at how the acceleration arises, you will need a technical language to describe rotational motion.

Angular displacement

Angular displacement is the angle through which an object moves in its circular motion. Angular displacement can be regarded as a scalar. Angular displacement can be measured in degrees (°) or in radians (rad). Radians are more commonly used than degrees in this branch of physics so, if you have not met radians before, read about the differences between radians and degrees in the *Tool for physics* section on page 335.

Angular speed and angular velocity

The term *speed* is usually used to refer to "linear speed"—motion in a straight line. When the motion is in a circle, there is an alternative: **angular speed**. This is given the symbol ω (the lower-case Greek letter, omega).

$$average\ angular\ speed = \frac{angular\ displacement}{time\ for\ the\ angular\ displacement\ to\ take\ place}$$

Figure 56 shows how the symbols are defined and you will see that, in symbols, the definition of angular speed becomes $\omega = \frac{\theta}{t}$, where θ is the angular displacement and t is the time taken for the angular displacement. In terms of the diagram:

$$\omega = \frac{\theta}{t_2 - t_1}.$$

You will also meet the term "angular velocity" in the IB Diploma Programme physics course. You will not have to treat it as a formal vector quantity.

ω, angular speed $= \dfrac{\theta}{t_2 - t_1} = \dfrac{\theta}{t}$

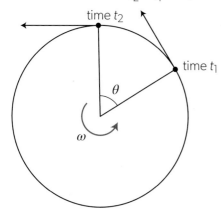

▲ Figure 56 Angular speed.

Angular speed or angular velocity?

You may be wondering about the distinction between angular speed and angular velocity, and whether angular velocity is a vector like linear velocity.

The answer is that angular velocity is a vector but with a surprising direction which is along the axis of rotation, as shown in Figure 57. In other words, through the centre of the circle around which the object is moving, and perpendicular to the plane of the rotation.

The direction follows a clockwise corkscrew rule so that, in this example, because the object is rotating clockwise, the direction of the angular velocity vector is into the plane of the paper.

In the IB Diploma Programme physics course, both the terms angular velocity and angular speed are used but they refer always to the magnitude of the angular velocity.

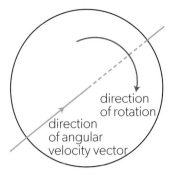

▲ Figure 57 Angular velocity direction.

Period and frequency

The time taken for a rotating point to go round its circle once is known as the **periodic time** or simply the **period** of the motion; it has the symbol T. In one time period, the angular distance travelled is 2π rad and therefore $T = \dfrac{2\pi}{\omega}$.

When T is in seconds, the units of ω are radians per second, abbreviated to rad s^{-1}.

If you have already studied waves in this course, you will have met the similar idea of **time period** as the time taken for one cycle of the wave.

Another quantity that is associated with T is **frequency**. Again, this quantity is common to both the physics of rotation and wave theory. Frequency is the number of times an object goes round a circle in unit time, and one way to express the unit of frequency would be in "per second" or s^{-1}. However, the unit of frequency is re-named after the 19th-century physicist Heinrich Hertz and is abbreviated to Hz. T is linked to f by: $T = \dfrac{1}{f}$

This leads to a link between ω and f: $\omega = 2\pi f$

Linking angular and linear speeds

Sometimes you know the linear speed around a circle and need the equivalent angular speed or vice versa.

This is straightforward. When the circle has a radius r, its circumference is $2\pi r$. T is the time taken to go around the circle once.

The linear speed of the object along the edge of the circle v is: $v = \dfrac{2\pi r}{T}$

Rearranging the equation gives: $T = \dfrac{2\pi r}{v}$

Also, $T = \dfrac{2\pi}{\omega}$, so equating the two equations for T gives $T = \dfrac{2\pi r}{v} = \dfrac{2\pi}{\omega}$.

Cancelling the 2π and rearranging gives $v = r\omega$.

Notice that both in this equation and in the earlier equation $s = r\theta$, the radius r multiplies the angular term to obtain the linear term. This is a consequence of our definition of angular measure.

Worked example 29

A large clock on a building has a minute hand that is 4.2 m long. Calculate:

a.   the angular speed of the minute hand

b.   the angular displacement, in radians, in the time periods:

  i.   12 noon to 12.20   ii.   12 noon to 14.30.

c.   the linear speed of the tip of the minute hand.

Solutions

a.   The minute hand goes round once (2π rad) every hour.
     One hour is 3600 s.

$$\text{Angular speed} = \frac{\text{angular displacement}}{\text{time taken}} = \frac{2\pi}{3600} = 0.001\,75\,\text{rad s}^{-1}$$

b.   i.   20 minutes is $\frac{1}{3}$ of 2π, so $\frac{2\pi}{3}$ rad

     ii.   2.5 h is $2\pi \times 2.5 = 5\pi$ rad

c. $v = r\omega = 4.2 \times 0.001\,75 = 0.007\,33\,\text{m s}^{-1} = 7.3\,\text{mm s}^{-1}$

Worked example 30

The International Space Station (ISS) moves in an approximately circular orbit, 420 km above Earth's surface, with a linear speed of 7650 m s^{-1}. The radius of Earth is 6370 km. Calculate the number of times the ISS orbits Earth each day.

Solution

The orbital radius of the ISS is 6370 + 420 = 6790 km. The period of one orbit is

$$\frac{\text{circumference}}{\text{linear speed}} = \frac{2\pi \times 6790 \times 10^3}{7650} = 5.58 \times 10^3\,\text{s}$$

The number of orbits in one day (24 × 60 × 60 seconds) is therefore $\dfrac{24 \times 60 \times 60}{5.58 \times 10^3} = 15.5$.

Practice question

41.   The motion of Earth around the Sun can be modelled as uniform circular motion around an orbit of radius 1.50×10^{11} m.

Calculate, for the orbital motion of Earth:

a.   the angular speed

b.   the linear speed.

42.   The blades of a toy helicopter complete 670 revolutions in one minute. Each blade is 16 cm long. Calculate:

a.   the angle, in radians, swept by a blade in one second

b.   the linear speed of the tip of a blade.

Centripetal acceleration

An object moving at a constant angular speed in a circle is being accelerated. Newton's first law tells us that, for any object in which the direction of motion or the speed is changing, there must be an external force acting. In circular motion, the direction of motion is constantly changing and so the object accelerates, and there must be a force acting on it to cause this to happen. The force that acts to keep the object moving in a circle is called the centripetal force. This force leads to a centripetal acceleration. (The word "centripetal" originates from two Latin words, *centrum* and *petere* — literally "to lead to the centre".)

The centripetal acceleration is directed inwards towards the centre of the circle.
The acceleration is always at 90° to the velocity vector.

The centripetal acceleration a is given by

$$a = \frac{v^2}{r} = \omega^2 r = v\omega$$

Worked example 31

Using the data provided in Worked example 30, calculate the centripetal acceleration of the International Space Station in its orbital motion.

Solution

$$a = \frac{v^2}{r} = \frac{7650^2}{(6370 + 420) \times 10^3} = 8.6\,\mathrm{m\,s^{-2}}.$$

The numerical value of the answer is slightly less than the acceleration of free fall near Earth's surface ($9.8\,\mathrm{m\,s^{-2}}$). This is not a coincidence! In Topic D.1 you will learn that all bodies moving in a gravitational field of a massive object such as Earth experience an acceleration due to gravity whose magnitude decreases with the distance from the centre of the object. At an altitude of 420 km, the gravitational acceleration is reduced to about 88% of its value on Earth's surface.

Worked example 32

A cyclist rides along a circular track of radius 25 m at a constant linear speed of 30 km h^{-1}. Calculate:

a. the angular speed of the cyclist

b. the centripetal acceleration.

Solutions

a. The linear speed should be converted to m s^{-1}, $v = \frac{30}{3.6} = 8.3\,\mathrm{m\,s^{-1}}$. Angular speed $\omega = \frac{v}{r} = \frac{30 \div 3.6}{25} = 0.333\,\mathrm{rad\,s^{-1}}$

b. Acceleration $a = \omega^2 r = 0.333^2 \times 25 = 2.8\,\mathrm{m\,s^{-2}}$

Practice questions

43. A paper disc of radius R rotates at a constant angular speed about an axis passing through the centre of the disc. P and Q are two particles of the disc. P is on the circumference of the disc and Q is at a distance of $\frac{R}{2}$ from the centre.

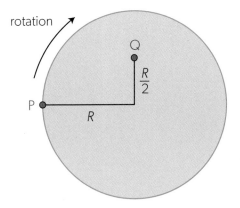

What is the ratio $\dfrac{\text{centripetal acceleration of P}}{\text{centripetal acceleration of Q}}$?

A. $\frac{1}{2}$

B. 1

C. 2

D. 4

44. Neutron stars, known for their relatively small size and fast rotation, are compressed remnants of exceptionally massive supergiant stars. A neutron star has a radius of 10 km and rotates at a frequency of 5 Hz. For a particle on the equator of the neutron star, calculate:

a. the linear speed

b. the centripetal acceleration.

45. A Ferris wheel rotating at a constant angular speed completes two revolutions per minute. Passengers riding on the Ferris wheel move at a linear speed of $3.0\,\text{m}\,\text{s}^{-1}$.

Calculate:

a.   the angular speed of the Ferris wheel

b.   the radius of the Ferris wheel

c.   the centripetal acceleration of a passenger taking the ride.

Investigating how *F* varies with *m*, *v* and *r*

- Tool 3: Determine the effect of changes to variables on other variables in a relationship.

- Inquiry 1: Identify and justify the choice of dependent, independent and control variables.

- Inquiry 3: Identify and discuss sources and impacts of random and systematic errors.

- Inquiry 3: Explain realistic and relevant improvements to an investigation.

This experiment tests the relationship $m\dfrac{v^2}{r} = Mg$

and uses the simple apparatus shown in Figure 58. To do this an object is whirled in a horizontal circle.

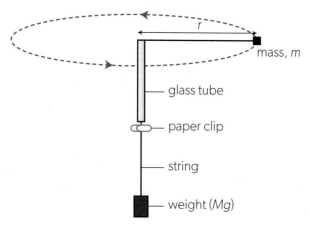

r

mass, *m*

glass tube

paper clip

string

weight (*Mg*)

▲ Figure 58   A simple experiment to confirm $F = m\dfrac{v^2}{r}$.

- An object (of weight *Mg*) hangs from one end of a string and a mass (a rubber bung of mass *m*) is attached to the other end of the string. A paper clip is attached to the string below a glass tube. The clip is used to ensure that the radius of rotation of the bung is constant. The bung should be rotated at a speed so that the paper clip stays just below the glass tube.

- The tension in the string is the same everywhere (whether below the glass tube or above in the horizontal part). This tension is equal to *Mg*, where *M* is the mass of the object that hangs vertically.

- Use an angular speed at which you can comfortably count the number of rotations of the bung in a particular time. From these data you can work out the linear speed *v* of the bung.

- To verify the equation, you need to test each variable against the others. There are several possible experiments. In each of these, one variable is held constant (a **control variable**), one is varied (the **independent variable**), and the third (the **dependent variable**) is measured. One example is:

Variation of *v* with *r*

- In this experiment, *m* and *M* must be unchanged. Move the clip to change *r* and, for each value of *r*, measure *v* using the method given above.

- Analysis: $\dfrac{v^2}{r}$ = constant and so a graph of v^2 against *r* ought to be a straight line passing through the origin. Alternatively, for each experimental run, you might simply divide v^2 by *r* and look critically at the result (which should be the same each time) to see if the value is constant. You should assess the errors in the experiment and put error limits on your value of $\dfrac{v^2}{r}$.

- What are other possible experimental tests?

- In practice, the string cannot rotate in the horizontal plane because of its own weight. How can you improve the experiment or the analysis to allow this?

Centripetal force

Newton's second law of motion in its simpler form tells us that $F = ma$ using the usual symbols. The second law applies to the force that provides the centripetal acceleration, so the magnitude of the force is:

$$ma = m\frac{v^2}{r} = m\omega^2 r = mv\omega$$

The question you need to ask for any situation involving circular motion is: what force provides the centripetal force in that situation? The direction of this force must be along the radial line between the object and the centre of the circle.

 ## Models—Centripetal or centrifugal?

When discussing circular motion, you will almost certainly have heard the term "centrifugal force"—probably everywhere except in a physics laboratory! However, so far this course has exclusively used "centripetal force". Why are two terms in use, and which is correct?

The alternative idea of centrifugal force comes from common experience. Imagine you are in a car going round a circle at high speed. You will undoubtedly feel that you are being "flung outwards".

One way to explain this is to imagine the situation from the point of view of a helicopter passenger hovering stationary above the circle around which the car is moving (Figure 59). From the helicopter you can see the passenger attempting to go in a straight line (Newton's first law). Nevertheless, the passenger is forced to move in a circle through friction forces acting between passenger and car seat. If the seat is frictionless and the passenger is not wearing a seat belt, then he or she will not get the "message" that the car is turning. The passenger will continue to move in a straight line eventually meeting the door that is turning with the car. If there is no door, what direction will the passenger take?

Another way to explain this is to imagine yourself in the car as it rotates. This is a rotating frame of reference that is accelerating and so does not obey Newton's laws of motion. You instinctively think that your rotating frame is stationary. Therefore, your tendency to go in what you believe to be a straight line feels like an outward force away from the centre of the circle (remember that the rest of the world now rotates round you, and the straight line you imagine that you are travelling along is actually part of a circle). Think about a cup of coffee sitting on the floor

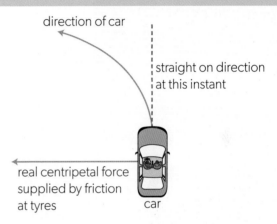

▲ **Figure 59** The centripetal force acting on a car seen from above.

of the car. When there is insufficient friction at the base of the cup, the cup will slide to the side of the car. In the inertial frame of reference (Earth) the cup is trying to go in a straight line. In your rotating frame of reference you have to "invent" a force acting outwards from the centre of the circle to explain the motion of the cup.

There are many examples of changing a reference frame in physics. Research the Foucault pendulum and perhaps go to see one of these fascinating pendulums in action or set one up in a high-ceilinged room in your school. Find out what is meant by the Coriolis force and how it affects the motion of weather systems in the northern and southern hemispheres.

Physicists often change reference frames. It is the Nature of Science to adopt alternative frames of reference to make explanations and theories more accessible.

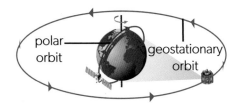

▲ Figure 60 Satellites in orbit. The polar orbit takes about 90 minutes. A geostationary satellite orbits once in 24 hours and so appears to sit at one point above the equator.

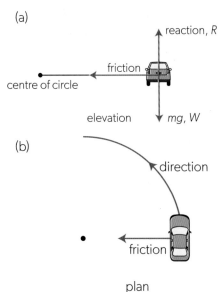

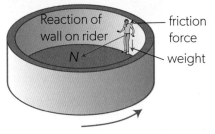

▲ Figure 61 The "rotor" fairground ride in action.

(a)

reaction, R

friction

centre of circle

elevation   mg, W

(b)

direction

friction

plan

▲ Figure 62 How friction between the road and the tyres enables a car to move around a circle.

Centripetal accelerations and forces in action

Satellites in orbit

Figure 60 shows satellites in a circular orbit around Earth. Why do satellites follow these paths? Gravitational forces act between the centre of mass of Earth and the centre of mass of the satellite. The direction of the force acting on the satellite is always towards the centre of the planet and it is the gravity that supplies the centripetal force. Topic D.1 returns to the subject of gravitational orbits and looks at them in a quantitative way.

Amusement park rides

Many amusement park rides take their passengers in curved paths. In the type of ride called a rotor (Figure 61), the riders are inside a drum that rotates about a vertical axis. When the rotation speed is large enough, the people are forced to the sides of the drum and the floor drops away. The people are quite safe because they are "held" against the inside of the drum as the reaction at the wall provides the centripetal force to keep them moving in the circle. The people in the ride feel the reaction between their spine and the wall. Friction between the rider and the wall prevents the rider from slipping down the wall.

Turning and banking

When a driver wants to make a car turn a corner, a resultant force must act towards the centre of the circle to provide a centripetal force. The car is in vertical equilibrium (the driving surface is horizontal) but not in horizontal equilibrium.

Turning on a horizontal road

For a horizontal road surface, the friction force acting between the tyres and the road becomes the centripetal force. The friction force is related to the coefficient of friction and the normal reaction at the surface where friction occurs (Figure 62).

The centripetal force required must be less than the frictional force if the car is not to skid:

$$m\frac{v^2}{r} < \mu_s mg$$

where μ_s is the static friction coefficient of friction. The expression can be rearranged to give a maximum speed before skidding occurs of

$$v_{max} = \sqrt{\mu_s gr}$$

for a circle of radius r.

Why is no work done on a body moving along a circular trajectory?

The expression for work done (= force × displacement in direction of force) has a vector nature that becomes apparent when dealing with motion in a circle. This is outlined in Topic A.4. The centripetal force that is required to maintain a circular trajectory acts towards the centre of the circle around which the object is moving. The displacement is zero in this direction as the circle radius is constant, and the work done is also zero. Of course, any force in the direction of travel—to overcome air resistance, for example—will require a transfer of energy in the usual way.

Banking

Tracks for motor or cycle racing, and even ordinary roads for cars are sometimes **banked** (Figures 63). The curve of the banked road surface is inclined at an angle, so that the normal reaction force contributes to the centripetal force that is needed for the vehicle to go round the track at a particular speed. Bicycles and motorcycles can achieve the same effect on a level road surface by "leaning in" to the curve. Tyres do not need to provide so much friction on a banked track compared to a horizontal road; this reduces the risk of skidding and increases safety.

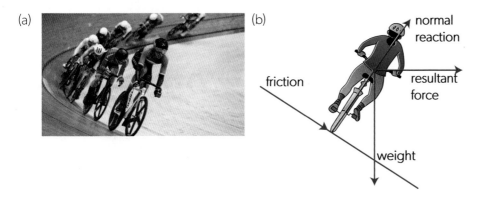

(a) (b)

▲ Figure 63 (a) A cycle velodrome is banked at varying angles to assist the cyclists. (b) The forces that act on a cyclist on a banked surface.

Figure 64 shows forces acting on an object rolling round a banked track. This is simplified to a point object moving in a circle to remove the complications of size and shape. A horizontal centripetal force directed towards the centre of the circle is needed for the rotation. The other forces that act on the ball are the force N normal to the surface (which is at the banking angle θ) and its weight mg acting vertically down. The vector sum of the horizontal components must equal the centripetal force.

The centripetal force is equal to $N \sin\theta$. The normal force resolved vertically is $N \cos\theta$ and is, of course, equal and opposite to mg. So,

$$F_{centripetal} = N \sin\theta = \left(\frac{mg}{\cos\theta}\right) \sin\theta = mg\tan\theta$$

As usual, $F_{centripetal} = \frac{mv^2}{r}$ and therefore $\tan\theta = \frac{v^2}{gr}$.

The banking angle is correct at a particular speed and a particular radius. Notice that it does not depend on the mass of the vehicle, so a banked road works for all the road users provided that they are going at the same speed. At speeds greater or less than this, there will need to be a horizontal component of friction supplied between the tyre and the road surface to prevent a slide. The direction of the frictional force can change: at speeds higher than the correct banking speed it is towards the centre of the circle, at lower speeds towards the outside of the bend.

Some more examples of banking

- Commercial airline pilots fly around a banked curve to change the direction of a passenger jet. When the angle is correct, the passengers will not feel the turn, they just feel a marginal increase in weight pressing down on their seat.

- Some high-speed trains tilt as they go around curves so that the passengers feel more comfortable.

Although you will not be asked to solve mathematical problems on this topic in your IB physics examination, you do need to understand the principles of banking.

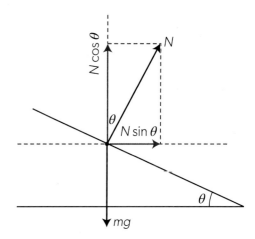

▲ Figure 64 The forces involved in banking.

Worked example 33

A cyclist riding at a speed of $11\,\text{m s}^{-1}$ attempts to make a sharp turn on a horizontal road. The turn can be modelled as a circle of radius $15\,\text{m}$.

a.   State the force that provides the centripetal acceleration of the bicycle.

The coefficient of static friction between the tyres of the bicycle and the road is $\mu = 0.75$.

b.   Deduce whether the cyclist will be able to turn without skidding.

c.   Outline how banking of the turn can increase the maximum safe speed of the cyclist.

Solutions

a.   The centripetal force is the friction between the tyres and the road. The direction of this force is towards the centre of the turn.

b.   The maximum centripetal force is μmg, where m is the combined mass of the cyclist and the bicycle. This can be equated to the expression $m\dfrac{v^2}{r}$ for the centripetal force, to give the condition for the maximum safe speed v of the bicycle. $m\dfrac{v^2}{r} = \mu mg$; hence $v = \sqrt{\mu gr} = \sqrt{0.75 \times 9.8 \times 15} = 10.5\,\text{m s}^{-1}$. The cyclist rides just above this speed, so he will start skidding during the turn.

c.   On a banked road, the normal reaction force has a horizontal component that adds to the frictional force, increasing the maximum centripetal force and hence the maximum centripetal acceleration.

Worked example 34

An amusement park ride called the rotor (see Figure 61) is a large cylinder rotating about the vertical axis. When the cylinder rotates fast enough, the floor drops out and the people taking the ride remain motionless against the inside of the cylinder.

a.   State the force that provides the centripetal acceleration of a passenger taking the rotor ride.

b.   Show that for the passenger to not slide down the inside of the cylinder, the centripetal acceleration a must satisfy the condition $a \geq \dfrac{g}{\mu}$, where μ is the coefficient of static friction between the passenger and the cylinder.

The coefficient of static friction is 0.36. The inner radius of the cylinder is $4.0\,\text{m}$.

c.   Determine the maximum period of rotation of the cylinder so that the passenger will not slide down the inside of the cylinder.

Solutions

a.   The centripetal force in this situation is the normal reaction force on the person from the inner surface of the cylinder.

b.   To prevent the passenger from sliding down, the static frictional force acting upwards must be equal to the passenger's weight, mg. The static frictional force cannot be greater than μN, where N is the normal force from the cylinder. This leads to the condition $\mu N \geq mg$. But N is the force that provides the centripetal acceleration a, so $N = ma$. Substitution into the previous inequality gives $\mu ma \geq mg$; hence $a \geq \dfrac{g}{\mu}$.

c.   It is convenient to first find the minimum angular speed ω of the cylinder.

$\omega^2 r \geq \dfrac{g}{\mu}$, so $\omega \geq \sqrt{\dfrac{9.8}{0.36 \times 4.0}} = 2.61\,\text{rad s}^{-1}$. The maximum period of rotation is then $T = \dfrac{2\pi}{\omega} = \dfrac{2\pi}{2.61} = 2.4\,\text{s}$.

This corresponds to about 25 rotations per minute!

Worked example 35

An aircraft flying at a constant speed makes a banked turn in a horizontal plane. The wings of the aircraft make an angle θ with the horizontal.

a. Draw a free-body diagram showing forces acting on the aircraft in the vertical plane.

b. The speed of the aircraft is $180\,\text{m s}^{-1}$ and the radius of the turn is $7400\,\text{m}$. Determine the banking angle θ.

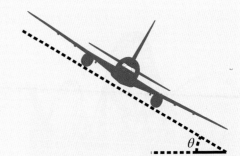

Solutions

a. The forces on the aircraft are the lift force (acting perpendicular to the plane of the wings, so at an angle θ to the vertical) and the weight. The diagram also shows the resultant force on the aircraft, which acts towards the centre of the turn and provides the centripetal acceleration. It is important to realize that this is not an independent force but rather the vector sum of the lift force and weight.

b. The centripetal acceleration of the aircraft is $a = \dfrac{180^2}{7400} = 4.38\,\text{m s}^{-2}$.

The lift force can be resolved into horizontal and vertical components. The horizontal component is equal to the centripetal force on the aircraft, ma, and the vertical component is equal to the weight, mg. Applying trigonometry to vector components leads to

$\tan\theta = \dfrac{\text{resultant force}}{\text{weight}} = \dfrac{ma}{mg} = \dfrac{a}{g}$. From here, $\tan\theta = \dfrac{4.38}{9.8} = 0.447 \Rightarrow \theta = 24°$.

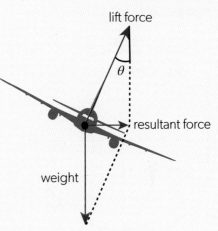

Practice questions

46. A small box is placed at the edge of a turntable of radius 30 cm that rotates about the vertical axis. The coefficient of static friction between the box and the turntable is 0.70.

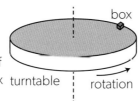

a. Determine the maximum angular speed of the turntable so that the box does not fall off.

b. Calculate the period of rotation of the turntable that corresponds to the angular speed you have found in part a.

47. Modern jet fighter aircrafts can structurally sustain accelerations of up to about $9g$ during manoeuvring ($1g$ is equivalent to the acceleration of free fall, $9.8\,\text{m s}^{-2}$).

A jet aircraft flying at a speed of $280\,\text{m s}^{-1}$ is to make a circular turn in a horizontal plane.

Estimate the minimum possible radius of the turn.

48. A small marble rolls around a horizontal circular path on the inner surface of a conical bowl. The surface of the bowl makes an angle θ with the vertical.

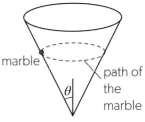

a. Draw a free-body diagram of the forces acting on the marble.

b. Show that the acceleration a of the marble is given by $a = \dfrac{g}{\tan\theta}$.

The speed of the marble is $1.5\,\text{m s}^{-1}$ and its mass is $3.0\,\text{g}$. The angle of the bowl is $\theta = 28°$.

c. Calculate:

i. the radius of the marble's path

ii. the magnitude of the reaction force on the marble from the bowl.

A small frictional force acts on the marble so that it gradually spirals down the bowl along a path that can be modelled as a circle of slowly decreasing radius.

d. Predict the effect this will have on the angular speed of the marble.

Moving in a vertical circle

The examples so far have been of motion around a horizontal circle. The amount of thrill from a ride such as that in Figure 65 depends on its height and speed, and the forces that act on the riders.

How must the horizontal situation be modified when the circular motion of the mass is in a vertical plane? What forces act when motion is in a vertical circle?

Imagine a mass on the end of a string that is moving in a vertical circle at constant speed.

Look carefully at Figure 66 and notice the way in which the tension in the string changes as the mass goes around in an anticlockwise direction.

Begin with the case when the string is horizontal, at point A. The weight acts downwards and the tension in the string is the horizontal centripetal force towards the centre of the circle.

The mass continues to move upwards and reaches the top of the circle at B. At this point, the tension in the string and the weight both act downwards. Thus,

$$T_{\text{down}} + mg = m\frac{v^2}{r}$$

and therefore

$$T_{\text{down}} = m\frac{v^2}{r} - mg$$

The weight of the mass combines with the string tension to provide the centripetal force and so the tension required is less than the tension T when the string is horizontal.

At C, the bottom of the circle, the tension and the weight both act vertically but in opposite directions, so

$$T_{\text{up}} = m\frac{v^2}{r} + mg$$

At the bottom, the string tension must include the weight and the required centripetal force.

As the mass moves around the circle, the tension in the string varies continuously. It has a minimum value at the top of the circle and a maximum at the bottom. The bottom of the circle is the point where the string is most likely to break. When the maximum breaking tension of the string is T_{break}, then, for the string to remain intact,

$$T_{\text{break}} > m\frac{v^2}{r} + mg$$

and the linear speed at the bottom of the circle must be less than

$$\sqrt{\frac{r}{m}(T_{\text{break}} - mg)}$$

If this seems to you to be a very theoretical idea without much practical value, think about a car going over a bridge (Figure 67). Assuming that its shape is part of a circle, then the bridge will have a radius of curvature r. What is the speed at which the car will lose contact with the bridge?

This is the case considered above, where the object, in this case, the car, is at the top of the circle. What is the "tension" (in this case, the force between car and road) when the car wheels lose contact with the bridge? To answer this question, you might begin with a free-body diagram. You should be able to show that the car loses contact at a speed equal to $\sqrt{gr}$.

▲ Figure 65 A theme park ride.

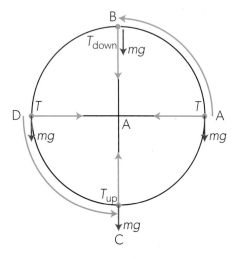

▲ Figure 66 Forces acting in circular motion in a vertical plane.

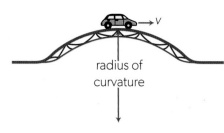

▲ Figure 67 A car going over a curved bridge.

Worked example 36

A hammer thrower in an athletics competition swings the hammer on its chain round 7.5 times in 5.2 s before releasing it. The hammer describes a circle of radius 2.1 m and has a mass of 4.0 kg. Assume that the hammer is swung in a horizontal circle and that the chain is horizontal.

a. Calculate, for the rotation:

   i. the average angular speed of the hammer

   ii. the average tension in the chain.

b. Comment on the assumptions made in this question.

Solutions

a. i. 7.5 revolutions $= 15\pi$ rad

     Angular speed $= \dfrac{15\pi}{5.2} = 9.1$ rad s^{-1}

  ii. Tension in the chain $=$ centripetal force required for rotation.

     Centripetal force $= mr\omega^2 = 4.0 \times 2.1 \times 9.1^2 = 690$ N

b. The thrower usually inclines the plane of the circle at about 45° to the horizontal in order to achieve maximum range. Even if the plane were horizontal, then the weight of the hammer would contribute to the system so that a component of the tension in the chain must allow for this. Both assumptions are unlikely.

Practice questions

49. A stone of mass 0.25 kg attached to the end of a string is moving in a vertical circle of radius 0.80 m at a constant speed. The string will break if the tension in it exceeds 10 N.

  a. Explain whether the string is more likely to break when the stone passes the lowest or the highest point of the path.

  b. Calculate:

    i. the minimum speed of the stone so that the string remains taut at the highest point

    ii. the maximum speed of the stone so that the string doesn't break.

50. A car of mass m moving at speed v goes over a bridge whose central part can be modelled as a section of a circle of radius r (see Figure 67).

  a. Derive an expression, in terms of m, v and r, for the magnitude of the normal reaction force between the car and the bridge as the car passes the top.

  It is given that $r = 60$ m and $m = 1400$ kg.

  b. Calculate:

    i. the normal reaction force on the car at the top of the bridge when the car moves at 50 km h^{-1}.

    ii. the speed that the car would have if the normal reaction force at the top of the bridge was reduced to 10% of its value on a horizontal road.

A.3 Work, energy and power

How are concepts of work, energy and power used to predict changes within a system?

How can a consideration of energetics be used as a method to solve problems in kinematics?

How can transfer of energy be used to do work?

Energy is a central concept throughout all of science. During an energy transfer, work can be done and the rate at which energy is transferred shows the power of the system that is doing the transfer. The law of conservation of energy predicts the overall transfers that occur when energy changes its nature. We can predict this change, either within a system or between systems. The fact that a conservation law exists underlines the importance of the energy concept in science.

So far in this theme, you have used derived equations (such as the *suvat* kinematic equations) and known physics laws (such as Newton's laws of motion) to predict the outcomes of mechanical changes. You can also calculate outcomes in mechanics by considering the energy changes. Indeed, in some circumstances, considering energy transfer can be the only reliable method possible.

Figure 1 shows the transfer of energy between many forms in a fountain. Kinetic energy is transferred to the water in this fountain as it rises into the air. As it does so, the kinetic energy is transferred to gravitational potential energy. The water does work against air resistance which causes energy to be transferred to thermal energy and the water will be marginally hotter when it returns to the ground. The lights also transfer energy (from electrical energy).

▲ Figure 1 A water fountain is an example of energy transfer.

In this topic, you will learn about:

- conservation of energy and work done by a force
- energy transfers
- kinetic energy, gravitational potential energy and elastic potential energy
- power as the rate of doing work and transferring energy
- efficiency
- Sankey diagrams.

Introduction

Topic A.3 examines the physics of energy transfer. The importance of energy only becomes clear when it moves between different forms. As this happens, we can make energy do useful work.

The study of energy has a long history stretching back to Aristotle (384–22 BCE). However, the energy concept that we use today stems from a proposal by Gottfried Leibniz (1646–1716) who suggested a "*vis viva*" (living force) which was *mass* × *speed*[2]. His "living force" was transformed into what we call "heat" as objects decelerate. The mathematician and natural philosopher Emilie du Châtelet (1706–49) is believed to have been the first person to recognise the concept of energy conservation in a closed system. By the middle of the 19th century, many people working in industry were examining the concept of energy in ways close to our present understanding. The idea that energy can be transferred was sealed by James Prescott Joule (1818–89) when he linked mechanical work to the production of heat in a quantitative way.

Energy forms and transfers

We now recognize that energy can be stored in many different forms. Some of the important ones are listed in Table 1.

You will learn more about the later developments of energy transfer—a study now known as thermodynamics—in Topic B.4.

Energy	Nature of energy associated with...	Note
kinetic	the motion of a mass	
(gravitational) potential	the position of a mass in a gravitational field	sometimes the word "gravitational" is not used
electric/magnetic	charge flowing	
chemical	atoms and their molecular arrangements	
nuclear	the nucleus of an atom	related to a mass change by $\Delta E = \Delta mc^2$
elastic (potential)	an object being deformed	The word "potential" is not always used.
thermal (heat)	a change in temperature or a change of state	A change of state is a change of a substance between phases, i.e. solid to liquid, or liquid to gas. The colloquial term "heat" is usually acceptable when referring to situations involving conservation of energy.
mass	conversion to binding (nuclear) energy when nuclear changes occur	
vibration (sound)	mechanical waves in solids, liquids or gases	The amount of sound energy transferred is almost always negligible when compared with other energy forms.
light	photons of light	another form of electric/magnetic energy, sometimes called "radiant energy"

▲ Table 1 The different forms of energy and what they are associated with.

Energy can be transferred between any of its forms, and it is during these transfers that you see the effects of energy. For example, water can fall vertically to turn the turbine of a hydroelectric power station and drive a generator. Many energy transfers occur in this apparently simple example. Water molecules are attracted gravitationally by Earth and accelerate downwards through a pipe. Their momentum is transferred to the blades of the turbine which rotates, gaining rotational energy, to turn the coils in the generator. As the coils turn, electrons are forced to move and there is an electric current in the coil. This chain of physical processes can be summed up as the transfer of gravitational potential energy of the water into electrical energy.

Another example of an energy transfer is when an animal converts stored chemical energy in its muscles into kinetic and gravitational potential energy. Some of this chemical energy is also transferred to frictional energy losses.

Other energy units

- Tool 3: Use of units (for example, eV, eVc⁻², ly, pc, h, day, year) whenever appropriate.

You will come across different energy units in some parts of science. These have usually arisen historically. For example, the electronvolt (eV), is the energy gained by an electron when it is accelerated through a potential difference of one volt. Another example is the calorie which is sometimes used to talk about the energy in food. One calorie (cal) is 4.2 J.

You will learn the detail about any special units used in the course in the appropriate place in this book.

You should learn to recognize the physical (and sometimes chemical) processes that occur in an energy transfer. It is easy to describe the changes in broad terms. Always try to explain the effects in terms of microscopic or macroscopic interactions.

Energy is a scalar quantity, which means it has no direction. Whatever the form of the energy transfer, we use a unit of energy called the joule (J). This is in honour of James Joule, the English scientist, who devoted his scientific efforts to studying energy and its transfers.

One joule is the energy transferred when a force of one newton acts through a distance of one metre.

In some applications, such as when discussing the output of power stations, the joule is too small a unit, so you will frequently see energies expressed in megajoules (MJ or 10^6 J) or even gigajoules (GJ or 10^9 J). You should become used to working both in powers of ten and with the SI prefixes when dealing with energy quantities.

How is the equilibrium state of a system, such as Earth's atmosphere or a star, determined?

Elsewhere in the course, equilibrium states, such as those in a star, are described in terms of a balance of forces. For the star, these are inward collapse due to gravitation and outwards expansion due to gas pressure. However, systems that are in equilibrium can also be described as having attained a minimum in their total energy. Imagine a cylinder of gas with a piston and a weight resting on the piston. The total energy of the system is the internal energy of the gas together with the gravitational potential energy of the weight. When the volume of the gas is too small for equilibrium, forces in the gas will push the piston and weight upwards. If this change occurs slowly, then the entropy of the system and surroundings (Topic B.4) will be constant. It is possible to show that when the force upwards on the piston (*pressure × piston area*) is equal to the weight acting downwards, the energy is minimised.

Worked example 1

Describe and explain the energy changes that occur when:

a. a balloon is inflated

b. the air is released from the balloon.

Solutions

a. Air molecules enter the balloon with a certain amount of kinetic energy. The newly added kinetic energy is distributed among the molecules of air through intermolecular collisions, and results in an increased pressure and increased force outwards on the skin of the balloon. As the balloon expands, the elastic potential energy stored in the skin of the balloon increases, until a new equilibrium is established between the tension in the skin and the atmospheric pressure.

b. The opposite process happens when the balloon is released and deflated. The elastic potential energy stored in the skin of the balloon is converted to kinetic energy of the air leaving the balloon.

Conservation of energy

When energy transfers from one form to another nothing is lost (providing that you take care to include every single form of energy involved in the transfers). This is known as the **principle of conservation of energy** which states that energy cannot be created or destroyed.

Physicists now recognize that mass must be included in any table of energy forms because if the mass changes the total energy also changes. For most changes, the mass difference between the beginning and end of a process is insignificant, but in a nuclear reaction it makes a major contribution to the transfer.

ATL Constructively assessing the contribution of peers

Emmy Noether (1882–1935) was a mathematician working in Germany in the early 20th century. In 1918, she published work, now referred to as Noether's theorem, which links the conservation of energy to the fact that the laws of physics do not change over time. A similar application to the conservation of momentum links this to the fact that the laws of physics do not change in different spaces.

At that time, it was unusual for women to enter higher education (she was one of only two female students out of almost one thousand) and even more unusual for women to teach at a university. To do this she had to lecture under the name of another mathematician and often was not paid for her work. However, she was held in high esteem by some of the most renowned mathematicians of the age and was always encouraged to persevere in her work.

When she died, Albert Einstein wrote in the *New York Times* that she was "the most significant creative mathematical genius thus far produced since the higher education of women began".

Laws of physics

The conservation of energy is a law of physics. As such it is empirical—that is, based on observation—rather than something that can be derived through mathematics or deduced with logic. Unlike a scientific theory, a scientific law makes no attempt to explain why the observations are as they are. In other words, the law of conservation of energy does not explain why energy is conserved. It merely states that it is.

Noether's theorem can be used to derive the conservation of energy from the fact that the laws of physics do not change over time. This merely moves the argument to the constant nature of the laws of physics—we have a law that they do not change over time; however, we cannot explain why.

Do laws of physics count as a valid way of knowing?

 Where do the laws of conservation apply in other areas of physics? (NOS)

The rule that energy is conserved underpins this entire topic. Without the conservation of energy and the concept of a conservative force, many of our theories would have no basis.

But you will see many conservation laws that apply throughout science. Topic B.5 shows that charge cannot be lost or destroyed. Several other properties of nuclear particles are also conserved (Theme E), so it is possible to predict the outcomes of nuclear interactions and events.

 How do travelling waves allow for a transfer of energy without a resultant displacement of matter?

Here in Theme A, kinetic energy is an attribute of a moving object. In Theme C, the assertion is that energy is transferred by the wave without displacement of the medium carrying it. This is a result of the cyclic nature of the particle motion in the medium. These particles undergo cycles of displacement and, as they move, are able to transfer energy in the propagation direction of the wave. When the wave has passed, the individual particles return to their equilibrium positions. They were the agents of energy transfer only while they were moving.

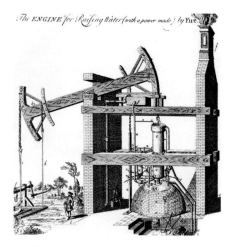

▲ Figure 2   A mine steam engine.

▲ Figure 3   Sand yachts.

Doing work

In 1826, Gaspard-Gustave Coriolis was studying the engineering involved in raising water from a flooded underground mine (Figure 2). He realized that energy was being transferred when the steam engines were pumping the water through a vertical distance. He described this energy transfer as "work done", and he recognized that the energy was transferred because the pumping engines exerted a force on a particular mass of water and lifted it from the bottom of the mine to the surface.

In other words:

work done (in J) = *force exerted* (in N) × *distance moved in the direction of the force* (in m)

When a weight of 5 N of water was lifted vertically through a height of 150 m, then the work done by the engine on the water was $5 \times 150 = 750$ J.

In the underground mine Gaspard studied, the force and the distance moved were in the same direction (vertically upwards), but in many cases this will not be the case. For example, in a sand yacht (Figure 3), the force F from the wind acts in one direction and the sail is set so that the yacht moves through a displacement s that is at an angle θ to the wind (Figure 4).

Because the distances moved in the direction of the force and by the yacht are not the same, you can use the force component in the direction of movement. In this case, *work done* = $F \cos \theta \times s$.

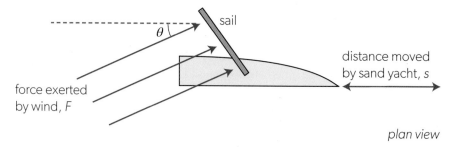

plan view

▲ Figure 4   The wind does work on the sand yacht, even though the wind direction and the direction in which the yacht travels are not the same.

Work done against a resistive force

Work is done when a resistive force is acting too. Consider a box being pushed at a constant speed in a horizontal straight line (Figure 5). For the speed to be constant, friction forces must be overcome. The force that overcomes the friction may not act in the direction of movement. Again, the work done by the force is *force acting* × *distance travelled* × $\cos \theta$.

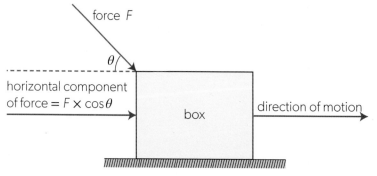

▲ Figure 5   Work done to overcome a resistive force.

Worked example 2

The thrust (driving force) of a microlight aircraft engine is 3.5×10^3 N. Calculate the work done by the thrust when the aircraft travels a distance of 15 km.

Solution

Work done = force × distance = $3500 \times 15\,000 = 5.3 \times 10^7$ J = 53 MJ

Worked example 3

A large box is pulled a distance of 8.5 m along a rough horizontal surface by a force of 55 N that acts at 50° to the horizontal. Calculate the work done in moving the box 8.5 m.

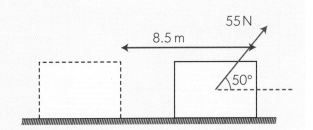

Solution

The component of force in the direction of travel is $55 \times \cos 50° = 35.4$ N.

The work done = this force component × distance travelled = $35.4 \times 8.5 = 301$ J.

Worked example 4

A cart rolls down a ramp that makes an angle of 25° with the horizontal. A horizontal force of magnitude 6.0 N acts on the cart.

The cart moves a distance of 0.75 m down the ramp. Calculate the work done by the horizontal force.

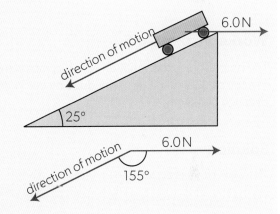

Solution

The angle between the force and the direction of motion is $180° - 25° = 155°$.

The work done = $6.0 \times 0.75 \times \cos 155° = -4.1$ J.

Note that the work done is negative, which indicates that the horizontal force has a component acting against the displacement of the cart. In other words, the force opposes the motion of the cart.

Practice questions

1. A particle travels a distance of 2.5 m along a straight line. A constant force of magnitude 0.60 N acts on the particle. Calculate the work done by the force on the particle, when the angle between the force and the displacement of the particle is:

   a. 60°

   b. 90°

   c. 160°.

2. A car moves at a constant speed of 50 km h⁻¹ on a horizontal road. The work done by the driving force of the car in one minute is 190 kJ.

   Calculate:

   a. the distance travelled by the car in one minute

   b. the magnitude of the resistive force acting on the car

   c. the work done by the resistive force.

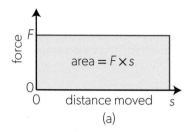

area = $F \times s$

force

0

0   distance moved   s

(a)

total area = (number of squares)
× (energy represented by
one square)

one square = $F \times s$

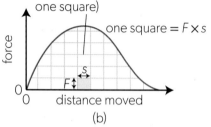

force

0

0   distance moved

(b)

▲ Figure 6   The work done in a process
can be determined by evaluating the area
under the graph of force–distance moved.

Force–distance graphs

In practice, it is rare for the force acting on a moving object to be constant. Real boxes or sand yachts are subject to air resistance and frictional losses. These lead to energy losses that vary depending on the speed of the object or the type of surface that the object runs over.

When you know how the force varies with distance, then you can use this to calculate the work done.

For example, when the force is constant (Figure 6(a)), the graph of force against distance will be a straight line parallel to the x-axis. The *work done* is the product of *force × distance* (we are assuming that $\theta = 90°$ in this case). This corresponds to the area under the graph of force against distance.

When the force is not constant with distance moved (Figure 6(b)), the work done is still the area under the line, but this time you must work a little harder and estimate the number of squares under the graph and equate each square to the energy that it represents. *Energy for one square × number of squares* will then give you the overall work done.

There is a further example of this type of calculation later in the topic on elastic potential energy.

Worked example 5

The graph shows the variation with displacement d of a force F that is applied to a toy car. Calculate the work done by the force F in moving the toy through a distance of 4.0 cm.

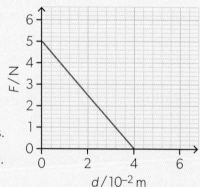

Solution

The work done is equal to the area of the triangle enclosed by the graph and the axis.

This is $\frac{1}{2}$ × base of the triangle × height of the triangle = $\frac{1}{2} \times 4.0 \times 10^{-2} \times 5.0 = 0.10\,\text{J}$.

Worked example 6

An initially stationary ice hockey puck is hit by a stick. The graph shows how the force exerted by the stick on the puck varies with the distance travelled by the puck.

The work done on the puck is 30 J. Determine the peak force F_{max} on the puck.

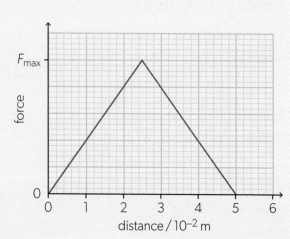

Solution

The area under the force-distance graph is $\frac{1}{2} \times 5 \times 10^{-2} F_{max}$. On the other hand, the area is equal to the work done on the puck.

$\frac{1}{2} \times 5 \times 10^{-2} F_{max} = 30$; hence $F_{max} = \frac{30 \times 2}{5 \times 10^{-2}} = 1200\,\text{N}$.

Practice questions

3. The graph shows how the driving force provided by the engine of a car varies with the distance moved by the car.

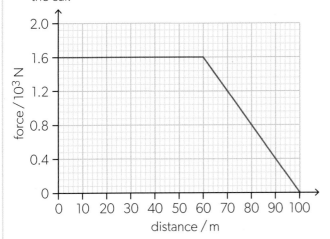

a. Calculate the work done by the driving force in moving the car through a distance of 100 m.

b. A constant force of 400 N opposes the motion of the car. The mass of the car is 1600 kg. Calculate the acceleration of the car when is has travelled a distance of:

   i. 50 m

   ii. 100 m.

4. The graph shows how a force exerted on an object in the direction of its motion varies with the distance travelled by the object.

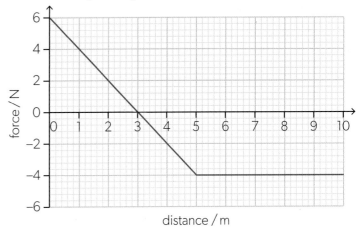

Calculate:

a. the work done by the force when the object has moved a distance of 10 m

b. the distance moved by the object when the work done by the force is zero.

Power

Imagine two people, Renee and Phillipe, with identical body weights (say, 650 N) who climb the same hill (70 m high). Because they have the same weight and climb the same vertical distance, they both gain the same amount of gravitational potential energy. However, Renee climbs the hill in 150 s whereas Phillipe takes 300 s.

Renee is gaining potential energy twice as fast as Phillipe because Renee's time is half that of Phillipe's. This is important when we compare machines or people taking different times to carry out the same amount of work.

The quantity **power** is used to measure the **rate of doing work**. In other words, the number of joules that can be converted every second. Power is defined as:

$$\text{power} = \frac{\text{energy transferred}}{\text{time taken for change}}$$

When the energy change is in joules, and the time for the transfer measured in seconds then the power is in watts (W).

$$1\,W \equiv 1\,J\,s^{-1}$$

In the example of Renee and Phillipe above, both did 45.5 kJ of work against gravity, but Renee's power in climbing the hill was $\frac{650 \times 70}{150} = 300\,W$ and Phillipe's was 150 W because Phillipe took twice as long in the climb.

 Which other quantities in physics involve rates of change?

The use of rate of change—the variation of a quantity with time—is often found in science. Topic A.1 shows the repeated use of rate of change of displacement leading first to velocity and then to acceleration. Power is the rate of transfer of energy which has a bearing both here and in the work on heat capacity and latent heat in Theme B. Similar arguments to those in mechanics occur in the wave theory of Theme C. Topic D.4 connects the generation of emf with rate of change of magnetic flux. Finally, the rate of change of activity of a sample of a radioactive nuclide is important in Topic E.3.

(ATL) Communication skills

▲ Figure 7 James Watt (1736–1819).

James Watt was a Scottish mechanical engineer who worked in the 18th and 19th centuries. He improved steam engine design and developed a way of making copies of paper documents used in offices until early in the 1900s.

Watt wanted to sell more engines and made money based on the amount of coal saved by using his more efficient engines. He came up with a unit called horsepower which enabled him to compare the power of his engines to machines previously run by horses. One horsepower is equal to about 750 W.

He was also careful to compare his engines in a favourable way. A horse can generate over ten horsepower for short amounts of time. But by taking an average over one day, Watt convinced potential buyers that his engine was more powerful than a horse.

Today, the unit of horsepower is still widely used by car manufacturers.

The equation *work done = force × distance* can be rearranged to give another useful expression for power.

$$\text{power} = \frac{\text{work done}}{\text{time}} = \frac{\text{force} \times \text{distance moved by force}}{\text{time taken}}$$

$$= \text{force} \times \frac{\text{distance moved by force}}{\text{time taken}}$$

This is the same as *power = force × speed*. The power required to move an object travelling at a speed *v* with a force *F* is *Fv*.

Maximum speed of a car

The maximum speed of any car is determined by several factors. There is a maximum force that the engine can exert through the tyres on the road surface. But, as with a sphere falling through a fluid, this is not the only force acting. There is a considerable drag force on the vehicle due to the air. This force increases significantly as the speed of the car becomes larger. Typically, when the speed doubles the drag force F_d will increase by at least a factor of four, in other words $F_d \propto v^2$.

There is a maximum power that the car engine can produce. When the car accelerates and the speed increases, the power dissipated in friction also increases. When the maximum energy output of the engine every second is completely used in overcoming the energy losses, then the car cannot accelerate further and has reached its maximum speed.

Worked example 7

A car is travelling at a constant speed of 25 m s⁻¹ and its engine is producing a useful power output of 20 kW. Calculate the driving force required to maintain this speed.

Solution

Driving force $= \dfrac{\text{power}}{\text{speed}} = \dfrac{20\,000}{25} = 800\,\text{N}$

Worked example 8

A cyclist rides up a 1.5 km long hill at a constant speed in a time of 350 s. The cyclist maintains a power of 240 W for the whole ride. The hill makes an angle of 3.0° with the horizontal. The mass of the cyclist and the bicycle is 80 kg.

Calculate:

a. the total work done by the cyclist

b. the forward force exerted on the system of the cyclist and the bicycle

c. the magnitude of the resistive force acting on the system.

Solutions

a. work = power × time = 240 × 350 = 84 kJ

b. force $= \dfrac{\text{work done}}{\text{distance}} = \dfrac{84 \times 10^3}{1.5 \times 10^3} = 56\,\text{N}$

c. The forces opposing the motion of the system are the component of the weight acting down the slope, equal to $80 \times 9.8 \times \sin 3° = 41\,N$, and the unknown resistive force. The cyclist rides at a constant speed; hence the net force on the system must be zero. Therefore the resistive force is $56 - 41 = 15\,N$.

Practice questions

5. A skier is towed at a constant speed of $1.5\,m\,s^{-1}$ by a ski lift whose cable makes an angle of $60°$ with the ski slope. The rate at which work is being done in towing the skier is $180\,W$.

    What is the tension force in the cable?

    A. $60\,N$ B. $120\,N$

    C. $180\,N$ D. $240\,N$

6. A railway locomotive is driven from rest along a horizontal track. The locomotive develops a constant power. What is correct about the acceleration of the locomotive?

    A. It is constant.

    B. It increases uniformly with time.

    C. It increases from zero to a maximum.

    D. It decreases from a maximum to zero.

7. A car of mass $1600\,kg$ is initially at rest and accelerates uniformly at $2.5\,m\,s^{-2}$. A constant force of $500\,N$ opposes the motion of the car.

    a. Calculate the driving force acting on the car.

    b. For the first $10\,s$ of the motion, calculate:

        i.   the total work done by the driving force

        ii. the average power developed by the car.

    c. Explain why the power developed by the car must increase to maintain a constant acceleration.

Kinetic energy (KE)

Kinetic energy E_k is the energy an object has because of its motion. Objects gain kinetic energy when their speed increases.

An object of mass m is at rest at time $t = 0$ and is accelerated by a force F for a time T. The kinematic equations and Newton's second law allow us to work out the speed of the object v at time $t = T$.

The acceleration a is $\dfrac{F}{m}$ (using Newton's second law of motion).

Therefore $v = 0 + \dfrac{F}{m}T$ (because the initial speed is zero). So $F = \dfrac{mv}{T}$.

The work done on the mass is the gain in its kinetic energy ΔE_k and is $F \times s$ where s is the distance travelled. So, $\Delta E_k = F \times s = \dfrac{mv}{T} \times \dfrac{vT}{2}$ because $s = \dfrac{(v + 0) \times T}{2}$.

The work done by the force is equal to the gain in kinetic energy and is:

$$\Delta E_k = \frac{1}{2}mv^2$$

Remember that this is the case where the initial speed was 0. When the object is already moving at an initial speed u, then the change in kinetic energy will be $\Delta E_k = \dfrac{1}{2}m(v^2 - u^2)$, as shown in Figure 8.

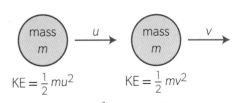

$KE = \dfrac{1}{2}mu^2$ $KE = \dfrac{1}{2}mv^2$

change in $KE = \dfrac{1}{2}m(v^2 - u^2)$

▲ Figure 8 The speed of an object of mass m increases from u to v. The change in kinetic energy, ΔE_k, is $\dfrac{1}{2}m(v^2 - u^2)$.

Kinetic energy equation

- Tool 3: Select and manipulate equations.

There is a subtle piece of notation here. When we talk about a value of kinetic energy, we write E_k, but when we are talking about a change in kinetic energy from one value to another then we should write ΔE_k where "Δ", as usual, means "the change in".

This equation for ΔE_k is one that needs a little care. Notice where the powers are: they are attached to each individual speed. This equation $\Delta E_k = \frac{1}{2}m(v^2 - u^2)$ is not the same as $\Delta E_k = \frac{1}{2}m(v - u)^2$.

Kinetic energy can also easily be linked to linear momentum p. Remember that $p = mv$ and that $E_k = \frac{1}{2}mv^2$. Therefore, E_k can be written as $E_k = \frac{1}{2}\frac{(mv)^2}{m} = \frac{p^2}{2m}$.

Worked example 9

A vehicle is being designed to capture the world land speed record. It has a maximum design speed of $1700\,\text{km h}^{-1}$ and fully fuelled mass of $7800\,\text{kg}$.

Calculate the maximum kinetic energy of the vehicle.

Solution

$1700\,\text{km h}^{-1} \equiv 472\,\text{m s}^{-1}$

$E_k = \frac{1}{2}mv^2 = 0.5 \times 7800 \times 472^2 = 8.7 \times 10^8\,\text{J} = 0.87\,\text{GJ}$

Worked example 10

A car of mass $1.3 \times 10^3\,\text{kg}$ accelerates from a speed of $12\,\text{m s}^{-1}$ to a speed of $20\,\text{m s}^{-1}$. Calculate the change in kinetic energy of the car.

Solution

$\Delta E_k = \frac{1}{2}m(v^2 - u^2) = 0.5 \times 1300 \times (20^2 - 12^2) = 1.7 \times 10^5\,\text{J}$

Worked example 11

An object of mass $0.80\,\text{kg}$ moving in a straight line has an initial kinetic energy of $24\,\text{J}$. Calculate:

a. the initial speed of the object

b. the distance in which the object will come to rest when a net force of $4.0\,\text{N}$ opposes the motion.

Solutions

a. $\frac{1}{2} \times 0.8 \times v^2 = 24 \Rightarrow v = \sqrt{\frac{24 \times 2}{0.8}} = 7.7\,\text{m s}^{-1}$

b. There must be $24\,\text{J}$ of work done to stop the motion of the object. The force acting is $4.0\,\text{N}$, so the stopping distance is $\frac{24}{4.0} = 6.0\,\text{m}$.

Practice questions

8. A railway truck of mass $2500\,\text{kg}$ is moving at an initial speed of $8.0\,\text{m s}^{-1}$. A forward driving force is applied to the truck. The graph shows how the driving force varies with the distance travelled. No other forces act in the direction of motion.

Calculate:

a. the work done on the truck by the driving force

b. the speed of the truck when it has travelled $100\,\text{m}$.

9. A cart of mass $1.2\,\text{kg}$ moves up a ramp with an initial speed of $2.7\,\text{m s}^{-1}$. The net force acting on the cart is constant. The kinetic energy of the cart is halved after $0.90\,\text{s}$.

Calculate, for the first $0.90\,\text{s}$ of motion:

a. the work done on the cart by the net force

b. the final speed of the cart

c. the distance travelled.

Gravitational potential energy (GPE)

Gravitational potential energy E_p is the energy an object has because of its position in a gravitational field. When a mass is moved vertically up or down in the gravity field of Earth, it gains or loses gravitational potential energy. Only the initial and final positions relative to the surface determine the change of GPE ΔE_p:

Conservative forces

There is an important difference between forces such as gravitational forces and frictional forces.

When an object is raised in a gravitational field, the work done is independent of the path. It depends *only* on the start height and end height of the motion. It does not depend on the *route* taken by an object to get from start to finish. The end position does not have to be vertically above the starting position. Gravitational force is said to be **conservative**. In other words, it conserves energy. We can recover all the energy by moving the object back to the starting point. When the object is moved in a closed path, the work done in a conservative field is zero.

In contrast, friction acts between a book and the surface of a table when a book is moved on the surface from one point to another. When the book goes directly from start to finish, a certain amount of energy will be required to overcome the friction. But if the book goes by a longer route, more energy is needed (*work done = friction force × distance travelled*). A force is said to be **non-conservative** when the exact route must be known to calculate the total energy conversion. If the book is moved back to its starting place, you cannot recover the energy in the way that you could when only a gravitational force acts.

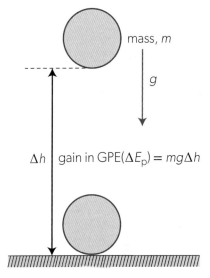

Calculating gravitational potential energy

The work done when an object is raised at constant speed through a change in height Δh is, as usual, equal to *force × distance* moved. In this case, the force required is mg and work done, $\Delta E_p = mg \times \Delta h$.

The value of g close to Earth's surface is $9.8\,\text{m s}^{-2}$, but this value becomes smaller when we move away from Earth.

▲ Figure 9 The gain in gravitational potential energy is equal to $mg \times \Delta h$.

 Why is the equation for the change in gravitational potential energy only relevant close to the surface of Earth? What happens when moving further away from the surface?

The value of g varies with the location where it is measured. This can be in terms of a position at sea level at different places on the planet, or it can be in terms of height above sea level. The theory behind this is explained in more detail in Topic D.1.

Gravitational field lines extend outwards from the centre of a sphere of uniform density. The gravitational force on a nearby object is directed along these field lines. With a large sphere the size of Earth these are (as far as we are concerned) locally at right angles to the surface. However, move away from Earth's surface and the lines begin to spread out indicating a reduction in the gravitational pull

and a smaller value for g. For the scale over which we can ignore the non-perpendicular nature of the field lines, we can use $mg\Delta h$. Otherwise, the full Newtonian law of gravitation in Topic D.1 must be used.

The value of g is also highly dependent on the density of the material, especially when there are density variations close to the surface. This is the basis of gravimetry where small local changes in the gravitational field at Earth's surface are measured by instruments on the surface and by those carried in orbiting satellites. The observed range is from $9.79\,\text{m s}^{-2}$ to $9.83\,\text{m s}^{-2}$.

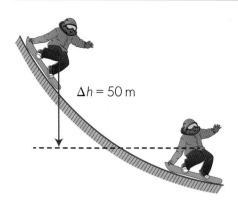

$\Delta h = 50\,\text{m}$

▲ Figure 10 The mechanics of snowboarding.

Energy transferring between GPE and KE

The kinetic energy and the gravitational potential energy of a body or system can be summed to give a new quantity known as the **mechanical energy** of the system. The mechanical energy of a system is constant provided that the system is isolated and is subject only to conservative forces. Thus, a satellite orbiting a planet with no atmosphere in an elliptical orbit is constantly transferring energy between the kinetic and gravitational potential forms with no overall change in the total mechanical energy available.

Sometimes the use of mechanical energy provides a neat way to solve a problem. For example, in Figure 10 a snowboarder is moving down a curved slope starting from rest ($u = 0$). The vertical change in height of the slope is $\Delta h = 50\,\text{m}$. What is the speed of the snowboarder at the bottom of the slope? Assume that we can ignore friction at the base of the snowboard and air resistance.

You must *not* use the kinematic (*suvat*) equations in this example. They do not apply here because the acceleration of the snowboarder is not constant. Although they give the correct numerical answer in this case, they should not be used because the physics is incorrect. The final answer is correct only because we use the start and end points and because the gravitational force is conservative. We are also using an average value for acceleration down the slope by assuming that the angle to the horizontal is constant. The kinematic equations would not give the correct answer if friction forces were involved.

Conservation of energy helps because (as friction loss is negligible) we know that the loss of gravitational potential energy as the snowboarder goes down the slope is equal to the gain in kinetic energy over the length of the slope (in other words, the mechanical energy is constant). Because we know that the GPE change depends only on the initial and final positions, then we do not need to worry at all about what is going on at the base of the snowboard.

So $\Delta E_{\text{p}} = m \times g \times \Delta h = \dfrac{1}{2} \times m \times v^2$. Rearranging this gives $v = \sqrt{2gh}$.

In this case, $v = \sqrt{2 \times 9.8 \times 50} = 31\,\text{m s}^{-1}$.

(This is a speed of about $110\,\text{km h}^{-1}$, which tells you that the assumption about no air resistance and no friction is a poor one!)

Notice that the answer does not depend on the mass of the snowboarder: the mass term cancels out in the equations. Again, including the effects of friction (depending on the snowboarder's mass) and air resistance (depending on the snowboarder's shape) will reduce the estimate of final speed.

Worked example 12

A ball of mass $0.35\,\text{kg}$ is thrown vertically upwards at a speed of $8.0\,\text{m s}^{-1}$. Calculate:

a.   the initial kinetic energy

b.   the maximum gravitational potential energy

c.   the maximum height reached.

→

Solutions

a. $E_k = \frac{1}{2}mv^2 = \frac{1}{2} \times 0.35 \times 8^2 = 11.2\,J$

b. At the maximum height, all the initial kinetic energy will have been converted to gravitational potential energy, so the maximum value of the GPE is also 11.2 J.

c. The maximum GPE is 11.2 J and this is equal to $mg\Delta h$, so $\Delta h = \frac{11.2}{mg} = \frac{11.2}{0.35 \times 9.8} = 3.3\,m$

Worked example 13

A pendulum bob is released from rest 0.15 m above its rest position. Calculate the speed as it passes through rest position.

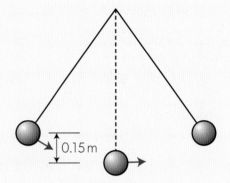

Solution

E_k at the rest position $= E_p$ at the release position so $\frac{1}{2}mv^2 = mg\Delta h$ which

rearranges to $v = \sqrt{2g\Delta h} = \sqrt{2 \times 9.8 \times 0.15} = 1.7\,m\,s^{-1}$.

Worked example 14

In ski jumping, competitors glide down a steep low-friction ramp before taking off. A ski jumper of mass 60 kg starts the in-run phase from rest and reaches a take-off speed of 90 km h^{-1} after travelling a vertical height of 40 m down the ramp.

a. Calculate the ratio $\dfrac{\text{kinetic energy gained by the ski jumper}}{\text{change in the gravitational potential energy}}$.

b. Calculate the work done on the ski jumper by any resistive forces.

The in-run ramp is 75 m long.

c. Estimate the magnitude of the average resistive force on the jumper.

Solutions

a. The take-off speed is 25 m s^{-1}.

$$\frac{\Delta E_k}{\Delta E_p} = \frac{\frac{1}{2}mv^2}{mg\Delta h} = \frac{0.5 \times 25^2}{9.8 \times 40} = 0.80.$$ It means that 80% of the GPE that the jumper had

on top of the ramp has been transferred to KE.

b. The GPE on top of the ramp is greater than the KE at take-off, and the work done by the resistive forces accounts for the difference.

work done by resistive forces = (GPE loss) − (KE gain) $= 60 \times 9.8 \times 40 - \frac{1}{2} \times 60 \times 25^2 = 4.8\,kJ$

c. Work = force × distance; hence the resistive force is $\dfrac{4.8 \times 10^3}{75} = 64\,N$. The force is

not necessarily constant during the in-run phase, so the answer represents the average value.

Worked example 15

A steel ball of mass m is attached to the end of a weightless string of length R and made to move in a vertical circle. The speed of the ball at the lowest point of the circle is u. The only forces acting on the ball are the weight and the tension in the string.

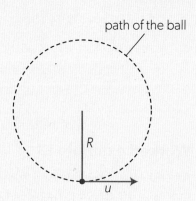

path of the ball

a. State the work done by the tension in the string during one full rotation.

b. Derive an expression, in terms of u and R, for the speed v of the ball at the highest point of the path.

c. Show that the ball will not reach the highest point of the path unless $u \geq \sqrt{5gR}$.

d. It is given that $R = 1.2\,\text{m}$. Calculate the minimum value of u so that the ball reaches the highest point of the path.

Solutions

a. The angle between the tension and the velocity is always $90°$, so the work done by the tension is zero.

b. The loss of KE of the ball between the lowest and the highest point of the path is equal to the gain of GPE.
$\frac{1}{2}mu^2 - \frac{1}{2}mv^2 = mg(2R)$, because the change in height is $2R$. The equation can be solved for the speed at the top of the circle: $v = \sqrt{u^2 - 4gR}$.

c. For the ball to reach the highest point and continue to move in a circular path, its instantaneous acceleration a must be related to the speed and radius through the equation $a = \frac{v^2}{R}$ (this is the centripetal acceleration formula that you met in Topic A.2). At the highest point, the net force on the ball is equal to or greater than the ball's weight (because weight and tension act in the same direction), so the smallest possible acceleration of the ball is g. This leads to the condition $\frac{v^2}{R} \geq g$, or $v^2 \geq gR$. Combining this inequality with the result of part b. gives $u^2 - 4gR \geq gR$, which results in a condition for the speed at the bottom of the path: $u \geq \sqrt{5gR}$.

d. Substitution into the inequality in part c. gives the minimum speed of the ball at the bottom of the path, $\sqrt{5 \times 9.8 \times 1.2} = 7.7\,\text{m s}^{-1}$.

Practice questions

10. An object of mass 1.5 kg slides down a 4.0 m long ramp that makes an angle of 30° with the horizontal.

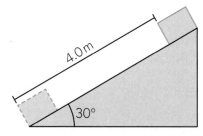

4.0 m

30°

What is the work done on the object by the gravitational force?

A. 3.0 J   B. 6.0 J   C. 30 J   D. 60 J

11. A pendulum bob is released from rest in the horizontal position. The only forces acting on the bob are the weight and the tension in the string.

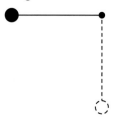

What is the acceleration of the bob when the string is vertical?

A. 0   B. g C. $2g$ D. $3g$.

12. A helicopter of mass 2900 kg takes off vertically and ascends to a height of 45 m above the ground. A constant lift force of 3.2×10^4 N acts on it.

    a. Calculate:

       i. the work done on the helicopter by the lift force

       ii. the change in the gravitational potential energy of the helicopter.

    b. Explain why the answers in a. i. and ii. are different.

    c. Determine the final vertical speed of the helicopter.

13. A block of wood of mass 250 g is suspended at rest from a string. An air rifle pellet of mass 1.80 g is fired horizontally at the block of wood at a speed of 200 m s^{-1}. The pellet becomes embedded in the block.

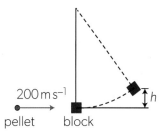

Determine the vertical height h through which the block of wood rises after the impact.

14. A tennis ball is served from a height of 2.50 m above the ground at an initial speed of 20.0 m s^{-1}. Air resistance is negligible.

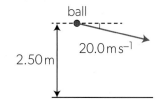

    a. Explain why the speed with which the ball hits the ground does **not** depend on the angle that the initial velocity of the ball makes with the horizontal.

    b. Calculate the speed with which the ball hits the ground.

Transferring GPE to KE

- Inquiry 1: Appreciate when and how to reduce friction.

- Inquiry 1: Design and explain a valid methodology.

- Inquiry 3: Compare the outcomes of an investigation to the accepted scientific context.

- Inquiry 3: Evaluate the implications of methodological weaknesses, limitations and assumptions on conclusions.

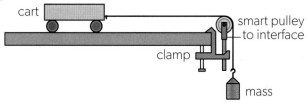

▲ Figure 11 The mass falling vertically transfers gravitational potential energy to the kinetic energy of *all* moving masses: cart, string, pulley and falling mass.

- Arrange a cart on a track and compensate the track for friction. This is done by raising one end of the track through a small distance so that the cart travels down the track at constant speed.

- Pass a string over the pulley. Attach a known mass to the other end of the string.

- Measure the mass of the cart.

- Devise a way to measure the speed of the cart. You could use a "smart pulley" that can measure the speed as the string turns the pulley wheel. Alternatively, use an ultrasound sensor, a data logger with light gates, or your mobile phone.

- When the mass is released, the cart gains speed, as its gravitational potential energy changes.

- Make measurements to assess the gravitational potential energy lost (you will need to know the vertical height through which the mass falls) and the kinetic energy gained (you will need the final speed). Notice that only the falling mass is losing GPE but both masses and the cart are gaining kinetic energy.

- Compare the two energies. Is the energy conserved? Where do you expect energy losses in the experiment to occur?

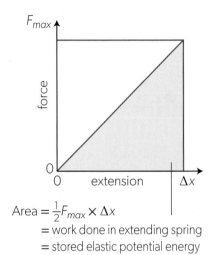

$$\text{Area} = \frac{1}{2}F_{max} \times \Delta x$$
= work done in extending spring
= stored elastic potential energy

▲ Figure 12 The force–extension graph for stretching a spring. The shaded area is the energy transferred to the spring as it stretches.

Elastic potential energy

As you saw in Topic A.2, when a force is applied, the shape of a solid can be changed. Some materials will be able to return the energy that has been stored in them when the force is removed. A metal spring, that obeys Hooke's law is a good example of this. Most springs are designed to store energy in this way in many different contexts. The materials that can return energy in this way have stored **elastic potential energy** E_H.

Hooke showed that, for small loads acting on a spring, the extension of the spring is directly proportional to the load. The graph of force F against extension Δx is a straight line going through the origin.

In symbols, this is $F \propto \Delta x$, or $F = k\Delta x$, where k is the spring constant as usual (see page 56). The gradient of the graph is equal to k.

You can now relate this graph to the work done in stretching the spring (Figure 12). The force is not constant (the bigger the extension, the bigger the force required) but we know how to deal with this. The work done on the spring is the area of the right-angled triangle under the $F–\Delta x$ graph. This is $E_H = \frac{1}{2}F_{max} \times \Delta x$.

Hooke's law is $F = k\Delta x$, so $k = \frac{F}{\Delta x}$ and $E_H = \frac{1}{2}k(\Delta x)^2$.

Worked example 16

A spring, of spring constant $48\,N\,m^{-1}$, is extended by $0.40\,m$. Calculate the elastic potential energy stored in the spring.

Solution

Energy stored $= \frac{1}{2}kx^2 = 0.5 \times 48 \times 0.40^2 = 3.8\,J$

Worked example 17

An object of mass $0.78\,kg$ is attached to a vertical spring of unstretched length $560\,mm$. When the object has come to rest, the new length of the spring is $620\,mm$. Calculate the energy stored in the spring as a result of this extension.

Solution

Change in length of spring, $\Delta x = 620 - 560 = 60\,mm$.

The final tension in the spring will be equal to the weight of the object $= mg = 0.78 \times 9.8 = 7.64\,N$.

The energy stored in the spring $= \frac{1}{2}F\Delta x = 0.5 \times 7.64 \times 0.06 = 0.23\,J$.

Practice questions

15. The elastic potential energy stored in a spring extended through a distance Δx from its relaxed length is E.

    How much work is done on the spring when its extension increases from Δx to $2\Delta x$?

    A. E B. $2E$

    C. $3E$ D. $4E$

16. A block of mass m is attached to the end of a weightless spring and placed on a frictionless horizontal surface. When the spring is compressed through a distance Δx from its unstretched length, it exerts a force F on the block.

    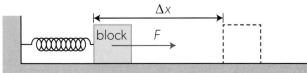

    The block is released from rest from the position shown in the diagram. What is the speed of the block when the spring returns to its unstretched length?

    A. $\sqrt{\dfrac{F\Delta x}{2m}}$ B. $\sqrt{\dfrac{F\Delta x}{m}}$

    C. $\sqrt{\dfrac{2F\Delta x}{m}}$ D. $2\sqrt{\dfrac{F\Delta x}{m}}$

17. A force of magnitude 60 N is needed to hold a spring that is extended through a distance of 3.0 cm from its unstretched length.

    a. Calculate:

        i.   the spring constant

        ii. the elastic potential energy stored in the spring when its extension is 3.0 cm.

    Work of 0.70 J is done to extend the spring further.

    b. Determine the new extension of the spring, relative to its unstretched length.

18. A block of mass 0.60 kg is dropped onto a vertical weightless spring. The spring is initially unstretched. The speed of the block just before it makes contact with the spring is 2.0 m s⁻¹.

    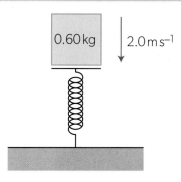

    The block instantaneously stops when the spring is compressed through a distance of 5.1 cm.

    a. Calculate:

        i.   the initial kinetic energy of the block

        ii. the work done on the block by the gravitational force, since the first contact with the spring until it stops

        iii. the elastic potential energy stored in the spring at the instant when the block is at rest.

    b. Hence, calculate the spring constant.

    c. Determine the acceleration of the block at the instant when the block is at rest.

Measuring the spring constant

- Tool 1: Recognize and address relevant safety, ethical or environmental issues in an investigation.

- Tool 3: Construct and interpret tables and graphs for raw and processed data including scatter graphs and line and curve graphs.

- Tool 3: Draw and interpret uncertainty bars.

- Tool 3: On a best-fit linear graph, construct lines of maximum and minimum gradients with relative accuracy (by eye) considering all uncertainty bars.

You must wear safety glasses for this experiment and ensure that your flying spring will not hit anyone.

- Get a spring and a length of wood (a wooden ruler would do).

- Make a notch in the top of the length of wood so that the end of the spring does not slip off (Figure 13).

- Mark the natural (unextended) length of the spring on the wood.

- Pull the spring so that it extends by 2 cm. Release it so that it flies vertically into the air.

- Measure the maximum height of the spring's motion. You should repeat this measurement three times.

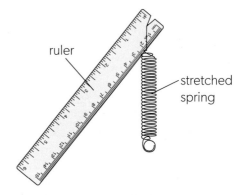

▲ Figure 13 Fire the spring vertically and estimate the maximum height reached.

- Repeat for five different extensions of the spring.

- Tabulate your values of extension and height. Take averages of the heights.

- By considering energy transfer, consider what you should plot on the x- or y-axes to give a linear graph. Plot this graph (use a computer spreadsheet to do it quickly). Is your graph linear?

- Use the variation in your repeats and the uncertainties to add error bars to your graph.

- Is it possible to draw a line of best fit that passes within the error bars?

- What does the gradient of your linear graph represent? How could you deduce the spring constant k of the spring from your gradient?

- What is the uncertainty of your gradient? Can you deduce the uncertainty in your measurement of k?

- Measure the spring constant in a different way (e.g. by considering Hooke's law). Try to evaluate the uncertainty in this measurement. Do your two values of k agree?

Efficiency

In most real experiments where gravitational potential energy is transferred to kinetic energy, some gravitational potential energy will not appear in the kinetic energy of the object. Some energy is lost:

- to internal energy, because of friction

- to elastic potential energy, etc.

We need to have a way to quantify these losses. One way is to compare the total energy put into a system with the useful energy that can be taken out. This is known as the **efficiency** of the transfer and can be applied to all energy transfers, whether carried out in a mechanical system, electrical system or other type of transfer. The definition can also be applied to power transfers because the energy change in these cases takes place in the same time for the total energy in and the useful work out.

$$\text{efficiency} = \frac{\text{useful work out}}{\text{total energy in}} = \frac{\text{useful power output}}{\text{total power input}}$$

Worked example 18

An electric motor raises a weight of 150 N through a height of 7.2 m. The energy supplied to the motor during this process is 3.5×10^3 J.

Calculate:

a. the increase in gravitational potential energy

b. the efficiency of the process.

Solutions

a. $\Delta E_p = 150 \times 7.2 = 1080$ J

b. efficiency
$$= \frac{\text{useful work out}}{\text{energy in}}$$
$$= \frac{1080}{3500} = 0.31 \text{ or } 31\%$$

Practice questions

19. A cyclist rides up a 50 m high hill in a time of 200 s. The average power developed by the cyclist is 270 W. The mass of the cyclist and the bicycle is 85 kg.

   Determine the efficiency with which the work done by the cyclist is transferred to the gravitational potential energy.

20. An electric car of mass 1600 kg accelerating on a horizontal road converts 65% of the electrochemical energy stored in the battery to kinetic energy.

   Calculate the energy transferred from the battery when the car accelerates from rest to a speed of 50 km h^{-1}.

Data-based questions

A ball is dropped from a fixed height of 1.00 m so that it bounces several times. The subsequent heights that it reaches are measured after each successive bounce. The results are given in the table.

- Calculate the energy of the ball on each bounce.

- Hence calculate the efficiency of the ball when it bounces.

- How could you show that these data display an exponential trend (once you have studied Topic E.3)?

- Identify other topics in this book where you find examples of exponential decay.

Number of bounces	Height / cm		
	Trial 1	Trial 2	Trial 3
0	100	100	100
1	94	93	91
2	85	84	82
3	77	78	76
4	69	73	71
5	64	61	66
6	59	58	59
7	53	55	54
8	49	50	50
9	46	45	44
10	42	40	42

Sankey diagrams

Sankey diagrams are visual representations of the flow of the energy in a device or in a process.

The rules to remember about using Sankey diagrams are:

- Each energy transfer in the process is represented by an arrow.
- The diagram is drawn to scale with the width of the arrow being proportional to the amount of energy transfer it represents.
- The energy flow is drawn from left to right.
- When energy is lost from the system, it moves to the top or bottom of the diagram.
- Power transfers as well as energy flows can be represented.

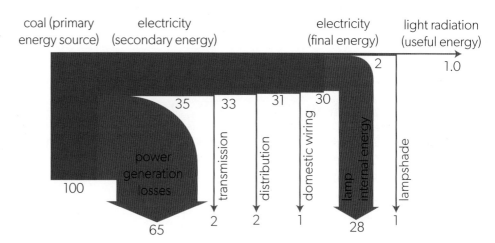

▲ Figure 14 A Sankey diagram for the energy transferred to an electric lamp.

Figure 14 is a Sankey diagram that shows the transfer of energy in an electric lamp. It shows the energy transfers that begin with the conversion of chemical energy from fossil fuels and end with light energy emitted from the filament of the lamp. Red arrows represent energy that is transferred from the system. In any process where there is an energy transformation, this energy is "lost" and is no longer available to perform a useful job. This **degraded energy** loss occurs in all energy transfers.

Of the original primary energy in Figure 14, only 35% appears as useful secondary energy. The remaining 65% is lost to the surroundings in the generation processes. The arrows that point downwards show this. There are losses involved in the transmission and distribution of the electricity, and losses in the house wiring. In the lamp itself, most of the energy (28% of the original) is transferred to the internal energy of the surroundings. Only 1% of the original primary energy is left as light energy for illumination.

A Sankey diagram is a useful way to visualize the energy consumption of nations. You can see many examples of this on the Internet. Figure 15 shows the energy flows associated with the US economy in 2017. Search the Internet to find the Sankey diagram for the energy demand of the country where you live.

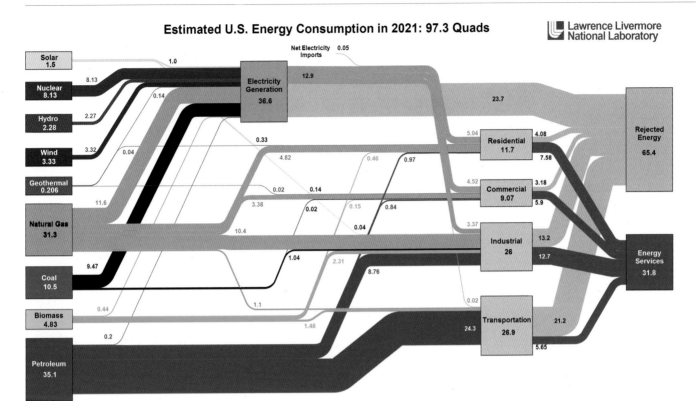

Estimated U.S. Energy Consumption in 2021: 97.3 Quads

▲ Figure 15 The energy flow for the USA in 2017. Lawrence Livermore National Laboratory and the US Department of Energy (March, 2022). One quad is equal to 2.9×10^{11} kW h (about 180 million barrels of petroleum, 39 million tons of coal, or 1000 billion cubic feet of natural gas).

Worked example 19

An electric kettle of rating 2.0 kW is switched on for 90 s. During this time 20 kJ of energy is lost to the surroundings from the kettle. Draw a Sankey diagram of this energy transfer.

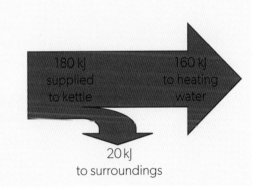

Solution

The energy supplied in 90 s is $2 \times 1000 \times 90 = 180$ kJ.

Percentage lost to the surroundings $= \dfrac{20}{180} \times 100 = 11\%$.

Worked example 20

In a petrol-powered car 34% of the energy in the fuel is converted into kinetic energy of the car. Heating the exhaust gases accounts for 12% of the energy lost from the fuel. The remainder of the energy is wasted in the engine, the gearbox and the wheels. Use these data to sketch a Sankey diagram for the car.

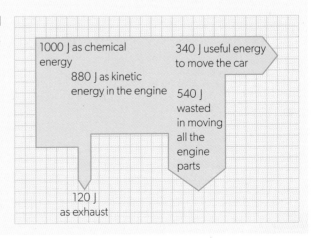

Solution

Energy lost in the engine and transmission
$= 100 - 34 - 12 = 54\%$.

A convenient way to draw the diagram is on squared paper. Use a convenient scale: 10% ≡ 1 large square is a reasonable scale here.

Practice question

21. The energy losses in a pumped storage power station are shown in the following table:

Source of energy loss	Percentage loss of energy
friction and turbulence of water in pipe	27
friction in turbine and ac generator	15
electrical heating losses	5

a.   Calculate the overall efficiency of the conversion of the gravitational potential energy of water in the tank into electrical energy.

b.   Sketch a Sankey diagram to represent the energy conversion in the power station.

Energy density

Much of the extraction of fossil fuels involves hard and dangerous work in mines or on oilrigs. The effort and risk of extracting fossil fuels does not seem to be justified when there are other sources of energy available. So why are fossil fuels still extracted? The answer becomes more obvious when we look at the energy available from the fossil fuel itself. Energy density is one way to quantify this.

Density is a familiar concept; it is the amount of quantity possessed by one cubic metre of a substance. **Energy density** is the number of joules that can be released from one cubic metre ($1\,m^3$) of a fuel. Table 2 shows the energy densities of some common fuels. Notice the wide range of values in this table.

Fuel	Energy density / $GJ\,m^{-3}$
uranium (nuclear fission)	1.3×10^9
coal	20–80
diesel	37
gasoline (petrol)	35
natural gas	0.036
hydrogen	0.01

▲ Table 2 The energy densities of some common fuels.

Worked example 21

A fossil-fuel power station burns coal. It has an efficiency of 25% and generates 1200 MW of useful electrical power. The energy density of the coal is 42 GJ m^{-3}. Calculate the volume of coal burnt every minute in the power station.

Solution

The fuel energy equivalent required every second is $\frac{1.2 \times 10^9}{0.25} = 4.8 \times 10^9$ J. In volume terms, this is $\frac{4.8 \times 10^9}{42 \times 10^9} = 0.114$ m^3. In one minute, the station will burn $0.114 \times 60 = 6.9$ m^3 of coal.

Worked example 22

A camping stove that burns gasoline (petrol) is used. 70% of the energy from the fuel reaches the cooking pot. The energy density of the gasoline is 35 GJ m^{-3}.

a. Calculate the volume of gasoline needed to raise the temperature of 1 litre of water from 10°C to 100°C. Assume that the heat capacity of the pot is negligible. The specific heat capacity of the water (see Topic B.1) is 4.2 kJ kg^{-1} K^{-1}.

b. Estimate the volume of fuel that a student should purchase for a weekend camping expedition.

Solutions

a. 1 litre of water has a mass of 1 kg so the energy required to heat the water is $4200 \times 1 \times 90$ which is 0.38 MJ. Allowing for the efficiency value, $\frac{0.38}{0.70} = 0.54$ MJ of energy is required, and this is a fuel volume of $\frac{0.54 \times 10^6}{3.5 \times 10^{10}} = 1.6 \times 10^{-5}$ m^3 or about 20 ml.

b. Assume that 2 litres of water are required for each meal, and that there will be five cooked meals during the weekend. So, 200 ml of fuel should be enough.

A.4 Rigid body mechanics

How can the understanding of linear motion be applied to rotational motion?

How does the distribution of mass within a body affect its rotational motion?

How is the understanding of the torques acting on a system used to predict changes in rotational motion?

This topic illustrates how scientists use one set of ideas in an altered context. The kinematics and mechanics of linear motion from Topics A.1 and A.2 are used as a model for rotational motion. On the face of it, the two types of movement could not be more different. One is a translational motion of a point object through space. The other is a motion of an object that has size and shape and which rotates around a fixed point. Nevertheless, all your understanding of quantities of force, energy and momentum can be extended to rotational concepts.

This extension requires the recognition that, for rotation, the distribution of an object's mass is important. The mass has a reduced relevance in rotational mechanics. The size,

shape and the arrangement of mass within a rigid body ultimately determine its response to changes in the forces acting.

However, it is no longer enough to relate the rotational response to the force alone. The force can act at any point on the body, so the response of the body will depend on where the force acts relative to the axis of rotation. This was not an issue with the translational and linear motion of the point object. You need to use a new concept—torque—the rotational equivalent of force. From this you can state the equivalents of Newton's laws of motion and the conservation rules already familiar from linear mechanics in the context of rotation.

In this additional higher level topic, you will learn about:

- torque and rotational acceleration
- rotational equilibrium
- angular displacement, angular velocity and angular acceleration
- the equations of motion for uniform angular acceleration

- the kinetic energy of rotational motion
- moment of inertia
- Newton's second law for rotation
- angular momentum and conservation of angular momentum
- angular impulse.

▲ Figure 1 An artist's impression of a black hole.

Introduction

There is an equivalent rotational quantity for each of the linear quantities used in Topics A.1 and A.2: mass, speed, acceleration, force, momentum and so on. There is a link between each of the quantities in linear mechanics and its equivalent in rotational terms.

Figure 1 shows an artist's impression of a black hole. Matter swirls around to form an accretion disk. Conservation of angular momentum means that, as the matter is pulled closer towards the black hole, the angular velocity increases. The gas close to a black hole is moving extremely quickly. It is extremely hot and emits large quantities of radiation. Once the matter falls inside the black hole, this radiation cannot escape. But, in falling, the release of energy is one of the most efficient processes in the universe. Up to around 40% of the mass–energy of the material that falls in is radiated away.

Angular acceleration

Just as in linear motion, a rotating object can speed up or slow down. To do this it must undergo an angular acceleration α, the change in angular velocity $\Delta\omega$ in a given time interval Δt. In equation form:

$$\alpha = \frac{\Delta\omega}{\Delta t} = \frac{\omega_2 - \omega_1}{t_2 - t_1}$$

where ω_1 is the initial angular speed at time t_1 and ω_2 is the final angular speed at time t_2. As you may expect, the units of α are rad s^{-2}.

> In Topic A.2, you were introduced to the terms angular displacement and angular velocity. Remember that angular velocity is treated as a scalar quantity.

 How are the laws of conservation and equations of motion in the context of rotational motion analogous to those governing linear motion?

Throughout this topic there are many parallels drawn between linear and rotational motion. Linking the descriptions of the two types of motion has lots of advantages. Not least, the common framework allows you to quick transfer the understandings you have mastered in one area to another area. Always look out for patterns and trends that link scientific ideas: this highlights the nature of science.

Rotational acceleration

Take care with acceleration in a rotational context. There are three accelerations to consider.

- Angular acceleration α: The rate of change of the angular velocity. The angular velocity itself is the rate at which the angular displacement changes.

- Centripetal acceleration a_c: The acceleration of the object directed towards the centre of the rotation. It is $a_c = \frac{v^2}{r}$ in our usual notation.

- Tangential acceleration a_t: This is a linear acceleration and is the instantaneous rate at which the object is changing its speed along the circumference of the circle. It is $a_t = \alpha r$ in our usual notation.

A constant angular velocity means that the angular acceleration is zero and the tangential acceleration is zero. There will still be a centripetal acceleration towards the centre of the rotation, however.

Worked example 1

A bicycle is placed on a repair stand and a mechanic spins its rear wheel from rest to a final angular velocity of 240 revolutions per minute in a time of $\Delta t = 4.0$ s.

a. Calculate the angular acceleration of the wheel, in rad s^{-2}, assuming that it is constant.

b. The radius of the wheel is 34 cm. Calculate the final linear speed of a point on the circumference of the wheel.

c. Outline how the centripetal acceleration of the point on the circumference of the wheel varies with time.

Solutions

a. The final angular velocity is equal to $\frac{240}{60} \times 2\pi = 25$ rad s^{-1}. The angular acceleration is therefore $\frac{25}{4.0} = 6.3$ rad s^{-2}.

b. The linear speed is $v = \omega r = 25 \times 0.34 = 8.5$ m s^{-1}.

c. The centripetal acceleration depends on the angular velocity of the wheel (alternatively, on the linear speed of the points on the circumference), as seen in the equations $a = \omega^2 r = \frac{v^2}{r}$. Since the velocity of the wheel increases, the centripetal acceleration of the points on the circumference increases too. Angular acceleration and centripetal acceleration are different quantities and should not be confused. Angular acceleration describes the rotation of the wheel as a whole, while centripetal acceleration refers to the circular motion of individual particles at a certain distance from the axis of rotation.

Symbol	Quantity
θ	angular displacement
ω_i	initial angular velocity
ω_f	final angular velocity
α	angular acceleration
t	time

▲ Table 1 Notation for rotational quantities.

Rotational equations of motion

In Topic A.1, we developed four kinematic equations of motion which can be applied to motion when its acceleration is linear and constant.

The definitions of angular displacement, angular velocity and angular acceleration are directly analogous to those of linear displacement, speed and acceleration. This means that an equivalent set of equations can be developed for rotational motion for all situations *when the angular acceleration is constant*. When acceleration is not constant, you must look for other methods to solve a particular problem.

Table 1 gives the notation used for some rotational quantities.

Therefore, the rotational kinematic equations are:

$$\omega_f = \omega_i + \alpha t$$
$$\theta = \omega_i t + \frac{1}{2}\alpha t^2$$
$$\omega_f^2 = \omega_i^2 + 2\alpha\theta$$
$$\theta = \left(\frac{\omega_f + \omega_i}{2}\right)t$$

Although these equations probably look strange to begin with, they can be used in exactly the same way as the set of linear equations.

Worked example 2

The angular acceleration of a wheel, rotated from rest, is $16\,\text{rad s}^{-2}$. The wheel accelerates for $5.0\,\text{s}$. Calculate:

a.   the final angular velocity

b.   the number of revolutions of the wheel in reaching this angular velocity.

Solutions

a. $\omega_f = \omega_i + \alpha t$ so $\omega_f = 0 + 16 \times 5 = 80\,\text{rad s}^{-1}$

b. $\theta = \dfrac{\omega_i + \omega_f}{2}t = \dfrac{(0+80)}{2} \times 5.0 = 200\,\text{rad}$

One revolution corresponds to the angular displacement of $2\pi\,\text{rad}$. Hence, the wheel makes $\dfrac{200}{2\pi} = 32$ revolutions during the acceleration.

Worked example 3

A spinning top is rotating with an initial angular velocity of $30\,\text{rad s}^{-1}$. It decelerates at a constant rate of $0.45\,\text{rad s}^{-2}$. The top falls over when its angular velocity has decreased to $5.0\,\text{rad s}^{-1}$. Calculate:

a.   the number of rotations the top makes before it falls over

b.   the time it takes for the top to fall over.

Solutions

a.   We use the equation $\omega_f^2 = \omega_i^2 + 2\alpha\theta$ to find the angular displacement θ. Note that the substituted value of the angular acceleration must be negative because the top is slowing down.

$5.0^2 = 30^2 - 2 \times 0.45\theta \Rightarrow \theta = 970\,\text{rad}$. The number of revolutions is $\dfrac{970}{2\pi} = 150$.

b. $5.0 = 30 - 0.45t \Rightarrow t = 56\,\text{s}$.

Worked example 4

A merry-go-round of radius 2.0 m is rotated from rest with a constant angular acceleration and makes two complete revolutions during a time of 16 s. Calculate:

a.   the angular acceleration

b.   the final linear speed of a point on the circumference of the merry-go-round

c.   the time it took to complete the first revolution.

Solutions

a.   Two revolutions correspond to the angular displacement of 4π. The initial angular velocity is zero and we use the equation $\theta = \frac{1}{2}\alpha t^2 \Rightarrow 4\pi = \frac{1}{2}\alpha \times 16^2$. From this, $\alpha = 0.098\,\text{rad s}^{-2}$.

b.   The final angular velocity can be calculated from $\omega_f = \alpha t = 0.098 \times 16$. The linear speed at the circumference is $\omega_f r = 0.098 \times 16 \times 2.0 = 3.1\,\text{m s}^{-1}$.

c.   The angular displacement for one revolution is 2π. $2\pi = \frac{1}{2} \times 0.098t^2 \Rightarrow t = 11\,\text{s}$.

Practice questions

1.   A wheel is rotating with an initial angular velocity of $5.0\,\text{rad s}^{-1}$. The angular velocity increases uniformly to $15\,\text{rad s}^{-1}$ during a time of 20 s. Calculate:

a.   the angular acceleration

b.   the number of revolutions the wheel makes during this time.

2.   The drum of a washing machine rotates from rest with a constant angular acceleration of $4.7\,\text{rad s}^{-2}$. The drum reaches its final angular velocity after a time of 20 s.

a.   Calculate the final angular velocity of the drum. Give the answer in revolutions per minute.

b.   Calculate the number of revolutions the drum makes during this time.

3.   A battery drill is switched on from rest and reaches its final angular velocity of 1800 revolutions per minute in a time of 0.10 s.

a.   Calculate the angular acceleration of the drill.

The drill, initially rotating at 1800 revolutions per minute, jams and comes to rest after making five complete revolutions.

b.   Calculate the time taken for the drill to stop, assuming a constant angular deceleration.

4.   The angular velocity of a flywheel is increased uniformly from $10\,\text{rad s}^{-1}$ to $50\,\text{rad s}^{-1}$. During the acceleration, the flywheel turns through 20 complete revolutions. Calculate:

a.   the angular acceleration of the flywheel

b.   the time to complete

i.   all 20 revolutions

ii.   the first 10 of the 20 revolutions.

Rotational mechanics graphs

You can also deduce the graphs showing the variation of angular quantities with time from their linear equivalents. Figure 2 shows how the second equation of rotational motion is derived from the graph of angular velocity against time.

The angular displacement is equal to the area below the line in a graph of angular velocity against time. You can derive an equation for angular displacement by adding together the two areas.

Angular displacement $\theta = $ total area $= \omega_i t + \frac{1}{2}\alpha t^2$

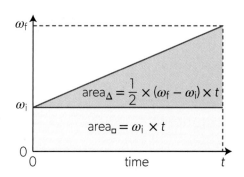

▲ Figure 2 The equation $\theta = \omega_i t + \frac{1}{2}\alpha t^2$ is derived by adding the two areas.

Using units in rotational mechanics

- Tool 3: Work with fundamental units.

Although the equations are similar to those for linear kinematics, think carefully about the units.

- *Time* should be in seconds. This is straightforward as the units are unchanged. Make sure that all values are in seconds before you do any calculations.

- *Angular velocity* must always be in radians per unit time. The time unit here must match the rest of the problem. Usually it will be rad s⁻¹.

- *Angular acceleration* must match the angular velocity and the time. The unit rad s⁻² is the most common. Try not to use mixed units such as rad s⁻¹ hour⁻¹. These are bound to give conversion difficulties at the end.

- *Angular displacement* can be the trickiest of all as there are several ways to specify it. Using radians is usually best for calculations. One revolution is equivalent to once round a circle, so one revolution $\equiv 2\pi$ rad.

For example, a flywheel is turning at 150 revolutions per minute. What is the angular velocity in rad s⁻¹?

In 1.0 s the flywheel must turn through $\frac{210}{60} = 3.5$ revolutions. This is $3.5 \times 2\pi = 22$ rad. The angular velocity is 22 rad s⁻¹.

Worked example 5

The diagram shows a hula-hoop (a large plastic ring) rolling with constant angular speed along a horizontal surface, A. It then rolls down a uniform inclined plane, B. When it reaches a second horizontal surface, C, it moves with a constant angular speed again. Sketch and explain graphs to show the variation with time of:

a.   the angular velocity

b.   the angular acceleration.

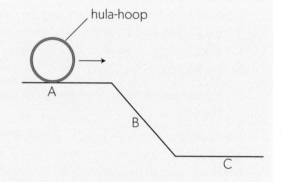

Solutions

a.   The graph shows the hula-hoop travelling with constant angular velocity along A and C. It has a greater value along C since it has now undergone angular acceleration. As B is of constant gradient, the angular acceleration is constant here.

b.   The second graph shows zero angular acceleration throughout A and C, and a constant angular acceleration along B.

Note: compare these graphs with the graphs for a point object moving along a frictionless surface.

When the hula-hoop is travelling at the higher angular velocity it covers the same distance in a shorter time

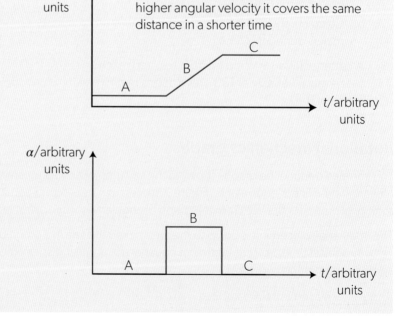

Worked example 6

The graph shows how the angular velocity of a rotating cylinder varies with time.

Calculate:

a. the angular acceleration of the cylinder during the first 4.0 s

b. the angular displacement of the cylinder from 0 to 10.0 s.

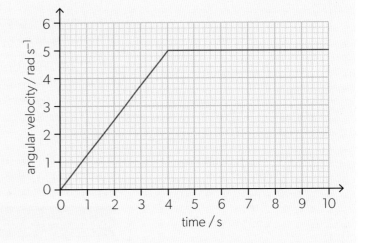

Solutions

a. The angular acceleration is the gradient of the graph: $\alpha = \dfrac{5.0}{4.0} = 1.25 \,\text{rad s}^{-2}$.

b. The angular displacement is the area under the graph: $\theta = \dfrac{1}{2} \times 4.0 \times 5.0 + 6.0 \times 5.0 = 40 \,\text{rad}$.

Moment of inertia

Mass played an important role in linear mechanics when you were treating masses as point objects. What is its equivalent in rotational mechanics?

It is not just the mass that is important but the way the mass is distributed about the centre of rotation. The inertial mass of an object is a measure of its opposition to changes in its linear motion. We need a rotational equivalent which gives a measure of how hard it is to change the rotational speed of an object. This is called the **moment of inertia**. The moment of inertia of an object is its resistance to a change in its rotational motion. The moment of inertia of an object depends on the axis about which it is rotated.

▲ Figure 3 Two flywheels both with a large mass placed as far from the rotation axis as possible.

For example, flywheels are designed to store rotational kinetic energy for machines. The two flywheels in Figure 3 are mounted on the same axle. Both are designed with large mass and these masses are arranged to be as far from the axis of rotation as possible. This makes it difficult to change the rotational speed and means that they have high moments of inertia.

Figure 4 shows a single point mass rotating in a circle of radius r. The axis of rotation is a line at 90° to the plane of the circle of rotation that goes through the centre of the circle. The moment of inertia I for this mass is given by

$$I = m \times r^2$$

The unit of moment of inertia is kg m^2 and it is a scalar quantity.

However, a single point mass rotating about a circle is of little practical use. You need to know how to treat objects that have more than one mass or a mass that is distributed (spread out) in space. When there is more than one mass the moment of inertia is calculated by adding together the moments of inertia for each point mass:

$$I = \sum m r^2$$

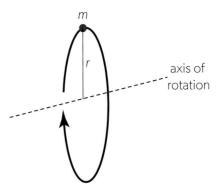

▲ Figure 4 A single mass m rotating in a circle of radius r has a moment of inertia of $m \times r^2$.

The symbol "$\sum$" here means "add up every mass × (its distance from the axis of rotation)²". For an arrangement of three masses, m_1, m_2, m_3 that are r_1, r_2 and r_3, respectively, from the axis, I is $m_1 r_1^2 + m_2 r_2^2 + m_3 r_3^2$.

Topic A.4 Rigid body mechanics

A simple dumbbell with a small mass at each end of a light rod of total length $2L$. The dumbbell is rotating about the midpoint of the light rod in a plane at 90° to the axis. Figure 5 shows that the moment of inertia is given by $2mL^2$.

When the mass is distributed, perhaps as a thin disc or as a pendulum, the problem becomes more complicated. Integral calculus or a numerical computation is used for such calculations.

Table 2 gives the equations for the moments of inertia of some common shapes that you may meet. (The mass is always m and the shapes are assumed to have a uniform density.) You do not need to learn these. In IB Diploma Programme physics examinations you will be provided with the equation for the moment of inertia of a particular shape distribution when you need it.

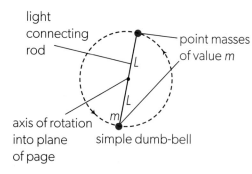

light connecting rod

point masses of value m

axis of rotation into plane of page

simple dumb-bell

▲ Figure 5 Two identical point masses connected by a light rod (a dumbbell shape) rotating in a circle of radius L have a moment of inertia I given by $m \times L^2 + m \times L^2 = 2mL^2$.

Shape and rotation axis	Moment of inertia
sphere of radius R rotating around a diameter	$I = \frac{2}{5}mR^2$
disc of radius R rotating about an axis perpendicular to and through the centre	$I = \frac{1}{2}mR^2$
rod of length L rotating about its centre perpendicular to the length	$I = \frac{1}{12}mL^2$
rod of length L rotating about one end perpendicular to the length	$I = \frac{1}{3}mL^2$
hoop of radius R rotating about its central axis	$I = mR^2$

▲ Table 2 Equations for the moment of inertia of different shapes.

 ## Models

Neutron stars are very dense objects created from the collapse of a massive star. A typical neutron star will have a mass of about 2.8×10^{30} kg and a radius of about 11 km. This would give it a moment of inertia of about 1.4×10^{38} kg m². This is important, as some neutron stars rotate with a time period of less than a second.

However, neutron stars are complex and, since they are almost impossible to investigate directly, somewhat mysterious. It is not easy to make measurements of their mass or radius and their rotational frequency is so small that it too can affect the radius and moment of inertia. Mathematical models are required to investigate how a neutron star might theoretically behave and these models are then compared with astronomical observations.

▲ Figure 6 An artist's impression of a neutron star.

Worked example 7

A bicycle wheel has a mass of 850 g and a radius of 34 cm. Estimate the moment of inertia of the wheel about the axis of rotation.

Solution

If we assume that all of the wheel's mass is concentrated in the rim, it can be modelled as a thin hoop. $I = MR^2 = 0.850 \times 0.34^2 = 0.098$ kg m². The actual moment of inertia is slightly less than the calculated value because some of the mass is in the hub and in the spokes, and hence closer to the axis of rotation than the wheel's radius.

Worked example 8

Two circular discs are cut from the same uniform metal plate. The moment of inertia of the first disc about its central axis is I. The radius of the second disc is twice the radius of the first disc.

Calculate, in terms of I, the moment of inertia of the second disc.

Solution

If M and R are the mass and the radius of the first disc, then $I = \frac{1}{2}MR^2$. The surface area and hence the mass of the second disc is four times greater than that of the first disc. The moment of inertia of the second disc is therefore $\frac{1}{2} \times 4M \times (2R)^2 = 16 \times \frac{1}{2}MR^2 = 16I$, which is sixteen times greater than that of the first disc.

Practice questions

5. Ball A is made of steel and ball B is made of lead. The density of lead is greater than that of steel. Outline which ball has a greater moment of inertia, when the balls have

    a. equal mass

    b. equal radius.

6. Two balls are made of the same material. The moment of inertia of the first ball is I. The radius of the second ball is half of the radius of the first ball. What is the moment of inertia of the second ball?

    A. $\dfrac{I}{32}$

    B. $\dfrac{I}{16}$

    C. $\dfrac{I}{8}$

    D. $\dfrac{I}{4}$

Torque—Newton's first and second laws of rotational motion

In linear mechanics, the net force acting on an accelerated object is $F = ma$ (Newton's second law), when we know m and a.

Moment of inertia I is the rotational equivalent of mass m; angular acceleration α is the rotational equivalent of linear acceleration a. The rotational equivalent of Newton's second law of motion, is:

$$\tau = I \times \alpha$$

This defines the **torque** τ acting on the object. The symbol used for torque is a Greek lower case tau τ.

The unit of torque is the newton-metre (N m).

You can use Figure 7 to define torque. A force F acts on a point P and causes a rotation about a point r away from P. The radius of the rotation is r. The force acts at angle θ to the line between the centre and P. Torque is then defined as:

$$\tau = Fr\sin\theta$$

This can be imagined also as $F \times (r\sin\theta)$ where the quantity in brackets is the perpendicular distance from the line of action of the force to the centre of rotation.

A given pair of F and r gives its maximum torque when $\theta = 90°$ ($\sin\theta = 1$). In this case, $\tau = Fr$.

A torque is also known as a "moment". However, you may want to avoid this term as it is easy to confuse with the word "momentum".

According to Newton's first law of motion, no resultant force acts on a body in translational equilibrium. The equivalent statement in rotational terms is that for an object in rotational equilibrium, no external resultant torque can act on it. Such an object continues at rest or rotating with a constant angular velocity.

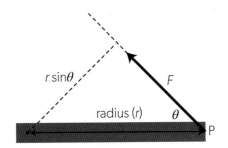

▲ Figure 7 The definition of torque in terms of the force F acting on a point P to move it about a radius r away from P.

Topic A.2 showed that a body in translational equilibrium does not accelerate. It remains either at rest or moving with a uniform velocity.

Formally, **Newton's first law for angular motion** may be stated as:

> **An object moves at a constant angular velocity (which may be zero) unless an external torque acts on it.**

For an object to be in **rotational equilibrium**, the total clockwise torque acting on the object must equal the total counter-clockwise torque acting. This statement is the **principle of moments**.

Direction rules

Strictly speaking, a torque is a type of vector with a direction at 90° to the plane of the circle in which the object rotates. The direction follows a right-hand corkscrew rule: imagine that your right hand grips the rotation axis so that the fingers curl round the axis in the direction of applied force; your thumb then points in the direction of the torque vector. Vector aspects of rotation are not used in IB Diploma Programme physics.

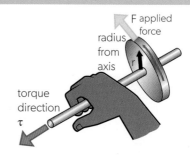

▲ Figure 8   Direction rules for rotational mechanics. The torque acting on an object follows a right-hand corkscrew rule.

Worked example 9

The picture shows a child leaning against a strong wind.

a. Draw a diagram showing the forces acting on the child. Assume that the effect of air resistance can be represented by a single force acting horizontally through the child's centre of mass.

b. The child remains in translational equilibrium. State the relationship between the magnitudes of the horizontal and the vertical forces acting on him.

c. To remain in rotational equilibrium, the child must lean at an angle of 65° to the horizontal. His mass is 42 kg.

   Determine the magnitude of the force of air resistance acting on the child.

Solutions

a. The forces acting through the centre of mass C are weight W and air resistance due to the wind, F_d. The forces applied to the point of contact with the ground are the static frictional force F_f and the normal reaction force N.

b. The net force must be zero; hence $F_d = F_f$ and $W = N$.

c. The net torque about the pivot point P must be zero. Friction and the normal force act through P and their torque is therefore zero. The air resistance force acts at an angle 65° to the line PC and provides a clockwise torque about P. The weight acts at an of angle 25° and provides an anticlockwise torque. The torques have equal magnitudes, so $mgr \sin 25° = F_d r \sin 65°$, where r is the distance between C and P. From this, $F_d = \dfrac{mg \sin 25°}{\sin 65°} = 190\,\text{N}$.

Practice questions

7. Two objects of masses 1.5 kg and m are attached to the ends of a weightless horizontal rod, as shown in the diagram. The rod is supported at point P.

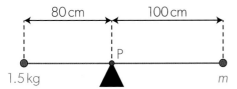

The rod remains in equilibrium. Calculate:

a. the mass m

b. the reaction force on the rod at P.

8. A ladder of mass 12 kg leans against a vertical frictionless wall. The ladder is at rest and makes an angle of 55° with the floor.

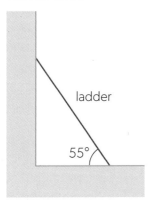

a. Draw a diagram showing the forces acting on the ladder.

b. Determine the magnitude of the reaction force between the wall and the ladder.

c. Calculate the minimum value of the coefficient of static friction between the floor and the ladder so that the ladder does not slide away from the wall.

9. A rod of mass 40 kg is attached to a vertical wall at point P. The other end of the rod is suspended from a string that makes a right angle with the rod. The rod remains in equilibrium.

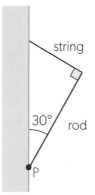

a. Determine the magnitude of the tension in the string.

b. Explain whether the contact force on the rod from the wall at P has a vertical component.

10. A uniform rod rests horizontally on two supports P and Q. A mass m is attached to the rod, at a distance of 10 cm from P and 40 cm from Q.

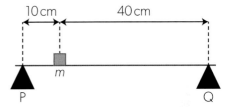

What is the increase in the reaction force on the rod from support P after the mass has been attached to the rod?

A. 0.25 mg B. 0.50 mg

C. 0.75 mg D. 0.80 mg

Couples

A common arrangement of forces that gives rise to a turning effect is known as a **couple**. A couple consists of two equal and opposite forces which do not act along the same straight line. Because the forces are offset, they produce a torque which causes the system to rotate. Because they are equal and opposite, they do not produce a linear acceleration. The system is in translational equilibrium but not rotational equilibrium.

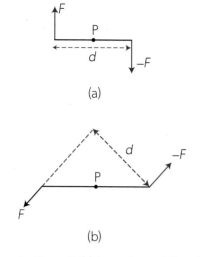

(a)

(b)

▲ Figure 9 (a) A couple consisting of two equal, parallel but opposite forces that will turn the rod about P but not translate it. (b) The line of action between the forces is still d (the distance between where the forces act is now greater) so the couple will be the same.

Figure 9(a) shows an example of a couple. The torque due to each force about the point P in the system of Figure 9(a) is $F \times \dfrac{d}{2}$. Both torques act in the same rotational direction (in this case, clockwise) so they add together making the total torque acting $2 \times F \times \dfrac{d}{2} = Fd$: that is, *the product of one of the forces, F, and the perpendicular distance d between the forces*.

To emphasise the importance of d being the perpendicular distance, consider the slightly different arrangement in Figure 9(b). Now the forces still act in opposite directions and are still offset by a distance d.

You must use the perpendicular distance between the **lines of action** of the forces (the construction with dashed lines shows this) and so the torque is still $F \times d$.

How does a torque lead to simple harmonic motion?

A Wilberforce pendulum is a mass–spring system that acts first as a vertical oscillator and later as a torsional oscillator, eventually reverting to the vertical motion again. Energy is being transferred between the two modes of motion. You can see many videos on the Internet of the pendulum in action.

In its torsional mode, the loaded cylinder, suspended on the spring, rotates about a vertical axis and the spring becomes twisted. This twist exerts a torque on the cylinder trying to restore the cylinder to its equilibrium position. For small rotations, the torque is directly proportional to the angular displacement and tends to return the spring to equilibrium; precisely the conditions required for simple harmonic motion (Topic C.1). This is a special case of resonance (Topic C.4); one mode can be regarded as providing the forcing oscillation for the other mode. Because the two modes have the same oscillation frequency, the system cannot make up its mind whether to oscillate vertically or in torsion. It moves between them one after the other!

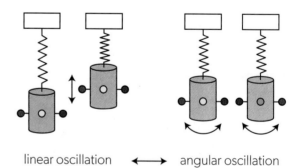

linear oscillation ⟷ angular oscillation

▲ Figure 10 The Wilberforce pendulum. The two modes of oscillation have the same time period and the system switches repeatedly between the two modes.

Worked example 10

Two parallel forces of magnitudes $F_1 = 3.0$ N and $F_2 = 4.0$ N are applied to a circular disc, as shown in the diagram.

a. Calculate the magnitude of:

    i.   the net force acting on the disc

    ii. the net torque about the centre P of the disc.

b. The moment of inertia of the disc is 0.080 kg m². Calculate the initial angular acceleration of the disc.

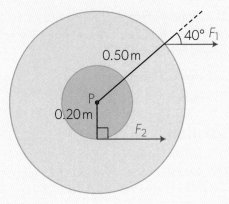

Solutions

a. i. The forces act in the same direction, so the net force is $4.0 + 3.0 = 7.0$ N.

    ii. The torque provided by F_1 is $\tau_1 = 3.0 \times 0.50 \times \sin 40° = 0.96$ N m and the torque provided by F_2 is $\tau_2 = 4.0 \times 0.20 = 0.80$ N m. τ_1 acts clockwise and τ_2 acts counter-clockwise, so the net torque is the difference $\tau = 0.96 - 0.80 = 0.16$ N m in the clockwise direction.

b. $\tau = I\alpha \Rightarrow \alpha = \dfrac{\tau}{I} = \dfrac{0.16}{0.080} = 2.0$ rad s⁻²

Worked example 11

A constant force of 25 N is applied tangentially to the sprocket of a bicycle wheel.

The wheel accelerates from rest to an angular velocity of 40 rad s⁻¹ in a time of 3.0 s. The radius of the sprocket is 4.0 cm.

Calculate the moment of inertia of the wheel.

Solution

The torque applied to the wheel is $\tau = 25 \times 0.040 = 1.0$ N m. The wheel's angular

acceleration is $\alpha = \dfrac{40}{3.0} = 13.3$ rad s⁻². Using Newton's second law for rotation, $I = \dfrac{\tau}{\alpha} = \dfrac{1.0}{13.3} = 0.075$ kg m².

Worked example 12

A constant frictional torque acting on a spinning top causes its angular velocity to decrease from 150 rad s⁻¹ to 80 rad s⁻¹. The top undergoes 300 revolutions while the angular velocity is changing. The moment of inertia of the spinning top is 5.0×10^{-3} kg m². Calculate the magnitude of the frictional torque acting on the top.

Solution

The angular acceleration can be found from $80^2 = 150^2 - 2\alpha \times 300 \times 2\pi \Rightarrow \alpha = 4.27$ rad s⁻². The torque acting on the top is therefore $\tau = 5.0 \times 10^{-3} \times 4.27 = 0.021$ N m.

Worked example 13

Two objects of masses 2.0 kg and 1.0 kg are attached to the ends of a weightless rod supported at point P as shown in the diagram. The rod is initially horizontal.

Determine the initial linear acceleration of each mass when the rod is released.

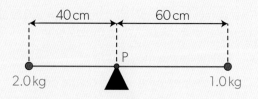

Solution

The moment of inertia of the system about P is $I = \sum mr^2 = 2.0 \times 0.40^2 + 1.0 \times 0.60^2 = 0.68\,\text{kg m}^2$. The weights of both masses are initially at right angles to the rod, so the initial torque about P is $\tau = 2.0 \times 0.40 - 1.0 \times 0.60 = 0.20\,\text{N m}$. The torque acting on the 2.0 kg mass is greater than that due to the 1.0 kg mass; hence the net torque is directed counter-clockwise. The initial angular acceleration is $\alpha = \dfrac{\tau}{I} = \dfrac{0.20}{0.68} = 0.294\,\text{rad s}^{-2}$ and is equal for both masses. The initial linear accelerations are different: $0.294 \times 0.40 = 0.12\,\text{m s}^{-2}$ downward for the 2.0 kg mass and $0.294 \times 0.60 = 0.18\,\text{m s}^{-2}$ upward for the 1.0 kg mass.

Practice questions

11. A couple consisting of two forces F acts on a system of two masses m connected by a weightless rod of length L.

    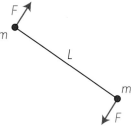

    What is the angular acceleration of the system about the midpoint of the rod?

    A. $\dfrac{F}{2mL}$ B. $\dfrac{2F}{mL}$ C. $\dfrac{F}{mL}$ D. $\dfrac{4F}{mL}$

12. A uniform cylinder of radius 4.0 cm and mass 1.2 kg can rotate freely around its central axis. A string is wrapped five times around the cylinder and a constant force of 1.8 N is applied to the string. The cylinder is initially at rest and the string unwinds without slipping.

    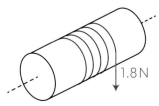

    Calculate:

    a. the angular acceleration of the cylinder

b. the final angular velocity of the cylinder when the string unwinds completely

c. the time taken to unwind the string.

13. A bicycle wheel has a moment of inertia of 0.090 kg m^2 and rotates at an angular velocity of 23 rad s^{-1}. When the brakes are applied to the wheel, it turns through one quarter of a revolution before coming to rest.

    a. Calculate the frictional torque acting on the wheel.

    The wheel is equipped with a disc brake system in which a force is applied to the disc by a pair of braking pads, one on each side of the disc, at a distance of 8.0 cm from the axis of rotation of the wheel. The coefficient of dynamic friction between braking pads and the disc is 0.85.

    b. Calculate the magnitude of the normal force acting between the disc and each braking pad.

14. A fan rotates at an initial rate of 320 revolutions per minute. When the motor is switched off, a resistive torque of 0.10 N m brings the fan to rest in a time of 8.0 s. Calculate:

    a. the moment of inertia of the fan

    b. the number of revolutions the fan rotates through before coming to rest.

Centre of mass

The centre of mass of an object is the point that moves as though the whole mass were concentrated there. An external force applied at the centre of mass causes linear but not angular acceleration. Figure 11 shows the centres of mass for some common shapes.

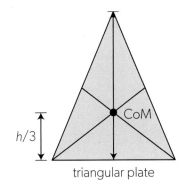

triangular plate

Consequences of centre of mass

- Inquiry 1: State and explain predictions using scientific understanding.

- Inquiry 2: Identify and record relevant qualitative observations.

- Inquiry 3: Compare the outcomes of an investigation to the accepted scientific context.

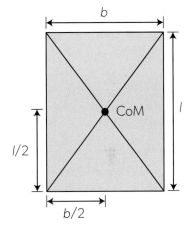

rectangular plate

- Set up the arrangement shown in the diagram, with two forks sticking into opposite ends of a cork and a match stuck in the middle of the cork.

- Consider where the centre of mass of this combination is.

- Investigate how you could find the centre of mass of this combination.

- The centre of mass of the combination of two forks and a cork lies somewhere in the space between the forks.

- Balance the fork, cork and match combination on the edge of a cup. The combined centre of mass is at the point where the match balances on the edge of the cup.

- Slightly wet the match above this balance point.

- Light the match and watch what happens. Consider why this is.

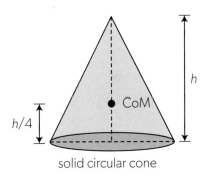

solid circular cone

When a single force acts on an object through the centre of mass, then the object will undergo linear acceleration (in a straight line). This is known as **translational acceleration**.

However, when the single force acts on an object in any other direction than through the centre of mass, two things happen. There is both a translational acceleration and a rotational acceleration. Figure 12 shows how this arises. The single force F acts near the top of the baseball bat. This has two effects on the bat:

- A force F acts at the centre of mass. This gives rise to a translational acceleration to the right.

- A couple acts on the bat producing a clockwise rotation. The couple has forces $+\frac{1}{2}F$ at the top of the bat and $-\frac{1}{2}F$ at the bottom.

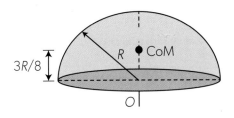

solid hemisphere

▲ Figure 11 The position of the centre of mass (CoM) for four objects.

When a force has acted on the bat, it will be moving at a constant linear velocity and rotating with a constant angular velocity.

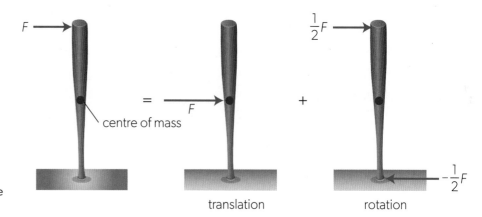

▶ Figure 12   The effects of a force acting on a baseball bat at a position other than the centre of mass.

translation   rotation

Centre of percussion

- Inquiry 1: Demonstrate creativity in the designing, implementation or presentation of the investigation

- Inquiry 1: Develop investigations that involve hands-on laboratory experiments, databases, simulations and modelling.

- Inquiry 1: Design and explain a valid methodology.

Anyone who plays a game that involves a bat will be aware that all bats and racquets have a single "sweet spot". This is the region on the bat where contact with a ball gives minimal jarring of the wrist(s) and produces the best result in terms of transferring energy to the ball.

This phenomenon is caused by the existence of a **centre of percussion**. Again, the example of a baseball bat shows how it occurs.

For the sake of simplicity, the bat is modelled as a horizontal block of wood of constant cross-section that is pivoted (hinged) at one end. The hinge corresponds to the player's wrist. When the ball impacts the bat at the centre of percussion, the wrist is not jarred.

When a force F is applied to the object there are always two effects:

- The force applies a rotational impulse about the centre of gravity (unless F acts exactly at the centre of gravity). This rotates the bat.

- The force applies a translational motion to the bat which moves it linearly.

Figure 13 shows these two effects in operation. When F acts below the centre of gravity, the rotation caused is counter-clockwise giving an impulse to the left at the hinge. The linear effect is to the right at the hinge. The centre of percussion is the point at which these two effects will have the same magnitude and be opposite in direction so that there is no net force acting at the hinge.

Devise an investigation along the lines of an internal assessment to determine the position of the centre of percussion for a bat used in sport.

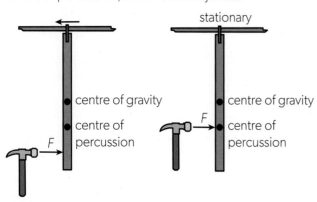

▲ Figure 13   Centre of percussion.

Newton's third law of rotational motion

The rotational equivalent of Newton's third law is straightforward. In rotational terms, the action torque and reaction torque are equal and opposite. The pair of torques, as in action–reaction pairs for linear motion, must act on different bodies.

When object A applies a torque to object B, then object B will apply an equal and opposite torque to object A.

Rotational kinetic energy

The work done by a torque when an object is rotated is an analogue of the work done by a force acting in a linear direction:

$$W = \text{torque} \times \text{angular displacement} = \tau \times \theta$$

and for power

$$P = \text{torque} \times \text{angular velocity} = \tau \times \omega$$

The kinetic energy of an object in linear motion is $E_k = \frac{1}{2}mv^2$. Therefore, the rotational kinetic energy of a rotating object is

$$E_k = \frac{1}{2}I\omega^2.$$

Worked example 14

A uniform disc of mass $0.25\,\text{kg}$ and radius $0.15\,\text{m}$ rotates about the central axis with an initial angular velocity of $8.0\,\text{rad}\,\text{s}^{-1}$. Calculate:

a. the initial rotational kinetic energy of the disc

b. the work done in increasing the angular velocity of the disc from $8.0\,\text{rad}\,\text{s}^{-1}$ to $16\,\text{rad}\,\text{s}^{-1}$.

Solutions

a. $I = \frac{1}{2}MR^2 = \frac{1}{2} \times 0.25 \times 0.15^2 = 2.8 \times 10^{-3}\,\text{kg}\,\text{m}^2$. The kinetic energy is $E_k = \frac{1}{2}I\omega^2 = \frac{1}{2} \times 2.8 \times 10^{-3} \times 8.0^2 = 0.090\,\text{J}$.

b. The work done is equal to the change in the rotational energy of the disc,

$$W = \frac{1}{2}I(\omega_f^2 - \omega_i^2) = \frac{1}{2} \times 2.8 \times 10^{-3} \times (16^2 - 8.0^2) = 0.27\,\text{J}.$$

Worked example 15

A stationary bicycle trainer has a flywheel with electronically controlled resistance to simulate different cycling conditions. When the cyclist stops pedalling, the flywheel comes to rest from an initial angular velocity of $180\,\text{rad}\,\text{s}^{-1}$ in a time of $9.0\,\text{s}$. The moment of inertia of the flywheel is $0.070\,\text{kg}\,\text{m}^2$. Calculate:

a. the resistive torque acting on the flywheel, assuming that it is constant

b. the work done by the resistive torque in stopping the flywheel

c. the power that the cyclist needs to transfer to the flywheel to maintain the constant angular velocity of $180\,\text{rad}\,\text{s}^{-1}$.

Solutions

a. The angular deceleration of the flywheel is $\alpha = \dfrac{180}{9.0} = 20\,\text{rad s}^{-2}$. The resistive torque is therefore
$\tau = I\alpha = 0.070 \times 20 = 1.4\,\text{N m}$.

b. The work done is equal to the change in rotational kinetic energy of the flywheel: $W = \dfrac{1}{2} \times 0.070 \times 180^2 = 1.1\,\text{kJ}$.

c. When the flywheel rotates with a constant angular velocity $\omega = 180\,\text{rad s}^{-1}$, the external torque applied to the flywheel by the cyclist must be equal to the resistive torque, and the power transferred by the cyclist is
$P = \tau\omega = 1.4 \times 180 = 250\,\text{W}$.

Worked example 16

A metre stick of mass 0.15 kg is suspended at one end and can rotate freely about the point of suspension. The metre stick is set in motion so that it passes through the vertical position with an angular velocity of $3.4\,\text{rad s}^{-1}$.

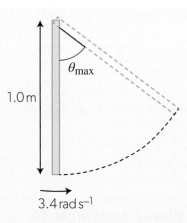

a. Calculate the rotational kinetic energy of the metre stick in the vertical position.

b. Explain why the torque acting on the metre stick increases with the angular displacement from the vertical position. State the direction of the torque.

c. Determine the maximum angle θ_{max} through which the metre stick rotates before it reverses the direction of motion.

Solutions

a. The moment of inertia of the metre stick through the fixed end is $I = \dfrac{1}{3}ML^2 = \dfrac{1}{3} \times 0.15 \times 1.0^2$
$= 0.050\,\text{kg m}^2$. The kinetic energy is therefore $E_k = \dfrac{1}{2} \times 0.050 \times 3.4^2 = 0.29\,\text{J}$.

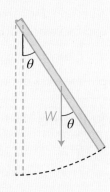

b. The torque is provided by the weight W, applied to the centre of mass of the metre stick at its midpoint.

Since $\tau = W\left(\dfrac{L}{2}\right)\sin\theta$, the torque increases with θ, and reaches the maximum value when the metre stick is horizontal. The torque is directed clockwise towards the vertical position, so against the angular displacement of the metre stick.

c. At the angular position θ_{max}, all of the kinetic energy of the metre stick has been converted to gravitational potential energy; hence $mgh = \dfrac{1}{2}I\omega_0^2$,

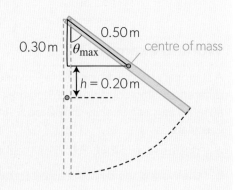

where h is the height through which the centre of mass of the metre stick has risen, and ω_0 is the angular velocity in the vertical position.
$0.15 \times 9.8h = \dfrac{1}{2} \times 0.050 \times 3.4^2 \Rightarrow h = 0.20\,\text{m}$. In the initial position, the centre of mass was 0.50 m below the fixed end so now it is $0.50 - 0.20 = 0.30\,\text{m}$ below the end. The angle of rotation can be determined using triangle trigonometry: $\cos\theta_{max} = \dfrac{0.30}{0.50} \Rightarrow \theta_{max} = 53°$.

Practice questions

15. A football can be modelled as a thin spherical shell of mass 0.45 kg and radius 0.11 m. The football is kicked so that it rotates with angular velocity 60 rad s^{-1}.

    a. Calculate the rotational kinetic energy of the football. The moment of inertia of a spherical shell of mass m and radius R is $\frac{2}{3}mR^2$.

    b. Determine the linear speed of the football if its translational and rotational kinetic energies are equal.

16. A wind turbine transfers kinetic energy of the wind to electrical energy at a rate of 1.5 MW. The turbine rotates at a constant angular velocity and makes one complete revolution in a time of 4.0 s. Calculate the torque acting on the turbine due to the wind.

17. Earth has radius 6.4×10^6 m and mass 6.0×10^{24} kg and rotates about its axis once every 24 hours.

    a. Calculate the rotational kinetic energy of Earth, assuming that it is a uniform sphere.

    b. The density of Earth is not uniform but increases towards the centre. Outline what effect this has on the energy calculated in a.

18. A merry-go-round rotates at an angular velocity of 0.60 rad s^{-1}. Assuming that the merry-go-round can be modelled as a ring of mass 150 kg and radius 2.5 m, calculate:

    a. the rotational kinetic energy of the merry-go-round

    b. the power required to stop it in a time of 5.0 s.

Angular momentum

Angular momentum L is the direct rotational equivalent of linear momentum. It is defined as the product of an object's moment of inertia and its angular velocity. It is a vector quantity but in the IB Diploma Programme physics you will only consider its sense (clockwise or counter-clockwise).

Using the symbols already defined:

$$L = I \times \omega$$

The units of angular momentum are kg m^2 [rad] s^{-1}. The radian is a ratio and so unitless. It is normally omitted.

Conservation of angular momentum

The total (linear) momentum of a system remains constant when no external forces act on the system. In rotational dynamics, a similar law of **conservation of angular momentum** applies:

The total angular momentum of a system remains constant providing no external torque acts on the system.

In equation terms, using the summation symbol ($\sum$) introduced earlier:

$$\sum(I_{initial} \times \omega_{initial}) = \sum(I_{final} \times \omega_{final}).$$

Imagine two co-axial flywheels (Figure 14) rotating in opposite directions at different angular velocities (I_1 and I_2) and with different initial angular velocities (ω_1 and ω_2). When these flywheels are suddenly clamped together so that they must rotate at the same speed Ω, then the equation that describes the conservation of angular momentum is $I_1\omega_1 - I_2\omega_2 = (I_1 + I_2)\Omega$.

The negative sign on the left-hand side is there because the flywheels are rotating in opposite directions. The final sign of Ω will show the direction in which the pair of flywheels rotate afterwards. If Ω is positive, the pair will rotate in the same direction as flywheel 1 initially. If Ω is negative, the pair will rotate in the same initial direction of flywheel 2.

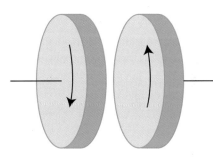

▲ Figure 14 Two flywheels rotating just before they are clamped together. Angular momentum must be conserved when this happens.

 Rotational momentum

Rotational kinetic energy can be written in terms of rotational momentum L: $E_k = \frac{L^2}{2I}$ to match the $E_k = \frac{p^2}{2m}$ formulation from linear mechanics.

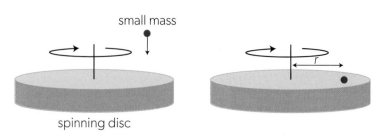

small mass

spinning disc

▲ Figure 15 Conservation of angular momentum as a small mass is dropped onto a turntable.

(a)   (b)

▲ Figure 16 The ice skater conserves angular momentum as she rotates (a). She pulls in her arms to decrease her moment of inertia (b). As a result, her angular velocity increases.

Figure 15 gives another example of angular momentum conservation when a small mass is dropped onto a disc that is spinning freely in a horizontal plane. The combined moment of inertia of the disc and mass is greater than the moment of inertia of the disc alone. Friction acts between the mass and the disc surface and this accelerates the mass. The reaction force to this friction decelerates the disc. The angular velocity of the system will decrease so that angular momentum is conserved.

Conservation of angular momentum is of great importance in sport. An ice skater can increase their angular velocity about a vertical axis by pulling their arms tightly into their body (Figure 16).

Worked example 17

A disc of radius 30 cm and mass 1.2 kg spins freely at an angular velocity of 5.0 rad s^{-1}. A small object of mass $m = 0.25$ kg is dropped onto the disc with a negligible initial velocity and comes to rest relative to the disc at a distance of 20 cm from the axis of rotation.

a. Calculate the final angular velocity of the system consisting of the disc and the mass, assuming that no external torques act on the system.

b. Determine the change in the rotational kinetic energy of the system.

c. Explain why the rotational kinetic energy has decreased.

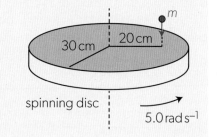

30 cm   20 cm m

spinning disc

5.0 rad s^{-1}

Solutions

a. The moment of inertia of the disc about the axis of rotation is $I_{disc} = \dfrac{1}{2} \times 1.2 \times 0.30^2 = 0.054$ kg m^2 and the moment of inertia of the mass is $I_{mass} = 0.25 \times 0.20^2 = 0.010$ kg m^2. Initially, the mass has zero angular velocity and it does not contribute to the angular momentum L of the system, $L = I_{disc}\omega_0 = 0.054 \times 5.0 = 0.27$ kg m^2 s^{-1}. When the mass has come to rest relative to the disc, the system rotates with a new angular velocity ω and with a combined moment of inertia $I_{disc} + I_{mass}$. The angular momentum is unchanged,

so $L = (I_{disc} + I_{mass})\omega = I_{disc}\omega_0$. From here, $\omega = \dfrac{I_{disc}\omega_0}{I_{disc} + I_{mass}} = \dfrac{0.27}{0.064} = 4.2$ rad s^{-1}.

b. The initial rotational energy is that of the spinning disc alone, $\dfrac{1}{2}I_{disc}\omega_0^2$. The final rotational energy

is $\dfrac{1}{2}(I_{disc} + I_{mass})\omega^2$. The change is therefore $\dfrac{1}{2} \times 0.064 \times 4.2^2 - \dfrac{1}{2} \times 0.054 \times 5.0^2 = -0.11$ J.

The negative sign indicates that the energy has decreased.

c. Before the mass has come to rest relative to the disc, a frictional force must have acted on the mass at the disc surface, slowing it down relative to the disc. The work done by the frictional force results in a decrease of the kinetic energy of the system by the amount equal to the work done.

Worked example 18

A rotating star collapses, decreasing its moment of inertia to $\dfrac{1}{1000}$ of the initial value. The initial angular velocity of the star is ω_0 and its initial rotational kinetic energy is E_0. For the star after the collapse, calculate:

a. the angular velocity

b. the rotational kinetic energy.

Solutions

a. The angular momentum is unchanged; hence $I_0\omega_0 = \dfrac{I_0}{1000}\omega$. From here, $\omega = 1000\omega_0$.

b. The new kinetic energy is $E_k = \dfrac{1}{2}I\omega^2 = \dfrac{1}{2}\dfrac{I_0}{1000}(1000\omega_0)^2 = 1000E_0$. The collapse of the star results in a thousandfold increase in both the angular velocity and the rotational kinetic energy. Where do you think this additional energy comes from?

Worked example 19

A child sits at the edge of a merry-go-round of radius $r = 2.5\,\text{m}$. The child throws a stone of mass $m = 0.50\,\text{kg}$ tangentially to the merry-go-round, with an initial speed of $v = 8.0\,\text{m s}^{-1}$.

a. Show that the angular momentum of the stone relative to the rotation axis of the merry-go-round is given by $L = mvr$.

The merry-go-round is initially at rest and can rotate without friction. The combined moment of inertia of the child and the merry-go-round is $I = 1400\,\text{kg m}^2$.

b. Calculate the final angular velocity ω of the merry-go-round.

Solutions

a. The initial moment of inertia of the stone is $I_s = mr^2$ and its initial angular velocity relative to the centre of the merry-go-round is $\omega_s = \dfrac{v}{r}$. Hence, $L = I_s\omega_s = mr^2\dfrac{v}{r} = mvr$.

b. The total angular momentum is zero; hence $I\omega - mvr = 0$. The minus sign means that the merry-go-round rotates away from the direction of the initial velocity of the stone.

$\omega = \dfrac{mvr}{I} = 7.1 \times 10^{-3}\,\text{rad s}^{-1}$. This corresponds to about $0.4°$ per second.

Practice questions

19. An ice skater doubles the angular velocity about the vertical axis of rotation by pulling in her arms close to her body. The initial rotational kinetic energy of the ice skater is E. No external torques act on the ice skater. What is the change in her rotational kinetic energy?

A. 0

B. E

C. $2E$

D. $3E$

20. A flywheel in the shape of a solid cylinder of mass $7.0\,\text{kg}$ and radius $0.10\,\text{m}$ rotates at an initial angular frequency of $80\,\text{rad s}^{-1}$. Another flywheel, of radius $0.25\,\text{m}$ and mass $3.0\,\text{kg}$, is initially at rest. The flywheels are coupled together. Calculate:

a. the final angular velocity of the system

b. the change in the rotational kinetic energy.

How does conservation of angular momentum lead to the determination of the Bohr radius?

Topic E.1 describes the important work of Niels Bohr who took empirical data for the hydrogen spectrum and linked it to the energy states of the atom. This work moved the description of the atom forward. Even though he used classical mechanics to describe the behaviour of what is a quantum-mechanical system, Bohr made predictions and hypotheses that were confirmed experimentally. He was able to calculate the radius of the electron orbit in the ground state using conservation of angular momentum.

 How does rotation apply to the motion of charged particles or satellites in orbit?

Kepler identified the three law of planetary motion which are described in Topic D.1 and are now named after him. The second law involves the line joining the centre of a planet to the centre of the Sun. The law suggests that this line sweeps out equal areas in equal times. This is clear for a circular orbit with the planet moving with a constant angular velocity. For comets where the orbit can be ellipses, then it is not so clear.

The conservation of angular momentum helps here. The angular momentum of the comet is equal to $m \times v \times r$, where m is the mass of the comet, v is its linear speed and r the radius of its orbit. When the distance of the comet from the Sun is large, its speed is small and vice versa. Analysing the orbit carefully and assuming conservation of angular momentum shows that the area the comet sweeps out is indeed constant, which verifies Kepler's second law.

Similar ideas apply to the rotation of a charged particle moving in a magnetic field. Topic D.4 contains the analysis which can be easily carried out in terms of rotational mechanics.

How can rotation lead to the generation of an electric current?

In Topic D.4, you will learn that a conducting coil rotating relative to a magnetic field will lead to the generation of an electromotive force across the terminals of the coil. There are direct links between the theory you meet in this topic and the later one which deals with the effects of changing a magnetic flux. Some of the mathematical ideas will appear again too.

▲ Figure 17 A gyroscope can be used to stabilize cameras because its large angular momentum means that a large torque is required to change its direction.

ATL Applying key ideas and facts in new concepts

A gyroscope consists of a disc that spins rapidly. Gyroscopes can be used to help keep things level and upright. For example, they can be used to aid navigation or to help stabilize cameras. In Figure 17, the camera operator is standing on a self-balancing transporter that uses gyroscopes to detect the tilt and adjusts the wheels to keep the rider in balance.

Because gyroscopes spin rapidly, they have a large angular momentum. The effects of this angular momentum can cause a gyroscope to have seemingly strange properties. Similar effects can be observed with other spinning objects such as bicycle wheels.

A fast-moving object with a large linear momentum requires a large force to stop it or to change its direction within a given timeframe—this is because of Newton's second law, $F = \dfrac{\Delta p}{\Delta t}$. The same effect applies to a gyroscope. Its large angular momentum requires a large torque to change the direction of rotation.

Investigating a spinning bicycle wheel

- Inquiry 2: Identify and record relevant qualitative observations.

- Inquiry 2: Interpret qualitative and quantitative data.

Imagine a bicycle wheel that is spinning counter-clockwise in a horizontal plane.

The wheel's angular momentum can be represented by a vector along its axis. In this instance, the direction of the vector would be vertically upwards.

Now imagine twisting the bicycle wheel by lifting the far side upwards and the near side downwards. This is represented by a torque with its vector direction acting along the axis of rotation to the right. This torque creates an angular impulse in the same direction.

The angular impulse gives the change in the angular momentum.

The axis of rotation has changed so that the left-hand side of the wheel rotates upwards and the right-hand side moves downwards. But this was not the direction in which the wheel was rotated.

- Try this with a bicycle wheel. Make it spin rapidly—you could add some mass evenly distributed around the rim to increase the moment of inertia of the wheel. This will make the effect more noticeable.

- Try standing on a rotating platform while you twist the bicycle wheel. What happens?

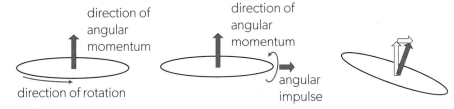

- If you cannot carry out the experiment, you could search for a video online—try searching for "angular momentum conservation with rotating bicycle wheel".

Behaviour of gyroscopes

▲ Figure 18 When spinning, a gyroscope can balance in a way that seems unexpected.

The behaviour of spinning objects such as gyroscopes can seem strange. They do not seem to follow the laws of physics. What is really happening is that we are making poor assumptions as to what the laws of physics predict.

Often, intuition is correct and can predict the outcome of events. Sometimes, however, it lets us down. Is intuition a valid way of predicting what will happen?

Global impact of science

Earth rotates about an axis which passes between the North and South poles. This axis is tilted, relative to the plane in which Earth rotates about the Sun, by an angle of about 23.5°. This tilt is responsible for the seasons—for example, when the North Pole is tilted towards the Sun, it is summer in the Northern Hemisphere.

The tilt varies slowly over a period of about 41,000 years. Additionally, the gravitational effect of the Sun and the moon create a torque which causes the direction of the tilt to wobble with a period of about 26,000 years.

The way in which the rotational motion of Earth varies over time causes variation in the seasons and in Earth's climate. These variations are called Milankovitch cycles

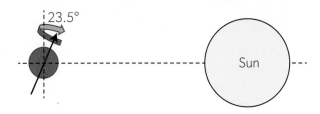

▲ Figure 19 The tilt of Earth is responsible for the seasons.

and they have caused periodic ice-ages in Earth's history. Understanding the rotational dynamics of Earth's orbit helps us to understand what effect these cycles currently have on Earth's climate and to evaluate the impact of humans on the planet.

Data-based questions

Angular momentum is important when considering the Solar System. Table 1 shows that the Sun accounts for approximately 99.9% of the Solar System's mass. However, the Sun does not account for the majority of the angular momentum in the Solar System. This is important when considering theories for how the Solar System formed.

The planets possess angular momentum because of their rotation about their axis (as a rotating sphere: $I = \frac{2}{5}MR^2$) as well as their orbit around the Sun (acting as a point mass: $I = MR^2$).

- Use the rotational period and radius to calculate the angular momentum of each of the solar system objects as they spin on their axis.

Two objects rotate in the opposite direction and so their rotational periods are expressed with a minus sign. You may wish to use a spreadsheet here.

- Use the orbital period and orbital radius to calculate the angular momentum of the planets as they orbit the Sun. The Sun has zero angular momentum from this motion.

- Hence calculate the total angular momentum of each object.

- Calculate the percentage of the total angular momentum of the solar system that is possessed by:

  a. the Sun

  b. Jupiter.

- Draw a pie chart to represent the distribution of the angular momentum of the Solar System objects.

Solar System object	Mass / $\times 10^{24}$ kg	Orbital period / $\times 10^6$ s	Orbital radius / $\times 10^9$ m	Rotational period / $\times 10^6$ s	Radius / $\times 10^6$ m
Sun	1 990 000	-	0	2.16	696
Mercury	0.330	7.60	57.9	5.07	2.44
Venus	4.87	19.4	108	−21.0	6.05
Earth	5.97	31.6	150	0.0860	6.38
Mars	0.642	59.4	228	0.0886	3.40
Jupiter	1900	374	779	0.0356	71.5
Saturn	568	929	1430	0.0385	60.3
Uranus	86.8	2640	2870	−0.0619	25.6
Neptune	102	5170	4500	0.0580	24.8

 Units of angular impulse

Because angular impulse is equal to the change in angular momentum, both quantities should have the same units. Earlier the units for angular momentum were stated as being $kg\,m^2\,rad\,s^{-1}$ or $kg\,m^2\,s^{-1}$. It is easy to show that the newton has units $kg\,m\,s^{-2}$ and so $(N \times m \times s)$ is $(kg\,m\,s^{-2} \times m \times s) \equiv kg\,m^2\,s^{-1}$.

Angular impulse

As you saw in Topic A.2, in linear mechanics, impulse J is the product of the *average resultant force acting on an object F* and the *contact time Δt over which the force acts*. In symbols, $J = F\Delta t$.

In rotational dynamics, the angular impulse ΔL is the product of the average torque τ and Δt:

$$\Delta L = \tau \times \Delta t$$

This can also be written as

$$\Delta(I \times \omega)$$

The units of angular impulse using $\Delta L = \tau\Delta t$ are $N\,m\,s$ (τ has the units $N\,m$).

Worked example 20

Two rocket thrusters T_1 and T_2 separated by a distance of 80 m are used to rotate a space station around its axis. Each of the thrusters provides a 1.0 kN force at a right angle to the line joining the thrusters. The moment of inertia of the space station about the point halfway between the thrusters is 2.8×10^8 kg m².

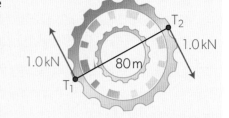

The space station rotates initially in a clockwise direction with a period of 20 minutes. The thrusters operate for 30 s. Calculate:

a. the initial angular momentum of the space station

b. the angular impulse delivered to the space station by the thrusters

c. the final rotational period of the space station.

Solutions

a. The initial angular velocity of the space station is $\omega_0 = \dfrac{2\pi}{T_0} = \dfrac{2\pi}{20 \times 60} = 5.24 \times 10^{-3}\,\text{rad s}^{-1}$.

   The angular momentum is $L_0 = I\omega_0 = 2.8 \times 10^8 \times 5.24 \times 10^{-3} = 1.47 \times 10^6\,\text{N m s}$.

b. The torque provided by the thrusters is $\tau = 1.0 \times 10^3 \times 80 = 8.0 \times 10^4\,\text{N m}$.
   The impulse delivered for 30 s is $\Delta L = \tau \Delta t = 8.0 \times 10^4 \times 30 = 2.40 \times 10^6\,\text{N m s}$.

c. The final angular momentum is $L = L_0 + \Delta L = 1.47 \times 10^6 + 2.40 \times 10^6 = 3.87 \times 10^6\,\text{N m s}$.
   This corresponds to the angular velocity $\omega = \dfrac{L}{I} = \dfrac{3.87 \times 10^6}{2.8 \times 10^8} = 1.38 \times 10^{-2}\,\text{rad s}^{-1}$,
   and the period of $\dfrac{2\pi}{1.38 \times 10^{-2}} = 455\,\text{s} = 7.6$ minutes.
   Solve the same task using the equations for uniformly accelerated motion, instead of angular momentum and impulse.

Worked example 21

A flywheel of the moment of inertia 0.070 kg m² is driven by an electric motor that provides a torque that varies with time t. The graph shows the variation with t of the net torque applied to the flywheel.

The flywheel is initially at rest. Calculate:

a. the angular momentum of the flywheel when $t = 5$ s

b. the angular velocity of the flywheel when $t = 10$ s

c. the average acceleration of the flywheel during the first 10 s.

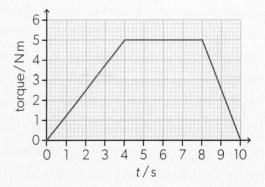

Solutions

a. The initial angular momentum is zero so the angular momentum at $t = 5$ s is equal to

   the area under the graph for $t \in [0,5]$ s. $L = \dfrac{1}{2} \times 4 \times 5 + 1 \times 5 = 15\,\text{N m s}$.

b. The angular momentum at $t = 10$ s is the total area under the graph, $L = 35\,\text{N m s}$.

   The angular velocity is $\omega = \dfrac{L}{I} = \dfrac{35}{0.070} = 500\,\text{rad s}^{-1}$.

c. $\alpha = \dfrac{\Delta \omega}{\Delta t} = \dfrac{500}{10} = 50\,\text{rad s}^{-2}$.

Practice questions

21. The motor of an electric saw delivers a torque of 40 N m to a circular cutting blade. The moment of inertia of the cutting blade about the rotation axis of the saw is 8.0×10^{-3} kg m^2. The cutting blade accelerates to its operating speed in a time of 0.10 s.

    a. Calculate the angular impulse delivered to the cutting blade in 0.10 s.

    b. Hence, calculate, in revolutions per minute (rpm), the final angular velocity of the cutting blade.

22. A golf ball can be modelled as a uniform solid sphere of radius 2.2 cm and mass 45 g. The ball is hit by a golf club and is launched with an initial angular velocity of 75 rad s^{-1}.

    a. Calculate the angular impulse delivered to the ball by the golf club.

    b. The ball is in contact with the golf club face for a time of 15 ms. Calculate the average torque exerted by the golf club on the ball.

23. The graph shows how the net torque acting on a turntable varies with time t. The turntable starts from rest and its moment of inertia about the rotation axis is 0.16 kg m^2.

    a. Determine the angular velocity of the turntable at $t = 1.0$ s.

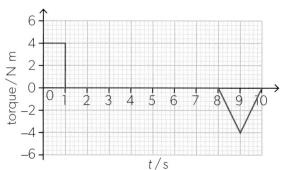

    b. Explain why the angular velocity remains constant between $t = 1.0$ s and 8.0 s.

    c. Calculate the number of revolutions the turntable makes between $t = 1.0$ s and 8.0 s.

    d. Explain why the turntable comes to rest at $t = 10$ s.

Change in angular momentum

As shown in Topic A.2, you can extend Newton's second law from the simple equation $F = ma$ to the more complex form $F\Delta t = \Delta(mv)$. This change expresses the link between change in momentum and impulse. This led to a second expression for Newton's second law as

$$F = m\frac{\Delta v}{\Delta t} + v\frac{\Delta m}{\Delta t}$$

The first term on the right-hand side of the equation is Newton's law in its simple form. The second term gives the dynamic contribution of the change of mass while the force acts to the overall change.

As you will expect, there is an equivalent statement for rotational dynamics:

$$\tau = I\frac{\Delta \omega}{\Delta t} + \omega\frac{\Delta I}{\Delta t}$$

It is clear from this equation that a torque is required to:

- change the angular velocity of an object that has a constant moment of inertia (e.g. accelerating a flywheel), or

- maintain, at a constant angular velocity, any object that has a changing moment of inertia (e.g. a figure skater).

Both of these conditions can change simultaneously.

Worked example 22

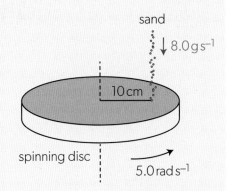

Sand is poured at a rate of $8.0\,\text{g s}^{-1}$ onto a spinning disc. The sand lands with a negligible vertical speed at an average distance of 10 cm from the centre of the disc. The moment of inertia of the disc is $0.040\,\text{kg m}^2$ and its initial angular velocity is $5.0\,\text{rad s}^{-1}$. No external torques act on the system consisting of the disc and sand.

a. Calculate the rate of change of the moment of inertia of the system.

b. Calculate the initial angular acceleration of the spinning disc.

c. Explain whether the acceleration will remain constant as the sand is being poured.

Solutions

a. $\dfrac{\Delta I}{\Delta t} = \dfrac{\Delta m}{\Delta t}R^2 = 0.0080\,\text{kg s}^{-1}\,(0.10\,\text{m})^2 = 8.0 \times 10^{-5}\,\text{kg m}^2\,\text{s}^{-1}$.

b. From $I\dfrac{\Delta \omega}{\Delta t} = -\omega\dfrac{\Delta I}{\Delta t}$, the angular acceleration is $\alpha = \dfrac{\Delta \omega}{\Delta t} = -\dfrac{\omega}{I}\dfrac{\Delta I}{\Delta t} = -\dfrac{5.0}{0.040} \times 8.0 \times 10^{-5} = -0.010\,\text{rad s}^{-2}$.

c. The angular velocity of the system decreases and the moment of inertia increases, so the magnitude of the angular acceleration will decrease, as given by $|\alpha| = \dfrac{\omega}{I}\dfrac{\Delta I}{\Delta t}$.

Rolling and sliding

It is important to distinguish between two motions: rolling and sliding. Rolling means that the object is rotating across a surface about an axis of rotation. Sliding means that the object is moving smoothly along the surface. When an object is moving on a perfectly frictionless surface, then it cannot roll—the only motion possible is sliding.

When there is friction between surface and object, the point of contact between the two is instantaneously at rest; this implies that the coefficient of static friction μ_s must be used in any calculation.

Figure 20 shows a rolling disc with its motion broken down into a linear, translational motion (Figure 20 (a)) and rotational motion (Figure 20 (b)). The disc of radius r has a linear motion of v_0 and an angular velocity ω. When the point of contact at the ground is instantaneously stationary, then v_0 must be equal to $r\omega$—which is the tangential velocity of each point on the edge of the disc. At the top of the disc the speed must be $v_0 + r\omega$, which is $2v_0$.

> There is more about coefficients of static friction in Topic A.2.

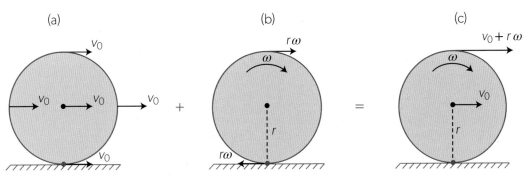

▲ Figure 20 When a disc or a sphere rolls, the point of contact with the ground has an instantaneous velocity of zero and the top of the object has a speed of $2v_0$, where v_0 is the linear speed of the centre of mass.

Rolling and slipping—An energy perspective

The total kinetic energy of a body of mass m and moment of inertia I rolling without slipping at linear speed v and angular velocity ω is $\frac{1}{2}mv^2 + \frac{1}{2}I\omega^2$. When the body rolls down a slope of vertical height Δh, it loses $mg\Delta h$ of gravitational potential energy. Applying conservation of total energy gives $mg\Delta h = \frac{1}{2}mv^2 + \frac{1}{2}I\omega^2$. By combining this equation with the moment of inertia for a particular shape, you can often eliminate ω and I, and use this to calculate quantities such as the velocity of the rolling object.

Worked example 23

A round object of mass M, radius R and moment of inertia I rolls without slipping down an inclined plane that makes an angle θ with the horizontal.

a. Draw a diagram showing the forces acting on the object.

b. Derive an expression for the linear acceleration of the object.

c. Calculate, in terms of θ, the linear acceleration of the object if the object is:

   i. a solid sphere   ii. a solid cylinder   iii. a thin hoop.

Solutions

a. The forces acting on the object are the weight, Mg, the normal reaction force, N, and the static frictional force F_f between the object and the plane. The frictional force provides the angular acceleration, because both Mg and N act through the centre of the object and their torque is zero. The diagram also shows the parallel and normal components of the weight.

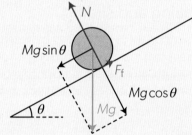

b. The object has both a linear acceleration a and an angular acceleration α. The net force acting on the object down the plane is $Mg\sin\theta - F_f$. Hence the linear acceleration is $a = g\sin\theta - \dfrac{F_f}{M}$. The torque provided by the frictional force is $F_f R$. Hence the angular acceleration is $\alpha = \dfrac{F_f R}{I}$. The object rolls without slipping so the linear and angular accelerations are related to each other, $a = \alpha R$. We eliminate α from the last two equations and get $F_f = \dfrac{aI}{R^2}$. This expression is substituted into the equation for the linear acceleration: $a = g\sin\theta - \dfrac{aI}{MR^2}$. From here, the acceleration is $a = \dfrac{g\sin\theta}{1 + \frac{I}{MR^2}}$.

c. i. For a solid sphere, $I = \frac{2}{5}MR^2$ and $a = \dfrac{g\sin\theta}{1 + \frac{2}{5}} = \frac{5}{7}g\sin\theta$.

   ii. For a cylinder, $I = \frac{1}{2}MR^2$ and $a = \dfrac{g\sin\theta}{1 + \frac{1}{2}} = \frac{2}{3}g\sin\theta$.

   iii. For a thin hoop, $I = MR^2$ and $a = \dfrac{g\sin\theta}{1 + 1} = \frac{1}{2}g\sin\theta$.

Of these three objects, the sphere is rolling down with the greatest acceleration, and the hoop with the smallest. Note that, for each object, the acceleration only depends on the angle of inclination, and not on the mass or the radius of the object.

Worked example 24

A solid cylinder rolls down an incline that makes an angle $\theta = 5.0°$ with the horizontal. The cylinder starts from rest and travels a distance of 1.0 m, measured along the plane.

a. Determine the final linear speed of the cylinder:

  i. by using the kinematic (*suvat*) equations for uniformly accelerated motion

  ii. by considering energy changes of the cylinder.

b. A solid sphere of the same mass as the cylinder in part a. rolls down the same inclined plane, starting from rest. Explain which of the objects has a greater rotational kinetic energy at the bottom of the plane.

Solutions

a. i. From worked example 23, the linear acceleration is $a = \frac{2}{3}g\sin\theta$. The final linear velocity can be calculated from $v^2 = 2as$.

$$v = \sqrt{2 \times \frac{2}{3} \times 9.8 \times \sin 5.0° \times 1.0} = 1.1\,\text{ms}^{-1}.$$

  ii. For rolling without slipping, the total kinetic energy is $E_k = \frac{1}{2}Mv^2 + \frac{1}{2}I\left(\frac{v}{R}\right)^2 = \frac{1}{2}\left(M + \frac{I}{R^2}\right)v^2$. In case of the cylinder, $I = \frac{1}{2}MR^2$ and $E_k = \frac{3}{4}Mv^2$. The increase in E_k is equal to the decrease in its gravitational potential energy, $\frac{3}{4}Mv^2 = Mgs\sin\theta$, where $s = 1.0\,\text{m}$ is the distance travelled along the plane. $v = \sqrt{\frac{4}{3} \times}$

$9.8 \times 1.0 \times \sin 5.0° = 1.1\,\text{ms}^{-1}$, in full agreement with the answer in i.

b. The sphere had a greater linear acceleration so its final speed and the translational kinetic energy is greater than that of the cylinder. On the other hand, both objects have moved through the same vertical distance and their total kinetic energies are therefore equal. This means that the cylinder must have a greater rotational energy than the sphere.

The total kinetic energy of each object is the same, but it is distributed in different proportions between translational and rotational components.

Practice questions

24. A solid cylinder and a thin hoop have the same mass and roll equal distances down the same inclined plane. The rolling occurs without slipping. Which is correct about the total kinetic energy and the translational kinetic energy of the objects?

	Total kinetic energy	Translational kinetic energy
A.	equal	equal
B.	equal	different
C.	different	equal
D.	different	different

25. A solid ball rolls down an inclined plane of length 1.8 m in a time of 2.9 s. The ball starts from rest and rolls without slipping.

  a. Calculate the linear acceleration of the ball.

b. Show that the linear acceleration satisfies the equation $a = \frac{5}{7}g\sin\theta$, where θ is the angle that the inclined plane makes with the horizontal.

c. Calculate θ.

26. A horizontal cylinder of radius R and moment of inertia I can rotate without friction around the central axis. A weightless thread is wrapped around the cylinder and an object of mass M is suspended from the thread. The object is released so that the system accelerates from rest. The thread unwinds from the cylinder without slipping.

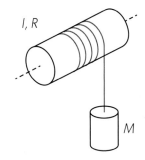

  a. State the force that provides the torque about the axis of rotation axis of the cylinder.

b. Show that the linear acceleration of the mass M is given by $a = \dfrac{g}{1 + \dfrac{I}{MR^2}}$.

It is given that $M = 0.400\,\text{kg}$, $I = 0.0200\,\text{kg m}^2$ and $R = 0.100\,\text{m}$.

c. Calculate the linear acceleration of the mass M.

d. Determine the rotational kinetic energy of the cylinder after 0.600 m of the thread has unwound from the cylinder.

27. A car of mass 1200 kg moves with a constant speed of $20\,\text{m s}^{-1}$. Each of the four wheels of the car can be modelled as a solid disc of mass 25 kg and radius 0.31 m.

a. Calculate the translational kinetic energy of the car.

b. Determine the ratio
$$\frac{\text{rotational kinetic energy of the wheels}}{\text{total kinetic energy of the car}}.$$

Verifying equations of rotational motion

- Tool 3: Select and manipulate equations.

- Tool 3: Derive relationships algebraically.

- Tool 3: Record uncertainties in measurements as a range (±) to an appropriate precision.

- Inquiry 3: Compare the outcomes of an investigation to the accepted scientific context.

For this experiment, you need a ramp and an object that will roll down it—for example, a sphere, a hollow cylinder or a solid cylinder (Figure 21).

The ramp should be about 1 m long and at an angle of about 10° to the horizontal. A longer ramp means that the object takes more time to reach the bottom, which reduces the uncertainty in the measurement of the time taken.

- Mark a suitable start line on your ramp so that you roll the object from the same place each time. Measure the height of the object when it is on the start line. Be careful to measure the difference in height of the object's centre of mass between the top and bottom of the ramp—in practice, this may mean measuring the change in height of the bottom of the object.

- Allow the object to roll down the slope and measure the time it takes to do so. Repeat this measurement three times.

- With no rotation you would expect the time taken for the object to move down the ramp to be given by $t^2 = \dfrac{2d^2}{gh}$. Use your values to calculate this predicted time.

- Compare your measured time with the time predicted by the equations.

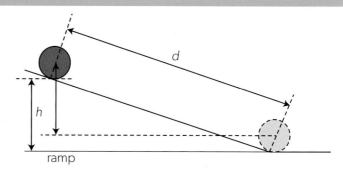

▲ Figure 21 The setup for the experiment.

- The time for the object to roll down the ramp should be given by $t^2 = \dfrac{2d^2}{gh}(1 + \alpha)$, where $\alpha = \dfrac{I}{MR^2}$. For a hollow cylinder $\alpha = 1$, for a solid cylinder $\alpha = \dfrac{1}{2}$ and for a solid sphere $\alpha = \dfrac{2}{5}$.

- Use your measurements to determine α for each type of rolling object and see whether it agrees with the theory.

- Use the variation in your time measurements to find the uncertainty. By considering this, and the other uncertainties in your experiment, find the uncertainty in your measurements of α. Can your experiment distinguish between a sphere and a cylinder? If not, consider how you might improve your experiment to do so.

- Finally, use the conservation of energy to derive the equations given above. Do not forget that the energy at the end of the ramp depends on the final velocity, while using your measured values of d and t will help you calculate the average velocity.

- This could be the basis for an internal assessment. How would you develop this experiment to answer a question about the design of a ball used in a sport?

🧪 Conserving energy—Determining the moment of inertia of a flywheel

- Tool 1: Understand how to measure mass, time and length to an appropriate level of precision.

- Tool 3: Derive relationships algebraically.

- Tool 3: Construct and interpret tables and graphs for raw and processed data including scatter graphs and line and curve graphs.

- Tool 3: Interpret features of graphs including gradient, changes in gradient, intercepts, maxima and minima, and areas under the graph.

It is possible to use the conservation of linear and rotational energies to estimate the moment of inertia of a flywheel.

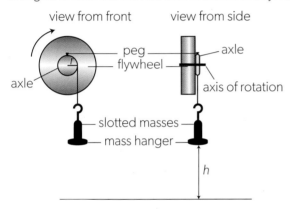

▲ Figure 22 Measuring the moment of inertia of a flywheel.

- Mount a flywheel on a horizontal axle so that it can rotate freely. (This experiment eliminates the effects of friction on the axle.) The total moment of inertia of both flywheel and axle is I.

- Attach a mass hanger of total mass m, to a string and wind it around the axle. Do not overlap the string. Ensure that it falls away from the axle just as the object reaches the floor.

- Release the object from rest. As it falls, the gravitational potential energy of the object is converted into its own linear kinetic energy + rotational kinetic energy of the flywheel + work done against the frictional forces.

- Conservation of energy shows that, just as the object reaches the floor:

$$mgh = \frac{1}{2}mv^2 + \frac{1}{2}I\omega^2 + n_1 W.$$

W is the work done against friction for each of the n_1 revolutions before the object reaches the floor.

- Once the string has disengaged, the axle will continue to turn and make n_2 extra revolutions before it stops moving (n_2 is determined by the friction acting on the axle).

- Make the following measurements:

  - the radius r of the axle

  - the mass m of the falling object

  - the initial height h of the falling object above the floor

  - the time t for the falling object to reach the floor

  - n_1 and n_2.

- Adjust the mass hanging from the string and then measure t, n_1 and n_2 again.

- Repeat several times with different masses and calculate the mean values.

- These data can be used graphically to find a value of I.

The rotational kinetic energy of the flywheel is transferred to the surroundings while it turns through the final n_2 revolutions, so $\frac{1}{2}I\omega^2 = n_2 W$, leading to

$$W = \frac{I\omega^2}{2n_2}.$$

Substituting this into the whole energy conservation energy gives

$$mgh = \frac{1}{2}mv^2 + \frac{1}{2}I\omega^2 + n_1\frac{I\omega^2}{2n_2},$$

or

$$mgh = \frac{1}{2}mv^2 + \frac{1}{2}I\omega^2\left(1 + \frac{n_1}{n_2}\right).$$

As usual, $v = r\omega$ and for the falling mass $\frac{v}{2} = \frac{h}{t}$ so that $\omega = \frac{2h}{rt}$.

This equation reduces to:

$$\frac{gt^2}{2h} = \frac{I}{mr^2}\left(1 + \frac{n_1}{n_2}\right) + 1 = \frac{I}{mr^2} \times \frac{n_1}{n_2} + \left(1 + \frac{I}{mr^2}\right).$$

- Compare this with $y = mx + c$. The variable you can alter is m. The dependent variables are t and n_2.

- Use this to plot a graph and determine I.

A.5 Galilean and special relativity

How do observers in different reference frames describe events in terms of space and time?

How does special relativity change our understanding of motion compared to Galilean relativity?

How are spacetime diagrams used to represent relativistic motion?

In 1905, a far-reaching paradigm shift took place in physics when Albert Einstein published the first of a series of scientific papers dealing with the dynamics of moving bodies. His new theories and his conceptual understanding of spacetime became famous as the special and (later) the general theories of relativity. Einstein and fellow scientists such as Albert Michelson and Hendrik Lorentz had overturned 200 years of Newtonian classical mechanics.

Isaac Newton based his theories of motion on an assumption of Galilean relativity. The term "relativity" has become associated with Einstein's work, but the concept itself is far older and is rooted in the idea of simultaneity—the question of whether two events take place at the same time. Humans (who move slowly compared with the speed of electromagnetic waves) assume that intervals of time observed in different frames are identical. Compare person A standing in an athletics stadium who is timing person B, an athlete running a 100 m race. We all assume that when A's stopwatch says 11.50 s, B's watch will also indicate 11.50 s. Special relativity overturns this. The fabric of space and time—spacetime—is more tightly knit than a simple equivalence of time intervals between different reference frames.

There are other key names in the history of special relativity. Einstein based his work on some suggestions by the Dutchman Hendrik Lorentz. Later developments by Hermann Minkowski, three years after Einstein's original papers, provided a visualization of reference frames. Minkowski (spacetime) diagrams owe their original existence to Minkowski's appreciation of the four-dimensional nature of spacetime. Is there conflict between the algebraic approach of Lorentz and the visualizations of Minkowski?

▲ Figure 1 Albert Einstein pictured in 1905—the year he published his theory of special relativity.

In this additional higher level topic, you will learn about:

- reference frames and inertial reference frames
- Galilean relativity and the Galilean transformation equations
- the two postulates of special relativity
- the Lorentz transformation equations
- the relativistic velocity addition equation

- invariant quantities such as the spacetime, the proper time interval and proper length
- time dilation and length contraction
- simultaneity
- spacetime diagrams
- muon decay experiments and the evidence they provide.

Introduction

On 8 December 1864, James Clerk Maxwell read a scientific paper to the Royal Society of London. It was entitled *A dynamical theory of the electromagnetic field*. In the paper, he presented four short equations that sum up the whole of electrical and magnetic theory. These had profound significance for the development of physics during the remainder of the 19th century and beyond.

Maxwell's four equations provoked Albert Einstein to think how it could be possible for Maxwell's conclusions to sit comfortably with the then agreed rules of classical physics. If Maxwell's theory were correct, the speed of an electromagnetic wave would be the same to any observer whatever the observer's motion relative to the source of the radiation. This flew in the face of the perceived wisdom of the day. Before Einstein it was believed that, when a moving observer approached a stationary light source, the speed of the light measured by the observer was the sum of the speed of the object and the speed of the light. Einstein showed that this could not be so and, in announcing his result, overturned 200 years of Newtonian mechanics—a mechanics based on simple assumptions about the independence of time and the existence of an absolute rest frame.

This topic introduces the physics of special relativity. The theory has profound implications for many aspects of life in the 21st century: magnetism, GPS, the properties of heavy metals, and the behaviour of light itself.

Spacetime

One concept underpins all the work in this topic: **spacetime**. Einstein realised that space and time together constituted a set of coordinates in just the same way that a 2D graph uses pairs of coordinates to display data. It is important not to regard space and time as separate when dealing with concepts in special relativity. A key element in understanding the special theory is the recognition that we move through the four dimensions of spacetime rather than the three dimensions of Euclidean space.

Much of our work in this topic uses just two of these dimensions: x (for space) and t (for time). However, all the work in this topic can be generalized into the four dimensions (t, x, y, z).

We will also use the term **event** extensively: an event is simply one set of spacetime coordinates that identifies a particular position in spacetime.

Reference frames

A **reference frame** allows us to refer to the position of a particle. A frame consists of an origin position together with a set of axes. In IB Diploma Programme physics, the Cartesian reference frame is used most often for this. Here, a position in space is defined using three distances measured along axes that are set at 90° to each other. For example, the axes of a three-dimensional graph make up what is known as a Cartesian reference frame (Figure 2). In this frame, the position of the point is specified by the three numbers (x_1, y_1, z_1).

Other frames are possible:

- Figure 2 specifies the same particle position using one distance r and two angles θ and ϕ referenced to the frame. Sailors use such angles as latitude and longitude; and, together with the distance of an object from the centre of Earth, they constitute a different but equivalent alternative to the Cartesian coordinates.

Theories—Validating a paradigm shift

The development of Einstein's ideas of relativity required a shift in the scientific view of the physical rules that govern the universe. How do scientists ensure that the need to shift perspectives is valid?

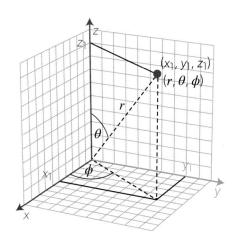

▲ Figure 2 A comparison between a Cartesian reference frame with three distances and a frame that uses spherical coordinates: two angles and a distance.

- Astronomers use angles to define the position of a star or galaxy that is being observed. Only two angles are required because the distance to the star (for observational purposes) is irrelevant.

For some frames of reference, one or more of Newton's laws of motion do not hold. Our own everyday frame, on the surface of a rotating planet, shows this. Everything off-planet appears to be spinning around us. Ancient civilisations factored this spin into their world view. This rotating frame has further implications. An object that is at rest relative to us on planet Earth cannot be moving at a constant velocity. This has the consequence, as we saw in Topic A.2, that we need to invent the fictitious "centrifugal" force when we are observing from the rotating frame. Another invented force, the Coriolis force, is sometimes used to "explain" the movement of weather systems.

Inertial frames of reference

It is inconvenient to have frames of reference in which forces need to be "invented" to explain physical effects. We define a concept called an **inertial frame of reference** to overcome this. There are several ways in which to define the inertial frame, but the one used here is that:

> An inertial frame of reference frame is a frame that is not accelerated.

All inertial frames are, by definition, moving at constant velocity (constant speed in a straight line) with respect to each other. The requirement that an inertial frame does not accelerate has wide implications. The lack of acceleration in one inertial frame means that Newton's first law is valid in that frame. Because all inertial frames are related by a constant relative velocity, then Newton's first law holds in every inertial frame.

Do inertial frames exist? The best way to find one is to take a spaceship out into deep space, far away from the gravitational effects of planets and stars, and then turn off the engines. No forces act from outside or inside the spacecraft and this will be a true inertial frame of reference.

Galilean relativity and Newton's postulates

There are an infinite number of inertial frames of reference in the universe and three ways to move between them (Figure 3):

- *Translation* from one frame to another frame. Figure 3(a) shows two Cartesian frames (x, y) and (x', y') that are offset in the x-direction by a constant distance X. Translation is the movement between these two frames.

- A *rotation* by a constant angle θ of one set of axes (x, y) to form another set (x', y') (Figure 3(b)). (Rotations are not discussed in IB Diploma Programme physics.)

- A *boost* from one frame (x, y) to another frame (x', y') that has a constant relative velocity (Figure 3(c)).

When an object is moving with a constant velocity in one reference frame, then, under any of these three conditions, it will also have a constant (but possibly different) velocity in the other reference frame.

You will see that the definition of an inertial frame excludes frames where, for example, a Coriolis force or a centrifugal force is required. Recall the discussion about a car turning in a circle on page 100 in Topic A.2. The car accelerates towards the centre of the circle even though it is travelling at constant speed (the velocity direction alone changes). The stationary helicopter above the car was in an inertial frame; the car itself was not. A passenger in the car needed to invoke a centrifugal force to explain the fictitious force that appeared to act.

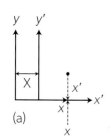
(a)

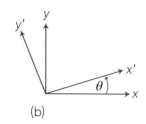
(b)

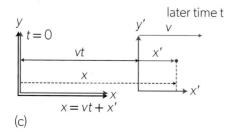
(c)

▲ Figure 3 Translations, rotations and boosts between reference frames.

The principle of relativity tells us about the nature of our universe.

- The translation rule is equivalent to saying that there is no special place in the universe; position is relative.

- The boost rule says that there is no special velocity. Being at rest (stationary) is not an absolute condition.

- The rotation rule says that there is no special direction; direction is relative.

For the translation in Figure 3(a), to convert a position in (x, y) to one in (x', y') you need to subtract X (the offset between frames) from x for it to become x' $(x - X, y)$.

A mathematical way to write this is as $x' \rightarrow x - X$. Here "$\rightarrow$" means "transforms to".

For a boost, when one inertial frame moves relative to the other by a constant relative velocity v along the x-axis, then the distance between the origins of the reference frames must be changing by v every second and this distance is $v \times t$, where t is the time since the frames coincided.

Imagine that the origins of the two reference frames S and S' (Figure 3(c)) were at the same position (that is, they coincided) at time $t = 0$. We say that clocks in the frames were synchronised so that $x = x' = 0$ when $t = t' = 0$. At a later time T, the origins of the reference frames will be separated by vT where v is the velocity of frame (x', y') relative to frame (x, y). Therefore a position X in frame (x, y) will be related to position X' in (x', y') by $X \rightarrow X' + vT$ or, alternatively, $X' \rightarrow X - vT$.

The velocities of an object in the two inertial frames also transform in an obvious way. When the velocity of an object in (x, y) is u and the velocity in (x', y') is u', then $u' \rightarrow u - v$. When the velocities u and v are not in the same direction, then they must be subtracted as vectors. An alternative is to treat u and u' as velocity components in the direction of the relative motion of the inertial frames.

These equations that link two reference frames by their relative velocity are known as the **Galilean transformations** (Figure 4).

▲ Figure 4 The travelling walkways at airports are good examples of Galilean transformations. They typically move at speeds of about $0.7 \, \text{m s}^{-1}$. A passenger standing on one walkway will see a passenger standing on the other walkway coming towards them at $1.4 \, \text{m s}^{-1}$. A stationary passenger by the side of the walkways will see each of the passengers moving at $0.7 \, \text{m s}^{-1}$ towards each other.

Observations in relativity

In one sense, as physicists working in a laboratory and taking measurements, we are all observers. However, the meaning of "observer" in special relativity is different from this. The observer now becomes the reference frame in which objects or, more usually, events are measured. Observers in inertial frames of reference cannot accelerate and are fixed within the frame.

The meaning of "observer" has continued to change in a subtle way throughout the development of special relativity. It is probably best to think not of a single observer, but of a team of observers each making observations in the local area. All members of the team have synchronized clocks so that they can report their findings to each other (with corrections arising from the finite speed of light).

This meaning of observer becomes crucial later in the discussion of simultaneity on page 182.

Observing relativity

The principle of relativity is often associated with Albert Einstein. In fact, Galileo was probably the first person to discuss the principle. He describes how, in a large sailing ship, butterflies in a cabin with no windows would be observed to fly at random whether the ship was moving at constant velocity or not. An observer in the cabin could not deduce, by an observation of the butterflies, any movement of the ship. The butterflies would not, for example, be pinned against the back wall of the cabin by the ship's forward motion!

The equation set when applied to two reference frames, initially coincident at time $t = 0$, moving apart in the x direction at speed v is:

- $x' = x - vt$

- $u' = u - v$ This is known as the **velocity addition equation** for the Galilean transformation.

The two postulates of special relativity

Newton developed Galileo's ideas further in his *Principia Mathematica* by suggesting two important postulates of special relativity. A postulate is an assertion or assumption that is not proved and acts as the starting point for a proof.

- Newton treated space and time as fixed and absolute. This is implied in our use of t in the equations above (t' does not appear, only t). A time interval between two events described in frame (x, y) is identical to the time interval between the same two events as described in frame (x', y'). The evidence of our senses seems to confirm this (but remember that we do not travel close to the speed of light in everyday life). This leads to a further transformation for time changes Δt: $\Delta t' \rightarrow \Delta t$.

- Newton recognized that two observers in separate inertial frames must make the same observations of the world. In other words, they will both arrive at the same physical laws that describe the universe. This is a direct consequence of Galilean relativity.

This basic principle is important:

> Galilean relativity means that Newton's laws of motion are the same in all inertial reference frames.

Einstein's great intellectual accomplishment was to recognize that, because Maxwell's four electromagnetic equations are true in all inertial frames (which had to be the case), then some modifications of Newton's postulates were needed.

Einstein's two postulates of **special relativity** were:

> The laws of physics are the same in all inertial frames of reference.

(Newton's second postulate is generalized to all scientific laws including those of optics and electromagnetism).

> The speed of light relative to any observer in an inertial frame is independent of the motion of the source of light relative to the inertial observer.

(This replaces the concept of absolute time and space in Newton's first postulate.)

Simultaneity

Do two events at different places occur at the same time? Are they simultaneous? This is the key issue in the concept of simultaneity. In Galilean relativity, Newton said that time was independent of the observer and simultaneity is absolute. So, if you synchronize two clocks in the same place and then move them into different frames, the clocks will keep time at the same rate as each other.

However, in Einstein's relativity the clocks will no longer measure identical time intervals for the same event as described by the observers in the different frames.

Later in this Topic (page 168), you will see that the transformation of time intervals in Einstein relativity must include an extra factor that is related to the speed of the inertial frame.

Theories

A different but equivalent version of Einstein's second postulate was later re-phrased to give an alternative version that you will often see:

> **The speed of light in free space (a vacuum) is the same in all inertial frames of reference.**

Hypotheses — The way Newton put it

These postulates in Galilean relativity were expressed by Newton in the *Principia Mathematica* — the book (in Latin) that he wrote to publish some of his discoveries. In translation his postulates were:

"I. Absolute, true, and mathematical time, of itself, and from its own nature, flows equably without relation to anything external.

II. Absolute space, in its own nature, without any relation to anything external, remains always similar and immovable."

AHL

Worked example 1

A bus moves at a constant velocity of $8.0\,m\,s^{-1}$ relative to the ground. A boy on the bus walks towards the rear of the bus at a constant velocity of $-1.5\,m\,s^{-1}$ relative to the bus.

a. Calculate the velocity of the boy relative to the ground.

The distance walked by the boy relative to the bus is $3.0\,m$.

b. Calculate the distance travelled during the same time by the boy relative to the ground.

Solutions

a. Relative to the ground, the velocity of the boy is smaller than the velocity of the bus: $8.0 - 1.5 = 6.5\,m\,s^{-1}$.

b. The time taken is $\dfrac{3.0}{1.5} = 2\,s$. During this time, the distance travelled relative to the ground is $6.5 \times 2.0 = 13\,m$.

Worked example 2

A cyclist rides a bicycle along a straight road, in the direction of the positive x-axis of the reference frame of the road. The position of the cyclist at $t = 0$ is $x = 0$ and the position at $t = 50$ is $x = 350\,m$. A car moving at a constant velocity of $12\,m\,s^{-1}$ relative to the road overtakes the cyclist. Calculate, relative to a reference frame in which the car is at rest:

a. the displacement of the cyclist from $t = 0$ to $t = 50\,s$

b. the average velocity of the cyclist.

Solutions

a. The displacement of the cyclist in the reference frame of the car can be found using the Galilean transformation: $\Delta x' = \Delta x - v\Delta t$, where $\Delta x = 350\,m$ is the displacement in the reference frame of the road, $\Delta t = 50\,s$ is the time taken and $v = 12\,m\,s^{-1}$ is the relative velocity of the reference frames. $\Delta x' = 350 - 12 \times 50 = -250\,m$. A negative displacement means that in the car's frame the cyclist moves towards the negative x'-axis.

b. average velocity $= \dfrac{\text{displacement}}{\text{time taken}} = \dfrac{-250}{50} = -5\,m\,s^{-1}$.

Changing perspectives

In Topic A.2, we began by describing Newton's second law of motion as: *force = mass × acceleration*.

Later, we showed that this was better expressed as: *force = rate of change of momentum*.

A similar change of expression is possible here: Newton's first law is usually given as a variant of "Every object continues in its state of rest or uniform motion unless net external forces act on it" (and you should continue to use this or a similar wording in your own work). But this is not the only possibility. A succinct and interesting way to express Newton's first law is "Inertial frames exist."

To what extent do concise forms of scientific laws help or hinder our understanding?

Measurement

James Clerk Maxwell established four equations which are today recognized as "Maxwell's equations". An important result of his work was that the speed of light (an electromagnetic wave) travels at a speed $c = \dfrac{1}{\sqrt{\varepsilon_0 \mu_0}}$. (You will meet the constants ε_0 and μ_0 in Topics D.2 and D.3.)

This result tells us that observers in different inertial frames observe the same value for the speed of light when they agree about physical laws such as the values of ε_0 and μ_0. This is directly contrary to the assumption of absolute time and absolute space as postulated by Newton, and as embodied in the Galilean transformations.

In 1887, two US physicists, Albert Michelson and Edward Morley, used an interferometer to observe light as it passed through different reference frames. Their apparatus was highly sensitive. It could send a beam of light over a distance of 11 m measuring changes to about 10^{-8} m in this distance. They mounted their apparatus on a massive stone slab floated on mercury so that it could be rotated. They expected to see a difference in the path of the light caused by the Earth's rotation about the Sun. However, despite their experiment being sensitive enough to measure the predicted change, they could not find any evidence for a difference. The experiment is sometimes referred to as a null experiment since they found no evidence of the effect that they were trying to measure. However, the result was an important confirmation that the speed of light is the same in all inertial reference frames.

Today, interferometers are still used as sensitive detectors. In 2015, the LIGO experiment (Laser-Interferometer Gravitational-Wave Observatory) used an interferometer to confirm one of the predictions of Einstein's theories: gravitational waves.

▲ Figure 5 Albert Einstein with Albert Michelson to his right during a visit to the Mt Wilson Observatory. Edwin Hubble is the second from the left of the picture.

Einstein first published his ideas of special relativity in 1905 in a paper whose title translates as "On the Electrodynamics of Moving Bodies": A. Einstein, Zur Elektrodynamik bewegter Körper, *Annalen der Physik* **17**, 891 (1905). Unusually, his paper contained no formal references, even for the Lorentz transformations which had already been published. However, he did mention Lorentz along with four other scientists: Newton, Maxwell, Hertz and Doppler.

Scientific historians have questioned whether Einstein worked independently and was unaware of the work of Lorentz and Henri Poincaré, a French mathematician and physicist. Others have asserted that Poincaré or Lorentz are the true founders of relativity. It is more likely that Lorentz and Poincaré had contributed many of the ideas of length contraction, time dilation, the Lorentz transformations and some of the mathematical framework, but that Einstein was the first to condense all these into the theory of special relativity.

Proper time interval

Some of the most dramatic differences between our everyday perceptions of space and time and the predictions made by special relativity concern the time and length differences that arise between observations made in different inertial reference frames.

Figure 6(a) shows a simple **light clock** that consists of two mirrors facing each other across a room. An observer is at rest in the room and watches light reflect between the mirrors. The distance across the room is L and so the time interval taken for the light to return to the first mirror is $\Delta t_0 = \dfrac{2L}{c}$. The symbol for the speed of light is c as usual.

(a)

(b)

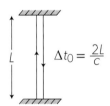

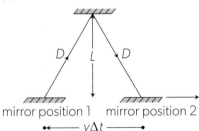

mirror position 1   mirror position 2

▲ **Figure 6** A light clock. Because the two references frames move relative to each other, the trip taken by the light differs according to two observers, one in each frame: (a) is in the frame of the mirrors, (b) is the frame for an observer moving to the left.

Another observer (Figure 6(b)) is moving to the left, parallel to the mirrors, at constant velocity v relative to the mirrors, watching the reflections. The diagram shows the bottom mirror at two positions as seen by this moving observer:

- position 1 when the light leaves the bottom mirror

- position 2 when the light returns to it.

In the frame of this moving observer, the light appears to travel to the right at an angle to the direction of motion (but, of course, at the same speed of light). The distance travelled by the light is now $2D$ and the time observed between reflections at the same mirror is now $\Delta t = \dfrac{2D}{c}$.

The distance travelled horizontally by the moving observer in time Δt is $v\Delta t$ and, by an application of Pythagoras' theorem: $D^2 = L^2 + \dfrac{v^2(\Delta t)^2}{4^2}$. We already know that $D = \dfrac{c\Delta t}{2}$ and rearranging for Δt gives $\Delta t = \dfrac{\frac{2L}{c}}{\sqrt{1 - \frac{v^2}{c^2}}}$;

so $\Delta t = \dfrac{\Delta t_0}{\sqrt{1 - \frac{v^2}{c^2}}}$, which reduces to $\Delta t = \gamma \Delta t_0$, where $\gamma = \dfrac{1}{\sqrt{1 - \frac{v^2}{c^2}}}$.

The time intervals in the two frames are very clearly different. The observer (clock) at rest in the inertial frame of the mirrors observes a time $\Delta t_0 = \dfrac{2L}{c}$ and the observer (a different clock) at rest in an inertial frame moving relative to the mirror (at constant speed v) observes a time $\Delta t = \dfrac{2L}{c} \times \dfrac{1}{\sqrt{1 - \frac{v^2}{c^2}}}$.

Because $\gamma = \dfrac{1}{\sqrt{1 - \frac{v^2}{c^2}}}$ is always greater than 1, this shows that the time interval for the journey of the light observed by the observer moving relative to the mirror is greater than the time interval for the observer who is stationary in the mirror frame. This is a general result: the time measured in the frame in which the event takes place is the shortest time observable for two events that have a time-like interval.

The time interval Δt_0 is known as the **proper time interval** (often shortened to proper time).

> **The proper time interval between two events is the time interval measured by an inertial observer at the place where the events occur.**

The meaning of the term "time-like interval" is explained on page 186.

Time dilation occurs in any other frame, with the time interval observed to be longer than the proper time interval.

How are equations of linear motion adapted in relativistic contexts?

The kinematic equations and other equations that arise throughout this theme were developed in Topics A.1 to A.3 in terms of Newtonian mechanics. A great deal of emphasis was placed on the importance of acceleration there, both in terms of its use to predict changes in velocity and displacement, and in terms of its relationship to applied force.

The discussion of special relativity in IB Diploma Programme physics deals exclusively with frames moving with constant relative velocities to each other. The twin paradox (see page 193) shows the difficulties that arise when acceleration takes an observer from one frame to another.

The concept of proper acceleration deals with this. Just as time intervals require a new definition—that of proper time—in relativity, so relative speeds and accelerations also require a new definition. Proper velocities and proper accelerations are all carefully defined to specify the reference frame in which measurements of displacement and time are taken. This was not required for Newtonian mechanics.

Patterns and trends — What Lorentz did

Initially, physicists tried to find a way round Maxwell's ideas within the context of classical physics. This led some to invent an "aether", a medium that was responsible for the transmission of electromagnetic radiation. However, the crucial experiment by Michelson and Morely showed that there was no aether and that, essentially, electromagnetic radiation can move through empty space. Lorentz and Fitzgerald independently conjectured that there must be a mysterious length contraction to explain this result. Later, Lorentz and others realized that such a length change implied a time change too. This resulted in the Lorentz transformations.

It was this work that sparked Einstein's great discovery. He was able to work from his two postulates to the same algebraic transformation as Lorentz without invoking any need for an aether.

The factor γ that appears in the time-dilation equation $\Delta t = \gamma \Delta t_0$ was first identified by Lorentz and is part of the Lorentz transformations that are the Einstein relativistic equivalents of the Galilean transformations. It is known as the **Lorentz factor**.

Figure 7 shows how the Lorentz factor changes with speed (or more properly the ratio $\frac{v}{c}$). The graph shows us that, for speeds up to 20% of that of light, the Lorentz factor remains close to 1. If the Galilean transformations were true for all speeds, then the graph would show a horizontal line at value 1 for all values of $\frac{v}{c}$. As v tends to c, γ tends to infinity.

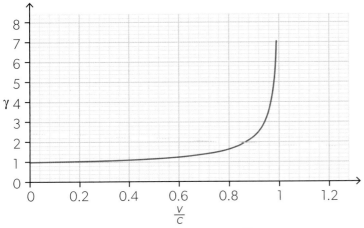

▲ Figure 7 Variation of the Lorentz factor γ with $\frac{v}{c}$.

Evidence and falsification

Are laws such as conservation of energy helpful to scientists? On the one hand they allow the prediction of, as yet, untested cases, but on the other hand they may restrict the progress of science. This can happen when scientists are not prepared to challenge the status quo.

In 2012, the results of an experiment suggested that neutrinos could travel faster than the speed of light in a vacuum. This flew in the face of the accepted science originally proposed by Einstein. Later investigations showed that small errors in the timings had occurred in the experiments. There was no evidence for faster-than-light travel. Were the scientists right to publish their results so that others could test the new proposals?

Intuition

Many aspects of physics can be described as intuitive — that applying a theory to a situation gives results that match our expectations. Intuition can be a helpful way of checking whether our answers are right. When they seem wrong, then it may be that we have made a mistake.

While our intuition often works well for areas of physics such as Newtonian mechanics, it does not help us with special relativity. The concepts of time dilation and length contraction are non-intuitive and seem hard to understand.

Does this mean that intuition is unhelpful? Or can intuition be a valuable way of knowing?

Models

In Topic E.1, you will meet the Bohr model for the atom. This model allows us to estimate the radius of electron orbitals in the hydrogen atom. The model predicts that the electron in the ground state of hydrogen is moving at $\frac{1}{137}$ of the speed of light. At this speed, the Lorentz factor is $\gamma = 1.00003$. Despite this small correction, its effect on the spectrum of hydrogen was first measured by Michelson and Morley in 1887. The fine structure constant, $\alpha = \frac{1}{137}$, has become an important quantity in atomic physics.

The electron orbitals of heavier elements are more complex as the outermost electrons are further from the nucleus. According to the Bohr model, the outer electrons move at a speed $\frac{Zc}{137}$. For atoms such as silver and gold, with atomic numbers $Z = 47$ and $Z = 78$, this gives Lorentz factors of $\gamma = 1.06$ and $\gamma = 1.22$. Silver and gold both have similar electron structures (they are in the same column of the periodic table) but the larger Lorentz

▲ Figure 8 Gold and silver have different colours because of special relativity.

factor for gold means that the electrons experience time dilation and length contraction on a greater scale. These relativistic effects on the outer orbitals change the energy gaps between them. This changes the wavelengths of light that are absorbed and reflected by gold. The result is that gold absorbs more blue light than silver. While silver reflects most wavelengths equally, giving it a silvery colour, gold gains its colour thanks to special relativity.

Worked example 3

Leah is flying a plane at $0.9\,c$. The landing lights on the plane flash every $2\,s$ as measured in the reference frame of the plane. Zosia watches the plane go by. Calculate the time between flashes as observed by Zosia.

Solution

When $v = 0.9c$, $\gamma = \dfrac{1}{\sqrt{1 - \frac{v^2}{c^2}}} = \dfrac{1}{\sqrt{1 - 0.9^2}} = 2.3$.

The time between flashes for Leah is $2\,s$. Therefore, the time between flashes for Zosia $= \gamma t = 2.3 \times 2 = 4.6\,s$.

Worked example 4

A spaceship travels at a constant speed from Earth to a space station. The space station is stationary relative to Earth. According to clocks on the spaceship, the journey takes 4.0 days and according to clocks on Earth and on the planet, it takes 5.0 days.

a. Outline which of these intervals is the proper time interval between the launch and the arrival of the spaceship.

b. Calculate the speed of the spaceship relative to Earth.

Solutions

a. The proper time interval is measured in a reference frame in which the launch and the arrival occur at the same position, so it is the interval measured by spaceship clocks.

b. 5.0 days $= \gamma \times 4.0$ days; hence $\gamma = \dfrac{5.0}{4.0} = 1.25$. The speed can be calculated

from the Lorentz factor, $\dfrac{1}{\sqrt{1 - \frac{v^2}{c^2}}} = 1.25 \Rightarrow \dfrac{v}{c} = \sqrt{1 - \frac{1}{1.25^2}} = 0.6$.

The speed of the spaceship is $0.6\,c$.

The Lorentz transformation

Hendrik Lorentz showed that the null result of Michelson and Morley (see page 166) could be avoided by using a set of transformation equations. Einstein proved that this Lorentz transformation could be derived assuming only Einstein's own modifications of Newton's postulates of special relativity.

In the Galilean transformation, lengths measured within one frame transform without change into the same length in any other frame. When length Δx is the difference between two positions in one frame and $\Delta x'$ is the difference in the other, then (using the Galilean transformations from earlier):

$$\Delta x' = x_1' - x_2' = (x_1 - vt) - (x_2 - vt) = x_1 - x_2 = \Delta x.$$

In Einstein relativity, this equality of Δx and $\Delta x'$ is no longer true and Lorentz proposed that the transformation must become $\Delta x' = \gamma(\Delta x - v\Delta t)$.

It is more useful to deal with a position x' in an inertial reference frame rather than two position determinations to give a length Δx, and therefore

$$x' = \gamma(x - vt)$$

gives the position x' of the object as observed in a reference frame moving at speed v relative to it.

The Lorentz transformation tells us about position measurements when one inertial reference frame is moving at a constant velocity relative to another. Imagine two observers, one in each frame, both making measurements of the same distance. The observer in one frame will not agree with the measurement made by the other observer. Space is no longer absolute.

Similarly, Lorentz realised that the transformation for time is: $t' = \gamma\left(t - \frac{vx}{c^2}\right)$.

With this transformation, time, like space, loses its Galilean property of being absolute. Time as measured in different reference frames differs when there is relative velocity between the frames. Also, terms in x now appear in the time equations and terms in t appear in the equations for x.

We have assumed so far that there is no relative motion between the frames in the y or z directions. When this applies, then there will be no relativistic changes in these directions either (this can be proved formally). The assumption of no motion in directions y and z along with the previous expressions lead to the complete set of Lorentz one-dimensional transformations which are compared in Table 1 with their Galilean equivalents.

The Lorentz transformations in Table 1 have been modified further so that the expression of time uses ct rather than t alone (this gives the time equation the dimensions of distance). A second change is to include the speed of light c twice in the distance equations. These changes make the equations appear more symmetrical and help to explain why later in this topic we use axes of ct and x to draw our spacetime diagrams.

Lorentz	Galilean
$x' = \gamma\left(x - \frac{v}{c} ct\right)$	$x' = x - vt$
$y' = y$ $z' = z$	$y' = y$ $z' = z$
$ct' = \gamma\left(ct - \frac{v}{c} x\right)$	$t' = t$

◀ Table 1 Lorentz one-dimensional transformations and their Galilean equivalents.

Inverse Lorentz transformations

The transformations given here allow you to transform from (x, ct) to (x', ct'). Sometimes the reverse change from the primed (') frame to the non-primed frame is required.

This is straightforward algebra. The Lorentz transformation for time gives $ct' = \gamma\left(ct + \frac{vx}{c}\right)$, which can be re-written with t as the subject:

$\frac{ct'}{\gamma} = ct - \frac{vx}{c}$, so that $ct = \frac{ct'}{\gamma} + \frac{vx}{c}$.

Substituting for x using

$x' = \gamma\left(x - \frac{v}{c} ct\right)$, which is equivalent

to $x = \frac{x'}{\gamma} + \frac{v}{c} ct$, gives $ct = \frac{ct'}{\gamma} + \frac{v}{c}$

$\left(\frac{x'}{\gamma} + \frac{v}{c} \times ct\right)$.

This expression can be simplified

to $ct = \gamma\left(ct' + \frac{vx'}{c}\right)$, remembering

that $1 - \left(\frac{v}{c}\right)^2 = \frac{1}{\gamma^2}$.

The equivalent inverse transformation from x' to x is

$x = \gamma\left(x' + \frac{v}{c} ct'\right)$.

These are known as the inverse Lorentz transformations.

Theories

The links between the x and t transformations mean that the coordinates for position and time have become entangled. The time interval between two events depends on the spatial separation of the events, and vice versa. This is another aspect of the paradigm shift that occurred when the special theory of relativity was accepted.

Observations

What does "observation" mean in special relativity? Are relativistic observations the same as the observations that we make in everyday life?

When we say "an observer observes a moving clock to be running slower", it does not mean that the observer can see this effect visually. The tick rate of the clock is revealed by measuring time and space coordinates within the reference frame of the observer.

In each case, when $v \ll c$, then $\gamma \approx 1$ and the Lorentz equations reduce to the Galilean equations.

The result from the thought experiment where light travels between mirrors (Figure 6) follows directly from the Lorentz transformations: the two events (the light leaving from, and then returning to, the bottom mirror) occur at t_1 and t_2. These events occur at the same place (the mirror), so we do not need to include the x terms in our proof (because $x_1 = x_2 = x$). The time interval between these events is $\Delta t = t_2 - t_1$. Therefore, in the observer frame,

$$\Delta t' = t_2' - t_1' = \gamma\left(t_2 - \frac{vx}{c^2} - \left(t_1 - \frac{vx}{c^2}\right)\right) = \gamma\Delta t \text{ and}$$

time interval in the observer frame = $\gamma \times$ time interval in the mirror frame

which is written $\Delta t = \gamma\Delta t_0$, as before.

This result shows that time measured in a frame moving relative to a clock is *always* longer than the time measured in a frame that is stationary relative to the clock. The effect is known as **time dilation** ("dilated" means "expanded"). When the moving observer in our example also has a clock, then an observer stationary with respect to the mirror frame observes the moving clock running slower than the mirror clock. The situation is symmetrical. We will discuss this later in this topic on page 187.

Why is the equation for the Doppler effect for light so different from that for sound?

In Topic C.5, you will study the Doppler effect for both sound and electromagnetic radiation. There are essential differences between them. In this topic, you see the importance of the relative difference in velocity between two observers in different inertial frames. Light does not require a medium for propagation and c is a limiting constant speed for the universe.

In the propagation of sound, movement of both source and observer relative to the medium is important and this leads to the difference between the two effects. There is a relativistic treatment of the Doppler effect for sound and this is discussed in more detail in Topic C.5.

Special relativity places a limit on the speed of light. What other limits exist in physics? (NOS)

There are limits on our knowledge. Sometimes these are technical limits. A measurement cannot be taken with sufficient precision at a particular time in scientific history. But these limits may be overcome in time or better methods may be developed.

There are philosophical limits to physics too. Science is built around the presumption of repeatability—do the same experiment twice and the same thing happens. When a reviewing scientist cannot replicate the result from another laboratory, problems arise in accepting the result. There is no obvious reason why the universe has to work in this way.

Finally, there are fundamental limits. The Heisenberg uncertainty principle predicts that we cannot know the precise position x and momentum p of a particle at the same time. Werner Heisenberg concluded, early in the history of quantum mechanics (Topic E.2), that $\Delta x \times \Delta p \geq \dfrac{h}{4\pi}$, where Δ stands for the "uncertainty in". This is just one of the fundamental limits that prevents us from knowing the exact state of the universe at any instant.

As you reflect on this course, what other limits to our knowledge can you identify? Are these limits shared with other areas of knowledge?

Worked example 5

A reference frame S' is moving at a velocity of $2.7 \times 10^8\,\mathrm{m\,s^{-1}}$ relative to a reference frame S, in the direction of the positive x-axis of S. The Lorentz factor γ for the relative speed of frames S and S' is 2.29. Spacetime coordinates of events A and B are measured in frames S and S'.

According to measurements in frame S, event A occurs at $x_A = 50\,\mathrm{m}$, $y_A = z_A = 0$, $t_A = 0.30\,\mu\mathrm{s}$, and event B occurs at $x_B = 80\,\mathrm{m}$, $y_B = z_B = 0$, $t_B = 0.40\,\mu\mathrm{s}$.
Calculate, as measured in S':

a.   the distance between events A and B

b.   the time interval between events A and B.

Solutions

a. $\Delta x' = x_B' - x_A' = \gamma((x_B - x_A) - v(t_B - t_A))$

   Substituting gives $\Delta x' = 2.29((80 - 50) - 2.7 \times 10^8(0.40 - 0.30) \times 10^{-6}) = 6.9\,\mathrm{m}$

b. $\Delta t' = t_B' - t_A' = \gamma\left((t_B - t_A) - \dfrac{v}{c}\dfrac{(x_B - x_A)}{c}\right) = 2.29\left((0.40 - 0.30) \times 10^{-6} - 0.9 \times \dfrac{80 - 50}{3 \times 10^8}\right) = 23\,\mathrm{ns}$

   This is a different time interval from that measured in S: $\Delta t = 0.1 \times 10^{-6}\,\mathrm{s} = 100\,\mathrm{ns}$. Time intervals are not absolute and depend on the relative motion of the observers!

Worked example 6

A rescue spaceship is sent from Earth to the site of a space accident. The spaceship moves in the direction of the positive x-axis of the reference frame of Earth, at a relative speed of $0.75c$. The clocks on Earth and in the spaceship are synchronized so that the launch of the spaceship occurs at $t = t' = 0$ and $x = x' = 0$. When the spaceship arrives at the site of the accident, the clock in the spaceship reads $8.0\,\mathrm{s}$.

a.   Explain why, according to Earth observers, the journey of the spaceship takes longer than $8.0\,\mathrm{s}$.

b.   Calculate, in the reference frame of Earth, the spacetime coordinates of the arrival of the spaceship at the site of the accident.

Solutions

a. $8.0\,\mathrm{s}$ is the proper time interval between the start and the end of the spaceship's journey, because it is measured in the rest frame of the spaceship. Earth is moving relative to the spaceship, so observers on Earth will measure a dilated (longer) time.

b.   In the reference frame of the spaceship, the coordinates are $x' = 0$ (because the spaceship is now at the site of the accident) and $t' = 8.0\,\mathrm{s}$. To find the coordinates (x, t) in the reference frame of Earth, we use the inverse Lorentz transformation:

$$\gamma = \frac{1}{\sqrt{1 - 0.75^2}} = 1.51$$

$$x = \gamma(x' + vt') = 1.51(0 + 0.75 \times 3 \times 10^8 \times 8.0) = 2.7 \times 10^9\,\mathrm{m}$$

$$t = \gamma\left(t' + \frac{v}{c}\frac{x'}{c}\right) = 1.51\left(8.0 + 0.75 \times \frac{0}{3 \times 10^8}\right) = 12\,\mathrm{s}$$

   According to observers on Earth, the spaceship has travelled a distance of $2.7 \times 10^9\,\mathrm{m}$ to the site of the accident, and the journey took $12\,\mathrm{s}$.

Practice questions

1. Clocks in two reference frames S and S' are synchronized so that $t = t' = 0$ when $x = x' = 0$. Frame S' has a speed of $0.300c$ relative to S. A lightning strike occurs at $x = 25.0$ km and $t = 0.150$ ms, as measured in S. Calculate the space and time coordinates of the lightning strike according to S', assuming that S' is moving:

   a. in the direction of the positive x-axis of S

   b. in the direction of the negative x-axis of S.

2. A space probe moves at a speed of $0.60c$ relative to the surface a planet. Clocks are synchronized so that $t = t' = 0$ when $x = x' = 0$.

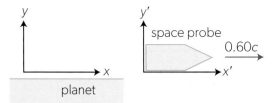

   Event P has coordinates $x' = 250$ m and $ct' = 400$ m in the reference frame of the space probe.

   Calculate the spacetime coordinates (x, ct) of P in the reference frame of the planet.

Proper length

The length of an object also changes when observed in frames that are moving relative to the object. When measuring a proper time interval, care is needed to specify that the position at which the times are measured is the same for both measurements (in the light clock on page 167 it was the position at which the light leaves the first mirror and to which it returns in the reference frame of the mirror). The **proper length** of an object (where the length is $x_2 - x_1$) must have x_1 and x_2 measured at the same time.

In reference frame S, x_1 and x_2 represent the ends of an object of length L_0. S is the frame in which the object is at rest. Frame S' is an inertial frame moving at speed v relative to S. In S', the positions become x_1' and x_2' with a length L'.

In S, $L_0 = x_2 - x_1$, which in S', using the Lorentz transformations, is equal to $\gamma(x_2' + vt_2') - \gamma(x_1' + vt_1')$.

The ends of the rod are measured at the same time and so t_1' and t_2' are equal.

Therefore $L_0 = \dfrac{L'}{\sqrt{1 - \dfrac{v^2}{c^2}}}$ and $L_0 = \gamma L'$. L_0 is the **proper length**.

Proper length is defined as the length of an object measured by an observer at rest relative to the object.

Both measurement events must be made at the same time. The proper length can also be regarded as the longest measured length that can be determined for an object. All other determinations of length made in a frame moving relative to the object frame will be shorter and are said to have undergone **length contraction** with a contracted length $L' = \dfrac{L_0}{\gamma}$.

Worked example 7

Priya and Orhan fly identical spacecraft that are 16 m long in their own frame of reference. Priya's spacecraft is travelling at a speed of $0.5\,c$ relative to Orhan's. Calculate the length of:

a. Priya's aircraft according to Orhan

b. Orhan's aircraft according to Priya.

Solutions

$\gamma = 1.15$ for this relative speed.

a. The length of Priya's aircraft is $\dfrac{16}{1.15}$ m according to Orhan. This is 13.9 m.

b. Because the situation is symmetrical, Priya will also think that Orhan's aircraft is 13.9 m long.

Worked example 8

Alpha Centauri is a nearby stellar system located at a distance of 4.4 light years (ly) from Earth, as measured in a reference frame in which Earth is at rest. A spaceship is sent from Earth to Alpha Centauri at a constant speed of 0.40c.

a. In the reference frame of the spaceship, calculate:

   i.   the distance between Earth and Alpha Centauri

   ii.   the time taken for the travel.

Immediately after the spaceship reaches the Alpha Centauri system, a radio message is sent to Earth.

b. Calculate, according to clocks on Earth, how long after the departure of the spaceship the radio message is received.

Solutions

a. i.   For $v = 0.40c$, $\gamma = 1.1$.
      The distance according to the spaceship is contracted and equals to $\dfrac{4.4}{1.1} = 4.0\,ly$

  ii.   The journey to Alpha Centauri would take $\dfrac{4.0\,ly}{0.40\,c} = 10$ years, according to clocks on the spaceship.

b. According to Earth observers, the journey takes $\dfrac{4.4\,ly}{0.40\,c} = 11$ years. The radio message travels at the speed of light and needs another 4.4 years to reach Earth. The message will be received $11 + 4.4 = 15.4$ years after the departure of the spaceship.

Practice questions

3. A rod of proper length 1.5 m moves at a speed of $0.8c$ relative to a laboratory. In the reference frame of the laboratory, the velocity of the rod is parallel to the rod. The rod passes through a light gate that is at rest relative to the laboratory.
Calculate, in the reference frame of the laboratory:

   a.   the length of the rod

   b.   the time taken for the rod to pass through the light gate.

4. Suppose that another spaceship is to be sent to the Alpha Centauri system, but it must reach the destination after 5.0 years, according to its own clocks.

   a.   Show, using the data in Worked Example 8, that the speed v of the spaceship relative to Earth satisfies the equation $\dfrac{\frac{v}{c}}{\sqrt{1 - \frac{v^2}{c^2}}} = 0.88$.

   b.   Hence, calculate v. Give the answer in terms of the speed of light.

Velocity addition

A spaceship is moving in frame A with a constant velocity u_A. Frame B is moving with a constant velocity v with respect to frame A. What do the Lorentz equations tell us about the velocity u_B of the spaceship when it is viewed by an observer in frame B?

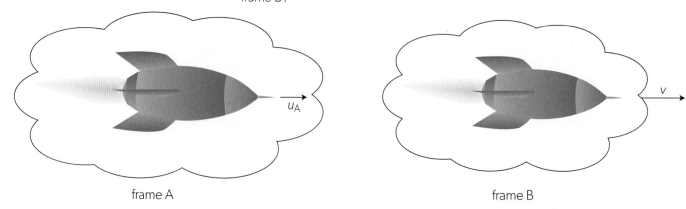

frame A   frame B

▲ **Figure 9** Relativistic velocity addition. The spaceship is moving in frame A. Frame B is moving relative to frame A. What is the speed of the spaceship in frame B?

Galilean relativity has no problem with this. The answer is simple: $u_B = u_A - v$. This cannot be correct for Einstein relativity, however. Suppose the rocket in frame A is moving at the speed of light. When the observer in frame B and the spaceship are moving in opposite directions relative to frame A, then v is negative and u_B will exceed c. This is not allowed by Einstein's second postulate.

We need to use the Lorentz equations. The speed u_A is equal to $\frac{x}{t}$ when viewed in frame A. Here, x is the distance moved from the origin in A together with the time t taken for the movement from the origin in A. Similarly, u_B is equal to $\frac{x'}{t'}$ when measured in frame B.

 Correct signs

One way to get the signs correct for a particular combination of frame velocities is to begin with the Galilean transformation (that is, when $v \ll c$) and work out what sign you expect. Then remember that the signs match at the top and bottom of the equation.

So $u_B = \dfrac{x'}{t'} = \dfrac{\gamma(x - vt)}{\gamma\left(t - \dfrac{vx}{c^2}\right)}$ using the Lorentz transformations.

You should satisfy yourself that substituting $x = u_A t$ into this expression gives:

$$u_B = \frac{u_A - v}{1 - \dfrac{u_A v}{c^2}}$$

This is the **relativistic velocity addition** equation.

Worked example 9

Jean and Phillipe are in separate frames of reference, neither of which is accelerating. Jean observes a spacecraft moving to his right at $0.8c$. Phillippe observes the same spacecraft moving to his left at $0.9c$. Calculate the velocity of Phillippe's frame of reference relative to Jean's.

→
Solutions

Relative to the spacecraft, Phillippe is moving at 0.9c to the right. The diagram shows the situation from the point of view of Jean's frame:

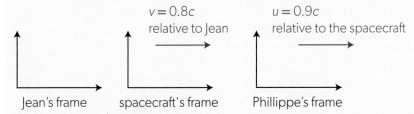

$v = 0.8c$
relative to Jean

$u = 0.9c$
relative to the spacecraft

Jean's frame   spacecraft's frame   Phillippe's frame

If the velocities were non-relativistic, we would expect Phillippe's velocity relative to Jean's frame to be $u + v$. Therefore, we use the "plus" sign in the relativistic velocity addition equation.

The velocity of Phillippe's frame relative to Jean's is $\dfrac{0.9 + 0.8}{1 + 0.9 \times 0.8}c = 0.988c$.

Practice questions

5. A probe moves at a speed of $0.15c$ away from a space station. A beam of electrons is emitted from the space station towards the probe, at a speed of $0.40\,c$ relative to the space station.

    Calculate, in the reference frame of the probe, the speed of the electrons according to:

    a. the Galilean transformation

    b. the Lorentz transformation.

6. Two spaceships A and B travel in opposite directions away from Earth. Relative to Earth, spaceship A moves at $0.40c$ and spaceship B moves at $0.60c$. Calculate, in the reference frame of spaceship A, the speed of:

    a. Earth

    b. spaceship B.

Invariant quantities and the spacetime interval

Einstein recognized that absolute time and absolute space are not invariant (unchanging) properties when moving from one inertial reference frame to another. However, there are some quantities that do not change between inertial frames. These quantities are said to be **invariant**.

Invariant quantities include the proper time interval and the proper length.

A third invariant quantity is the **spacetime interval**. This arises because of the deduction from Einstein's second postulate that the speed of light is universal. In Galilean relativity, time intervals are invariant and cannot change between reference frames. In Einstein relativity, it is no longer true that $\Delta t' = \Delta t$. The time interval must be replaced by the spacetime interval to reflect the unification of space and time as four coordinates.

We expressed time in the table of Lorentz transformations not as plain t but as the product of the speed of light and time, ct. This quantity has the dimensions of length and leads to the **spacetime interval** Δs that is defined for motion in the x-direction as

$$\Delta s^2 = (c\Delta t)^2 - \Delta x^2$$

When the terms are written as $(ct)^2 - x^2 = \text{constant}^2$, the equation is known as the **invariant hyperbola** (the equation here assumes that one pair of x and t are zero). The importance and use of this curve are discussed on page 184.

Another name used for Δs is the **invariant interval**. (You may see Δs defined in some books as $\Delta x^2 - (c\Delta t)^2$, in other words, as the negative of our definition.)

The spacetime interval is invariant because comparing two different frames of reference and using the ct formulation of the Lorentz transformations,

$$c\Delta t = c(t_2 - t_1) = \gamma\left(ct'_2 + \frac{v}{c}x'_2\right) - \gamma\left(ct'_1 + \frac{v}{c}x'_1\right) = \gamma\left(c\Delta t' + \frac{v}{c}\Delta x'\right)$$

and

$$\Delta x = (x_2 - x_2) = \gamma\left(x'_2 + \frac{v}{c}ct'_2\right) - \gamma\left(x'_1 + \frac{v}{c}ct'_1\right) = \gamma\left(\Delta x' + \frac{v}{c}c\Delta t'\right)$$

Therefore

$$\Delta s^2 = (c\Delta t)^2 - \Delta x^2 = \gamma^2\left(c\Delta t' + \frac{v}{c}\Delta x'\right)^2 - \gamma^2(\Delta x' + v\Delta t')^2$$

$$= \gamma^2(c^2 - v^2)\Delta t'^2 - \gamma^2\left(1 - \frac{v^2}{c^2}\right)\Delta x'^2 = \gamma^2\left(1 - \frac{v^2}{c^2}\right)c^2\Delta t'^2 - \gamma^2\left(1 - \frac{v^2}{c^2}\right)\Delta x'^2$$

So $\Delta s^2 = (c\Delta t')^2 - \Delta x'^2 = \Delta s'^2$ because, as $\gamma^2 = \dfrac{1}{\left(1 - \frac{v^2}{c^2}\right)}$, the terms in γ cancel.

This is obviously identical to the original definition using the same quantities (and no others) but is measured in the other reference frame.

In three dimensions, the spacetime interval becomes $\Delta s^2 = (c\Delta t)^2 - \Delta x^2 - \Delta y^2 - \Delta z^2$.

Worked example 10

Events A and B are two lightning strikes. In the reference frame of a ground observer, A and B have coordinates $x_A = 2.5 \times 10^3$ m, $t_A = 0$ and $x_B = 8.0 \times 10^3$ m, $t_B = 1.2 \times 10^{-5}$ s. A rocket flies above the ground at a speed of $0.75c$ towards the positive x-axis of the ground observer.

a. Calculate the spacetime interval between events A and B, using the coordinates of:

    i.   the frame of reference of the ground observer

    ii. the frame of reference of the rocket.

b. Discuss whether it is possible that A and B occur at:

    i.   the same position

    ii. the same time relative to some other inertial reference frame.

Solutions

a. i.   Relative to the ground, time and space differences between A and B are $\Delta t = 1.2 \times 10^{-5}$ s and $\Delta x = 5.5 \times 10^3$ m. The spacetime interval is therefore $(\Delta s)^2 = (3 \times 10^8 \times 1.2 \times 10^{-5})^2 - (5.5 \times 10^3)^2 = -1.73 \times 10^7$ m².

    ii. Time and space differences relative to the rocket can be calculated using the Lorentz transformation. For $v = 0.75c$, $\gamma = 1.51$.

$$\Delta x' = 1.51(5.5 \times 10^3 - 0.75 \times 3 \times 10^8 \times 1.2 \times 10^{-5}) = 4.23 \times 10^3 \text{ m}$$

$$\Delta t' = 1.51\left(1.2 \times 10^{-5} - 0.75 \times \frac{5.5 \times 10^3}{3 \times 10^8}\right) = -2.65 \times 10^{-6} \text{ s}$$

$(\Delta s')^2 = (3 \times 10^8 \times (-2.65 \times 10^{-6}))^2 - (4.23 \times 10^3)^2 = -1.73 \times 10^7 \, m^2$. This is the same value as in part i., demonstrating that the spacetime interval is invariant. Note that $\Delta t' < 0$, and hence in the frame of the rocket, lightning B struck before lightning A!

b. i.   In a reference frame in which the lightning strikes are at the same position, $\Delta x = 0$ and the spacetime interval between A and B would have to be greater than zero, because, in this case, $(\Delta s)^2 = (c\Delta t)^2$. But we know from part a.i. that the interval is negative. Hence, no such frame exists. This is an example of a space-like interval (see page 185). A and B are spatially separated in every inertial frame of reference, regardless of its relative velocity.

ii.   The negative value of the spacetime interval between A and B implies that there is a reference frame in which the events happen at the same time. In this particular frame, $\Delta t = 0$ and $(\Delta s)^2 = 0^2 - (\Delta x)^2 = -(\Delta x)^2$ which is a negative quantity as required by a.i. Note that the separation Δx in this equation is the *proper distance* between the lightning strikes, which is greater than that measured in any other reference frame.

Muon decay in the upper atmosphere

There is direct experimental evidence for both time dilation and length contraction (which are two sides of the same coin).

Muons are particles that can be created either in high-energy accelerators or in the upper atmosphere when cosmic rays strike air molecules. These muons have short mean lifetimes of about 2.2 μs. When travelling at 0.98c, the distance the muon will travel in one mean lifetime is roughly 660 m. This distance is far less than the height of 10 km above Earth's surface where the muons are created. Based on Newtonian physics, very few muons would be expected to reach the surface as the time to reach it is about 15 mean lifetimes. Nevertheless, a considerable number of muons are detected at the surface. Many more than would be expected.

The presence of muons at the surface is due to time dilation (or length contraction, whichever viewpoint you choose). At a speed of 0.98c,

$\gamma = \dfrac{1}{\sqrt{1-\dfrac{v^2}{c^2}}} = \dfrac{1}{\sqrt{1-0.98^2}} = 5.0$. So, in the reference frame of Earth, the mean

lifetime becomes 11 μs. The time to travel 10 km at 0.98c is 33 μs, so a significant number of muons remain undecayed at the surface.

In the frame of reference of the muon, the 10 km from atmosphere to Earth's surface (as measured by an observer on the Earth) is only $\dfrac{10}{\gamma}$ km (as far as the muon is concerned). This is 2.0 km in the muon's rest frame corresponding to a travel time of about 3 mean lifetimes, which allows many more muons to reach the surface than Galilean relativity would suggest.

Thus, depending on the viewpoint of the observer, either time dilation or length contraction can be used to explain the observed large number of muons at the surface.

Data-based questions

In 1940, B. Rossi and D. Hall measured the decay of atmospheric muons. Their experiments were improved in 1963 by D. Frisch and J. Smith who detected muons both in Cambridge, Massachusetts and on Mount Washington which is 1907 m higher in altitude than Cambridge. Their results are shown in the table.

Run	Number of muons detected in one hour	
	on Mt Washington	at Cambridge
1	568	412
2	554	403
3	582	436
4	527	395
5	588	393
6	559	

Data taken from D. H Frisch, J. H. Smith, Measurement of the Relativistic Time Dilation using μ-mesons, *Physical Review* 64 (7–8), 199–201 (1963).

- Calculate the average number of muons detected in each location and give an uncertainty with your value.

- The muons were measured to have a speed of 0.9952c.

  - Calculate the time taken, in Earth's rest frame for a muon to travel 1907 m.

  - Calculate the value of γ for a speed of 0.9952c.

  - Hence calculate the time taken in the muon's rest frame to travel this distance.

- $\lambda = 4.55 \times 10^5\,\text{s}^{-1}$ is the decay constant of muons. Use the equation $N = N_0 e^{-\lambda t}$ (see Topic E.3) (where N_0 is the number of muons detected on Mt Washington) to calculate the number of muons (N) that are expected to be detected at Cambridge:

  - without time dilation

  - when the effects of time dilation are included.

- Calculate an uncertainty in your two answers.

- Compare the measured value of the number of muons at Cambridge with your predicted values.

Practice questions

7. An unstable particle is produced in a high-energy collision in a particle accelerator. The particle travels a distance of 2.4 cm at a speed of 0.85 c, both measured relative to the accelerator, before it decays. Calculate the proper lifetime of the particle.

8. Charged pions are particles present in cosmic ray bursts in Earth's atmosphere. The mean lifetime of a charged pion is 2.60×10^{-8} s, in a reference frame in which the pion is at rest. The mean lifetime of charged pions in a particular cosmic ray burst is estimated as 1.00×10^{-7} s in the reference frame of Earth. Calculate:

   a. the speed of the charged pions in the cosmic ray burst

   b. the mean distance travelled by the charged pions relative to Earth.

 Global impact of science—GPS

Satellite navigation units (satnavs) in cars and other devices use global positioning systems to pinpoint a position on Earth's surface to within a few metres. Several satellites are always above the horizon anywhere on Earth. Inside each satellite is an atomic clock, accurate to about ±1 ns, which controls the transmission of a signal to the receivers on Earth.

These receivers triangulate the signals from satellites above the local horizon to arrive at a positional fix to within metres in a few seconds. Wait a little longer with some special GPS receivers and this precision can rise to orders of millimetres. A satnav in a moving vehicle can show the speed and heading in real time.

The design of both the satellite transmitters and the GPS receivers need to take account of relativity. The atomic clocks are adjusted so that once in orbit they run at the same rate as Earth-bound clocks. The GPSs contain computers that carry out the calculations required to make the relativistic corrections.

Special relativity predicts that the clocks on the satellites fall behind ground clocks by about 7 μs in every 24 hours due to their relative motion with respect to the surface. General relativity, Einstein's later theory that links gravitational effects and spacetime, predicts that the satellite clocks should advance compared with the surface clocks by about 45 μs every day.

The net result of the two relativistic corrections is that the clocks in the GPS satellite gain on clocks back on Earth by about 1.6 μs every hour. This factor swamps the 20 ns accuracy required of Earth-bound GPS receivers.

When relativistic effects are not considered, then the position errors become serious after about 100 s and accumulate at a rate of tens of kilometres every day. This is unacceptable for navigation. The GPS receivers in our cars and on our mobile phones are constantly carrying out relativistic corrections to adjust for the unavoidable time changes due to relativistic effects.

Spacetime diagrams and worldlines

In 1908, Hermann Minkowski introduced a way to visualize the concept of spacetime. Minkowski represented the four-dimensional nature of spacetime using a graph known as a **spacetime diagram** (sometimes known as a Minkowski diagram).

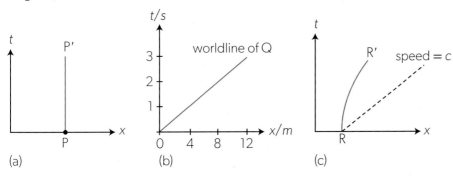

(a) (b) (c)

▲ **Figure 10** Spacetime diagrams. (a) P is stationary in the inertial frame of the diagram. (b) Q is moving at a constant velocity relative to the diagram frame. (c) R is accelerating relative to the diagram frame.

Spacetime diagrams (Figure 10) show the position of an object in one dimension (x) at a time (t) in an inertial frame. The axes themselves constitute the inertial frame. The diagram resembles (but should not be confused with) the ordinary distance–time graphs which are familiar from Topic A.1. One difference is that time is plotted on the y-axis and position on the x-axis.

Figure 10(a) shows the spacetime diagram for a particle that is stationary with respect to the inertial frame represented by the diagram. At $t = 0$, the particle P is on the x-axis. As time increases, because the object is stationary, it does not change its position (x) in the reference frame. Line PP' shows the **trajectory** of the particle through spacetime and is known as the **worldline** of the particle.

Figure 10(b) shows a different particle Q moving at a constant velocity in the reference frame of the same spacetime diagram. At $t = 0$, Q is at the origin of the diagram ($x = 0$) and it is moving at $4\,\text{m s}^{-1}$ to the right. Each second after $t = 0$, Q is 4 m further to the right and so its worldline in the spacetime diagram is a line at an angle to the axes. When another particle R is accelerating relative to the reference frame of the diagram, then the worldline of R is a curve (Figure 10(c)).

There must be a limit to the gradient of the R worldline because nothing can exceed the speed of light in free space. The gradient of the dashed line on Figure 10(c) shows the maximum limiting speed of R. This dashed line also represents the worldline of a photon in the diagram (the minimum gradient of RR').

We can combine two separate spacetime diagrams for different inertial frames moving at constant speed relative to each other. This combination of axes is

Models

The algebraic approach of the Lorentz transformations and the geometric approach devised by Minkowski are complementary. They describe the same phenomena, one in a visual way, one using algebra. They give the same results and yield the same conclusions.

Neither approach is more fundamental than the other. Indeed, it was the geometry of the Minkowski diagram that led Einstein, who had earlier been Minkowski's student, to extend his work from special to general relativity.

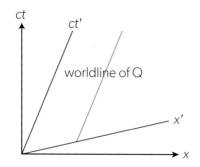

▲ Figure 11 The spacetime diagram can show the inertial frame for Q.

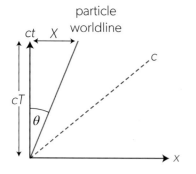

▲ Figure 12 The angle θ between the worldline and the ct-axis is a measure of the velocity of the particle.

of help later in this topic when we resolve some of the paradoxes that special relativity appears to create.

Figure 11 shows spacetime diagram for particle Q which is again moving at constant velocity relative to the frame of P (P is not shown). Q is displaced from the origin of its own reference frame and remains there. Notice also how the time axes have changed. They are now not *t* but have been changed to *ct*. This is how you will see the axes written from now on. It is a convenient convention because it means that:

- both *x*- and *ct*-axes have the dimensions of length
- the worldline of a photon must be at 45° to both *x*- and *ct*-axes.

In Figure 11, the Q axes (*x'*–*ct'*) are the rest frame for Q. They have rotated away from *ct* and *x* and become closer together. The *x'*-axis is now at the same angle as the previous Q worldline in the P reference frame. Q is moving along a worldline parallel to the time axis of its own reference frame. In other words, Q is at a constant position along its own *x'*-axis.

An additional convention is that sometimes physicists define *c* to be equal to 1 so that, in calculations, large values for the answers do not trouble them. Equally, expect to see speeds quoted as, for example, 0.95*c* meaning 95% of the speed of light in free space ($2.85 \times 10^8 \, \text{m s}^{-1}$).

Some simple geometry (Figure 12) shows that when we are using (*x*–*ct*) axes, then the angle θ between the particle worldline and the *ct*-axis is given by:

$$\tan \theta = \frac{\text{opposite}}{\text{adjacent}} = \frac{X}{cT} = \frac{1}{c} \times \frac{X}{T} = \frac{v}{c}.$$

or

$$\theta = \tan^{-1}\left(\frac{v}{c}\right).$$

When $v = c$, then $\theta = 45°$ (because $\tan 45° = 1$) and

the worldline for a photon starting at the origin of the spacetime diagram is a line at 45° to both ct- and x-axes.

Simultaneity

Simultaneity in Galilean relativity is absolute. All frames in Galilean relativity share a universal, or absolute, time. Newton assumed that both time and place were absolute and independent. For Newton, time was independent of the observer, so that direct comparisons between different frames were possible because the "flow" of time proceeded at the same rate in every frame. An alternative way to imagine this is to think of a series of clocks that can be synchronized across all frames. After being brought together for synchronization, the clocks can be moved to different frames. Under Galilean relativity, the clocks in their different frames will keep time at the same rate as each other. This is because every time interval Δt transforms (in the Galilean invariance) into $\Delta t'$, where $\Delta t = \Delta t'$.

An important axiom of Einstein's relativity is that the clocks, after moving to different reference frames, will no longer measure identical time intervals for the same event as described by the observers in the different frames.

There are significant changes to our ideas about the order in which things happen or whether two events happen simultaneously under Einstein's relativity compared with Galilean relativity. This is because the speed of light is always observed to have the same value by observers in different frames.

The classic "thought" experiment to illustrate this is the example of a train carriage moving at constant velocity past an observer standing on a station platform (Figure 13). A person in the carriage (Jack) switches on a lamp that hangs from the centre of the ceiling. Jack observes that the light from the lamp reaches the two end walls of the carriage (R and L) at the same moment.

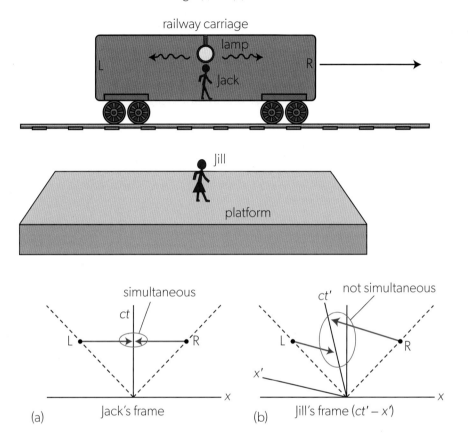

▲ Figure 13 The lamp in the carriage is switched on as Jack, in the train, passes Jill who is on the platform. (a) Jack thinks the light reaches each end of the carriage simultaneously. (b) Jill thinks that the light reaches the left-hand end of the carriage first.

Jill, who is on the platform, does not agree with this observation. The light from the lamp moves to both ends of the carriage at the same speed c. However, while the light is moving to the ends, the left-hand end of the carriage moves towards the light and the right-hand end moves away. The consequence is that the light (according to Jill) hits the left-hand end first (event L) before the right (event R).

This result becomes clear in a spacetime diagram. In the reference frame of Jack (x–ct), the events R and L occur at the same instant because they are on a line parallel to the x-axis and are at the same ct-coordinate (Figure 13(a). In Jill's frame (x'–ct'), plotted on Figure 13(b), L occurs before R when you consider these events in terms of the ct'-axis.

It is possible to misunderstand and to think that the loss of simultaneity is to do with the transmission of the information. In other words, that this difference of opinion between Jack and Jill arises because the light travels through different distances from the ends of the carriage to their eyes. That is not the explanation of what is happening. The lack of simultaneity arises because the speed of light is always constant even when a particular observer is moving relative to the light source. As far as Jack is concerned, he is *always* midway between the carriage end walls. As far as Jill is concerned, once the photons have left the lamp, then they travel at c and the carriage will continue to move while the photons themselves are in transit.

As mentioned earlier in this topic, one of the reasons for this confusion is the use of the term "observer", which is universal in books and articles about relativity. We often think of an observer as being located at one point in the inertial reference frame. This is not the true meaning. It is better to think of the observer as being in overall charge of an (infinitely) large number of clocks and rulers that are located throughout the observer's frame. Jack (the stationary observer in this case) can take a reading at the instant when the light hits the end wall of the carriage without having to worry about the time taken for this information to travel from the carriage to his position. Another way to think about the observer is as a whole team of observers with each one able to make measurements of his or her immediate region of the reference frame.

The distinction between Jack's and Jill's observations can also be explained in terms of the Lorentz transformations. Imagine that two events are simultaneous in one frame of reference so that $\Delta t = 0$. The full transformation for the time interval between the events is is $\Delta t' = \gamma\left(\Delta t - \frac{v\Delta x}{c^2}\right)$, with the usual notation. When $\Delta t = 0$, then $\Delta t'$ can only be zero when $\Delta x = 0$. Two events can only be simultaneous *in both frames* when they occur at the same position (so, essentially, they must be the same event).

The invariant hyperbola

Earlier we wrote the equation for the spacetime interval as $(ct)^2 - x^2 = \pm(\text{constant})^2$ by taking one of the values of t in $\Delta t = (t_2 - t_1)$ to be 0. This equation has a geometric meaning in terms of the spacetime diagram.

Time-like intervals
Begin with the equation with a positive right-hand side:
$$(ct)^2 - x^2 = +(\text{constant})^2$$

Figure 14 shows two frames (x, ct) and (x', ct') where the observer in the second frame is moving at $0.5c$ relative to the first. For convenience, we will call (x, ct) the laboratory frame and (x', ct') the moving frame (relative to the lab). Added to the diagram are three lines that correspond to the upper branches of the invariant hyperbolas where the constant in the equation is equal to $+1$, $+2$, or $+3$. (Hyperbolas below the x-axis are equivalent but relate to events in the past, that is, before $ct = 0$; these are not shown on the diagram.)

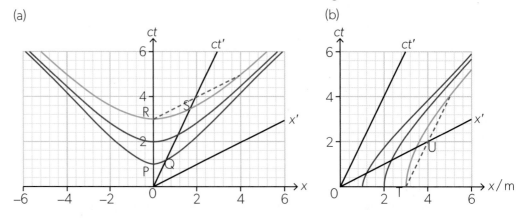

▲ Figure 14 The invariant hyperbola gives information about (a) time-like and (b) space-like intervals.

When the sign on the right-hand side of the equation is positive, then it is possible for x to be equal to 0 but $c(\Delta t)$, and therefore Δt, cannot equal zero. The events represented by the set of hyperbolas are separated by a non-zero time intervals (three intervals are shown on Figure 14(a)). These are therefore known as **time-like intervals**.

Imagine that a first event occurs at the origin of the spacetime diagram. This occurs at the same point in both frames and so both observers will agree about its position and time. A second event P then occurs at the position $x = 0$ and at the time $ct = 1$ in the lab frame (the red curve in Figure 14(a)). The invariant hyperbola that intersects P shows what happens in the moving frame. P has the same time coordinate in frame S as Q has in frame S'. In other words, intersections of the hyperbola with the ct' axis sets the scale for the axis. Another event at R in the lab frame is similarly related to S in the moving frame. The observer in the lab frame sees P and R as occurring in the same place but at $ct = 3$ (the green curve), which is not true for the moving observer.

The invariant hyperbolas allow another question to be answered: is a succession of events separated in the lab frame by equal times also separated by equal times in the moving frames? In other words, can the moving-frame axes be calibrated? The red, blue and green hyperbolas cross the lab ct-axis at equal times ($ct = 1$, 2 and 3). From the perspective of the lab frame, this is not true for the moving frame because the distances along the ct'-axis decrease with increasing ct'. An additional question is: are these crossing times spaced equal distances apart in the moving frame of the observer?

To answer this, look at the red dashed line on Figure 14(a); it is parallel to the x'-axis. This line passes through event R ($x = 0$, $ct = 3$). It intersects the ct'-axis at ($x' = 0$, cT'), where (using the Lorentz time transformation) $cT' = \gamma\left(ct - \frac{vx}{c}\right)$. We know that this intersection is simultaneous with R in the moving frame and so $cT' = \gamma\left(3 - \frac{v \times 0}{c}\right)$, leading to $cT' = 3\gamma$. As ct has the value 3 this means that the ct'-axis must be scaled by the factor γ. As γ itself depends only on v and c, we can infer that the scaling on the ct'-axis is uniform and does not change with time.

Space-like intervals

Similar arguments apply for the set of hyperbolas shown in Figure 14(b). These are the curves for $(ct)^2 - x^2 = -(\text{constant})^2$. Again, the first event is at the origin (0, 0) for both frames, but we focus on the position, not the timing, of subsequent events.

For this set of curves, Δx cannot be zero (but Δt can). The intervals in this case are **space-like**. Events that are space-like can be observed to take place at the same time although never at the same position.

Again, a first event happens at the origin of the frames and both observers agree about the time and the position. Events also occur simultaneously in the lab frame at distances 1 m, 2 m, and 3 m from the origin. The intersections of the hyperbolas with the x'-axis indicate the position and time at which the moving observer thinks that these other events occur. Again, the scaling on the x'-axis is not the same as that on the x-axis.

The red dashed line on Figure 14(b) is parallel to the ct'-axis and passes through the event T at $ct = 0$ for $X = 3$ m. The intersection at U on the x'-axis is at $ct' = 0$ (because U lies on the x'-axis) and so $X' = \gamma\left(x - \frac{v}{c}ct\right)$. Cancelling c gives $x' = (x - vt)$. As $x = 3$, $X' = 3\gamma$. Once again, the factor γ acts as a scaling factor between the x-axes for the two frames.

The reason why ct cannot be zero is that, if it is, then x must be imaginary (a complex number) as opposed to a real value. Complex numbers are treated in some parts of the IB Diploma Programme mathematics, so you may meet them there. A complex number contains both a real and an imaginary part. The imaginary part arises from the solution of an equation such as $x^2 = -1$. You will recognize that when $c(\Delta t) = 0$, then this must be the case here for $-x^2 = + (\text{constant})^2$.

What about time travel?

The spacetime interval has a bearing on the cause-and-effect relationship between two objects or events. Intervals can be classified as space-like, time-like, or light-like depending on the value of Δs^2 for the two events that make up the interval.

Figure 15 shows space-like and time-like regions for the invariant hyperbolas. These regions are separated by the photon worldlines. A time interval on the photon worldlines is light-like.

Light-like intervals

When Δs^2 is zero, the spatial distance and the time interval are the same because $(c\Delta t)^2 = \Delta x^2$. Such events are linked by a photon travelling at the speed of light. Suppose event S is the eruption of a solar flare on the Sun and W is the event of an astronomer on Earth witnessing the eruption. The photons from the eruption travel directly from event S to event W (along the photon worldline) and because this world line is common to both reference frames (Sun and astronomer), both observers will agree about the order of the events and their light-like separation.

On the invariant hyperbola (in this case a straight line) the information about the event moves along the photon world line.

Space-like intervals

For an interval to be space-like, Δs^2 must be < 0 (that is, negative). This corresponds to $(c\Delta t)^2 - \Delta x^2 = -$(constant)^2, so $\Delta x^2 > (c\Delta t)^2$. This means that the distance between the events is too great for light from one event (or anything travelling slower) to have any effect on the other event. They do not occur in each other's past or future. Although there is a reference frame in which they occur at the same *time*, there is no reference frame in which the events can occur at the same *place*.

Suppose the Earth-bound astronomer turns on a light 5 minutes after the flare occurs on the Sun. Call this event L. (The distance from the Sun to Earth is 8 light minutes.) L occurs 3 minutes before the astronomer sees the flare. To get from event S to event L, it will be necessary to travel faster than the speed of light. There is a space-like interval between event S and event L.

Time-like intervals

Finally, when Δs^2 is > 0 (that is, positive), then $(c\Delta t)^2 > \Delta x^2$. There can be a cause-and-effect relationship between the events because the time

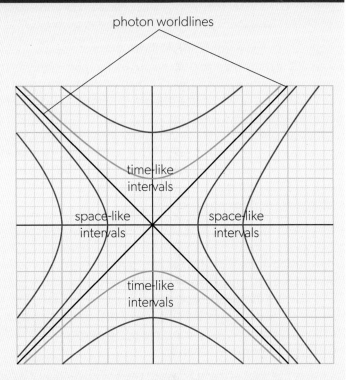

▲ Figure 15 A set of invariant hyperbolas together with photon worldlines.

part of the spacetime interval is greater than the spatial separation. Time-like intervals lie in the regions defined by photon worldlines that include the ct-axis.

Event L and event W occur at the same place (the astronomer turns on the light and sees the solar flare without moving). These events are time-like. They would also be time-like if they occurred at different places if the astronomer was moving because the astronomer cannot travel faster than c.

Another observer travelling in a different reference frame from either the Sun or the astronomer (an astronaut, say) will disagree with the astronomer about the times and positions at which the events occur. The astronaut may well disagree about the order in which events L and S occur because these are space-like separated. However, both astronaut and astronomer will agree about the spacetime intervals so that, for example, the interval Δs^2 between events S and W is zero for all observers.

Time travel has always been a fascination of science-fiction authors. The spacetime interval can tell us the extent to which two events in space and time can affect each other.

To what extent do fictional works that you know mirror scientific truth?

Time dilation and length contraction re-visited

The effect of time dilation can be visualized in spacetime diagrams.

Figure 16(a) shows two frames, S (x, ct) and S' (x', ct'), in the usual way. A clock is at rest in S' and ticks once at the origin $(ct' = 0)$ and later at event E. The green curve in Figure 16 is the invariant hyperbola for this event. Event G in frame S and event E in frame S' both lie on the invariant hyperbola and so have the same time coordinate. This means that an identical clock placed in S would also tick at the origin and G. The dashed line represents simultaneity in frame S. Hence, according to S, events F and E have the same time coordinate. F is later than G according to S, so the time of event E is greater (dilated) than the time observed in S'.

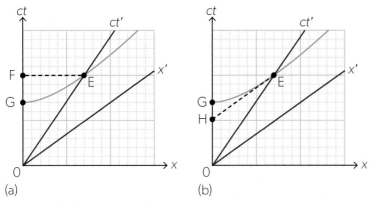

(a)   (b)

▲ **Figure 16** Space diagrams with invariant hyperbolas can be used to explain time dilation when (a) a clock is at rest in frame S' and (b) the clock is at rest in frame S.

In Figure 16(b), a different clock is at rest in S and ticks at H. This time, the dashed line shows simultaneity in frame S'. The events H and E have the same time coordinate in frame S' because they lie on the same line parallel to the x' axis. This time coordinate is the same as the time coordinate of event G in frame S (because E and G remain joined by the invariant hyperbola). G is later than H, so S' measures a longer time for event H than observer S. The time dilation effect is symmetrical between the two frames.

Length contraction can also be visualized in spacetime.

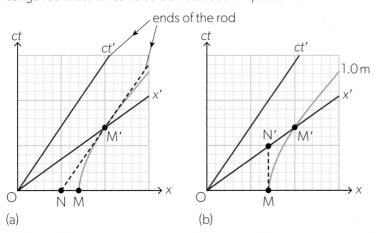

(a)   (b)

▲ **Figure 17** Using the invariant hyperbola to explain length contraction when a rod is at rest in (a) frame S' and (b) in frame S.

Figure 17(a) shows a rod at rest in S'. The length of the rod in S' is OM'. An identical rod in S has the length OM. M and M' therefore both lie on the same (green) spacelike invariant. However, the length of the moving rod according to S is ON, because the position of both ends of the rod must be measured at the same time in S. ON is less than OM and so the measurement in S shows a length contraction.

Figure 17(b) shows the spacetime diagram for a rod of length 1.0 m (the spacelike hyperbola is labelled with this value). This time, the rod is at rest in S with the end labelled M as before. The argument is left to you to show that there is length contraction according to frame S′ when the observer compares a 1.0 m rod with the frame S rod.

Remember that OM′ is a scaled length and does not indicate the true calibration of the x′-axis from the point of view of the S′ observer.

Worked example 11

In the distant future a network of four warning beacons W, X, Y, and Z is set up to warn spaceship commanders of the approach lanes for planet Earth. The beacons flash in sequence. The spacetime diagram shows the reference frame in which the beacons are at rest and one cycle of the sequence. The worldline for a spaceship is also shown.

a. Determine the order in which the four beacons flash according to:

   i. an observer stationary in the frame of the beacons

   ii. an observer on the spaceship.

b. Determine the order in which the observer on the spaceship sees the beacons flash.

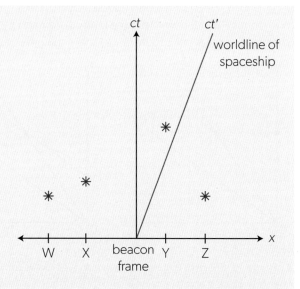

Solutions

a. i. The spacetime diagram in the frame of the beacons indicates the chronological order in which the beacons flash: W and Z simultaneously, then X, and then Y.

   ii. The order of the flashes in the spaceship frame has to be obtained from constructing lines parallel to the x′-axis. In this frame, Z is observed to flash first, then W and X flash simultaneously. Finally, Y flashes.

b. To decide on the arrival of the light from the beacon, it is necessary to add the photon worldlines to the diagram. These are lines that begin at the beacon flash and travel at 45° to the axis. The intersection of the photon worldline with the ct′-axis gives the arrival time at the spacecraft. The order is Z, Y, X, W.

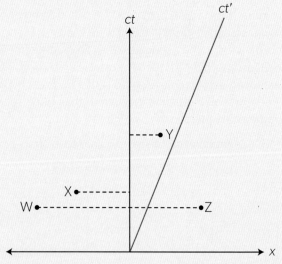

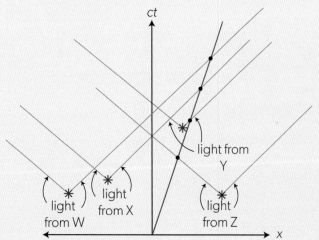

Worked example 12

A train moves at a constant velocity relative to a station. When the train passes the station, the train's front and rear lights are switched on simultaneously in the train's reference frame. Event F is the train's front lights switched on and event R is the train's rear lights switched on. The spacetime diagram shows event F and the worldlines of the front and rear of the train in the frame of reference of the station.

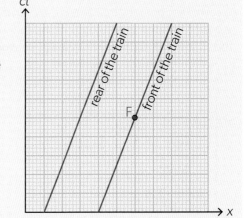

a. i. Calculate the speed of the train relative to the station.

ii. Construct event R on the spacetime diagram.

iii. Explain which event, F or R, happened first according to an observer at rest relative to the station.

The proper length of the train is 100 m.

b. Calculate, in the frame of reference of the station:

i. the length of the train

ii. the time elapsed between events F and R.

Solutions

a. i. The ratio $\frac{v}{c}$ is equal to the tangent of the angle θ between either of the worldlines of the train and the ct-axis and can be estimated from the diagram.

$$\frac{v}{c} = \tan \theta = \frac{\text{opposite}}{\text{adjacent}} = \frac{2}{5} = 0.4$$

The speed of the train is $0.4c$.

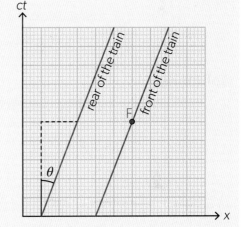

ii. R and F are simultaneous in the train's frame, and all events simultaneous with F are represented by a line through F that makes an angle θ to the x-axis. The diagram shows how this simultaneity line for F can be constructed. R is at the intersection of the simultaneity line and the worldline of the rear of the train.

iii. From the diagram, the ct-coordinate of event R in the reference frame of the station is less than the ct-coordinate of event F. Hence, R happens before F according to measurements done in this frame.

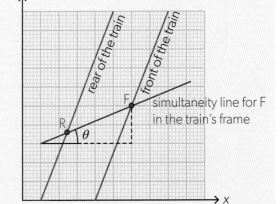

b. i. The Lorentz factor for the speed is $0.4c$ is $\gamma = \frac{1}{\sqrt{1-0.4^2}} = 1.09$. The length of the train in the frame of the station is shorter than the proper length by this factor.

$$L = \frac{100}{1.09} = 91.7 \, \text{m}.$$

ii. The time difference between R and F in the frame of the station can be calculated using the inverse Lorentz transformation:

$t_F - t_R = \gamma\left((t'_F - t'_R) + \frac{v}{c^2}(x'_F - x'_R)\right)$. From the simultaneity of R and F in the train's frame, we have

$t'_F - t'_R = 0$ and $x'_F - x'_R = L_0$. Hence $t_F - t_R = 1.09 \times \frac{0.4}{3 \times 10^8} \times 100 = 1.45 \times 10^{-7} \, \text{s} = 145 \, \text{ns}$.

Worked example 13

A rod of proper length 2.0 m is moving at a constant velocity relative to a laboratory reference frame (x, ct). In the laboratory frame, the velocity of the rod is parallel to the rod and directed towards the positive x-axis. The ct'-axis in the spacetime diagram represents the worldline of the left-hand end of the rod.

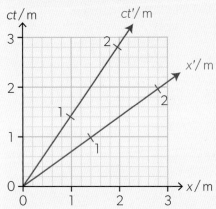

a.   Construct, on the spacetime diagram, the worldline of the right-hand end of the rod.

b.   Estimate the length of the rod in the laboratory reference frame:

    i.   by making an appropriate coordinate measurement on the diagram

    ii.   by using the length contraction equation.

Solutions

a.   The worldline of the right end of the rod is parallel to the ct'-axis and crosses the x'-axis at $x' = 2$ m.

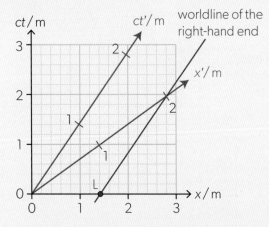

b.   i.   The measurement of the length of the rod in the laboratory frame involves subtracting the x-coordinates of the ends of the rod measured simultaneously at the same value of ct. When $ct = 0$, the rod extends from the origin of the coordinate system to the point labelled L on the diagram. Hence, the length of the rod in the laboratory frame is equal to the x-coordinate of L, which is approximately 1.4 m.

    ii.   To use the length contraction equation, we need to estimate the speed of the rod relative to the laboratory. The method explained in worked example 12.a.i. gives $v = 0.7c$. The length of the rod in the laboratory frame is therefore

$$\frac{2.0}{\gamma} = 2.0 \times \sqrt{1 - 0.7^2} = 1.4 \text{ m}.$$ This is, of course, consistent with the answer to part b.i.!

Practice questions

9. The spacetime diagrams show coordinate axes of reference frames $S(x, ct)$ and $S'(x', ct')$. For each diagram, state which event is simultaneous with event P, according to measurements done in the reference frame:

   i.   S

   ii. S'.

   a)

   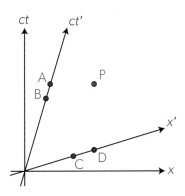

   b)

   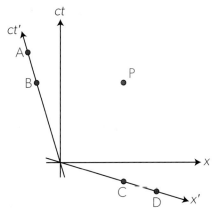

10. The spacetime diagram shows events P, Q and R in coordinate axes of reference frames $S(x, ct)$ and $S'(x', ct')$.

    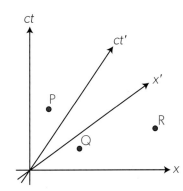

    What is the order of these events, from earliest to latest, according to an observer at rest with respect to frame S'?

    A. P, Q, R   B. P, R, Q

    C. R, Q, P   D. Q, R, P

11. A space probe moves at a constant velocity. Two photons are emitted in opposite directions towards the probe and arrive at the probe simultaneously. The diagram shows the emission event and the worldline of each photon and the event of their common arrival at the probe. ct' is the worldline of the probe and x' is the space axis of its reference frame.

    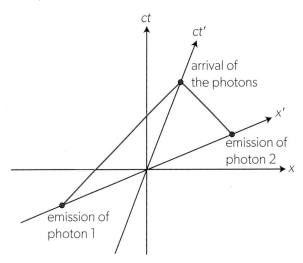

    Which statement is correct in the reference frame of the space probe?

    A. Photon 1 travels a longer distance than photon 2.

    B. Photon 1 travels for a longer time than photon 2.

    C. Photon 2 travels at a higher speed than photon 1.

    D. Both photons travel an equal distance.

12. A spaceship moves away from Earth at a constant velocity. On the spacetime diagram, the ct-axis is the worldline of Earth and the ct'-axis is the worldline of the spaceship.

    a. Calculate the speed of the spaceship relative to Earth.

    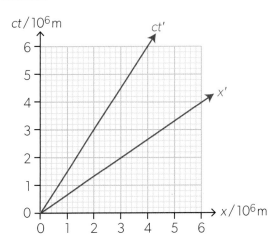

A radio pulse is emitted from the spaceship towards Earth. In the reference frame of Earth, the emission occurs at a distance of 2.00×10^6 m from Earth. The emission of the pulse is event P and its detection is event Q.

b.   Construct events P and Q on the spacetime diagram.

c.   Calculate:

   i.   the proper time interval between the launch of the spaceship and event P

   ii.   the coordinates (ct', x') of event Q in the reference frame of the spaceship.

d.   State the distance travelled by the radio pulse according to:

   i.   a spaceship observer

   ii.   an observer on Earth.

13.   A spaceship is sent from Earth towards a distant planet. The diagram shows the spacetime axes (x, ct) of the reference frame of Earth and the worldline ct' of the spaceship. The spaceship leaves Earth at $t = 0$ and moves at a constant velocity relative to Earth, in the direction of the negative x-axis. Coordinate axes are scaled in light years (ly). When $t = 2.00$ years, a communication signal travelling at the speed of light is sent from Earth towards the spaceship.

a.   Copy the spacetime diagram and draw on it:

   i.   the space axis x' for the reference frame of the spaceship

   ii.   the worldline of the communication signal.

b.   Estimate, using the diagram, the time at which the communication signal is received:

   i.   according to the clock in the spaceship

   ii.   according to the clock on Earth.

c.   The speed of the spaceship is $0.500\,c$. Determine, using the Lorentz transformation, the spacetime coordinates in the reference frame of the spaceship of the event when the communication signal is received.

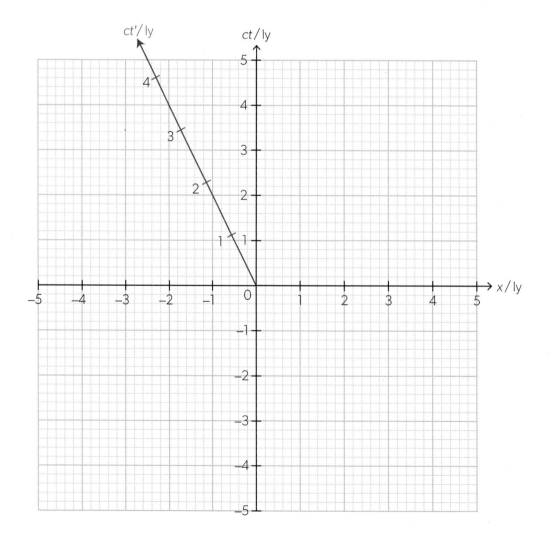

Theories — The twin paradox

Many ideas in this topic lead to paradoxes in which deductions using special relativity are at odds with everyday existence. The twin paradox is the most famous of these and is simply stated:

Malik and Maha are twins. Malik journeys to a distant star at a high speed (Lorentz factor equal to γ) taking a time T in Maha's frame of reference, Malik returns and his return journey also takes time T (according to Maha). Maha has aged $2T$ in her frame but, to her, Malik has only aged by $\frac{2T}{\gamma}$.

This is time dilation — so where does the paradox arise? Malik sits in his spacecraft and watches Maha move away at the same high speed so why is Maha not younger than him on his return? We expect symmetry between the frames.

In fact, there is no symmetry between the two cases. Maha remains in an inertial frame throughout Malik's journey. He accelerated four times: at the start of the trip, when slowing down at the star, when accelerating back to top speed towards Maha, and, finally, when he decelerates to arrive home. Leaving an inertial reference frame even once breaks the symmetry. This is why Malik and Maha age at different rates relative to each other.

Figure 18(a) shows what happens. Maha's frame is (x, ct), Malik's is (x', ct'). Maha remains at the origin of her frame moving along the ct-axis. Malik moves along his worldline at his origin $x' = 0$ or at $x = vt$ in Maha's frame (v is Malik's speed relative to Maha). Malik reaches the star at event P and lines of simultaneity are given for Malik and Maha. Maha thinks that Malik arrives at the star at Q. R is when Malik thinks that Maha observes his arrival at the star. They disagree about the simultaneity of Q and R as we expect. At this stage, both Malik and Maha think that the other is younger by a factor of γ — as predicted by symmetrical time dilation.

Malik must change velocity to return. This is not necessary if, when he reaches the star, he synchronizes his clock with another clock on a spacecraft carrying Jay. Jay is already on his way to Earth (and therefore Maha) with the Lorentz factor γ. Figure 18(b) includes the worldline for Jay. When Jay leaves the star, he thinks that Maha is at S. Were Malik to decelerate, turn round, and

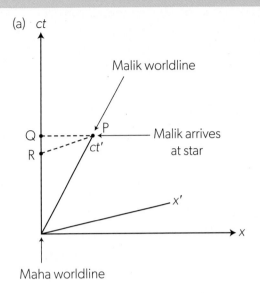

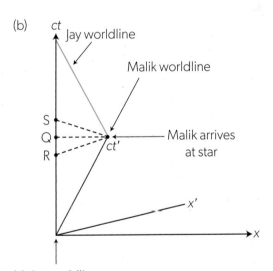

▲ Figure 18 A spacetime view of the twin paradox for (a) up to the instant at which Malik arrives at the star and (b) the whole journey as Malik/Jay return to Earth.

go back to Earth, the acceleration would make Maha appear to age rapidly from R to S.

This prediction has been used as a test of the special theory of relativity using high-speed aircraft. This was just one of a number of tests that have been used to confirm the theory and to justify the paradigm shift it involved.

Theme A — End-of-theme questions

1. A student strikes a tennis ball that is initially at rest so that it leaves the racquet at a speed of $64\,m\,s^{-1}$. The ball has a mass of $0.058\,kg$ and the contact between the ball and the racquet lasts for $25\,ms$.

   a. Calculate:

      i. the average force exerted by the racquet on the ball

      ii. the average power delivered to the ball during the impact.

   b. The student strikes the tennis ball at point P. The tennis ball is initially directed at an angle of $7.00°$ to the horizontal.

   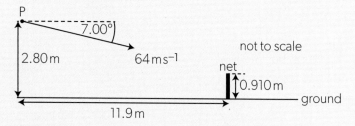

      i. Calculate the time it takes the tennis ball to reach the net.

      ii. Show that the tennis ball passes over the net.

      iii. Determine the speed of the ball as it strikes the ground.

2. A company designs a spring system for loading ice blocks onto a truck. The ice block is placed in a holder H in front of the spring and an electric motor compresses the spring by pushing H to the left. When the spring is released, the ice block is accelerated towards a ramp ABC. When the spring is fully decompressed, the ice block loses contact with the spring at A. The mass of the ice block is $55\,kg$.

   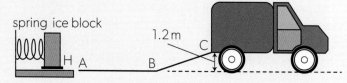

   Assume that the surface of the ramp is frictionless and that the masses of the spring and the holder are negligible compared to the mass of the ice block.

   a. i. The block arrives at C with a speed of $0.90\,m\,s^{-1}$. Show that the elastic energy stored in the spring is $670\,J$.

      ii. Calculate the speed of the block at A.

   b. Describe the motion of the block:

      i. from A to B with reference to Newton's first law

      ii. from B to C with reference to Newton's second law.

   c. Sketch a graph to show how the displacement of the block varies with time from A to C. (You do not have to put numbers on the axes.)

   d. The spring decompression takes $0.42\,s$. Determine the average force that the spring exerts on the block.

3. A company delivers packages to customers using a small unmanned aircraft. Rotating horizontal blades exert a force on the surrounding air. The air above the aircraft is initially stationary.

   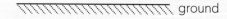

   The air is propelled vertically downwards with speed v. The aircraft hovers motionless above the ground. A package is suspended from the aircraft on a string. The mass of the aircraft is $0.95\,kg$ and the combined mass of the package and string is $0.45\,kg$. The mass of air pushed downwards by the blades in one second is $1.7\,kg$.

   a. i. State the value of the resultant force on the aircraft when hovering.

      ii. Outline, with reference to Newton's third law, how the upward lift force on the aircraft is achieved.

      iii. Determine v. State your answer to an appropriate number of significant figures.

   b. The package and string are now released and fall to the ground. The lift force on the aircraft remains unchanged. Calculate the initial acceleration of the aircraft.

4. A small ball of mass m is moving in a horizontal circle on the inside surface of a frictionless hemispherical bowl.

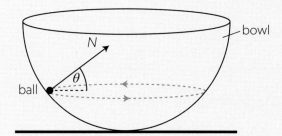

The normal reaction force N makes an angle θ to the horizontal.

a. i. State the direction of the resultant force on the ball.

   ii. On the diagram, construct an arrow of the correct length to represent the weight of the ball.

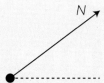

   iii. Show that the magnitude of the net force F on the ball is given by the equation $F = \dfrac{mg}{\tan\theta}$

b. The radius of the bowl is 8.0 m and $\theta = 22°$. Determine the speed of the ball.

c. Outline whether this ball can move on a horizontal circular path of radius equal to the radius of the bowl.

d. A second identical ball is placed at the bottom of the bowl and the first ball is displaced so that its height from the horizontal is equal to 8.0 m.

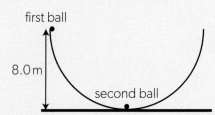

The first ball is released and eventually strikes the second ball. The two balls remain in contact. Determine, in m, the maximum height reached by the two balls.

5. A constant force of 50.0 N is applied tangentially to the outer edge of a merry-go-round. The following diagram shows the view from above.

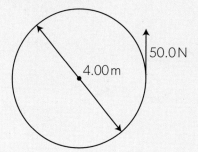

The merry-go-round has a moment of inertia of 450 kg m² about a vertical axis. The merry-go-round has a diameter of 4.00 m.

a. Show that the angular acceleration of the merry-go-round is 0.2 rad s⁻².

b. The merry-go-round starts from rest and the force is applied for one complete revolution. Calculate, for the merry-go-round after one revolution:

   i. the angular speed

   ii. the angular momentum.

A child of mass 30.0 kg is now placed onto the edge of the merry-go-round. No external torque acts on the system.

c. Calculate the new angular speed of the rotating system.

d. The child now moves towards the centre.

   i. Explain why the angular speed will increase.

   ii. Calculate the work done by the child in moving from the edge to the centre.

6. a. Explain what is meant by the statement that the spacetime interval is an invariant quantity.

   b. Observer A detects the creation (event 1) and decay (event 2) of a nuclear particle. After creation, the particle moves at a constant speed relative to A. As measured by A, the distance between the events is 15 m and the time between the events is 9.0×10^{-8} s. Observer B moves with the particle. For event 1 and event 2:

   i. Calculate the spacetime interval.

   ii. Determine the time between them according to observer B.

   c. Outline why the observed times are different for A and B.

B The particulate nature of matter

Introduction

Look at a polished stone, an electron microscope image of a plastic, or an X-ray diffraction pattern. The structure of matter and its intrinsic beauty springs out at you immediately. Matter does not have the uniform, homogeneous nature that we assumed in Theme A. In this theme we recognise the differences between solids, liquids, and gases – and we acknowledge the different descriptions needed to explain their observed properties. These explanations range from simple models for the flow of electric charge through to a complex multi-factorial description of the origin of the weather.

One concept that links these descriptions is that of the **particle**. In Theme B a particle is taken to be an atom, molecule or – in the case of electric current in Topic B.5 – an electron. The very name "atom" itself comes from the Greek language of 400 BCE where the word ἄτομος (atomos) was coined by Democritus for something that could not be subdivided further. Atoms and molecules can be combined in different ways to form the four phases of matter. These combinations and their consequences are discussed in Topic B.1.

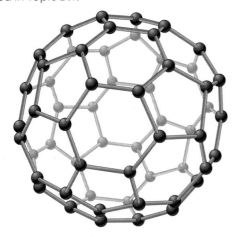

The particle concept becomes apparent as we alternate between macroscopic and microscopic views of matter. This is of particular importance in Topics B.3 and B.4 where the observable macroscopic properties of gases are explained using microscopic descriptions of the interacting particles of the gas. In Topic B.5, electric current is described through the collisions between electrons and positive ions in a conductor – both entities treated as particle-like.

Our over-arching concept of **force** has its part to play too. The link between the macroscopic and particulate descriptions of a gas is provided by the momentum conservation at the walls of the gas container and the force, and hence pressure, to which it leads.

The concept of particle merges with our over-arching theme of **energy** too. In Topic B.2, the impact of particle behaviour on the Earth's climate is considered. Topic B.4 discusses the impact of energy change within the context of individual particle behaviour on a thermodynamic system.

Throughout the theme we **model** phenomena. Establishing a **theory** is an essential part of the Nature of Science. This modelling is backed up by that other essential of the subject: empirical (**experimental**) results. As you study this theme you will become aware of the importance within this area of physics of **shared endeavour** effort by many scientists and engineers. Work undertaken over decades and centuries that provides us with useful and accurate models of the behaviour of the everyday material world.

How do macroscopic observations provide a model of the microscopic properties of a substance?

How is energy transferred within and between systems?

How can observations of one physical quantity be used to determine the other properties of a system?

The transfer of energy between systems affects the behaviour of the particles which make up the systems. At the simplest level, transferring energy into a system increases the kinetic energy of the particles that make up the system. They will move faster, and one could say that the material is "hotter". There are also issues of energy transfer throughout the system. Interactions between the particles govern this. Our models must provide a mechanism for energy transfer between systems. These models must also provide mechanisms to explain how this energy can spread through a whole system of particles. Above all, macroscopic observations and microscopic properties must align so that changes to one give the observed response in the other. This link between observation and explanation is an important aspect of 21st-century science.

How do the systems of particles respond to the transfer of energy into them? What are the links between the observed macroscopic effects and the energy changes at the microscopic level? The quantitative thermal properties of materials have been well studied since the 1700s, but qualitative explanations for the properties are more recent and are underpinned by our knowledge of the nature of the microscopic particles that make up our world.

▲ Figure 1 The energy transferred from the Sun, and the subsequent energy transfers as it heats the oceans and the land, form an important environmental system on Earth.

In this topic, you will learn about:

- molecular theory for solids, liquids and gases
- density
- the Celsius temperature scale
- the Kelvin temperature scale as a measure of the average kinetic energy of particles
- the internal energy of a system
- phase changes
- specific heat capacity of a substance
- specific latent heat of fusion and vaporization of substances
- conduction, convection and thermal radiation as energy transfer mechanisms
- thermal conductivity
- the Stefan–Boltzmann law
- apparent brightness and luminosity
- the emission spectrum of a black body
- Wien's displacement law.

Introduction

In Topic B.1, you will meet types of matter that have different structures: solids, liquids and gases. Some materials conduct energy well; others do not. Materials transfer thermal energy at different rates and in different ways. Models are used to explain these differences in terms of the microscopic and macroscopic changes that occur when energy is transferred to and from materials.

Particles

Describing material structure requires a new technical language. One important concept is that of the **particle**. Two hundred years ago, particles were taken to be individual minute point objects. We now understand that materials consist of atoms. For some purposes, we still describe them as small impenetrable objects. We use further descriptions of ions and free electrons when, for example, we deal with electrical conduction. These atoms (or molecules when chemically combined) interact with each other to yield the macroscopic behaviour that we see in our everyday objects.

Phases of matter

You will probably be familiar with the three phases of matter: solid, liquid and gas. Each state is modelled as an arrangement of particles and as the movement of the particles relative to each other. The movement arises because, at any temperature above the lowest, each particle has kinetic energy and therefore some degree of random motion.

The three states are:

- **Solid.** Solids have a fixed shape, meaning they must have a fixed volume. Each particle vibrates about a given fixed position. The arrangement of these fixed positions may determine other features of the solid too: for example, whether it is a crystal or a more amorphous material such as a plastic. Solids tend to have higher densities than gases.

- **Liquid.** Liquids do not have a fixed shape, but they do have a fixed volume at a particular temperature. The particles also vibrate relative to each other as in a solid, but this time the particles of the liquid have more freedom. They can move around relative to their nearest-neighbour particles, perhaps exchanging positions with a neighbour from time to time. Liquids and solids have similar densities and nearest-neighbour distances.

- **Gas.** A gas has neither a fixed shape nor a fixed volume. It completely fills its container. For most of the time the gas particles move in straight lines until they collide with another particle or with the container wall. Gases typically have a low density but they can be compressed over a wide range of pressures to give variable densities.

Water, like many substances, moves between the three phases (ice, water and steam) depending on the amount of kinetic energy available to its particles and the external conditions. When there are large amounts of kinetic energy, then particles can break apart from each other and behave as a gas.

Temperature, energy transfer and internal energy

Before atoms were discovered, scientists did not describe solids, liquids and gases in terms of particle vibration and movement. What they observed were transitions between phases as energy is transferred to and from the substance.

You will use the concept of a point particle when you model the behaviour of gases as a series of particle interactions in Topic B.2.

Models—Macroscopic and microscopic

The models in this topic describe materials in either macroscopic or microscopic terms.

- **Macroscopic** means that we view the materials and objects largely, ignoring internal structure, much as in Theme A.

- **Microscopic** means that the objects are modelled at the atomic or molecular level and that the interactions between the basic building blocks become the focus of attention.

There was a paradigm shift when scientists realized that matter consists of atoms. Later it was recognized that the atoms themselves have a structure with even smaller particles whose properties determine behaviour. This internal structure to the atom is the focus of Theme E.

A fourth phase

There is a further phase known as a **plasma**, which is a high-temperature phase where the atoms are completely ionized and act as an ensemble of high-speed charged particles. This phase is not discussed in the IB Diploma Programme physics course.

From the 17th century onwards, scientists thought that a substance called *phlogiston* (or sometimes *caloric*) flowed from hot objects to cold objects rather like water being poured from one container to a lower one. Even today, a common (though not SI) unit of the energy transferred by food is the "calorie".

Another concept that scientists developed gradually, and that we still use today, is the description of the "degree of hotness" of an object using the term **temperature** or, more properly, a **temperature scale**.

▲ Figure 2 All four states of matter are visible in this picture. The air is transparent, but its presence can be seen in the colour of the sky. (The Sun is a plasma – the fourth state of matter.)

Measurements—Comparing temperature scales

A temperature scale has two fixed points. The choice of fixed points and the number of degrees between them defines the scale. Temperature scales used today are the Celsius, Kelvin and the Fahrenheit scales.

The Celsius (formerly called Centigrade) and Kelvin scales are compared in Figure 4. In the Kelvin scale, the lower temperature is absolute zero (0 K), a theoretical temperature which is the lowest attainable. The upper fixed point is the triple point of water: a unique temperature at which ice, water and water vapour all co-exist at a pressure of 611 Pa and a temperature of 273.16 K. This pressure is about 0.6% of atmospheric pressure. Small changes in the conditions will make the substance change into one of the three phases.

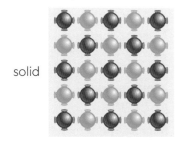

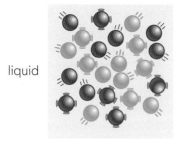

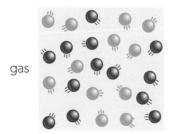

▲ Figure 3 Motion of molecules in three of the four phases of matter.

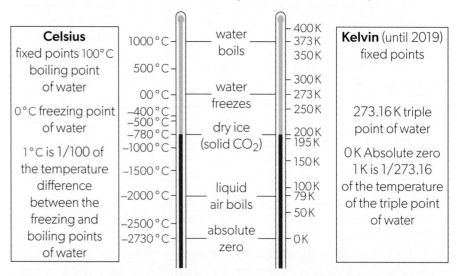

▲ Figure 4 Two temperature scales, Celsius and Kelvin, compared.

Figure 5 shows how a mercury-in-glass thermometer can be calibrated by marking the lengths reached by the mercury when the thermometer is in an ice-water mixture and in steam at 100 °C. The distance between the two lengths is divided into 100 divisions to give the degree Celsius intervals.

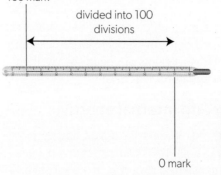

◀ Figure 5 A mercury-in-glass thermometer is made by marking the 100 °C and the 0 °C point and dividing the length between the marks into 100 divisions.

Calibrating a thermocouple

- Tool 1: Understand how to accurately measure temperature and electric potential difference to an appropriate level of precision.

- Tool 3: Construct and interpret tables and graphs for raw and processed data including scatter graphs and line and curve graphs.

- Inquiry 2: Assess accuracy, precision, reliability and validity.

A thermocouple is a temperature sensor that uses the potential difference between two different metals that are joined together. This pd varies with the temperature difference between the metals.

- You need a voltmeter sensitive enough to measure to the nearest 0.1 mV. The most sensitive scale on a multimeter is usually sufficient. Two lengths of wire made of copper with crocodile clips attached and one length of another type of wire are also required. Nichrome wire (a nickel–chromium alloy) is suitable for the single length.

- One of the junctions must be at a reference temperature. A mixture of ice and water can give a reference of 0 °C.

- Attach the nichrome wire to the other copper wires that lead to the voltmeter with the clips (Figure 6). Put one junction in the ice water and the other in a beaker of warm water. Use a mercury-in-glass thermometer to measure the temperature of the warm water and record this temperature as well as the reading on the voltmeter.

- Vary the temperature of the warm water and record your voltmeter readings. Plot a calibration graph of your results to show the recorded voltage against thermometer temperature.

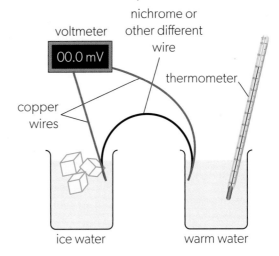

▲ Figure 6 The Seebeck effect allows two metal junctions to act as a thermometer.

- Use your calibrated thermocouple to measure the temperature of another object—perhaps a freezer compartment or something recently removed from the freezer. How accurate do you think your thermocouple is?

- This may be a temperature measurement technique that you will need in your IA. How would the thermometer calibration fit into your inquiry cycle (page 702)?

ATL Communication skills—Using terminology

Temperature and temperature conversions

Temperature is measured in units of kelvin (K) or degrees Celsius (°C). Note that there is *no* degree symbol for the K of kelvin but the symbol *must* be used for the Celsius measure.

Temperature difference is measured in units of kelvin or "degrees" (deg). Technically, Celsius, Fahrenheit and any other scale (other than the Kelvin scale) cannot be used to specify a temperature change in SI.

Only Celsius and Kelvin scales are regularly used in scientific work. A change in temperature is the same in Kelvin and Celsius so that the conversion between them is straightforward.

To go from a Celsius temperature to a Kelvin temperature, add 273. Thus, $T/K = \theta/°C + 273$.

To go from Kelvin to Celsius, subtract 273. Remember that expressions such as $\theta/°C$ in this context mean: θ "measured in" degrees C.

Worked example 1

The temperature of a metal sample is increased from 100 °C to 500 °C. Calculate, in K:

a. the initial absolute temperature of the sample

b. the temperature change of the sample.

Solutions

a. $100 + 273 = 373 K$

b. 400 K. The temperature change is the same expressed in kelvin as in degrees Celsius.

Absolute zero

The choice of the unattainable 0 K (absolute zero) for a temperature scale seems odd. The point is that it is never required in a practical way to calibrate the thermometers. In fact, the thermometer used for these purposes is a gas thermometer which requires only one calibration measurement at the ice or triple point.

This occurs elsewhere in physics. In field theory (Theme D), a point called "infinity" is imagined as a point so far away that no force acts.

How can readings or positions that cannot be attained be useful?

Practice questions

1. Several physical properties vary with temperature.
   a. State an example of a physical property that varies with temperature.
   b. Outline how this property can be used to measure temperature.

When a cold object is in contact with a hot object with no other energy supplied, the two will eventually come to a common "degree of hotness". At this point they are said to be in **thermal equilibrium**.

All objects (at a temperature above absolute zero) possess internal kinetic energy due to the motion of their particles. The higher the temperature of the object, the greater the internal energy associated with the particles.

Internal energy

The phase of a substance determines the freedom of movement of its particles. As energy is transferred to the particles, two processes can take place:

- Particles can move further apart from each other, increasing the stored potential energy of the system of the particles.

- Particles can move faster, increasing the kinetic energy of the system.

Taken together, this means that for a substance:

> **The internal energy of the system is the sum of the total intermolecular potential energy arising from the forces between the molecules and the total kinetic energy of the molecules arising from their random motion.**

The particles of a gas are, on average, far apart and the stored potential energy must be small, meaning that, **for a gas, the internal energy is almost completely made up of kinetic energy**. For a solid, the sizes of the stored potential energy and the kinetic energy are about the same. The internal energy of a gas is an important quantity and is discussed further in Topic B.3.

Given the opportunity, energy always transfers spontaneously from a high-temperature region to one at a low temperature. "Heat", we say rather loosely, "flows from hot to cold". On pages 214–224, we will look at the ways in which energy can transfer between and within objects due to temperature differences.

Patterns and trends — Changes in stored potential energy

It may seem to be a contradiction that, as energy is transferred to a substance, its intermolecular potential energy can increase but the gas phase (when most energy has been accumulated) has effectively zero intermolecular potential energy.

This is because the stored potential energy for solids and liquids is negative. These are similar ideas to those of binding energy in nuclear physics (Theme E) and the negative gravitational potential energies of bound planet–satellite systems in Theme D.

The solid atoms and molecules are said to be in a potential well, the bottom of which is a negative energy. As energy is transferred to the system, the potential energy increases, getting nearer and nearer to the zero value for a gas. The stored potential energy is a measure of the strength of the links between atoms. As atoms get further and further apart, these links become weaker and weaker.

Regarding the internal energy as made up of two parts helps to explain phase change. As energy is transferred to a solid, both the random kinetic part and the intermolecular potential part increase. The former increases the vibrational speed, the latter the distance between the molecules. But, as the molecules move apart, the force of attraction between two molecules is reduced enough for the inter-particle bonds to be broken. Eventually, groups of molecules become free. We say that the solid has melted. During this phase of energy transfer, the intermolecular potential energy has been rising, but the kinetic energy has stayed the same. This means that the temperature is constant. The solid is melting at constant temperature (it is at the melting point). As more energy is supplied, the bulk of the energy now begins to increase the random kinetic energy of the molecules with only a fraction going to intermolecular potential energy. The temperature is increasing again.

Eventually, the molecules have so much random kinetic energy that they begin to break apart (to overcome the intermolecular forces) and become unbound from their nearest neighbours. They are now free and act as a gas. Again, while this occurs, the incoming energy is transferred to the potential form not the kinetic. When all molecules are free, then the temperature begins to increase again, and other parameters of the gas can also change (pressure, density, volume).

Average kinetic energy and the Boltzmann factor

Increased temperatures are linked to increases in the internal energy for a substance. However, we have not made this idea quantitative. The quantity, now known as the Boltzmann constant, was first introduced by Max Planck in a theory that described black-body radiation (you will meet this theory in Topic B.2). However, Planck also linked the Kelvin temperature T of a gas to the average translational kinetic energy E_k of a gas particle by

$$E_k = \frac{3}{2} k_B T$$

The value of k_B is $1.381 \times 10^{-23}\ J\,K^{-1}$.

One way to think of the Boltzmann constant is as the conversion factor between Kelvin temperature and average kinetic energy of a gas. It reminds us that temperature is the macroscopic measure that we use to assess the amount of kinetic energy in a substance.

Energy terminology

Heat and heat transfer are the terms used for the energy being transferred into or away from a system through mechanisms such as radiation, conduction or convection. The transfer must involve the surroundings. Therefore, heat is not a property of a single system.

Internal energy can be regarded as the energy difference between the present state of a system and a reference state. The reference state we assume is absolute zero (0 K). With this assumption, the internal energy of a system is the total of the kinetic energies of the entities in the system (the atoms and molecules) and the potential energies of the entities.

Thermal energy is a loose term that is generally taken to relate to the movements of the atoms or molecules within an object or system. These movements can be translational, vibrational or rotational.

How can observations of one physical quantity allow for the determination of another? (NOS)

The Boltzmann factor k_B provides a vital link—both numerical and conceptual—between the average kinetic energy of the particles and temperature. Observing one quantity determines the other.

The factor k_B itself is relatively recent and was introduced by Max Planck in 1900. The ideal gas constant R (which you meet in Topic B.3) was all that could be used to link the macroscopic quantities used to describe a gas.

If you study Topic B.4, you will find another interpretation for k_B which follows the link that Boltzmann himself made (even though he never used a fundamental constant in his ideas). The link here is between entropy and probability. So, again, k_B provides the means to determine the value of one quantity when another has been measured.

Worked example 2

A sample of a gas is heated from an initial temperature of 20 °C. The average kinetic energy of the molecules of the sample is doubled during this heating. Calculate, in °C, the final temperature of the sample.

Solution

The average kinetic energy is directly proportional to the absolute temperature; hence the absolute temperature must have doubled during the heating. The initial absolute temperature is $273 + 20 = 293$ K and the final temperature is $2 \times 293 = 586$ K. This is equivalent to $586 - 273 = 313$ °C.

Worked example 3

Samples of two gases A and B are kept at the same temperature. Molecules of gas A have a greater mass than molecules of gas B. Compare the average speed of the molecules of gas A with that of gas B.

Solution

The average kinetic energy of the molecules depends on the temperature only; hence it is equal for both samples. From $E_k = \frac{1}{2}mv^2$, molecules of sample B are moving at a greater average speed.

Practice questions

2. A sample of a gas is kept at a temperature of 100 °C. The average kinetic energy of the particles of the sample is E. The temperature of the sample is increased to 470 °C. Which is the best estimate of the change in the average kinetic energy of the particles?

   A. E B. $2E$ C. $4E$ D. $5E$

3. Air is a mixture of molecular oxygen (O_2) and molecular nitrogen (N_2). The ratio of molecular masses of O_2 to N_2 is approximately $\frac{8}{7}$.

   What is $\dfrac{\text{average velocity of } O_2 \text{ molecules}}{\text{average velocity of } N_2 \text{ molecules}}$ at a given temperature?

   A. $\dfrac{7}{8}$ B. $\sqrt{\dfrac{7}{8}}$ C. $\sqrt{\dfrac{8}{7}}$ D. $\dfrac{8}{7}$

Measuring energy transfers in temperature and phase changes

When identical amounts of energy are transferred to equal masses of two different solids it is unlikely that the same temperature change will occur in both.

To measure these energy transfers we use two quantities:

- **Specific heat capacity** when a substance changes its temperature but not its phase.

- **Specific latent heat** when a substance changes its phase at constant temperature. Specific latent heat is itself divided by the type of phase change:

  - **Specific latent heat of fusion** (melting) when the substance is changing between a liquid and a solid.

  - **Specific latent heat of vaporization**, when the substance is changing between a liquid and a gas.

Historical terminology

Some of the terms used in this area of physics come from the English language of two centuries ago and need some explanation. "Specific" has an exact meaning in science that is not now used in everyday language. It means "per unit mass". So "specific heat capacity" means the heat capacity of an object that has a mass of one kilogramme. It makes sense to refer the energy transfer to a standard mass for a particular material.

In the same way, the specific energy of a fuel is the energy transferred per unit mass of fuel.

"Latent" is another old word that means "hidden". The energy transfer to a substance as it changes phase is hidden because it does not appear as a temperature change.

The term "heat capacity" is used in this course, but some authors prefer to use the term "thermal capacity". This makes a more direct link between energy transfer and changing temperature. You can expect to see either "specific heat capacity" or "specific thermal capacity" used in books. They mean the same thing.

Specific heat capacity

Transfer 1000 J of energy to 1 kg of water and the temperature will rise by about $\frac{1}{4}$ K. Transfer the same amount of energy to 2 kg of copper and the temperature increase will be roughly 1.3 K. These two materials have different **heat capacities.** When the masses are the same (at 1 kg each), then the temperature change for copper will be about 2.5 K. To make the comparison easily:

The specific heat capacity of a substance is defined as the energy transfer required to raise the temperature of 1 kg of the substance by 1 K.

Algebraically, $c = \dfrac{Q}{m \times \Delta T}$, where Q is the energy transferred, m is the mass of the substance, ΔT is the temperature change and c is the specific heat capacity.

The units of c are $J\,kg^{-1}\,K^{-1}$ or $J\,kg^{-1}\,deg^{-1}$.

This equation is commonly written as $Q = mc\Delta T$

Using copper and water as examples, the energy, mass and temperature change above mean that the specific heat capacity for water c_{water} is about $\dfrac{1000}{1 \times 0.25}$, which is $4000\,J\,kg^{-1}\,K^{-1}$ and the value for copper c_{copper} is $\dfrac{1000}{2 \times 1.3}$, which is $380\,J\,kg^{-1}\,K^{-1}$. Table 1 shows typical values of specific heat capacity for some solids and liquids.

Units

You should avoid writing $J\,kg^{-1}\,°C^{-1}$ because a Celsius temperature is a point on the temperature scale not a temperature difference. Determinations of specific heat capacity always involve temperature change.

Substance	Specific heat capacity / $J\,kg^{-1}\,K^{-1}$
copper	380
iron	410
diamond	510
glass	840
wood	~2000
ethanol	2400
water	4200

▲ Table 1 Typical values of the specific heat capacity of some substances.

Patterns and trends — Water

Water has one of the highest specific heat capacities of all common liquids and solids (ammonia is another notable anomaly). This has a significant impact on the water-based life forms of Earth. A large specific heat capacity implies that a large energy is required to change the temperature of the organism by a small amount. This large value helps animals to maintain a more-or-less constant temperature to survive.

In a similar way, the large quantities of water in liquid that exist in the atmosphere can buffer temperature changes because large amounts of energy are required to change the air temperature significantly.

The value of c for water is large due to an unusual bonding between a hydrogen atom in one water molecule and the oxygen atom in another molecule. This linking is known as a hydrogen bond.

▲ Figure 7 Despite being almost in the Arctic Circle, Iceland has a relatively mild climate due to it being surrounded by ocean water with a high specific heat capacity.

Measuring the specific heat capacity of a solid

- Inquiry 1: Appreciate when and how to maintain constant environmental conditions of systems.

- Inquiry 2: Collect and record sufficient relevant quantitative data.

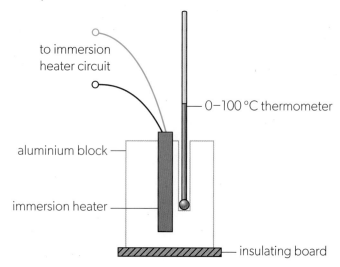

▲ **Figure 8** Measuring the specific heat capacity of an aluminium block.

Energy is transferred to a metal block (in this case, aluminium) from a coil of resistance wire (an immersion heater) inserted into it. When the current in the wire is I and potential difference across the wire is V, the energy supplied every second to the wire (the power) is VI. Energy is supplied to the coil for a time t and this causes

the temperature of the block to rise. This temperature change $\Delta\theta$ is measured with a thermometer. This is shown as a mercury-in-glass thermometer inserted into the block here, but it could also be a data-logging temperature sensor.

The analysis is straightforward:

- Energy transferred to the block (and the thermometer and heater) $= VIt$.

- Energy absorbed by block $= mc\Delta\theta$, where c is the specific heat capacity of the aluminium.

- Therefore, $c = \dfrac{VIt}{m\Delta\theta}$.

There are some errors and assumptions to think about.

- The energy does not spread instantly from heater to block. It is important to wait until the recorded temperature is at its greatest. (You can tell when the value starts to drop.)

- The block is placed on an insulating board to prevent energy loss to the surface below the block.

- You can insulate the sides of the block to prevent energy transfer to the air surrounding the block.

- If you know the specific heat capacity and the mass of the thermometer and the heater, then you can allow for the energy transfer to them too. Normally, however, these are ignored with this apparatus.

 How can the phase change of water be used in the process of electricity generation?

When energy is transferred to water, first its temperature increases and then it changes into the vapour phase. The steam can do useful work by rotating a turbine linked to an electrical generator (see Topics B.3 and B.4). The use of steam was historically important both as a stimulus to scientists and engineers investigating steam engines and also as a driver of societal change.

B. The particulate nature of matter

Measuring the specific heat capacity of a liquid

- Tool 1: Understand how to accurately measure mass and temperature to an appropriate level of precision.

- Inquiry 1: Pilot methodologies.

- Inquiry 3: Identify and discuss sources and impacts of random and systematic errors.

You can use a technique called the **method of mixtures** to measure heat capacities, as shown in Figure 9. Two substances both of known mass but at different initial temperatures are mixed. The final temperature after the mixing is determined. When the specific heat capacity of one substance is known, the energy transferred to the other substance can be determined.

A known mass of water is placed in a container together with a thermometer. A block of metal of known mass m_m and specific heat capacity c_m is placed in boiling water for long enough to ensure that the entire block is at 100 °C. The metal is then removed from the boiling water, quickly dried on paper tissue to remove droplets of water and then immediately transferred to the water in the container. The thermometer is read when the temperature of the mixture has reached its maximum.

The calculation is straightforward:

energy transferred *from* the block = energy transferred *to* the water

$$m_m \times c_m \times (100 - T_{final}) = m_w \times c_w \times (T_{final} - T_{initial})$$

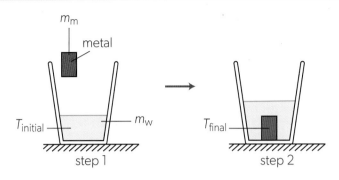

▲ Figure 9 The method of mixtures with a solid and a liquid.

leading to $c_w = \dfrac{m_m \times c_m \times (100 - T_{final})}{m_w \times (T_{final} - T_{initial})}$

where $T_{initial}$ is the starting temperature of the water before the addition of the block and T_{final} is the final temperature of the water and block together.

However, the estimation of T_{final} is not so easy because, in practice, energy is transferred to the surroundings and the container. One way to allow for this is to design the experiment so that $T_{initial}$ begins as far below room temperature as it ends above it. Then, to a fair approximation, the energy transferred into the mixture in the first half of the experiment is equal to the energy transferred out in the second half. A preliminary run is normally required to estimate what m_w needs to be.

Measuring the efficiency of a kettle

- Tool 1: Understand how to accurately measure mass, time, temperature, electric current and electric potential difference to an appropriate level of precision.

- Tool 3: Select and manipulate equations.

- Inquiry 2: Collect and record sufficient relevant quantitative data.

You need a mains power meter that can measure the current and that attaches to a plug socket and an electric kettle (or a portable electric hob and a saucepan).

- Measure the mass of a quantity of water (about 1 kg would be suitable) and put it in the kettle.

- Use a thermometer to measure the initial temperature of the water.

- Switch the kettle on until the water boils. Measure the time that this takes. While the water is boiling, note the energy being transferred to the kettle.

- Once the water has boiled, measure the final temperature of the water.

- Determine the energy supplied from the mains. Use the temperature difference, the mass of water and the specific heat capacity of water ($c = 4200$ J kg^{-1} K^{-1}) calculate the thermal energy transferred to the water. Hence calculate the efficiency of the kettle.

Note: This experiment could also be carried out using a microwave and a bowl of water. The microwave should have a power rating (there may be different settings with different powers). Compare this rated power to the measured heating of the water.

Worked example 4

In the following examples take the specific heat capacity of water to be $4200\,\mathrm{J\,kg^{-1}\,K^{-1}}$.

1. Energy is transferred from an electric heater to a metal sample of mass 1.2 kg. The graph shows how the temperature change ΔT of the sample varies with the energy Q absorbed by the sample.

    a. Determine the specific heat capacity of the sample.

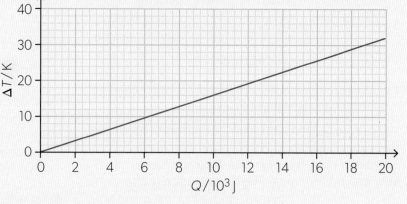

    When the metal sample reaches the temperature of 100 °C, it is transferred to a thermally insulated container filled with 5.0 kg of water at an initial temperature of 20 °C.

    b. Determine the final equilibrium temperature of the system.

Solutions

a. From $Q = mc\Delta T$, the gradient of the line is $\dfrac{\Delta T}{Q} = \dfrac{1}{mc}$. The value of the gradient can be obtained directly

from the graph, $\dfrac{\Delta T}{Q} = \dfrac{32}{20 \times 10^3} = 1.6 \times 10^{-3}\,\mathrm{K\,J^{-1}}$.

$\dfrac{1}{mc} = 1.6 \times 10^{-3}\,\mathrm{K\,J^{-1}} \Rightarrow c = \dfrac{1}{1.2 \times 1.6 \times 10^{-3}} = 520\,\mathrm{J\,kg^{-1}\,K^{-1}}$

b. Energy lost by the metal is equal to energy gained by the water.

$1.2 \times 520 \times \Delta T_{metal} = 5.0 \times 4200 \times \Delta T_{water} \Rightarrow \Delta T_{metal} = 33.6 \times \Delta T_{water}$.

On the other hand, $\Delta T_{metal} + \Delta T_{water} = 80\,\mathrm{K}$. Combining the equations gives $33.6\Delta T_{water} + \Delta T_{water} = 80\,\mathrm{K}$

$\Rightarrow \Delta T_{water} = \dfrac{80}{34.6} = 2.3\,\mathrm{K}$. The temperature of the water has increased by 2.3 K, so the final equilibrium

temperature is 22.3 °C.

Worked example 5

Energy is supplied to a mass of 300 g of water at a constant rate of 600 W. The temperature of the water increases by 25 K in 1.0 minute.

Calculate:

a. the increase in internal energy of the water

b. the average power transferred by the water to the surroundings.

Solutions

a. The increase in internal energy is proportional to the increase in temperature.
    $Q = mc\Delta T = 0.300 \times 4200 \times 25 = 31.5\,\mathrm{kJ}$

b. The power required to increase the internal energy of the water is $\dfrac{31.5 \times 10^3}{60} = 525\,\mathrm{W}$. This is less than the

    total power supplied to the water, and the remaining power is transferred to the surroundings:
    $600 - 525 = 75\,\mathrm{W}$.

Practice questions

4. a. Discuss, with reference to molecular behaviour, the process of mixing two liquids of different initial temperatures.

    b. A piece of metal of mass 100 g at a temperature of 90 °C is dropped into a thermally insulated flask containing 150 g of a liquid at a temperature of 20 °C. The liquid is stirred until it reaches the final temperature of 30 °C.

    What is $\dfrac{\text{specific heat capacity of the liquid}}{\text{specific heat capacity of the metal}}$?

    A. 2   B. 4   C. 6   D. 9

5. Two samples of the same material are allowed to reach thermal equilibrium without exchanging thermal energy with the surroundings. The graph shows how the temperature of each sample varies with time.

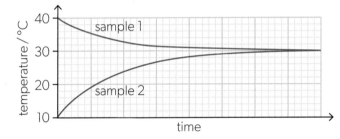

    The mass of sample 2 is 600 g. What is the mass of sample 1?

    A. 200 g   B. 300 g   C. 900 g   D. 1200 g

6. A block of metal of mass 300 g transfers 8.3 kJ of thermal energy to the surroundings. The temperature of the block decreases by 60 K. Calculate the specific heat capacity of the block.

7. Two metal samples A and B of equal masses in contact are allowed to reach thermal equilibrium. The following data are given:

    Specific heat capacity of sample A = 920 J kg^{-1} K^{-1}
    Initial temperature of sample A = 20 °C
    Initial temperature of sample B = 250 °C
    Final equilibrium temperature of the system = 87 °C
    Thermal energy losses to the surroundings are negligible.
    Calculate the specific heat capacity of sample B.

8. 640 J of thermal energy is transferred to a lead ball of mass 80.0 g at an initial temperature 22.0 °C. The specific heat capacity of lead is 127 J kg^{-1} K^{-1}.

    a. Calculate the final temperature of the lead ball.

    The ball is now transferred to a thermally insulated container filled with 160 g of cold water. The ball and the water reach thermal equilibrium.

    b. Estimate the ratio $\dfrac{\text{temperature change of the water}}{\text{temperature change of the lead ball}}$

9. An electric kettle contains 0.90 kg of water at an initial temperature of 20 °C. The kettle supplies a power of 2.2 kW to the water.

    a. Calculate the time required to increase the temperature of the water to 95 °C. Ignore any thermal energy losses.

    b. The actual time taken to increase the temperature of the water to 95 °C is 150 s. Calculate the power transferred by the water to the surroundings.

10. Hot water enters a wall-mounted radiator at a temperature 65 °C and leaves it at a temperature 30 °C. The mass of water flowing through the radiator each minute is 0.50 kg. Determine the power that the radiator transfers to the room.

Specific latent heat

The **specific latent heat** L for a phase change is the amount of energy Q transferred when changing one kilogramme of the substance from one phase to another. Algebraically:

$$L = \frac{Q}{m}$$

where m is the mass of the substance.

The units of specific latent heat are $J\,kg^{-1}$. Notice that there is no reference to K in the unit because there is no temperature change.

This equation can also be written as $Q = mL$

To help you to think about all these energy-transfer quantities, imagine a small block of ice removed from a freezer. Energy is transferred to the block at a constant rate until the ice has changed first into liquid water and then into water vapour. Figure 10 shows how the temperature of the block changes with time.

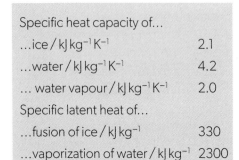

Specific heat capacity of…	
…ice / $kJ\,kg^{-1}\,K^{-1}$	2.1
…water / $kJ\,kg^{-1}\,K^{-1}$	4.2
… water vapour / $kJ\,kg^{-1}\,K^{-1}$	2.0
Specific latent heat of…	
…fusion of ice / $kJ\,kg^{-1}$	330
…vaporization of water / $kJ\,kg^{-1}$	2300

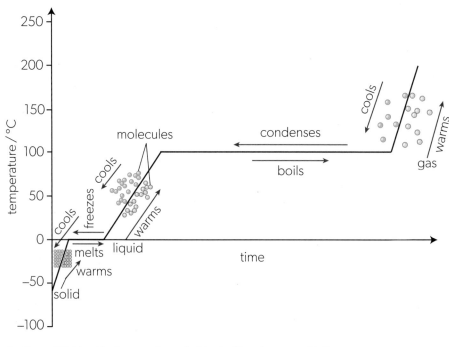

▲ Figure 10 How the temperature of a block of ice changes with time.

Energy is being transferred to the water at a constant rate, so the x-axis could also be "energy supplied". There would be no change in the shape of the graph.

The horizontal parts of the graph correspond to the temperatures at which latent heat (phase) changes occur. Even though energy is being supplied, the temperature is constant:

- from the instant the ice begins to form water until it is all completely melted

- from the instant the water begins to boil until it is completely vaporized.

The relative lengths of these sections give you an indication of the comparative sizes of the specific latent heats of fusion and vaporization.

The sections of the graph that have positive gradients correspond to heat-capacity changes where temperature is increasing without change of state. Again, the value of the gradient gives a relative sense of the values of the three specific heat capacities. Study the graph and link the data to the various parts of the graph.

Measuring the specific latent heat of the fusion of ice

- Tool 3: Use basic arithmetic and algebraic calculations to solve problems.

- Tool 3: Select and manipulate equations.

- Inquiry 2: Collect and record sufficient relevant quantitative data.

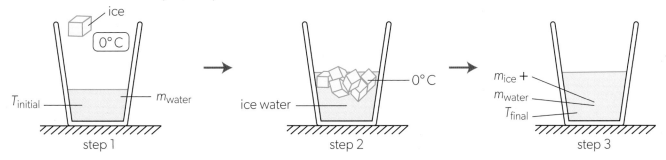

▲ Figure 11 Using the method of mixtures to measure a specific latent heat.

This is another method-of-mixtures where a substance with an unknown specific latent heat (ice, in this case) is added to a substance with a known specific heat capacity (here, water). The basic procedure is shown in Figure 11:

- Step 1. A known mass of ice at 0 °C is added to a known mass of water at a higher temperature.

- Step 2. The ice melts at a temperature of 0 °C. The energy required to melt the ice is transferred from the original water, the temperature of which decreases.

- Step 3. The original water then continues to transfer energy to the melted ice until all the water (original and melted ice) in the container is at the same temperature. This will be somewhere between the initial water temperature and 0 °C.

The water is placed in a container and its initial temperature is measured. The container should have a small mass so as not to absorb too much energy itself. A paper cup is ideal. The ice is dried (on a paper tissue) and added to the container, stirring the original water all the time. The mass of the ice is determined at the end of the

experiment by knowing the difference between the initial mass of water and the final mass of water plus melted ice.

This time the algebra is more complicated as there are two contributions for the ice:

- The energy transferred to the ice to melt it, $m_{ice} \times L$.

- The energy transferred to heat the ice from 0 °C to the final temperature, $T_{final} = m_{ice} \times c_{water} \times (T_{final} - 0)$.

- The original water has energy transferred from it equal to $m_{water} \times c_{water} \times (T_{initial} - T_{final})$ as it cools from $T_{initial}$ to T_{final}.

Equating the energy transfers gives

$$m_{ice} \times L + m_{ice} \times c_{water} \times (T_{final} - 0) = m_{water} \times c_{water} \times (T_{initial} - T_{final})$$

This leads to $L = \left[\dfrac{m_{water}}{m_{ice}} \times c_{water} (T_{initial} - T_{final}) \right] - c_{water} T_{final}$,

although it is easier to evaluate the energy contributions separately rather than use this equation.

How is the understanding of systems applied to other areas of physics?

In physics the word "system" is applied to any self-contained group of interacting objects. In this theme, "system" often applies to a body of gas sealed in a container. The properties of this group of gas particles can be predicted by assuming that they obey the ideal-gas equation (Topic B.3).

We talk about the Solar System, meaning the group of objects that constitute the Sun and all its satellites. There are gravitational interactions between these objects the effects of which can be predicted using Newton's laws of motion and his law of gravitation and the Keplerian laws, as outlined in Topic D.1.

One interpretation of "system" is of a complete entity that self-interacts but that has only limited interaction with the environment beyond it. An example in Topic B.4 is the distinction made between the system under consideration, the surroundings and the universe.

Worked example 6

In an experiment to determine the specific latent heat of fusion of ice, an ice cube of mass 15 g at a temperature 0 °C is dropped into an insulated container filled with 210 g of water at a temperature of 25 °C. The ice melts completely and the final temperature of the system is 18 °C.

a. Calculate the thermal energy:
   i.   transferred from the original water as it cools from 25 °C to 18 °C
   ii. transferred to the molten ice to heat it from 0 °C to 18 °C.
b. Hence, determine the specific latent heat of fusion of ice.

Solutions

a. i. $0.210 \times 4200 \times (25 - 18) = 6.17\,kJ$
    ii. $0.015 \times 4200 \times (18 - 0) = 1.13\,kJ$
b. The energy transferred to melt all the ice is equal to the difference between the energies calculated in part a.

$$L = \frac{Q}{m} = \frac{6.17 - 1.13}{0.015} = 340\,kJ\,kg^{-1}$$

Worked example 7

A piece of ice of mass 40.0 g and temperature −10.0 °C is dropped into 300 g of water at a temperature 18.0 °C.
    Specific heat capacity of ice = 2100 J kg⁻¹K⁻¹
    Specific latent heat of fusion of ice = 334 kJ kg⁻¹

Predict the final equilibrium temperature of the system, ignoring any energy transfers to or from the surroundings.

Solution

The energy required to increase the temperature of the ice to 0 °C and melt it completely is $2100 \times 0.040 \times 10.0 + 334 \times 10^3 \times 0.040 = 14.2\,kJ$

The temperature change of the original water resulting from this energy transfer is $\Delta T = \dfrac{14.2 \times 10^3}{0.300 \times 4200} = 11.3\,K$. Once the ice has melted, the system can be thought of as consisting of 300 g of water at a temperature 18.0 − 11.3 = 6.7 °C and 40 g of water at a temperature 0 °C. The original water will still transfer energy to water that was originally ice, until equilibrium is reached. Energy lost by the original water is equal to energy gained by the original ice,

so $40(T_{final} - 0) = 300(6.7 - T_{final})$. From this, the final temperature is $T_{final} = \dfrac{300 \times 6.7}{300 + 40} = 5.9\,°C$.

Worked example 8

Energy is supplied at a constant rate of 750 W to a sample of ammonia of mass 0.51 kg that is initially in liquid phase. The graph shows how the temperature of the sample varies with time. The sample begins to boil during heating.

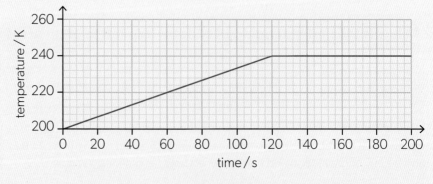

a. State, in K, the boiling temperature of ammonia.
b. Determine the specific heat capacity of ammonia in its liquid phase.
c. The specific latent heat of vaporization of ammonia is 1400 kJ kg⁻¹. Determine the mass of the sample that remains in the liquid phase at a time of 200 s.

Solutions

a. 240 K

b. Thermal energy delivered to the sample during the first 120 s is $750 \times 120 = 90$ kJ and the temperature change is 40 K.

$$90 \times 10^3 = 0.51 \times c \times 40 \Rightarrow c = \frac{90 \times 10^3}{0.51 \times 40} = 4400 \, \text{J kg}^{-1} \text{K}^{-1}$$

c. After 120 s, the sample is at the boiling point and a further $750 \times (200 - 120) = 60$ kJ of thermal energy is delivered to it. The mass of the sample that has boiled away is $\frac{60 \times 10^3}{1400 \times 10^3} = 0.04$ kg. The mass remaining in the liquid phase is $0.51 - 0.04 = 0.47$ kg.

Practice questions

11. In cold climates, the decrease of air temperature during the winter is often slower near lakes and other large bodies of water. Discuss this phenomenon.

12. A mass of 50 g of crushed ice at an initial temperature 0 °C is added to a thermos flask containing water at a temperature of 20 °C.

 a. Calculate the energy required to melt the ice.

 b. Determine the minimum mass of water that must be present in the flask to melt the ice completely.

 c. Describe the final state of the system if the flask initially contains **less** water than the value you have calculated in b.

13. A thermally insulated container has 0.50 kg of water at a temperature 30 °C. Determine the mass of ice that must be added to the container to decrease the temperature to 10 °C, when the ice is initially at:

 a. 0 °C

 b. −18 °C.

14. Tin–lead alloy used for soldering has the following properties:

> Melting temperature = 190 °C
> Specific heat capacity = 210 J kg^{-1} K^{-1}
> Specific latent heat of fusion = 52×10^3 J kg^{-1}

 A soldering iron has a power of 45 W. Calculate the minimum time required to melt a 5.0 g sample of tin–lead solder that is initially at 20 °C.

15. A metal sample is heated in a furnace. The graph shows how the temperature of the sample varies with time t. The sample starts melting at $t = 10$ s.

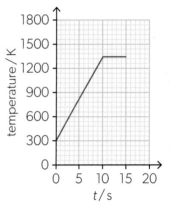

 The following data are given:

> Power of the furnace = 1000 W
> Melting temperature of the metal = 1340 K
> Specific heat capacity of the metal = 126 J kg^{-1} K^{-1}

 a. Estimate the mass of the metal in the furnace.

 b. The sample melts completely when $t = 15$ s. Calculate the specific latent heat of fusion of the metal.

 c. The heating of the sample continues. The specific heat capacity of the metal in the liquid phase is greater than the specific heat capacity of the metal in the solid phase.

 Outline how the temperature of the sample is changing for $t > 15$ s.

Cooling

This is a term that needs care. Throughout this discussion of specific heat capacity and specific latent heat there is a careful distinction between temperature change (heat capacity) and constant temperature conditions (phase change). The term "cooling" only has the meaning of a temperature decrease in physics. In everyday life we are less careful in using the term. On a hot day you might say that you need to "cool down" but what you mean is that you need to transfer more energy from your body to maintain the correct temperature even though you feel "hot". This is a physiological, not a physical, effect.

When water is in the process of freezing to ice, it is not cooling. Its temperature is constant (at 273 K) and energy is being transferred from it.

Should language vary between scientific practice and everyday life?

Absolute temperature

Absolute temperature is equivalent to the average translational kinetic energy of the molecules of a gas using the conversion

$$\overline{E_k} = \frac{3}{2} k_B T.$$

You study electrical conduction in Topic B.5.

Thermal energy transfer

Earlier in this topic you saw that an object with a temperature above absolute zero possesses internal energy that is due to two contributions:

- the random motion of its atoms and molecules
- their intermolecular potential energy.

The higher the temperature of the object, the greater the internal energy associated with the molecules.

Energy spontaneously transfers from a region at a high temperature to a region at a low temperature.

There are three ways in which this **thermal energy transfer** can be achieved:

- conduction
- convection
- thermal radiation.

All are important to us on both an individual level and in global terms.

▲ Figure 12 Ice is often put into drinks. When the ice starts at a temperature of less than 0 °C, then adding more ice to a drink will result in less ice melting.

Thermal conduction

Conduction can occur in a thermal sense (thermal conduction) and in an electrical sense (electrical conduction), but it is normally just shortened to 'conduction'.

Everyone has experienced practical conduction in some way. Burning a hand on a camping stove, plunging a hot metal into cold water which then boils, or melting ice in the hand all give the experience of energy moving by conduction from a hot source to a cold sink.

Metals are excellent thermal conductors, just as they are also good electrical conductors. Poor thermal conductors such as glass or some plastics also conduct electricity poorly. This suggests similarities between the mechanisms that lead to both types of conduction. However, you should note that there are still considerable differences in scale between the best metal conductors (copper, gold) and the worst metals (brass, aluminium). There are many laboratory experiments that you can carry out to determine the different thermal properties of good and poor conductors. Figure 13 shows just two demonstrations of conduction.

In conduction processes, energy transfers through the bulk of the material without any large-scale relative movement of the atoms that make up the solid. Thermal conduction and electrical conduction are collectively known as **transport phenomena**.

vibrating ions transfer
energy to free electrons

free electrons transfer
energy through the metal

hot

cold

○ atom ● electron

▲ Figure 14 A good conductor heated to a high temperature at one end soon transfers energy along its length.

Atomic vibration occurs in all solids, both metals and non-metals. At all temperatures above 0 K, the ions in the solid have an internal energy. They are vibrating about their average fixed position in the solid. The higher the temperature, the greater is their average kinetic energy, and therefore the higher their mean speed.

Imagine a metal rod heated at one end and cooled at the other (see Figure 14). At the high-temperature end, the ions have a larger average kinetic energy than at the low-temperature end. The ions transfer energy to the free electrons, which re-distribute this energy along the rod. The ion with the smaller energy tends to gain energy, and the other one loses energy in the electron collisions. There is a transfer of internal energy along the metal rod until the whole of the metal rod is at the same temperature.

Conduction can occur in gases and liquids as well as solids. However, the inter-atomic connections are weaker and the gas atoms are about ten times further apart than in solids and liquids. Thus, conduction is much less important than convection in many fluids.

The use of the words "source" and "sink" link to the same usage in Topic B.4 where they have distinct meanings: source means a provider of energy capable of transferring energy to an absorber (sink) of the energy at a lower temperature.

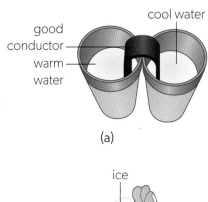

cool water

good
conductor

warm
water

(a)

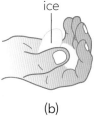

ice

(b)

▲ Figure 13 Examples of conduction.
(a) A good conductor is able to transfer energy from a hot object to a cold one.
(b) Ice held in the hand soon chills the fingers.

Models

This section on conduction moves between microscopic and macroscopic descriptions of the phenomenon. Both sets of descriptions are equally valid. The concept of energy transfer and the idea of thermal equilibrium with a balance of incoming and outgoing energy is shared by both models.

 ## What role does the molecular model play in understanding other areas of physics? (NOS)

Although thermal conduction by atomic vibration is universal in solids, there are other conduction processes that vary in importance depending on the type of solid. This is one example of a physical model that occurs in several areas of physics. We use the atoms and ions of the substance in discussing thermal conduction. The agent for electrical conduction is different but the ideas lying behind the model are the same.

Electrical conductors have a covalent (or metallic) bonding that releases **free electrons** to form what is essentially an electron gas filling the whole of the interior of the solid. These free electrons are in thermal equilibrium with the positive ions that make up the atomic lattice of the solid. The electrons can interact with each other and the energy from the high-temperature end of the solid "diffuses" along the solid by interactions between these electrons. When an electron interacts with an atom, energy is transferred back into the atomic lattice to alter the vibrational state of the atom. This free-electron mechanism for conduction depends critically on the numbers of free electrons available to the solid. Good electrical conductors, where there are many charge carriers (free electrons) available per unit volume, are likely also to be good thermal conductors. For example, in copper there is one free electron per atom. You should be able to use Avogadro's number, the density of copper and its relative atomic mass, to show that there are 8.4×10^{28} electrons in one cubic metre of copper.

Analogies are often used in science to aid our understanding of phenomena. Electrical and thermal conduction are also closely linked in terms of their mathematical descriptions. You can read more about this in Topic B.5 on page 300.

 ## Data-based questions

Since free electrons have a role in thermal conductivity, it should not surprise you that thermal conductivity is related to electrical conductivity. The table below gives the thermal conductivity k and resistivity ρ of some metals.

You will meet resistivity in Topic B.5. For this question, you only need to appreciate that a lower resistivity means that the metal is a better electrical conductor.

Metal	$\rho / 10^{-8}\,\Omega\,m$	$k / W\,m^{-1}\,K^{-1}$
aluminium	2.7	240
copper	1.7	400
gold	2.2	320
lead	21	35
magnesium	4.4	160
platinum	11	72
silver	1.6	430
titanium	42	22

- It is suggested that $\rho \propto k^{-1}$. Plot a suitable graph to verify whether metals obey this relationship.

- The values given in the table are to the nearest 10%. Add appropriate error bars to your graph.

- Use your graph to determine the constant of proportionality. Include an uncertainty with your constant.

Note that this relationship only applies to metals. Diamond, for example, not only has a very high thermal conductivity ($k \approx 2000\,W\,m^{-1}\,K^{1}$) but a high resistivity as well ($\rho \geq 10^{13}\,\Omega\,m$).

Thermal conductivity

Conduction is an important factor in the design of many engineering projects. It is important to minimize the transfer of energy from buildings in parts of the world with cold winters. Equally, a good design will maximize the transfer of energy in heat exchangers for power stations of all types. Engineers must be able to quantify the amount of energy conducted through different materials. This is done by defining a quantity known as **thermal conductivity** which is a measure of how good a thermal conductor is at transferring energy through itself when in steady state. **Steady state** is when the temperature at any point in the slab does not change with time. The defining word equation is

$$\text{thermal conductivity} = \frac{\text{rate of energy transfer}}{\text{area of material} \times \text{temperature gradient across conductor}}$$

or *rate of energy transfer = thermal conductivity × area of material × temperature gradient across conductor*, which in symbols (see Figure 15(a)) is

$$\frac{\Delta Q}{\Delta t} = k \times A \times \frac{\Delta T}{\Delta x}$$

where:

- an energy ΔQ is transferred across the material in a time Δt

- through an area A

- when there is a temperature difference $\Delta T = T_1 - T_2$ across the conductor that has a length Δx.

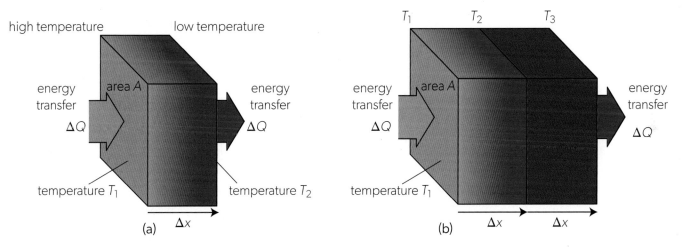

▲ Figure 15 Energy flow through a conductor of area A and length Δx. The energy transfer is ΔQ in time Δt and leads to a temperature difference of ΔT.

It can be seen intuitively that this equation makes sense because, if you:

- double the area of the conductor, there will be twice the energy transferred in time t for the same temperature difference ΔT and Δx

- double the time for the transfer, then the total energy transfer will double (everything else remains unchanged).

Imagine one slab joined to another identical slab (Figure 15(b)) with the same energy transfer through both. Then the temperature difference ΔT is the same individually across each slab, which leads to a temperature difference of $2\Delta T$ across a width of $2\Delta x$. In other words, $T_1 - T_2 = T_2 - T_3$.

Units for *k*

- Tool 3: Work with fundamental units.

Take care when writing the units of thermal conductivity. Writing W/mK with a solidus and the m close to the K means something quite different.

Other technical language is possible: $\frac{\Delta Q}{\Delta t}$ is the rate of energy transfer, in other words, the power input to the slab. For a steady state, this is also the power output from the slab. This means that there must be no energy transfer out of the sides of the slab. (This has important implications for experiments involving thermal-conductivity measurements later.)

Re-writing the definition leads to

$$k = \frac{\Delta Q}{\Delta t} \times \frac{\Delta x}{A \Delta T}$$

The units of *k* are therefore $W\,m^{-1}\,K^{-1}$ (W comes from $J\,s^{-1}$ as usual).

Thermal conductivity values vary widely as you might expect given the range of behaviour from the best conductors to the worst insulators. Table 2 gives you an idea of the range involved.

Substance	$k\,/\,W\,m^{-1}\,K^{-1}$
silver	430
copper	400
iron	80
glass	0.5–0.8
water	0.55–0.7 over range 0–100 °C
air	0.026

▲ Table 2 Thermal conductivity values for a range of substances. These are measured at room temperature, except where specified.

Worked example 9

A glass window of surface area $1.2\,m^2$ is made of a single glass pane of thickness 5.0 mm and thermal conductivity $0.50\,W\,m^{-1}\,K^{-1}$. The outside temperature is +5 °C and the inside temperature is +20 °C.

a. Calculate the rate of energy transfer through the window.

b. In cold climates, windows are usually made of two or more glass panes separated by an air space. Outline what effect the air space has on the rate of energy transfer through the window.

Solutions

a. The temperature difference is 15 K. $\frac{\Delta Q}{t} = 0.50 \times 1.2 \times \frac{15}{5.0 \times 10^{-3}} = 1.8\,kW$

b. Air has a much lower thermal conductivity than glass. Therefore, the air space between the glass panes greatly reduces the rate of energy loss through the window.

Worked example 10

A layer of ice of a uniform thickness 7.0 cm has formed on the surface of a lake. The temperature of the air above the ice is −12 °C and that of the water below the ice is 0 °C.

a. Calculate the rate of thermal energy transfer per unit area through the ice. The thermal conductivity of ice is $2.1\,W\,m^{-1}\,K^{-1}$.

b. Calculate the mass of water that freezes during one hour below one square metre of the ice. The specific latent heat of fusion of ice is $334\,kJ\,kg^{-1}$.

c. Hence calculate, in mm per hour, the rate of change of thickness of the ice. The density of ice is $920\,kg\,m^{-3}$.

Solutions

a. $\frac{\Delta Q}{\Delta t} = 2.1 \times \frac{12}{7.0 \times 10^{-2}} = 360\,W\,m^{-2}$.

b. The energy transferred from the water through one square metre of the ice in one hour is $360 \times 60 \times 60 = 1.3 \times 10^6\,J$. This loss of thermal energy causes water to freeze, and the mass frozen in one hour can be calculated using the equation $Q = mL$.

$$m = \frac{Q}{L} = \frac{1.3 \times 10^6}{334 \times 10^3} = 3.9\,kg$$

c. Let *d* be the thickness of the additional ice that forms during one hour and $A = 1.0\,m^2$, the surface area under consideration.

$$\text{density} = \frac{\text{mass}}{A \times d} \Rightarrow d = \frac{\text{mass}}{A \times \text{density}} = \frac{3.9}{1.0 \times 920} =$$

$4.2 \times 10^{-3}\,m$. The ice is being formed at a rate of 4.2 mm per hour.

Note that, assuming constant air temperature, the rate of ice formation decreases with time, because as the thickness of the ice grows, the temperature gradient through it decreases.

Practice questions

16. On a winter day, the outside temperature is −5 °C and the temperature inside a house is 20 °C. Calculate the rate of thermal energy loss per unit area of a window, when the window is made of:

    a. a single glass pane of thickness 4.0 mm and thermal conductivity 0.50 W m⁻¹ K⁻¹.

    b. two parallel panes separated by 1.0 cm of air of thermal conductivity 0.023 W m⁻¹ K⁻¹.

    Assume that conduction is the only energy transfer mechanism and that the glass surfaces in contact with the air have temperatures −5 °C and 20 °C.

17. One end of a thin cylindrical rod made of pure aluminium is kept in a mixture of water and ice at 0 °C and the other end is maintained at a constant temperature of 300 °C. The length of the rod is 60 cm and its diameter is 2.0 cm. The rod is insulated and energy losses through its cylindrical surface are negligible.

    a. Calculate the rate at which thermal energy is transferred through the rod. The thermal conductivity of aluminium is 240 W m⁻¹ K⁻¹.

    b. Calculate the mass of the ice that melts during one minute.

Measuring thermal conductivity—A study in experimental design

- Tool 3: Determine rates of change.
- Inquiry 1: Appreciate when and how to insulate against heat loss or gain.

The way in which k is measured for a material depends on whether the material is a good or poor conductor. k must be measured when the material is in a steady state. The energy transfer rate $\left(\frac{\Delta Q}{\Delta t}\right)$ and the temperature gradient $\left(\frac{\Delta T}{\Delta x}\right)$ need to be measured with as small a fractional uncertainty as possible.

For a *good conductor*, the temperature gradient across it is likely to be small. This means that, to be able to measure a reasonable temperature difference, x must be large. It also helps when A is small $\left(\text{because } k = \frac{\Delta Q}{\Delta t} \times \frac{\Delta x}{A\Delta T}\right)$. The optimum shape for the conductor is therefore a long, thin

cylinder. This shape has a large surface area, which means that energy will transfer easily through the sides of the cylinder unless this is prevented. This leads to a method first developed by Searle and known as the Searle's bar experiment.

The criteria are different for a *poor conductor*. Here the power transfer will be small, and is maximized by having a large area for the conductor. Even a thin specimen has a large temperature change across it, but a thin poor conductor will not be prone to energy loss from the sides. This leads to a method for measuring k for a poor conductor developed by Lee and known as the Lee's disc experiment.

Notice how both scientists used good experimental design to arrive at different solutions that were matched to the experiments they were trying to perform.

Searle's bar—k for a good conductor

- Tool 3: Calculate areas and volumes for simple shapes.
- Tool 3: Carry out calculations involving fractions.
- Inquiry 2: Collect and record sufficient relevant quantitative data.
- Inquiry 3: Discuss the impact of uncertainties on the conclusions.

The energy flowing through the bar is estimated by knowing the power supplied to the water-cooled coil at the right-hand end. This is $m'c_{water}(T_3 - T_4)$ where m' is the mass of water entering the cooled coil every second. The temperature gradient of the material is estimated from the difference between the thermometers. This is $\frac{(T_1 - T_2)}{L}$.

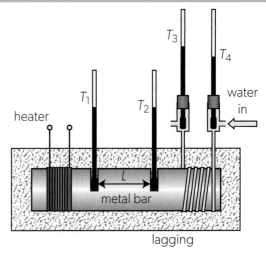

▲ Figure 16 Searle's bar apparatus.

You should be able to show that

$$k = m' c_{water} (T_3 - T_4) \times \frac{1}{A} \times \frac{L}{(T_1 - T_2)}$$

where A is the area of the bar.

The apparatus is set up and allowed to reach steady state. Take all the readings when they have reached constant values.

Energy is lost from cylindrical sides of the bar is reduced by lagging the bar.

•   The bar diameter should be measured carefully using vernier callipers.

• m' should be measured over a long time period to reduce its fractional uncertainty.

Lee's disc — k for a poor conductor

•   Tool 3: Determine rates of change.
•   Tool 3: Construct and interpret tables and graphs for raw and processed data including scatter graphs and line and curve graphs.
•   Inquiry 2: Carry out relevant and accurate data processing.

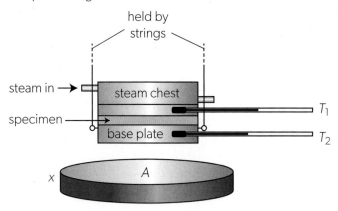

▲ Figure 17 Lee's disc method for a poor conductor.

The specimen has a disc shape of thickness x and area A. The energy transfer through the poor conductor is hard to estimate and is usually evaluated graphically.

Steam is passed through a steam chest at 100 °C. The whole apparatus is allowed to reach a steady state, shown by unchanging thermometer readings, to give the temperature difference across the specimen and the temperature gradient of $\frac{(T_1 - T_2)}{x}$.

The specimen is then removed so that the steam chest sits on top of the base plate with nothing between them. The temperature of the base plate is then allowed to rise well above T_2. The steam chest is removed and replaced by an insulating felt pad that prevents energy loss from the upper surface of the plate. A graph of temperature against time is recorded from just above to just below the original temperature T_2 of the base plate. This enables the rate of temperature change $\frac{\Delta T}{\Delta t}$ to be calculated at T_2 and hence (using its specific heat capacity) the energy transfer rate through the disc.

The thermal conductivity equation becomes

$$mc_{plate} \frac{\Delta T}{\Delta t} = kA \frac{(T_1 - T_2)}{x}$$

leading to

$$k = \frac{x}{(T_1 - T_2)} \times \frac{mc_{plate}}{A} \times \frac{\Delta T}{\Delta t}$$

Practice questions

18.   The thermal conductivity of a metal bar is investigated using the apparatus shown in Figure 16. Water flows through the cooling coil at a steady rate of 0.300 kg per minute. Water enters the coil at a temperature of 11.0 °C and leaves it at a temperature of 17.2 °C.

   a.   Calculate, in W, the rate of energy transfer from the metal bar to the water.

   The cross-sectional area of the bar is $1.25 \times 10^{-3} \, m^2$. The steady-state temperatures of the bar, measured at two points separated by 0.100 m, are 76.5 °C and 49.2 °C.

   b.   Calculate the thermal conductivity of the metal bar.

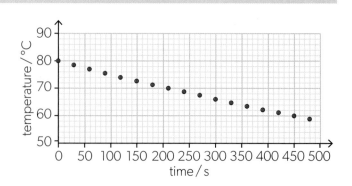

19. Consider the apparatus shown in Figure 17. A thin specimen of cardboard is placed between the metal plates. In a steady state, the plates reach temperatures $T_1 = 91\,°C$ and $T_2 = 70\,°C$. When the cardboard is removed, the temperature of the base plate quickly rises above the steady-state value. The steam chest is then removed and the base plate is allowed to cool. The graph shows how the temperature of the base plate varies with time.

a. Estimate the gradient of the temperature–time graph, at the instant when the temperature of the base plate is $70\,°C$.

b. Hence, calculate the rate at which the thermal energy is removed from the base plate. The mass of the base plate is 0.55 kg and its specific heat capacity is $380\,J\,kg^{-1}\,K^{-1}$.

c. Determine the thermal conductivity of the specimen of cardboard. The specimen has thickness 2.0 mm and surface area $8.0 \times 10^{-3}\,m^2$.

Convection

Convection is the movement of groups of atoms or molecules within fluids (liquids and gases) because of variations in density. Unlike conduction, which involves the microscopic transfer of energy, convection is a bulk property and is described in macroscopic ways. Convection cannot take place in solids. An understanding of convection is important in many areas of physics, astrophysics and geology. In some hot countries, houses are designed to take advantage of natural convection to cool them down in hot weather.

Examples of convection

Figure 18 shows three experiments that involve convection. In all three cases, energy is supplied to a fluid. In Figure 18(a), a candle heats the air underneath a tube (a chimney) that leads vertically out of the box. The air molecules immediately above the flame move further apart decreasing the air density in this region. With a reduced density compared with the surrounding air, these molecules experience an upthrust and move up through the left-hand chimney.

This upward air movement reduces the pressure in the box slightly and causes cooler air to be pulled down the right-hand chimney. Further heating of the air above the flame leads to a continuous current of cold air down the right-hand chimney and hot air up the left-hand tube. This is a **convection current**.

Similar currents can be demonstrated in liquids. Figure 18(b) shows a small crystal of a soluble dye (potassium permanganate, $KMnO_4$) placed at the bottom of a beaker of water. When the base of the beaker is heated gently near to the crystal, water at the base heats, expands becoming less dense, and rises.

There is also a convection current in Figure 18(c) where a glass tube, in the shape of a rectangle, again with a small soluble coloured crystal in the tube, can sustain a convection current that moves all around the tube.

A convection current is the mechanism through which all the water heated in a saucepan eventually reaches a uniform temperature. There are many examples of convection in action. Figure 19 shows examples from the natural world. There are many others.

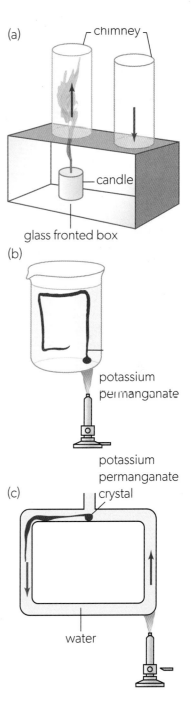

(a)

(b)

(c)

▲ Figure 18 Convection in a gas and a liquid.

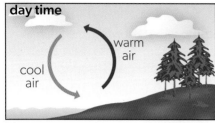

(a)

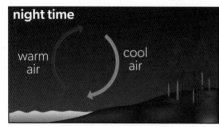

(b)

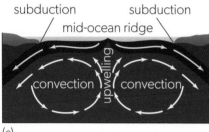

(c)

▲ Figure 19 (a) and (b) Changes in land and sea temperatures account for on- and off-shore convected breezes at different times. (c) Convection currents in Earth's mantle.

Sea breezes

The direction of the breeze near an ocean changes during a 24-hour period. During the day, breezes blow on-shore from the ocean to the land. At night, the direction is reversed, and the breeze blows off-shore.

Convection effects explain this (Figure 19(a) and (b)). During the day, the land is warmer than the sea. The air above the land expands so that the fluid density of the air decreases compared with that of the cooler air above the ocean. This warm air rises over the land mass, pulling in cooler air from above the ocean. At night, the land cools down much more quickly than the sea (the variation in sea temperature over one day is less). Now the warmer air rises from the sea, so the wind blows off-shore. (You might like to use your knowledge of specific heat capacity to explain why the sea temperature varies much less than that of the land.)

Convection in Earth

At the bottom of the Atlantic Ocean, and in other places on the planet, new crust is being created (Figure 19(c)). This is due to convection effects that occur below the surface. Earth's core is at a high temperature and drives convection effects in the upper mantle just below the surface. Two convection currents operate and drive material in the same direction. Material is upwelling at the top of these currents to reach the surface of Earth at the bottom of the ocean. This creates new land that is forcing the Americas, Europe and Africa apart at the rate of a few centimetres every year. In other parts of the world, convection currents are pulling material back down below the surface (subduction). These convection currents, over time, have produced the continental drift that has shaped the continents that we know today.

Why winds blow

Winds are driven by uneven heating of Earth's surface by the Sun. This differential heating can be due to many causes including geographical factors and the presence of cloud. However, where the land or the sea heat up, the air just above them rises and creates an area of low pressure. Conversely, where the air is falling, a high-pressure zone is set up. The air moves from the high- to the low-pressure area and this is what we call a wind. The wind velocity also interacts with the rotation of Earth (through an effect known as the Coriolis force—another fictitious force like those in Topic A.2). This leads to rotation of the air masses such that air circulates clockwise around a high-pressure region in the northern hemisphere but counter-clockwise around a high-pressure area in the southern hemisphere.

 ### Theories—Modelling convection

Faced with a hot cup of morning coffee and little time to drink it, most of us blow across the liquid surface to cool it more quickly. This increases the rate at which energy is transferred from the liquid and the temperature of the liquid drops more quickly too. This is an example of **forced convection**—when the convection cooling is aided by a draught of air.

→ Newton stated an empirical law for cooling under conditions of forced convection. He suggested that the rate of change of the temperature of the cooling body $\frac{d\theta}{dt}$ was proportional to the temperature difference between the temperature of the cooling body θ and the temperature of the surroundings θ_s. In symbols, $\frac{d\theta}{dt} \propto (\theta - \theta_s)$.

Newton's law of cooling leads to a half-life behaviour in just the same way that radioactive half-life (Topic E.3) follows from the radioactive equation $\frac{dN}{dt} \propto N$, where N is the number of radioactive atoms in a sample.

We can use the mathematics of radioactivity for cooling too. The cooling half-life is the time for the *temperature excess* over the surroundings to halve, and this is always the same for a particular situation of hot object and its surroundings.

Two scientists Shigenao Maruyama and Shuichi Moriya repeated Newton's 1692–3 experiments involving forced convection in 2020. They allowed for effects that Newton could not have begun to imagine and found that, even so, his measurements were "quite accurate".

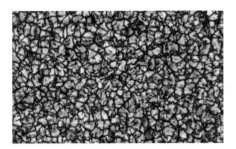

▲ Figure 20 This is a picture of the Sun's surface taken by the Inouye Solar Telescope. The Sun's surface is divided into regions, an effect called granulation. Each of these regions is a convective cell. Hotter plasma rises to the surface in the centre of each cell. At the edges, the cooler plasma sinks back into the Sun.

Worked example 11

Explain the role played by convection in the flight of a hot-air balloon.

Solution

The air in the gas canopy is heated from below and as a result of convection currents its temperature increases. The hot air in the balloon expands and its density decreases below that of the cold air outside the gas envelope. There is therefore an upwards force on the balloon. If this exceeds the weight of the balloon (plus basket and occupants), then the balloon will accelerate upwards.

Worked example 12

Suggest two reason why covering the liquid surface of a cup of hot chocolate with marshmallows will slow down the loss of energy from the hot chocolate.

Solution

The marshmallows, having air trapped in them, are poor conductors, so they allow only a small flow of energy through them. The upper surface of marshmallows will be at a lower temperature than the lower surface. This reduces the amount of convection occurring at the surface, as the convection currents that are set up will not be so strongly driven because the density differential will not be as great.

Thermal radiation

Thermal radiation is the transfer of energy by means of electromagnetic radiation. This radiation travels as a wave but does not need a medium in which to move (propagate). We receive energy from the Sun even though it has passed through about 150 million km of vacuum to reach us. Radiation is different from conduction and convection, which require a bulk material to carry the energy from place to place.

Thermal radiation has its origins in the random thermal motion of atoms. These contain charged particles and when these charges are accelerated, they emit electromagnetic radiation. There is more information on page 401.

Electromagnetic radiation acting in a wave-like way is described in more detail in Topic C.2.

Black and white surfaces

- Inquiry 1: Justify the range and quantity of measurements.
- Inquiry 1: Demonstrate independent thinking, initiative, or insight.
- Inquiry 2: Collect and record sufficient relevant quantitative data.
- Tool 3: Construct and interpret tables and graphs for raw and processed data including scatter graphs and line and curve graphs.

Use this method or something similar to investigate the impact of black and white surfaces on cooling.

- Take two identical tin cans and cut out a lid for each one from thick card. Make a hole in each lid for a thermometer. Paint one can completely with matt black paint. Paint the other shiny white.
- Fill both cans with the same volume of hot water at the same temperature. Replace the lids and place the thermometers in the water.
- Position the cans so that radiation from one cannot be incident on the other.
- Collect data to enable you to plot a graph to show how the temperature of the water in each can varies with time. This is called a cooling curve.

- You could also consider doing the experiment in reverse, beginning with cold water and using a radiant heater to provide energy for the cans. In this case, you must make sure that the heater is the same distance from the surface of each can and that the shiny can is unable to reflect radiation to the black one.
- Experiments such as this suggest that matt black surfaces are good at both radiating and absorbing energy. The opposite is true for white or shiny surfaces. These reflect rather than absorb energy, and are poor at radiating energy. Containers used to store or heat hot drinks are often shiny — it helps them to retain the energy.

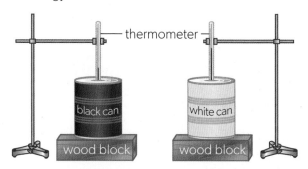

▲ Figure 21 The black can transfers more energy in a given time to the environment than the white can.

▲ Figure 22 A radiator, used circulate hot water to heat a room.

What's in a name—Radiator or not?

In many parts of the world, buildings need to be heated during all or part of the year. One way to achieve this is to circulate hot water from a boiler through a thin hollow panel often known as a "radiator" (Figure 22). But is this the appropriate term?

The outside metal surface of the panel becomes hot because energy is *conducted* from the hot water through the metal.

The air near the surface of the panel becomes hotter and less dense. It rises, setting up a *convection current* in the room.

There is some *thermal radiation* from the surface but, as its temperature is close to that of the room, the net radiation is low—certainly lower than the contributions from convection.

Should the radiator be called a radiator? How important is the accurate use of scientific language in everyday life?

Global impact — Making a saucepan

Pans are needed for cooking across the world. What is the best strategy for designing a saucepan?

The pan will be placed on a flat hot surface heated either by flame, through solar energy, or radiant energy from an electrically heated filament or plate. The energy conducts through the base and heats the contents of the pan. The base of the pan needs to be a good conductor to allow a large rate of energy transfer into the pan. The walls of the saucepan need to withstand the maximum temperature at which the pan will be used but should not lose energy if possible. Giving them a shiny silver finish reduces this energy loss.

The handle of the pan needs to be a poor conductor or a good insulator so that the pan can be lifted easily and safely. Make the handle strong and easy to hold but as thin as possible (giving a small A in the thermal conductivity equation).

Conclusion: a good pan will have a thick copper base (a good conductor), handle and sides made from stainless steel (a relatively poor conductor for a metal) and the overall finish will be polished and silvery.

Black-body radiation

The observation that black surfaces are poor reflectors of thermal energy leads to an important idea in the theory of thermal radiation: that of the **black-body radiator**. A black body is one that absorbs all the wavelengths of electromagnetic radiation that fall on it. Equally, a black body is a perfect emitter of radiation. Like some other concepts developed in physics, the black body is an idealization that cannot be realized in practice – although there are radiators and emitters that are close to the ideal.

One way to produce a good approximation to a black body is to make a small hole in the wall of an enclosed container (known as a cavity) and to paint the interior of the container matt black. The interior of the container when viewed through the hole will look very black inside.

Some of the first experiments into the physics of the black body were made by Lummer and Pringsheim in 1899 using a porcelain enclosure made from fired clay. When such enclosures are heated to high temperatures, radiation emerges from the cavity. The radiation appears coloured depending on the temperature of the enclosure. At low temperatures the radiation is in the infrared region, but as the temperature rises, the colour emitted is first red, then yellow, eventually becoming white when the temperature is high enough.

- The intensity of the radiation coming from the hole or cavity is higher when the cavity is at a higher temperature.

- The radiation emitted from the hole is not dependent on the material from which the cavity is made.

This can be seen in the picture of the interior of a steel furnace (see Figure 23). In the centre of the furnace at its very hottest point, the colour appears white. At the edges the colour is yellow. At the entrance to the furnace where the temperature is very much lower, the colour is a dull red. The scale in the figure will allow you to estimate the temperatures inside the furnace.

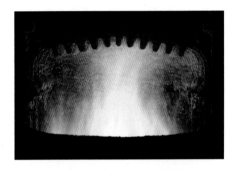

colour and temperature / K

	1000
	2000
	2500
	3200
	3300
	3400
	3500
	4500
	4000
	5000

▲ Figure 23 The interior of a furnace and a temperature scale to allow the furnace temperature to be estimated.

Global impact of science — The potter's kiln

Fabrication of ceramic objects was an early technology developed by humans.

A potter needs to know the temperature of the inside of a kiln while the clay is being "fired" to transform it into porcelain. Some potters simply view the interior of the kiln through a small hole. They can tell by experience what the temperature is from the emitted colour of the pots inside. Other potters use an instrument called a pyrometer. A tungsten filament is placed at the entrance to the kiln between the kiln interior and the potter's eye. An electric current is supplied to the filament, and this is increased until the filament disappears by merging into the background. At this point the filament is at the same temperature as the interior of the kiln. The filament system will have previously been calibrated so that the current required for the filament to disappear can be equated to the filament temperature.

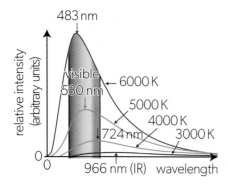

▲ Figure 25 Black-body spectra for other temperatures.

The emission spectrum from a black body

Although there is a predominant colour to the radiation emitted from a black-body radiator, this does not mean that only one wavelength emerges. To study the whole of the radiation that the black body emits, an instrument called a spectrometer is used. It measures the intensity of the radiation at a particular wavelength across a range of wavelengths.

Intensity is the power emitted per square metre. As an equation, this is

$$I = \frac{P}{A}$$

where I is the intensity, P is the power emitted and A is the area on which the power is incident. The units of intensity are $W\,m^{-2}$ or $J\,s^{-1}\,m^{-2}$.

A typical intensity–wavelength graph is shown in Figure 24 for a black body at the temperature of the visible surface of the Sun, about 5800 K. The Sun can be considered as a near-perfect black-body radiator. The graph shows how the relative intensity of the radiation varies with the wavelength of the radiation at which the intensity is measured. No scale is given on either axis. The graph shows relative values of intensity.

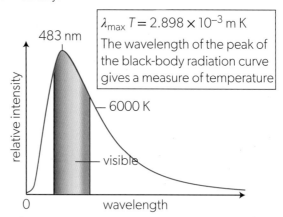

▲ Figure 24 The spectrum of the Sun assuming that it is a black-body radiator.

Important features of this graph:

- There is a peak value at about 500 nm (somewhere between green and blue light to our eyes). Is it a coincidence that the human eye has a maximum sensitivity in this region or is this biological evolution at work?

- There are significant radiations at all visible wavelengths.

- There is a steep rise from zero intensity. Notice that the line does not quite go through the origin.

- At large wavelengths, beyond the peak of the curve, the intensity falls to low levels and approaches zero asymptotically. Figure 25 shows the graph when curves at other temperatures are added. This gives some further perspectives on the emission curves.

As before, the units are arbitrary, meaning that the graph shows relative and not absolute changes between the curves at the four temperatures.

This family of curves tells us that, as **temperature increases**:

- at each wavelength, the **overall intensity increases** (because the curve is higher)
- the **total power emitted per square metre increases** (because the total area under the curve is greater)
- the curves skew towards shorter wavelengths (higher frequencies)
- the peak of the curve moves to shorter wavelengths.

The next step is to focus on the exact changes between these curves.

Wien's displacement law

In 1893, Wilhelm Wien was able to deduce the way in which the shape of the black-body emission graph depends on temperature. He showed that the height of the curve and the overall width depends on temperature alone. His full law allows predictions about the height of any point on the curve, but you will only use it to predict the peak of the intensity curve.

Wien's displacement law states that the wavelength at which the intensity is a maximum λ_{max} is related to the absolute temperature of the black body T by:

$$\lambda_{max} T = b$$

where b is known as **Wien's displacement constant** which has the value 2.9×10^{-3} m K.

Stefan–Boltzmann law

The scientists Stefan and Boltzmann independently derived an equation that predicts the total power radiated from a black body at a particular temperature. The law applies across all the wavelengths that are radiated by the black body. Stefan derived the law empirically in 1879 and Boltzmann produced the same law theoretically five years later.

The **Stefan–Boltzmann law** states that the total power (luminosity) L radiated by a black body is given by the equation

$$L = \sigma A T^4$$

where A is the total surface area of the black body and T is the absolute temperature of the surface. The constant σ is known as the Stefan–Boltzmann constant and has the value 5.67×10^{-8} W m^{-2} K^{-4}. The law refers to the total power radiated by the object, but this is the same as the energy radiated per second. It is easy to show that the energy radiated each second by one square metre of a black body (so $A = 1$) is σT^4. This variant of the full law is known as Stefan's law.

The unit of L is the watt (W $\equiv$ J s^{-1}).

Units for *b*

Notice that the unit for b is metre kelvin, m K, and must be written with a space between the symbols. Take care not to write it as mK which means millikelvin.

What applications does the Stefan–Boltzmann equation have in astrophysics and in the use of solar energy?

You will meet the Stefan–Boltzmann equation a number of times in this course. It is used in astrophysics for the calculation of the properties of individual stars. When applied to our Sun, it allows us to reach conclusions about the energy reaching the top of Earth's atmosphere and therefore make climate models of our planet.

Worked example 13

A student uses a prism spectrophotometer to investigate how the intensity of light from two incandescent lamps A and B varies with wavelength. The graph shows the results obtained.

a. Estimate the temperatures T_A and T_B of the light-emitting wire filaments of each lamp.

b. The surface area of the filament of lamp A is the same as that of lamp B.

Calculate, using the Stefan–Boltzmann law, the ratio $\dfrac{L_A}{L_B}$ of the total powers radiated by the lamps.

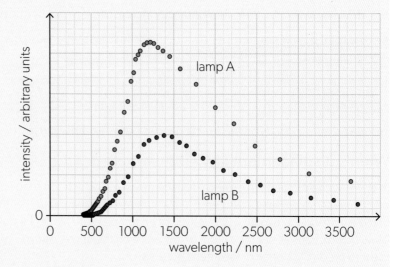

Solutions

a. We assume that both lamps are black-body radiators. The peak of the intensity curve of lamp A occurs at a wavelength of about 1200 nm. The temperature of the filament is therefore $T_A = \dfrac{2.9 \times 10^{-3}}{1200 \times 10^{-9}} \simeq 2400\,\text{K}$. Lamp B has the peak of intensity at about 1400 nm. $T_B = \dfrac{2.9 \times 10^{-3}}{1400 \times 10^{-9}} \simeq 2100\,\text{K}$.

b. Since the surface areas of the filaments are equal, the ratio of the radiated powers depends on the temperatures only: $\dfrac{L_A}{L_B} = \left(\dfrac{T_A}{T_B}\right)^4 = \left(\dfrac{2400}{2100}\right)^4 = 1.7$.

Worked example 14

A hot plate has a surface area of $0.025\,\text{m}^2$ and a constant temperature of $150\,°\text{C}$. Its surroundings are kept at a temperature of $20\,°\text{C}$. We assume that the hot plate behaves like a black body and that thermal radiation is the only energy transfer mechanism. Determine the net power exchanged by the exposed face of the hot plate with the surroundings.

Solution

The absolute temperatures of the hot plate and the surroundings are $T_{\text{hot plate}} = 423\,\text{K}$ and $T_{\text{surroundings}} = 293\,\text{K}$. The net power is the difference between the power radiated by the hot plate and the power it absorbs from the surroundings:

$P_{\text{net}} = \sigma A(T^4_{\text{hot plate}} - T^4_{\text{surroundings}}) = 5.67 \times 10^{-8} \times 0.025(423^4 - 293^4) = 35\,\text{W}$

Worked example 15

The Sun has a diameter of $1.4 \times 10^9\,\text{m}$ and a surface temperature of 5800 K. Calculate:

a. the power radiated by one square metre of the surface of the Sun

b. the total energy radiated by the Sun in one day.

Solutions

a. $P = \sigma T^4 = 5.67 \times 10^{-8} \times 5800^4 = 6.4 \times 10^7\,\text{W}$

b. The surface area of the Sun is $4\pi\left(\dfrac{1.4 \times 10^9}{2}\right)^2 = 6.2 \times 10^{18}\,\text{m}^2$.

The energy radiated in one day is $6.4 \times 10^7 \times 6.2 \times 10^{18} \times 24 \times 60 \times 60 = 3.4 \times 10^{31}\,\text{J}$

Practice questions

20. Visible light has wavelengths in the range from about 400 nm to about 700 nm. Calculate the minimum and the maximum temperature of a black body that has the peak intensity of its radiation within the visible part of the spectrum.

21. The coiled metal filament in a light bulb has a diameter of 0.050 mm and an uncoiled length of 0.30 m. The light bulb radiates a power of 45 W.
    a. Determine the absolute temperature of the filament.
    b. Calculate the peak wavelength of the radiation emitted by the light bulb.

22. A hot plate of surface area 0.025 m^2 is placed in an environment of temperature 20 °C.
    a. The hot plate exchanges energy with the environment at a net rate of 50 W. Calculate the temperature of the hot plate.
    b. Determine the additional power that must be delivered to the hot plate to increase its temperature to 200 °C.

Science as a shared endeavour — Building a theory

By the end of the 19th century, the graph of radiation intensity emitted by a black body as a function of wavelength was well known. Wien's equation fitted the experiments but only at short wavelengths. Rayleigh attempted to develop a new theory based on classical physics. He suggested that charges oscillating inside the cavity produce standing electromagnetic waves (see Topic C.4) as they bounce backwards and forwards between the cavity walls. Standing waves that escape from the cavity produce the observed black-body spectrum. Rayleigh's model fits the observations at long wavelengths, but predicts an "ultraviolet catastrophe" with an infinitely large intensity at short wavelengths.

Max Planck varied Rayleigh's theory slightly. He proposed that the standing waves could not carry all possible energies, but only certain quantities of energy E given by nhf where n is an integer, h is a constant (Planck's constant) and f is the frequency of the allowed energy. Planck's model fitted the experimental results at all wavelengths and thus, in 1900, a new branch of physics was born: quantum physics. Planck limited his theory to the space inside the cavity, he believed that the radiation was continuous outside.

Some years later, Einstein realized that the photons outside the cavity also had discrete amounts of energy. Planck was the scientific referee for Einstein's paper, and it is to Planck's credit that he recognized the value of Einstein's work and accepted the paper for publication even though it overturned some of his own ideas.

Observational astronomy and black-body radiation

Given that most stars are black-body radiators to a very close approximation, the Stefan–Boltzmann law and Wien's displacement law have crucial importance for astronomers. Indeed, the Stefan–Boltzmann law is sometimes known as the **luminosity law** in astronomy because the radiated power L is known as **luminosity** when used in the context of stars (hence the symbol L).

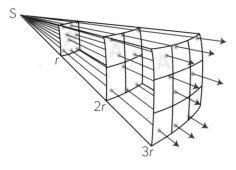

▲ Figure 26 As radiation spreads out from a point source S it covers an area that increases as the square of the distance from the source.

The luminosity of a star allows comparisons between stars of similar ages or sizes. It allows astronomers to apply the properties and distances of a known star to those of a newly discovered star.

When a star radiates a power P, at a large distance from the star the radiation can be imagined as being emitted by a point source. Imagine the surface of a sphere of radius d, centred on a star S (Figure 26 shows part of the sphere). The **intensity** I is the power that falls on unit area of the sphere surface. When the sphere has double the radius, $2d$, then the intensity at the new surface falls to $\frac{1}{4}$. This is an **inverse-square law**:

$$I = \frac{P}{4\pi d^2}$$

The units of intensity are watt metre^{-2} (W m^{-2} ≡ J s^{-1} m^{-2}).

In astrophysics, luminosity can be quoted relative to the luminosity of the Sun. The Sun's luminosity is often given the symbol $L_\odot$ so that a star with an radiated power output that is 20 times that of the Sun will have a luminosity of $20L_\odot$. The ☉ notation can also apply to the mass of the Sun $M_\odot$ and the surface temperature of the Sun $T_\odot$. As an example, the star Betelgeuse is the tenth-brightest star in the sky with a mass of $17\,M_\odot$, a temperature of $0.62\,T_\odot$ and a luminosity of $130\,000\,L_\odot$. This means that its mass is $(2.0 \times 10^{30} \times 17 =)\ 3.4 \times 10^{31}$ kg, the surface temperature is $(0.62 \times 5800 =)\ 3600$ K and the luminosity is $(3.8 \times 10^{26} \times 130\,000 =)\ 4.9 \times 10^{31}$ W.

 Where do inverse-square relationships appear in other areas of physics? (NOS)

The relationship between the intensity of a wave and the distance that the wave has travelled from the source is met frequently in physics. You will meet it again in the wave theory of Theme C. It also describes the variation of intensity of the radiation with distance from a point source of gamma photons.

Apparent brightness

Using the notation for luminosity introduced earlier ($L \equiv P$), then the intensity b falling on the surface of the sphere at a distance d from the star is $b = \frac{L}{4\pi d^2}$.

Astronomers call b the apparent brightness of the star rather than its intensity.

The **apparent brightness** is the power received at a telescope or other detector at Earth. A knowledge of the luminosity of the star and details of its spectrum (the peak wavelength λ_{max}, for example) allows astronomers to deduce the distance of the star from Earth.

 **How has international collaboration helped to develop the understanding of the nature of matter?**

▲ Figure 27 A photograph of the first Solvay conference in 1911. Einstein (standing second from the right) was the second youngest participant. Max Planck can be seen standing second from the left and Wilhelm Wien is sitting third from the right.

Scientists regularly hold conferences to share ideas and promote research. The first Solvay conference in 1911 was titled "The Theory of Radiation and the Quanta". It was one of the first conferences bringing physicists from different countries together. Hendrik Lorentz chaired the conference and impressed all the participants with his careful chairing.

Today, the Solvay Institute still hosts international conferences. The aim of all physics conferences is to promote the sharing of ideas so that progress can be made.

Observations — Discovering galaxies

The concept of a galaxy did not exist until the early 20th century. In 1925, Edwin Hubble deduced that M31 was 900 000 light years away and therefore well beyond the edge of our own galaxy. This enabled Hubble to use the apparent brightness of M31 to deduce that its mass was about $3.5 \times 10^9 \, M_\odot$ and its luminosity about $7 \times 10^8 \, L_\odot$.

It was later discovered that M31 (the Andromeda galaxy) is in fact 2.5 million light years away. Its greater distance means that it is much brighter than Hubble thought. The luminosity of the andromeda galaxy is now thought to be $2.6 \times 10^{10} \, L_\odot$.

Combining the Stefan–Boltzmann law with the equation for the area of a sphere of radius R, assuming that these stars are spherical in shape, gives $L = \sigma 4\pi R^2 T^4$. The luminosity of a star depends on its temperature and its radius.

▲ **Figure 28** The M31 galaxy.

Topic E.5 takes this relationship further to discuss an important classification tool in stellar astronomy — the Hertzsprung–Russell diagram.

Practice questions

23. The star Antares has a luminosity of $76\,000 \, L_\odot$ ($L_\odot = 3.83 \times 10^{26} \, W$) and is at a distance of $5.2 \times 10^{18} \, m$ from Earth.

    Calculate the apparent brightness of Antares as seen from Earth.

24. The minimum intensity of light that can be detected by unaided human eye is approximately $10^{-10} \, W\,m^{-2}$.

    Estimate the distance at which the Sun could just be seen by unaided eye.

25. A sphere of radius R has an absolute temperature T and it radiates power P. A second sphere has a radius $2R$. The spheres can be assumed to be black-body radiators. Determine:

    a. the power radiated by the second sphere, when its absolute temperature is T

    b. the absolute temperature of the second sphere, when it radiates power P.

How does the greenhouse effect help to maintain life on Earth and how does human activity enhance this effect?

How is the atmosphere as a system modelled to quantify the Earth–atmosphere energy balance?

A concern of both national governments and world citizens is the future of Earth's climate. Science can model the behaviour of the atmosphere and the results of this modelling are discouraging. There need to be urgent efforts to reduce the rate at which we release greenhouse gases into the atmosphere to avoid catastrophic temperature increases on Earth. But we rely on the greenhouse effect to prevent the temperature of the planet from plunging. What is the greenhouse effect and how has our impact modified its effects? All these questions rely on the development and understanding of climate models.

Conditions in our Sun dictate the energy that arrives at Earth. Physics tells us about the balance of wavelengths emitted from the star and inform us about the interaction of these wavelengths with the atmosphere. The Sun's conditions therefore ultimately influence the emission of radiation by Earth and the relative intensities of the wavelengths in this radiation.

▲ Figure 1 The atmosphere is responsible for creating the possibly unique conditions that have allowed life to develop on Earth.

In this topic, you will learn about:

- energy conservation
- emissivity
- albedo and the albedo of Earth
- the solar constant and the intensity of solar radiation at Earth's surface
- the main greenhouse gases and how they absorb infrared radiation
- the greenhouse effect and the enhanced greenhouse effect.

Introduction

On the face of it, modelling Earth's atmosphere should be straightforward. Three gases, nitrogen, oxygen and argon make up all but 0.04% of the total. But it is the last one part in 2500 that makes a profound difference to life on Earth. Remove the ozone and the damaging ultraviolet radiations from the Sun will destroy much of life on Earth. Increase the amounts of carbon dioxide, methane, nitrous oxide and water vapour by too much and the greenhouse effect on which we rely becomes enhanced. This enhancement leads to dramatic changes in the climate and, specifically, in the mean temperature of the atmosphere with consequent changes in sea level and climate. A good understanding of the relationship between climate change and the composition of the atmosphere is vital for all concerned citizens of this planet not just physicists, chemists and climatologists.

Grey bodies and emissivity

In Topic B.1, you were introduced to the concept of a black body. The spectrum emitted by a black body has a maximum of intensity at a characteristic wavelength λ_{max} for a particular temperature T of the body. Two mathematical laws are associated with black bodies:

- Wien's displacement law that links λ_{max} and T: $\lambda_{max} T = 2.9 \times 10^{-3}$ m K

- the Stefan–Boltzmann law that links the total power output of a black body to its surface area A and the absolute temperature T: $P = \sigma A T^4$.

In practice, objects can approximate to a black body without being 100% perfect in the way they behave. These approximations to black bodies are called grey to account for this. A **grey body** at a particular temperature emits less energy per second than a perfect black body of the same dimensions at the same temperature. How much less is shown by the quantity **emissivity** e which is the measure of the ratio between these two powers:

$$e = \frac{\text{power emitted by a radiating object}}{\text{power emitted by a black body with the same dimensions and at the same temperature}}$$

Emissivity has no units because it is a ratio.

For an object with emissivity e, the Stefan–Boltzmann law becomes

$$P = eA\sigma T^4$$

and

$$e = \frac{P}{A} \times \frac{1}{\sigma T^4}$$

The emissivity is

$$e = \frac{\text{power radiated per unit area}}{\sigma T^4}$$

- A perfect black body has an emissivity value of 1.

- An object that completely reflects radiation without any absorption at all has an emissivity of 0.

- All real objects have an emissivity somewhere between these two values.

Table 1 shows typical values of emissivity at visible wavelengths for some substances. It is important to note that emissivity values such as these are a function of the wavelength of the radiation. It is surprising that snow, although apparently white and reflective, is an effective emitter (and absorber) at infrared wavelengths.

Substance	Emissivity
brick	0.90
glass	0.95
ice	0.97
polished silver	0.02
snow	0.8–0.9

▲ Table 1 Typical values of emissivity at visible wavelengths.

Worked example 1

The data points in the diagram show experimental results of the variation with wavelength of the intensity of radiation emitted by a metal sample. The solid curve corresponds to a black body of the same shape and at the same temperature as the metal sample.

a. Calculate the temperature of the sample.

b. Estimate the emissivity of the sample at the wavelength corresponding to the peak of the intensity curve.

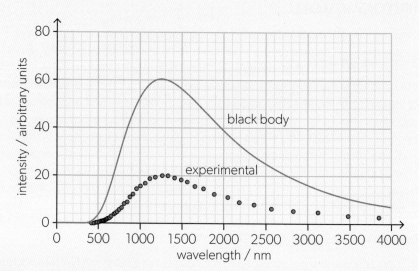

Solutions

a. λ_{max} is approximately 1250 nm. $T = \dfrac{2.9 \times 10^{-3}}{\lambda_{max}} = \dfrac{2.9 \times 10^{-3}}{1250 \times 10^{-9}} = 2300\,K$

b. Emissivity is the ratio of the intensity emitted by the sample to that emitted by the black body. At 1250 nm, the ratio is $e = \dfrac{20}{60} = 0.33$. Note that the ratio would be slightly different at different wavelengths, indicating that the emissivity varies with the wavelength!

Worked example 2

A body of surface area $0.50\,m^2$ and temperature 400 K radiates energy at a rate of 580 W. Calculate the emissivity of the body.

Solution

$$e = \frac{\text{power radiated per unit area}}{\sigma T^4} = \frac{580}{0.5 \times 5.67 \times 10^{-8} \times 400^4} = 0.80$$

Worked example 3

Human skin can be considered a grey body of emissivity 0.97. An adult person of normal body weight has a total surface area of about $1.8\,m^2$.

a. Calculate the total energy radiated by the human body during one hour. Assume that the body temperature is 37 °C.

b. If the temperature of the surroundings is 25 °C, calculate the net energy transferred from the human body as radiation during one hour.

Solutions

a. The absolute temperature of the human body is $273 + 37 = 310\,K$.

Energy = power × time = $0.97 \times 5.67 \times 10^{-8} \times 1.8 \times 310^4 \times 3600 = 3.3\,MJ$

b. The temperature of the surroundings is $273 + 25 = 298\,K$.

The net energy transfer is therefore $0.97 \times 5.67 \times 10^{-8} \times 1.8 \times (310^4 - 298^4) \times 3600 = 480\,kJ$.

Practice questions

1. A body of surface area $1.4\,m^2$ and emissivity 0.90 radiates energy at a constant rate of 1.1 kW. Calculate the absolute temperature of the body.

2. A sphere of radius 0.12 m and surface temperature 55 °C emits thermal radiation at a rate of 100 W. Calculate the emissivity of the sphere.

3. A cube of side length 0.15 m, initial temperature 0 °C and emissivity 0.75 is placed in an environment of constant temperature 50 °C.

   a. Calculate the net power exchanged by the cube with the environment as thermal radiation.

   b. The mass of the cube is 28 kg and its specific heat capacity is $380\,J\,kg^{-1}\,K^{-1}$. Estimate, using the answer in a., the initial rate of change of the temperature of the cube. State the answer in $K\,s^{-1}$.

   c. The actual temperature of the cube increases at a higher rate than calculated in b. Suggest a possible reason for this.

The solar constant and our Sun

The Sun emits large amounts of energy every second because of nuclear fusion reactions. As Earth is small and a long way from the Sun, only a small fraction of this energy arrives at the top of Earth's atmosphere. A black body at the temperature of the Sun has just under half of the energy of its radiation in our visible region, roughly the same amount in the infrared, and about 10% in the ultraviolet.

The energy gained by Earth from the Sun every second is the overall difference between the powers of the incoming solar radiation and the radiation that Earth subsequently emits back into space. This energy gained by the planet is used by plants in photosynthesis. It also drives the changes in the world's oceans and atmospheres. This energy from the Sun is crucial to life on the planet.

The intensity I of radiation is defined as the incident power P arriving per unit area A, so that $I = \dfrac{P}{A}$. The intensity of the radiation that arrives at the top of the atmosphere is known as the **solar constant** S. A precise definition of S is that:

The solar constant is the intensity of solar radiation across all wavelengths that is incident at the mean distance of Earth from the Sun on a plane perpendicular to the line joining the centre of the Sun and the centre of Earth.

One way to evaluate the solar constant is to imagine that the energy from the Sun, at Earth's orbit, is spread over the area of an imaginary sphere that has a radius equal to Earth–Sun distance (this area is A in the intensity equation). Earth is roughly 1.5×10^{11} m from the Sun and so the surface area of this sphere is $4\pi r^2 = 4\pi \times (1.5 \times 10^{11})^2 = 2.8 \times 10^{23}\,m^2$.

The Sun emits about 3.8×10^{26} J of electromagnetic radiation every second. The energy incident in one second on one square metre at the distance of Earth from the Sun is therefore $I = \dfrac{P}{A} = \dfrac{3.8 \times 10^{26}}{2.7 \times 10^{23}} = 1400$ J. This answer is quoted to 2 s.f., which is a reasonable precision for this estimate which uses 2 s.f. data. It represents a fraction of about 5×10^{-10} of the entire power output of the Sun.

When data with more precision are used, the solar constant is $1360\,W\,m^{-2}$ to 3 s.f.

Details of the nuclear fusion that is taking place in the Sun and other stars are given in Topic E.5.

235

The astronomical unit

The mean distance from Earth to the Sun is 1.50×10^{11} m. This is known as the astronomical unit (symbol: AU). This is an important length both for climatologists and for astronomers. Astronomers rely on the distance across Earth's orbit (2 AU) for their baseline in the determination of distances to stellar objects, as you see in Topic E.5.

The value of the solar constant varies periodically for several reasons:

- The output of the Sun varies by about 0.1% during its principal 11-year sunspot cycle.

- Earth's orbit is elliptical with Earth slightly closer to the Sun in January compared with July; this accounts for a difference of about 7% in the solar constant. (This difference is *not* the reason for a January summer in the southern hemisphere. Seasons occur because the axis of rotation of Earth is not perpendicular to the plane of its orbit around the Sun.)

- Other longer-period cycles are believed to occur in the Sun's luminosity and Earth's orbit. These have periods ranging from roughly hundreds to thousands of years.

Worked example 4

The distances between some of the planets and the Sun are given.

Venus–Sun distance = 0.72 AU

Saturn–Sun distance = 9.6 AU

Calculate the intensity of the solar radiation at the position of:

a. Venus

b. Saturn.

Solutions

The intensity at the location of Earth is $S = 1.36 \times 10^3$ W m^{-2}. The intensity at the location of a planet can be expressed in relative terms:

$$\frac{I_{planet}}{S} = \frac{1}{d^2_{planet}} \div \frac{1}{d^2_{Earth}} \Rightarrow I_{planet} = \left(\frac{d_{Earth}}{d_{planet}}\right)^2 \times S$$

a. $I_{Venus} = \dfrac{1}{0.72^2} \times 1.36 \times 10^3 = 2.6 \times 10^3$ W m^{-2}.

This is nearly twice the solar constant.

b. $I_{Saturn} = \dfrac{1}{9.6^2} \times 1.36 \times 10^3 = 15$ W m^{-2}.

This is a little more than 1% of the solar constant.

Practice questions

4. The Earth–Sun distance changes due to the elliptical shape of Earth's orbit. The minimum Earth–Sun distance is 0.9833 AU, in early January each year. The maximum distance is 1.0167 AU, in early July.

    Show that the changes in the Earth–Sun distance result in a difference of about 7% in the intensity of the solar radiation received by Earth.

5. A space observatory in an Earth orbit is powered by an array of solar cells of total surface area 16 m^2. The efficiency of solar to electrical energy conversion in the solar array is 0.10.

    Calculate the power output of the solar array when sunlight is incident at right angles to it.

Energy balance in Earth's surface — Atmosphere system

The solar constant of 1360 W is the power incident per square metre at the top of the atmosphere. It is not the average power arriving at the surface.

The energy from the Sun falls on only *half* the area of the sphere at any one time. The energy is then transferred to the *whole* of Earth's surface (Figure 2). The energy arriving comes through a disc of radius R, where R is the radius of Earth.

This disc area is πR^2. This energy is then spread over the whole of the sphere surface (day side and night side) which has a total area of $4\pi R^2$.

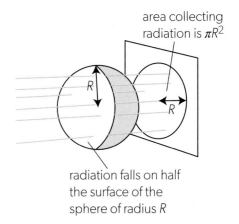

area collecting radiation is πR^2

radiation falls on half the surface of the sphere of radius R

The average incident intensity $\bar{I}_{surface}$ throughout the course of one day (24 h) at any point on the surface must be $\dfrac{\text{power arriving through the disc}}{\text{total surface area of Earth}} = \dfrac{S \times \pi R^2}{4\pi R^2} = \dfrac{S}{4}$.

Therefore $\bar{I}_{surface} = \dfrac{1360}{4} = 340\,\text{W m}^{-2}$ to 2 s.f.

This is not the end of the story because we have neglected the effect of Earth's atmosphere. As the radiation from the Sun enters and travels through the atmosphere, it is subject to losses that reduce the energy arriving at the surface. Radiation is absorbed and scattered by the atmosphere. The degree to which this absorption and scattering occur depends on the position of the Sun in the sky at a particular place. When the Sun is lower in the sky (at dawn and sunset and near the poles), its radiation passes through a greater thickness of atmosphere and thus more scattering and absorption takes place. This also gives rise to the colours in the sky at dawn and dusk.

The inverse-square law used here is discussed in Topic B.1.

The energy arrives at ground level and is incident on the surface. The surface of Earth is not a black body and it scatters some energy back up towards the atmosphere. The extent to which a particular surface can scatter energy is known as its **albedo** (from the Latin word for "whiteness").

Albedo is given the symbol a:

$$a = \frac{\text{energy scattered by a given surface in a given time}}{\text{total energy incident on the surface in the same time}}$$
$$= \frac{\text{total scattered power}}{\text{total incident power}}$$

Like emissivity, albedo is a ratio and has no units. It varies from 0 for a surface that scatters no energy (a black body) to 1 for a surface that absorbs no radiation at all. Unless stated otherwise, the albedo in Earth system is normally quoted for visible light.

The average annual albedo for the whole of Earth is about 0.3, so that, on average, about 30% of the intensity from the Sun that reaches the ground is scattered. Thus about 70% of the energy is absorbed.

This value for a of 0.3 is an average because albedo varies depending on several factors:

- It varies daily and with the seasons, depending on the amount and type of cloud cover (thin clouds have albedo values of 0.3–0.4, thick cumulo-nimbus cloud can approach values of $a = 0.9$).

- It depends on latitude.

- It depends on the terrain and the material of the surface.

Using the mean average intensity at the surface together with the average albedo shows that the average intensity absorbed by Earth's surface is

$$(1-a) \times \frac{S}{4} = 0.7 \times \frac{1360}{4} = 238\,\text{W m}^{-2}.$$

The factor $(1-a)$ corresponds to the ratio $\dfrac{\text{total absorbed power}}{\text{total incident power}}$.

Table 2 gives typical albedo values for some common land and water surfaces.

Surface	Albedo
ocean	0.06
fresh snow	0.85
sea ice	0.60
ice	0.90
urban areas	0.15
desert soils	0.40
pine forest	0.15
deciduous forest	0.25

▲ Table 2 Typical albedo values.

Models—Albedo vs emissivity

Why do physicists use two quantities, albedo and emissivity, when they appear to be related?

The answer is that both albedo and emissivity depend strongly on wavelength but are different quantities. Emissivity is the ratio of the radiation *emitted* by a grey body compared to that emitted by a black body. Albedo is the fraction of the radiation that is *reflected* by a surface compared to the radiation incident on it.

When considering the energy balance of Earth, the incoming radiation from the Sun has a black-body temperature of about 5700 K and a peak wavelength in the visible part of the spectrum. Earth itself has a black-body temperature of 288 K and a peak wavelength that is in the infrared. Earth's surface will absorb and reflect light differently in these different regions of the electromagnetic spectrum.

There are many other effects to consider. For example, the Sun's light is not incident at 90°. This affects the albedo of the surface.

When radiation is incident at right angles to the surface and the black-body temperature of the incident radiation and the surface are the same, then there will be a

relationship between albedo and emissivity so that $e = 1 - a$. This is shown in Figure 4.

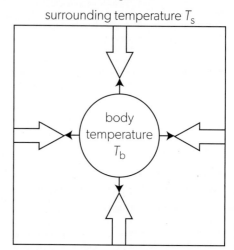

▲ **Figure 4** Radiation incident at right angles to the surface.

The grey body *radiates* at a rate $P_{rad} = e\sigma A T_b^4$. The grey body *absorbs* a fraction $(1 - a)$ of the incident radiation at a rate $P_{abs} = (1 - a)\sigma A T_s^4$.

The system will reach equilibrium when $P_{rad} = P_{abs}$. This will happen when $T_s = T_b$ and so $e = 1 - a$.

The importance of albedo will be familiar to anyone who lives where snow is common in winter. Fresh snow has a high albedo and scatters most of the radiation that is incident on it. The snow stays frozen for a long time when the temperature remains low. However, sprinkle some earth or soot on the snow and, as the sun shines, the snow soon disappears because the dark surface material absorbs energy. The radiation provides the latent heat energy needed to melt the snow.

Science as a shared endeavour—Developments in climate science

In 2021, Syukuro Manabe and Klaus Hasselmann were awarded part of the Nobel Prize in Physics for their work on climate change. It is remarkable that a study of the unpredictable weather systems on Earth can yield firm predictions about climate change. Manabe was one of the first to develop climate models in terms of the interaction between radiation balance and the movement of air masses. This was the basis for our present-day

models. Hasselmann was able to identify the imprint of both natural phenomena and human activity on climate, leading to an understanding that changes in atmospheric temperature are due to human activities.

Although the two scientists did not work together directly, it is the impact of one scientist's influence on another's work that makes science such a powerful tool.

Figure 5 shows the correlation between climate changes over the past hundred years and both natural and human events.

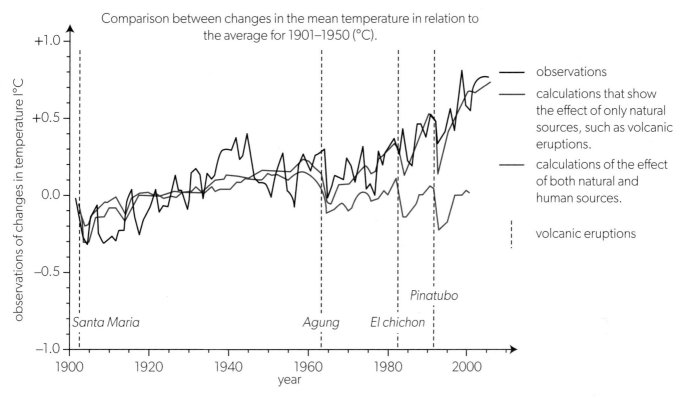

▲ Figure 5 The impact of human activity on Earth's temperature over the past hundred years, as deduced by Hasselmann. Klaus Hasselmann developed methods for distinguishing between natural and human causes (fingerprints) of atmospheric heating.

Worked example 5

Four habitats on Earth are: forest, grassland (savannah), the sea and an ice cap. Discuss which of these have the greatest and least albedo.

Solution

A material with a high albedo scatters the incident visible radiation. Ice is a good reflector and consequently has a high albedo. On the other hand, the sea is a good absorber and has a low albedo.

Worked example 6

The data give details of a model of the energy balance of Earth. Use the data to calculate the albedo of Earth that is predicted by this model.

Incident intensity from the Sun = $340\,\mathrm{W\,m^{-2}}$

Scattered intensity at surface = $100\,\mathrm{W\,m^{-2}}$

Radiated intensity from surface = $240\,\mathrm{W\,m^{-2}}$

Solution

$$\text{Albedo} = \frac{\text{power scattered by a given surface}}{\text{total power incident on the surface}}$$

In this case, the value is $\dfrac{100}{340} = 0.29$.

Worked example 7

Mercury is a planet in the Solar System. The average Mercury–Sun distance is 0.40 AU. Mercury has no atmosphere.

a. Calculate the average intensity of solar radiation incident at any point on the surface of Mercury.

b. The albedo of Mercury is 0.11. Calculate the average intensity of solar radiation absorbed by the surface of Mercury.

Solutions

a. The intensity of solar radiation at the location of Mercury is $\left(\dfrac{d_{Earth}}{d_{Mercury}}\right)^2 \times S = \dfrac{1.36 \times 10^3}{0.40^2} = 8.5 \times 10^3 \, W\,m^{-2}$. The

exposed area of Mercury is $\dfrac{1}{4}$ of its total surface area, so the average incident intensity is $\dfrac{8.5 \times 10^3}{4} = 2.1 \times 10^3 \, W\,m^{-2}$.

b. Absorbed intensity = (1 − albedo) × (incident intensity) = (1 − 0.11) × 2.1 × 10³ = 1.9 × 10³ W m⁻².

Practice questions

6. The Moon has no atmosphere and an average albedo of 0.12. Calculate the average intensity of solar radiation absorbed by the surface of the Moon.

7. Sunlight of intensity 900 W m⁻² is incident at right angles on a roof of surface area 64 m². The radiant power absorbed by the roof is 43 kW.

   Calculate the albedo of the roof.

8. Eris is a dwarf planet in the Solar System with an unusually high albedo of 0.96. Its moon, Dysnomia, has an albedo of 0.04.

   Calculate the ratio

   $$\dfrac{\text{average intensity of sunlight absorbed by Dysnomia}}{\text{average intensity of sunlight absorbed by Eris}}.$$

Models—Greenhouses

The glass in a greenhouse, like the one in Figure 6, allows the Sun's light through but blocks infrared radiation from the ground. The environment in the greenhouse remains warm so that crops grow in cooler climates. Greenhouses work in a similar way to the greenhouse effect in the atmosphere. However, a greenhouse also traps warmer air by preventing convection losses. Perhaps a greenhouse is not the best example of the greenhouse effect after all!

▲ Figure 6 The greenhouse effect in the atmosphere works in a similar way to a greenhouse.

The greenhouse effect and temperature balance

Earth and the Moon are the same average distance from the Sun, yet the average surface temperature of the Moon is 255 K, while that of Earth is about 288 K. The discrepancy is due to Earth's atmosphere because the Moon has effectively no atmosphere.

The difference is due to a phenomenon known as the **greenhouse effect** in which certain gases in Earth's atmosphere trap energy within the atmospheric system and produce a consequent rise in Earth's average temperature. The most important gases that cause this effect are carbon dioxide (CO_2), water vapour (H_2O), methane (CH_4) and nitrous oxide (dinitrogen monoxide; N_2O), all of which occur naturally in the atmosphere. Ozone (O_3), which has natural and man-made sources, also contributes to the greenhouse effect.

It is important to distinguish between:

- the "natural" greenhouse effect that is due to the naturally occurring levels of the responsible gases, and

- the **enhanced greenhouse effect** in which increased concentrations of these gases lead to further increases in Earth's average temperature and therefore to climate change.

The principal gases in the atmosphere are nitrogen, N_2, and oxygen, O_2 (roughly, 80% and 20% by weight). Both gases are made up of tightly bound molecules and, because of this, do not absorb energy from sunlight. They make little contribution to the natural greenhouse effect. The 0.04% of the atmosphere that is made up of CO_2, H_2O, CH_4 and N_2O has a much greater effect.

The structure of greenhouse-gas molecules means that they absorb ultraviolet and infrared radiation from the Sun as it travels through the atmosphere. Visible light on the other hand is not so readily absorbed by these gases and passes through the atmosphere to be absorbed by the land and water at the surface. As a result of this absorption, the temperature of the surface rises.

Earth then re-radiates just like any other hot object. The temperature of Earth's surface is far lower than that of the Sun, so the wavelengths radiated from Earth peak in the long-wavelength infrared. The absorbed radiation from the Sun was mostly in the visible region of the electromagnetic spectrum. Just as the gases in the atmosphere absorb the Sun's infrared radiation on its way in, now they absorb energy in the infrared region being radiated by Earth. The atmosphere then re-radiates the energy yet again, this time in all directions, meaning that some returns to the surface. Therefore, energy is trapped in the complex system that consists of the surface of Earth and its atmosphere.

The whole system is in a dynamic equilibrium for which:

$$\text{total energy incident on the system from the Sun} = \text{total energy being radiated away by Earth}$$

The enhanced greenhouse effect begins when there are increased levels of absorbing gases. These increased levels mean that more energy is being retained and so a greater energy must be radiated away to arrive at a new balance. The amount of radiated energy depends on temperature according to the Stefan–Boltzmann law ($P = \sigma A T^4$). The temperature of Earth has to rise to achieve this increased amount of radiation.

This increase continues until the balance of incoming and outgoing energies is reached again. Of course, this balance was originally established over billions of years. It has also varied steadily as the composition of the atmosphere and the albedo have changed with variations in vegetation, and with continental drift and other geological processes.

Observations — Other worlds, other atmospheres

The dynamic equilibrium in our climate has been important for the evolution of life on Earth. The planets Venus and Mars evolved very differently from Earth. We have been able to make observations of other worlds through the development of spacecraft sophisticated enough to travel to harsh environments.

Venus has similar dimensions to Earth but is closer to the Sun with an albedo of about 0.76. This higher albedo is caused by a thick cloud layer. Without its atmosphere, the Venusian surface is thought to have an albedo of about 0.6 and this would give Venus a surface temperature of about 260 K. However, the thick atmosphere is composed almost entirely of carbon dioxide and the surface temperature reaches 730 K. A runaway greenhouse effect acts on the planet.

Mars has very little atmosphere. The pressure at its surface is about 0.6% that of Earth's air pressure. Its surface temperature is about 215 K and the greenhouse effect only contributes about 5 K to this. It is thought that Mars had a thicker atmosphere in its past and that at this time, the greenhouse effect could have raised its surface temperature sufficiently to have liquid water and possibly even life.

ATL Social skills — Reducing greenhouse emissions

Reducing the emission of greenhouse gases is a crucial step to limiting climate change. Much of the progress in reducing the emission of greenhouse gases must be made by industries, governments and large organizations. However, individual actions are also important—particularly when made by people who live in wealthy countries. The lifestyles of the richest 10% of the world's population are estimated to be responsible for 50% of all greenhouse-gas emissions.

People try to reduce their greenhouse emissions by:

- travelling less, particularly air travel

- eating less meat

- insulating houses (in cooler countries)

- reducing their electricity consumption, particularly where this electricity is generated from burning hydrocarbons.

What other actions can individuals take to reduce greenhouse emissions?

▲ Figure 7 Cattle farming is one industry which releases greenhouse gases. Some people try to reduce their intake of milk and beef to reduce greenhouse emissions.

→

Venus and Mars are clear reminders to us of the fragility of a planetary climate.

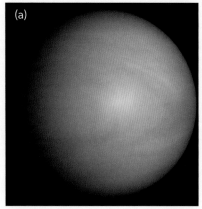 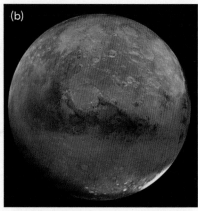

▲ Figure 8 Venus and Mars. (a) Venus has such a thick atmosphere that its surface cannot be seen. (b) Mars, on the other hand has very little atmosphere.

Why greenhouse gases absorb energy

Both ultraviolet and long-wavelength infrared radiations are absorbed by Earth's atmosphere.

Photons in the ultraviolet region of the electromagnetic spectrum are energetic and have enough energy to break the bonds within the gas molecules. This leads to the production of ionized materials in the atmosphere. A good example is the reaction that leads to the production of ozone from the oxygen atoms formed when oxygen molecules are split apart by photons with ultraviolet frequencies.

The energies of infrared photons are much smaller than those of ultraviolet photons and are not sufficient to break molecules apart. When the frequency of a photon matches the frequency of a vibrational state in a greenhouse-gas molecule, then an effect called **resonance** occurs. The vibrational states and resonance in carbon dioxide are described here, but similar effects occur in all the greenhouse-gas molecules.

In a carbon dioxide molecule, the oxygen atoms at each end are attached by double bonds to the carbon in a linear arrangement. The bonds resemble springs in their behaviour.

The molecule has four vibrational modes, as shown in Figure 9. The first of these modes—a linear symmetric stretching mode—does not cause infrared absorption, but the remaining three motions do. Each one has a characteristic frequency. When the frequency of the incident radiation matches this frequency, then the molecule is stimulated into vibrating at the matching mode. The energy of the vibration comes from the incident radiation. This leads to vibrational absorption at infrared wavelengths of $2.7\,\mu m$, $4.3\,\mu m$ and $15\,\mu m$.

The effects of these absorptions can be clearly seen in Figure 10, which shows part of the absorption spectrum of carbon dioxide.

Mechanical resonance is discussed in more detail in Topic C.4 (page 449).

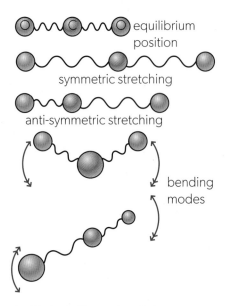

equilibrium position

symmetric stretching

anti-symmetric stretching

bending modes

▲ Figure 9 The modes of vibration in a carbon dioxide molecule.

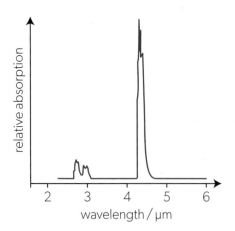

relative absorption

wavelength / µm

▲ Figure 10 Part of the absorption spectrum for carbon dioxide. The peaks indicate wavelengths at which significant absorption occurs.

What relevance do simple harmonic motion and resonance have to climate change?

Electromagnetic radiation, as its name implies, consists of an electric field and a magnetic field that propagate in the same direction but are at 90° to each other. The fields alternate in field strength and direction.

The waves interact with a greenhouse-gas molecule and drive a forced oscillation of the molecule. The molecules have a polarity so that one part of the molecule is slightly positive with respect to another part. As the wave passes through the gas, the alternation in field direction rotates the molecule in a cycle: first in one direction and then the opposite way, at the frequency of the electric field. This is what stimulates the absorption of the radiation at resonance frequencies.

As the molecule continues to move, charged regions of the molecule oscillate and this movement of charge gives rise to both electric and magnetic fields—another electromagnetic wave. This time, however, the wave will not be moving in the original direction. The molecule can be thought of as a point source that radiates equally in all directions.

Imagine the molecule to be a simple system consisting of three or more atoms connected by springs. Provide energy to it and the spring system will oscillate in a complex way. Provide this energy at exactly the correct frequency and the system will resonate.

What limitations are there in using a resonance model to explain the greenhouse effect?

Oxygen and nitrogen are abundant gases in the atmosphere but do not contribute to the greenhouse effect even though they have vibrational modes. This is because oxygen and nitrogen are so symmetrical that even when vibrating their charges do not become "lopsided". It is the absence of symmetry in the charge positions that causes greenhouse gases to absorb and then re-radiate the energy from the Sun.

Practice questions

9. Explain, with reference to greenhouse gases, why some of the thermal radiation emitted by the land and water is returned to Earth's surface from the atmosphere.

10. Explain why the symmetric stretching mode of oscillation of a carbon dioxide molecule (see Figure 9) does not cause absorption of infrared radiation, but the anti-symmetric stretching mode does.

Modelling climate balance

Knowledge of the present energy balance leads to a simple model for the climate balance of Earth. Taking everything into account, the average intensity arriving at the surface is 238 W m^{-2}. The knowledge of this emitted power allows us to predict the temperature T of a black body that will emit an intensity of 238 W m^{-2}. Using the Stefan–Boltzmann law, $238 = e\sigma T^4$ and, assuming a value for the emissivity of 0.9, this gives $T = \sqrt[4]{\dfrac{238}{0.9 \times 5.67 \times 10^{-8}}} = 261$ K. This is −12 °C

and is close to the value for the average temperature of the Moon's surface. An improved estimate can be made by accounting for the fact that the Moon has a lower albedo than Earth. As a result, less of the incident radiation from the Sun is reflected and the average intensity arriving at the Moon's surface is greater than 238 W m^{-8}. We need to investigate why the mean temperature of Earth is about 27 K greater than this.

▲ Figure 11 With very little atmosphere, but the same distance from the Sun, the Moon shows us the effect that Earth's atmosphere has on its surface conditions.

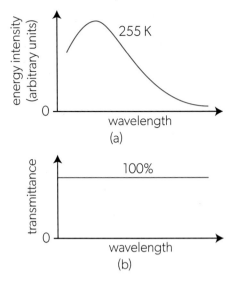

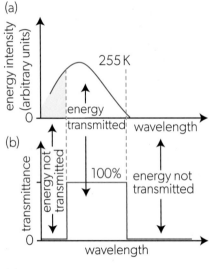

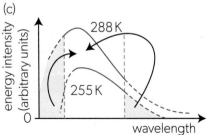

▲ Figure 12 (a) The intensity–wavelength graph for a black-body at 255 K. (b) The variation of transmittance with wavelength for a transparent atmosphere. Combining this with (a) does not change the emitted radiation profile.

▲ Figure 13 When the transmittance is zero for some wavelength ranges the emitted energy profile changes.

We made the assumption that Earth emits 238 W m⁻² $^{238\,W\,m^{-2}}$, and that this energy leaves the surface and the atmosphere completely. This would be true for an atmosphere that is completely transparent at all wavelengths, but Earth's atmosphere is not transparent because of the absorbing effects of the greenhouse gases.

Figure 12(a) shows the intensity–wavelength graph for a black body at 255 K. As expected, the total area under this curve will be $238\,W\,m^{-2}$ because it represents the predicted emission from Earth's surface assuming no atmospheric absorption.

Figure 12(b) shows the transmittance of the atmosphere as a function of wavelength. Transmittance is the measure of how well the atmosphere transmits a particular wavelength. The value of 100% means that all energy is completely transmitted at the wavelength; 0% means that no energy is transmitted at the wavelength. With a completely transparent atmosphere (or no atmosphere at all), all the black-body radiation leaves Earth because the transmittance equals 100% for all wavelengths in this model. When Figures 12(a) and 12(b) are combined, they indicate the overall intensity radiated from the planet (at the top of the atmosphere) allowing for atmospheric effects. The overall radiated intensity is identical to the emitted intensity because a transparent atmosphere has no effect.

A more realistic model uses the fact that the atmosphere absorbs energy in both the infrared and ultraviolet regions. In this model the transmittance remains at 100% for the visible wavelengths but is reduced to zero above and below the visible wavelengths. Figure 13(b) shows the modified transmittance–wavelength graph after this change.

The infrared and ultraviolet wavelengths (the yellow shaded areas in Figure 13(a)) will be absorbed by the atmosphere. The total area under the overall emission curve will now be less than $238\,W\,m^{-2}$ because of this absorption. The infrared and ultraviolet radiations are absorbed by the atmosphere rather than being radiated away into space. The atmosphere then re-radiates these wavelengths in all directions and so some energy returns to the surface.

For the system to reach equilibrium again, the temperature of the emission curve must be raised enough to compensate for the energy that has been trapped. The total area under the emission curve must return to $238\,W\,m^{-2}$ to match the incoming energy from the Sun. The only way that this can happen is for Earth's temperature to increase so that the energy deficit is included in the emission curve. As the curve changes with the increase in temperature, the area under the curve increases too. The calculation of the temperature change required is difficult and not given here. However, for the emission from the surface to equal the incoming energy from the Sun, allowing for the absorption, the surface temperature must rise to about 288 K. The net effect is shown in Figure 13(c) where the emission curve is raised and its peak shifts to shorter wavelengths to compensate exactly for the energy that was not transmitted through the atmosphere because of absorption.

The suggestion that the atmosphere completely removes wavelengths above and below certain wavelengths is, of course, an over-simplification. Figure 14 shows the complicated transmittance pattern in the infrared and indicates the molecules responsible for some of the absorption regions.

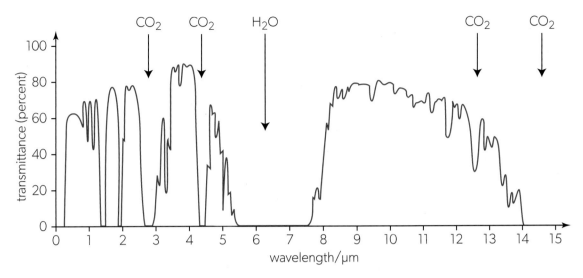

▲ Figure 14 Transmittance of Earth's atmosphere in the infrared.

The energy balance of Earth

The surface–atmosphere energy balance system is very complex. Figure 15 is a diagram showing the basic interactions and you should study it carefully.

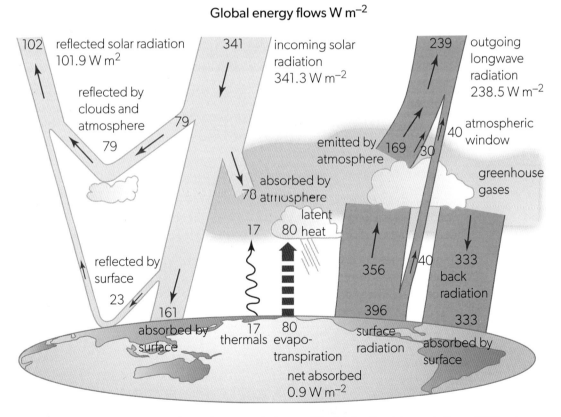

Global energy flows W m^{-2}

102 reflected solar radiation 101.9 W m^2

341 incoming solar radiation 341.3 W m^{-2}

239 outgoing longwave radiation 238.5 W m^{-2}

reflected by clouds and atmosphere 79

79

emitted by atmosphere 169

30

40 atmospheric window

absorbed by 78 atmosphere

greenhouse gases

latent 17 80 heat

reflected by surface 23

161 absorbed by surface

17 thermals

80 evapo-transpiration

356

40

333 back radiation

396 surface radiation

333 absorbed by surface

net absorbed 0.9 W m^{-2}

▲ Figure 15 The factors that make up the energy balance of Earth (after Stephens and others, 2012 An update on earth's energy balance in light of the latest global observations. *Nature Geoscience*.)

245

Global warming

There is no doubt that climate change is occurring on our planet. There is a significant warming that will ultimately lead to changes in sea level and climate across the world. The fact that there is change should not surprise us. We have recently (in geological terms) been through several Ice Ages and we are thought to be in an interstadial phase (between Ice Ages) now. In the 17th century, a "Little Ice Age" covered much of northern Europe and North America. The River Thames, in London, regularly froze, and the citizens held fairs on the ice. In 1608, the Dutch painter Hendrick Avercamp painted a winter landscape showing the typical extent and thickness of the ice in Holland (Figure 16).

Many models have been suggested to explain global warming, they include:

- changes in the composition of the atmosphere (and specifically the greenhouse gases) leading to an enhanced greenhouse effect

- increased solar flare activity

- cyclic changes in Earth's orbit

- volcanic activity.

Scientists now recognize that climate change is due to the burning of fossil fuels, which has gone on at increasing levels since the Industrial Revolution in the 18th century. There is much evidence for this. Table 3 shows some of the changes in the principal greenhouse gases over the past 250 years.

Gas	Pre-1750 concentration / ppb	Recent concentration / ppb	% increase since 1750
carbon dioxide	280 000	410 000	46
methane	700	1900	170
nitrous oxide	270	330	20

ppb = parts per billion

▲ Table 3 Changes in greenhouse gases over the past 250 years.

The recent values in this table have been collected directly in many parts of the world. There is a long-term study of the variation of carbon dioxide in Hawaii where, recently, carbon dioxide levels exceeded 400 ppm for the first time (Figure 17).

These changes become even more stark when viewed over the longer term. Figure 18 shows the carbon dioxide concentration going back roughly 0.8 million years. These data were obtained from the analysis of Antarctic ice cores.

▲ Figure 16 Winter landscape with skaters (1608), Hendrick Avercamp.

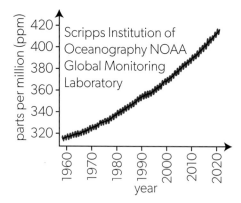

▲ Figure 17 Data from Mauna Loa, Hawaii, showing recent changes in carbon dioxide concentration.

▶ Figure 18 Long-term change in carbon dioxide concentration in the atmosphere. (Lüthi, D., M. Le Floch, B. Bereiter, T. Blunier, J.-M. Barnola, U. Siegenthaler, D. Raynaud, J. Jouzel, H. Fischer, K. Kawamura, and T.F. Stocker. 2008. High-resolution carbon dioxide concentration record 650,000–800,000 years before present. Nature, Vol. 453, pp. 379–382, 15 May 2008.)

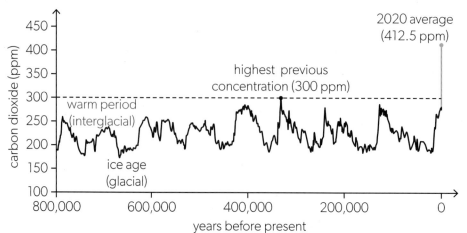

The cores are extracted from deep in the ice and yield data for the composition of the atmosphere during the era when the snow originally fell on the continent.

The enhanced greenhouse effect results from changes in concentration of the greenhouse gases. As the amounts of these gases increase, more absorption occurs both when energy enters the system and when the surface re-radiates. For example, in the transmittance–wavelength graph for a particular gas, when the concentration of the gas rises, the absorption peaks increase too. The surface temperature increases to emit sufficient energy at sea level so that emission of energy by Earth from the top of the atmosphere will equal the incoming energy from the Sun.

Global warming is likely to lead to other mechanisms that will themselves make global warming increase at a greater rate:

- The ice and snow cover at the poles will melt. This decreases the average albedo of Earth and increases the rate at which energy is absorbed by the surface.

- Increased water temperatures in the oceans reduce the extent to which CO_2 is dissolved in seawater leading to a further increase in atmospheric concentration of the gas.

Other human-related mechanisms also drive global warming as the amount of carbon fixed in plants is reduced. This is a problem that must continue to be addressed at both an international and an individual level. The world needs greater efficiency in power production and a major review of the types of fuel used. As individuals we need to be aware of our personal impact on the planet. We should all encourage the use of non-fossil-fuel methods and be conscious of our carbon footprint. Nations can capture and store carbon dioxide and agree to increase the use of combined heating and power systems. Everyone agrees that doing nothing is no longer an option.

 How do different methods of electricity production affect the energy balance of the atmosphere?

At present, much of the electrical energy production around the world comes from the generation of steam that is used to power turbines that rotate electrical generators. Much of the steam generation comes from the burning of fossil fuels and this has a major impact on the atmosphere and hence on global warming.

However, the use of nuclear fission to generate the steam leads to no large-scale release of carbon dioxide. The same goes for renewable generation methods such as wind turbines and solar panel farms.

Climate modelling and the physics of energy conversion both impact the future of the planet.

You learn more about electrical energy generation in Topic D.4.

Science and society

Opinions about climate change vary with age, nationality, gender and education. A 2021 UN study found that:

- In the US and Canada, women and girls were more than 10% more likely than men and boys to believe in the climate emergency, whereas the opposite was true in Vietnam and Nigeria.

- The highest proportion of people who believed that climate change is a global emergency was in Italy and the UK (81%).

- High-income countries (72%) and small-island developing states (74%) had a higher belief in the emergency than the least-developed countries (58%).

More and more people believe that climate change is caused by human activities. However, there are still discrepancies between the number of people who believe in the human cause of climate change and the scientific community. Studies of peer-reviewed scientific articles concluded that more than 99% of scientists believed that humans were the cause of climate change.

Surely just being a scientist does not guarantee that you are right? Or have scientists failed to be persuasive?

Science as a shared endeavour — An international perspective

There have been a number of international attempts to reach agreement over the ways forward for the planet. These have included:

- the Kyoto Protocol that was originally adopted by many (but not all) countries in 1997 and later extended in 2012

- the Intergovernmental Panel on Climate Change

- the Asia–Pacific Partnership on Clean Development and Climate

- various other United Nations Conventions on Climate Change, e.g. Cancùn 2010, Paris 2015, Glasgow 2021.

Internet research will show you what agreements are in force at present between governments.

How are developments in science and technology affected by climate change?

You should understand the impact that climate change is likely to have on both the planet and the plants and animals that inhabit it. Will this affect developments in science and technology in years to come? The answer must be: almost certainly. Already we see the development of electric-vehicle technology as industry and governments attempt to remove the effects of unnecessary fossil-fuel consumption. Work continues to make nuclear fusion (Topic E.5) a reality as an alternative source of energy.

All this work is driven by a desire to back away from the climate catastrophe that will affect future generations. What other changes do you expect to see in the development of science and technology?

Data-based questions

Climate scientists rely heavily on models to predict how changes to the atmosphere may affect Earth's energy balance, and how this in turn might change the average equilibrium temperature of Earth's surface.

The table on the right shows some data from a model of how the change in the temperature of Earth's surface depends on the concentration of carbon dioxide in the atmosphere C measured in parts per million (ppm).

The model is logarithmic:
$\Delta T = a \times \ln C - b$, where a and b are constants.

C / ppm	$\Delta T / °C$ ($\pm 0.20 °C$)
250	0.46
300	0.33
350	0.98
400	1.56
450	2.06
500	2.51
550	2.92

- Deduce the units of a and b.

- Tabulate values of $\ln C$

- Plot a graph of ΔT against $\ln C$. Include error bars on your graph.

- Use the graph to find the values of a and b.

- Use maximum and minimum gradients to find uncertainties in your values for a and b.

- The Paris Agreement in 2021 aimed to limit global warming to below 2 °C and ideally to 1.5 °C. Assuming that no other factors contribute to global warming, use your values of a and b to deduce the limits for the amounts of carbon dioxide in the atmosphere in order to achieve these targets. Give an uncertainty with your answer.

To find out more about this sort of model, research "radiative forcing". This is a concept used by climate scientists to measure the imbalance of Earth's energy. It is often assumed that the increase in global temperature will be directly proportional to the increase in radiative forcing. The constant of proportionality is often called "the climate-sensitivity parameter". Modelling the values of these parameters is important in climate science.

Using models such as these enables climate scientists to compare the effects of other greenhouse gases to that of carbon dioxide. As a result, the amount of carbon dioxide in the atmosphere has become a unit of measurement for climate impact. Hence, the idea of carbon offsetting when engaging in one activity which acts to reduce global warming is matched against an activity which acts in the reverse way. The activities do not necessarily require carbon dioxide to be released or absorbed. Rather, the amount of carbon dioxide in the atmosphere is used as a conversion factor or a currency in trading between the two.

Worked example 8

The diagram shows a simplified model of the energy balance in Earth surface–atmosphere system. The arrows represent the intensities of radiation. The average intensity of the radiation entering the system is equal to the average intensity of the radiation leaving it.

a.   The albedo of Earth is $a = 0.300$. Calculate the intensity I_1 of the outgoing radiation emitted by Earth.

Earth's surface may be assumed to be a black body of absolute temperature T that emits thermal radiation of intensity I_2. A fraction k of this intensity is re-radiated by the greenhouse gases in the atmosphere back towards the surface.

b.   Show that $(1 - k)\sigma T^4 = (1 - a)\dfrac{S}{4}$

c.   Explain why Earth surface–atmosphere system can be considered a grey body of emissivity $1 - k$.

d.   The present average temperature of Earth is close to 288 K. Determine k.

e.   Outline what effect the enhanced greenhouse effect has on the value of k and therefore, on the global average temperature.

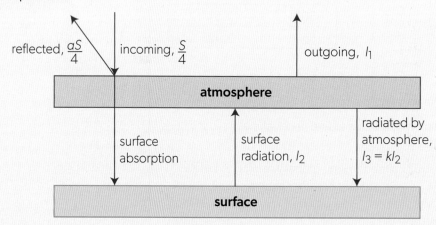

Solutions

a.   The intensity leaving the system is the sum of the intensities of the reflected solar radiation and the outgoing thermal radiation of Earth. Energy balance of Earth as a whole implies that this sum must be equal to the incoming intensity, $\dfrac{aS}{4} + I_1 = \dfrac{S}{4}$.

Hence, $I_1 = (1 - a)\dfrac{S}{4} = (1 - 0.300) \times \dfrac{1.36 \times 10^3}{4} = 238\ \mathrm{W\,m^{-2}}$.

b.   We have $I_2 = \sigma T^4$ and $I_3 = kI_2 = k\sigma T^4$. On average, the energy entering the atmosphere must be equal to the energy leaving it, so $\sigma T^4 = k\sigma T^4 + I_1$. Combining this equation with the result of part a.,

we get $(1 - k)\sigma T^4 = I_1 = (1 - a)\dfrac{S}{4}$.

c.   emissivity $= \dfrac{\text{intensity radiated by Earth as a whole}}{\sigma T^4} = \dfrac{I_1}{\sigma T^4} = 1 - k$.

d.   Rearranging the equation from part b. gives $k = 1 - \dfrac{238}{5.67 \times 10^{-8} \times 288^4} = 0.39$. This means that, according to the model, 39% of the longwave radiation emitted by Earth's surface is returned to it by the greenhouse gases in the atmosphere.

e.   The increased concentration of the greenhouse gases in the atmosphere leads to a greater fraction of the radiation returned to the surface, so k would increase. The energy balance equation in part b. requires that the average global temperature T increases, too.

Practice questions

11. Consider the energy-balance model given in Worked example 8.

    a. Suppose that the fraction of the thermal radiation returned to the surface increases to 0.40 because of the enhanced greenhouse effect.

       Use the model to predict the change in the average global temperature of Earth.

    b. There is evidence that the increased global average temperature results in melting of sea ice and increased cloud cover. Discuss, by reference to albedo, how each of these changes modifies the effects of global warming.

12. a. The average temperature of Earth's surface is 288 K. Calculate the intensity of thermal radiation emitted by Earth's surface.

    A more detailed energy balance model of Earth considers additional ways of energy transfer in the Earth surface–atmosphere system.

    b. Outline two mechanisms other than thermal radiation by which thermal energy may be transferred from Earth's surface to the atmosphere.

    c. The average intensity of the solar radiation absorbed by Earth's surface is 161 W m^{-2}. The intensity of radiation returning to the surface from the atmosphere is 333 W m^{-2}. Estimate the intensity that must be transferred from Earth's surface to the atmosphere by other means than thermal radiation.

How are macroscopic characteristics of a gas related to the behaviour of individual molecules?

What assumptions and observations lead to universal gas laws?

How can models be used to help explain observed phenomena?

We can measure the bulk properties of a gas: its pressure, volume, temperature and the mass of the sample. A gas consists of many particles all interacting on a microscopic level. How can all these individual interactions be incorporated in a model that allows us to move from a microscopic description of a gas to a prediction of its macroscopic bulk quantities?

The behaviour of a gas can be modelled using the mechanics of a single moving particle that describe the effects that occur when the particle strikes the wall of its container. This is straightforward and follows directly from the work in Theme A. Then we can consider the effects of many particles moving in many different directions and at many different speeds — an averaging process over a very large scale. When these parameters are linked to the bulk properties, the behaviour of the gas in our theoretical interpretation should match the experimental observations if our model is to be confirmed.

This mechanical model is based on assumptions about the gas particles: the nature of their interaction with each other and with the walls of their container, and the nature of their motion. The model also paints a picture of the gas — one that we can visualize, but not image or observe directly. It aids our understanding of the nature of a gas and helps us to link our observations of bulk gas behaviour to the reality of gas behaviour at the microscopic level.

▲ Figure 1 The pressure and temperature of the atmosphere govern the existence of life on Earth. Should humans want to explore outside Earth's atmosphere, they must replicate these conditions.

In this topic, you will learn about:

- pressure and how it arises at a microscopic level

- the mole and the Avogadro constant

- how ideal gases approximate the behaviour of real gases

- the ideal gas law equation

- how pressure is related to the average translational speed of the molecules of a gas

- how the internal energy of a gas is related to its Kelvin temperature

- approximating a real gas from an ideal gas.

Introduction

The atmosphere has a mass of 5×10^{18} kg and contains roughly 10^{44} molecules. It reaches over 100 km above our heads and produces a pressure at sea level of about 10^5 Pa. But how does this pressure arise?

This topic examines the origin of gas pressure and links it to other properties of a gas. Macroscopic and microscopic concepts are connected using ideas from Theme A and Topics B.1 and B.2. This topic also looks at ways in which the microscopic model fails and where it cannot match the behaviour of a real gas. Recognizing the limitations of a model is as important to a scientist as the construction of the model itself.

Topic A.2 contains an explanation of why a buoyancy force acts on a solid floating in a liquid. This explanation involved the concept of the pressure exerted by a liquid on the sides of the floating solid.

Pressure in solids, liquids and gases

Pressure P is always defined as the force F per unit area that acts perpendicular to the surface of the object:

$$P = \frac{F}{A}$$

where A is the area on which the force acts. This allows pressure to be defined for solids, liquids and gases.

- **Solid**. The weight W of a solid acts vertically downwards (Figure 2(a)) onto a horizontal surface on which it rests. The contact area of the solid on the surface is A. The solid pressure is $P = \frac{W}{A}$.

- **Liquid**. The pressure P in a liquid at a depth h depends on the gravitational field strength g and the liquid density ρ (Figure 2(b)). The pressure definition leads to $P = \rho g h$, as in Topic A.2.

- **Gas**. The pressure is caused by the force exerted by the gas molecules on the wall of the gas container when they transfer momentum as they rebound (bounce) at the wall (Figure 2(c)). The pressure of the gas depends on other quantities such as the number of gas particles in the container, the volume of the container and the absolute temperature (in K) of the gas.

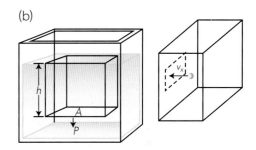

▲ Figure 2 Pressure in (a) solids, (b) liquids and (c) gases.

Worked example 1

A cyclist applies a force of 800 N to a pedal of a bicycle. The pedal can be modelled as a rectangle of dimensions 6.5 cm × 2.5 cm. Calculate the pressure applied between the shoe and the pedal.

Solution

The contact area is $0.065 \times 0.025 = 1.6 \times 10^{-3}$ m². The pressure is

$P = \frac{F}{A} = \frac{800}{1.6 \times 10^{-3}} = 4.9 \times 10^5$ Pa. This is about five times greater than

the standard atmospheric pressure of about 10^5 Pa.

Worked example 2

The quietest sound a young person with normal hearing can hear corresponds to an air-pressure variation of about 2×10^{-5} Pa. The diameter of the human eardrum is about 9 mm. Estimate the amplitude of the force exerted on the eardrum by the quietest detectable sound.

Solution

The surface area of the eardrum is $\pi\left(\dfrac{9 \times 10^{-3}}{2}\right)^2 = 6 \times 10^{-5}$ m^2. The variation of the force is of the order of $P \times A = 2 \times 10^{-5} \times 6 \times 10^{-5} \simeq 10^{-9}$ N. Such a small force is sufficient to produce a vibration of the eardrum that the human ear can detect!

Practice questions

1. An ice skater pushes off the ice by applying a force of 1.2 kN to the skate. The contact area between the blade of the skate and the ice is 7.0×10^{-4} m^2. Calculate the pressure that the skater exerts on the ice.

2. A push pin is pressed with a finger into a cork board by applying a force of 2.5 N to its flat head, which has a diameter of 8.0 mm.

    a. Calculate the pressure between the head of the pin and the finger.

    b. The pressure at the sharp tip of the pin is 300 MPa. Estimate the diameter of the tip of the pin.

ATL Working collaboratively — Hot-air balloons

Even though hot-air balloons were used by the Chinese in about 300 CE for signalling, much of the history of ballooning is linked to the development of the gas laws. This work was centred around France in 1783. In June of that year, the Montgolfier brothers demonstrated a hot air balloon at Annonay. They repeated the demonstration in September at the Palace of Versailles, this time carrying a duck, a cockerel and a sheep. Following the success of this experiment, they demonstrated the first piloted flight later that year.

The principle of hot-air ballooning is that hotter air expands and becomes less dense. The hot air trapped in the balloon displaces cooler, denser surrounding air. When the resulting buoyancy force exceeds the weight of the balloon and its load, then the balloon rises.

Jacques Charles (after whom Charles's law is named) was also interested in ballooning. Rather than using the hot-air principle, he used hydrogen to provide buoyancy and demonstrated a balloon in Paris in August 1783. The balloon took four days to fill and the demonstration attracted large crowds. By December, he had launched a piloted balloon in front of a crowd of 400 000.

Although balloons filled with hydrogen were more popular at first, they have not proved so resilient. There was a vogue for commercial passenger flights using hydrogen-filled airships at the start of the 20th century.

However, a series of catastrophic crashes, in which the hydrogen in the balloon ignited, stopped the development of this form of transport. Nowadays, tethered helium balloons are used to take passengers above a viewpoint.

Although the first of the gas laws, Boyle's law, was published in 1662, Charles's law and the pressure law were not published until 1802 by Joseph Gay-Lussac who stated that Jacques Charles had observed the effect of Charles's law 15 years earlier but had not published his results.

▲ Figure 3 Hot-air balloons.

Avogadro's number and the mole

The **mole** is the SI measure of quantity or amount of a substance.

The mole is defined to be $6.022\,140\,76 \times 10^{23}$ particles. The abbreviation for it is **mol**.

The recent redefinition of the mole is helpful as it reminds us that "mole" is simply a collective name for a certain number of something (just as we say a "dozen" meaning twelve). The number itself is known as the **Avogadro number**, and has the symbol N_A.

$$\text{The Avogadro number, } N_A = 6.022\,140\,76 \times 10^{23}$$

You can specify 3 mol of electrons (meaning $3 \times 6.022\,140\,76 \times 10^{23}$ electrons) or 0.5 mol of water (meaning $0.5 \times 6.022\,140\,76 \times 10^{23}$ H_2O molecules). The quantity of the 0.5 mol of water and its chemical formula together tell you straight away that there are $6.022\,140\,76 \times 10^{23}$ hydrogen atoms and $3.011\,070\,38 \times 10^{23}$ oxygen atoms making a total of about 10^{24} atoms in each half mole of water.

Molar mass

The use of the mole to yield numbers of atoms and molecules leads to **molar mass**. Nitrogen gas normally has two atoms and is written as N_2. One mole of nitrogen has 12.044×10^{23} atoms. The mass of one mole of nitrogen atoms is 14.01 g and so the mass of one mole of nitrogen molecules must be 28.02 g.

Measurements

Up until 2019, the mole was defined as the amount of substance in 12 g of carbon-12, the isotope of carbon that has 6 protons and 6 neutrons in one nucleus. This was one of the changes made to the SI to improve the basis for scientific measurement. Some of the other changes made in 2019 are discussed elsewhere in this book.

Worked example 3

A bottle contains 0.500 kg of water. The molar mass of water is 18.0 g mol⁻¹. Calculate:

a. the quantity of water, in mol, in the bottle
b. the number of water molecules in the bottle
c. the mass, in kg, of one molecule of water.

Solutions

a. $n = \dfrac{500}{18.0} = 27.8\,\text{mol}$
b. $N = 6.02 \times 10^{23} \times 27.8 = 1.67 \times 10^{25}$ molecules
c. One mole of water has a mass of 0.0180 kg and contains N_A molecules.

The mass of one molecule is therefore $\dfrac{0.0180}{6.02 \times 10^{23}} = 2.99 \times 10^{-26}\,\text{kg}$.

Worked example 4

A single serving of espresso typically contains about 2.5×10^{20} molecules of caffeine, which has a molar mass of 194 g mol⁻¹. Calculate the total mass, in mg, of caffeine in one serving of espresso.

Solution

The number of moles in one serving is $n = \dfrac{2.5 \times 10^{20}}{6.02 \times 10^{23}} = 4.15 \times 10^{-4}\,\text{mol}$. The total mass of caffeine is $4.15 \times 10^{-4} \times 194 = 8.1 \times 10^{-2}\,\text{g} \simeq 80\,\text{mg}$.

Practice questions

3. A ring contains 12.0 g of gold and 4.0 g of copper. The molar mass of gold is 197 g mol⁻¹ and that of copper is 63.5 g mol⁻¹. Calculate the number of atoms of each element in the ring. Are there more gold or copper atoms in the ring?

4. One egg yolk contains about 1 µg of vitamin D_3, an important dietary component. The molar mass of

vitamin D_3 is 385 g mol⁻¹. Calculate the number of molecules of vitamin D_3 in the egg yolk.

5. A typical human body contains about 8×10^{14} atoms of radioactive carbon-14, naturally absorbed from the environment and replenished by breathing and eating. Calculate the mass, in µg, of carbon-14 present in the body.

Models — The ideal gas law

The properties of gases were investigated experimentally over a period of about 150 years starting in the mid-17th century. In those days, the experiments were carried out with gases at about atmospheric pressure and temperatures around 300 K. We will see later that under these conditions a gas can be regarded as an **ideal gas**. An ideal gas always obeys the gas laws. In fact, no real gas is ideal at high pressures and high temperatures. When the pressure is roughly atmospheric, the approximation to ideal behaviour is good.

You can see how real gases deviate from the ideal gas on page 270 of this topic.

The gas laws

Boyle's law was the first of the laws. Robert Boyle, an Irish physicist, carried out the work in 1662.

He showed that:

> **The pressure *P* of a fixed quantity of gas (that is, a constant number of molecules) is inversely proportional its volume *V* when the temperature does not change.**

In fact, the requirement that the temperature should be constant was due to the French scientist Edmé Mariotte who repeated Boyle's experiment in 1679.

The law can be written algebraically in two ways:

$$P \propto \frac{1}{V} \quad \text{or} \quad PV = \text{constant (at constant temperature)}$$

The law leads to two graphs (shown in Figure 4) of the variation of pressure with (a) volume and (b) $\frac{1}{\text{volume}}$.

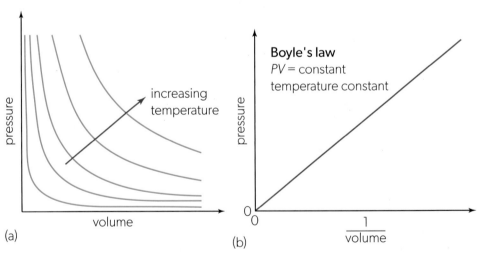

(a)

(b)

▲ **Figure 4** Boyles law gives the variation of pressure with volume when the temperature of a fixed mass of gas is constant.

The inverse nature of the relationship between *P* and *V* leads to the *P–V* graph being curved for each temperature. A fixed quantity of gas gives a different curve at each different temperature (Figure 4(a)). These curves are known as **isothermals** (the name is from Greek words "isos" meaning equal and "therme" meaning heat).

The alternative graph (Figure 4(b)) is more useful as it allows easier predictions of gas behaviour when conditions change. As well giving *P* for a change in $\frac{1}{V}$, it is possible to predict the line:

- for a higher temperature and the same mass of gas (the gradient of the line will increase)

- for a smaller mass of gas at the original temperature (the gradient of the line will decrease).

Charles's law is named for the French experimenter Jacques Charles who, around 1787, repeated the earlier experiments of Guillaume Amontons. In these experiments the temperature of the fixed mass of gas was varied and the volume change measured; the pressure was held constant. Charles found that every time the temperature changed by 1 K, the volume of the gas changed by $\frac{1}{273}$ of the volume at 273 K (0 °C). This implies that when the temperature has dropped to 0 K (−273 °C) the volume will be zero, assuming that the properties of the gas do not change with temperature.

Although Charles did not publish this work, it was announced in 1802 by Joseph Gay-Lussac. In words, the law becomes:

The volume of a fixed quantity of gas at a constant pressure is directly proportional to the absolute temperature (T)

Mathematically,

$$V \propto T \quad \text{or} \quad \frac{V}{T} = \text{constant (at constant pressure)}$$

We now have two laws that connect the three quantities P, V and T.

The origin of the third **pressure law** (or Gay-Lussac's law) is obscure. Amontons investigated the variation of pressure with temperature, but his equipment was not sensitive enough to reach a firm conclusion. The law in words is:

The pressure of a fixed quantity of gas at a constant volume is directly proportional to the absolute temperature

or

$$P \propto T \quad \text{or} \quad \frac{P}{T} = \text{constant (at constant volume)}$$

The final strand to this experimental work on gases is due to the Italian scientist Count Amadeo Avogadro who hypothesized in 1811 that all gases at the same temperature and pressure contain equal numbers of particles per unit volume. This follows from the knowledge that gases expand by equal amounts for equal temperature increases. This has become known as **Avogadro's law** and can be stated as:

The quantity of gas (in mol) at constant temperature and pressure is directly proportional to the volume of the gas.

Once again this leads to a mathematical relationship:

$$n \propto V \quad \text{or} \quad \frac{n}{V} = \text{constant (at constant temperature and pressure)}$$

(Incidentally, each of the four relationships leads to a constant quantity, but these quantities are different for each equation and different gas samples.)

The **ideal gas equation,** also known as the **equation of state for an ideal gas,** can be derived using all four equations met so far. They combine to give

$$\frac{PV}{nT} = R \quad \text{or} \quad PV = nRT$$

Theories — Law or theory?

The three rules developed empirically to predict the behaviour of a gas are known as "laws". These rules do not attempt to explain the behaviour of the gases. They simply report what most gases do when the pressure is small at around room temperature. Move away from these limits and the "laws" do not apply anymore.

Laws are statements that allow scientists and engineers to make predictions without requiring an explanation.

Are there other examples of such laws?

An interpretation of *R*?

In Topic B.1, the specific heat capacity was the energy transfer required to change the temperature of a unit mass of a substance by 1 K. The units of specific heat capacity are $J\,kg^{-1}\,K^{-1}$. That should remind you of the unit of *R*. The only difference is the replacement of kg by mol. You can regard *R* as a type of specific heat capacity. It will also remind you that $P \times V$ has the units of energy.

This is, of course, the same constant from Topic B.1 where the average kinetic energy of particles $\bar{E}_k$ is given by $\bar{E}_k = \frac{3}{2}k_B T$.

You should be able to recognize that each of the four laws is contained in this over-arching equation.

The constant *R* is the gas constant and it applies for any gas that can be treated as ideal. When pressure is measured in pascal, volume in cubic metres and temperature in kelvin, then *R* takes the value $8.31\,J\,mol^{-1}\,K^{-1}$.

There is one final modification to make to the equation and that is to convert from an equation in terms of moles to one that contains the number of molecules *N* in the gas.

One mole of the gas contains N_A molecules (using the Avogadro number from earlier in this topic), so it is convenient to define a new constant $\frac{R}{N_A}$. Now *n* can be replaced by the total number of molecules if we also replace *R* with a new quantity k_B, the **Boltzmann constant**. This is defined as $k_B = \frac{R}{N_A}$ and leads to the ideal gas equation in an alternative form:

$$\frac{PV}{NT} = k_B \quad \text{or} \quad pV = Nk_B T$$

The Boltzmann constant has the value

$$k_B = \frac{R}{N_A} = \frac{8.31}{6.02 \times 10^{23}} = 1.38 \times 10^{-23}\,J\,K^{-1}.$$

Verifying the gas laws

- Tool 3: Construct and interpret tables and graphs for raw and processed data including scatter graphs and line and curve graphs.
- Tool 3: Extrapolate and interpolate graphs.
- Inquiry 1: Justify the range and quantity of measurements.
- Inquiry 1: Appreciate when and how to maintain constant environmental conditions of systems.

Boyle's law

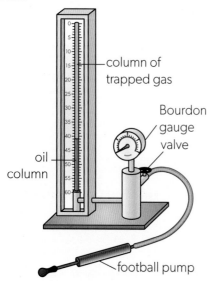

There are many forms of apparatus used to verify Boyle's law. They all require measurements of the pressure of a fixed mass of gas at a known volume under conditions in which the temperature will remain constant. The apparatus shown in Figure 5 achieves that well, provided some precautions are observed.

- The fixed mass of gas is trapped in a transparent column by a tube of oil. Oil can be forced into the tube using a bicycle or football pump. This compresses the gas in the tube, decreasing its volume and increasing its pressure. The pressure is read directly using a Bourdon gauge.

- It is important to carry out the change slowly and then to wait for several seconds because compressing the gas increases its temperature. The wait allows the gas to return to room temperature.

- Once a range of pressure values have been obtained for various volumes, the graph of *P* against $\frac{1}{V}$ can be plotted. A straight-line graph like the one in Figure 4(b) will verify the law.

▲ Figure 5 Apparatus to verify Boyle's law.

Charles's law

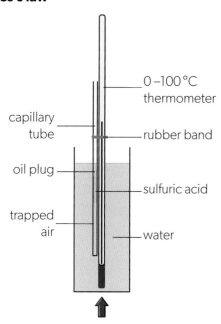

▲ Figure 6 Apparatus to verify Charles's law.

The aim here is to keep the pressure constant. Air is trapped between the sealed end of a glass capillary tube and a length of sulfuric acid in the tube (the acid keeps the air dry). The top end of the tube is open to the atmosphere, meaning that the pressure in the gas is equal to atmospheric pressure plus the pressure due to the length of the acid.

- The temperature of the gas is changed by heating or cooling the liquid. Sufficient time needs to be left to allow the gas inside the glass to be in thermal equilibrium with the water. The process of thermal transfer by conduction through the glass takes time.

- The volume of the gas is directly proportional to the length of the gas column. This assumes that the internal diameter of the capillary tube is constant.

- Readings of the length of the gas column should be made over as wide a temperature range as possible, recording both length and temperature.

- Plot length against temperature. There is no need to convert to gas volume. The graph should be a straight line that can be extrapolated back to zero length to give an estimate of absolute zero.

Pressure (Gay-Lussac's) law

The final experiment involves a constant volume of gas with readings of the pressure and temperature. Again, there are many forms of this apparatus (Figure 7).

- The fixed mass of gas is trapped in a round-bottomed flask. The flask is connected to a Bourdon gauge. Notice that a significant error here is caused by the tube connecting the gauge and the flask. The gas inside it is not at the temperature of the water bath. This is a reason to keep the tube volume small compared with the volume of the flask.

- The temperature of the water bath is varied and, for a wide range of temperatures, the pressure is read directly. Again, time must be allowed for the system to reach thermal equilibrium.

- A plot of pressure against temperature should be a straight line and again you can use the graph to estimate a value for absolute zero.

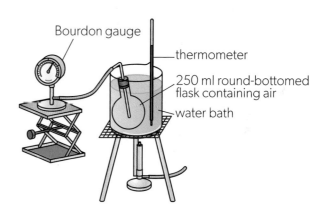

▲ Figure 7 Apparatus to verify the pressure (Gay-Lussac's) law.

Worked example 5

A sample of an ideal gas expands at constant pressure from an initial volume of 80 cm³ and temperature 13 °C to a final volume of 120 cm³. Calculate, in °C, the final temperature of the sample.

Solution

For constant pressure, $\frac{T}{V} = \text{const.}$

$$\frac{273 + 13}{80} = \frac{T_f}{120} \Rightarrow T_f = \frac{120}{80} \times 286 = 429\,K = 156\,°C$$

Note that each of the gas laws requires the use of the *absolute* temperature of the gas. Since the calculation involves the ratio of the volumes, it is *not* necessary to convert the volume to m³.

Worked example 6

A sample of ideal gas at a temperature of 280 K and pressure 9.0×10^4 Pa is compressed from an initial volume of 240 cm^3 to a volume of 60 cm^3. Its final pressure is 5.4×10^5 Pa. Calculate the final temperature of the gas.

Solution

We use the ideal gas equation, written in the form $\dfrac{PV}{T} = \text{const.}$

$$\frac{9.0 \times 10^4 \times 240}{280} = \frac{5.4 \times 10^5 \times 60}{T} \Rightarrow T = \frac{5.4 \times 10^5 \times 60}{9.0 \times 10^4 \times 240} \times 280 = 420 \text{ K}$$

Worked example 7

A sample of argon undergoes compression in a sealed container. The following data are given:

  initial volume $= 35.0 \text{ cm}^3$
  initial pressure $= 9.60 \times 10^4$ Pa
  final volume $= 6.00 \text{ cm}^3$
  final pressure $= 5.60 \times 10^5$ Pa

a. Show that the initial and final temperatures of the gas are the same.

b. The mass of the sample is 58.0 mg. The molar mass of argon is 39.9 g mol^{-1}. Determine the temperature of the sample.

Solutions

a. $P_i V_i = 9.60 \times 10^4 \times 35.0 \times 10^{-6} = 3.36 \text{ Pa m}^3$; $P_f V_f = 5.60 \times 10^5 \times 6.00 \times 10^{-6} = 3.36 \text{ Pa m}^3$

   The product *pressure × volume* has the same value in the initial and final states, indicating that the temperature of the gas is also the same.

b. The amount of the gas in the sample is $n = \dfrac{58.0 \times 10^{-3}}{39.9} = 1.45 \times 10^{-3}$ mol. Using the ideal gas equation,

   $$T = \frac{PV}{nR} = \frac{3.36}{1.45 \times 10^{-3} \times 8.31} = 278 \text{ K}$$

Worked example 8

A fixed mass of 19 mg of an ideal gas is kept in a container of a constant volume $5.0 \times 10^{-5} \text{ m}^3$. The gas is heated. The graph shows how the pressure of the gas varies with temperature.

a. Calculate the number of molecules of the gas in the container.

b. Calculate the molar mass of the gas.

c. The experiment is repeated with half of the original amount of the gas in the container. Sketch the variation of pressure with temperature for this new experiment.

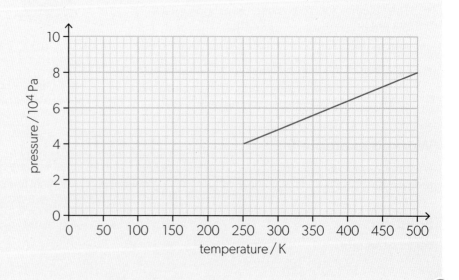

Solutions

a. From the ideal gas equation, $N = \dfrac{P}{T} \times \dfrac{V}{k_B}$,

where $\dfrac{P}{T} = \dfrac{4.0 \times 10^4}{250} = 160\,\text{Pa}\,\text{K}^{-1}$ is

the slope of the pressure–temperature

graph,

$N = 160 \times \dfrac{5.0 \times 10^{-5}}{1.38 \times 10^{-23}} = 5.8 \times 10^{20}$

b. The number of moles of the gas in the

container is $\dfrac{N}{N_A}$. Hence the molar mass is

$\dfrac{m \times N_A}{N} = \dfrac{19 \times 10^{-3} \times 6.02 \times 10^{23}}{5.8 \times 10^{20}}$

$= 20\,\text{g}\,\text{mol}^{-1}$

c. The graph will be directly proportional, and the slope will be reduced to one half of the original value.

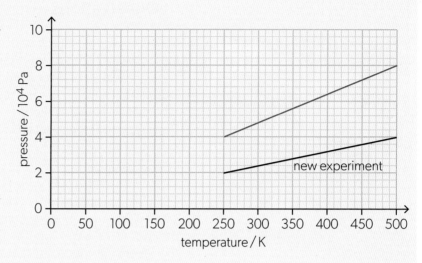

Practice questions

6. The temperature of a fixed amount of an ideal gas changes from 100 °C to 200 °C at a constant volume. The initial pressure of the gas is 2.0 MPa. What is the final pressure of the gas?

   A. 1.0 MPa   B. 1.6 MPa   C. 2.5 MPa   D. 4.0 MPa

7. A fixed amount of an ideal gas is kept at a constant pressure P. The graph shows how the volume of the sample varies with the absolute temperature.

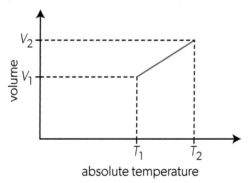

   What is the quantity of gas in the sample?

   A. $\dfrac{T_2 - T_1}{V_2 - V_1} \times \dfrac{R}{P}$

   B. $\dfrac{T_2 - T_1}{V_2 - V_1} \times \dfrac{P}{R}$

   C. $\dfrac{V_2 - V_1}{T_2 - T_1} \times \dfrac{R}{P}$

   D. $\dfrac{V_2 - V_1}{T_2 - T_1} \times \dfrac{P}{R}$

8. A sample of diatomic nitrogen (molar mass 28 g mol⁻¹) occupies a volume of 150 cm³ at a temperature 25 °C and pressure 1.0×10^5 Pa. The sample expands to a final volume of 180 cm³ and its temperature increases to 150 °C. Calculate:

   a. the final pressure of the sample

   b. the mass of the sample.

9. A container of constant volume $2.0 \times 10^{-3}\,\text{m}^3$ is filled with air at a pressure of 1.0×10^5 Pa and temperature 22 °C.

   a. Calculate, in mol, the quantity of air in the container.

   The air in the container is cooled to 4.0 °C.

   b. Calculate the final pressure of the air in the container.

   The container has a lid of surface area 80 cm². The pressure outside of the container is 1.0×10^5 Pa.

   c. Determine the force needed to open the lid.

10. The gravitational wave detector Virgo near Pisa in Italy is one of the largest vacuum installations in the world. Its vacuum system has the total volume of 7000 m³ and is kept at a residual pressure of about 10^{-7} Pa. Assume that the temperature of the gas is 290 K.

    Calculate:

    a. the quantity of the gas, in mol, in the Virgo's vacuum system

    b. the number of gas particles in one cubic centimetre of the vacuum.

Observations

Robert Boyle was one of the giants of European science. During his lifetime the "scientists" of his day resorted to philosophical reasoning rather than experimentation. Boyle preferred the latter. He described his techniques carefully so that they could be repeated by others. This was in contrast to the other scientists who worked independently and guarded their work closely. Boyle reported his results and conclusions swiftly so that others could make progress too.

Scientifically speaking, the 17th century was a different time from ours. Many who we now hold in the highest regard were alchemists as well as scientists—Isaac Newton is a famous example. They believed that if they could discover a way to transmute metals—to turn base metal in gold—then they would make their fortunes. This is why they generally kept their results to themselves. Of course, true transmutation of one element into another had to wait 200 years until nuclear fission and fusion were studied.

A microscopic interpretation of gases

Despite the suggestion by Democritus, a Greek philosopher from about 400 BCE, the existence of atoms has only been recognized by scientists over the past century or so. In the 19th century, a chemist named John Dalton realized that elements combine in fixed ratios. He proposed atoms to account for this. However, this was not direct evidence and relied on careful measurement of the products of chemical reactions. Even today, we cannot form direct images of gas molecules in motion. We have indirect evidence, however, from the effects of diffusion and the phenomenon known as Brownian motion.

Cook food on a barbecue and the smell soon reaches all parts of the garden, even on a windless day. Air molecules produce a random motion in the vapours from the food. This is known as diffusion. Figure 8 shows a demonstration of this. The gas jars on the right contain bromine gas (bottom jar) and air (top jar). These are prevented from mixing by a piece of glass between the jars. Remove the glass, keeping the jars together, and the bromine gas gradually diffuses up into the upper jar.

The air molecules are colliding at random with the bromine molecules and, by chance, some are knocked upwards into the jar that was initially full of air.

Robert Brown, in 1827, first observed the motion of small particles that were suspended in water.

▲ Figure 8 A demonstration of diffusion.

these specks of light are the smoke particles scattering light

the random motion of a smoke particle showing how it moves linearly in between collisions with air molecules

▶ Figure 9 The cell contains smoke particles that can be seen to be buffeted by invisible, fast-moving air molecules.

Some particles of smoke are introduced into a cell that contains air (Figure 9). The space is illuminated, and the interior of the cell, which has transparent walls, is viewed with a microscope. Small specks of light are seen. These are smoke particles and they have a curious random motion in which they move in straight lines but then change direction abruptly. Some of the invisible air molecules are moving at large speeds and when they collide with a smoke particle there is a transfer of momentum. This deflects the smoke particle giving the random motion that Brown saw.

However, the explanation had to wait 80 years until a young Albert Einstein, making one of his first scientific contributions, explained the effect. He went on to analyse the statistics of the motion of the smoke particles with great success. Another statistical model explaining Brownian motion was developed independently by a Polish physicist Marian Smoluchowski.

Today we are clear that gases consist of molecules or atoms in constant motion. We use this as a basis for a theoretical model of the kinetics of a gas.

Empirical or theoretical modelling

At this point, we move from empirical models of gas behaviour to a theoretical model based on a set of assumptions. Both types of model have their place in science.

In the 17th century, the understanding of the microscopic composition of a gas was not sufficient to allow Boyle and Charles and their collaborators to develop a theory. Nevertheless, the rules of gas behaviour that they produced from experiments allow a prediction of the bulk behaviour of a gas. We assume that a gas under the same conditions of an experiment will always behave in same way.

Theoretical models are based on a set of assumptions about the system they model. Theoretical models are, superficially, very attractive because they appear to allow predictions for all circumstances. But this is an illusion. Even today, attempts to produce a unified theory for all gases under all conditions have not been successful. An important piece of work by the Dutch scientist Johannes van der Waals (mentioned later in this topic) requires individual sets of empirical constants for each individual gas for his theory to work. We still need empirical results when the assumptions of a model break down.

Data-based questions

Brownian motion can be modelled by a random-walk process. A computer was used to model the data below. A particle starts at the origin (0, 0). It then takes a step of length 1 unit in a random direction. The computer repeats this process N times.

The model tracks 10 particles, determining the average distance d of a particle from the origin after N steps and the uncertainty in d. The results are given in the table.

N	d	Uncertainty in d
10	2.97	0.48
20	2.96	0.68
50	5.52	0.82
100	7.54	0.93
200	12.7	1.8
500	26.6	5.0
1000	18.0	3.2
2000	46.6	7.4
5000	62.6	8.5
10 000	92.0	11
20 000	137	17
50 000	213	34
100 000	325	24

- Theory suggests that $d^2 \propto N$. Plot a graph of $\log d$ against $\log N$ and explain whether your graph confirms the relationship.

- Plot a graph of d^2 against N.

- Calculate the uncertainties in d^2 and add suitable error bars to your graph.

- Find the gradient. Use maximum and minimum gradients to find the uncertainty in your value for the gradient.

Kinetic model of an ideal gas

The gas laws outlined above were based solely on experimental investigations that involved work by different scientists that stretched over 150 years. It is important to recognize that they are **empirical** results.

It is also possible to construct a **theoretical model** of an ideal gas that begins with a series of assumptions and uses mechanics principles developed in Theme A to arrive at a description of the gas. We will then compare the results of these two routes to see the extent to which they complement each other.

The ideal gas model we develop is known as the **kinetic model of an ideal gas** and it is based on a set of assumptions. These are:

1. A gas consists of many identical particles in a container. They have the same mass as each other. (We now know that these particles are the atoms or molecules of the gas.)

2. These particles are in constant random motion.

3. The total volume of the particles is negligible compared with the total volume of the gas. (This is the same as saying that the average distance between particles is much greater than their average size.)

4. The particles collide elastically with each other and with the walls of their container.

5. Intermolecular forces between the particles and the walls can be ignored except during collisions. (This means that the energy in the gas can be considered as entirely kinetic with no potential-energy contribution.)

6. The time for a collision between particles, and the time for a collision between a particle and the wall are negligible compared with the time between collisions.

7. External forces (such as gravity) are ignored.

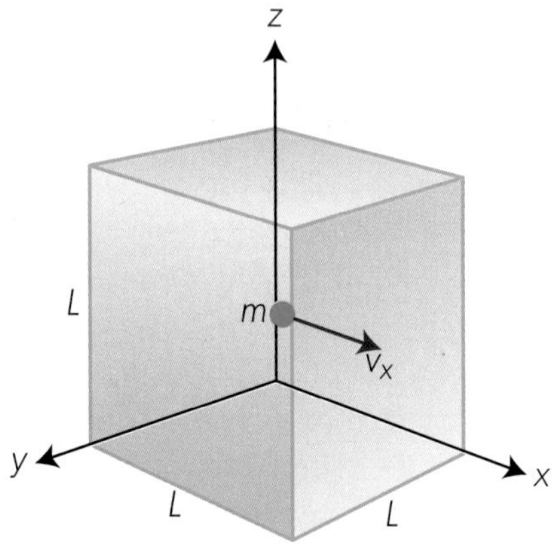

The particle is moving at velocity v as shown which can be resolved into three components v_x, v_y and v_z.

▲ Figure 10 The particle strikes the right-hand wall at 90° and then retraces its path to the far wall. There is a transfer of momentum between the wall and the particle.

	Assumption
Our model begins with the single particle of mass m shown in Figure 10 moving inside a cube of wall length L. This box has a volume V, where $V = L^3$.	1
The particle strikes the cube wall at right angles with speed v_x and it is the x-direction on which we initially focus.	
The particle collides elastically with the wall and rebounds with a velocity that is equal and opposite to its original value.	4
The momentum of the particle before the collision is mv_x and afterwards is $-mv_x$, giving an overall momentum change of $-mv_x - (mv_x)$, which is $-2mv_x$ (the signs are important).	5
The particle will then travel to the opposite wall and back again taking a time T to do so. The average force F_x that the wall exerts on the particle over the whole of this motion is the change of momentum divided by T and this is $-\dfrac{2mv_x}{T}$. This time T can be expressed as	6
	7

$$\frac{\text{total distance travelled as the particle crosses the box}}{\text{x-component of speed}} = \frac{2L}{v_x}.$$

The force F_x is therefore $\dfrac{-2mv_x}{\left(\dfrac{2L}{v_x}\right)} = -\dfrac{mv_x^2}{L}$.

This tells us that the particle exerts a force of $\dfrac{mv_x^2}{L}$ on the right-hand wall of the box (this is, by Newton's third law of motion, in the opposite direction to the force exerted by the wall on the particle).

For a gas, the box will be filled with N gas particles. One of them is shown in Figure 11. These particles have a range of velocities and therefore a range of speed components. The first particle has three speed components $(v_{x_1}, v_{y_1}, v_{z_1})$ in the axis directions x, y and z. The second particle has components $(v_{x_2}, v_{y_2}, v_{z_2})$ and so on up to the final particle $(v_{x_N}, v_{y_N}, v_{z_N})$.

In terms of the original wall, each of the particles collides with it to give an averaged-out force on the wall, given by

$$F_x = \frac{m\left(v_{x_1}^2 + v_{x_2}^2 + \cdots + v_{x_N}^2\right)}{L}.$$

This averaging can be taken further by using the mean of the squared velocities, in other words, $\overline{v_x^2} = \dfrac{\left(v_{x_1}^2 + v_{x_2}^2 + \cdots + v_{x_N}^2\right)}{N}$. The quantity $\overline{v_x^2}$ is known as the "mean square speed" of the x-components. When the square root of it is taken, in other words, $\sqrt{\overline{v_x^2}}$, this is the "root mean square speed of the x" components.

Replacing the individual speed components with their mean square speed, the average force on the wall becomes

$$F_x = \frac{Nm}{L}\,\overline{v_x^2}$$

For an individual particle, its actual speed v is the combination of the three components, given by $v^2 = v_x^2 + v_y^2 + v_z^2$.

Applying this to the mean square speeds means that $\overline{v^2} = \overline{v_x^2} + \overline{v_y^2} + \overline{v_z^2}$.

The quantity $\overline{v^2}$ is the **mean square speed of the molecules.**

The line above the symbol extends over the ^{2} to remind you that the speeds are first squared and then averaged.

When N is very large (and when there is one mole of gas in the box, there will be around 10^{24} individual atoms), then the gas will look the same in whatever direction we observe it. This means that the magnitudes of mean components of velocity must be the same:

$$\overline{v_x^2} = \overline{v_y^2} = \overline{v_z^2} \text{ and so } \overline{v^2} = 3\overline{v_x^2} \text{ or } \overline{v_x^2} = \frac{\overline{v^2}}{3}.$$

Replacing v_x by v leads to a total force F on the wall of

$$F = \frac{1}{3}\frac{Nm}{L}\,\overline{v^2}$$

The pressure on the wall is $P = \dfrac{\text{force acting on wall}}{\text{area of wall}}$, and because the area of the wall is L^2, the pressure on a wall is

$$P = \frac{1}{3}\frac{Nm}{L} \times \frac{1}{L^2} \times \overline{v^2} \quad \text{or} \quad P = \frac{1}{3}\frac{Nm}{V}\,\overline{v^2}$$

Assumption

2

1

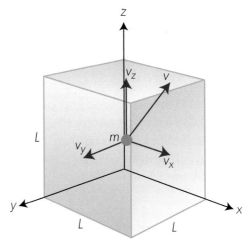

▲ Figure 11 The particle is moving with a random velocity v that has components (v_x, v_y, v_z)

Large numbers

The statistical idea of an average or a mean value works well for large numbers. The numbers of atoms or molecules of a gas are usually very large indeed. Despite the random motion of the particles, the overall effect is very unlikely to deviate from the behaviour of the model.

Where else do large numbers lead to increased confidence in knowledge?

Although you need to understand the physics that underlies the derivation of $P = \dfrac{1}{3}\dfrac{Nm}{V}\,\overline{v^2}$, you do not need to recall the proof that leads to the result.

How does the concept of force and momentum link mechanics and thermodynamics?

How does a consideration of the kinetic energy of molecules relate to the development of the gas laws?

The kinetic theory is an analysis of gas behaviour at the microscopic level. It begins with a consideration of the kinematics of a single gas particle moving in a highly constrained way and gradually broadens the analysis until it encompasses all the particles in the ensemble. The theory applies the familiar concepts of force and rate of change of momentum from Theme A to link the particle motion to the pressure that the whole gas exerts on the internal walls of its container.

Coupled to this kinematic approach is the underlying property that each particle has a kinetic energy due to its motion and that therefore the whole ensemble of atoms has a total kinetic energy. This can also be analysed to give an average energy for all the atoms in the gas.

The gas laws arise empirically. The kinetic theory arises theoretically. They both meet in the equation that links the internal energy of the gas to the amount of gaseous substance in the container and its temperature.

The kinetic theory of an ideal gas was itself extended later. Four giants of 20th-century physics, Fermi and Dirac, and Einstein and Bose used the kinetic theory as a basis for a model of the behaviour of other particles such as electrons, neutrons and photons.

This is also the pressure of the gas itself because pressure acts equally all directions.

We will now drop the line that is printed over the v^2, but you should remember that, from here on, v is the square root of the average (mean) translational speed^2 of a gas particle. The equation becomes $P = \dfrac{1}{3}\left(\dfrac{Nm}{V}\right)v^2$. This square root of the mean squared speed is usually called the **root mean square** (rms) value.

The quantity $N \times m$ is the *number of molecules × mass of one molecule*. In other words, it is the total mass of the gas in the box. The term in brackets is the total mass divided by the gas volume: the gas density. A simpler form of this equation is therefore

$$P = \frac{1}{3}\rho v^2$$

where ρ is the density of the gas.

 Root mean square

- Tool 3: Carry out calculations involving decimals, fractions and exponents.

Root mean square (rms) speeds are not the same as mean speeds or mean velocities. Squaring the magnitude of the velocity remove direction.

Imagine that a gas has four particles and at one instant these particles have velocities of $+1\,m\,s^{-1}$, $-3\,m\,s^{-1}$, $+5\,m\,s^{-1}$ and $+7\,m\,s^{-1}$.

- The mean speed of the four particles is $\dfrac{(1 + 3 + 5 + 7)}{4} = \dfrac{16}{4} = 4.0\,m\,s^{-1}$.

- The mean velocity of the four particles is $\dfrac{(1 - 3 + 5 + 7)}{4} = \dfrac{10}{4} = 2.5\,m\,s^{-1}$.

- The mean square speed is $\dfrac{(1^2 + (-3)^2 + 5^2 + 7^2)}{4} = \dfrac{84}{4} = 21\,m^2\,s^{-2}$.

- The root mean square speed is $\sqrt{21} = 4.6\,m\,s^{-1}$.

Quantities with rms values occur elsewhere in physics, most notably in alternating current (AC) theory where the rms value of a current has the same heating effect as a direct current with the same value.

 Thinking skills—Credibility

The kinetic model of an ideal gas is successful as it simplifies a complicated situation. The application of simple physics leads to predictions which match experimental results.

It is important to consider the credibility of any model and whether it has limitations. Make some estimates of the air in your room and answer the following questions.

Practice questions

11. The kinetic model assumes that the total volume of gas particles is much smaller than the volume of the box.

    a. Estimate the volume of the room.

    b. Estimate the number of gas molecules in the room.

    c. The diameter of a nitrogen molecule is about 3×10^{-10} m. Estimate the total volume of the gas molecules in the room.

12. The kinetic model assumes that we can ignore the effects of external forces such as gravity.

    a. Estimate the change in gravitational potential energy (Topic D.2) as a gas molecule travels from the floor to the ceiling.

    b. What is this change in terms of a percentage of the molecule's kinetic energy?

13. The kinetic model assumes that gas molecules collide elastically with each other and with the walls of the container. What would an inelastic collision of gas molecules be like? Is this a good assumption?

Worked example 9

A beam of electrons is formed in an electron tube. The cross-sectional area of the beam is $1.0 \, \text{mm}^2$. Electrons are emitted at a rate of 9.0×10^{14} per second and travel at a speed of $1.0 \times 10^7 \, \text{m s}^{-1}$. The mass of an electron is 9.11×10^{-31} kg. The electrons are incident normally on the inner wall of the tube and are absorbed by the wall. Determine the pressure exerted by the beam on the wall.

Solution

The force exerted by the electrons on the wall is equal to the rate of change of momentum of the electrons.

$$F = \frac{\Delta m}{\Delta t} v = 9.0 \times 10^{14} \times 9.11 \times 10^{-31} \times 1.0 \times 10^7 = 8.2 \times 10^{-9} \, \text{N}.$$

The pressure is $P = \dfrac{F}{A} = \dfrac{8.2 \times 10^{-9}}{1.0 \times 10^{-6}} = 8.2 \times 10^{-3} \, \text{Pa}.$

Worked example 10

Dry air at room temperature and a pressure of 100 kPa has a density of $1.16 \, \text{kg m}^{-3}$. Calculate the root mean square speed of the air molecules in these conditions.

Solution

$$P = \frac{1}{3}\rho v^2 \Rightarrow v = \sqrt{\frac{3P}{\rho}} = \sqrt{\frac{3 \times 100 \times 10^3}{1.16}} = 509 \, \text{m s}^{-1}$$

Note that this value represents the average over different gases that make up the air. The air is a mixture of molecular nitrogen, oxygen and other gases, which all have different root mean square speeds depending on their molecular mass. See Figure 13 later in this chapter and question 3 in Topic B.1.

Worked example 11

Nitrogen makes up about 75.5% of dry air by mass. The rms speed of nitrogen molecules at room temperature is $517 \, \text{m s}^{-1}$. Calculate the pressure due to nitrogen molecules alone.

Solution

The density of nitrogen in air is $0.755 \times 1.16 = 0.876 \, \text{kg m}^{-3}$, using the data from Worked example 10.

$$P = \frac{1}{3} \times 0.876 \times 517^2 = 78.0 \, \text{kPa}.$$

This is known as partial pressure—exerted by the molecules of just one constituent gas in a mixture. Other atmospheric gases are responsible for the remaining 22.0 kPa of the atmospheric pressure.

Practice questions

14. A balloon is filled with 240 kg of helium at a pressure of 1.1×10^5 Pa. The volume of the balloon is 1300 m³. Calculate the root mean square speed of helium atoms in the balloon.

15. Oxygen makes up 23.1% of air by mass. At room temperature, the molecules of atmospheric oxygen exert a pressure of 20.8 kPa. The density of air is 1.16 kg m⁻³.
Calculate the root mean square speed of oxygen molecules.

16. The atmosphere of the planet Venus has an average pressure of 9.0×10^6 Pa and an average density of 5.2 kg m⁻³.
Calculate the root mean square speed of the molecules that make up the atmosphere of Venus.

17. A sealed container of volume 5.0×10^{-4} m³ is filled with 2.2×10^{-4} kg of a gas at a temperature of 300 K and pressure 2.5×10^4 Pa.
Calculate:
   a. the density of the gas
   b. the root mean square speed of the gas molecules
   c. the molar mass of the gas.

How can gas particles of high kinetic energy be used to perform useful work?

Gas particles with high speed and therefore large kinetic energies can do work by acting as sources of energy that is transferred to energy sinks. This transfer is the basis of a heat engine (Topic B.4).

A heat engine is a general concept, not confined simply to the internal combustion engine. Any process where work is done as a result of energy transfer from a hot to a cold body can be regarded as a heat engine—including biological systems such as animals.

Interpreting temperature

The empirical ideal gas equation can be linked to this theoretical kinetic model. The expression for the pressure in a gas, $P = \dfrac{1}{3}\dfrac{Nm}{V}\overline{v^2}$, can be rewritten as $PV = \dfrac{Nm}{3}\overline{v^2}$. This leads to a new equation that uses elements of both the empirical and theoretical equations:

$$PV = \frac{Nm}{3}\overline{v^2} = Nk_B T$$

A further rearrangement of $\dfrac{Nm}{3}\overline{v^2} = Nk_B T$ and the introduction of a numerical factor $\dfrac{3}{2}$ gives

$$\frac{3}{2}\frac{Nm}{3}\overline{v^2} = N \times \frac{1}{2}m\overline{v^2} = \frac{3}{2}Nk_B T$$

Therefore, $N \times \dfrac{1}{2}m\overline{v^2} = \dfrac{3}{2}Nk_B T$ and

$$\frac{1}{2}m\overline{v^2} = \frac{3}{2}k_B T$$

This equation has a very specific physical meaning. The left-hand side of the of the equation is $\dfrac{1}{2} \times mass \times speed^2$ and is therefore the translational kinetic energy of the "average" gas molecule in the box. The right-hand side of the equation is $1.5 \times Boltzmann\ constant \times absolute\ temperature$.

This is an important result. The kinetic theory links the macroscopic quantity *temperature* to the microscopic quantity *kinetic energy of the "average" molecule*. This interpretation is discussed in Topic B.1 (page 203)

Recall that an ideal gas has no long-range intermolecular forces (assumptions 5 and 7) and therefore no potential energy contribution to the total energy. The total energy of the gas relies solely on the kinetic energy. Given that there are N molecules in the gas, then

$$\text{total internal energy of an ideal gas} = \frac{3}{2}Nk_B T$$

This is an important step in thinking about the properties of an ideal gas. Its total internal energy is directly proportional to the Kelvin temperature. The units of k_B (J K⁻¹, as we saw earlier) reflect this and will help you to remember the importance of this relationship.

Worked example 12

Calculate the root mean square speed of water vapour molecules at a temperature of 300 K. The mass of a water molecule is 3.0×10^{-26} kg.

Solution

We assume that water vapour behaves like an ideal gas.

$$\frac{1}{2}mv^2 = \frac{3}{2}k_B T \Rightarrow v = \sqrt{\frac{3k_B T}{m}} = \sqrt{\frac{3 \times 1.38 \times 10^{-23} \times 300}{3.0 \times 10^{-26}}} = 640 \, \text{m s}^{-1}$$

Worked example 13

A sample of 2.0 mol of a monatomic ideal gas is kept at a constant volume. Calculate the energy transferred to the sample when its temperature increases from 30 °C to 100 °C.

Solution

The energy transferred to the sample is equal to the increase in its internal energy:

$$\Delta U = \frac{3}{2}Nk_B \Delta T = \frac{3}{2}nR\Delta T = \frac{3}{2} \times 2.0 \times 8.31 \times (100 - 30) = 1.7 \, \text{kJ}$$

Practice questions

18. A container of constant volume is filled with an unknown quantity of a monatomic ideal gas. When 50 J of thermal energy is transferred to the container, the temperature of the gas increases by 85 K.

    a. Calculate, in mol, the quantity of gas in the container.

    The molar mass of the gas is 40 g mol^{-1}.

    b. Calculate:
       i. the mass, in kg, of the sample.
       ii. the specific heat capacity, in J kg^{-1} K^{-1}, of the gas when it is kept at constant volume.

19. Helium is a monatomic noble gas. The mass of one atom of helium is 6.65×10^{-27} kg. Thermal energy of 80.0 J is transferred to a sample of helium of mass 2.50×10^{-3} kg, without changing the volume of the sample.

    Calculate:

    a. the number of atoms in the sample
    b. the change in the temperature of the sample.

Ideal and real gases

It is important to remember that the kinetic model derived above applies to a monatomic (single atom) gas—not a gas where the molecules have two or more atoms. For molecules the theory must be adapted.

Some of the behaviours that are shown by real gases should not occur if the assumptions made in the kinetic theory are correct.

An example of non-ideal behaviour is that gases can be liquefied (turned from a gas into a liquid). This should not be possible with an ideal gas. Liquefaction was studied by Thomas Andrews, an Irish chemist of the mid-19th century. He showed that there is a critical temperature above which a gas cannot be liquefied; for carbon dioxide this is 31 °C. Once the gas has become a liquid, then, of course, it cannot be compressed.

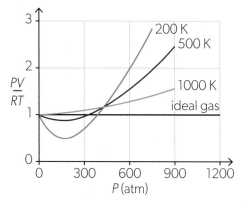

▲ Figure 12 The behaviour of a real gas showing the deviation from ideal-gas behaviour.

 Hypotheses

Van der Waals won the Nobel Prize in 1910 for his work on modifying the ideal gas equation. In its modified form his equation is

$$\left(P + \frac{n^2a}{V^2}\right)(V - nb) = nRT.$$ The $\frac{n^2a}{V^2}$

term that modifies P is to account for the intermolecular forces of attraction. As the number of molecules increases (increase in n) or as the volume decreases, the molecules become closer together. The forces between them can no longer be ignored and the pressure term is effectively increased.

Similarly, the volume available for molecular movement itself is reduced as n increases. This accounts for the change to the V term and for its negative sign.

The values of a and b depend on the gas under consideration so that van der Waals produced a hybrid empirical–theoretical model.

A graph of $\frac{PV}{RT}$ against P for one mole of a real gas hints at this behaviour. The quantity $\frac{PV}{RT} = n$ and so an ideal gas has a constant value on the y-axis. This is not found for real gases at low temperatures and high pressures (Figure 12). For an incompressible material we would expect the line to rise vertically, parallel to the y-axis. Look at the lowest temperature here (the green line) and you can see that the behaviour of the gas is beginning to approach this at high pressures and low temperatures.

The best approximations to ideal gas behaviour occur when a gas is at a high temperature and low pressure.

 Applying a theory — The Maxwell–Boltzmann distribution and the atmosphere

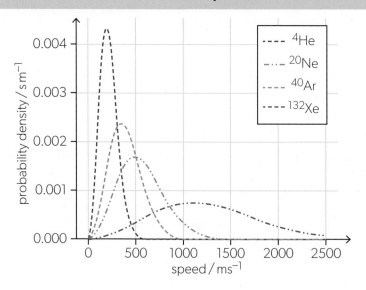

▲ Figure 13 The distribution of speeds of four gases at room temperature.

In principle, Newton's laws of motion can be used to predict the behaviour of the collection of molecules that make up a gas. In practice, this is not possible. The number of particles is far too great even for the most powerful computer. In 1860, Maxwell produced arguments resembling our derivation of the kinetic model but he used ranges of molecular speeds rather than the mean square speed we used in our proof.

In 1868, Boltzmann improved Maxwell's argument and his work led to an understanding of the molecular-speed distribution that is shown in Figure 13. This shows the speed distribution for four gaseous elements, all for the same temperature. Helium, the least massive molecule shown here, has the largest tail and therefore more molecules with the highest speeds. Hydrogen would have speeds even higher in its tail. This is one reason why the atmosphere contains little helium and hydrogen. The escape speed for Earth is 11 km s⁻¹ and for hydrogen there is a significant number of molecules with this speed. These molecules are lost from the atmosphere. The speeds are then re-distributed through collisions so that another group of molecules gains the highest speeds and the loss of hydrogen continues.

Your work on gravitational fields (Topic D.1) links to the kinetic theory in this phenomenon.

Data-based questions

The speed of sound in a gas is closely related to the typical speeds of the molecules. As a consequence, the speed of sound in air varies with temperature. The table shows the speed of sound at various temperatures. The uncertainty in each speed is $\pm 1\,\text{m s}^{-1}$.

T / °C	Speed of sound / m s⁻¹
−20	319
−10	325
0	331
10	337
20	343
30	349
40	355

- Plot a graph of the speed of sound vs temperature. Include the uncertainties in the speed.

- Find the gradient of your graph and use the maximum and minimum gradients to find the uncertainty in your gradient.

- Although the data seem to follow a linear trend very well, theory would predict that $v^2 = bT$, where T is the absolute temperature (in kelvin) and b is a constant. Plot a graph of v^2 against T in kelvin.

- Add error bars to your graph.

- Use the gradient of your graph to determine b and determine the uncertainty in b.

- At what temperature would a measurement of the speed of sound have to be made in order to distinguish between these two trends?

What other simplified models do we rely on to communicate our understanding of complex phenomena? (NOS)

You will come across many examples of complex phenomena that are explained using simple models. Examples include:

- the motion of electrons through a metallic lattice (Theme B.5)

- the behaviour of single waves as they diffract through apertures and multiple coherent waves as they interfere (Theme C)

- the behaviour of the air in a pipe when a standing sound wave is propagated in it (Theme C)

- the use of field theory for the description of gravitational, electric and magnetic fields (Theme D)

- the energy levels in atoms (Theme E.1).

What other simplified models are used in this course?

B.4 Thermodynamics

How can energy transfers and storage within a system be described and analysed?

How can the future evolution of a system be determined?

How is entropy fundamental to the evolution of the universe?

Society uses machines that transfer heat into mechanical energy. Thermodynamics is the branch of physics that models the conversion of the internal energy of a system into mechanical work. It also answers the question of the extent to which such a conversion is possible.

Inevitably, in physics, these systems are large. Even $\frac{1}{1000}$ of a mole will contain 10^{21} particles. This means that statistics can apply to the system in the same way that the averaging of a collection of molecules allowed us to develop the kinetic theory of Topic B.3. Thermodynamics looks at the wholesale transfer of energy between energy reservoirs, not at the movement of individual particles, each carrying a minuscule amount of energy.

The tools of thermodynamics allow future predictions of behaviour too. They can describe the likely (though not completely certain) outcome of a transfer, or whether it will even occur or not. Again, this is a statistical judgment. The sheer number of microscopic particles in a system means that only average macroscopic behaviour is likely to be seen. This is itself driven by the assumed randomness of these large systems. A measure of this randomness, called entropy, is introduced towards the end of this topic. This quantity allows a measure of prediction about the possible outcomes of an energy transfer.

Our knowledge of the universal nature of thermodynamics allows very long-term predictions. Entropy is a measure of randomness and, as we allow systems to interact, the amount of randomness increases with time. One way to interpret any change is as a transfer of energy from a

▲ **Figure 1** A fire releases thermal energy and this energy is dissipated to the cooler surroundings. The ordered structure of the wood with its complex molecules storing energy, is a useful source of fuel. Once this thermal energy is distributed to the environment in a disordered way, it is less useful and impossible to put back.

hot body (at a high temperature) to a cold body (at a low temperature). The statistics of thermodynamics predict that transfers are always in this direction except in the very rarest cases. Essentially, everything in the universe is moving towards a uniform temperature. This so-called "heat death" is one way to describe the fate of the universe.

In this additional higher level topic, you will learn about:

- the first law of thermodynamics
- modelling isovolumetric, isobaric, isothermal and adiabatic processes
- entropy and the degree of disorder of a system
- entropy and the number of possible microstates of a system

- the second law of thermodynamics
- entropy changes in a real isolated and non-isolated systems
- irreversibility
- cyclic heat engines and their efficiency
- the Carnot cycle and its theoretical efficiency.

Introduction

The ideal gas and how the behaviour of real gases differs from it occupied Topic B.3. This topic is also largely concerned with gas behaviour but in a different way. It examines changes in the state of gases and other substances, not in isolation, but in the context of their surroundings. This is much closer to our real interactions with matter. We expand and compress gases, we allow them to act as energy sinks and then we use the stored energy to perform useful work for us. It is interactions like this that are examined in this topic.

In earlier topics, our interests concerned the overall state before and after a change to the gas state. Now we are also concerned with the way in which the change is made. What can we say about the consequences of the rate at which the changes happen? And what models emerge from these considerations?

The modelling used in this topic again involves the assumption that the molecules in a solid, gas or liquid can be treated as single entities.

System and surroundings

The Guiding questions for this topic use the term "system". This word needs some explanation as the term has an exact meaning in thermodynamics. Figure 3 shows the connection between a system, its surroundings and the universe. The **system** is the body or bodies that we are considering. Normally, this system can interact with the **surroundings** through the transfer of energy. The system together with the surroundings constitute the **universe**. The total energy within the universe is assumed to have a constant value.

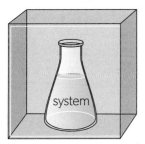

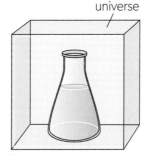

the flask represents the boundary in this case

▲ **Figure 3** In thermodynamics, there is a system under consideration and its surroundings. Taken together, these are the universe.

Sometimes the system is said to be a **closed system**. This means that the entities (often the particles of a gas) inside the system cannot vary in number. A phrase such as "a fixed mass of an ideal gas" indicates this—because when the mass is fixed, so is the number of particles in the gas which, in this case, is ideal and will follow the kinetic gas model and the gas laws. However, energy can flow into and out of a closed system. When neither matter nor energy can enter or leave, then the system is said to be **isolated**.

Much of the work on thermodynamics originated from scientists trying to make steam engines more efficient. Throughout the 18th and 19th centuries, steam engines were able to do more work, using less fuel. They became more useful, not just for transport, but for driving machinery in factories. This was the industrial revolution—the nature of society changed as fewer people were needed for farming and more people worked in factories. The industrial revolution began to increase life expectancy, improve living conditions and raise wages, although some of these improvements did not appear straight away. However, the industrial revolution also marked the beginning of population increase, more pollution and a reliance on fossil fuels.

▲ **Figure 2** A steam locomotive hauling a passenger train. These locomotives are direct descendants of the steam engines originally developed in the early 1800s.

Hypotheses — Joule, work and temperature

The unit of energy was named after James Joule, the first scientist to show that doing work on a system is equivalent to transferring thermal energy to it.

He showed this using apparatus in which weights fall (transferring gravitational potential energy to kinetic energy) and turn paddles which stir the water in a container (Figure 4). The temperature of the water increases as a result. This result seems obvious to us today, but in the 1840s this conversion was not yet fully understood.

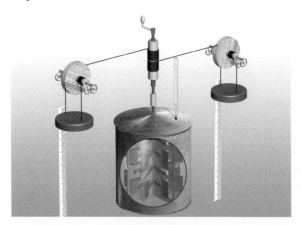

▲ **Figure 4** Joule's apparatus used to find the mechanical equivalent of heat.

Joule tried other more direct experiments, too. He took a pair of sensitive thermometers with him on his honeymoon and placed one at the top and one at the bottom of a tall waterfall in Switzerland. He hoped to find a temperature difference (higher at the bottom because energy was transferred to the falling water). Unfortunately, because most waterfalls mix the water with the surrounding air very effectively, the experiment did not give the result he expected (Figure 5).

▲ **Figure 5** Even the highest waterfall in the world — Angel Falls in Venezuela with a single drop of over 800 m — would only have a theoretical temperature difference of 1.9 °C between the top and the bottom. In reality, the surrounding air is likely to differ in temperature by more than this between the top and bottom, and the water mixes with the air and its surroundings.

This conversion of "work to heat" was a hypothesis by Joule. He designed an ingenious and successful experiment to verify his idea (Figure 4). We can interpret the waterfall test as a falsification of the hypothesis. However, further examination of the conditions of this experiment show that it is not a good test of Joule's idea.

The first law of thermodynamics

The internal energy of a system changes when:

- the system does work, or has work done on it

- energy is transferred into or out of the system when there are temperature differences between the system and the surroundings.

This is a statement of **conservation of energy** and leads to the **first law of thermodynamics**. There are several ways to express this law; the version used in IB Diploma Programme physics is

$$Q = \Delta U + W$$

where Q is energy transferred, ΔU is the change in internal energy of the system and W is the work done by the system.

It is important to use a consistent sign convention with this equation.

- A *positive* Q means that heat energy is transferred *from* the surroundings *to* the system

- A *positive* ΔU means that the internal energy of the system *increases.*

- A *positive* W means that the system does work *on* the surroundings

This is the **Clausius sign convention**.

ATL Using terminology consistently—What's in a sign?

There is another sign convention in common usage, so it is important to be clear about which one is used in a particular case. The second convention is that of the International Union of Pure and Applied Chemistry (IUPAC) where the first law is written as $\Delta U = Q + W$. Although this looks similar at first glance, when the IUPAC equation is rearranged in the same way as the IB Diploma version, it becomes $Q = \Delta U - W$. So in the IUPAC formulation, work done *on* the system *by* the surroundings is treated as positive.

It is important to stick to the IB Diploma version. However, the two equations always lead to the same answers when used consistently.

There are many other cases where you must be consistent in defining what is a positive quantity. In mechanics, for example, when considering projectile motion, you may decide that the upwards direction is positive. However, taking the downwards direction as positive should still lead to the same answers. Can you find any other examples where a sign convention is important?

Worked example 1

80 J of work is transferred to a system. The internal energy of the system increases by 60 J.

a. State, referring to the first law of thermodynamics, the magnitude and the sign of:

    i. W ii. ΔU.

b. Calculate the energy transferred between the system and the surroundings. State whether the energy is transferred into or out of the system.

Solutions

a. i.   The work is done by the surroundings on the system, so W is negative. $W = -80$ J.

    ii.   The internal energy has increased, so ΔU is positive. $\Delta U = 60$ J.

b. $Q = \Delta U + W = 60 + (-80) = -20$ J. A negative Q indicates that thermal energy is transferred from the system to the surroundings.

Worked example 2

An ideal gas in a sealed container with a piston expands without change in temperature. Explain why thermal energy must be transferred to the gas.

Solution

Since the temperature of the gas does not change, its internal energy remains constant: $\Delta U = 0$. The first law of thermodynamics becomes $Q = W$. The gas does positive work on the surroundings when it expands. Hence, an equal amount of thermal energy must be supplied to the gas.

Worked example 3

A cylinder with a piston contains 0.025 mol of monatomic ideal gas at a temperature 10 °C. Thermal energy of 12 J is supplied to the gas. The gas expands and does a work of 5.0 J on the piston.

Calculate:

a. the change in the internal energy of the gas

b. the final temperature of the gas.

Solutions

a. In this situation, $Q > 0$ and $W > 0$. $\Delta U = Q - W = 12 - 5.0 = 7.0$ J.

b. For an ideal gas, the change in temperature is proportional to the change in the internal energy.

$$\Delta T = \frac{\Delta U}{\frac{3}{2}nR} = \frac{7.0}{\frac{3}{2} \times 0.025 \times 8.31} = 22 \text{ K. The final temperature is } 10 + 22 = 32 \,^{\circ}\text{C.}$$

Practice questions

1. The work done to compress a sample of a gas is 50 J. During the compression, an energy of 30 J is transferred to the gas from the surroundings. What is the change in the internal energy of the gas?

   A. −80 J B. −20 J C. +20 J D. +80 J

2. A quantity of 0.060 mol of monatomic ideal gas has a fixed volume. 40 J of energy is supplied to the gas.

   a. State the work done on the gas.

   b. Calculate the change in the temperature of the gas.

3. When 0.030 mol of monatomic ideal gas is compressed, the temperature of the gas increases from 20 °C to 90 °C. The gas does not exchange thermal energy with the surroundings. Calculate the work done in compressing the gas.

4. A quantity of 8.0×10^{-3} mol of a monatomic ideal gas in a cylinder is compressed by a piston. The work done on the gas is 7.0 J and the temperature of the gas increases by 50 K.

   a. Calculate:

      i. the change in the internal energy of the gas

      ii. the energy transferred between the gas and the surroundings during compression.

   b. The gas remains compressed at a constant volume and its temperature increases by another 50 K. State the energy transferred to the gas.

Using pressure–volume diagrams

The work on the three gas laws in Topic B.3 leads to three graphs: pressure–volume, pressure–temperature, and volume–temperature. These are all graphs that represent gas processes (that is, changes in the conditions of a gas). The first of these (P–V) has a particular use because it can lead us directly to the work that a gas is doing on its surroundings or the work that is being done on the gas.

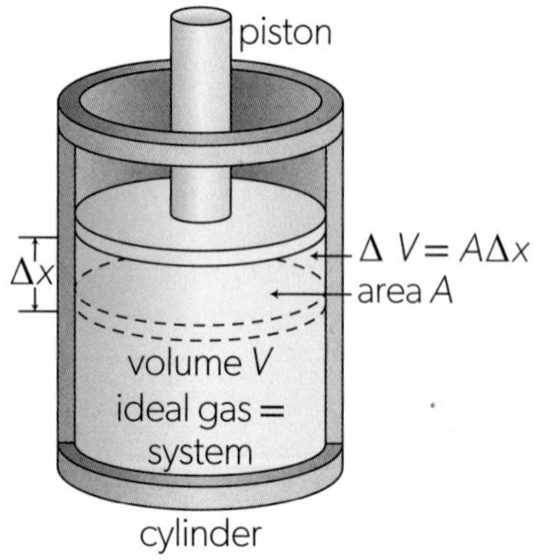

An ideal gas at a pressure P is trapped in a cylinder of cross-sectional area A by a piston (Figure 6). In this example, the gas constitutes the system. The gas is allowed to expand moving the piston upwards by a distance Δx. The gas volume was initially V and it increases by $\Delta V = A \times \Delta x$ because of this movement. The distance Δx is taken to be small, so that the change to the volume is also small.

Some energy is now transferred from the surroundings to the gas in a way that ensures that the gas pressure remains constant. The surroundings in this case are the cylinder and the atmosphere outside it. We ignore temperature and energy changes to the cylinder walls and the piston. The volume of the gas increases; the top of the piston moves upwards and so compresses the atmosphere. Work is being done on the atmosphere.

▲ Figure 6 When a gas expands, it pushes back against the atmosphere through the piston. This means that the gas does work on the atmosphere.

The work done by the gas on the surroundings during the expansion is equal to the force on the piston (which is constant at $P \times A$). The work done is therefore

$$F \times \Delta x = (P \times A) \times \Delta x = P \times (A\Delta x) = P \times \Delta V$$

Indicator diagrams — Visualizing the work done

P–V graphs are commonly used in engineering physics when dealing with the performance of steam engines or internal combustion engines. In this practical context, the graph is known as an **indicator diagram**. The diagram was used very early in the development of steam engines by James Watt and his associate John Southern to study the efficiency of the engines. The P–V graph itself was first used by Émile Clapeyron in 1834 to illustrate the Carnot cycle that you will meet later in this topic.

To what extent do graphs such as this represent the scientific fact itself?

In terms of the first law of thermodynamics, $W = +P\Delta V$. This is positive because the gas (the system) is doing work on the surroundings. Figure 7 shows the change X to Y for a gas expanding at constant pressure. The work done by the gas is the area below the line that joins X and Y on the graph. Notice the arrow on this line: it shows the direction of the change (expanding gas in this case) and shows also that the work done by the gas is positive. When the arrow is in the other direction, then the work done is by the surroundings on the gas and, using our sign convention, is negative.

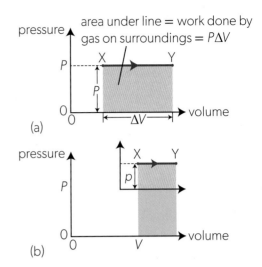

area under line = work done by gas on surroundings = $P\Delta V$

▶ Figure 7 (a) The work done by a gas as it expands from X to Y is the area under the line XY. When the change is in the opposite direction work is done on the gas. (b) The graph sometimes has a false origin. Always make sure that you calculate the total area under the line.

Graph origins and areas

- Tool 3: Use basic arithmetic and algebraic calculations to solve problems.
- Tool 3: Interpret features of graphs including gradient, changes in gradient, intercepts, maxima and minima, and areas under the graph.

Look closely at the graph in Figure 7(a). There is an origin marked as (0, 0). You will not always have a real origin; sometimes the graph will be the small graph (top-right in Figure 7(b)) which has a false origin. Don't forget to include the green area in your calculation for the total work done.

When the P–V graph is not straight, then the task becomes one of counting squares or estimating the area in some other way. You must use the technique of dividing the area below the line into strips or other convenient shapes. Figure 8 shows you how. Simply add up the values for each approximate rectangle. In other words,

$$\text{work done} = P_1\Delta V_1 + P_2\Delta V_2 + \cdots + P_N\Delta V_N$$

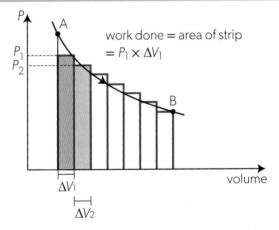

work done = area of strip = $P_1 \times \Delta V_1$

▲ Figure 8 When the variation of P with V is a curve, use thin rectangular strips or counting squares to evaluate the work done.

Make sure that the strips are thin enough to give a reasonable answer but not so thin that the problem takes a long time to complete.

Worked example 4

A monatomic ideal gas is compressed along the path AB as shown.

a. Calculate the work done during the compression.

The temperature of the gas at A is 480 K.

b. Calculate the temperature of the gas at B.

c. Determine:
   i. the change in the internal energy of the gas
   ii. the energy transferred between the gas and the surroundings.

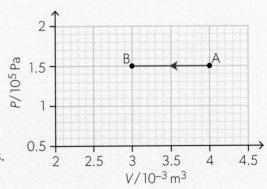

Solutions

a. The gas is compressed at a constant pressure; hence $W = P\Delta V = 1.5 \times 10^5 \times (3 - 4) \times 10^{-3} = -150\,J$. The value is negative because the work is done on the gas to compress it.

b. For an ideal gas at a constant pressure, $\frac{T}{V} = $ constant, so $\frac{T_B}{V_B} = \frac{T_A}{V_A}$.

$\frac{T_B}{3 \times 10^{-3}} = \frac{480}{4 \times 10^{-3}}$ and $T_B = 360\,\text{K}$.

c. i.   The ideal gas equation implies that $nR\Delta T = P\Delta V$. Hence, the change in the internal energy is

$$\Delta U = \frac{3}{2} nR\Delta T = \frac{3}{2} P\Delta V = \frac{3}{2} \times (-150) = -225\,\text{J}.$$

    ii. From the first law of thermodynamics, $Q = \Delta U + W = -225 - 150 = -375\,\text{J}$. The energy is transferred out of the gas.

Worked example 5

The $P\text{–}V$ graph shows two possible processes that can change the conditions of a gas from the initial state A to the final state B.

Deduce which of the processes results in a greater amount of:

a. work done by the gas

b. thermal energy transferred to the gas.

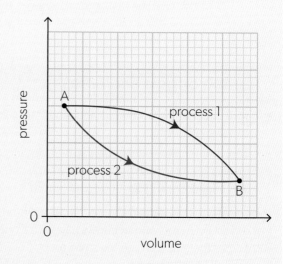

Solutions

a. The change in the volume of the gas is the same for both processes. Process 1, on average, has a greater gas pressure than process 2. From $W = P\Delta V$, the work done by the gas during process 1 is greater.

b. The change in the internal energy of the gas depends only on the temperature difference between the states A and B. Hence, ΔU is the same for both processes. From $Q = \Delta U + W$, the heat transferred to the gas is greater during process 1, because the gas must do more work for the same change in the internal energy.

Worked example 6

An ideal gas expands along the path AB shown in the $P\text{–}V$ diagram.

a. Calculate the work done by the gas.

b. Show that the initial and the final temperature of the gas is the same.

c. Hence, state the energy supplied to the gas.

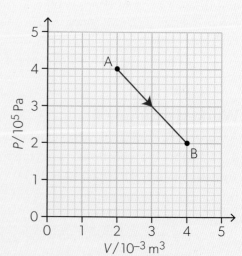

Solutions

a. The work is equal to the area under the path AB. The gas expands hence the work is positive. $W = \dfrac{2 \times 10^5 + 4 \times 10^5}{2} \times 2 \times 10^{-3} = 600\,\text{J}.$

b. $P_A V_A = 4 \times 10^5 \times 2 \times 10^{-3} = 800\,\text{Pa\,m}^3$ and $P_B V_B = 2 \times 10^5 \times 4 \times 10^{-3}$ $= 800\,\text{Pa\,m}^3$. From the ideal gas equation, since $P_A V_A = P_B V_B$, the temperature at A and at B must be the same.

c. Equal temperature means that the gas has the same internal energy in both states, so $\Delta U = 0$. The first law of thermodynamics gives $Q = W = 600\,\text{J}.$

Practice questions

5. A monatomic ideal gas expands at a constant pressure of 1.2×10^5 Pa. The initial temperature of the gas is 300 K and the initial volume is 1.5×10^{-4} m^3.

    a. Calculate the number of moles of the gas.

    The work done by the gas during the expansion is 6.0 J.

    b. Determine:

        i. the final volume of the gas

        ii. the final temperature

        iii. the energy transferred to the gas during the expansion.

6. A cylinder with a moveable piston contains an ideal gas. The cylinder is placed in a cold environment and thermal energy is removed from the gas. The volume of the gas decreases from 1.8×10^{-3} m^3 to 1.4×10^{-3} m^3 at a constant atmospheric pressure of 1.0×10^5 Pa.

    a. Calculate the work done.

    The final temperature of the gas is 260 K.

    b. Calculate the initial temperature.

    The energy removed from the gas is 100 J.

    c. Calculate the change in the internal energy of the gas.

An ideal gas and the first law of thermodynamics

P–V diagrams help to visualize four common types of change that occur in a gas:

- **isobaric**—a change carried out at constant pressure

- **isovolumetric**—a change carried out at constant volume

- **isothermal**—a change carried out at constant gas temperature which is constant internal energy

- **adiabatic**—a change carried out with no energy transferred to or from the system.

Isobaric change

Isobaric changes occur at constant pressure. (The word "isobaric" comes from two Greek words "iso" meaning "the same" and "baros" meaning "weight".)

This case was considered in Figure 7(a) where the P–V diagram shows a line parallel to the V-axis and, in this case, the work transferred was easy to calculate.

For an isobaric change, the first law of thermodynamics can be written as

$$Q = \Delta U + P\Delta V$$

The equation of state for the gas, in this case, is

$$\frac{V}{T} = \text{constant}$$

Isovolumetric change

An isovolumetric change occurs when the volume of the gas is constant. Therefore the equation of state is

$$\frac{P}{T} = \text{constant}$$

Figure 9 shows the P–V diagram for this case. The single line parallel to the pressure axis makes it clear that the change in volume is zero ($\Delta V = 0$).

The first law becomes

$$Q = \Delta U + P\Delta V = \Delta U + P \times 0$$

This leads to $Q = \Delta U$. All the heat energy transferred into the system appears as internal energy. Conversely, when the first law is written with a negative sign as $Q = -\Delta U$, then all the energy is removed from the system and transferred from the internal-energy store of the gas.

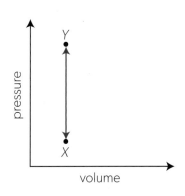

▲ Figure 9 An isovolumetric change shown on a P–V graph. No work is done because the area under the line is zero.

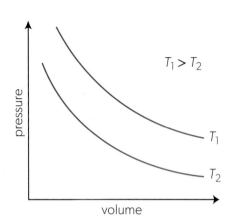

▲ Figure 10 Two isotherms for an ideal gas.

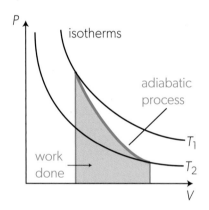

▲ Figure 11 An adiabatic process compared with two isothermal changes. The gas in the adiabatic change moves between the two temperatures T_1 and T_2.

Why $\frac{5}{3}$?

The exponent $\frac{5}{3}$ only applies to monatomic gases. For molecules with two or more atoms, the value changes. Calculations that involve nitrogen (N_2) with two atoms in the molecules take the value $\frac{7}{5}$. Often the adiabatic equation is written as $PV^\gamma =$ constant, where γ is a constant that depends on the number of atoms in a gas molecule.

Isothermal change

The following two equivalent statements describe an isothermal change.

- The internal energy of the system remains constant, and therefore

- the temperature of the system remains unchanged.

This time, the first law $Q = \Delta U + W$ becomes $Q = W$ because $\Delta U = 0$. All the thermal energy is transferred into the system and appears as work done by the gas.

- When the work done is $+W$, then the gas in its container is expanding against the atmosphere (surroundings).

- When the work done is $-W$, then energy is transferred away from the gas and work is done by the surroundings on the gas which is being compressed.

The equation of state for this change is $PV =$ constant. The P–V diagram for two different temperatures T_1 and T_2 is shown in Figure 10. Each line is known as an **isotherm**. T_1 is greater than T_2 here. The larger temperature is always further from the origin.

An important question is whether an isothermal change is ever possible. The answer is that, in practical terms, it is not. The energy must transfer through the boundary between the system and the surroundings without changing the temperature of the gas. This can only happen when the energy transfer happens very slowly through a boundary that is itself a good thermal conductor. An isothermal change will take an infinite time to happen. However, in practice, a slow change can be a good *approximation* to an isothermal change.

Adiabatic change

When no energy is transferred between the surroundings and the system, the change is said to be **adiabatic**. This is achieved when there is an insulating boundary between the system and the surroundings. (They can in principle be at different temperatures because of this insulation.)

The first law of thermodynamics tells us that, for adiabatic conditions, $Q = 0$. Therefore, $Q = \Delta U + W$ becomes:

- $0 = -\Delta U + W$, so $\Delta U = W$ (work done on the system with an increase in internal energy), or

- $0 = \Delta U - W$, so $\Delta U = -W$ (work done by the system with a decrease in internal energy).

As always, you should note the use of the signs here.

For an ideal monatomic gas, the equation for an adiabatic change is $PV^{\frac{5}{3}} =$ constant. You do not need the proof for this equation, but it follows from a combination of the first law and the general gas equation.

At constant temperature, $P_1V_1^{\frac{5}{3}} = P_2V_2^{\frac{5}{3}}$, and when this is divided by $\frac{P_1V_1}{T_1} = \frac{P_2V_2}{T_2}$, then $T_1V_1^{\frac{5}{3}} = T_2V_2^{\frac{5}{3}}$ for conditions where the pressure is constant.

Figure 11 shows a single adiabatic curve (green) on a P–V diagram together with lines (black) for two isothermal changes. The gradient of the adiabatic curve is always greater than that of the isothermals because the exponent of V in the adiabatic equation is greater than one.

Is an adiabatic change possible? As for the isothermal case, the answer strictly is no. There must be no opportunity for any energy to transfer between the system and the surroundings. The change in internal energy of the system (that is, its temperature change) must be the work being done. Any transfer of energy through the system boundary will make the process non-adiabatic.

This leads to the (perhaps surprising) result that an adiabatic change can be approximated by a very rapid change from one gas state to another. The essential point is that there must be no time for energy to transfer through the system boundary (which, in practice, is the wall of the gas container).

Worked example 7

An ideal gas undergoes a cyclic process along the loop ABCA, as shown in the P–V graph. The change AB is isothermal at a temperature of 900 K.

a. Calculate the temperature of the gas at C.

The work done by the gas during the change AB is 330 J.

b. Determine the net work done in one cycle.

c. Explain whether thermal energy is transferred to the gas or from the gas during each of the changes AB, BC and CA.

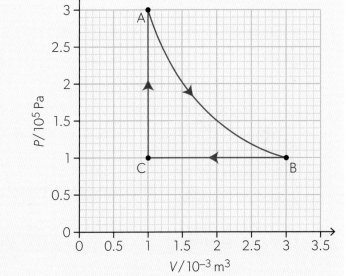

Solutions

a. The change CA is isovolumetric, so $\frac{P}{T}$ = constant.
$$\frac{1 \times 10^5}{T_C} = \frac{3 \times 10^5}{900} \Rightarrow T_C = 300 \text{ K}.$$

b. The work done in compressing the gas during the isobaric change BC is $P\Delta V = 1 \times 10^5 \times (-2 \times 10^{-3}) =$ −200 J. No additional work is done during the final change CA because the volume remains constant. The net work is therefore $W = 330 - 200 = 130$ J. The sign is positive; hence the work is done by the gas on the surroundings. Note that this work is equal to the area enclosed by the cycle.

c. During change AB, the internal energy remains constant, $\Delta U = 0$. The first law of thermodynamics gives $Q = W$, and since $W > 0$ (work done by the gas), we must have $Q > 0$ (energy transferred to the gas).

During change BC, we have $W < 0$ and $\Delta U < 0$, because the temperature of the gas decreases. From $Q = \Delta U + W$, it can be deduced that $Q < 0$ (energy transferred from the gas to the surroundings).

During change CA, $W = 0$ and $\Delta U > 0$, because the gas returns to its original temperature. In this case, $Q = \Delta U$ and therefore $Q > 0$ (energy transferred to the gas).

Worked example 8

A monatomic ideal gas at an initial temperature 300 K, pressure 1.00×10^5 Pa and volume 3.60×10^{-4} m³ is compressed adiabatically to a new volume of 1.20×10^{-4} m³. The compression is represented by the change AB in the P–V graph. The gas is then allowed to cool at constant volume to the original temperature, as represented by the change BC. The dashed line is the isothermal at 300 K.

a. Calculate, for the gas in state B:

   i.   its pressure

   ii. its temperature.

b. Determine the change in the internal energy of the gas during the compression. Give the answer to the nearest joule.

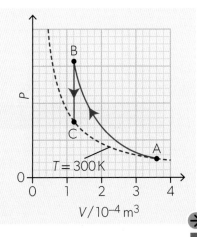

[object Object]

c. Hence, state:

i.   the work done during change AB

ii. the energy transferred during change BC.

Solutions

a. i.   We use the equation of an adiabatic change, $PV^{\frac{5}{3}} = \text{constant}$, so $P_B V_B^{\frac{5}{3}} = P_A V_A^{\frac{5}{3}}$.

$P_B (1.2 \times 10^{-4})^{\frac{5}{3}} = 1.00 \times 10^5 (3.6 \times 10^{-4})^{\frac{5}{3}}$. From here, $P_B = 6.24 \times 10^5$ Pa.

ii. For an ideal gas, $\dfrac{PV}{T} = \text{constant}$, so $\dfrac{P_B V_B}{T_B} = \dfrac{P_A V_A}{T_A}$.

$$\frac{6.24 \times 10^5 \times 1.20 \times 10^{-4}}{T_B} = \frac{1.00 \times 10^5 \times 3.60 \times 10^{-4}}{300} \Rightarrow T_B = 624 \,\text{K}.$$

b. The change in the internal energy is $\Delta U = \dfrac{3}{2} nR(T_B - T_A)$. We find the quantity (number of moles) of the gas by

applying the ideal gas equation to state A. $n = \dfrac{P_A V_A}{RT_A} = \dfrac{1.00 \times 10^5 \times 3.6 \times 10^{-4}}{8.31 \times 300} = 1.44 \times 10^{-2}\,\text{mol}.$

$\Delta U = \dfrac{3}{2} \times 1.44 \times 10^{-2} \times 8.31(624 - 300) = 58\,\text{J}.$

c. We use the first law of thermodynamics, $Q = \Delta U + W$, to answer both parts of this question.

i.   The change AB is adiabatic. Hence, $Q = 0$. The increase in the internal energy is solely due to the work done on the gas, $0 = \Delta U + W \Rightarrow W = -\Delta U = -58\,\text{J}.$

ii. The gas does not change volume from B to C. Hence, $W = 0$. The gas returns to the original temperature, so $\Delta U = -58\,\text{J}$. An equal amount of energy is removed from the gas: $Q = \Delta U = -58\,\text{J}.$

Practice questions

7. An ideal gas undergoes a cyclic process that consists of an adiabatic compression AB, an isovolumetric process BC and an isobaric expansion CA.

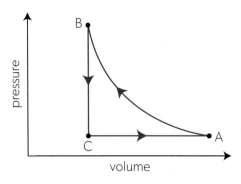

volume

The table contains some of the data about the process. Q is the energy transferred to the gas, ΔU is the change in the internal energy, and W is the work done by the gas.

	Q/J	$\Delta U/J$	W/J
AB			−970
BC		−1570	
CA	1000		

a. Complete the table by filling in the missing quantities, including the appropriate sign.

b. Determine the net energy that leaves the gas during one cycle.

8. A monatomic ideal gas is compressed at a constant temperature of 346 K from an initial state A of volume $5.00 \times 10^{-3}\,\text{m}^3$ and pressure $2.00 \times 10^5\,\text{Pa}$ to a new state B of volume $2.20 \times 10^{-3}\,\text{m}^3$. The isothermal compression is followed by an adiabatic expansion to a final state C of volume $5.00 \times 10^{-3}\,\text{m}^3$.

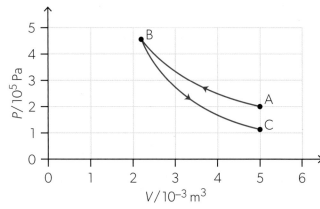

Calculate:

a. the pressure at B

b. the temperature at C.

9. An ideal gas undergoes adiabatic compression AB, isobaric expansion BC and isovolumetric process CA. The volume at A is $6.00 \times 10^{-4}\,m^3$, the pressure at A is $2.00 \times 10^5\,Pa$ and the pressure at B is $6.00 \times 10^5\,Pa$.

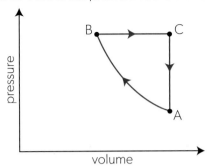

a. Calculate the volume at B.

b. Explain why the temperature at B is greater than at A.

c. Calculate the work done during process BC.

d. The work done on the gas during process AB is 99 J. Calculate the net work done in one cycle.

e. The energy that leaves the gas during process CA is 360 J. Calculate the energy transferred to the gas during process BC.

Heat cycles and engines

The conversion of internal energy into work is an important one for society because it is the basis of a motor or engine.

A device that performs useful work by *continuously* converting energy to work is known as a **heat engine**. The principle behind any heat engine (Figure 12) is that energy Q_h is transferred into the engine at a high temperature T_h and that energy Q_c is rejected by the engine at a lower temperature T_c. The energy difference $(Q_h - Q_c)$ is used for work. Any engine that is to work continuously must eventually be returned to its initial state and thus work in a cycle. There must be some way to return the heat engine to its original situation.

The thermal efficiency η of the heat engine is

$$\eta = \frac{\text{useful work output}}{\text{input energy}} = \frac{Q_h - Q_c}{Q_h}$$

The first description of a cyclic heat engine was given by the French engineer and physicist Nicolas Léonard Sadi Carnot in 1824. He described a cycle, as shown in Figure 13, in which an ideal gas is carried around four processes in sequence: two isothermal and two adiabatic. The moving parts of the heat engine are assumed to be completely frictionless. This is known as the **Carnot cycle**. You should compare Figure 13 with Figure 11.

The steps in the cycle are:

*   Step A→B. The gas undergoes an isothermal expansion when an energy Q_h is supplied to it at high temperature T_h. As the temperature is constant, the internal energy of the gas is unchanged. All the energy absorbed (Q_h) does work on the surroundings through the expansion of the gas.

*   Step B→C. The gas expands adiabatically. No energy is absorbed or rejected by the gas ($Q = 0$) but because this is an expansion, the internal energy of the gas falls and its temperature decreases to T_c. The gas has transferred internal energy into work done on the surroundings, $-\Delta U = W$.

*   Step C→D. The gas returns to its original state (A) and it does this in two stages. First, there is an isothermal compression in which energy Q_c is rejected to the surroundings. This is entirely work done on the gas as its internal energy is unchanged and it is compressed.

*   Step D→A. Finally, there is a further adiabatic compression during which the work done on the gas increases the internal energy as the gas returns to temperature T_h.

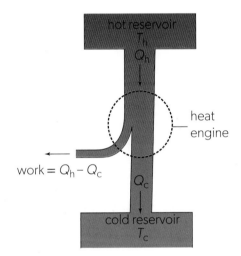

▲ Figure 12 A representation of a heat engine. This works between a hot source at temperature T_h and a cold sink at temperature T_c. Energy Q_h is removed from the source while Q_c is transferred to the sink. The difference $Q_h - Q_c$ is used to perform work.

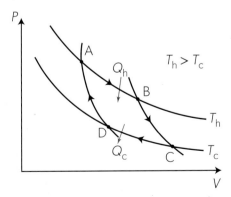

▲ Figure 13 A Carnot cycle in which an ideal gas is taken through two isothermal and two adiabatic processes ABCD. The area bounded by the cycle is work done by the gas when the arrow directions are clockwise.

AHL

How are efficiency considerations important in AC/DC motors and generators?

Discussions in physics textbooks are usually in terms of gases and processes in which gases are moved around a cycle of changes. But this tends to conceal the importance to an engineer of the efficiency of any process that involves a cycle.

Take, for example, the electrical generators of Topic D.4. Such generators are often fed by turbines that are driven by steam. The steam will be superheated (that is, will be above 100 °C) when it enters the turbine and will leave at a lower temperature when it has expanded and driven the turbine around. The difference between these temperatures is crucial to the overall efficiency. The designers of the power station will maximize the steam temperature and minimize the final temperature at which it is rejected to improve the efficiency. This may involve a two-stage cooling where the steam is returned to a heat exchanger to decrease its temperature even further. The energy from this cooling will be used to pre-heat the water that is about to be boiled to form steam. This second stage of temperature reduction improves the overall efficiency of the station.

The cooling towers shown in Figure 14 are common to many forms of power station whether they use fossil or nuclear fuels.

A heat engine that performs the Carnot cycle is **reversible**.

> A reversible process is one in which a system can be returned to its previous state with only an infinitesimal change to the properties of the system or its surroundings.

Another interpretation is that:

> A reversible process operates continuously in a quasi-static state. Quasi-static means that the system and its surroundings are always in a state of thermodynamic equilibrium.

These definitions of reversibility imply that a reversible change must be carried out infinitely slowly so that the system can return to its exact initial state at the end of the cycle.

For a Carnot cycle, the thermal efficiency η_{Carnot} is given by

$$\eta_{Carnot} = \frac{\text{useful work output}}{\text{input energy}} = \frac{Q_h - Q_c}{Q_h} = \frac{T_h - T_c}{T_h} = 1 - \frac{T_c}{T_h}.$$

where T_h and T_c are absolute (Kelvin) temperatures.

When the process is **irreversible** (not reversible), some of the energy will be lost to non-useful processes such as friction or turbulence and the difference $Q_h - Q_c$ will not all be useful work output.

This equation implies that to increase the efficiency of a Carnot cycle (and in principle any heat engine), T_h should be as large as possible and T_c should be as small as possible.

Modelling a real heat engine

Do not think that heat engines are theoretical devices invented by physicists. Such engines can be very real. Take a locomotive steam engine, for example. High temperature superheated steam is allowed to expand in a cylinder, pushing back a piston. As a result, the steam–air mixture cools and is ejected from the piston at a much lower temperature. The original internal energy of the steam is greatly reduced. Energy is transferred to the piston and, eventually, to the driving wheels and kinetic energy of the locomotive.

Worked example 9

A heat engine is modelled as a Carnot cycle whose hot and cold reservoirs are maintained at constant temperatures of 900 K and 350 K, respectively.

a. Calculate the efficiency of this cycle.

The energy is transferred from the hot reservoir into the engine at a rate of 720 W.

b. Calculate:

   i. the useful power developed by the engine

   ii. the rate at which waste thermal energy is rejected to the cold reservoir.

Solutions

a. $\eta = 1 - \dfrac{350}{900} = 0.61$

b. i. useful power = (input power) × (efficiency) = 720 × 0.61 = 440 W

   ii. 720 − 440 = 280 W

Worked example 10

An ideal gas operating in a Carnot cycle absorbs 1.50 kJ of energy per cycle from a hot reservoir of a constant temperature of 490 °C. The gas transfers energy to the low-temperature reservoir at a rate of 0.63 kJ per cycle.

Calculate:

a. the efficiency of the cycle

b. the temperature of the cold reservoir.

c. Explain why a practical heat engine operating between the same temperatures as a Carnot cycle is likely to require a higher energy input from its hot reservoir to do the same useful work.

▲ **Figure 14** This nuclear power station uses nuclear fuel to power a turbine. The cooling towers in this nuclear power station act to keep the temperature T_c as small as possible, so that the efficiency is improved.

Solutions

a. The useful work done by the gas is 1.50 − 0.63 = 0.87 kJ. The efficiency is $\dfrac{0.87}{1.50} = 0.58$.

b. The absolute temperature of the hot reservoir is $T_h = 490 + 273 = 763$ K.

$0.58 = 1 - \dfrac{T_c}{763} \Rightarrow T_c = 320\,\text{K} = 47\,°\text{C}$

c. For given hot and cold temperatures, the efficiency of a practical engine is always lower than that of an ideal Carnot cycle.

Since input energy $= \dfrac{\text{useful work}}{\text{efficiency}}$, the lower efficiency results in a higher input energy required.

Achieving optimum efficiency

The Carnot cycle represents the optimum efficiency of a heat engine. However, since the isothermal stages in the cycle must be carried out infinitely slowly, the power output of a Carnot engine is zero.

The consequence is that a compromise must be made between power output and efficiency.

Is scientific knowledge limited when it is only relevant to an unattainable reality?

Practice questions

10. The work done per cycle of a reversible heat engine is 5.8 kJ. The engine rejects energy at a rate of 2.4 kJ per cycle into a cold reservoir kept at a constant temperature of 100 °C.

   a. Calculate the energy transferred to the engine from the hot reservoir during one cycle.

   b. Calculate the efficiency of the engine.

   c. Assuming that the engine can be modelled as a Carnot cycle, calculate the temperature of the hot reservoir.

11. The diagram shows a Carnot cycle for an ideal gas.

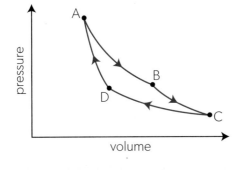

a. Identify the part of the cycle during which the energy is:

   i. transferred to the gas

   ii. transferred from the gas.

The efficiency of the cycle is 0.40. The work done during one cycle is 960 J.

b. Calculate the energy:

   i. transferred to the gas during one cycle

   ii. removed from the gas during one cycle.

12. A heat engine of efficiency 0.30 can be modelled as a Carnot cycle. The engine rejects waste thermal energy into a cold reservoir of temperature $T_c = 150\,°\text{C}$. An engineer wants to improve the efficiency of the engine to 0.50 without changing the temperature of the **hot** reservoir. Suggest how this can be achieved.

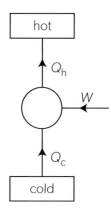

▲ Figure 15 The energy transfer diagram for a refrigerator or heat pump. The aim now is to use energy to heat up the hot reservoir at the expense of the cold reservoir (which becomes even colder). External work must be supplied to do this.

Refrigerators and heat pumps

The energy transfers in an ideal heat engine can be reversed. A net amount of work can be done on the gas or other fluid in the system to transfer energy from the cold to the hot reservoir. As before, the engine works between hot and cold reservoirs, but the direction of energy transfer is now as shown in Figure 15.

Such a device can be used either as a:

*   **refrigerator**, where as much energy as possible is transferred from the cold reservoir for each joule of work done, or a

*   **heat pump**, where as much energy as possible is transferred to the hot reservoir for each joule of work done.

The coils of a refrigerator contain a fluid known as the refrigerant (Figure 16). The properties of a good refrigerant include:

*   low boiling point

*   high specific latent heat of vaporization

*   low specific heat capacity of liquid

*   low vapour density

*   easy to liquefy at moderate pressure and temperature.

A compressor (the point at which work is done on the system) raises the pressure and temperature of the refrigerant, which then flows into a set of coils on the outside of the refrigerator. At this stage, the refrigerant is a gas. The coils are hot compared with the temperature of the kitchen and so the gas cools down (heating the room) and condenses into a liquid. The internal energy of the liquid is rejected into the kitchen as latent heat.

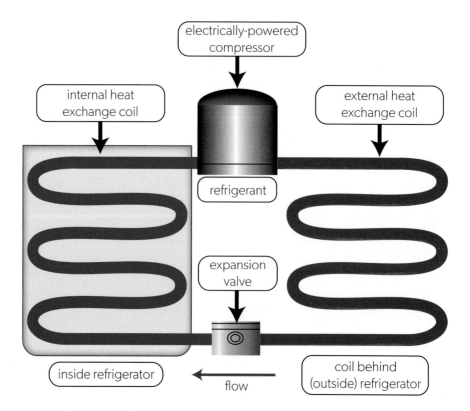

▲ Figure 16 A schematic diagram for a refrigerator.

The fluid then passes through a thermal expansion valve and—as its name suggests—the liquid refrigerant expands and becomes a gas again in the coils inside the refrigerator. To do this, the fluid requires latent heat and this can only come from the interior of the refrigerator, which loses energy and cools. The gas travels through these coils and back to the compressor where its state of pressure and temperature are changed again. The cycle then continues.

The principle of a heat pump is identical to that of a refrigerator. This time, the evaporator coils are in the air outside the house or in the ground nearby. The condenser (the equivalent of the coil outside the refrigerator) is inside the house so that the work done on the system is transferring energy to make the outside become colder and the inside of the house hotter. The benefit of using a heat pump for heating a house is that the amount of heat energy Q_h provided to the house is greater than the work done, since $Q_h = W + Q_c$.

Data-based questions

The power usage of a refrigerator was measured at different room temperatures. The data are shown in the table below.

The internal temperature of the refrigerator was $3 \pm 1\,°C$.

It is suggested that $P \propto \Delta T^2$, where P is the average power and ΔT is the difference in temperature between the inside and outside.

$T/°C\ (\pm 1\,°C)$	Average power $/\,W\ (\pm 2\,W)$
8	8
15	13
20	18
25	25
30	35
35	46

- Tabulate values of ΔT and ΔT^2. Calculate the uncertainties in these values.
- Plot a graph of P against ΔT^2.
- Find the gradient of your line of best fit.
- By considering maximum an minimum gradients, establish the uncertainty in your gradient.
- Write an equation to express the relationship between P and the ambient temperature.

Worked example 11

The working substance of a refrigerator is an ideal gas that undergoes a Carnot cycle in reverse. During one cycle, the refrigerator extracts energy $Q_c = 5.0\,J$ from a cold reservoir of temperature $2\,°C$ and transfers energy Q_h to a hot reservoir of temperature $25\,°C$.

Calculate:

a. Q_h

b. the work done by the refrigerator.

Solutions

a. The Carnot cycle equation is satisfied also if the cycle is taken in reverse.

$$\frac{Q_h - Q_c}{Q_h} = 1 - \frac{T_c}{T_h}. \quad \frac{Q_h - 5.0}{Q_h} = 1 - \frac{273 + 2}{273 + 25} \Rightarrow Q_h = 5.42\,J.$$

b. $W = Q_h - Q_c = 5.42 - 5.0 = 0.42\,J.$

Global impact of science — Real heat engines

There are several types of internal combustion engines used for public and private transport, including the diesel engine and the four-stroke petrol (Otto cycle) engine.

The German engineer Nicolaus Otto developed the internal combustion (petrol) engine in the mid 19th century.

In the theoretical Otto cycle, the gas steps are:

- A–B: an adiabatic compression—a constant entropy change because $\Delta Q = 0$ (see page 280)

- B–C: thermal energy is supplied in an isovolumetric change

- C–D: an adiabatic expansion—again at constant entropy

- D–A: energy rejection from the system (a cooling) under isovolumetric conditions.

In this theoretical cycle (which ignores the intake and exhaust processes), the work done by the engine is, as usual, the area ABCD enclosed by the loop. Try to match the four strokes of the practical Otto cycle to Figure 17.

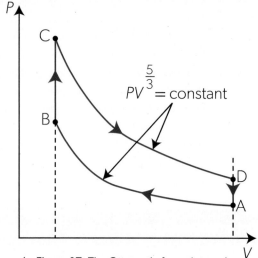

▲ **Figure 17** The Otto cycle for an internal combustion engine.

It is the aim of automotive engineers to maximize the efficiency of these engines by choosing ideal running conditions for real engines.

Why is there an upper limit on the efficiency of any energy source or engine?

For a heat engine to operate at an efficiency of 100%, the sink temperature must be 0 K. This is not a practical proposition, so this 100% is not a possible efficiency. The upper limit is set by the Carnot cycle. However, two parts of this cycle involve changes that can only take place infinitely slowly. The Carnot cycle is certainly not a practical proposition either!

Any real engine or energy source must be less efficient that the Carnot cycle (because of non-conservative losses).

Zeroth law of thermodynamics

The term "zeroth law of thermodynamics" appears to come from the 1930s when Sir Ralph Fowler, a British scientist, was discussing the (then) recent work of Meghnad Saha and Bishwambhar Nath Srivastava. These were two in a long line of scientists who realized that there is a basic, almost unspoken, requirement before the laws of thermodynamics can be expressed. The zeroth law essentially says that temperature scales must exist and be equivalent to each other. Another way to state this is to imagine three objects, A, B and C. When A and B are in thermal equilibrium, and B and C are in thermal equilibrium, then A and C must also be in thermal equilibrium. We make this basic assumption every time we use a thermometer.

To what extent does a failure to note the basis of a theory invalidate the results of the theory?

Worked example 12

A heat engine is modelled by the cycle ABCDA shown in the P–V diagram. The working substance is 7.00×10^{-2} mol of a monatomic ideal gas. Change AB is an adiabatic compression to $\frac{1}{4}$ of the original volume. During change BC, the pressure of the gas increases by a factor of 3 under a constant volume. Change CD is adiabatic and change DA is isovolumetric. The pressure of the gas at A is 1.00×10^5 Pa.

a. Calculate the pressure at B.

The temperature of the gas at B is 866 K.

b. Calculate the temperature at C.

c. Calculate the energy transferred to the gas during the change BC.

The energy transferred from the gas to the environment during the change DA is 600 J.

d. Calculate:

i. the work done by the gas during one cycle

ii. the efficiency of the cycle.

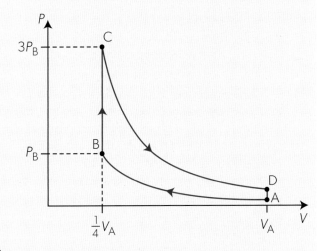

Solutions

a. $P_B = P_A \left(\dfrac{V_A}{V_B}\right)^{\frac{5}{3}} = (1.00 \times 10^5)(4)^{\frac{5}{3}} = 1.01 \times 10^6$ Pa.

b. For an isovolumetric change, the temperature is directly proportional to the pressure; hence $T_C = 3T_B = 3 \times 866 = 2598$ K.

c. Since $W = 0$, we have $Q = \Delta U = \frac{3}{2}nR\Delta T = \frac{3}{2}(7.00 \times 10^{-2})(8.31)(2598 - 866) = 1.51 \times 10^3$ J.

d. i. After one full cycle, the internal energy of the gas returns to the original value. Hence, the work done is equal to the net energy transferred to the gas. The adiabatic processes AB and CD transfer no energy. Hence, $W = Q_{BC} - Q_{DA} = 1510 - 600 = 910$ J.

ii. $\eta = \dfrac{910}{1510} = 0.603$.

Practice questions

13. A heat engine whose working substance is a monatomic ideal gas operates on the reversible cycle shown in the P–V diagram. The change AB is an isothermal compression, BC is an isovolumetric heating and CA is an adiabatic expansion. The volume of the gas at B is $\frac{1}{3}$ of the volume at A.

    At A the pressure is 9.00×10^4 Pa, the volume is 1.20×10^{-3} m^3 and the temperature is 400 K.

    a. Calculate:

       i. the pressure at B

       ii. the temperature at C

       iii. the thermal energy added to the gas during change BC.

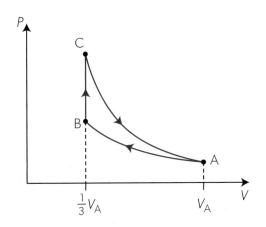

The work done on the gas during the isothermal compression is 119 J.

    b.   Determine:

        i.   the net work done during one cycle

        ii. the efficiency of the heat engine.

14. Initially, a monatomic ideal gas has a temperature of 320 K and a pressure of 1.0×10^5 Pa, and it occupies a volume of 8.0×10^{-4} m^3. The gas undergoes a cycle that consists of three changes: an adiabatic compression to a volume of 3.0×10^{-4} m^3, an isothermal expansion to the initial volume and cooling at a constant volume to the initial pressure.

    a.   Calculate, at the end of the adiabatic compression:

        i.   the pressure

        ii. the temperature.

    b.   Calculate the pressure of the gas at the end of the isothermal expansion.

    c.   Sketch the cycle in the following coordinate grid.

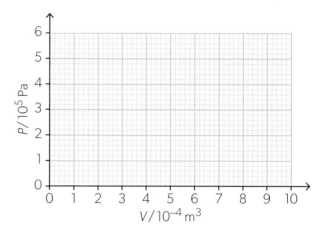

The energy supplied to the gas during the isothermal expansion is 150 J and the energy removed from the gas during the isovolumetric cooling is 110 J.

    d.   Calculate the efficiency of the cycle.

The second law of thermodynamics

The first law of thermodynamics is a statement of the conservation of energy, but it does not give any clue as to the preferred direction of energy transfer. The second law considers the direction in which energy transfer can occur. There are several different ways to express the second law. Two are outlined here. A third way reveals another way to think about energy transfer, in terms of the disorder that occurs because of the transfer. This involves a new concept called entropy that is discussed later in this topic.

One version of the second law is due to Clausius:

Energy cannot be transferred from a body at a lower temperature to a body at a greater temperature unless work is done on the system.

In other words, energy will not move spontaneously from a low-temperature object to a high-temperature object. Imagine that a cup containing a hot drink is placed in a room where the temperature is lower than that of the drink. It is impossible for the drink to become hotter at the expense of the temperature (internal energy) of the room — which would then cool down. The domestic refrigerator is a heat engine that warms a room at the expense of the cold interior of the refrigerator, but an input of energy (to the compressor) is required to achieve this.

Another version of the second law is due to Kelvin and Planck (sometimes attributed just to Kelvin):

Energy cannot be extracted from a hot object and transferred entirely into work.

This means that Q_c, the rejected energy, can never be zero for a heat engine and therefore $\eta = \dfrac{Q_h - Q_c}{Q_h}$ can never be equal to 1. The consequence is that a heat engine must always be less than 100% efficient because there must always be some Q_c rejected to the sink at the lower temperature T_c.

In any heat engine or other change at the microscopic level, we are transforming the internal energy of an object (in other words, the random motion of the molecules) into work. This must cool down the object giving the collection of molecules a smaller internal energy, effectively making them more ordered.

Entropy—A macroscopic interpretation

The idea of order in cool objects and disorder in hot objects leads us directly to another way to think about the second law of thermodynamics. This involves the concept of entropy. We discuss changes in entropy from the macroscopic view of a thermodynamic system and then from a microscopic viewpoint.

A **change in entropy** S can be defined on a macroscopic scale for a reversible change as

$$\Delta S = \frac{\Delta Q}{T}$$

where ΔS is the *increase* in entropy, ΔQ is the energy transferred *into* a system and T is the temperature at which the transfer occurs.

Entropy is a scalar quantity. Its units are joule per kelvin ($J\,K^{-1}$). Entropy is a property of a system, like temperature and (therefore) internal energy.

When a process is reversible, then there is no change in entropy of the system. In symbol terms, $\Delta S = 0$.

As an example, imagine a gas flowing along a pipe that has a constriction (a narrowing in the pipe). Before the constriction, the state of the gas is P_1, V_1, and T_1—these states completely define the gas. At the point of constriction, the state is P_2, V_2 and T_2. But imagine also that no energy flows in or out through the pipe walls or that there is no turbulent flow anywhere in the gas. This means that, beyond the constriction, the state returns to P_1, V_1 and T_1 and, because no energy has entered or left the gas, $\Delta Q = 0$ as well. From the definition of entropy, there has been no change in entropy during the gas flow. If, on the other hand, there had been friction at the walls of the pipe, then there would have been a transfer of energy from the gas to the pipe and hence to the surroundings. The gas could not then have returned to the initial state after passing the constriction. Because energy ΔQ flows out of the gas through the walls of the pipe, this implies that the gas temperature must be higher than the surroundings. This is therefore the simple case of energy being transferred from a hot body to a cold body where $T_{gas} > T_{surroundings}$. The change in entropy of the universe is

$$\Delta S = -\frac{\Delta Q}{T_{gas}} + \frac{\Delta Q}{T_{surroundings}} = \Delta Q\left(\frac{1}{T_{surroundings}} - \frac{1}{T_{gas}}\right)$$

making ΔS positive. The overall entropy of the universe has increased.

This is a general rule:

Entropy formulation of the second law of thermodynamics

The entropy of the universe always increases during an irreversible change.

(Remember that the universe is system + surroundings.) This is the third way to express the second law of thermodynamics.

 ΔS when the temperature changes

The definition $\Delta S = \dfrac{\Delta Q}{T}$ only applies for a change where the temperature is constant or does not change appreciably. When there is a temperature change, then the definition becomes a calculus equation: $dS = \displaystyle\int \frac{1}{T} dQ$. Calculations of entropy that involve calculus will not be tested in IB Diploma Programme physics.

The laws of physics

The laws of physics can never be proved. They can only be tested through experiments which may support the laws or may falsify them. Laws of physics can have exceptions. They may apply only to certain situations (for example, Ohm's law or Hooke's law), or they may be universal (laws of thermodynamics and the conservation of energy, for example).

Scientists often seek theories that are simple and that can be applied to a wide variety of situations. The laws of thermodynamics fit this principle well. As such, they have always been held in high esteem.

Einstein is quoted as saying:

"A theory is the more impressive the greater the simplicity of its premises, the more different kinds of things it relates, and the more extended is its area of applicability. Therefore, the deep impression which classical thermodynamics made upon me. It is the only physical theory of universal content concerning that, I am convinced within the framework of the applicability of its basic concepts, will never be overthrown."

Sir Arthur Eddington said in 1927:

"The law that entropy always increases—the Second Law of Thermodynamics—holds, I think, the supreme position among the laws of Nature. If someone points out to you that your pet theory of the universe is in disagreement with Maxwell's equations—then so much the worse for Maxwell's equations. If it is found to be contradicted by observation—well these experimentalists do bungle things sometimes. But if your theory is found to be against the second law of thermodynamics I can give you no hope; there is nothing for it but to collapse in deepest humiliation."

Can the laws of thermodynamics ever be disproved?

Worked example 13

a. A sample of 0.15 kg of water at 0 °C freezes. The specific latent heat of fusion of water is $3.3 \times 10^5 \, \text{J kg}^{-1}$. Calculate the change in entropy of the water during freezing.

b. The freezing takes place outdoors on a winter day when air temperature is −10 °C. Calculate the change in entropy of the water–air system as the water freezes.

c. Comment on your answers to a. and b. with reference to the second law of thermodynamics.

Solutions

a. The energy transferred to the surroundings is $mL = 0.15 \times 3.3 \times 10^5 = 5.0 \times 10^4 \, \text{J}$.

$\Delta S_{water} = \dfrac{\Delta Q}{T} = -\dfrac{5.0 \times 10^4}{273} = -180 \, \text{J K}^{-1}$. The change is negative (the entropy decreases) because the energy is *removed* from the water.

b. The energy gained by the air is equal to the energy lost by the water. The air temperature does not change by any appreciable amount, so the energy transfer to the air happens at a nearly constant temperature of 273 − 10 = 263 K.

The change in entropy of the water–air system is $\Delta S = \Delta S_{water} + \Delta S_{air} = -\dfrac{5.0 \times 10^4}{273} + \dfrac{5.0 \times 10^4}{263} = 7.0 \, \text{J K}^{-1}$.

c. The entropy of the water decreases, but the entropy of the surroundings (cold air) increases by a greater amount. The total entropy of the universe has increased in agreement with the second law of thermodynamics, indicating that freezing is an irreversible change.

Worked example 14

The diagram shows a Carnot cycle for an ideal gas.

a. Describe the change in entropy taking place during each part of the cycle.

b. State the change in entropy during once complete cycle ABCDA.

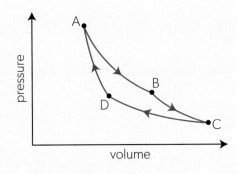

Solutions

a. The entropy is constant during adiabatic changes BC and DA.
$\Delta Q = 0$; hence $\Delta S = 0$.

The entropy increases during isothermal expansion AB because energy is added to the gas.
$\Delta Q_{AB} > 0$; hence $\Delta S_{AB} > 0$.

The entropy decreases during isothermal expansion CD because energy is removed from the gas.
$\Delta Q_{CD} < 0$; hence $\Delta S_{CD} < 0$.

b. The gas returns to the original state, so the net change in entropy is zero.

Entropy—A microscopic interpretation

Entropy can be defined on a microscopic scale too. The molecules in a gas constantly explore all the alternative speeds and positions available to them. When there are Ω different arrangements for a group of molecules, then the entropy of these molecules is defined to be $S = k_B \ln \Omega$. Notice that this is the *value* of S not a *change in the value* of S, unlike the macroscopic definition.

The individual arrangements of the same group of molecules are known as **microstates**. Each microstate is different from the others, but each is equally likely to be observed. The atoms in a solid could be perfectly ordered with no gaps in the lattice (there is only one way to achieve this), but once some atoms are removed to leave gaps, then there are a many different possibilities for the positions of the gaps. Many possibilities means many microstates and an increase in entropy when the ions are removed.

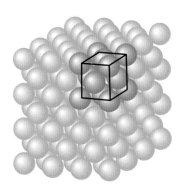

▲ Figure 18 A perfect crystal of a solid can have only one possible arrangement. Remove just one atom. There are many ways in which that this can be done. The entropy has increased with this removal.

As an example, imagine a copper crystal that initially has a perfect lattice arrangement (Figure 18). Suppose that one copper atom is removed at random from the lattice. There are as many ways to do this as there are atoms, but only one arrangement can happen in practice. The total number of possible arrangements of the lattice with one atom missing are the microstates. The removal has changed Ω from 1 to a large number and this increases the entropy (the disorder) in the system using $S = k_B \ln \Omega$.

The alternative view of entropy says that energy ΔQ has been required to remove the atom from the lattice and that this has been carried out at a temperature T, leading to the change in entropy, $\Delta S = \dfrac{\Delta Q}{T}$.

It may not be immediately clear why a definition of change of entropy $\Delta S = \dfrac{\Delta Q}{T}$ and a definition in terms of numbers of arrangements available to gas molecules ($S = k_B \ln \Omega$) are equivalent.

To understand the link between entropy and randomness, imagine two boxes. One box initially contains six counters, labelled 1 to 6. The other box is empty (Figure 19).

Throw a six-sided die and move the counter with the number that landed upwards into the other box. Then throw again, and so on. Whenever the number of a counter occurs, then that counter moves to the other box. Figure 19 shows the first few moves.

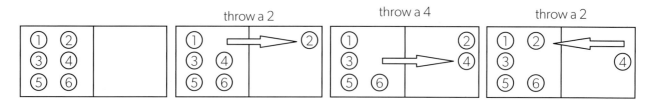

▲ Figure 19 A game that explores all the states of a system. When a six-sided die is thrown, the number that falls uppermost dictates which counter is moved. This counter is moved to the other box irrespective of which box it was in originally.

As you play this game, record the average number in each box. You will find that this average is three when you play the game for long enough. Even though the system is exploring all the possibilities, this value of three in each box occurs quite often because there are so many ways in which three counters in the box can result.

Eventually, if you play the game long enough, you will observe every possible arrangement (every possible microstate) of the counters. How many are there altogether? The answer is $2 \times 2 \times 2 \times 2 \times 2 \times 2 = 64 = 2^6$. Each counter can be placed in one of two ways, independently of the others, so the probabilities multiply. Notice that the chance of all counters returning to one box is 1 in 64, so, if you are lucky, you may see this occur.

The configuration where there are three counters is one of six **macrostates** for this system. We do not know which particular counters make up the three. However, we know that there are 20 ways to achieve three in each box. This is the most common macrostate of the six possible arrangements. Figure 20(a) shows a histogram of the number of microstates for each macrostate of the six-counter system.

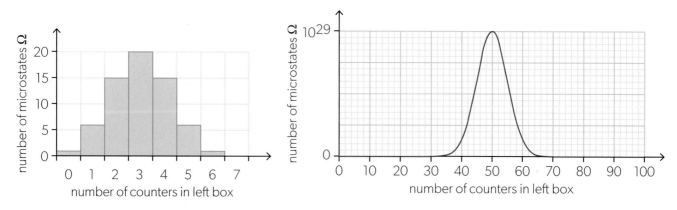

▲ Figure 20 The number of ways to arrange counters in the box when there are (a) six counters, and (b) 100 counters.

Now suppose you played the game with 100 counters with suitable computer software to generate the moves. There are now 2^{100} possible alternatives for the counter arrangements, and all counters will appear in one box on average only once in 2^{100} throws — roughly one billion throws. At one throw per second, this will take 34 years to achieve. Finally, imagine one mole of counters: the number of alternatives is now $2^{6\times10^{23}}$ and this number, when written out with digits, is greater than the radius of the Solar System.

Figure 20(b) shows the variation of number of microstates for the macrostates when there are 100 counters. When the system spends an equal time in each microstate, it is overwhelmingly more likely that the preferred macrostate is 50 counters in each box.

The game models the expansion of a gas from one container into a container of twice the volume. When the gas molecules explore a volume twice as large, there are more possible arrangements for the molecules than before. For the six-counter game, the ratio $\dfrac{\text{number of arrangements with volume } 2V}{\text{number of arrangements with volume } V} = \dfrac{\Omega'}{\Omega} = \dfrac{2^6}{1}$ $= \dfrac{64}{1}$. Remember that $\dfrac{\text{larger volume}}{\text{original volume}} = \dfrac{V'}{V} = 2$, for this case.

The general case of N counters with a volume change from V_1 to V_2, a ratio of $\dfrac{V_2}{V_1}$, leads to a ratio for $\dfrac{\Omega'}{\Omega}$ of $\left(\dfrac{V_2}{V_1}\right)^N$. There is only one arrangement possible for Ω (all molecules in the same volume), so that the number of arrangements is $\Omega' = \left(\dfrac{V_2}{V_1}\right)^N$. The definition of entropy is that $S = k_B \ln \Omega$, so the change in entropy is $\Delta S = k_B N(\ln V_2 - \ln V_1)$. This can be written as $\Delta S = k_B N \Delta(\ln V)$ and, because $\Delta(\ln V) = \dfrac{\Delta V}{V}$, $\Delta S = k_B N \dfrac{\Delta V}{V}$.

The maximum amount of work that a gas can deliver when it expands at constant temperature is $P\Delta V$. We know from earlier that $PV = Nk_B T$, so $P\Delta V = Nk_B T \dfrac{\Delta V}{V}$ (dividing both sides by V and multiplying both sides by ΔV).

This rearranges to $P\Delta V = T \times \left(Nk_B \dfrac{\Delta V}{V}\right)$, where the expression in brackets is just ΔS. Finally, this shows that $P \times \Delta V = T \times \Delta S$, where $P \times \Delta V$ is, as usual, the energy change of the gas when expanding at constant temperature. This leads to the first definition of entropy in terms of Q and T:

$$Q = T \times \Delta S \quad \text{or} \quad \Delta S = \dfrac{Q}{T}$$

This non-rigorous proof shows that the concept of entropy as a measure of randomness in a system is equivalent to the concept that emerges from the energy transfer at a particular temperature.

Worked example 15

A box contains ten identical particles that move randomly and have an equal probability of being found in the left-hand half and in the right-hand half of the box. A, B and C are three configurations of the system, with zero, two and five particles in the left half of the box.

configuration A

a. State the number of microstates of the system in configuration A.

b. Show that the number of microstates of the system in configuration B is 45.

configuration B

c. Calculate the number of microstates of the system in configuration C.

d. All microstates of the system are equally probable. Explain why the system is more likely to be found in a state with particles evenly divided between the two halves of the box.

configuration C

e. Calculate, in terms of k_B, the value of $S_C - S_A$, where S_C and S_A are the entropies of the system in configurations C and A. Comment on the answer with reference to the second law of thermodynamics.

Solutions

a. There is only one microstate with all of the particles in the right-hand half of the box.

b. Suppose that the particles are assigned to the left-hand half of the box one at a time. The first particle can be selected in ten ways (any of the ten particles can be the first one to be placed in the left-hand half), which leaves nine choices to pick the second particle. There will be $10 \times 9 = 90$ arrangements of two particles in the left-hand half of the box. But the particles are indistinguishable, so this value must be divided by the number of ways in which the two particles can be ordered, $1 \times 2 = 2$. The number of microstates is $\dfrac{10 \times 9}{1 \times 2} = 45$.

c. We generalize the procedure of assigning one particle at a time to the left-hand half of the box.
$$\frac{10 \times 9 \times 8 \times 7 \times 6}{1 \times 2 \times 3 \times 4 \times 5} = 252.$$
If you take Mathematics AA in the Diploma Programme, you will have noticed that the numbers of microstates follow the pattern of the binomial coefficients $^{10}C_0$, $^{10}C_2$ and $^{10}C_5$.

d. The number of microstates in configuration C is greater than in any uneven configuration, so the probability of finding the system in this configuration is greatest.

e. $S_C - S_A = k_B(\ln 252 - \ln 1) = 5.5\,k_B$. The entropy of configuration C is greater than that of A. From the second law of thermodynamics, a system initially in configuration A would spontaneously evolve towards a more disordered configuration C.

ATL Social — Resolving conflicts

Ludwig Boltzmann was an Austrian physicist who developed our understanding of entropy. The constant k_B was named after him by Max Planck who developed the equation $S = k_B \log \Omega$.

However, Boltzmann found it hard to persuade others of his ideas. His theory relied on the concept of atoms which could not be directly observed at the time. As a result, not all scientists of the time believed that atoms existed. While Boltzmann was lecturing in Vienna about atoms, Ernst Mach is reputed to have called out "Have you ever seen one?" Boltzmann moved away from Vienna, partly to get away from his rival, Mach. He only returned to the city when Mach retired.

Sadly, Boltzmann struggled with his mental health and eventually committed suicide. The equation for entropy appears on his gravestone.

▲ Figure 21 The gravestone of Boltzmann. Even at the time of his death, some scientists did not believe in the existence of atoms.

What are the consequences of the second law of thermodynamics to the universe as a whole?

The implications of the thermodynamic processes discussed here are that — for the universe as a whole — entropy is increasing and temperature differences are being smoothed out. These temperature differences drive the processes that occur both on the small scale (in car engines) and on the largest scales (events in stars and galaxies). When the differences have disappeared and entropy reaches its maximum value, then the processes cannot operate any longer.

This state is sometimes called "the heat death of the universe" and is a hypothesis about the fate of the universe. Because everything is in thermodynamic equilibrium, heat engines cannot operate.

This is, in one sense, the ultimate hypothesis because we cannot test this theory. Lord Kelvin postulated the idea of a universe heat death in the 1850s while he was attempting to disprove a suggestion that the universe was infinitely old and would last for an infinite time.

Kelvin's idea cannot be called a theory as it cannot be falsified.

What paradigm shifts enabling change to human society, such as harnessing the power of steam, can be attributed to advancements in physics understanding? (NOS)

Don't think of paradigm shifts as changes in understanding shared amongst a small number of scientists. Such shifts can lead to very real changes for everyone.

There has always been a close historical link between advancements in physics and developments in engineering. These paradigm shifts have led to changes in human society and to the way in which we live our lives. There are many examples of this in the history of science. Changes associated with Theme B are obvious:

- The European industrial revolution was largely driven by the development of steam-powered devices which themselves arose from the work carried out by the early workers in thermodynamics. Before the change, many societies were based on agricultural practice, with artefacts made individually by artisan craftspeople. After the revolution people moved away from the countryside to congregate in large towns and cities. From then on, their lives were based around an industrial lifestyle.

- The development of electrical devices throughout the 19th and 20th centuries has had a profound impact on all human activity. These changes range from the extension of the working day through electric lighting, to the proliferation of mobile devices in the 21st century—expected to exceed 20 billion worldwide during the lifetime of this book.

What other advances in physics have resulted in paradigm shifts that had have led to societal change? Try to find one from each of the other Themes in this course.

Global impact of science—Does life break the second law of thermodynamics?

Living plants and animals develop large amounts of order as they grow. As a result, it can appear that they break the second law of thermodynamics by decreasing entropy. However, the second law does not prohibit parts of a system from developing large-scale order provided that the overall system has an entropy increase. The Sun provides the energy source for life on Earth. The system under consideration is the Sun and Earth (to a good approximation, this system is closed). Entropy increase in the Sun can allow for life to develop on Earth.

◀ **Figure 22** A plant growing on Earth develops ordered structures that seem to contradict the idea that entropy must increase.

Science as shared endeavour—Disorder or random?

The concept of entropy is not confined to thermodynamics or even physics. It is used in all sciences and engineering. Many scientists and engineers talk and write about entropy as the degree of disorder in a system. By this they mean that increasing the amount of disorder in a system increases the number of possible arrangements and therefore an increase in the system's entropy.

Information theory makes use of the idea of entropy. In a communication system, a transmitter uses a communication channel to send a message to a receiver. Because noise occurs in the system, the received message becomes unpredictable compared with the transmitted, pure, message. The entropy of the system is the average level of uncertainty in the information that makes up the message. In 1948, Claude Shannon was able to use the result of Harry Nyquist and Ralph Hartley to predict the maximum rate at which a message can be sent over a channel with a particular bandwidth and still be 100% recovered. Although this may seem a long way from a gas in a box, the mathematics turns out to be very similar.

How do charged particles flow through materials?

How are the electrical properties of materials quantified?

What are the consequences of resistance in conductors?

The scientific study of electricity stretches back to the earliest days of science. Electrical phenomena, ranging from lightning storms down to the simple electrostatic attraction between small objects, have always fascinated scientists.

Although the effects of charge flow have been known for thousands of years, an understanding of the exact mechanisms that allow charges to move through materials is relatively recent. In this topic we concentrate on the motion of electrons through solid conductors – which are usually metals, although some non-metals are also conductors. But charge flow through other phases is also possible. Chemical electrolysis and the flow of ions in gases are important in industrial chemistry and in a description of some natural phenomena. The description here of electrons flowing around the fixed positive ions of a metal can easily be broadened to cases where positive charges are also mobile or absent. The charged particles are subject to the electric fields that you meet in detail in Theme D and these electric fields can be created using familiar laboratory items such as cells and power supplies.

Variations in the density of electrons and the other microscopic constituents of the solid material imply variations in the conduction properties of materials. To quantify these, we require a vocabulary that helps us to identify the factors that alter the charge flow in a material. This is where electrical current (a rate of charge flow) and potential difference (a measure of energy transfer) arise.

Using these two quantities leads us to the resistance of a conductor. However, resistance depends on the size and shape of the conductor. Resistivity, which follows from a definition of resistance, describes the response of a particular material to electric charge flow rather than the response of an individual specimen.

Finally, we examine the consequences of resistance, both in terms of how resistors can be combined in various configurations and how they can be used to provide us with variations in electrical charge flow.

▲ **Figure 1** The developments in thermodynamics in Topic B.4 are associated with the industrial revolution. The developments in our understanding of electricity in this topic led to the technological revolution.

In this topic, you will learn about:

- electric cells as a source of emf that has internal resistance
- chemical cells and solar cells
- circuit diagrams
- charge carriers and direct current
- electrical conductors and insulators
- electric potential difference as the energy transfer per unit charge

- electrical resistance and electrical resistivity
- Ohm's law
- ohmic and non-ohmic behaviour
- electrical power
- combining resistors in circuits
- variable resistors and their uses.

Introduction

Take a plastic comb, pull it through your hair and the comb can pick up small pieces of paper. Look closely and you may see paper being repelled just after it touches the comb. This is because charges can be separated by friction. This discovery is attributed to the Greek scientist Thales who lived about 2600 years ago. Then, silk was spun on amber spindles and it became charged and attracted to the amber. The ancient Greek word for "amber" is ηλεκτρον (electron).

In the 1700s, du Fay found that both materials could be "electrified" and that there were two opposite "electrifications". Gradually, scientists developed the idea that there were positive and negative charges. The American physicist, Benjamin Franklin, carried out a series of experiments in which he flew kites during thunderstorms. He named the charge on a glass rod rubbed with silk as "positive electricity". The charge left on materials such as ebonite when rubbed with animal fur was called "negative".

At the end of the 19th century, J. J. Thomson detected a small particle that he called the electron. Later experiments showed that all electrons have identical charge and that atoms contain electrons. Atoms also have protons with the same magnitude of electronic charge as electrons but an opposite charge sign. Now we assign negative charge to the electron and positive to the proton, unlike Franklin. Only these two species of charge are known.

Figure 3 shows a shuttling ball, charged at one plate, that is repelled to strike the other. The ball then gains the opposite charge to be repelled again. The ammeter shows an electric current in the metal wires of the circuit. This links Franklin's static charges to the moving charges of current electricity.

Conduction in metals

The metal atoms in a solid are bound together by metallic bonds. The full details of the bonding are complex. However, a simple model of the interior of a metal solid suggests that the atoms form a regular lattice arrangement. The details of the lattice vary, but the feature common to all metals is that one electron is donated from the outer shell of each atom to a common sea of electrons that occupies the entire volume of the metal.

Figure 4 shows the model. The positive ions occupy fixed lattice positions. These are positive ions because each atom has lost an electron. Of course, at all temperatures above absolute zero the ions vibrate at these positions. Around the ions is the sea of free electrons or **conduction electrons**; these are responsible for electrical conduction.

The conduction electrons interact with the ions and transfer kinetic energy to them. It is this transfer of energy from electrons to ions that accounts for the phenomenon that we call "electrical resistance".

The energy transfer arises as follows:

In the absence of an electric field, the free electrons move and interact with the ions in the lattice. This is a random process. However, when an electric field is present (Figure 5), then an electric force acts on the electrons because they are charged. The direction of an electric field is defined as the direction in which a positive charge moves (Topic D.2),

▲ Figure 2 Amber—fossilized tree resin—is a material that becomes charged through friction with materials such as silk.

The term **field** is used in physics for cases where two objects, not in contact, exert forces on each other. We say that, in the case of the comb picking up paper, the paper is in the **electric field** due to the comb. Field theory is studied in more detail in Theme D.

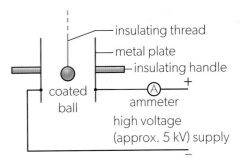

▲ Figure 3 A table-tennis ball coated in graphite will shuttle indefinitely between charged plates due to charge transfer.

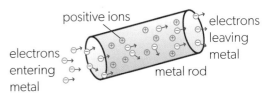

▲ Figure 4 A simple model for conduction by free electrons in a metal.

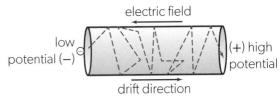

▲ Figure 5 The electrons drift in the opposite direction to that of the electric field.

Conduction in gases and liquids

Electrical conduction is also possible in gases and liquids which contain free ions because of their chemical bonding. When an electric field is applied to these materials, ions move: positive charges in the direction of the field; negative charges the opposite way. When this happens, an electric current is observed.

so the force on the electrons will be in the opposite direction to the electric field in the metal.

In the presence of an electric field, the negatively charged electrons drift along the conductor. The electrons are known as **charge carriers**. One way to imagine the movement of the charge carriers is as a colony of ants carried along a moving walkway. Each ant moves at random with an overall drift along the walkway due to its motion.

While this model of conduction applies to all metallic conductors, there are some circumstances in which we ignore it. A particular case is when we deal with the connecting wires used to link practical circuits. Such wires are deliberately made from metals such as copper, which provide little resistance to the flow of the charge carriers. Using the language of later parts of this topic, the electrical resistance of these connecting wires is taken to be zero and they are assumed to have zero potential difference between their ends.

Models of conduction

The simple model given here is of a fluid-like flow of free electrons through a solid, a liquid or a gas. But this is not the end of the story. There are other, more sophisticated models of conduction in solids that can explain the differences between conductors, semiconductors and insulators. This simple flow model cannot do this. These more advanced models involve the electronic band theory which arises from the interactions between the electrons within individual atoms and between the atoms themselves.

This band model (which comes from the atomic models discussed in Topic E.1) suggests that the electrons adopt different energies within the substance. This leads to certain ranges of energy levels (called band gaps) that

are not available to the electrons. Where these band gaps are wide, electrons cannot easily move from one set of levels to another, and this makes the substance an insulator. Where the band gap is narrow, adding energy to the atomic structure allows electrons to jump across the band gap and conduct more freely. This is what gives a semiconductor its curious properties. You will later see that a property of semiconductors is that, by adding internal energy to them (for example, by raising their temperature), free electrons are released so that there is better conduction. In conductors, the band gap is of less relevance because the electrons have many energy states available to them and so conduction happens very readily. You can find out more about band theory on the Internet.

point P

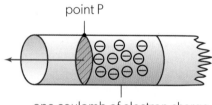

one coulomb of electron charge

▲ **Figure 6** Electric current is the flow of electric charge.

▲ **Figure 7** A lightning strike can transfer about 15 C of charge or more.

Electric current and charge

When charge flows in a material, we say that there is an electric current in it. The unit of current is the **ampère**, the symbol for this is A. Often, in the English-speaking world, the accent is omitted.

Current is linked to flow of charge carriers in a simple way.

Figure 6 shows a block of electrons with a total charge of one coulomb moving to the left along a conductor. An observer at point P watches these electrons move. When all the electrons in the block move past P in one second, then the current is defined to be one ampere. When it takes twice as long (2 s) for the electrons to pass, then the current is half what it was before and is 0.5 A. When the block takes only 0.1 s to pass the observer, then the current is 10 A.

Mathematically:

$$\text{electric current, } I = \frac{\Delta q}{\Delta t}$$

where Δq is the charge and Δt is time.

The ampere is a **fundamental unit** and is defined as part of the SI. The definition of the ampere was changed in 2019 to reflect the link between charge and current. The ampere is now defined as one coulomb flowing for one second. Implicit in this definition is the knowledge that the magnitude of the charge of one electron is $1.602176634 \times 10^{-19}$ C, which links one coulomb to the number of electrons that must flow past a point in the one-second time interval.

The coulomb is a large unit. When you run a comb through your hair, a charge of somewhere between 1 pC and 1 nC can be transferred to it.

Units

The unit of charge, unlike the ampere, is not a fundamental unit, so when you are asked for the fundamental units of a quantity involving the coulomb you should immediately change C into A s.

How are the fields in other areas of physics similar to and different from each other?

You meet the concept of the electric field both in this topic and, extensively, in Theme D. In the present topic, we are concerned with the way in which an electric field influences the behaviour of conduction electrons. In Theme D, electric, magnetic and gravitational fields are studied to reveal similarities and differences between them. Electric and gravitational fields are linked by the inverse-square relationship between their strength and distance from a point charge or mass. However, they differ because of the existence of positive and negative charge, whereas there is only positive mass. Magnetic field strengths vary with distance in a different way to electric and gravitational fields because magnetic monopoles are not observed. The field strength–distance relationship for a dipole is not inverse-square.

Worked example 1

A conducting ball suspended from a long insulating thread (Figure 3) moves between the two charged plates at a frequency of 0.67 Hz. The ball carries a charge of magnitude 72 nC each time it crosses from one plate to the other. Calculate:

a. the average current in the circuit
b. the number of electrons transferred each time the ball touches one of the plates.

Solutions

a. The time interval between the ball hitting the same plate $= \dfrac{1}{f} = \dfrac{1}{0.67} = 1.5$ s. The time to transfer

72 nC is therefore half of that: 0.75 s. So, current $= \dfrac{7.2 \times 10^{-8}}{0.75} = 96$ nA

b. The charge transferred is 72 nC $= 7.2 \times 10^{-8}$ C. Each electron has a charge of -1.6×10^{-19} C, so the number of electrons involved in the transfer is $\dfrac{7.2 \times 10^{-8}}{1.6 \times 10^{-19}} = 4.5 \times 10^{11}$.

Worked example 2

a. Calculate the current in a wire through which a charge of 25 C passes in 1500 s.
b. The current in a wire is 36 mA. Calculate the charge that flows along the wire in one minute.

Solutions

a. Current, $I = \dfrac{\Delta q}{\Delta t} = \dfrac{25}{1500} = 17$ mA

b. $\Delta q = I \Delta t = 3.6 \times 10^{-2} \times 60 = 2.2$ C

Practice questions

1. A lightning strike lasts for 2.0 ms and carries an average current of 8.0 kA. Calculate:
    a. the charge
    b. the number of electrons transferred between the thundercloud and the ground.

2. Calculate the time needed for a charge of 5400 C to flow through a wire if the current in the wire is 1.2 A.

Hypotheses — Making the invisible imaginable

The link between charge movement and current is a crucial one. Electric current is a macroscopic quantity. Transfer of charge by electron flow is a microscopic phenomenon in every sense of the word. The link between the flow of charge and the existence of an electric current is another example of a link in physics between macroscopic observations and inferences about what is happening on the smallest scales. It was the lack of knowledge of what happens inside conductors at the atomic scale that meant that scientists, up to the end of the 19th century, had to develop concepts such as current and field to hypothesize the "invisible" effects they were observing.

Modelling — Charge carrier speeds in a conductor

Turn on a lighting circuit at home and the lamp lights almost immediately. Does this tell us that the electrons are moving very quickly around the connecting wire? In fact it does not, because the electrons travel in a conductor at speeds of a few millimetres per second. They do not (as is often imagined) travel through the wires at velocities close to the speed of light. This slow speed at which the ions move along the conductor is known as the **drift speed**.

A mathematical model of conduction confirms this.

Figure 8 shows a cylindrical conductor carrying an electric current I. We assume that there are n charge carriers in 1 m^3 of conductor — this quantity is known as the **charge density**.

In one second, a volume Av of charge carriers passes P. The total number of charge carriers in this volume is nAv and therefore the total charge in the volume is $nAvq$. This is the total charge that passes point P in one second, in other words, the electric current, which leads to $I = nAvq$.

The slow drift speed in metals, even for large currents, poses the question of how a lamp turns on so quickly when there is a significant cable run between the switch and the lamp. The information that the switch has closed travels much more quickly — close to the speed of light as an electromagnetic wave. All the free electrons in the circuit begin to drift virtually simultaneously as the wave propagates. The lamp turns on almost instantaneously, even though, for direct current, it may take an individual electron many minutes to reach the lamp from the power supply.

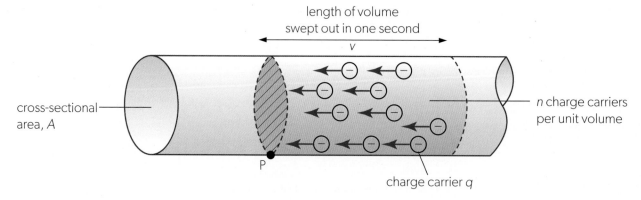

▲ Figure 8 A mathematical model for electrical conduction.

Potential difference

Electric cells and power supplies provide the electric field and transfer energy to the electrons. As the electrons move through the conductors, they collide with the positive ions in the lattice and transfer the energy they have gained from the field to the ions. This is not a transfer of kinetic energy as the electrons do not speed up or slow down in the conductors. We shall see later that the electric current in a series circuit is the same everywhere.

In a situation where a field is acting, physicists use two quantities called **potential** and **potential difference** to deal with the energy transfers. Potential difference (often abbreviated to "pd") is a measure of the electrical potential energy transferred to or from an electron when it is moving between two points in a circuit. However, given the small amount of charge possessed by each electron, this amount of energy is also very small. It is better to use the much greater energy transfer associated with one coulomb of charge.

Potential difference is defined as the work done (energy transferred) W when one unit of positive charge q moves between two points along the path of the current.

$$\text{potential difference, } V = \frac{W}{q}$$

The symbol given to potential difference is V. Its unit is JC^{-1} (joules per coulomb) and is named the volt (symbol: V) after the Italian scientist Alessandro Volta who was born in the middle of the 18th century and who made early discoveries about electricity.

In fundamental units, the volt is $kg\,m^2\,s^{-3}\,A^{-1}$.

The potential difference between two points is one volt (1 V) when one joule (1 J) of energy is transferred per coulomb of charge passing between the two points.

The simple circuit in Figure 9 illustrates these ideas.

An electric cell is connected to a lamp via a switch and three leads. Figure 9 shows a picture of the circuit as set up on a bench.

When the switch is closed, electrons flow round the circuit. The diagram also shows the direction of a **conventional current** and the electronic current. The two directions are opposite: in this case, clockwise for the electron flow and counter-clockwise for the conventional current. The reason for this difference is explained later. You need to take care with this difference, particularly when using some of the direction rules that are introduced later in this topic and in Theme D.

What happens to an electron as it goes round the circuit once? The electron gains electric potential energy as it moves through the cell. The electron then leaves the cell and moves through the connecting lead and switch. The potential differences across the leads and the switch are small because the passage of one coulomb of charge through them will not result in much energy transfer to their metal lattices.

After moving through another connecting lead, the electron reaches the lamp. The pd across the lamp will be large because it is deliberately designed to transfer electrical potential energy from the electrons as they pass through it. The metal lattice in the filament of the lamp gains energy. The filament in the lamp glows brightly at its increased temperature. We say that "the lamp is lit".

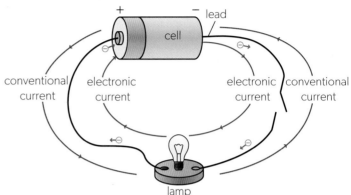

▲ Figure 9 Conventional current in a circuit is from the positive terminal to the negative terminal outside the cell. Electronic current is in the opposite direction.

(ATL) Thinking skills

In an electric current, the transfer of negative charge (electrons) in one direction is the same as considering the transfer of positive charge in the opposite direction.

There are many situations where there is an equivalence between transferring a quantity of something in one direction and transferring the opposite quantity in the opposite direction. For example, banks and financial institutions trade debt. Since a debt is a lack of money, moving a debt away from a bank is equivalent to moving money into the bank.

Try to think of other examples where a negative quantity is moved in one direction instead of a positive quantity in the opposite direction.

Hypotheses—Conventional and electron currents

Scientists use working hypotheses to explain their discoveries. Sometimes they make the wrong decisions.

In early studies of current electricity, the idea emerged that there was a flow of "electrical fluid" in wires and that this flow was responsible for the observed effects of electricity. This is where much of the language of "current and charge flow" arose. At first, the suggestion was that there were two types of fluid known as "vitreous" and "resinous". Benjamin Franklin proposed that there was only one fluid but that it behaved differently depending on the circumstances. He was also the first scientist to use the terms "positive" and "negative".

Then scientists assigned a positive charge to the "fluid" thought to be moving in the wires. This positive charge

was said to flow out of the positive terminal of a power supply (because the charge was repelled from the supply terminal) and went around the circuit, and then re-enter the power supply through the negative terminal. This is what we now term the **conventional current**. You should not confuse this with the electron current.

You may ask: why do we now not simply drop the conventional current and talk only about the **electronic current**? The answer is that other rules in electricity and magnetism were set up on the assumption that charge carriers are positive. All these rules would need to be reversed to take account of our later knowledge. It is better to leave things as they are.

Worked example 3

A high efficiency LED lamp is lit for 2 hours. Calculate the energy transfer to the lamp when the pd across it is 240 V and the current in it is 50 mA.

Solution

2 hours is $2 \times 60 \times 60 = 7200$ s

The charge transferred, $\Delta q = I \Delta t$
$= 7.2 \times 10^3 \times 50 \times 10^{-3} = 360$ C

Work done = charge × pd
$= 360 \times 240 = 86\,400$ J

Worked example 4

A cell has a terminal voltage of 1.5 V and can deliver a charge of 460 C before it becomes discharged.
a. Calculate the maximum energy that the cell can deliver.
b. The current in the cell never exceeds 5 mA. Estimate the lifetime of the cell.

Solutions

a. Potential difference, $V = \dfrac{W}{q}$, so $W = qV = 460 \times 1.5 = 690$ J

b. The current of 5 mA means that no more than 5 mC flows through the cell at any time. So $\dfrac{460}{0.005} = 92\,000$ s, which is about 26 hours.

Practice questions

3. The potential difference between a thundercloud and the ground is 3.2×10^7 V. A charge of 16 C is transferred during a lightning strike. Estimate the energy released during the lightning strike.

4. A fully charged mobile phone battery stores 60 kJ of energy. During recharging, a current of 1.5 A flows through the battery at a potential difference of 5.0 V. Assume that the recharging process is 100%

efficient. In other words, all the work done in moving the charge through the battery is transferred into electrochemical energy of the battery. The battery is initially at 10% of its full capacity. Calculate:
a. the total amount of charge that flows through the battery during recharging
b. the time needed to fully recharge the battery.

Effects of electric current

Electric currents can produce different effects when charge flows. These include:

- a **heating effect** when energy is transferred to a resistor as internal energy

- a **magnetic effect**, when a current produces a magnetic field, or when magnetic fields change near conductors to induce an emf in the conductor

- a **chemical effect**, when chemicals react together to alter the energy of electrons and to cause them to move in a cell or a battery of cells, or when electric current in a material causes chemical changes.

Electromotive force (emf)

An important concept is that of **electromotive force** (usually written as "emf" for brevity). Emf describes the energy transfers in a power source rather than the "force" that is implied in its name.

When charge flows electrical energy can go *into* another form such as internal energy (through the heating effect), or it can be converted *from* another form (for example, light (radiant energy) in solar (photovoltaic) cells).

The term emf is used when energy is transferred *to* the electrons in, for example, a cell that is using a chemical effect. Other devices can also convert energy into an electrical form via magnetic effects. Examples include microphones, electrical generators and dynamos.

The term *potential difference* will be used when the energy is transferred *from* the electrical form. Examples of this are from the electrical form into heat and light (resistors and lamps), or from electrical into kinetic energy (motors).

▲ Figure 10 Electric currents can cause heating effects, such as in the heating element of this toaster; magnetic effects, which cause this fan to work; or chemical effects such as the chemical reaction that occurs in a cell or battery.

 In what ways can an electrical circuit be described as a system like the Earth's atmosphere or a heat engine?

A heat engine (Topic B.4) transfers energy from a hot source to a cold sink and performs useful work during the process. Heat engines work in a cycle.

Similarly, there are sources that transfer energy into the atmosphere and sinks that absorb the energy. The atmosphere also operates in a cyclical way.

A complete electric circuit has a cyclic character too. This is because, while the switch in the circuit is closed, electrons will continue to move around the circuit transferring energy from sources of emf to sinks of resistance where the energy appears in a thermal form. When a direct current (dc) circuit is switched on for a long time, one electron can complete the circuit many times.

Another similarity between the circuit and the engine and the atmosphere is that the circuit will stop when the chemical sources in the cell have all been converted into their discharged form. While a heat engine continues to run, a temperature difference exists between source and sink. The atmosphere will continue to generate convection currents and winds for as long as there are temperature differences to drive them.

This highlights a difference: temperature differences drive a heat engine and the atmosphere. Chemical changes drive a chemical cell.

Power, current and pd

We can now answer the question of how much energy is transferred to a conductor by the electrons as they move through it. Suppose there is a conductor with a potential difference V between its ends when a current I is in the conductor.

In time Δt the charge q that moves through the conductor is $q = I \times \Delta t$.

The energy W transferred to the conductor from the electrons is $q \times V$ which is $(I\Delta t) \times V$.

There is also the rate of electrical energy transfer to consider. The energy transferred in time Δt is $W = IV\Delta t$. The electrical power being supplied to the conductor is $\dfrac{\text{energy transferred}}{\text{time taken to transfer}} = \dfrac{W}{\Delta t}$ and therefore

electrical power supplied, $P = IV$

Alternative forms of this expression that you will find useful are

$$I = \frac{P}{V} \quad \text{and} \quad V = \frac{P}{I}.$$

The term "potential difference" implies that there is a physical quantity, known as "potential", which can differ from point to point. This is indeed the case. Potential is a concept from field theory described in detail in Topics D.1 and D.2. Potential differences exist between areas of high potential and low potential. Positive charges—when free to do so—move from regions of high to low potential. Negative charges move from low to high potential.

The unit of power is the watt (W). One watt (1 W) is the power developed when 1 J is converted in a time interval of 1 s. This is the same definition in both mechanics and electricity. Another way to think of the volt is as the power transferred per unit current in a conductor.

Worked example 5

A 3 V, 1.5 W filament lamp is connected to a 3 V battery. Calculate:

a. the current in the lamp

b. the energy transferred in 2400 s.

Solutions

a. Electrical power $P = IV$, so $I = \dfrac{P}{V} = \dfrac{1.5}{3} = 0.5\,A$

b. The energy transferred every second is 1.5 J, so in 2400 s the energy transferred is $1.5 \times 2400 = 3600\,J$.

Worked example 6

An electric motor that is connected to a 12 V supply is able to raise a 0.10 kg load through a distance of 1.5 m in 7 s. The motor is 40% efficient. Calculate the average current in the motor while the load is being raised.

Solution

The energy gained $= mg\Delta h = 0.10 \times 9.8 \times 1.5 = 1.47\,J$

The power output from the motor must be $\dfrac{1.47}{7} = 0.21\,W$

The current $= \dfrac{P}{V} = \dfrac{0.21}{12} = 17.5\,mA$. Since the motor is 40% efficient, the current in the motor will be 44 mA.

Practice questions

5. A cordless drill develops a power of 270 W. A work of 8100 J is done when a charge of 450 C flows through the motor of the drill. What is the current in the motor?

    A. 9 A   B. 12 A   C. 15 A   D. 18 A

6. The temperature of 1.2 kg of water in an electric kettle is raised by 25 °C in one minute. The specific heat capacity of water is 4200 J kg^{-1} K^{-1}.

    a. Calculate the power developed by the kettle.

    b. The kettle is connected to a 230 V source. Calculate the current in the kettle's heating element.

7. The terminal potential difference of the battery of an electric car is 450 V. When the car drives at a constant speed of 80 km h^{-1}, the average power transferred from the battery is 16 kW. The total energy stored in the battery is 3.6×10^{8} J.

    a. Calculate the average current from the battery.

    b. Estimate, in km, the range of the car.

    The car's battery is being recharged from a domestic 230 V wall socket. The current from the socket is limited to 25 A.

    c. Calculate the maximum power available for recharging.

    d. The recharging process is 95% efficient. Estimate the minimum time needed to fully recharge the car's battery.

Drawing circuits

A set of agreed electrical symbols has been devised so that all physicists understand what is represented in a circuit diagram. They are shown on page 5 of the IB DP Physics Data Booklet. Ensure that you can draw and identify all of them accurately. There are conventions for drawing, interpreting, and using circuit diagrams. When the value of a particular component is important for the operation of the circuit, it is normal to write its value alongside it.

There are separate symbols for cells and batteries. Most people use these two terms interchangeably, but there is a difference. A battery is a collection of cells arranged positive terminal to negative terminal—the diagram for the battery shows how they are connected. A cell only contains one source of emf.

Care needs to be taken when drawing one connecting lead over another. The convention is that, if two leads cross and are joined to each other, then a dot is placed at the junction, as shown in Figure 11. When there is no dot, then the leads are not considered to be connected to each other.

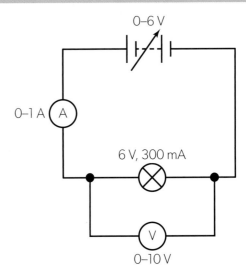

▲ Figure 11 Important values for the components in an electrical circuit can be written near the component. Connections are always shown with a "dot" where the wires connect.

What are the advantages of cells as a source of electrical energy?

Electric cells were one of the first laboratory sources of electrical energy. They provided the early scientists with reliable and relatively stable electrical supplies for their experiments. The work of scientists such as Georges Leclanché in the 1880s paved the way for portable torches and radio batteries in the early part of the 20th century.

▲ Figure 12 A solar-powered lantern.

For example, Figure 12 shows a solar-powered lamp. It has a cell charged by light from the Sun. Such a lamp makes an incredible difference to the lives of people in remote areas of Africa. Children can continue to study when night has fallen and it enables their parents to extend their working day with all the economic benefits that this can bring.

The advantage of all cells is their flexibility. Some cells, such as the rechargable lead–acid cell, can provide short bursts of large current. Button cells used in wrist watches and similar devices can provide very small currents of around 1 mA or less and sustain these currents for years before being completely discharged.

See page 347 in the *Tools for physics* section for more on constructing circuits from circuit diagrams.

Circuit troubleshooting

Circuit troubleshooting is a useful skill to have. It is an art in itself and comes with experience. A possible sequence is as follows.

- Check the circuit: is it really set up as in your diagram?

- Check the power supply (try it with another single component such as a lamp that you know is working properly).

- Check that all the leads are correctly inserted and that there are no loose wires inside the connectors.

- Check the pd across individual components using an extra multimeter with insulated test leads. This will indicate whether the live circuit is working as designed.

- Check that the individual components are working by substituting them into an alternative circuit that is known to be working.

- Check that the fuses in multi-meters are not blown when the meters are used in series to measure current.

▲ Figure 13 A conventional laboratory analogue multimeter and a digital multimeter.

Worked example 7

The current in a component is 5.0 mA when the pd across it is 6.0 V. Calculate:

a. the resistance of the component

b. the pd across the component when the current in it is 150 μA, assuming that the resistance of the component is the same as in (a).

Solutions

a. $R = \dfrac{V}{I} = \dfrac{6}{5 \times 10^{-3}} = 1.2\,k\Omega$

b. $V = IR = 1.5 \times 10^{-4} \times 1.2 \times 10^{3}$
$= 0.18\,V$

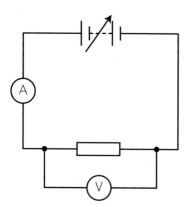

▲ Figure 14 Ammeters are connected in series with the component whose current they are measuring; voltmeters are connected in parallel.

Measuring current and potential difference

We often need to measure the current in a circuit and the pd across components in the circuit. This can be achieved with the use of meters or sensors connected to computers (data loggers). There are many types and varieties of meters. For example, an analogue meter uses a mechanical system with a coil and magnet to move a pointer on a scale, whereas a digital meter converts a reading into a digital display.

Ammeters measure the current in the circuit. As we want to know the size of the current in a component, the ammeter needs to be in **series** with the circuit or component. An ideal ammeter will not take any energy from the electrons as they flow through it, otherwise it would disturb the circuit it is trying to measure. An ideal ammeter has zero resistance. Figure 14 shows where the ammeter is placed to measure the current.

Voltmeters measure the energy converted per unit charge that flows in a component or components. You can think of a voltmeter as needing to compare the energy of the electrons before they enter a component to when they leave it, rather like the turnstiles (baffle gates) to a railway station that count the number of people (charges) going through as they give a set amount of money (energy) to the rail company. To do this the voltmeter must be placed across the terminals of the component or components whose pd is being measured. This arrangement is called **parallel**. In an ideal world, the voltmeter will not require any energy. An ideal voltmeter has an infinite (large) resistance and no charge flows through it. Figure 14 shows how to connect voltmeter.

Electrical resistance

As an electron moves through a metal, it interacts, at the microscopic level, with the positive ions and transfers energy to them. At the macroscopic level, an electric current is observed to have a heating effect.

However, simple comparisons between different conductors show that the amount of energy transferred varies greatly from metal to metal or even between different shapes of the same metal. When there is the same current in wires of similar size made of tungsten or copper, the tungsten wire will heat up more than the copper. We need to take account of the fact that some conductors can achieve better energy transfers than others. The concept of **electrical resistance** is used for this.

> The resistance of a component is defined as
> $$\dfrac{\text{potential difference across the component}}{\text{current in the component}}.$$

The symbol for resistance is R and the definition leads to a well-known equation:

$$R = \frac{V}{I}$$

The unit of resistance is the ohm (symbol Ω; named after Georg Simon Ohm, a German physicist). In terms of its fundamental units, the $\Omega \equiv kg\,m^2\,s^{-3}\,A^{-2}$. Using the ohm as a unit is much more convenient!

Alternative forms of the equation are

$$V = IR \text{ and } I = \frac{V}{R}$$

When both the pd across a component and the current in the component are known, then it is possible to calculate the resistance of the component for that current.

How does a particle model allow electrical resistance to be explained? (NOS)

How can the heating of an electrical resistor be explained using other areas of physics?

Electrical resistance is explained through the interaction of the electrons with the bulk of the solid—the lattice. Initially, an electron gains energy from the electric field that exists across a conductor because of the emf from the energy source (the cell). Topic D.2 goes into more detail about the way in which charges interact with an electric field. The electron interacts with the metal lattice and its atoms, transferring energy to them. This energy appears within the material as atomic vibrations. The more energy that electrons transfer, the greater the energy stored in atomic vibrational states and therefore the greater the average speeds of the atoms and the greater the temperature. It is the effectiveness of the energy transfer that we call resistance.

The nature and regularity of a metallic lattice lead to the existence of quantized acoustic waves that travel within the lattice. You will meet the idea of quanta in Theme E.

Practice questions

8. The heating element of a hair dryer has a resistance of 35 Ω. The hair dryer is connected to a 230 V power source. Calculate, for the hair dryer:
   a. the current
   b. the power developed.

9. A potential difference of 4.5 V is applied to a coil of wire. The coil dissipates an energy of 60 J in a time of 20 s. Calculate:
   a. the current in the coil
   b. the resistance of the coil
   c. the charge flowing through the coil in 20 s.

Ideal and non-ideal meters

An **ideal ammeter has zero resistance**—clearly not attainable in practice as the coils or circuits inside the ammeter have resistance.

An **ideal voltmeter has an infinite resistance**—again, not a practical situation.

Some modern digital meters can get close to these ideals of zero resistance for ammeters and infinite resistance for voltmeters. Digital meters are used more and more in modern science.

In questions, you can assume that a voltmeter or ammeter is ideal unless otherwise stated. As you will see later in this topic, you can still do calculations as long as you know the resistance of the meter.

Measuring the resistance of a metal wire

- Tool 1: Understand how to accurately measure electric current to an appropriate level of precision.

- Tool 3: Express quantities and uncertainties to an appropriate number of significant figures or decimal places.

- Tool 3: Calculate mean and range.

- Take a piece of metal wire (an alloy called constantan is a good one to choose) and connect it in the circuit shown in Figure 14.

- Use a power supply with a variable output rather than a cell so that you can alter the pd across the wire easily. Your teacher will suggest suitable power supplies and meters for your experiment.

- If your wire is long, coil it around an insulator (perhaps a pencil) and ensure that the coils do not touch.

- Take readings of the current in the wire and the pd across it for a range of currents. Your teacher will tell you an appropriate range to use to avoid changing the temperature of the wire.

- For each pair of readings divide the pd by the current to obtain the resistance of the wire in ohm.

- Calculate an average value for the resistance, and round it to the correct number of significant figures based on the significant figures of your data.

Global impact of science—Edison and his lamp

▲ **Figure 15** One of the first lamps developed by Thomas Edison.

The conversion of electrical energy into internal energy was one of the first uses of distributed electricity. Thomas Edison was an inventor and entrepreneur who worked in the US in the second half of the 19th century. He pioneered electric lighting, the earliest forms of which were provided by producing a current in a metal or carbon filament which then glowed (Figure 15). Early lamps were primitive, but produced a revolution in the way that homes and public spaces were lit. The development continues today as inventors and manufacturers seek more and more efficient electric lamps such as light-emitting diodes (LED). More developments in lighting will undoubtedly occur during the lifetime of this book.

Science as a shared endeavour—Ohm and Barlow

Ohm's law has its limitations because it only tells us about a material when the physical conditions do not change. It did not find immediate favour with the scientific community. In contrast, Barlow was an English scientist who was highly respected for his earlier work and had recently published an alternative theory on conduction. People simply did not believe that Barlow could be wrong.

This immediate acceptance of one scientist's work over another would not necessarily happen today. Scientists use a system of peer review. Work published by one scientist or scientific group must be set out in such a way that other scientists can repeat the experiments or collect the same data to check that there are no errors in the original work. Only when the scientific community has verified the data is new work accepted as scientific "fact".

Ohm's law

Lots of information can be obtained from a graph of the variation of pd with current for a component. Such a graph is known as a *V–I* graph. Figure 16 shows the results from the plot of a graph of *V* against *I* for a metal wire.

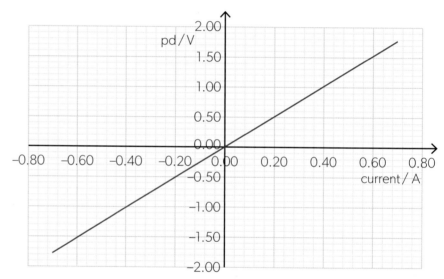

▲ **Figure 16** A typical *V–I* graph for a metal conductor at constant temperature.

A straight line of best fit has been drawn through the data points. For this wire, the resistance is the same for all values of current measured. Such a resistor is called an **ohmic conductor**. An equivalent way to say this is that the potential difference and the current are directly proportional (the line is straight and goes through the origin). In the experiment carried out to obtain these data, the temperature of the wire did not change.

This behaviour of metallic wires was first observed by Georg Simon Ohm in 1826. It leads to a rule known as **Ohm's law**.

> Ohm's law states that the potential difference across a metallic conductor is directly proportional to the current in the conductor providing that the physical conditions of the conductor do not change.

By physical conditions we mean the temperature (the most important factor as we shall see) and all other factors about the wire. But the temperature factor is so important that the law is sometimes stated with the term "temperature" replacing the words "physical conditions".

Worked example 8

The graph shows how the current *I* varies with the potential difference *V* for an ohmic conductor R and a non-ohmic component S.

a. Calculate the resistance of R.

b. Outline how the resistance of S varies with the potential difference across it.

c. Calculate the resistance of S when the potential difference is 6.0 V.

d. Determine the current in S when it dissipates a power of 3.6 W.

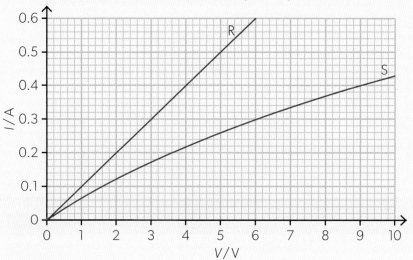

Solutions

a. $R = \dfrac{V}{I} = \dfrac{6.0}{0.6} = 10\,\Omega$. The ratio is constant and does not depend on the particular point chosen to read off the values of *V* and *I*.

b. The ratio $\dfrac{V}{I}$ increases with *V* (a straight line joining the origin with an arbitrary point on the curve becomes less steep as *V* increases). Hence the resistance of S increases with the potential difference.

c. $R = \dfrac{6.0}{0.30} = 20\,\Omega$.

d. We need to identify *I* and *V* so that *IV* = 3.6 W. By examining the curve, we find that the correct combination is *V* = 9.0 V and *I* = 0.40 A.

Ohm's law

This statement attributed to Ohm is always called a law—but is it? In reality, it is an experimental description of how a group of materials behave under rather restricted conditions.

Does that make it a law?

🧪 Resistance of a lamp filament

- Tool 3: Construct and interpret tables and graphs for raw and processed data including scatter graphs and line and curve graphs.

- Inquiry 2: Collect and record sufficient relevant quantitative data.

Use the circuit (Figure 14) that you used to calculate the resistance of a metal wire but replace the wire by a filament lamp.

- Your teacher will suggest the range of currents and pds to use.

- Do the experiment twice, the second time with the charge flowing through the lamp in the opposite direction to the first. There are two ways to achieve this. The first is to reverse the connections to the power supply, also reversing the connections to the ammeter and voltmeter (if the meters are analogue). The second way is easier. Simply reverse the lamp and call all the readings negative because the currents are in the opposite direction through the lamp.

- Plot a graph of *V* (*y*-axis) against *I* (*x*-axis) with the origin in the centre of the paper. Figure 17 shows an example of a *V–I* graph for a lamp.

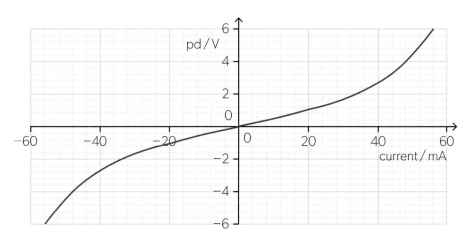

▲ Figure 17 A lamp filament does not have a constant temperature as the current in it increases. The *V–I* graph which is a curve going through the origin shows this.

Non-ohmic behaviour

The graph in Figure 17 is not straight (although it goes through the origin), so V and I are not proportional to each other unlike the metal wire at constant temperature. The lamp does not obey Ohm's law and it is said to be **non-ohmic**.

When the resistance is calculated using the data points it is not constant either. Table 1 shows the resistance values for six data points.

The data show that resistance of the lamp increases as the current increases. At large currents, greater changes in pd are required to change the current by a fixed amount. This is exactly what you can predict. Each electron transfers the same amount of energy to the filament. As the current increases, the drift speed of the electrons increases and so more energy is transferred from each electron every second. The energy goes into increasing the kinetic energy of the lattice ions and therefore the temperature of the bulk material rises too. But the more the ions vibrate in the lattice, the more the electrons collide with them so at higher temperatures even more energy is transferred to the lattice by the moving charges.

Current / mA	Resistance / Ω
20	50
34	59
41	73
47	85
52	96
55	109

▲ Table 1 How the resistance of a lamp filament changes with current.

Other non-ohmic conductors include semiconducting diodes and thermistors. These are devices made from a group of materials known as semiconductors.

Data-based questions

Table 1 shows how the resistance of a lamp's filament changes with current. Use these data to complete the following tasks.

It is suggested that the resistance varies according to the relationship $R - R_0 = kI^n$, where R is the resistance of the filament, R_0 is the resistance at small currents (approaching zero), I is the current, k is a constant and n is an integer.

- Plot a graph of R against I and show that R_0 is about $47\,\Omega$.
- Tabulate values of $\log(I)$ and $\log(R - R_0)$ using $R_0 = 47$.

- Plot a graph of $\log(R - R_0)$ against $\log(I)$. Hence, show that a suitable value for n is 3.
- The uncertainties in the values of I can be assumed to be $\pm 1\,\text{mA}$. Tabulate values of I^3 and the uncertainties in I^3.
- Plot a graph of R against I^3. Include uncertainties on your graph (the uncertainty in the values of R may be assumed to be $\pm 1\,\Omega$).
- Add a line of best fit to your graph of R against I^3. Use this to find a value of the constant k and an improved estimate of the value of R_0.

Ohm's law and the definition of resistance

Notice that the resistances in the table were calculated for each individual data point using $R = \dfrac{V}{I}$. They were *not* evaluated using the tangent to the graph at a particular current. In other words, the definition of resistance is $\dfrac{V}{I}$ *not* in terms of $\dfrac{\Delta V}{\Delta I}$, which is the value of the tangent.

Another aspect of Ohm's law is also misunderstood.

Our definition of resistance is that $R = \dfrac{V}{I}$ or $V = IR$.

However, Ohm's law states that:
$$V \propto I$$
and, including the constant of proportionality k,
$$V = kI.$$
Even though R is defined in the same way as k, the definition of resistance does not correspond to Ohm's law (which just states proportionality). Therefore, $V = IR$ is not a statement of Ohm's law.

Resistivity

The resistance of a sample of a material depends not only on what it is made of, but also on the physical dimensions (the size and shape) of the sample itself.

The resistance of a conductor is:

- proportional to its length L, $R \propto L$

- inversely proportional to its cross-sectional area A (or diameter^2, d^2),
  $R \propto \dfrac{1}{A}$ or $\dfrac{1}{d^2}$.

Combining these two results suggests that $R \propto \dfrac{L}{A}$. This means that $\dfrac{RA}{L}$ is constant for a given material and leads to a definition of a new quantity called resistivity.

$$\text{Resistivity } \rho \text{ is defined as } \rho = \frac{R \times A}{L}.$$

The unit of resistivity is the ohm-metre (symbol $\Omega\,m$).

Resistivity is a useful quantity. The electrical resistance of an object depends not only on what it is made from, but also the shape of the sample. Even a constant volume of a material will have values of resistance that depend on the shape. However, **the value of the resistivity is the same for all pure samples of the material.**

Resistivity is independent of shape or size just like quantities such as *density* (where the value is mass per unit volume) or *specific latent heat* (where the value is related to unit mass of the material).

 Units of resistivity

Take care here: the resistivity unit is ohm metre. It is *not* ohm metre^{-1}—a mistake frequently made by students. The meaning of ohm metre^{-1} is the resistance of one metre length of a particular conductor, which is a relevant quantity to know, but is not the same as resistivity.

Worked example 9

A uniform wire has a radius of 0.16 mm and a length of 7.5 m. Calculate the resistance of the wire when the resistivity of the metal is $7.0 \times 10^{-7}\,\Omega\,m$.

Solution

Unless told otherwise, assume that the wire has a circular cross-section. So,
area of wire $= \pi(1.6 \times 10^{-4})^2 = 8.04 \times 10^{-8}\,m^2$

$$\rho = \frac{RA}{L}, \text{ so } R = \frac{\rho L}{A} = \frac{7.0 \times 10^{-7} \times 7.5}{8.04 \times 10^{-8}} = 65\,\Omega$$

Worked example 10

Calculate the resistance of a block of copper that has a length of 0.012 m with a width of 0.75 mm and a thickness of 12 mm. The resistivity of copper is $1.7 \times 10^{-8}\,\Omega\,m$.

Solution

The cross-sectional area of the block is $7.5 \times 10^{-4} \times 1.2 \times 10^{-2} = 9.0 \times 10^{-6}\,m^2$

The relevant dimension for the length is 0.012 m,

$$\text{so } R = \frac{\rho L}{A} = \frac{1.7 \times 10^{-8} \times 0.012}{9.0 \times 10^{-6}} = 0.023\,m\Omega$$

Practice questions

10. Two wires X and Y are made of the same material.

    Wire Y has half the length of wire X and twice the diameter. What is $\dfrac{\text{resistance of X}}{\text{resistance of Y}}$?

    A. 1   B. 2   C. 4   D. 8

11. The filament of a light bulb is a coil of tungsten wire of length 1.5 m and diameter 4.0×10^{-5} m. The resistivity of tungsten at the operating temperature of the filament is $7.4 \times 10^{-7}\ \Omega\,m$.

    a. Calculate the resistance of the filament.

    b. The light bulb is connected to a 230 V source. Calculate:

       i. the current in the filament

       ii. the power of the light bulb.

12. A wire of length 5.0 m and a uniform cross-sectional area 7.9×10^{-7} m^2 carries a current of 1.5 A. The potential difference across the wire is 0.25 V. Calculate:

    a. the resistance of the wire

    b. resistivity.

13. The resistance of a copper wire of length 12 m and uniform diameter is $0.10\ \Omega$. The resistivity of copper is $1.7 \times 10^{-8}\ \Omega\,m$. Determine the diameter of the wire.

14. The graph shows how the resistance R of a wire varies with the length L of the wire. The wire has a uniform diameter of 0.75 mm.

    Determine the resistivity of the material of the wire.

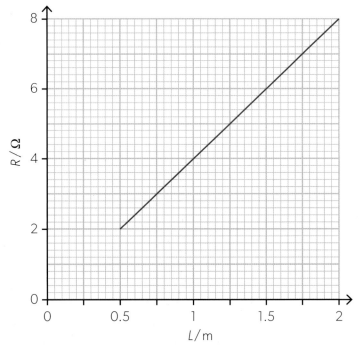

What are the parallels in the models for thermal and electrical conductivity? (NOS)

There is a simple analogy between thermal and electrical effects.

Compare thermal conduction with electrical conduction for a wire of cross-section area A and length ΔL.

Thermal

The rate of transfer of internal energy Q is proportional to temperature gradient $\dfrac{\Delta T}{\Delta x}$:

$\dfrac{Q}{\Delta t} = -\dfrac{A}{K}\dfrac{\Delta T}{\Delta x}$, where K is the thermal resistivity of the material. This follows from the equation for thermal energy transfer in Topic B.1. Notice that the constant K is the reciprocal of k, not the same as in the earlier topic.

Electrical

The rate of transfer of charge q (in other words, the current) is proportional to potential gradient: $\dfrac{\Delta q}{\Delta t} = -\dfrac{A}{\rho}\dfrac{\Delta V}{\Delta L}$, where ρ is the electrical resistivity of the material.

The thermal resistivity K of a material and the electrical resistivity ρ are analogous. The temperature gradient and the electric potential gradient are equivalent in the expressions. The rate at which both internal energy and charge are carried through the wire is related to the presence of free electrons in the metal. This is more than a similarity between macroscopic equations: similar physics is involved at the microscopic scale.

Are good electrical conductors also good thermal conductors? See the data-based question on page 216, or you could compare the values of ρ and $\dfrac{1}{k}$ for different metals using a book of data or values from the internet.

Resistivity of pencil lead

- Tool 3: Select and manipulate equations.

- Inquiry 1: Demonstrate creativity in the designing, implementation or presentation of the investigation.

- Inquiry 2: Collect and record sufficient relevant quantitative data.

Graphite is a semi-metallic conductor and is a constituent of the lead in a pencil. Another constituent in the pencil lead is clay. It is the ratio of graphite to clay that determines the "hardness" of the pencil. This experiment enables you to estimate the resistivity of the graphite.

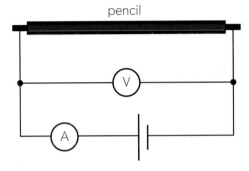

▲ Figure 18 A circuit to estimate the resistivity of graphite.

- Take a B grade pencil (also known as 1 grade in the US) and remove about 1.5 cm of the wood from each end leaving a cylinder of the lead exposed. Attach a crocodile (alligator) clip firmly to each end.

- Connect the pencil in a circuit to measure the resistance of the lead. Expect the resistance of the pencil lead to be about $1\,\Omega$ when choosing your power supply and meters.

- Determine the resistance of the lead.

- Measure the length of pencil lead between the crocodile clips.

- Measure the diameter of the lead using a micrometer screw gauge or digital callipers to enable you to calculate the area of the lead.

- Use your data to calculate the resistivity of the lead. The accepted value of the resistivity of graphite is about $30\,\mu\Omega\,m$ but you will not expect to get this value because the clay in the lead changes the value.

You can take the experiment one step further with this challenge:

- Use your pencil to shade a 10 cm by 2 cm area on a piece of graph paper as uniformly as possible. This will make a graphite resistance film on the paper.

- Devise a way to attach the graphite film to a suitable circuit. Then measure the resistance of the film. Knowing the resistance and the dimensions of your shaded area should enable you to work out how thick the film is.

(*Hint:* in the resistivity equation, the length is the *distance across the film*, and the area is *width of the film* × *thickness of the film*.)

Combining resistors

Electrical components can be linked together in two ways in an electrical circuit:

- **in series**, where the components are joined one after another like the ammeter, the cell and the resistor in Figure 14, or

- **in parallel** like the resistor and the voltmeter in the same figure.

Components connected in series (the power supply and the ammeter in Figure 14) have the same current in each. The number of free electrons leaving the first component must equal the number entering the second component. If free electrons remained in the first component, then it would become negatively charged and would repel further electrons, preventing them from entering it. The flow of charges would rapidly stop.

In series the potential differences (pds) add. As the charge travels through two components, the total energy lost is equal to the sum of the two separate amounts of energy in the components. Because the same charge flows through both (they are in series), the sum of the pds is equal to the total drop in pd across them.

Components connected in parallel (the voltmeter and the resistor in Figure 14), on the other hand, have the same pd across them, but the currents in the components differ when their resistances are different.

Consider two resistors of different resistance values, in parallel with each other and connected to a cell with no resistance of its own. When one of the resistors is temporarily disconnected, then the current in the remaining resistor is given by $\frac{\text{emf of the cell}}{\text{resistance}}$. This will also be true for the other resistor when connected alone. When both resistors are now connected in parallel with each other, both resistors have the same pd across them because a terminal of each resistor is connected to one of the terminals of the cell. The cell will have to supply more current than when either resistor was there alone. To be precise, it supplies the sum of the separate currents.

	Currents...	Potential differences...
In series	... are the same	... add
In parallel	... add	... are the same

▲ Table 2 A summary of currents and pds in series and parallel.

These rules about currents and potential differences in series and parallel components are important for you to understand and to be able to use.

Combining resistors in series and parallel

- Inquiry 1: Demonstrate independent thinking, initiative, or insight.

- Inquiry 2: Collect and record sufficient relevant quantitative data.

- Inquiry 3: Relate the outcomes of an investigation to the stated research question or hypothesis.

For this experiment you need six resistors, each one with a tolerance of ±5%. Two of these resistors should be the same. The tolerance figure means that the manufacturer only guarantees the value to be within 5% of the nominal value. "Nominal" means the value marked on the resistor. Your resistors may have the nominal value written on them or you may have to use the colour code printed on them. The code is easy to decipher (Figure 19).

You also need a multimeter set to measure resistance directly and a way to join the resistors together and connect them to the multimeter.

- Measure the resistance of each resistor alone and record this in a table.

- Take the two resistors that have the same nominal value and connect them *in series*. Measure the resistance of the combination. Can you see a rule for the combined resistance of two resistors?

- Repeat with five of the possible combinations for connecting resistors *in series*.

- Now measure the combined resistance of the two resistors with the same nominal value when they are in parallel. Is there an obvious rule this time?

- One way to express the rule for combining two resistors R_1 and R_2 in parallel is that the combined resistance is $\frac{R_1 R_2}{R_1 + R_2}$. Test this relationship for five combinations of parallel resistors.

- Test your two rules together by forming combinations of three resistors such as:

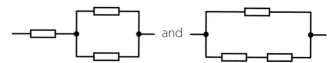

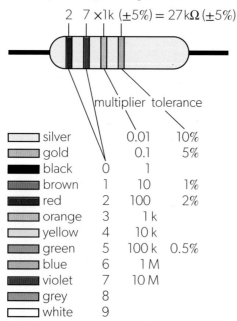

$$2 \quad 7 \times 1k \ (\pm 5\%) = 27\,k\Omega \ (\pm 5\%)$$

multiplier tolerance

silver	0.01	10%
gold	0.1	5%
black	0	1
brown	1	10	1%
red	2	100	2%
orange	3	1 k
yellow	4	10 k
green	5	100 k	0.5%
blue	6	1 M
violet	7	10 M
grey	8	
white	9	

▲ Figure 19 The colour code used to mark resistors.

Resistors in series

Three resistors R_1, R_2 and R_3 are in series in Figure 20.

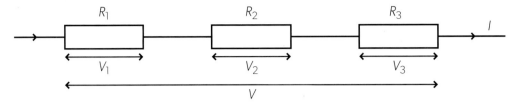

▲ Figure 20 A single resistor is to replace the three series resistors. The value of this single resistor must be equal to the sum of the three resistance values.

What is the resistance of the single resistor that can replace them so that the resistance of the single resistor is equivalent to the combination of all three?

The resistors are in series and therefore the current I is the same in each resistor. The definition of resistance tells us that the pd across each resistor, V_1, V_2 and V_3 is $V_1 = IR_1$, $V_2 = IR_2$ and $V_3 = IR_3$.

The single resistor with a resistance R must be indistinguishable from the three in series. In other words, when the current through this single resistor is the same as that through the three, then it must have a pd V across it such that $V = IR$. As the three resistors make up a series combination, the potential differences add, so $V = V_1 + V_2 + V_3$. Therefore $IR = IR_1 + IR_2 + IR_3$ and so

in series

$$R_s = R_1 + R_2 + \cdots + R_n$$

When resistors are combined in series, the resistances add to give the total resistance.

Resistors in parallel

Three resistors in parallel (Figure 21) have the same pd V across them.

The current in the connecting lead is equal to the sum of the currents in the three separate resistors. Therefore $I = I_1 + I_2 + I_3$. Each current can be written in terms of V and R using the definition of resistance: $\dfrac{V}{R} = \dfrac{V}{R_1} + \dfrac{V}{R_2} + \dfrac{V}{R_3}$. Finally, both sides of the equation are divided by V, so

in parallel

$$\frac{1}{R_p} = \frac{1}{R_1} + \frac{1}{R_2} + \cdots + \frac{1}{R_n}$$

In parallel combinations of resistors, the reciprocal of the total resistance is equal to the sum of the reciprocals of the individual resistances.

The parallel equation needs some care in calculations. The steps are:

• Calculate the reciprocals of each individual resistor.

• Add these reciprocals together.

• Take the reciprocal of the answer.

A frequent error is to ignore the last step; Worked example 11 shows the correct approach.

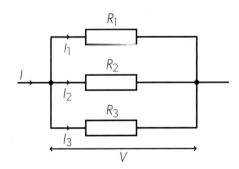

▲ Figure 21 The single resistance value that can replace the three resistors in parallel is the reciprocal of the sum of the individual reciprocal resistances.

Worked example 11

Three resistors of resistance $2.0\,\Omega$, $4.0\,\Omega$ and $6.0\,\Omega$ are connected. Calculate the total resistance of the three resistors when they are connected

a. in series

b. in parallel.

Solutions

a. In series, the resistances are added together, so $2 + 4 + 6 = 12\,\Omega$.

b. In parallel, the reciprocals are used:

$$\frac{1}{R} = \frac{1}{R_1} + \frac{1}{R_2} + \frac{1}{R_3} = \frac{1}{2} + \frac{1}{4} + \frac{1}{6} = \frac{6+3+2}{12} = \frac{11}{12}$$

The final step is to take the reciprocal of the sum, so $R = \frac{12}{11} = 1.1\,\Omega$.

When the networks of resistors are more complicated, then the individual parts of the network need to be broken down into the simplest form. Do this in the following order: parallel then series.

Worked example 12

$2.0\,\Omega$, $4.0\,\Omega$ and $8.0\,\Omega$ resistors are connected as shown. Calculate the total resistance of this combination.

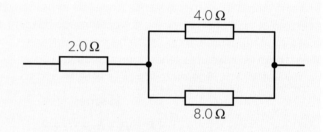

Solution

The two resistors in parallel have a combined resistance of

$$\frac{1}{R} = \frac{1}{R_1} + \frac{1}{R_2} = \frac{1}{4} + \frac{1}{8} = \frac{3}{8}. \text{ So } R = \frac{8}{3} = 2.67\,\Omega.$$

This $2.67\,\Omega$ resistor is in series with $2.0\,\Omega$, so the total combined resistance is $2.67 + 2.0 = 4.7\,\Omega$.

Worked example 13

Four resistors, each of resistance $1.5\,\Omega$, are connected as shown. Calculate the combined resistance of these resistors.

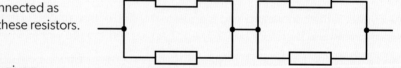

Solution

Two $1.5\,\Omega$ resistors in parallel have a resistance given by $\frac{1}{R} = \frac{1}{1.5} + \frac{1}{1.5} = \frac{2}{1.5}$. So $R = 0.75\,\Omega$.

Two $0.75\,\Omega$ resistors in series have a combined resistance of $0.75 + 0.75 = 1.5\,\Omega$.

Worked example 14

Three resistors of resistances $100\,\Omega$, $200\,\Omega$ and $300\,\Omega$ are connected in series to a cell of emf $12\,V$. Calculate:

a. the current in each of the resistors

b. the potential difference across the $100\,\Omega$ resistor.

Solutions

a. The total resistance of the circuit is $100 + 200 + 300 = 600\,\Omega$. The current is the same in each resistor and equal to the overall current in the circuit, $I = \frac{12}{600} = 0.020\,A$.

b. $V = IR = 0.020 \times 100 = 2.0\,V$. Note that the overall potential difference of $12\,V$ is divided between the individual resistors in the proportion of their resistances. $100\,\Omega$ is $\frac{1}{6}$ of the combined resistance, and therefore the pd across the $100\,\Omega$ resistor is $\frac{1}{6}$ of the emf of the cell.

Practice questions

15. Calculate the combined resistance of each of the following arrangements of resistors.

    a.
    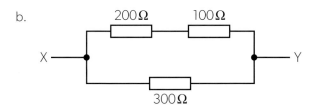

    b.
    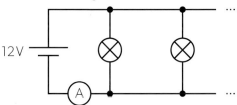

16. A potential difference of 6.6 V is applied between points X and Y in each of the arrangements in question 15.

    Determine, for each arrangement, the current in the 100 Ω resistor and the potential difference across it.

17. Ten identical lamps of resistance 30 Ω each are connected in parallel to a 12 V voltage source as shown in the diagram.

    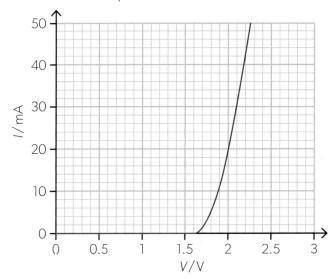

    a. Calculate:
       i.   the current in each of the lamps
       ii. the total resistance of the circuit
       iii. the reading of the ammeter.

    b. One of the lamps fails. Deduce, without any further calculation, how the reading of the ammeter will change.

18. A resistor of resistance R and a light emitting diode are connected in series to a 3.0 V cell, as shown.

    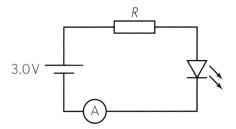

    The diagram shows the variation of the current I in the diode with the potential difference V across it.

    a. The current in the circuit is 20 mA. State the potential difference across:
       i. the diode   ii. the resistor.

    b. Calculate R.

Worked example 15

A 250 Ω resistor is connected in series with a 500 Ω resistor and a 6.0 V battery.

a. Calculate the pd across the 250 Ω resistor.

b. Calculate the pd that will be measured across the 250 Ω resistor if a voltmeter of resistance 1000 Ω is connected in parallel with it.

Solutions

a. The pd across the 250 Ω resistor $= \dfrac{V \times R_1}{(R_1 + R_2)} = \dfrac{6 \times 250}{(250 + 500)} = 2.0\,V$.

b. When the voltmeter is connected, the resistance of the parallel combination is $R = \dfrac{R_1 R_2}{(R_1 + R_2)} = \dfrac{250 \times 1000}{1250} = 200\,\Omega$.

   The total resistance is now 700 Ω, so the pd across the parallel combination is $V = \dfrac{200 \times 6}{700} = 1.7\,V$

Worked example 16

An ammeter with a resistance of 5.0 Ω is connected in series with a 3.0 V cell and a lamp rated at 300 mA, 3 V. Calculate the current that the ammeter will measure.

Solution

Resistance of lamp $= \dfrac{V}{I} = \dfrac{3}{0.3} = 10\,\Omega$. Total resistance in circuit $= 10 + 5 = 15\,\Omega$.

So current in circuit $= \dfrac{V}{R} = \dfrac{3}{15} = 200\,\text{mA}$. This assumes that the resistance of the lamp does not vary between 0.2 A and 0.3 A.

Variable resistors

Some resistors are designed so that their resistance can be changed to a required value. This can be done by changing the dimensions of the resistor or by using a property of the material that it is made from that varies in a predictable way (as in a thermistor or a light-dependent resistor).

Thermistors

Thermistors (as their name implies) are devices whose electrical resistance varies with temperature. They are made from one of two chemical elements that are electrical semiconductors: silicon and germanium. There are several types of thermistor, but we will only consider the negative temperature coefficient type (ntc). As the temperature of an ntc thermistor increases, its resistance falls. This is the opposite behaviour to that of a metal.

Semiconductors have many fewer free electrons per cubic metre than metals. Their resistances are typically 10^5 times greater than those of metals with similar sizes. However, unlike a metal, the charge density in semiconductors depends strongly on the temperature. The higher the temperature of the semiconductor, the more charge carriers are made available in the material.

As the temperature rises in a semiconductor:

- The lattice ions have an increased vibration and impede the movement of the charge carriers more strongly. This is the same effect as in metals and leads to an *increase* in resistance.

- However, more and more charge carriers become available to conduct because the increase in temperature provides them with enough energy to break away from their atoms. This is not the case in a metal. This leads to a large *decrease* in resistance.

- The second effect is much greater than the first and so the net effect is that conduction increases (resistance falls) as the temperature of the semiconductor rises.

Light-dependent resistors

Light dependent resistors (LDRs) are devices that, like thermistors, are made from semiconductor materials. This time, however, the LDRs are affected not by temperature but by light. Photons that arrive at the surface of the LDR transfer energy to the material and release electrons from the lattice in a similar way to the thermal release of electrons in the thermistors.

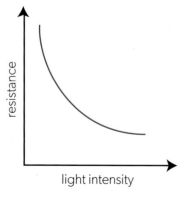

▲ Figure 22 A graph of the variation of resistance with light intensity for a light-dependent resistor.

The greater the intensity of light falling on the surface of the LDR, the smaller the resistance of the LDR. Figure 22 shows the typical variation of resistance with light intensity for an LDR.

Investigating the dependence of resistance in a metal on the size and shape of the conductor

- Inquiry 1: Demonstrate independent thinking, initiative, or insight.

- Tool 3: Construct and interpret tables and graphs for raw and processed data including scatter graphs and line and curve graphs.

- Inquiry 1: Demonstrate creativity in the designing, implementation or presentation of the investigation.

- Inquiry 1: Develop investigations that involve hands-on laboratory experiments, databases, simulations and modelling.

You need a set of wires made from the same metal or metal alloy. The wires should have a circular cross-section and be available in a range of different diameters. You will also need to devise a way to connect the wires into the circuit and a way to vary the length of one of the wires. Use the circuit in Figure 14 as a guide.

- How does the resistance R of **one** of the wires vary with length l?

- How does the resistance R of the wires vary with diameter d when the wires all have the same length?

Try to make things easy for your analyses. In the first investigation, begin by doubling and halving the length of the wire to see what difference this makes to the resistance. Is there an obvious relationship? When you think you know what this is, plot a graph on suitable axes to test your idea.

In the second investigation, the diameter of the wire may be more difficult to test in this way, but a graph of resistance against diameter should give you an immediate clue.

You may decide that the best way to answer these questions is to plot graphs of R against l and R against $\frac{1}{d^2}$.

Variable resistors, potential dividers and potentiometers

Variable resistor

A variable resistor circuit is shown in Figure 23(a). It consists of a power supply, an ammeter, a variable resistor, and a fixed resistor.

When the variable resistor is set to its minimum value of zero, then there will be a pd of 2 V across the fixed resistor and a current of 0.2 A in the circuit.

When the variable resistor is set to its maximum value, 10 Ω, then the total resistance in the circuit is 20 Ω, and the current is 0.1 A.

This means that with 0.1 A in the 10 Ω fixed resistor, only 1 V is dropped across it. Therefore, the range of pd across the fixed resistor can only vary between 1 V and 2 V — half of the available pd that the power supply can, in principle, provide. You should be able to predict the range across the variable resistor too.

This reduced range of pd is a significant limitation in the use of a variable resistor. To achieve a better range, we could use a variable resistor with a much higher range of resistance. To get a pd of 0.1 V across the fixed resistor, the resistance of the variable resistor must be about 200 Ω. But when the fixed resistor has a much greater resistance, then the variable resistor needs an even higher value and this limits the current available from the circuit.

Potential divider

The arrangement known as the **potential divider** (Figure 23(b)) allows a much greater range of pd to a component than a variable resistor in series with the component .

A potential-divider arrangement uses a piece of equipment known as a **potentiometer**; this involves a three-terminal variable resistor such as the one

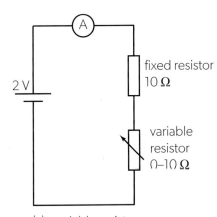

(a) variable resistor

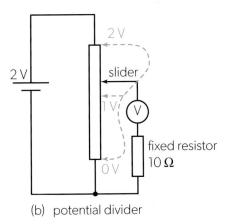

(b) potential divider

▲ Figure 23 A comparison between (a) a variable resistor arrangement and (b) a potential divider.

▲ Figure 24 A variable resistor; also called a rheostat. It has three terminals: one at each end of the resistance coil and one connected to a slider that touches the resistance coil.

shown in Figure 24. One lower terminal of the variable resistor is connected to one side of the cell in Figure 23(b). The other lower terminal is connected to the other terminal of the cell. The upper terminal attached to the slider contact is connected to the rest of the circuit.

The potential at any point along the resistance winding depends on the position of the slider (or wiper) that can be swept across the windings from one end to the other. Typical values for the potentials at various points on the windings are shown for the three blue slider positions on Figure 23(b). The component that is under test (again, a fixed resistor) is connected in a secondary circuit between one terminal of the resistance winding and the slider on the rheostat. When the slider is positioned at one end, the full 2 V from the cell is available to the resistor under test. When at the other end, the pd between the ends of the resistor is 0 V (the two leads to the resistor are effectively connected directly to each other).

Variable resistor or potentiometer?

- Set up the two circuits shown in Figure 23. Match the value of the fixed resistor to the variable resistor. They do not need to be exactly the same but should be reasonably close.

- Add a voltmeter connected across the fixed resistor to check the pd that is available across it.

- Make sure that the maximum current rating for the fixed resistor and the variable resistor cannot be exceeded.

- Check the pd available in the two cases and convince yourself that the potentiometer gives a wider range of voltages.

Worked example 17

A light sensor consists of a 6.0 V battery, a 1800 Ω resistor and a light-dependent resistor in series. When the LDR is in darkness, the pd across the 1800 Ω resistor is 1.2 V.
a. Calculate the resistance of the LDR when it is in darkness.
b. When the sensor is in the light, its resistance falls to 2400 Ω. Calculate the pd across the LDR.

Solutions

a. As the pd across the 1800 Ω resistor is 1.2 V, the pd across the LDR must be $6 - 1.2 = 4.8$ V. The current

in the circuit is $I = \dfrac{V}{R} = \dfrac{1.2}{1800} = 0.67$ mA. The resistance of the LDR is $\dfrac{V}{I} = \dfrac{4.8}{0.67 \times 10^{-3}} = 7200\,\Omega$.

b. The ratio of $\dfrac{\text{resistance across LDR}}{\text{resistance across } 1800\,\Omega} = \dfrac{2400}{1800} = 1.33$. This is the same value as $\dfrac{\text{pd across LDR}}{\text{pd across } 1800\,\Omega}$.

For the ratio of pds to be 1.33, the pds must be 2.6 V and 3.4 V with the 3.4 V across the LDR.

Worked example 18

A thermistor is connected in series with a fixed resistor and a battery. Describe and explain how the pd across the thermistor varies with temperature.

Solution

As the temperature of the thermistor rises, its resistance falls. The ratio of the pd across the fixed resistor to the pd across the thermistor rises too because the thermistor resistance is dropping. As the pd across the fixed resistor and thermistor is constant, the pd across the thermistor must fall.

The change in resistance in the thermistor occurs because more charge carriers are released as the temperature rises. Even though the movement of the charge carriers is impeded at higher temperatures, the release of extra carriers means that the resistance of the material decreases.

Practice questions

19. A circuit consists of a fixed 10.0 kΩ resistor and a light-dependent resistor (LDR) connected in series with a cell of emf 5.00 V. When the LDR is in darkness, it has a resistance of 1.20 MΩ.

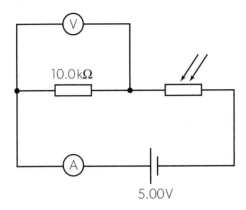

a. Calculate the minimum current in the ammeter.
b. Calculate the reading on the voltmeter when the LDR is in darkness.
c. When the LDR is in daylight, the voltmeter reads 4.96 V. Calculate:
    i.   the resistance of the LDR in daylight
    ii. the reading on the ammeter.

20. A fixed resistor R and a thermistor are connected in series with a cell of a constant terminal potential difference. The resistance of the thermistor decreases with temperature.

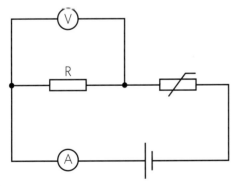

What is the change in the voltmeter reading and in the ammeter reading when the temperature of the thermistor is increased?

	Voltmeter reading	Ammeter reading
A.	increases	increases
B.	increases	decreases
C.	decreases	increases
D.	decreases	decreases

21. A fixed 30 Ω resistor is in a circuit with a potentiometer AB and a cell of emf 9.0 V.

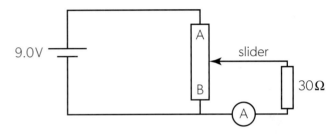

Determine the range of currents that the ammeter will measure as the slider of the potentiometer is moved from A to B.

Heating effect equations

The power P dissipated in a component is related to the pd V across it and the current I in it. The use of V = IR gives two other useful equations:

$$P = IV = I^2 \quad \text{and} \quad R = \frac{V^2}{R}$$

The energy E converted in time Δt is E = IV Δt. When either V or I are unknown, then two more equations become available: $E = IV\,\Delta t = I^2$ and $R\Delta t = \frac{V^2}{R}\,\Delta t$.

These equations will allow you to calculate the energy converted in electrical heaters and lamps. Applications include heating calculations and determining the consumption of energy in domestic and industrial situations.

Worked example 19

Calculate the power dissipated in a 250 Ω resistor when the pd across it is 10 V.

Solution

$$P = \frac{V^2}{R} = \frac{10^2}{250} = 0.40\,\text{W}$$

Worked example 20

A 9.0 kW electrical heater for a shower is designed for use on a 250 V mains supply. Calculate the current in the heater.

Solution

$P = IV$, so $I = \dfrac{P}{V} = \dfrac{9000}{250} = 36\,A$

Worked example 21

Calculate the resistance of the heating element in a 2.0 kW electric heater that is designed for a 110 V mains supply.

Solution

$P = \dfrac{V^2}{R}; R = \dfrac{V^2}{P} = \dfrac{110^2}{2000} = 6.1\,A$

Practice questions

22. A 150 Ω resistor is connected to a battery. The resistor dissipates 32 J of thermal energy in 60 s. Calculate the emf of the battery.

23. A floor heating system consists of a resistive wire that transfers electrical energy at a rate of 800 W when the current in it is 3.5 A. Calculate:
    a.   the resistance of the wire
    b.   the potential difference across it.
    The wire has a length of 48 m and is made of a metal of resistivity $1.5 \times 10^{-6}\,\Omega\,m$.
    c.   Determine the diameter of the wire.

24. A radiant heater dissipates a power of 1200 W when connected to 230 V mains supply. Calculate the power dissipated by the same heater when connected to 110 V mains supply.

25. A wire of length L and a uniform cross-sectional area converts electrical energy at a rate of 200 W when connected across a certain potential difference. The wire is cut in half and one of these pieces of length $\dfrac{L}{2}$ is connected across the same potential difference. What is the rate of energy conversion in the shorter wire?
    A. 50 W   B. 100 W   C. 400 W   D. 800 W

26. Circuit 1 is formed by connecting two identical resistors in parallel with a cell. Circuit 2 is formed by connecting the same two resistors in series with the cell.

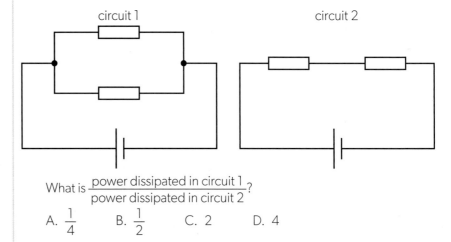

What is $\dfrac{\text{power dissipated in circuit 1}}{\text{power dissipated in circuit 2}}$?

    A. $\dfrac{1}{4}$ B. $\dfrac{1}{2}$ C. 2   D. 4

Cells and batteries

Electric currents can produce a **chemical effect**. This has great importance in chemical industries as it can be a method for extracting ores or purifying materials. However, in this course we do not investigate this aspect of the chemical effect. Our emphasis is on the use of an electric cell to store energy in a chemical form and then release it as electrical energy to perform work elsewhere.

Cells operate as direct-current (dc) devices which means that the cell drives charge in one direction. The electrons (the charge carriers) leave the negative terminal of the cell. After passing around the circuit, the electrons re-enter the cell at the *positive* terminal. The positive terminal has a higher potential than the negative terminal. So electrons appear to gain energy (whereas positive charge carriers would appear to lose it).

The chemicals in the cell react as charge flows and, because of this reaction, energy is transferred to the electrons moving through the cell.

Chemical cells and solar cells as energy sources

Chemical cells

Many of the portable devices we use today, such as torches and music players, can operate with internal cells, either singly or in batteries. In some cases, the cells are used once until they are completely discharged and are then thrown away. These are **primary cells**. The chemicals have completely reacted; the cell cannot be recharged.

Other devices such as laptops use rechargeable cells. When the chemical reactions have finished, the original chemicals can be re-formed by sending charge through the cell in the reverse direction. The cell is then again available as a chemical-energy store. Rechargeable cells are **secondary cells**. Secondary cells include lead–acid accumulators, nickel–metal-hydride (NiMH), and lithium–ion cells. These can be recharged many times. As the recharge is cheap, the overall cost is lower than that of buying many primary cells.

Solar cells

The first solar "photocells" were developed around the middle of the 19th century by Alexandre-Edmond Becquerel (the father of Henri, the discoverer of radioactivity). For a long time, solar cells, based on the element selenium, were only used in photography. However, when semiconductor technology was developed, it led to the invention of photovoltaic (solar) cells to power everything from calculators to satellites. In many parts of the world, solar panels are mounted on the roofs of houses. These panels not only supply energy to the house, but excess energy transferred during sunny days is often sold to the local electricity supply company.

The photovoltaic materials in the solar panel convert electromagnetic radiation from the Sun into electrical energy. When a photon is incident on the cell, electrons are released and gain energy to move. The electrons transfer this energy into the external circuit and do useful work. A full explanation of the way in which this happens goes beyond the IB syllabus.

One single cell has a small emf of about 1 V (this is determined by the nature of the semiconductor) and so banks of cells are manufactured to produce usable currents on both a domestic and commercial scale.

The efficiencies of present-day solar cells are about 25% or a little higher. However, extensive research and development are being carried out in many countries and it is likely that these efficiencies will rise significantly over the next few years.

Comparing energy sources

Different energy sources have different advantages and disadvantages:

- Primary chemical cells. These can be cheap and with a small mass. However, they can only be used once.

- Secondary chemical cells. These can also be small mass for some applications but when high currents are required, then batteries of the cells need a large mass and large volume. This has disadvantages for applications involving transport. The cells can be recharged many times, although not indefinitely, and the replacement cost can be high.

- Solar cells transfer solar energy so rely on sunlight for their operation. The cost of the panels can be high, although it is dropping steadily. The lifetime of the cells can be limited. Solar cells often require a battery of secondary cells for their use to be effective at times when it is cloudy or at night.

Global impact of science

Rechargeable batteries are used in many devices, from mobile phones to electric cars. These batteries are often made using lithium. However, the world's reserves of lithium are limited. As our use of renewable resources increases, the need for batteries to store this energy will also increase.

▲ Figure 25 A lithium–ion battery for a mobile phone.

Internal resistance and emf of a cell

The materials from which cells are made have electrical resistance in just the same way as the metals in the external circuit. This **internal resistance** has an important effect on the total resistance and current in the circuit. Figure 26(a) shows a simple model for a cell.

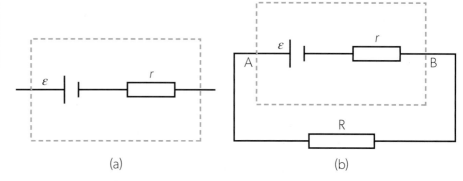

(a)   (b)

▲ **Figure 26** (a) A simple model of an electrical cell with a constant source of emf in series with a fixed internal resistance. (b) The cell model connected in series with a resistor R.

Inside the dotted box is an "ideal" cell that has no resistance of its own. Also inside the box is a resistance symbol that represents the internal resistance of the cell. The two together make up our model for a real cell. The model assumes that:

- the internal resistance is constant (for a practical cell it varies with a number of factors)

- the emf is constant (which can also vary for a real cell).

Our model cell has an emf ε and an internal resistance r in series with an external resistance R. The current in the circuit is I.

We can apply conservation of energy to this circuit:

- The emf of the cell supplying energy to the circuit $= \varepsilon$.

- The sum of the pds $= IR + Ir$.

- This means that $\varepsilon = IR + Ir$.

- Or more simply, $\varepsilon = I(R + r)$.

- When the pd across the external resistor is V, then $\varepsilon = V + Ir$ or $V = \varepsilon - Ir$.

It is important to realize that V, which is the pd across the external resistance, is equal to the terminal pd across both the ideal cell and the internal resistance in series in our model (in other words, between A and B).

> The emf of a cell is the open circuit pd across the terminals of a power source—in other words, the terminal pd when no current is supplied.

The pd between A and B is less than the emf unless the current in the circuit is zero. The difference between the emf and the **terminal pd** (the measured pd across the terminal of the cell) is sometimes referred to as the "lost pd" or the "lost volts". These lost volts represent the energy required to drive the charge carriers—the electrons—through the cell itself. Once the energy has been used in the cell in this way, it cannot be available for conversion in the external circuit. You may have noticed that, when a cell is being charged, or when a cell is discharging at a high current, it becomes warm. The energy required to raise the temperature of the cell has been transferred in the internal resistance.

Worked example 22

A cell of emf 6.0 V and internal resistance 2.5 Ω is connected to a 7.5 Ω resistor. Calculate:
a.   the current in the cell
b.   the terminal pd across the cell
c.   the energy lost in the cell when charge flows for 10 s.

Solutions

a.   Total resistance in the circuit is 10 Ω, so current in circuit $= \dfrac{6}{10} = 0.60\,A$

b.   The terminal pd is the pd across cell $= IR = 0.6 \times 7.5 = 4.5\,V$

c.   In 10 s, 6 C flows through the cell and the energy lost in the cell is 1.5 J C⁻¹. The energy lost is 9.0 J.

Worked example 23

A battery is connected in series with an ammeter and a variable resistor R. When $R = 6.0\,\Omega$, the current in the ammeter is 1.0 A. When $R = 3.0\,\Omega$, the current is 1.5 A. Calculate the emf and the internal resistance of the battery.

Solution

Using $V = \varepsilon - Ir$ and knowing that $V = IR$ gives two equations:

$6 \times 1 = \varepsilon - 1 \times r$ and $3 \times 1.5 = \varepsilon - 1.5 \times r$.

These can be solved simultaneously to give $(6 - 4.5) = 0.5r$ or $r = 3.0\,\Omega$ and $\varepsilon = 9.0\,V$.

Practice questions

27. A 100 Ω resistor is connected to a cell of emf 5.0 V. The potential difference across the resistor is 4.9 V.
    a.   Calculate the internal resistance of the cell
    b.   Calculate the terminal pd across the cell when the resistor is replaced by one of resistance 20 Ω. Assume that the internal resistance of the cell is constant.

28. Which statement is correct about the terminal pd across a real cell?
    A.   It is constant.
    B.   It increases with the load resistance.
    C.   It increases with the current in the cell.
    D.   It is zero when no current is in the cell.

29. A variable resistor is connected to a cell as shown in the diagram. The ammeter and the voltmeter are ideal.

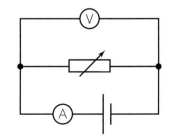

The graph shows how the potential difference V across the resistor varies with the current I in the ammeter.

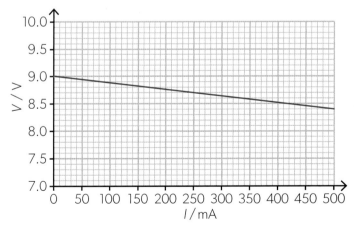

a.   State the emf of the cell.
b.   Determine the internal resistance of the cell.
c.   Calculate the resistance of the variable resistor when the current is 500 mA.

Measuring the internal resistance of a fruit cell

- Tool 1: Understand how to accurately measure electric current and electric potential difference to an appropriate level of precision.

- Tool 3: Construct and interpret tables and graphs for raw and processed data including scatter graphs and line and curve graphs.

- Inquiry 2: Collect and record sufficient relevant quantitative data.

The method given here works for any type of electric cell. However, for novelty, a citrus fruit cell (orange, lemon, lime, etc.) or even a potato can be used for the measurement. The ions in the flesh of the fruit react with two different metals to produce an emf. With an external circuit that only requires a small current, the fruit cell can discharge over surprisingly long times.

- To make the cell, take a strip of copper foil and a strip of zinc foil, both about 1 cm by 5 cm, and insert these, about 5 cm apart, deep into the fruit. You may need to use a knife to make an incision unless the foil is stiff. (Other metals can be used, for example the zinc foil can be replaced with some magnesium ribbon or even an iron nail. A copper coin can be used instead of copper foil.)

- Connect the circuit shown in Figure 27 using a suitable variable resistor. (Ideally the variable resistor should have a range up to about 100 kΩ.)

- Measure the terminal pd across the fruit cell and the current in the cell for the largest range of pd you can manage.

- Compare the equation $V = \varepsilon - Ir$ with the equation for a straight line $y = c + mx$.

Terminal pd / V	Current / A
1.13	0.05
1.01	0.10
0.89	0.15
0.77	0.20
0.65	0.25
0.53	0.30
0.41	0.35
0.29	0.40
0.17	0.45
0.05	0.50

- A plot of V on the y-axis against I on the x-axis should give a straight line with a gradient of $-r$ and an intercept on the V-axis of ε.

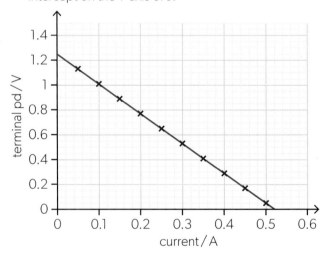

▲ Figure 28 A graph of V against I for a cell. The intercept for this graph, and therefore the emf of the cell, is 1.25 V. The gradient is −2.4 which gives an internal resistance for the cell of 2.4 Ω. You should check these results for yourself.

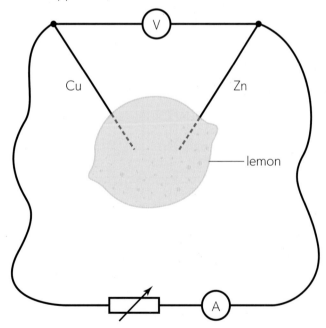

▲ Figure 27 A circuit to measure the internal resistance and emf of a fruit cell.

Power supplied by a cell

The total power supplied by a non-ideal cell is equal to the power delivered to the external circuit plus the power wasted in the cell.

Algebraically, $P = I^2 R + I^2 r$ using the notation used earlier.

The power delivered to the external resistance is $\dfrac{\varepsilon^2}{(R + r)^2} \times R$

Figure 29 shows how the power delivered to the external circuit varies with R.

The peak of this curve is when $r = R$, in other words, when the internal resistance of the power supply is equal to the resistance of the external circuit. The load and the supply are "matched" when the resistances are equal in this way. This matching of supply and circuit is important in several areas of electronics.

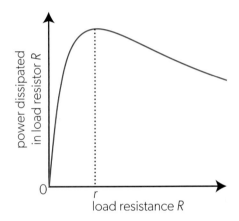

▲ Figure 29 The power supplied by a cell is at a maximum value when the external resistance and the internal resistance of the cell are equal.

Worked example 24

A battery of emf 9.0 V and internal resistance 3.0 Ω is connected to a load resistor of resistance 6.0 Ω. Calculate the power delivered to the external load.

Solution

Using the equation $\dfrac{\varepsilon^2}{(R + r)^2} \times R$ leads to $\dfrac{9^2}{(6 + 3)^2} \times 6 = 6.0 \, \text{W}$

Theme B — End-of-theme questions

1. The Sun has a radius of 7.0×10^8 m and is a distance 1.5×10^{11} m from Earth. The surface temperature of the Sun is 5800 K.

    a. Show that the intensity of the solar radiation incident on the upper atmosphere of Earth is approximately 1400 W m^{-2}.

    b. The albedo of the atmosphere is 0.30. Deduce that the average intensity over the entire surface of Earth is 245 W m^{-2}.

    c. Estimate the average surface temperature of Earth.

2. The average temperature of ocean surface water is 289 K. Oceans behave as black bodies.

    a. Show that the intensity radiated by the oceans is about 400 W m^{-2}.

    b. Explain why some of this radiation is returned to the oceans from the atmosphere.

    c. The intensity in b. returned to the oceans is 330 W m^{-2}. The intensity of the solar radiation incident on the oceans is 170 W m^{-2}.

        i. Calculate the additional intensity that must be lost by the oceans so that the water temperature remains constant.

        ii. Suggest a mechanism by which the additional intensity can be lost.

3. A mass of 1.0 kg of water is brought to its boiling point of 100 °C using an electric heater of power 1.6 kW. A mass of 0.86 kg of water remains after it has boiled for 200 s.

    a. i. Estimate the specific latent heat of vaporization of water. State an appropriate unit for your answer.

        ii. Explain why the temperature of water remains at 100 °C during this time.

    b. The heater is removed and a mass of 0.30 kg of pasta at −10 °C is added to the boiling water. Determine the equilibrium temperature of the pasta and water after the pasta is added. Other heat transfers are negligible.
        Specific heat capacity of pasta = 1.8 kJ kg^{-1}K^{-1}
        Specific heat capacity of water = 4.2 kJ kg^{-1}K^{-1}

    c. The electric heater has two identical resistors connected in parallel.

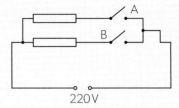

The circuit transfers 1.6 kW when switch A only is closed. The external voltage is 220 V.

    i. Show that each resistor has a resistance of about 30 Ω.

    ii. Calculate the power transferred by the heater when both switches are closed.

4. a. State what is meant by the internal energy of an ideal gas.

    b. A quantity of 0.24 mol of an ideal gas of constant volume 0.20 m^3 is kept at a temperature of 300 K.

        i. Calculate the pressure of the gas.

        ii. The temperature of the gas is increased to 500 K. Sketch a graph to show the variation with temperature T of the pressure P of the gas during this change.

    c. A container is filled with 1 mole of helium (molar mass 4 g mol^{-1}) and 1 mole of neon (molar mass 20 g mol^{-1}). Compare the average kinetic energy of helium atoms to that of neon atoms.

5. An ideal gas consisting of 0.300 mol undergoes a process ABCD. AB is an adiabatic expansion from the initial volume V_A to the volume $1.5V_A$. BC is an isothermal compression. The pressures at C and D are the same as at A.

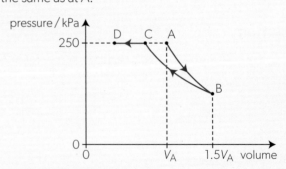

The following data are available:

Pressure at A = 250 kPa

Volume at C = 3.50×10^{-3} m^3

Volume at D = 2.00×10^{-3} m^3

    a. i. Show that the pressure at B is about 130 kPa.

        ii. Calculate the ratio $\dfrac{V_A}{V_C}$.

AHL

b. The gas at C is further compressed to D at a constant pressure. During this compression the temperature decreases by 150 K.

For the compression CD:

　i. Determine the thermal energy removed from the system.

　ii. Explain why the entropy of the gas decreases.

　iii. State and explain whether the second law of thermodynamics is violated.

6. A photovoltaic cell is supplying energy to an external circuit. The photovoltaic cell can be modelled as a practical electrical cell with internal resistance.

The intensity of solar radiation incident on the photovoltaic cell at a particular time is at a maximum for the place where the cell is positioned.

The following data are available for this particular time:

Operating current = 0.90 A
Output potential difference to = 14.5 V
external circuit
Output emf of photovoltaic cell = 21.0 V
Area of panel = 350 mm × 450 mm

a. Explain why the output potential difference to the external circuit and the output emf of the photovoltaic cell are different.

b. Calculate the internal resistance of the photovoltaic cell for the maximum intensity condition using the model for the cell.

c. The maximum intensity of sunlight incident on the photovoltaic cell at the place on Earth's surface is 680 W m⁻². A measure of the efficiency of a photovoltaic cell is the ratio

energy available every second to the external circuit
───
energy arriving every second at the photovoltaic cell surface

Determine the efficiency of this photovoltaic cell when the intensity incident upon it is at a maximum.

7. A lighting system consists of two long metal rods with a potential difference maintained between them. Identical lamps can be connected between the rods as required.

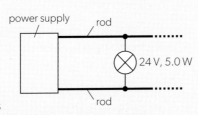

The following data are available for the lamps when at their working temperature.

Lamp specifications	24 V, 5.0 W
Power supply emf	24 V
Power supply maximum current	8.0 A
Length of each rod	12.5 m
Resistivity of rod metal	7.2 × 10⁻⁷ Ω m

a. Each rod is to have a resistance no greater than 0.10 Ω. Calculate, in m, the minimum radius of each rod.

b. Calculate the maximum number of lamps that can be connected between the rods. Neglect the resistance of the rods.

c. One advantage of this system is that if one lamp fails then the other lamps in the circuit remain lit. Outline one other electrical advantage of this system compared with one in which the lamps are connected in series.

8. The graph shows how current I varies with potential difference V across a component X.

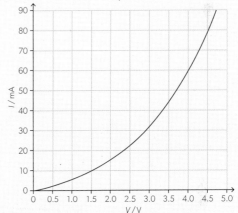

a. Outline why component X is considered non-ohmic.

b. Component X and a cell of negligible internal resistance are placed in a circuit. A variable resistor R is connected in series with component X. The ammeter reads 20 mA.

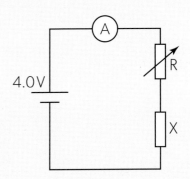

Determine the resistance of the variable resistor.

c. Calculate the power dissipated in the circuit.

331

Tools for physics

Introduction

The nature of science is that it seeks through experiment and theory to link the observed behaviour of the universe to a set of principles. In order to do this, physicists require a set of tools, both experimental and mathematical. This section of the book sets out some of these principles.

Science is inquiry-based. This is another aspect to its nature. The inquiry process itself is described at the end of this book. But access to this inquiry process requires scientific skill (Figure 1). Some of these skills you have when you begin the course. Others will be developed as you study IB Diploma physics.

▶ Figure 1 The skills required for the inquiry process are in three groups: mathematical techniques, experimental techniques and technology skills.

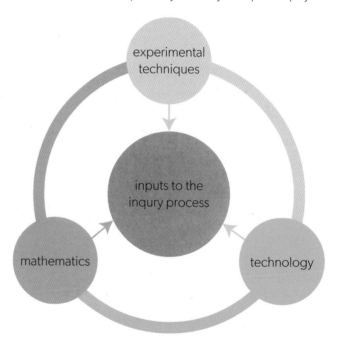

This chapter contains three sections. These outline the skills required for the inquiry process:

- Mathematical tools for physics

- Experimental tools for physics

- Handling data and modelling physics

The whole chapter is a series of short sections arranged for reference purposes. You may wish to refer to it repeatedly as you carry out practical and arithmetic work and develop your skills during the course.

The Inquiry process and internal assessment (IA) chapter at the end of the book suggests how you can best demonstrate your understanding of these skills in the internal assessment which you undertake towards the end of your IB Diploma Programme physics course.

Mathematical tools for physics

- quantities and variables, both discrete and continuous

- units and the unit system used in science

- dimensional analysis

- significant figures and orders of magnitude

- estimation and approximations

- correlation and proportionality

- vectors and scalar arithmetic.

You will be provided with an IB Physics data booklet in your examinations. This contains a list of the fundamental constants, conversions between units, and equations. So you do not need to learn these.

Symbols, units and numerical values

Quantities

Physicists use measurable quantities such as time, length and mass. These quantities are linked by relationships such as speed $= \dfrac{\text{distance travelled}}{\text{time taken}}$. Algebra transforms this to $v = \dfrac{s}{t}$, where v, s and t are the speed, distance travelled and time taken. Quantities also require units that can be used for the measurements: metres (m) and seconds (s), so speed is in metres per second (m s^{-1}).

Quantities are written using *italic* symbols (for example, s for distance rather than s for time), whereas a unit is written using Roman (upright) symbols. It is important to identify symbols from your textbooks correctly. When m appears in this book and in your examinations it represents "mass", but m stands for the unit "metre".

It is important to be clear about the context in which a symbol is being used. For example, d can mean a distance in Theme A, but it means slit separation in Topic C.3.

Variables and constants

Fundamental constants have fixed values. The speed of light in a vacuum c is an example. The value of c does not change.

Other constants include numbers that occur as a result of the way we define the shape of the world and our mathematical system. The irrational number pi (π) is an example of this, so is e, the base of natural logarithms.

A third set of constants arise from physical theory and represent the properties of individual substances or the natural world. Unlike fundamental constants they vary from place to place or material to material. Examples here include:

- g, the acceleration due to gravity at Earth's surface. This varies with position on the surface due to local density variations in Earth and variations in Earth's radius.

- ρ, electrical resistivity. This changes from material to material and is temperature dependent. However, it does not depend on the dimensions of the material and is therefore more useful to an engineer than the electrical resistance of a particular sample.

Continuous and discrete variables

Quantities that can change in physics are known as **variables**. The variables you use in your practical and theoretical work fall into two categories.

- **Continuous variables** are data that can take any value. The temperature of a room is an example of continuous data when a thermometer is used to record the variation throughout the day.

- **Discrete variables** are data that can only take certain values. The number of electrons emitted in a photoelectric experiment, for example, is discrete because only integer numbers of emitted electrons are possible. Half an electron is never observed.

Discrete data are often displayed using bar charts, pie charts or histograms. Histograms are explained more on page 353.

Système Internationale d'Unités (SI)

Science needs a common language for its communication. To aid this, there is an internationally agreed system of units called, for short, the SI.

The SI defines **units** for quantities that are either:

- base (fundamental) units

- derived units, that are combinations of base units

- supplementary units, such as the radian (rad).

Base units

SI gives seven **base units**.

- second (s)

- meter (m)

- kilogram (kg)

- ampere (A)

- kelvin (K)

- mole (mol)

- candela (not used in this course)

The way that some base units were defined underwent an important change in May 2019. Books older than this may use the old definitions. The change in 2019 began with the creation of defined constants, the values of which were based on the best estimate at the time.

These base units are defined in terms of seven fundamental constants (one is not given here):

- the speed of light in a vacuum: $c = 299\,792\,458\,\mathrm{m\,s^{-1}}$

- the Planck constant: $h = 6.626\,070\,15 \times 10^{-34}\,\mathrm{J\,s}$

- the charge on the electron: $e = 1.602\,176\,634 \times 10^{-19}\,\mathrm{C}$

- the Boltzmann constant: $k = 1.380\,649 \times 10^{-23}\,\mathrm{J\,K^{-1}}$

- the Avogadro constant: $N_A = 6.022\,140\,76 \times 10^{23}\,\mathrm{mol^{-1}}$

- the hyperfine transition frequency of cesium-133: $\Delta \nu_{Cs} = 9192\,631\,770\,\mathrm{Hz}$.

Each base unit is defined in terms of one or more fundamental units so that, for example,

- the metre is defined to be the distance travelled by light in a vacuum in $\dfrac{1}{299\,792\,458}$ th of a second, and
- the kilogram is defined to be the mass whose rest energy is equal to the energy of a collection of photons of a combined frequency of $1.356\,392\,489\,652 \times 10^{50}$ Hz. This definition uses $E = mc^2 = hf$ leading to an energy equivalence to mass of $\dfrac{h \times \Delta v_{Cs}}{c^2}$.

Prefixes

Units can be modified by the use of the prefixes allowed in SI (Table 1).

Derived units

Most units are known as derived units because they are combinations of the seven base units. For example, the equation for kinetic energy is $E_k = \dfrac{1}{2}mv^2$, where m is the mass of the moving object and v is its speed. This can be written in unit terms as $mass \times speed^2$; the $\dfrac{1}{2}$ is dropped here because it is a scale factor without units.

Speed has the units $\dfrac{distance}{time}$, so the whole unit for E_k is $mass \times speed^2$ which is $mass \times \left(\dfrac{distance}{time}\right)^2$ which, reverting to the unit symbols, is $kg\,m^2\,s^{-2}$. This is the unit for energy—the joule (J)—expressed in base units.

It is important to develop the skill of converting between derived units and their fundamental equivalents.

Non-SI units

Not all units that you will meet are part of the SI. Some used in the course are allowed because of their convenience to physicists working in particular fields. For example, the kilowatt hour (kW h) and the electronvolt (eV). Astronomy, in particular, uses non-SI units, for example, light year (ly), the astronomical unit (AU) and the parsec (pc)) because the distances are so large. Even Celsius (°C) for temperature is not strictly part of the SI and kelvin (K) should properly be used as much as possible.

Radians and degrees

Calculations of circular motion involve the use of angles. In the science you studied before starting this course, you will almost certainly have measured all your angles in degrees. In some areas of physics (including circular motion), there is an alternative measure of angle that is much more convenient: the **radian**. Radians are based on the geometry of the arc of a circle.

Prefix	Abbreviation	Value
peta	P	10^{15}
tera	T	10^{12}
giga	G	10^{9}
mega	M	10^{6}
kilo	k	10^{3}
hecto	h	10^{2}
deca	da	10^{1}
deci	d	10^{-1}
centi	c	10^{-2}
milli	m	10^{-3}
micro	μ	10^{-6}
nano	n	10^{-9}
pico	p	10^{-12}
femto	f	10^{-15}

▲ Table 1 Prefixes for SI units.

See page 115 in Topic A.3 for more about the equation for kinetic energy.

You can see another example of unit conversion on page 518. There are others in the book.

See page 519 in Topic D.2 for more about the electronvolt (eV) and page 693 in Topic E.5 for a discussion of units in astronomy.

The kilogram (kg) is the SI unit of mass, but in atomic and nuclear physics it is convenient to express mass in the unified atomic mass unit (u) or in the units of $MeV\,c^{-2}$. See page 596 and worked examples in Topic E.3.

Conversions between these and other units are provided in the IB Physics data booklet.

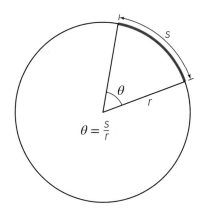

▲ Figure 1 Definition of the radian.

An angle measured in radians (Figure 1) is defined as the circumference s of an arc of a circle divided by the radius r of the circle. In symbols θ (in radians) $= \dfrac{s}{r}$.

The radian is abbreviated as "rad" in SI. However, as it is a ratio, it is strictly speaking not a unit.

Going around a circle once means travelling all the way round the circumference. This is a distance of $2\pi r$. The angle θ in radians subtended by the whole circle is $\dfrac{2\pi r}{r} = 2\pi$ rad.

So $360° \equiv 2 \times \pi = 2 \times 3.14... = 6.28$ rad and 1.00 rad $= 57.3°$.

Sometimes, the radian numbers are left as fractions, for example $90° = \dfrac{\pi}{2}$.

To convert other values for yourself, use the equation:

$$\frac{\text{angle in degrees}}{360} = \frac{\text{angle in radian}}{2\pi}$$

Finally, a practical point: scientific and graphic calculators work happily in either degrees or radians, but the calculator must know what to expect! Always check that your calculator is set to work in radians if that is what you want, or in degrees when those are the units you are using.

Dimensional analysis

Dimensional analysis is a technique that you may find useful when planning your internal assessment and when checking expressions. The principle is to use a knowledge of the dimensions (units) of a quantity to establish its algebraic relationship to other variables. The dimensions are often written in square brackets to make it clear that these are not variables in their own right. Thus, speed will be written as [length] [time]$^{-1}$ and force will be written as [mass] [length] [time]$^{-2}$ (using Newton's second law of motion). A dimensional analysis cannot give you any constants of proportionality. They need to be determined by experiment.

Revisiting the simple pendulum

You already know the equation for the time period of a pendulum from Topic C.1 but it is possible to derive part of it using dimensional analysis.

Some factors that you might expect to affect the time period of a pendulum are the length l, the mass of the pendulum bob m and the gravitational field strength g. These lead to the equation

[time]^1 = [length]^x × [gravitational field strength]^y × [mass]^z.

The term [gravitational field strength] needs to be split into its separate dimensions [length × time^{-2}] so that (collecting terms)

[time]^1 = [length]$^{x+y}$ × [time]$^{-2y}$ × [mass]^z.

Looking at each dimension separately gives $z = 0$, $y = -\dfrac{1}{2}$ and $x = +\dfrac{1}{2}$. The predicted equation for the time period is $T \propto \sqrt{\dfrac{l}{g}}$. As before, this analysis cannot give the 2π constant.

You can now carry out an experiment to confirm the constant of proportionality. Vary l and measure T. Find g using an independent method. Careful measurements can produce a value that is less than 1% away from the actual value.

Worked example 1

A student wants to know how the speed v of a surface water wave in the deep ocean varies with other variables.

Step 1. What could the wave speed depend on? Possible variables could be density of water ρ, wavelength of the wave λ, depth of the ocean D, acceleration due to gravity g, water temperature, salinity of the water etc.

The water temperature and salinity can be eliminated because the wave speed must eventually have the dimensions of [length] [time]$^{-1}$ and these variables have the dimensions of [temperature] and [quantity], respectively. The ocean depth is unlikely to be important because a deep ocean will probably not influence the behaviour of a wave at the surface. Therefore, the student can assume that the relationship between wave speed and its variables is $v = \rho^x \times g^y \times \lambda^z$ where they wish to determine x, y and z.

Step 2. Write this expression in full with all the dimensions to give
[length]^1 [time]$^{-1}$ = [mass]^x [length]$^{-3x}$ × [length]^y [time]$^{-2y}$ × [length]^z.

Step 3. The dimensions must match on each side of the equation. Therefore, collecting powers together for each dimension separately:

[mass]: $0 = x$ [length]: $1 = -3x + y + z$ [time]: $-1 = -2y$

These can be solved as three simultaneous equations. Clearly, $x = 0$ and so the density is not part of the relationship. Then $y = \frac{1}{2}$ from the third equation, leading to $z = \frac{1}{2}$ also from the second equation. The final relationship is that $v \propto g^y \lambda^z = g^{\frac{1}{2}} \lambda^{\frac{1}{2}} = \sqrt{g\lambda}$. This suggests that the wave speed depends only on the wavelength and acceleration due to gravity.

Worked example 2

The distance s travelled in time t by an object accelerating uniformly from rest is given by $s = \frac{1}{2} at^2$, where a is the acceleration. A cart starts from rest and travels 1.45 m in the first 0.90 s of motion. Calculate the acceleration of the cart, stating the answer to an appropriate number of significant figures.

Solution

$a = \frac{2s}{t^2} = \frac{2 \times 1.45}{(0.90)^2} = 3.5802... \, \text{m s}^{-2}$.

The distance is given to 3 sf but the time of travel is only to 2 sf. The result of the calculation should be rounded to the lowest number of significant figures in the input data, which in this case is 2 sf. Therefore, $a = 3.6 \, \text{m s}^{-2}$.

Significant figures

Often numerical results from calculators or spreadsheets will appear as a long string of numbers. It is important only to quote the numbers that have meaning in the context of the problem or experiment. These are called the **significant figures**, often abbreviated to "s.f." or "sf".

When you write a number, such as 1.602, "1" is the most significant digit, "2" is the least significant digit. When the number is between 0 and 1, such as 0.001 352, the same is true, so that "1" is the most significant digit.

The following are rules for the use of significant figures.

- A non-zero digit is always significant: 789 is 3 s.f.

- A zero between two non-zero digits is always significant: 709 is 3 s.f.

- Leading zeros are never significant: 0.0123 is 3 s.f.

- Trailing zeros are not significant when there is no decimal point: 54 800 is 3 s.f.

- Zeros to the left of a decimal point are significant: 54 800. is 5 s.f.

- Zeros to the right of a decimal point and to the right of a non-zero digit are significant: 0.009 is 1 s.f; 0.023 is 2 s.f.; 0.175 is 3 s.f.; 0.175 00 is 5 s.f.

As examples of the use of significant figures, compare Worked examples 3 and 4 in Topic D.1. Worked example 3 has data all to 3 sf. The final answer from a calculator is 9.81344... but is quoted only to 3 sf here. Had one piece of data (say, the radius of Earth) only been given to 2 sf, then 2 sf would be appropriate for the final answer.
In Worked example 4, all the data and the final answer are given to 2 sf.

Practice questions

Calculate the following quantities, giving the answer to an appropriate number of significant figures:

1. The speed v of a football of mass $m = 0.450\,kg$, when the kinetic energy of the football is $E_k = 35\,J$. Use the equation $E_k = \frac{1}{2}mv^2$.

2. The volume V of an ice cube of mass $m = 1.00 \times 10^2\,g$. The density of ice is $\rho = 0.917\,g\,cm^{-3}$. Use the equation $\rho = \frac{m}{V}$.

3. The current I in a resistor of resistance $R = 1.00 \times 10^2\,\Omega$ when the resistor is dissipating a power of $P = 1.35\,W$. Use the equation $P = I^2R$.

When you carry out a multi-step calculation, avoid rounding off intermediate values and only round the final answer to the required number of s.f. Worked example 15 in Topic A.1 illustrates a typical multi-step calculation that requires intermediate values to be kept to a higher precision than the input data.

Practice questions

Consider these estimates. Don't use any measuring instruments.

4. A girl jumps off a wall. Estimate the force that she exerts on the ground as she lands.

5. What is the total weight of the house or apartment block in which you live?

6. How many key depressions were involved in typing out the manuscript for this book?

7. What is the total floor area of your school?

8. What is the total length of wire in a grand piano?

Whenever you do a calculation (in a lesson, exam or experiment), carefully choose how many significant figures to write each number to using the following rules.

- When all the data are supplied to 3 s.f. then give your answer to 3 s.f.

- If there are mixed s.f., usually this will be 2 or 3, then quote your answer to the lowest number of s.f.

- Sometimes in nuclear masses or binding energies you will be given up to 5 or 6 s.f. in some data because there are subtractions involved in the calculations that reduce the number of s.f. In this case, it is safe to quote to a level of 3 s.f.

- Finally, there is a group of problems in which you can be asked to "Show that" a particular answer is correct. In this case, you must quote your answer to one s.f. better than the s.f. used in the question. When the data are quoted to 3 s.f., give your answer to 4 s.f. or more. This is to show that you have carried through all the steps of the problem, including the final calculation.

Orders of magnitude

Whenever you arrive at a final numerical answer at the end of a calculation or an experiment, take a moment to think about the result. Is it credible? Any result that indicates that the mass of person is 2000 kg is unlikely to be correct. A temperature of 10 million K is only found in the interior of a star. Answers should be of the right order of magnitude. If they are not, check your data and working carefully.

An answer is given as an **order of magnitude** when it is expressed to the nearest power of ten. A sweep second hand of an analogue clock takes $60\,s$ — roughly $10^2\,s$ — to go round once. A feature film lasts about two hours – $7200\,s$ or roughly $10^4\,s$. The orders of magnitude of these two time periods are 2 and 4. The difference in the order of magnitude is $4 - 2 = 2$.

Approximation and estimation

It is helpful to estimate values when it is not possible to know the exact value or when a rough value is all that is required. Estimation may provide an answer to better than an order of magnitude, but this depends on the particular estimation that is being carried out.

Estimation can be practical or theoretical. For example:

- Using a lens to magnify a scale. A metre ruler is calibrated to the nearest millimetre, but a magnifying glass will enable you to make an estimate of a length to the nearest tenth of a millimetre where necessary. This can be done with most analogue scales.

- Estimating the area under a graph. A reasonable count of the number of grid squares allowing for incomplete squares is required. When this is combined with the known size of one grid square, a surprisingly close value to the true value can usually be obtained.

- Planning an experiment. Theoretical estimates are of great value for giving you a rough idea of the magnitudes of the measurements that you will need to make. In making such estimates, always use rounded data: human beings are about 2 m tall, a car has a mass of about 1000 kg and room temperatures are about 300 K.

In making estimates, you often need to make assumptions because some effects may not make a large difference to the answer. For example, air resistance is often ignored in mechanics problems when speeds are small (of the order of $10^0 \, \text{m s}^{-1}$). At high speeds (around $10^3 \, \text{m s}^{-1}$), this would be an unreasonable assumption and the frictional drag would need to be included in an estimate.

Correlation

An aim of physics experimentation is to confirm correlations between variables. This means to establish the extent to which one quantity varies with another.

Always remember that correlation does not prove causality. Just because a graph of x against y is a straight line going through the origin, it does not mean that the change in x is necessarily the cause of the change in y. There are many examples in medical and other sciences where correlation and cause have been confused. When there is a physical reason why the increase in one quantity causes a change in another quantity, there is said to be a **causal relationship**.

Correlations can be:

- **positive**; meaning that as the independent variable increases, the dependent variable tends to increase

- **negative**; meaning that as the independent variable increases, the dependent variable tends to decrease.

Direct and inverse proportionality

There are two types of correlation that have particular importance in science: direct and inverse proportionality. Both types are easily identified on graphs or by calculation.

Two variables x and y have a correlation that is **direct proportion** when $y = kx$, where k is a constant. This will be represented on a graph of the variation of y with x by a straight line with gradient k that goes through the origin (Figure 2).

The two variables x and y are in **inverse proportion** when $y = \dfrac{k}{x}$ and k is a constant. This time the graph of y against x is no longer a straight line (Figure 2), but the relationship can be identified relatively easily because $x \times y = k$. A series of data pairs when multiplied together should give a product that does not change (within experimental error).

Scalars and vectors

Scalars are quantities with magnitude only. They obey the rules of algebra and can be added, subtracted, multiplied and divided as with normal numbers. Scalars retain their units under addition and are combined when multiplied. A distance of 50 m added to a distance of 30 m is 80 m and when it takes 20 s to cross this distance, the speed is $4.0 \, \text{m s}^{-1}$.

Vectors have magnitude and direction. Vector algebra is required so that both elements of the vector can be combined correctly. Vectors are sometimes given their own name to distinguish them from scalar equivalents: distance and displacement (Topic A.1) are examples of this.

You can multiply or divide a vector by a scalar. Suppose the displacement 80 m due north takes 20 s, then the velocity is $4.0 \, \text{m s}^{-1}$ due north. The direction is unaffected when multiplying or dividing by a scalar quantity. The scalar scales the vector quantity in a ratio-like way.

Worked example 16 in Topic A.1 shows how the distance travelled can be estimated using the area under a speed–time graph.

Fermi questions

One type of estimate is known as a Fermi question; the physicist Enrico Fermi was well known for his ability to make estimates. He based his estimates on either little or no data, or on a method that seemed remote from the problem in hand. A famous example is Fermi's estimate of the energy released in an explosion. He did this by releasing pieces of paper as the air blast from the explosion reached him and then observing the distances travelled horizontally by the blast. His estimate was within a factor of 2 of the true result. In every Fermi estimate, the scientist needs to make an educated guess at the value of one or more variables in the calculation or measurement.

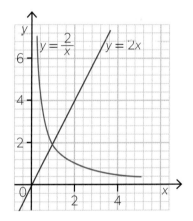

▲ Figure 2 The graph shows a direct proportional relationship (straight red line) and an inverse proportional relationship (blue line).

Scalar quantity	Symbol
distance	s, d, l
speed	u, v
time	t
mass	m
volume	V
temperature	T, θ
density	ρ
pressure	P
energy	E
power	P
electric current	I
resistance	R
gravitational potential	V_g
electric potential	V_e
magnetic flux	Φ

▲ Table 2 Scalar quantities and symbols.

Vector quantity	Symbol
displacement	s
velocity	u, v
acceleration	a
momentum	p
force	F
gravitational field strength	g
electric field strength	E
magnetic field strength	B
area	A

▲ Table 3 Vector quantities and symbols.

5.0 N
this vector can represent a force of 5.0 N in the given direction (using a scale of 1 cm representing 1 N, it will be 5 cm long)

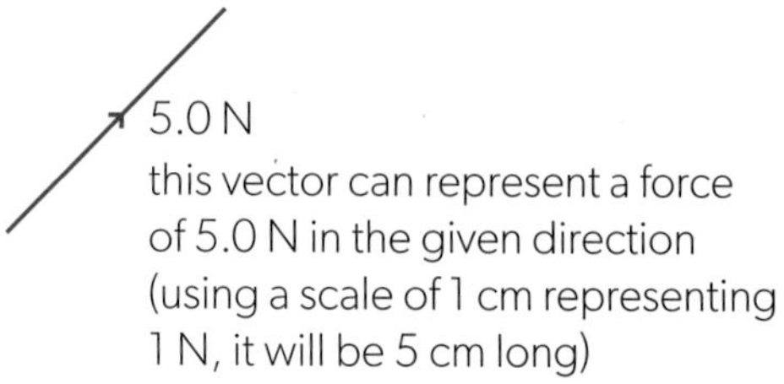

▲ Figure 3 This vector is drawn at half-scale. It represents a force of 5.0 N in the given direction. 1 cm scale represents 0.50 N.

Tables 2 and 3 give some of the scalars and vectors used in this course together with their usual symbols.

Drawing and representing vectors

A vector can be represented on a scale diagram using a line with a direction given by an arrow (Figure 3).

Relative to other vectors in the diagram:

* The length of the line gives the size (magnitude) of the vector.

* The direction gives the direction of the vector (its **line of action**).

The starting point of the vector is the point at which it acts: its **point of application**.

Vector addition and subtraction

Collinear addition

When only one dimension is involved, the problem of vector combination whether addition or subtraction is straightforward. The example is of subtraction.

Figure 4 shows an object suspended by a spring (which is not shown for clarity).

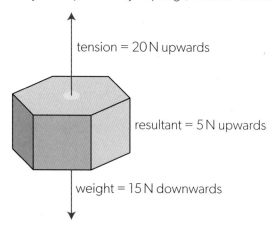

tension = 20 N upwards

resultant = 5 N upwards

weight = 15 N downwards

▲ Figure 4 When vectors are collinear, the overall direction does not change. For subtraction the magnitudes are subtracted using normal arithmetic.

The weight of the object is 15 N downwards. The upwards force on the object from the spring is 20 N. The point of application of both vectors is at the centre of mass of the object and the vectors act in opposite directions.

Although a vector diagram can be drawn here, it is simple to imagine a line down of scale length 15 and a line upwards of scale length 20. The sum of these two forces is 5 length units upwards. In other words, the subtraction of the vectors gives a force of 5 N upwards.

Addition when non-collinear

When vectors are not **collinear**, then the problem is slightly more complicated. Figure 5 shows two vectors not acting along the same line.

Figure 5(a) shows two vectors that need to be added. The steps to achieve this by scale drawing are as follows.

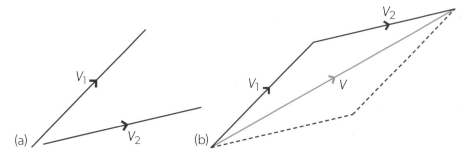

▲ **Figure 5** Two non-collinear vectors can be added using a scale drawing.

- Choose a suitable scale for the vectors so that they fit on the drawing page.

- Draw the vectors at their scaled length and in their correct direction so that the point of application of one vector (V_2 in this case) is at the end of the other vector. This forms two sides of a parallelogram (Figure 5(b)).

- The vector sum is the diagonal of the parallelogram that stretches from the point of application of V_1 to the end of the V_2 vector. Construct this line which is the resultant vector of $V_1 + V_2$.

Subtraction when non-collinear

To subtract two vectors, all that is needed is to form the negative of one of them. The negative of a vector is simply a vector with the same magnitude, but opposite in direction.

Figure 6 shows two vectors V_1 and V_2. We want to find the vector $V_2 - V_1$.

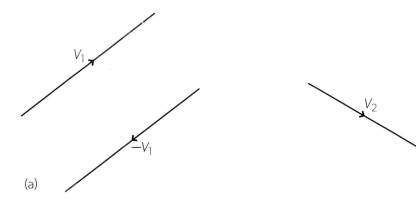

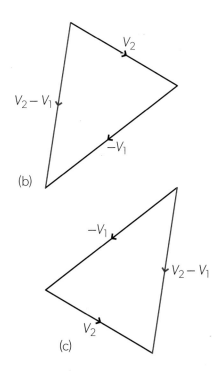

(a)

▲ **Figure 6** Subtracting two vectors.

- Form $-V_1$ by reversing the arrow on the vector (Figure 6(a)).

- Use the earlier technique to add the new vector to V_2 (Figure 6(b)).

- Notice that the order in which the vectors are added does not matter (Figure 6(c)). This is because vector addition is commutative, so that $a + b = b + a$.

You can be asked to add up to three vectors all in the same plane. This can be achieved by drawing the sum of two of the vectors and then redrawing with the third vector added to the sum. Again, the order in which this is done does not matter.

> Table 3 uses the notation introduced on page 550 where the symbols for vector quantities are emboldened. Generally, in this book, there is no distinction made in equations between vector and scalar quantities.

> There is another method of linking vectors shown on page 53. In this approach, the vectors are "slid" along the page until they link nose to tail. In the work on page 53 the vectors form a closed loop indicating that they sum to zero.

Manipulating vectors algebraically

Two perpendicular vectors

This is a common situation (often there is a horizontal and a vertical vector to be added) and is a straightforward application of Pythagoras' theorem.

Vectors V_1 and V_2 are to be added (Figure 7(a)).

Our drawing method would give a solution that looks like Figure 7(b). Because the angle between the vectors is a right angle, the parallelogram is a rectangle.

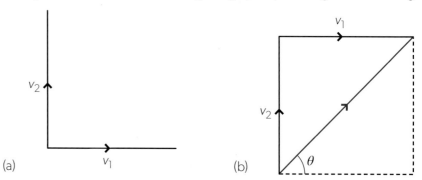

(a)   (b)

▲ **Figure 7** Pythagoras' theorem gives the magnitude (length) of the summed vector while trigonometry gives θ.

Therefore, the resultant vector has:

- a magnitude $\sqrt{V_1^2 + V_2^2}$

- an angle θ to vector V_1 that is given by $\tan \theta = \left(\dfrac{V_2}{V_1}\right)$, so that $\theta = \tan^{-1}\left(\dfrac{V_2}{V_1}\right)$.

Resolving vectors

Sometimes a single vector needs to be deconstructed into two component vectors at right angles to each other. This is called **resolving a vector**. The single vector is the **resultant** of the two new vectors. Resolution into two perpendicular components is helpful because in many situations the two vectors can be treated independently. This is used in projectile motion in Topic A.1.

The work on adding vectors shows that this is vector addition in reverse (Figure 8).

Vector F has to be resolved horizontally and vertically. It is at an angle θ to the horizontal.

In trigonometry, $\sin = \dfrac{\text{opposite side}}{\text{hypotenuse}}$ and $\cos = \dfrac{\text{adjacent side}}{\text{hypotenuse}}$ so that:

- The horizontal component $= F \times \cos \theta$.

- The vertical component $= F \times \sin \theta$.

As a check, F should equal $\sqrt{F^2 \sin^2 \theta + F^2 \cos^2 \theta}$, which it does because $\sin^2 \theta + \cos^2 \theta = 1$.

Worked example 6 in Topic D.2 shows an application of this technique to electric fields.

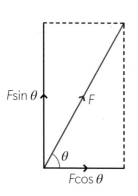

▲ **Figure 8** Vector F is resolved into two components at right angles to each other. The enclosed angle θ gives the $\cos \theta$ component. The other angle is $90 - \theta$ and $\cos(90 - \theta)$ is $\sin \theta$.

Resolving vectors into perpendicular components is essential in many mechanics problems, such as Worked examples 6, 13 and 25 in Topic A.2.

Experimental tools for physics

In this section, you will learn about:

- measuring and controlling variables
- using data loggers

- safe working in a laboratory
- practical electrical work.

Measuring variables

Measurements of physical quantities need to be made to an appropriate level of **precision**. Precise measurements have very little spread about the mean value and the degree of precision usually depends on the type of instrument used for the measurement.

It is important to use the right measuring tool for the job. Imagine measuring the floor area of your laboratory. The appropriate instrument here (Figure 1) is a measuring tape. In principle, this can measure to about ±2 mm (remembering that there is 1 mm error at each end of the tape). A single metre ruler would be inappropriate because it would need to be moved several times during the measurement with a consequent error at each replacement. Something like a vernier calliper would be completely inappropriate because it is not designed for linear measurements of this sort.

An alternative to the measuring tape would be a digital rangefinder in which the time taken for a laser beam to reach a distant wall and return is converted into a distance using the speed of light.

Table 1 gives some of the common quantities you may need to measure in your laboratory and some of the alternative techniques that you can use. Your teacher will tell you which instruments are available.

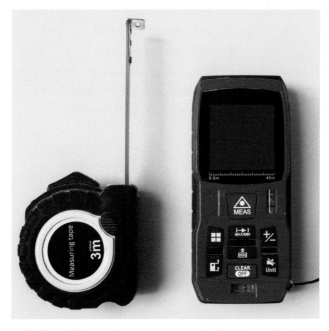

▲ Figure 1 A metal tape measure and a digital rangefinder.

Quantity	Measurement instrument or technique	Comments
mass	analytical (chemical) balance	reads to a few mg
	digital kitchen scale	reads to 1 g or 0.1 g
	analogue or digital bathroom scales	reads to 0.1 kg
	digital floor scale	up to 3000 kg to a resolution of 0.5 kg
time	electronic laboratory timer or mobile phone	reading to 1 ms; can be connected electronically to avoid human reaction time errors
	mechanical stop watch	to 0.05 s but subject to human reaction time
	clock or wrist watch	to 0.5 s
volume	calculation from measured lengths	most suitable for regular solids; appropriate instruments should be chosen
	displacement methods	most suitable for irregular solids
temperature	thermometer examples of different types: • liquid-in-glass thermometer • digital thermometer • thermocouple (temperature-varying emf between dissimilar metals) • resistance thermometers (variation of resistance with temperature)	several types can be linked to a data logger to collect temperature data automatically or quickly
force	simple spring balance (newton meter)	
	strain gauge	in which the compression/extension of a metal wire changes its electrical resistance
	load cell	in which compression or extension of the device changes another electrical property
angle	plastic or engraved metal protractor	normally to 0.5°
	digital angle rule	0.1°
	digital rotation sensor	for example, simple electrical potentiometer, Hall effect sensor that responds to magnetic field, rotary encoders
sound intensity	analogue microphone and cathode-ray oscilloscope or digital oscilloscope	for absolute intensity measurements, careful calibration is often required; relative intensities are more straightforward
	digital sound sensor and data logger	
light intensity	direct reading analogue light meter	
	digital light sensor	
electric potential difference and electric current	these are discussed in a separate section — see Topic B.5.	

▲ Table 1 Techniques for measuring quantities.

Using data loggers

Put in its simplest terms, a data logger is a computer with the specialist function of recording and storing data. It can be programmed to read one or more series (or channels) of data in various ways:

- as one-off measurements or one or more quantities at the press of a button — although this is not usually effective as the set-up time for the apparatus and the data logger could be prohibitive

- as a series of measurements either with regular time intervals between them or with the reading triggered by external events such as a light switching on or a temperature change.

The main advantages of data logging are:

- reduction in error while taking readings

- ability to capture data at both very fast and very slow rates that are outside the normal range of human capability

- data storage both in terms of accuracy of storage and volume of material

- near-simultaneous logging of several pieces of data

- ability to capture data in remote or dangerous situations, a cold freezer or dark environments, for example.

The only real disadvantage of a data logger is the risk that the equipment may develop a fault which may not be apparent until data are analysed.

Many types of data logger are available — your teacher will help if you need to use a particular data logger for experimental work or for your internal assessment. Modern data loggers are either:

- completely self-contained so that they have their own purpose-designed computing hardware and software, or

- designed to connect to an existing computer and to act as an interface between sensors and the computer.

Self-contained data loggers in schools are often designed to plot the data graphically as a function of time or as variable against variable.

It is unusual for data loggers to have particular sensors attached to them. You will need to be able to select and attach the sensors to the data logger correctly.

An important aspect of sensor use is the possible need for sensor calibration. A preliminary experiment may need to be performed to adjust the output of the sensor to suit a particular set or sets of stimuli. The sensor instructions will state the need for this and show how it should be performed.

Sensors divide into two broad groups:

- **Analogue** sensors where a varying voltage or current is output by the sensor in response to changes in the experiment. An example is a current sensor which can be as simple as a calibrated resistor. When charge flows through the resistor, the pd across it can be measured and transferred to the data logger which reads the pd at predetermined intervals.

Using a smartphone as a data logger

Some of the data-based questions in this book use experimental data from a smartphone data logger. You may decide to use a smartphone data logger in your own IA.

Data loggers can have many advantages but the quantity of data that they can record may be overwhelming unless you design your experiment and set up the logger carefully.

Try an experiment with a smartphone sensor. Measure the light intensity at varying distances from a light bulb (or other light source). Can you show that the intensity obeys an inverse square law? (See Topics B.2 and E.5). Consider how the sensor readings can be used to give an average intensity as well as the uncertainty at each value.

Another example is the data-based question on page 451. You could try to reproduce this experiment yourself. If you allow the data logger to take data for 30 s, how will you process the data?

- **Digital** sensors undertake some pre-processing of the signal and output a digital code that is detected by the logger. A simple example is a push switch that is either on or off and this can be sent as a binary 1 or 0 to the logger, perhaps to pause the data-logging operation. More sophisticated digital devices include digital temperature sensors that output the temperature value in digital code. These can be more accurate and have higher resolution than analogue temperature sensors such as thermistors.

Sensors are available for many quantities. The most important for your use include light sensors, including light gates; electrical sensors: current, pd, power; sound sensors; pressure and stress sensors; temperature sensors; accelerometers; magnetic sensors (using the Hall effect); load cells (for force measurements); strain sensors.

Your mobile phone is often a convenient source of sensors for data logging. There are a number of apps available for both iOS and Android devices that will exploit the built-in phone sensors to make logging measurements over time. Typical sensors found in mobile phones detect motion: tilt, shake, rotation; environmental change: temperature, moisture, pressure; position: accelerometer, geomagnetic field; light; proximity; gyroscope; compass.

You may also own a graphical calculator that can accept the input from a sensor. These can be used with a wide variety of sensors in physics and other sciences.

Using a data logger

One example of the use and convenience of data logging is that of plotting the electrical characteristic (the *IV* curve) for a filament lamp.

To perform this experiment without a data logger, you would have to reset the position of the potential divider by hand and manually note down the current and pd readings shown on the meters. Figure 2 shows an arrangement that will allow automated recordings of the data.

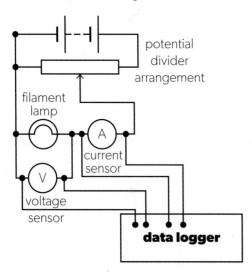

▲ Figure 2 The two sensors act as inputs to the data logger. It is also possible for a change in one sensor input to trigger the beginning of the data-acquisition sequence.

- Connect the two sensors, voltage and current, into the circuit. The instructions for your sensors will indicate how this should be done.

- The data logger can be set up so that any change in the sensed voltage will start the data collection. All you then need to do is slide the wiper on the potential divider from one end to the other.

- Before beginning the experiment, you will need to decide how many readings to take. Usually, you can either specify a total number or you can program the data logger with a time interval between readings. This is your choice, but a decision to take as many readings as possible may not be sensible. Data loggers have finite memory storage for data and you must take account of this—though this may be more important when the readings are taken over long time periods of days or weeks.

- Having varied the pd across the lamp, you should now have data waiting in the data logger. You can then choose to download the data to a computer or use the data logger itself to display the characteristic in a way that you choose. You may be able to program the data logger to calculate resistance values or plot a graph of the variation of resistance against current automatically.

Working safely

It is important to protect yourself and others around you when working in a laboratory. There are three areas that you must consider:

- **Safety from accident.** There are many potential hazards in a physics laboratory, including high temperatures, electrical accidents, falling weights and toxic substances. It is everyone's duty to protect themselves and others while experiments are being carried out. If you are unsure about a procedure and its safety, you must ask for your teacher's advice.

- **Environmental safety.** In a similar way, it is everyone's responsibility to protect the environment. Do not use materials needlessly and thoughtlessly. Are your experiments planned so that they minimize their impact on the local and global environment?

- **Ethical working.** Ethics should be in your mind during the planning and research stages of an investigation. It links to the previous bullet point. Is what you are doing ethically appropriate?

Constructing electrical circuits from a diagram

Wiring up a circuit is an important skill. If you are organized, then you should have no problems with even complex circuits.

This is how you might tackle the potential-divider circuit from page 321 that is used to vary the pd across a component, in this case, a lamp.

Look carefully at the circuit diagram (Figure 3a). It is really two smaller circuits linked together: the top sub-circuit with the cell and potential divider, and the bottom sub-circuit with the lamp and the two meters. The bottom circuit itself consists of two parts: the lamp/ammeter link together with the voltmeter loop.

The potential divider needs care. In one form it has three terminals (Figure 3b), and sometimes the three terminals are in an arc of a circle. The linear device has a terminal at one end of a rod with a wiper that touches the resistance windings. The other two terminals are at each end of the resistance winding.

- Begin with your circuit diagram. Get your teacher to check it if you are not sure that it is correct.

- Lay out the circuit components on the bench as they appear on the diagram.

- Connect up one loop of the circuit at a time.

- Check that components are set to give minimum or zero current when the circuit is switched on.

- Do not switch the circuit on until you have checked everything.

Figure 3c shows the sequence for setting up the circuit step-by-step.

Global impact of science

Scientists have a responsibility to ensure that their experiments are safe. In 2008, a lawsuit was brought against the Large Hadron Collider at CERN on the basis that it might create a black hole and destroy Earth. Both the lawsuit and a subsequent appeal were rejected.

Your own experiments are unlikely to pose such hazards but you must be aware of laser beams, falling or sharp objects, high voltages and high temperatures in your work.

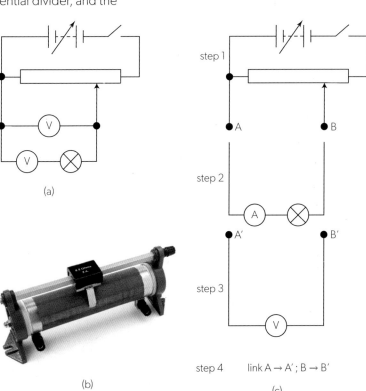

(a)

(b)

(c)

step 1

step 2

step 3

step 4   link A → A'; B → B'

▲ **Figure 3** Setting up an electrical circuit. Set out the components without connecting them. Connect the power supply section without turning on the supply. Arrange the series part of the circuit and then add the parallel components in the correct places.

Handling data and modelling physics

In this section, you will learn about:

- errors in measurement, how to estimate and combine them
- displaying data effectively
- using graphs and charts
- modelling physical phenomena using software.

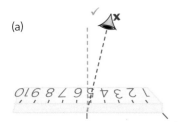

(a)

(b)

(c)

▲ Figure 1 (a) Parallax error and its cause. (b) Analogue and digital meters. (c) A digital vernier calliper.

Suppose the manufacturer of your digital meter describes it as having a resolution of $3\frac{1}{2}$ digits and an accuracy at full scale deflection of 2.00 V of 25 ppm.

- The meter can show up to 1999 on its scale so that its maximum resolution on the 2.0 V scale is 1.999 V. The left-most digit can only display 0 or 1. If this pd rises by 0.002 V, then the best the display can do is 02.00 V as you will need to choose the 20 V range for the display.

- The accuracy of the meter at 2.0 V is $\pm 2.00\,V \times \dfrac{25}{1\times 10^6}$
$= \pm 50\,\mu V$

Errors in measurement

Measurements of data have implicit errors. These errors come in two forms.

- **Random error.** All measurements are prone to this error. It occurs because of small variations in the factors that go into making the measurement. One example is the parallax error that occurs when reading a scale. When the observer's eye is not exactly at right angles to the plane of the scale and there is some distance between the pointer and the scale, then the reading will be in error (Figure 1).

  In repeating the reading an observer is very unlikely to place their head in exactly the same place so that the observation position is random.

  Another example is the effect of reaction time on the start and stop of a time measurement using a stop watch. Although reaction time is, broadly speaking, constant for an individual, there will be some differences between the two measurements due to anticipation.

  Random errors are dealt with by making a series of repeat measurements under the same experimental conditions. Then find the mean (average) of all the measurements. This is the result to use in further analysis. The variation of the readings also gives an estimate of the absolute uncertainty in the mean value. This is covered later.

  Digital instruments can be more difficult. Digital electrical meters (Figure 1(b)) in particular often appear to give an unchanging reading and it is tempting not to take repeat readings. The answer here, for accurate work, is to look at the calibration data supplied with the instrument. This will give you an indication of the accuracy of the meter and its true **resolution**.

  Readings that have a small spread about the mean are said to be **precise**.

- **Systematic error**. These can be more difficult to cope with. They are characterized by constant offsets to the true reading caused by the instrument or some aspect of the measurement technique. A simple example is that of an analogue electrical meter where the point does not indicate zero when the meter is not in a circuit (Figure 1(b)). The best approach here is to adjust the meter so that zero input gives a zero on the scale. If this cannot be done, then you must be careful to add or subtract the offset from each data point you measure.

Another example is that of a micrometer screw gauge or a vernier calliper where the instrument can give a non-zero reading when the jaws are fully closed (Figure 1(c)). All such instruments have a method for resetting the zero. It is important to check the zero every time that you use these devices. Systematic uncertainties can cause a measurement to be different from the true value. We call such a measurement inaccurate.

Handling errors

When a quantity is measured, there is an implicit error in the measurement. This should always be quoted when discussing the quantity and its value. The usual forms for writing the value and the error are:

$$\text{length} = 2.4 \text{ cm} \pm 0.2 \text{ cm} \quad \text{OR} \quad (2.7 \pm 0.2) \text{ cm}$$

value   absolute
uncertainty
in value

The **absolute uncertainty** of the value has the same units as the quantity and is an estimate of the range of the value within which the true result for the quantity lies. When the true result and the measured result are close, the measurement result is said to be **accurate**.

There are statistical methods to determine the absolute uncertainty in a measurement but, for your purposes, a quicker method may be good enough.

To obtain your estimate of the absolute uncertainty, look at the range of values from which you established the mean value. Then take the difference between the greatest and smallest values that you observed. Halve this difference to arrive at the absolute uncertainty. Here is an example.

- You measure six speeds:

  6.78 m s^{-1}, 6.56 m s^{-1}, 6.92 m s^{-1}, 6.42 m s^{-1}, 6.54 m s^{-1} and 6.48 m s^{-1}.

- The mean of these six data points is 6.62 m s^{-1} (rounded to 3 s.f.).

- The greatest and least values are 6.92 m s^{-1} and 6.42 m s^{-1}. This is a range of $6.92 - 6.42 = 0.50 \text{ m s}^{-1}$.

- Half of this range is 0.25 m s^{-1} which leads to an absolute uncertainty of $\pm 0.25 \text{ m s}^{-1}$.

- The speed should be quoted as $(6.6 \pm 0.3) \text{ m s}^{-1}$. The uncertainty is usually rounded up to 1 s.f. and the mean value of the data is given to the same precision as the uncertainty.

- Although, exceptionally, you may wish to write this as $6.6(2) \text{ m s}^{-1} \pm 0.2(5) \text{ m s}^{-1}$ to emphasize the sensitivity of the measurement.

There are other ways of expressing experimental uncertainty too.

The **fractional uncertainty** is used later when we combine uncertainties in particular ways. The **percentage uncertainty** is an excellent way to get a feeling for the size of the error estimate in relation to the value being measured.

Admitting fault

Sometimes students confuse the terms error and uncertainty. Although the terms are often used interchangeably, the use of the word "uncertainty" reminds us that it is not necessarily the fault of the experimental equipment or the experimenter.

Is confessing to uncertainty an admission of fault, or is the greater error to underestimate uncertainty? Can there ever be a measurement with no uncertainty?

- **To calculate the fractional uncertainty.** Calculate:

$$\frac{\text{estimate of the absolute uncertainty in a value}}{\text{estimate of the value}}.$$

  In the earlier example, the fractional uncertainty is $\dfrac{0.25}{6.62} = 0.038$.

  Fractional uncertainty has no units; it is a ratio of values with identical units.

- **To calculate the percentage uncertainty.** Calculate the fractional uncertainty and multiply by 100. In the earlier example, the percentage uncertainty is $\dfrac{0.25}{6.62} \times 100 = 3.8\%$.

Combining uncertainties

Practise combining uncertainties using the experiment on page 65 in Topic A.2.

You will frequently need to combine uncertainties in your practical work. This technique is known as the **propagation of uncertainty** and is needed when there are uncertainties that have to be combined to form a further value.

Addition and subtraction of uncertainty

> **When values are added or subtracted, the absolute uncertainties of the values are added.**

Two quantities a and b are added to give a third value c: $c = a + b$. The values of these with their absolute uncertainties are $a \pm \Delta a$ and $b \pm \Delta b$.

The absolute uncertainty of c is $\Delta c = \Delta a + \Delta b$.

When a and b are subtracted to give a different value d: $d = a - b$, then the absolute uncertainty of d is still the sum of the absolute uncertainties $\Delta d = \Delta a + \Delta b$.

To show this, a student measures the length of a piece of card with a metre ruler. The two measurements (Figure 2) are 195.0 mm and 118.5 mm. Each of these measurements has an absolute uncertainty of ± 0.5 mm.

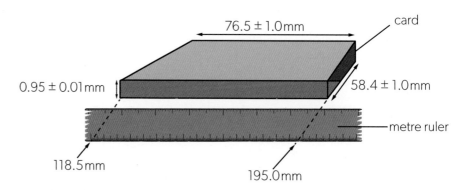

▲ **Figure 2** The dimensions of the card (not to scale).

The length of the card is $195.0 - 118.5 = 76.5$ mm.

The absolute uncertainty of the length of the card is $0.5 + 0.5 = 1.0$ mm.

The card length should be quoted as (77 ± 1) mm or (76.5 ± 1.0) mm.

The percentage uncertainty in the length estimate is $\dfrac{1.0}{76.5} = 1.3\%$.

Multiplication and division of uncertainty

The position is more complicated when quantities are being multiplied or divided; in other words, when $c = a \times b$ or $d = \dfrac{a}{b}$. Again, a and b have absolute uncertainties of Δa and Δb, respectively. This time, the fractional uncertainties are added: $\dfrac{\Delta c}{c} = \dfrac{\Delta a}{a} + \dfrac{\Delta b}{b}$.

When one quantity is divided by another, the fractional uncertainties are still added $\dfrac{\Delta d}{d} = \dfrac{\Delta a}{a} + \dfrac{\Delta b}{b}$. As with the absolute uncertainties earlier, the fractional uncertainties are added and never subtracted.

> **When values are multiplied or divided, the fractional uncertainties of the values are added.**

Worked example 1

The student measuring the card length earlier went on to determine the width and thickness of the card using a metre ruler and a vernier calliper. The three values are:

> length = (76.5 ± 1.0) mm; width = (58.4 ± 1.0) mm; thickness = (0.95 ± 0.01) mm

a. What is the volume of the card?

b. What are the fractional and absolute errors in the volume?

The card has a mass of (1.1 ± 0.1) g.

c. What is its density?

Solutions

a. The volume is $76.5 \times 58.4 \times 0.95 = 4240$ mm³.

b. The fractional errors are $\dfrac{1.0}{76.5}, \dfrac{1.0}{58.4}$ and $\dfrac{0.01}{0.95}$. This leads to a fractional error in the volume of $\dfrac{1.0}{76.5} + \dfrac{1.0}{58.4} + \dfrac{0.01}{0.95} = 0.0407$. The absolute error in the volume is $0.0407 \times 4240 = 172$ mm³.

The volume is (4240 ± 180) mm³, where the absolute error is rounded upwards. Rounding down in this case would make the estimate in the absolute error appear too small.

c. The density ρ of the card is $\rho = \dfrac{\text{mass}}{\text{volume}}$ and has the value (259 ± 35) kg m⁻³. You should be able to verify this answer. Don't forget to add the fractional uncertainties.

Raising quantities to a power

Remember that a number cubed x^3 is simply $x \times x \times x$. This leads to the result that, when $y = x^n$, then $\dfrac{\Delta y}{y} = \left| n \times \dfrac{\Delta x}{x} \right|$. Here the modulus sign ("| |") means that the result can be positive or negative, but we must treat it as positive.

The fractional uncertainty when a quantity is raised to a power is equal to the power multiplied by the fractional uncertainty of the quantity, ignoring the sign.

Uncertainly in experiments

Test your ability to handle uncertainties in this experimental task.

Measuring the weight of paper

Paper can be specified by its area density—the mass in grams of one square metre (g m⁻² or gsm). Terms such as paper weight, grammage, or basis weight are often used for this.

Typical office printer paper is marked as 80 gsm. Paper marked as having a basis weight of 20 lb will be equivalent to 75 gsm.

Take a single sheet of printer paper, preferably from a pack of paper which states the basis weight or gives the gsm. Design and carry out suitable measurements to establish a measurement of the actual gsm of the paper and the uncertainty in your value.

Careful measurements may show that your observed paper weight is less than the value declared on the packet. This is usually because, by convention, paper weight can be determined before the paper is fully dried during manufacture.

Would using more than one sheet of paper when determining the mass improve the experiment?

Many of the data-based questions involve relationships that are power laws (i.e. $y = kx^n$ where n is an integer). See the data-based questions on page 26 and page 263 for examples.

Worked example 2

A student is calculating the time period of a pendulum using $T = 2\pi\sqrt{\dfrac{l}{g}}$, where l is the length of the pendulum and g is the acceleration due to gravity.

The percentage uncertainty in l is 12% and the percentage uncertainty in g is 3%.

What is the absolute uncertainty in the time period when $l = 1.9\,\text{m}$ and $g = 9.8\,\text{m s}^{-2}$?

Solution

The two constants 2 and π have no error. $\dfrac{\Delta T}{T} = \dfrac{1}{2}\times\dfrac{\Delta l}{l} + \dfrac{1}{2}\times\dfrac{\Delta g}{g}$.

(Notice that, although the expression gives $g^{-\frac{1}{2}}$, the minus sign is ignored.)

$\dfrac{\Delta T}{T} = \dfrac{1}{2}\times 0.12 + \dfrac{1}{2}\times 0.03 = 0.075$

$T = 2\pi\sqrt{\dfrac{1.9}{9.8}} = 2.77\,\text{s}$.

The absolute uncertainty is $2.77\times 0.075 = 0.21\,\text{s}$. The time period can be quoted as $(2.8 \pm 0.2)\,\text{s}$ or $(2.7(7) \pm 0.2(1))\,\text{s}$.

Look at the data-based question on page 125. Use the table of data, or carry out the experiment yourself.
Find the ratio of the height of a bounce to the height of the previous bounce. Use your knowledge of uncertainties to establish a mean value and an uncertainty in this mean value.

Displaying data

Tables can be used to display data in a report and can be useful in making a point. Every row or column in a table should contain a heading with the quantity and its unit. Do not put a unit with each value in the table.

Consider the number format. Scientific notation ($1.2 \times 10^2\,\text{m s}^{-1}$ for a speed of $120\,\text{m s}^{-1}$) may be best. The table containing this number would begin:

Speed / $10^2\,\text{m s}^{-1}$
1.2

Data tables can be used to test relationships such as $y = \dfrac{k}{x}$, $y = kx^2$, $y = \dfrac{k}{x^2}$, $y = e^{-kx}$, relationships including $\sin x$ and $\cos x$ and so on. Combine the variables in relationship in a way that should yield a constant.

For example, in Boyle's law (Topic B.3), the prediction is that pressure $= \dfrac{k}{\text{volume}}$ $\left(P = \dfrac{k}{V}\right)$. Therefore $P \times V$ should be a constant. This can be demonstrated easily in a table. Pressure and volume are the variables obtained from an experiment, a third column gives the product PV.

P / kPa	V / m³	PV / kJ
115	15.6	1794
105	17.2	1806
95	18.8	1786
85	21.0	1785
80	22.3	1784

The question is whether the values in the third column are close enough to confirm the relationship. In this case, they are with a mean and uncertainty for PV of $(1790 \pm 10)\,\text{kJ}$.

A data-table test for exponential change is also possible. The damped response of an oscillator is an example of this (Topic C.4). This test uses the knowledge that amplitude ratios over identical time intervals (or numbers of oscillations in this case) are the same. The amplitude of the oscillation is measured every tenth oscillation.

amplitude of oscillation / cm	6.1	3.8	2.4	1.5	0.95	0.60
number of oscillations	10	20	30	40	50	60
$\dfrac{\text{this amplitude}}{\text{previous amplitude}}$		0.62	0.63	0.63	0.63	0.63

Using graphs and charts

Charts

The following diagrams can be seen in physics texts.

- *Pie charts* (see, for example, the origins of background radioactivity on page 655) are useful for showing the relative contribution of the components of a quantity. Taking 360° as equivalent to 100% allows the relative percentage of a particular contribution to be shown as an angle within a circle (Figure 3).

- *Bar charts*. The height of the bar or column represents the value of the quantity as distributed between different components.

- *Histograms* are superficially the same as bar charts, but have a different purpose being intended to show the frequency density of a variable (not be confused with wave frequency in Theme C). Their use is best shown by an example.

Continent	% area
Africa	20.4
Asia	29.2
North America	16.5
Antarctica	9.2
South America	12
Europe	6.8
Australia	5.9

(a)

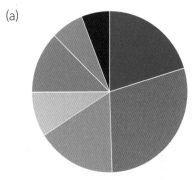

- Africa ▪ Asia ▪ North America ▪ Antarctica
- South America ▪ Europe ▪ Australia

(b)

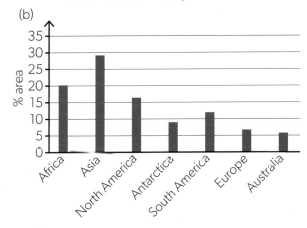

▲ **Figure 3** A pie chart (a) and bar chart (b) showing the relative areas of the seven continents.

ATL Communication skills

Graphs provide a visual representation of the data and reveal trends. However, your choice of graph must be appropriate to demonstrate this.

Compare the graphs on pages 666 and 683. These are the same as Figure 5 overleaf, however they have been plotted with different axes. On page 666 a false-origin is used on the *y*-axis so that the data for high nucleon numbers can be seen clearly. The graph on page 683 does not use a false-origin but uses a logarithmic scale on the *x*-axis so that the elements with a small nucleon number are clearer.

Consider the data-based question on page 668. Which type of graph is most appropriate here? Since the total of all the probabilities must add to 1, a pie chart could be drawn. This would demonstrate the most common number of neutrons released in a fission reaction. Would it demonstrate the mathematical trend?

Worked example 3

A radioactivity counter is used to detect the time of arrival of the next event when measuring the background count in a laboratory. On four occasions out of 30, the first event happened within 0.20 s of switching the counter on. On nine occasions the first event happened between times of 0.20 s and 0.35 s, and so on.

The complete data are:

Time t / s	$0 < t \leq 0.20$	$0.20 < t \leq 0.35$	$0.35 < t \leq 0.50$	$0.50 < t \leq 0.60$	$0.60 < t \leq 0.80$
Frequency	4	9	6	6	5

Solution

The first step is to calculate the frequency density, which is $\dfrac{\text{frequency}}{\text{width of time period}}$ ($\dfrac{4}{0.2} = 20$ for the first group). The complete table is:

Time / s	$0 < t \leq 0.20$	$0.20 < t \leq 0.35$	$0.35 < t \leq 0.50$	$0.50 < t \leq 0.60$	$0.60 < t \leq 0.80$
Frequency	4	9	6	6	5
Time width / s	0.20	0.15	0.15	0.10	0.20
Frequency density	20	60	40	60	25

Figure 4 shows the histogram. Notice that the widths of the columns vary and that the frequency density (as opposed to the frequency alone) allows for this.

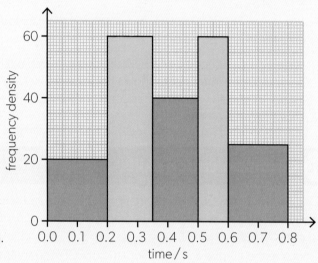

▶ **Figure 4** An example of a plotted histogram.

• *Scatter graphs.* These are plots on *x*–*y*-axes, but do not necessarily show a causal relationship. A good example in physics is the plot of binding energy per nucleon against nucleon number from Topic E.3, which is shown again here (Figure 5). It is inappropriate to draw a line of best fit for this descriptive chart.

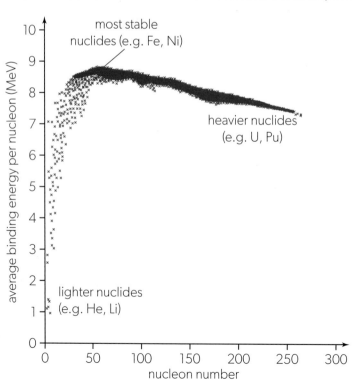

▶ **Figure 5** The chart of average binding energy per nucleon against nucleon number for the elements.

Throughout this book, you will see two types of graph.

- **Sketch graphs**, often without data points, that are meant to illustrate a point in physics or the theoretical variation of y as x changes.

  Sketch graphs should:

  - be drawn neatly in pencil

  - show the relationship between y and x clearly — keeping the shape of the curve accurate

  - have axes labels — there may not be a need to include axis scales or units on sketch graphs

  - include any known data points on the graph — in this case, scales and units will be required.

  There are some simple sketch graphs that it is useful to be able to draw freehand quickly and accurately. These include:

  $y = \dfrac{k}{x}$, $y = kx^2$, $y = \dfrac{k}{x^2}$, $y = \pm \sin x$, $y = \pm \cos x$, and (for AHL) $y = ke^{-cx}$, where k and c are constants.

- **Drawn graphs**, with data points, that represent the outcome of data collection in a laboratory experiment. You will typically use data that you alone, or with a group of fellow students, have collected. Such graphs must always have axes with scales, labels and units.

Graphs are a highly convenient way to display data and information. Drawing them well is a traditional skill of the physicist and, even though spreadsheets and other software can produce graphs quickly, it is important that you can produce clear and accurate graphs by hand under examination conditions and perhaps for your internal assessment.

Whichever spreadsheet program you use, you should be confident that you can use it to produce all the features that you need: **axes + labels**, **plotted points**, **error bars**, **line of best fit** and so on. It takes just as much skill to produce a good computer-derived graph as a clear hand-drawn one.

Patterns of results become very clear on a graphical plot. Some students plot a rough graph as an experiment progresses. This enables them to see whether there are any values omitted that need to be filled in before the apparatus is dismantled.

Graphical analysis

A curved line on a graph shows the basic variation of one variable with another.

A straight line on a graph helps you to see exactly how one variable varies with another. With a straight line, precise mathematical statements can be made about the variation of the y-axis quantity with the x-axis quantity. Always manipulate data so that it produces a straight line, if possible. Follow these steps:

- Find a physics equation that models how the variables behave.

- Manipulate the algebraic equation to the form $y = mx + c$.

- Plot the graph.

Figure 6 shows a $y = mx + c$ graph to remind you of the meaning of the symbols.

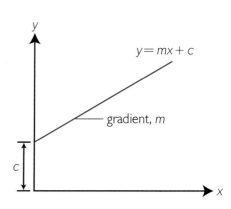

▲ **Figure 6** The equation of a straight line with gradient m and an intercept on the y-axis of c is $y = mx + c$.

Using logarithms to give a straight-line graph

(a) When the relationship is exponential

This occurs in the analysis of radioactive decay in Topic E.3. Using the notation of that topic, $N = N_0 e^{-\lambda t}$, where N is the number of nuclei remaining at time t when the initial number was N_0 and the decay constant for the nuclide is λ.

Figure 7(a) shows the variation of N with t. However, a better plot can be obtained by taking $\log_e$ of both sides of the equation: $\ln N = \ln N_0 - \lambda t$. Comparing this with $y = mx + c$ suggests that a plot of $\ln N$ against t will be a straight line (Figure 7(b)). The gradient of the line is $-\lambda$, leading to the half-life:

$$T_{\frac{1}{2}} = \frac{\ln 2}{|\text{gradient}|}.$$

(b) When the relationship may be a power law

This is a technique to use when you hypothesize that the relationship between y and x is of the form $y = kx^n$, where k and n are unknown. This time you take logs of both sides of the equation:

$$\log y = \log k + n \log x$$

$$y = c + mx$$

A plot of $\log y$ against $\log x$ gives a straight line and a comparison with the straight-line equation shows that the gradient is n and the intercept of the graph when $\log x = 0$ is $\log k$.

The data in Figure 8 show a gradient of 1.5 and an intercept of about 0.43, giving $k = 10^{0.43} = 2.7$. The equation for the relationship is $y = 2.7x^{\frac{3}{2}}$.

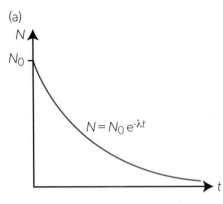

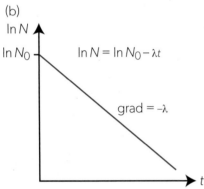

▲ **Figure 7** (a) A plot of N against t gives a curve from which a radioactive half-life can be obtained directly. (b) A better method (because it averages all the data points) is to plot $\ln N$ against t to give the line of best fit and to obtain λ from the gradient.

See the data-based questions on page 263 and page 688 for examples of using a log-log graph to find the power of a power law relationship.

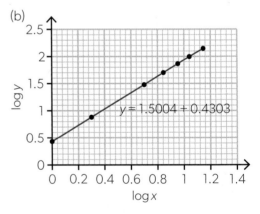

(a)	x	y	log x	log y
	1	2.7	0.000	0.431
	2	7.6	0.301	0.881
	5	30	0.699	1.477
	7	50	0.845	1.699
	9	73	0.954	1.863
	11	99	1.041	1.996
	14	140	1.146	2.146

▲ **Figure 8** (a) The data table contains the original data and the logarithms to base 10 of the same data. (b) The log data plotted onto a conventional squared grid.

Plotting points and drawing graphs

The steps that you should go through in plotting a graph are as follows:

- **Plan your scales for both axes.** You will need a minimum of six data points if at all possible. Look at the range of values for the two variables. These ranges and the grid on which you are plotting will determine the scales. Sensible scale ratios are 1:2, 1:5 and 1:10. Poor scale ratios are 1:3, 1:6, 1:7, and 1:9 — you should **not** use these. The scale 1:4 is intermediate — it can be used, but needs care.

  Your scale should be such that at least half the grid is filled with data points. If this appears difficult to do, consider using a **false origin**, where the bottom left-hand corner of the graph is not (0, 0). When you anticipate using a logarithmic graph or a graph with the reciprocal of a quantity, then you should take care when choosing your data ranges. Such scales tend to compress the points at one end or the other. Figure 8(b) shows this.

- **Label the axes correctly** with quantity + power of ten + unit. A solidus (/) should separate the quantity and the unit: thus, speed $/ 10^3 \, \text{m s}^{-1}$. The power of ten indicates that each number on the axis must be multiplied by 1000.

  The reason for the quantity / unit notation is because this is really $\dfrac{\text{quantity}}{\text{unit}}$ written on one line. The SI rule is that the number on the graph is just a number and must not have units attached to it. Writing in the form quantity / unit makes the data plotted on the axis unitless, like a ratio.

- **Mark the data points on the grid consistently and accurately.** Acceptable ways to show the point include: ×, + and ⊗ and, for digital plots, a small solid circle •. The symbol ⊙ is not so good because, when a data point lies on a grid line, the exact point may not be clear.

- **Draw lines and points with a sharp pencil**, never ink (in case you need to erase a mark).

- **Add error bars.** You should have determined the absolute error in your data. This can be incorporated into a graph using error bars that are the length of the absolute error above and below the point (in the case of vertical error bars) or from side to side for horizontal bars.

- **Use a ruler to draw any straight lines.** A transparent plastic ruler is best.

- **Use free hand for curves.** Draw the curve in one flowing movement. Practise the curve a few times without letting the pencil touch the paper. Keep your drawing hand inside the curve.

- When drawing the line (straight or curved), **get a balance of points each side of your line.** Minimize the distance from each point to the line.

- **Do not force the line to go through the origin** unless you have strong evidence to do so. This is a common error. Surprisingly few graphs of experimental data go through (0, 0).

ATL Communication skills — Using digital media for communication

The suggestions for graph plotting given here apply to both hand-drawn and computer-plotted graphs.

The advantage of using software to plot a graph is that it will plot the points and choose some axes for you. However, it is important to check these for yourself. For example, software is unlikely to label the axes for you automatically. It may not select the scales that you would like.

Computer programs often plot a line of best fit. However, they are usually unable to identify anomalous data points. Software usually weights all points equally and calculates the trend based on the values. It may be that a computer-generated line of best fit does not pass through all the error bars even though you would be able to draw such a line.

If you use graph-plotting software, you will need to ensure that you know how to adjust the axes, how the data points are chosen, how the error bars are constructed and how the line of best fit is selected.

The rule of thumb used by examiners who mark practical graphs is that the lines you draw should be thinner than the thickest grid line on the paper.

See page 45 in Topic A.2 for a Data-based question using error bars.

Figure 9 shows a graph complete with error bars and line of best fit.

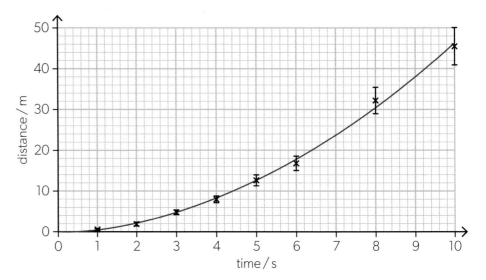

▲ **Figure 9** A graph that contains all the essential elements: clear plots with error bars, a line of best fit and axes that are sensible and labelled.

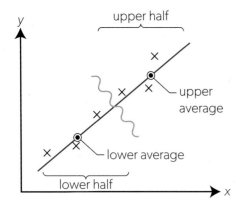

▲ **Figure 10** A quick way to compute the line of best fit is to group the data points into lower and upper groups. Find the mean of each group and the line of best fit is the line joining these two mean points.

Nowadays, it is easy to determine the exact position of a line of best fit using a graphical calculator or a computer spreadsheet. Use your calculator for this task, for speed and accuracy. However, if this is not possible, an approximate way to find the position of a line of best fit is as follows.

- Divide the graph data into a lower and an upper half. With an odd number of points, one group will have to be one larger than the other.

- Find the mean of all the x-values in the lower set. Do the same for the y-values. Plot this new average point on the graph using a different symbol from your main plots.

- Do the same thing for the x mean and the y mean of the upper set. Plot this point too.

- The line joining the two mean plotting points will be a good approximation to the line of best fit (Figure 10).

Uncertainty in the gradient of a line of best fit

The error bars attached to the x- and/or y-values for the coordinates on a graph give rise to uncertainty in the true position of the line of best fit. This leads to uncertainty in the gradient, which can be estimated easily.

Figure 11 shows a graph with error bars for the y-values. This is the outcome of an experiment in which a student drew five circles of known radius and estimated the area of each circle by drawing a grid on each circle. The graph shows the variation of estimated area with circle radius^2. The gradient should be π. The error that the student estimated in each area is shown as an error bar.

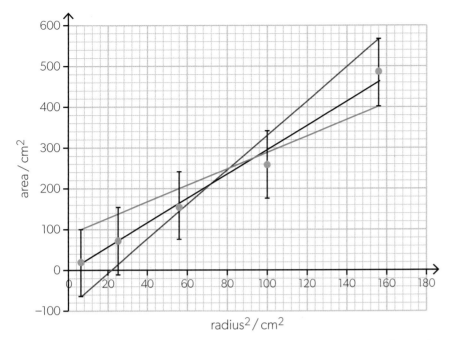

area / cm²

radius² / cm²

▲ Figure 11 A graph with error bars and the line of best fit, and maximum gradient and minimum gradient lines. A calculation of the minimum and maximum gradients leads to an estimate of the absolute uncertainty in the gradient.

Three lines are shown.

- The black line is the line of best fit as judged by the student. It has a gradient of 3.04.

- The red line is the steepest line that can be drawn with the line remaining within every error bar.

- The green line is the flattest line that can be drawn within every error bar.

The gradient of the red line is the maximum gradient. It has the value

$$\frac{560 - (-10)}{156 - 20} = \frac{570}{136} = 4.19.$$

The gradient of the green line is the minimum gradient. It has the value

$$\frac{395 - 130}{156 - 20} = \frac{265}{136} = 1.95.$$

This has a range of 2.24 and a half range of 1.12.

The student should quote the answer for π as 3 ± 1.

 Measuring the density of air

Many smartphones have a built-in pressure sensor. You should be able to find a suitable app that will allow you to access the readings from this sensor.

Place the smartphone on the floor and measure the pressure. It is possible that the pressure will fluctuate. In this case, you might choose to stop the sensor read-out and record the value multiple times. This will give an average measurement and its range. Alternatively, you might use the smartphone as a data logger and record a set of values.

Now take measurements of the pressure P at different heights h above the floor. A staircase may be useful to do this.

Plot a graph of h on the x-axis and P on the y-axis. Find the gradient of your graph and use your error bars to find the uncertainty in your gradient.

Using the equation $P = \rho g h$, determine a value for the density of air and give an uncertainty with your value. (You can choose to take the local accepted value of g or use the smartphone to provide a measurement of g using one of its sensors.)

Practice question

1. A student investigates how current I in a metal wire depends on the length L of the wire, when the potential difference across the wire is kept constant. The theoretical prediction for the relationship of I and L is given by $I = \frac{k}{L}$, where k is a constant that depends on the potential difference and the properties of the wire.

   The graph shows the variation of I with $\frac{1}{L}$ and the line of best fit.
   a. Suggest whether the data supports the theoretical prediction.

   b. Estimate the value of k.

   c. On the graph, sketch the maximum and minimum gradient lines for the data.

   d. Hence, state the value of k with its absolute uncertainty.

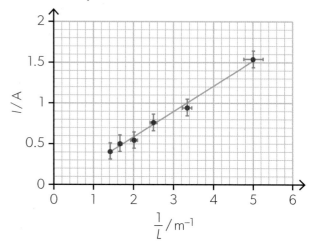

Dealing with data using graphs

Interpolation

When you have drawn the graph and line of best fit for your experimental data, it is straightforward to obtain values within the graphical range. Simply read off y-values for particular x-values that you need. A classic example of this is the use of a V–I graph to give resistance values for particular currents.

Figure 12 shows such a graph with a trend line drawn. The resistance at a current of 1.15 A is $R = \frac{V}{I} = \frac{4.9}{1.15} = 4.3\,\Omega$.

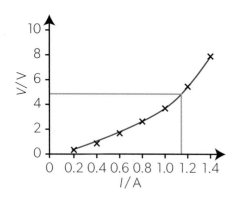

▲ Figure 12 Interpolation within a graphed range involves careful read-offs of the (x, y)-coordinate.

Extrapolation

Things are a little more difficult when working outside the range of your graph. You have to make assumptions about the behaviour of y as x varies and these may not necessarily hold. You are unlikely to be asked to extrapolate in a theory paper except for the estimation of an intercept from a straight line. However, you may need this skill for your own practical work.

Gradients from a straight line

Calculating the gradient of a straight-line graph is an essential skill and you should be able to do this both manually and using graphical calculators or spreadsheets.

Figure 13 shows a typical graph with a straight line. The steps to calculate the gradient are as follows.

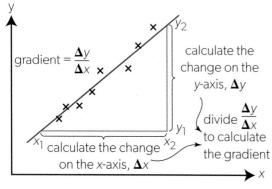

▲ Figure 13 The steps to calculating a gradient of a line of best fit.

- **Draw a large triangle** that occupies at least half the length of the trend line. Ideally, the ends of the triangle should lie at data values that are easy to read: near grid intersections, for example. This is one reason why your axis scales should not be awkward fractions, such as 1:7.

- Read off the intersection coordinates (x, y) where your triangle meets the line. Only use actual data points from your experiment when they lie *exactly* on the trend line.

- Calculate the change on the y-axis, $\Delta y = y_2 - y_1$ and the change on the x-axis, $\Delta x = x_2 - x_1$.

- Evaluate $\dfrac{\Delta y}{\Delta x}$. This is the value of the gradient.

- In physics, it is normal to include the unit of the gradient too. Strictly, the gradient itself has no units because the y-values and x-values are unitless as a result of the way the axis labels are written. This is a subtle point but it is more convenient to re-introduce the gradient value to its unit as soon as possible.

Gradients from a curve

The gradient to a curve gives the instantaneous value of $\dfrac{\Delta y}{\Delta x}$ at the x-value chosen.

- Draw a tangent to the curve at the point concerned. If you find this difficult, use a plane mirror as shown in Figure 14.

- When the curve appears to be continuous, your mirror is at 90° to the tangent. Draw along the mirror and then use a set square to construct the tangent.

- You can now continue as though for a straight-line gradient.

Intercepts

Intercepts on either axis are usually just a straight read-off where your line cuts the axis.

However, if you have used a false origin, then things may be more difficult. You will need to calculate the intercept using either trigonometry or the equation of a straight line ($y = mx + c$). When you know m then a read-off of a point (x, y) on the line will give the y-intercept c_y from $c_y = y - m \times x$ and the x-intercept c_x from $c_x = -\dfrac{c_y}{m}$.

A quick way when you have a drawn graph in front of you is shown in Figure 15.

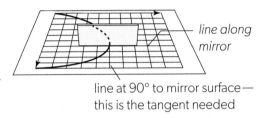

line along mirror

line at 90° to mirror surface — this is the tangent needed

▲ Figure 14 When you need a tangent to a graph curve, use a mirror to ensure that your tangent is along the curve.

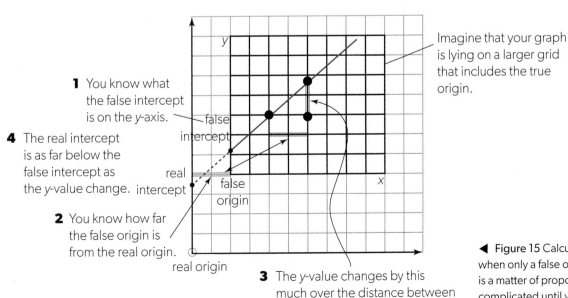

1 You know what the false intercept is on the y-axis.

4 The real intercept is as far below the false intercept as the y-value change.

2 You know how far the false origin is from the real origin.

3 The y-value changes by this much over the distance between the false origin and the real origin.

Imagine that your graph is lying on a larger grid that includes the true origin.

◀ Figure 15 Calculating a real intercept when only a false origin is available is a matter of proportion. This looks complicated until you have practised it a few times.

Area under a graph line

This is a technique used in Topic A.1 and elsewhere in this book.

For a curved line:

- Calculate the area of one grid rectangle.

- Count up the total number of grid rectangles between the x limits for which the area is required. Usually, some estimation is needed to cope with partly filled squares.

- Multiply the area of the single rectangle by the number of rectangles to give the area.

- The units of the area are (units of y-axis) $\times$ (units of x-axis).

For a straight line:

- Divide the area into a rectangle and a right-angled triangle.

- Use the equations for the areas of these shapes to calculate the total area.

Worked examples 8 and 16 in Topic A.1 illustrate the use of area under a graph to estimate distances.

Physics modelling and simulation

Throughout the course, there are examples of the use of modelling software. This is an important tool for physicists who can use it to answer many questions about the behaviour of physical systems. Such studies are called modelling or simulations.

Modelling software can be used to:

- indicate the underlying processes that control a system's behaviour

- investigate the past or future behaviour of a system

- show how to influence the behaviour of the system.

An advantage of a simulation is that it partially removes the need for expensive experimentation. However, it cannot be used to eliminate experiments completely and is, ultimately, an addition to the normal inquiry process, not a replacement for part of it.

There are a number of ways to simulate the behaviour of a scientific system. These include using:

- spreadsheets to plot the known outcomes of a model as functions of time

- spreadsheets to solve the physical equations using iterative methods to simulate behaviour

- software that is purpose-built to model physical systems.

Using spreadsheets

You can use a spreadsheet to plot a graph of a function. For example, Figure 16 shows a graph plotted for variation of displacement with time in simple harmonic motion. Such plots can help to visualize the behaviour of a system. However, this does not allow easy prediction of future variations unless the mathematical function of this variation is known.

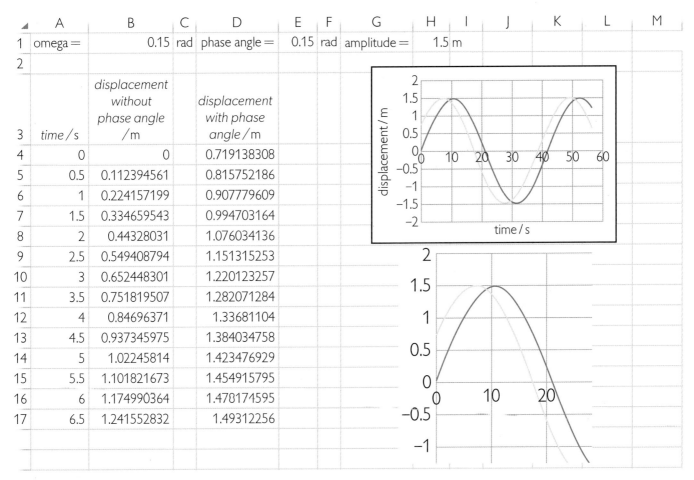

	A	B	C	D	E	F	G	H	I
1	omega =	0.15	rad	phase angle =	0.15	rad	amplitude =	1.5	m
2									
3	time /s	displacement without phase angle /m		displacement with phase angle /m					
4	0	0		0.719138308					
5	0.5	0.112394561		0.815752186					
6	1	0.224157199		0.907779609					
7	1.5	0.334659543		0.994703164					
8	2	0.44328031		1.076034136					
9	2.5	0.549408794		1.151315253					
10	3	0.652448301		1.220123257					
11	3.5	0.751819507		1.282071284					
12	4	0.84696371		1.33681104					
13	4.5	0.937345975		1.384034758					
14	5	1.02245814		1.423476929					
15	5.5	1.101821673		1.454915795					
16	6	1.174990364		1.478171595					
17	6.5	1.241552832		1.49312256					

▲ **Figure 16** Part of a spreadsheet that graphs the variation of displacement with time for simple harmonic motion, at two phase angles. The function used in cell D4 is "=$H$1*SIN(($B$1*A4)+$E$1)".

It is also possible to solve equations iteratively using a spreadsheet.

An example of this is the free fall of an object released close to Earth's surface. The spreadsheet is set up to compute, step-by-step, the successive positions of an object as time goes on.

Over the time interval Δt, the speed of the object changes from v_{old} to v_{new} by $v_{new} = v_{old} + a \times \Delta t$, where a is the acceleration.

This is entirely equivalent to the definition of acceleration as $a = \dfrac{\Delta v}{\Delta t}$. Similarly, the change in displacement is from x_{old} to x_{new} through $x_{new} = x_{old} + v \times \Delta t$.

The object is released 100 m above Earth's surface and air resistance is ignored. Figure 17 shows part of a simulation of this using a spreadsheet. The model predicts that the object will reach the surface after a time of about 4.5 s. This time is confirmed by the use of kinematic equations (Topic A.1). The values in this standard calculation from Topic A.1 are shown to the right of the graph.

 Cell references

The use of dollar ($) signs indicates to the program that, when the cell is replicated, the cell references do not change. You can use this to copy formulae down a column or along a row but keep some cell references the same.

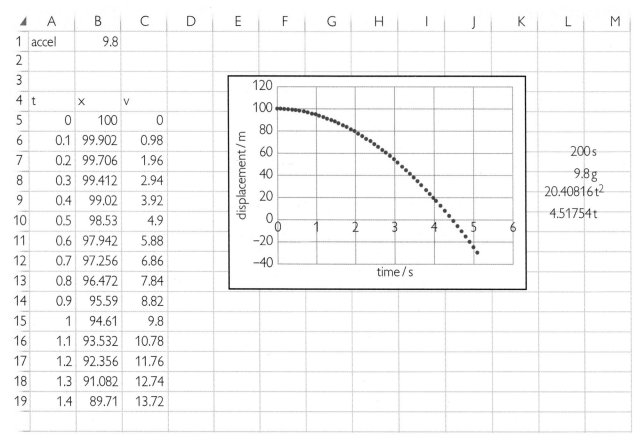

▲ **Figure 17** Part of the spreadsheet to simulate falling from rest under gravity. The function in cell B6 is "=B5-0.1*C6" so that the new x is the old x (vertically above it on the spreadsheet) less the time interval (0.1 s) multiplied by the speed in the right-hand cell. The speed is computed from the fixed value of g in cell $B$2 and from the old speed immediately above it: for cell C6, this is "=C5+$B$1*0.1".

The simulation shown in Figure 17 is based on a numerical approach called Euler's method. You may have met Euler's method as a part of your HL Mathematics AA course.

It is possible to simulate the effects of air resistance on this model by modifying the formulae slightly. Such a model would make a good addition to an experimental internal assessment that investigated the effects of friction or air resistance on motion.

Important considerations in designing such a model include the size of the time interval Δt. The acceleration is assumed constant during each Δt and, if Δt is too large, it will cause errors in the modelling. Such models drift away from reality as the number of data points increases and errors gradually creep into the simulation. You should check the outcomes of all models for this behaviour and modify your approach if necessary.

Using modelling software

There are many free and commercial software packages available for physics modelling, such as *ModellusX*, *Desmos* and *Geogebra*. These packages are free and have extensive video and written tutorials. Any sophisticated software package needs practice, but these three are easy to use and understand.

Many graphical calculators can also solve differential equations and be used for modelling. The instruction book for your calculator will explain how to enter and run the simulation.

ModellusX has the advantage that the models are written in a script that does not necessarily require detailed mathematical knoweldge. For example, the model for motion under gravity with air resistance used earlier in *ModellusX* is shown in Figure 18.

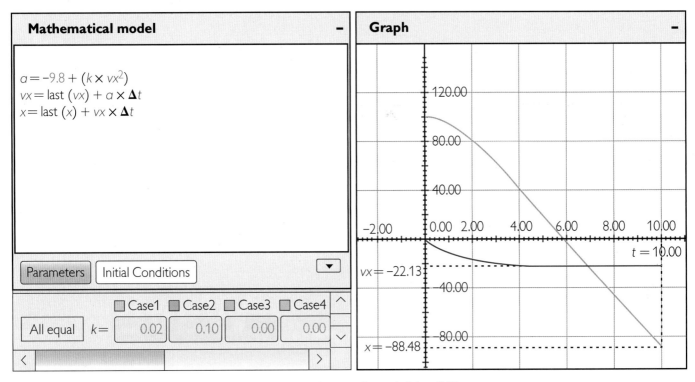

Mathematical model

$a = -9.8 + (k \times vx^2)$
$vx = \text{last } (vx) + a \times \Delta t$
$x = \text{last } (x) + vx \times \Delta t$

Parameters | Initial Conditions

☐ Case1 ☐ Case2 ☐ Case3 ☐ Case4
All equal $k=$ 0.02 0.10 0.00 0.00

Graph

120.00
80.00
40.00
−2.00   0.00 2.00   4.00   6.00   8.00   10.00
$t = 10.00$
$vx = -22.13$
−40.00
−80.00
$x = -88.48$

▲ **Figure 18** A *ModellusX* model for acceleration under gravity from rest from a height of 100 m.

The script for the model is shown on the left, and the graphical outcome is on the right. The blue line is the distance–time graph and the red line is the speed–time graph. The first line of the model sets the acceleration as being a negative g (9.8 m s^{-2} downwards) plus the air resistance which is proportional to *speed*^{2} (upwards). The second line computes the new speed as the old speed plus the change in speed, which is equal to the current (*acceleration* × *time increment*). The third line computes the new height above the surface from the old height plus the (negative) change in distance. The parameter shown is a constant of proportionality for the *speed*^{2} term. The initial conditions are a speed of zero and a distance (height) of 100 m. The model predicts that the object hits the ground after 6 s of travel at a speed of 22 m s^{-1}.

Similar models can be constructed within the *Geogebra* and *Desmos* environments. These operate within a more mathematical framework than *ModellusX*.

All the themes in this book contain opportunities to write physics simulations. Some of these could provide the starting point for an IA. Examples you could consider as simulations include:

- the motion of a parachutist from the instant of leaving the plane until landing (Theme A)
- the emf induced in a wire or coil as it moves towards a long straight wire carrying a current (Theme B and D)
- the thought experiment of balls moving between boxes to illustrate ideas in entropy (Theme B)
- diffraction or interference of waves (Theme C)
- refraction effects (Theme C)
- the parameters that affect criticality in nuclear fusion, eg moderator mass, amount of control material (Theme E)

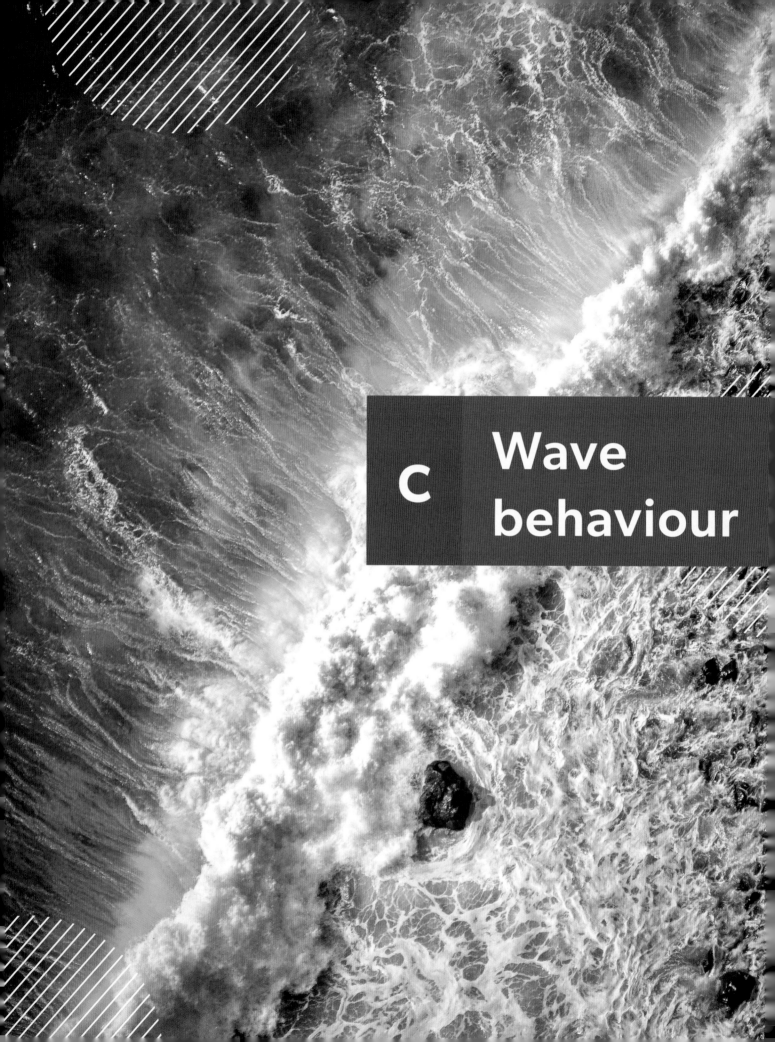

C Wave behaviour

Introduction

Every year, a human heart beats more than 50 million times – an oscillation that leads to the movement of blood around our bodies. Motions such as this and the beating of a hummingbird's wings are periodic. The pattern of the motion repeats again and again, sometimes with the same fixed time interval between each repeat. This theme deals with the physics of such a motion; known as an oscillation.

We begin in Topic C.1 with a detailed analysis of one important type of oscillation: simple harmonic motion. Simple harmonic motion has a fundamental importance. Complex oscillations can be described as the combined sum of many simple harmonic motions. This summation is important in many fields of science and engineering.

Oscillations lead to the production and transmission of mechanical waves. Waves come in many forms: Sound waves transmit through all materials and enable us to hear. Earthquake waves travel through the Earth. Our knowledge of wave theory enables us to understand and predict the behaviour of many man-made and natural phenomena.

Topic C.2 begins the work on waves themselves with the description of a model for wave motion. Topic C.3 looks at the effects that occur when waves interact with different media and with each other. Topic C.4 continues with a

description of the mechanisms that lead to standing waves – an important part of the production of sound in musical instruments. The theme ends with Topic C.5 that deals with the Doppler effect when waves are emitted and detected by sources and observers moving relative to each other.

The concepts of **particles** and **energy** are inextricably linked throughout Theme C. Waves transfer energy but the medium that carries the wave is undisturbed when the wave has gone through.

A mechanical wave is made up of the movement of particles. The particles are the medium for the wave. However, electromagnetic waves do not have a particulate nature and do not require a medium. The physics of this wave transfer is significantly different from that of mechanical wave motion. These differences have led to profound changes in our understanding of spacetime.

Two features of physics that have underpinned the theory of oscillations are **observations** and **measurements**. Galileo is said to have used his own pulse to time the slow swings of the huge candelabra in the cathedral of Pisa. He recognised that, whatever the amplitude of the swing, the period was constant. To what extent would we regard these as reliable observations today?

What makes the harmonic oscillator model applicable to a wide range of physical phenomena?

Why must the defining equation of simple harmonic motion take the form it does?

How can the energy and motion of an oscillation be analysed both graphically and algebraically?

Simple harmonic motion is an oscillation with an unchanging amplitude and frequency and which never ends. No energy is transfering from the oscillating system. It may seem strange to learn about such a specific type of motion, but there is a good reason. Joseph Fourier showed that any periodic motion could be regarded mathematically as a sum of individual simple harmonic motions. Study simple harmonic motion and you have studied more complex oscillations too. However, many oscillations are either purely or approximately simple harmonic. A buoy floating in the sea, a mass oscillating on a spring and a pendulum are just three common examples of this motion.

The motion itself is characterized by a simple defining equation. The acceleration of a system is directly proportional to the displacement of the system and acts opposite to the displacement direction. The equation contains only three quantities, including a constant of proportionality, but the way in which these interact generates oscillations. The constant of proportionality tells us about the time taken to complete one oscillation. The statement about direction is crucial too. It says that the further the object is from an equilibrium position, then the larger is the acceleration back towards the equilibrium point. This already suggests an oscillation of some kind.

The oscillation trades displacement for velocity, and potential energy for kinetic energy. When the system is far from equilibrium it is travelling slowly. Around the equilibrium point it is moving quickly so that its momentum carries it through equilibrium to the other half of the cycle. At this point, the force on the system (and therefore the acceleration) reverses direction, once more acting towards the equilibrium point.

Our defining equation also leads to sets of equations linking the displacement, velocity and acceleration of the oscillating system with time. This means that we can go on to use knowledge from Theme A to describe the energy transfers in the oscillating system too. These can also be expressed in terms of time and distance.

Finally, graphical representations of energy–time and displacement–time can be linked to real examples of simple harmonic motion. This allows us to confirm that our equation for harmonic motion and the predictions it makes are a good fit to the real oscillations that we observe in a practical context.

In this topic, you will learn about:

- oscillations and simple harmonic motion

- the defining equation of simple harmonic motion

- the conditions for simple harmonic motion

- displacement, amplitude, time period, frequency, angular frequency and equilibrium position

- phase angle   **AHL**

- the mass–spring system and the simple pendulum

- energy changes during an oscillation

- kinematic and energy calculations involving simple harmonic motion

- kinematic and energy calculations involving simple harmonic motion.   **AHL**

Introduction

In this topic, you will meet the language of oscillation and consider the harmonic oscillator, usually referred to as simple harmonic motion. True simple harmonic motion can only be obtained in some systems under certain limited conditions, such as small displacements. Nevertheless, you can still use simple harmonic motion as a model in these systems, if you accept the conditions and the limitations they impose.

(ATL) Drafting, revising and improving academic work

Joseph Fourier was a French mathematician and physicist who lived from 1768 to 1830. In 1807, he read a paper to the Paris Institute *"On the Propagation of Heat in Solid Bodies"*. In it he used a mathematical method to reduce a complicated oscillation to a series of sine waves.

You can try this for yourself. Use a graphical calculator or a spreadsheet to help plot the function $y = \sin x + \frac{1}{3}\sin 3x + \frac{1}{5}\sin 5x + \cdots$. You can add further terms of $\frac{1}{n}\sin nx$ for odd values of the integer n. It does not require very many terms in the series to show that the series approaches a square-wave. You could also try the even terms to see what happens.

However, Fourier's paper did not convince everyone in the audience. He had relied on intuition in places and there were some gaps in his logic. His mathematical method also contradicted some of the work of one of the examiners in the audience—Joseph-Louis Lagrange.

To settle the matter, a prize problem was set in 1810 and Fourier submitted his original paper along with some new work. There was only one other paper, and Fourier won the competition. But the feedback (possibly from Lagrange) was not entirely favourable, and the result was that Fourier's work was not published until 1822.

Fourier's method of splitting a signal into sinusoidal waves of different frequencies is widely used today and is the principle behind the spectral analysis of sound.

▲ Figure 1 Knowledge of simple harmonic motion led to the development of the pendulum clock. For about 300 years, pendulum clocks were the most precise clocks available. This is a sidereal clock used to help make astronomical observations.

Oscillations

Many oscillations in science and engineering are **isochronous**. This means that the oscillation repeats, taking the same repetition time irrespective of its size. This is important because, unless energy is transferred to them, real oscillating systems "run down" and eventually stop. The **amplitude**—the maximum displacement—of the system decreases when it transfers energy to the environment.

▲ Figure 2 A swinging pocket watch is an example of a simple pendulum oscillating with approximate simple harmonic motion.

Technology for timing

Galileo is reputed to have first observed that the time period of a simple pendulum did not depend on its amplitude (provided that the amplitude remained small). The story is that when he was about 17 years old, he was bored during a service in Pisa Cathedral and observed the way that the chandelier swung as the wind blew it. He compared the time for the swings with his pulse. Sometimes the wind blew the chandelier into large oscillations and sometimes the oscillations were small. However, the number of oscillations in a certain number of pulse beats was always the same.

While it is likely that other scientists may have observed that a pendulum's period does not change with amplitude, Galileo was perhaps one of the first to use the pendulum in experiments. As a result, scientists could now measure time and hence other quantities such as speed. Without this timing mechanism, experiments in mechanics would have been impossible. The importance of experimental evidence in scientific knowledge was still a relatively new concept at this time, and the increased ability to conduct experiments increased the importance of this evidence.

How else has technology affected the value we place on different forms of knowledge?

Figure 2 shows a pocket watch oscillating about its centre (equilibrium) position from the maximum position on one side to the other. The watch is illuminated with a flash that occurs every 0.25 s and so it takes 1.0 s for the watch to complete each oscillation (to go from one side to the other and back again). A simple pendulum only performs approximate simple harmonic motion which changes at large amplitudes of swing. Nevertheless, a timepiece can be governed to make it into an isochronous oscillator.

Defining periodic motion

Before we can develop the mathematics of simple harmonic motion, we need a technical language.

To illustrate the terms we use, imagine an experiment with a mass hanging at the end of a spring (Figure 3(a)). The position of a small card attached to the mass is detected by a motion sensor on a data logger that produces a graph of displacement against time for the mass (Figure 3(b)).

- The mass with its card is shown on the left in its **equilibrium position**. This is the position it adopts when at rest.

- The mass–spring system oscillates when displaced vertically and released (it takes both a spring and a mass to oscillate; hence the word "system").

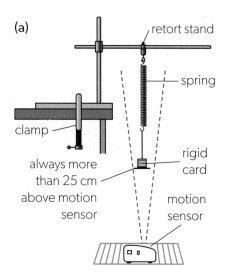

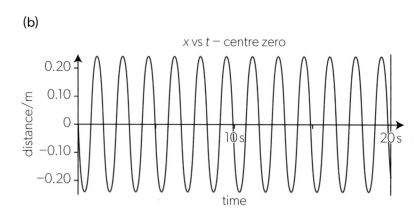

▲ Figure 3 (a) The experimental arrangement and (b) the resulting displacement–time graph for an illustration of simple harmonic motion.

The position of the oscillator at any moment in time is known as the **displacement** x. As in Theme A, displacement is a vector that can be positive (when the mass is above the equilibrium position here) or negative (when it is below). Once the positive direction has been chosen as upwards, then it must be used in a consistent way for all vector quantities in the oscillation, including forces, velocities, displacements and accelerations.

- The maximum displacement of the oscillator is known as the **amplitude** x_0. This amplitude is measured from the equilibrium position to the extreme (largest) displacement. It is *not* the distance from one extreme to the other. Amplitude does not have a sign and is not a vector quantity.

- One complete **cycle** of the oscillation occurs when the mass (in this situation) goes from one position in the motion through the extreme position on the opposite side, back to the other extreme, finally moving through the original position *in the original direction*. It is easiest to understand this for the mass–spring system by starting at the equilibrium position. The mass goes down to the bottom, back through the equilibrium, moving upwards, and to the top. Then it goes down through the equilibrium again. The cycle only ends with this second transit through the equilibrium. Trace this motion out on the graph (Figure 3(b)). There are six cycles in 10 s.

- The time taken to complete one cycle is known as the **time period**, T. For the isochronous mass–spring system, the time period (often shortened to **period**) does not depend on where the cycle starts or on the amplitude.

- The **frequency** f of the oscillation is the number of cycles that the system goes through in one second. Thus

$$f = \frac{1}{T}$$

The unit of frequency is the hertz (Hz), which is the same as s^{-1}.

Practice questions

1. Which of the following quantities describing an oscillation can be negative?

    A. displacement   B. amplitude   C. period   D. frequency

2. A mosquito flaps its wings at a frequency of 580 Hz. Calculate the period of mosquito's flaps.

3. An object undergoes simple harmonic motion with a period of 0.40 s. The distance between the extreme positions of the object is 6.0 cm. Calculate:

    a. the frequency

    b. the amplitude.

Applying the definitions

These definitions apply to many repetitive phenomena such as the rhythm of a human heart. Figure 4 shows the electrocardiograph of a healthy heart that is beating at 65 beats per minute, a frequency of $\frac{65}{60} = 1.08$ Hz. This means that T for the graph is $\frac{1}{f} = \frac{1}{1.08} = 0.92$ s. The overall height of the voltage spike from 0 V, shown as A in Figure 4, is the amplitude signal output by this sensor.

Worked example 1

The pendulum of a wall clock completes 25 oscillations in 30 s. Calculate:

a. the period

b. the frequency of the oscillations.

Solutions

a. $T = \dfrac{30}{25} = 1.2$ s

b. $f = \dfrac{1}{T} = \dfrac{1}{1.2} = 0.83$ Hz

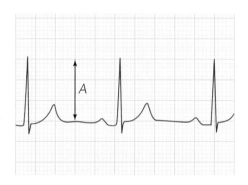

▲ Figure 4 The normal heart rhythm of an adult male. The graph shows the pd measured using a voltage sensor attached to the chest wall.

Investigating a mass—spring system

- Tool 2: Use sensors.
- Tool 2: Represent data in a graphical form.
- Inquiry 1: Develop investigations that involve hands-on laboratory experiments, databases, simulations and modelling.
- Inquiry 1: Design and explain a valid methodology.

In this investigation, an ultrasound motion sensor is used to monitor the position of a mass suspended from the end of a long spring. The data logger software processes the data to produce a graph showing the variation of displacement with time.

- Arrange the apparatus as shown in Figure 3(a). The system needs to have a period of at least 1.0 s. Avoid reflections from the surroundings by keeping objects well away from the apparatus.
- Put the mass into oscillation by displacing it vertically and then releasing it.

- Set up the data logger so that it is triggered to start reading at a given displacement value.
- Use software to plot graphs of velocity and acceleration (in addition to displacement) against time.
- Devise an investigation to find out how the time period of the oscillation varies with:
  - spring constant k
  - mass m on the spring.
- You may wish to carry out a preliminary set of runs to get an idea of the relationships between k and T, and between m and T. Try doubling the mass or quadrupling it to see the effect on T. Two or more identical springs can be joined together in series or in parallel to vary k. (Hint: look at page 58 to remind yourself how k depends on the arrangement of springs).
- You can also perform similar investigations with other oscillations, such as a mass swinging from side to side at the end of a long string—a simple pendulum.

Global impact of science

Heinrich Hertz, for whom the frequency unit was named, was a German physicist working in the mid-19th century. He demonstrated the existence of electromagnetic radiation in the radio wavelengths and (famously) suggested that his work had no future application! Within 15 years, the Italian nobleman Count Marconi had sent messages across the Atlantic Ocean using radio waves. Hertz, unfortunately, never lived to see the application of radio waves, as he died in 1894 aged 36.

There is a direct link between the frequency of simple harmonic motion and the frequencies of the electromagnetic radiation that Hertz identified. His waves consisted of oscillating electric and magnetic fields that are modelled as sinusoidal variations just like those of an oscillating spring.

Simple harmonic motion

The variation with time of the displacement of the mass–spring system shown in Figure 3(b) is regular and simple. This is a negative sine curve (making the mass go upwards first will make this a positive sine curve). Oscillations that follow this model with a sinusoidal displacement–time graph are undergoing **simple harmonic motion**.

There are two requirements for motion to be simple harmonic. Both relate to the restoring force (and therefore the acceleration) acting on the system.

- The size (magnitude) of the force (acceleration) must be proportional to the displacement of the object from a fixed point.
- The direction of the force (acceleration) must be towards the fixed point.

Newton's second law of motion links acceleration and force in these statements.

At the equilibrium position, the weight of the mass is equal and opposite to the tension in the spring (assuming that the spring has negligible mass).

When the spring obeys Hooke's law (Topic A.2, page 57), then $F = -kx$, where F is the restoring force on the spring, k is the spring constant and x is the spring extension.

Substituting for F means that, for simple harmonic motion:

$$a = -(\text{constant})^2 \times x$$

You can find out more details of electromagnetic radiation in Topic C.2.

The constant is squared. This forces it to be positive, so that the minus sign always indicates that the displacement and acceleration vectors are in opposite directions. As a result, this equation now agrees with both of the requirements for simple harmonic motion.

In simple harmonic motion, the system is always accelerated towards the centre of the motion—the equilibrium position. When the mass is moving away from the equilibrium position, the system is slowing the mass down, accelerating it towards the equilibrium position. When the mass has reached the extreme of the motion, the system still accelerates it towards the equilibrium position, but now the speed of the motion increases until it reaches a maximum in the motion's centre.

This is summed up in Figure 5, which shows the variation of acceleration with displacement for any simple harmonic motion, not just the mass–spring system here. The gradient of the graph is negative as expected.

We need to know more about the constant in the defining equation. It is often written as

$$a = -\omega^2 \times x$$

with the constant as ω. This makes an important link between simple harmonic motion and the circular motion of Topic A.2.

Angular frequency

The oscillation of the pendulum can be compared with circular motion using the apparatus shown in Figure 6.

Two metal spheres are used, one acting as the mass for the pendulum. The other sphere is mounted on a horizontal turntable that rotates at a constant angular speed. The length of the string is adjusted so that the time period T of the simple harmonic motion oscillation is the time taken for the turntable to rotate once. When the arrangement is illuminated from the side, the two spheres move together and are synchronized on the screen. The circular motion is projected onto a vertical plane (the screen) and has the same pattern of movement as a pendulum when viewed in the same vertical plane.

The angular speed of the rotating sphere is

$$\frac{\text{angular displacement in radians}}{\text{time for one rotation}} = \frac{2\pi}{T}$$

In Topics A.2 and A.4, the quantity angular speed was given the symbol ω and therefore

$$\omega = \frac{2\pi}{T}$$

Putting this all together gives

$$T = \frac{1}{f} = \frac{2\pi}{\omega}$$

The same is true for ω in the simple harmonic motion equation, but here the quantity is known as **angular frequency** because it has the unit s^{-1} equivalent to the hertz (Hz). As before, although this is rad s^{-1}, the radian is ignored because it is a unitless ratio.

Because ω is linked to T, which depends only on the properties of the harmonic oscillator, it also links the magnitude of the acceleration of the oscillator to its displacement. To show this link in more detail, we will look at two oscillators in detail: the mass–spring system and the simple pendulum.

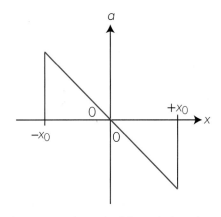

▲ Figure 5 A graph of the variation of acceleration with displacement for simple harmonic motion. The graph is a straight line of negative gradient going through the origin.

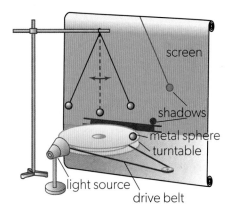

▲ Figure 6 The projection of a ball moving in a horizontal circle onto a vertical plane gives the same motion as a simple pendulum performing simple harmonic motion.

 How can circular motion be used to visualize simple harmonic motion?

The demonstration above shows the close link between circular motion and simple harmonic motion. One can be regarded as a one-dimensional projection of the other. The link extends beyond the purely practical, however, as the mathematics of circular motion from Topic A.2 and the mathematics of simple harmonic motion are themselves closely related. Similar physical quantities are defined in the same way in both.

How can the understanding of simple harmonic motion apply to the wave model? (NOS)

Topics C.2 and C.3 deal with waves — periodic movements of interconnected individual particles that transfer energy. There must be an agent that generates the waves and this could easily be an object moving in a circle. The waves in deep oceans are linked to a circular motion of water that becomes an up-and-down motion of the surface. The mathematics developed in this Topic applies to the wave motions later.

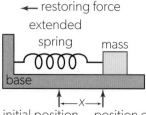

▲ Figure 7 Waves on the surface of the ocean are caused by the circular motion of the water.

▲ Figure 8 A mass–spring system.

The mass–spring system

The mass–spring system here is a mass on a horizontal frictionless surface oscillating at the end of a spring. This is known as "exact simple harmonic motion" when the spring obeys Hooke's law. The horizontal case is easier to analyse than when the spring hangs vertically. (You can analyse the vertical case for yourself, remembering to include the weight of the mass as part of the net force that acts on the spring.)

The force F_H acting on the spring is directly proportional to its extension x: $F_H = -kx$ (from Topic A.2) and acts to return the spring to its equilibrium position. Therefore $ma = -kx$. When the positive direction is defined to be to the right and the mass is displaced to the right, the force must be directed to the left. The negative sign shows this.

This equation rearranges to $a = -\left(\dfrac{k}{m}\right)x$ and shows the shape of the simple harmonic motion equation with its negative sign and positive constant inside the brackets.

Therefore, $\omega^2 = \dfrac{k}{m}$ and $\omega = \sqrt{\dfrac{k}{m}}$, leading to

$$T = 2\pi\sqrt{\dfrac{m}{k}}$$

as the equation for the time period of a mass–spring system.

The simple pendulum

A simple pendulum consists of an object on the end of a string of negligible mass that is swinging in a vertical plane. The pendulum obeys simple harmonic motion provided that the angle of swing from the vertical is small (<10°).

The string has a length l and is displaced with its bob of mass m through a vertical angle θ (Figure 9). When released, the bob moves with time period T.

The restoring force that pulls the bob back to the equilibrium position is $-mg\sin\theta$. The negative sign is because θ is measured to the right (anticlockwise on the diagram), but the restoring force is to the left (clockwise).

So $-mg\sin\theta = ma$, leading to $a = -g\sin\theta$.

The length of the arc from the equilibrium position to the bob is x, so

$$\theta = \frac{x}{l} \text{ and } a = -g\sin\left(\frac{x}{l}\right)$$

giving

$$a = -\frac{g}{l}x$$

providing that $\theta < 10°$.

You can check, using your calculator, that when $\theta < 12°$ (about 0.2 rad), then $\sin\theta$ and θ are within 1% of each other when calculated using radian measure.

Thus, $\omega^2 = \dfrac{g}{l}$ and $\omega = \sqrt{\dfrac{g}{l}}$, with

$$T = 2\pi\sqrt{\frac{l}{g}}$$

which is the equation for the time period of a simple pendulum.

Analyses such as this can be carried out for many more types of oscillator too, including floating cylinders bobbing up and down on the flat surface of a lake .

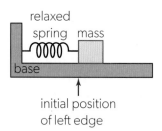

relaxed
spring   mass

base

initial position
of left edge

← restoring force

extended
spring   mass

base

← x →

initial position   position of left edge
of left edge   when spring extended

Data-based questions

As you have seen, a simple pendulum obeys simple harmonic motion (i.e. it is isochronous) provided that the amplitude is small (less than 10°). What happens if the pendulum swings through a larger amplitude?

The graph shows the variation of the time period T with angle θ for a pendulum of length 1.8 m.

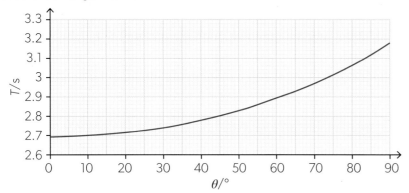

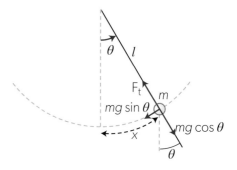

▲ Figure 9 A simple pendulum.

- Use the graph to estimate the percentage difference in T when the pendulum swings with $\theta = 80°$ and when $\theta = 10°$.

- A student measures the time period of the pendulum by using oscillations with $\theta = 10°$. Explain why it would not be appropriate to give this measurement to 3 decimal places.

- You are asked to design an experiment to confirm that T changes between a 10° and a 45° amplitude. You have a stopwatch which reads to the nearest 0.01 s. Assume that your reaction time is 0.1 s. You decide to time the pendulum over several oscillations and then divide the total time by the number of oscillations to arrive at T. How many oscillations would you need to measure to verify that T is longer at 45° than at 10°?

▲ Figure 10 Many mechanical objects can be approximated as either a mass on a spring or a pendulum. This picture shows a car's suspension which consists of a spring to absorb the shocks from bumps in the road. The car behaves like a mass on a spring and will have a time period for its oscillations.

Worked example 2

The graph shows how the acceleration a of an object varies with the displacement x.

a. Outline why the object performs simple harmonic motion.

b. State the amplitude of the oscillations.

c. Determine the period.

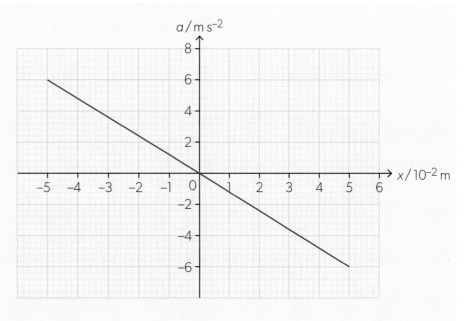

Solutions

a. The graph is a straight line with a negative slope through the origin. Hence, the acceleration is proportional to negative displacement and satisfies the defining equation of simple harmonic motion, $a = -\omega^2 x$.

b. The amplitude is equal to the maximum displacement, 5.0 cm.
c. The period is related to the angular frequency ω, which can be determined from the slope of the graph.

$$\text{Slope} = -\omega^2 = -\frac{6.0}{5.0} \Rightarrow \omega = \sqrt{\frac{6.0}{5.0}} = 1.1 \text{ rad s}^{-1}. \text{ From here, } T = \frac{2\pi}{\omega} = \frac{2\pi}{1.1} = 5.7 \text{ s}.$$

Worked example 3

A mass of 0.045 kg oscillates simple harmonically at the end of a spring of spring constant 1.3 kN m^{-1}. Calculate the frequency of the oscillations.

Solution

$$T = 2\pi\sqrt{\frac{m}{k}} = 2\pi\sqrt{\frac{0.045}{1.3 \times 10^3}} = 3.7 \times 10^{-2} \text{ s}.$$

$$f = \frac{1}{T} = \frac{1}{3.7 \times 10^{-2}} = 27 \text{ Hz}.$$

Worked example 4

An object of mass 2.1 kg attached to a spring undergoes simple harmonic motion on a horizontal frictionless surface. The period of oscillations is 1.8 s and the amplitude is 0.25 m.

Calculate:

a. the angular frequency
b. the maximum force acting on the object
c. the spring constant.

Solutions

a. $\omega = \dfrac{2\pi}{T} = \dfrac{2\pi}{1.8} = 3.5 \text{ rad s}^{-1}.$

b. From the defining equation of simple harmonic motion, the maximum acceleration of the object is $a_{max} = \omega^2 x_0$, where x_0 is the amplitude of oscillation. The maximum force is therefore

$$F_{max} = m a_{max} = m\omega^2 x_0 = (2.1)\left(\frac{2\pi}{1.8}\right)^2 (0.25) = 6.4 \text{ N}.$$

c. $k = \dfrac{F_{max}}{x_0} = \dfrac{6.4}{0.25} = 26 \text{ N m}^{-1}.$

Practice questions

4. A force F acting on a point mass depends on the displacement x of the mass. Which of the relationships between F and x leads to simple harmonic motion?

   A. $F = -x^2$ B. $F = -2x$ C. $F = 3x$ D. $F = 4x^2$

5. Calculate:
   a. the period of a simple pendulum whose length is 0.80 m
   b. length of a simple pendulum whose period is 2.4 s.

6. An object of mass 0.45 kg is attached to a spring with spring constant 12 N m^{-1}. The object undergoes simple harmonic motion with an amplitude of 0.15 m. Calculate:
   a. the period of oscillation
   b. the maximum force acting on the object from the spring.

7. A mass–spring system undergoes simple harmonic oscillations of a frequency 0.58 Hz. The mass is 0.90 kg. Calculate the spring constant.

8. A weightless spring of spring constant $k = 2.9\,\text{N}\,\text{m}^{-1}$ hangs vertically with a mass $m = 0.050\,\text{kg}$ attached to its free end. When the mass is in the equilibrium position, the spring extends by a distance L_0 relative to the unstretched length.

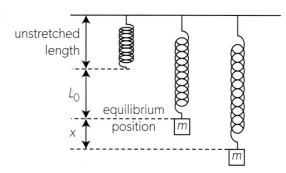

a. Calculate L_0.

The mass is displaced vertically from the equilibrium position by a distance x and released.

b. Draw a free-body diagram for the mass at the displaced position.

c. Show that the magnitude of the net force acting on the mass is kx.

d. Compare the period of the vertical mass–spring system to that of a horizontal system, if the mass and the spring are the same in both systems.

e. Calculate the period of the oscillations.

Energy changes during simple harmonic motion

One way to interpret simple harmonic motion is in terms of energy transfer.

Figure 11 shows the transfers that occur in the horizontal mass–spring system.

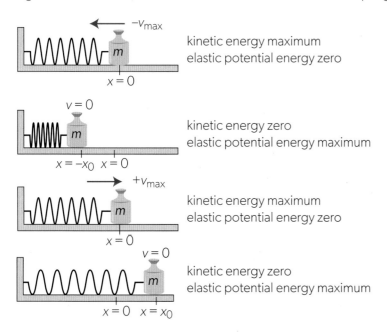

▲ Figure 11 The energy transfers that occur in simple harmonic motion for a mass–spring system.

The mass is oscillating between $-x_0$ and $+x_0$. The amplitude of the motion is x_0. At each extreme, the speed of the mass is zero, so the kinetic energy is also zero. At this point, all the energy is in the form of stored elastic potential energy. At the centre of the motion the spring is at its natural (unextended) length and the mass is moving at its fastest, so the kinetic energy is also at a maximum with no energy stored in the form of elastic potential energy.

During one cycle of the oscillation, there are two kinetic-energy maxima because there are two velocity maxima, one in each direction when the mass is at the equilibrium position. In the same way, there are two maxima of elastic potential energy. The frequency of the energy transfers is double that of the frequency of the oscillation itself. Conversely, the time period for one energy cycle is half that of the time period for the simple harmonic motion.

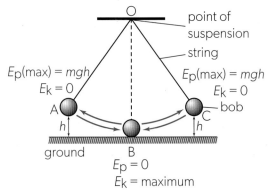

$E_p(\text{max}) = mgh$
$E_k = 0$

$E_p(\text{max}) = mgh$
$E_k = 0$

bob

A

C

h

h

ground

B

$E_p = 0$
$E_k = \text{maximum}$

▲ Figure 12 The energy analysis is similar for the simple pendulum. The transfers between gravitational potential energy and kinetic energy for the pendulum bob are shown here.

Worked example 5

A body undergoes simple harmonic motion of a frequency 20 Hz. How many times during one second is the kinetic energy of the body zero?

Solution

The KE is zero twice during one oscillation; hence $2 \times 20 = 40$ times per second.

Figure 12 shows the energy transfers for the simple pendulum.

For both oscillators, there is a continuous transfer between the kinetic and potential energies. When there are no energy losses from a system, such as those due to air resistance or friction, then the total energy in the system must be constant.

Figure 13 shows three graphs for the variation with time of the kinetic E_k, potential E_p and total energies E_{tot} for simple harmonic motion. It also shows how the displacement varies with time, so that the difference between the period of energy transfer and the period of simple harmonic motion is clear.

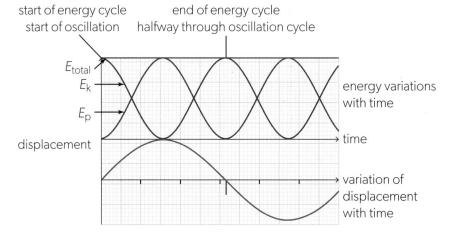

▲ Figure 13 The variations of kinetic and potential energies in simple harmonic motion with time. The total energy in the system is constant.

Worked example 6

The graph shows how the potential energy of a simple pendulum varies with time.

a. Identify the first time when:
   i. the pendulum passes through the equilibrium position
   ii. the kinetic and the potential energies are equal.

b. State the period of oscillations.

c. Draw a graph of the variation of the kinetic energy of the pendulum with time.

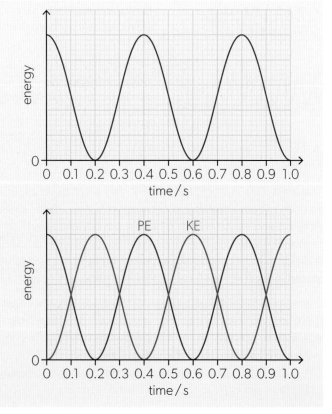

Solutions

a. i. In the equilibrium position the potential energy is zero. This happens for the first time at 0.2 s.
   ii. The potential energy must decrease to one half of its maximum value. This happens at 0.1 s.

b. It takes 0.4 s to move from one extreme position (of maximum amplitude and potential energy) to the other. This is one half of the complete oscillation. The period is therefore $2 \times 0.4 = 0.8$ s.

c. The KE is a maximum when the PE is zero, and vice versa.

Energy loss and simple harmonic motion

Strictly speaking, once resistive losses of any sort occur for an oscillating system, then the oscillation is no longer simple harmonic. True simple harmonic motion never stops. The graphs for the variation with time of

displacement/velocity/acceleration and the energy–time graphs have constant amplitudes as there are no resistance or energy losses to reduce the amplitude.

How can greenhouse gases be modelled as simple harmonic oscillators?
What physical explanation leads to the enhanced greenhouse effect? (NOS)

Topic B.2 gives the absorption mechanisms of electromagnetic radiation by molecules of the greenhouse gases. These molecules have vibrational states that are excited by the radiation. This leads to the temporary storage of the electromagnetic energy with subsequent re-radiation in different directions. There are links here both to the work in this topic but also to the resonance effects discussed in more detail in Topic C.4.

When changes to the atmosphere occur, then the levels of radiation absorption reflect the change. With greater concentrations of greenhouse gases, the absorption and re-radiation increases, leading to climate change.

How does the creation of links within physics enable scientists to develop greater understanding of the linked topics?

How can the understanding of simple harmonic motion apply to the wave model? (NOS)

There are strong links from Topic C.1 to the physics of Topics C.2 and C.3. Wave motion is a common phenomenon and a working knowledge of the mathematics of simple harmonic motion helps our understanding of wave behaviour and vice versa.

One way to describe the motion of a particle in a wave is in terms of a vector of constant length that rotates at a

constant speed. Such a vector is known as a "phasor". This is the function of the red arrow in Figure 14. The arrowhead of the phasor traces out the motion of the wave particle. Wave motion and simple harmonic motion are closely interlinked, with the same terms and quantities being used in both.

Do links such as these give us further insights into the physical world?

Linking circular motion and simple harmonic motion

When a circular motion in a horizontal plane is projected onto a vertical plane as in Figure 6, it is equivalent to a motion that is simple harmonic (Figure 14).

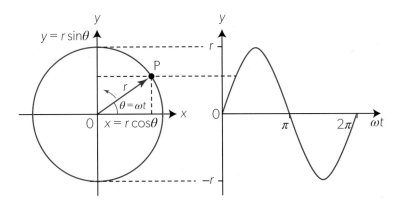

▲ Figure 14 Projecting circular motion onto a y-axis.

The y-axis point P is moving around the circle in Figure 14 at a constant angular speed ω.

AHL

Two equations relate x and y to the circle of radius r and the angle θ between P and the x-axis. These are:

- $x = r\cos\theta$

- $y = r\sin\theta$.

The angle θ between the arrow and the x-axis is known as the **phase angle**.

As $\theta = \omega t$, the equations for the projection of P onto the diameter of the circle along the x-axis become $x = r\cos\omega t$ or $y = r\sin\omega t$. The radius of the circle is the amplitude of the simple harmonic motion so $r = x_0$ and we obtain the simple harmonic motion equations:

- $x = x_0\cos\omega t$ for simple harmonic motion that begins at the extremes

- $x = x_0\sin\omega t$ for simple harmonic motion that begins in the centre.

Two further equations also follow from the definition of simple harmonic motion and from $x = x_0\sin\omega t$:

The velocity $v = \dfrac{dx}{dt} = \omega x_0\cos\omega t$ and the acceleration $a = \dfrac{dv}{dt} = -\omega^2 x_0\sin\omega t$.

Notice that, because $x = x_0\sin\omega t$, then $a = -\omega^2(x_0\sin\omega t) = -\omega^2 x$. Our solution for the simple harmonic motion equation that arises from the projected circular motion satisfies the defining equation.

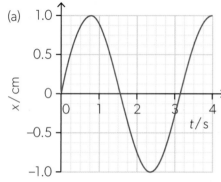

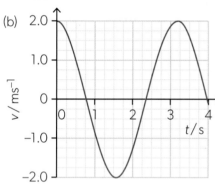

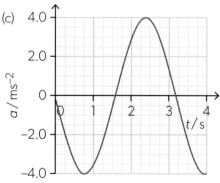

(a)

(b)

(c)

The three equations lead to three graphs.

Figure 15 shows the variations with time of (a) displacement, (b) velocity and (c) acceleration for the case where the motion starts at the centre. The displacement graph (a) is a sine curve, (b) is a cosine curve and (c) is a negative sine curve. (For motion starting at the positive extreme, they will be respectively (a) cosine, (b) −sine and (c) −cosine.)

The gradient at a particular time for the velocity–time graph gives the acceleration at that instant, and, similarly, the gradient of the displacement–time graph yields the velocity at that moment. This is easy to see at the extremes when the motion is momentarily at rest ($v = 0$).

There is another equation for the velocity that you will find useful because it does not contain t. Using the identity, $\sin^2\theta + \cos^2\theta = 1$, so that $\cos\theta = \pm\sqrt{1 - \sin^2\theta}$ and substituting this into the speed equation gives $v = \pm\omega x_0\sqrt{1 - \sin^2\theta}$. However, $\sin\theta = \dfrac{x}{x_0}$. To see why, look at Figure 15 and notice that $\sin\theta$ is the ratio of the displacement of P (which is at x) to the radius of the circle (which corresponds to the amplitude x_0). This gives $v = \pm\omega x_0\sqrt{1 - \dfrac{x^2}{x_0^2}}$, which rearranges to

$$v = \pm\omega\sqrt{x_0^2 - x^2}$$

The $\pm$ sign reminds us that the object can be travelling in either direction at a particular x. As you can see, this is a useful equation when you know the amplitude and displacement of an object but do not know the time at which the displacement occurs.

▲ Figure 15 Variation with time of (a) displacement, (b) velocity and (c) acceleration. These graphs all assume that the motion starts at the equilibrium position.

Displacement	$x = x_0\sin\omega t$
Velocity (x unknown)	$v = \omega x_0\cos\omega t$
Velocity (t unknown)	$v = \pm\omega\sqrt{x_0^2 - x^2}$
Acceleration	$a = -\omega^2(x_0\sin\omega t) = -\omega^2 x$

▲ Table 1 The four equations for simple harmonic motion.

Worked example 7

The graph shows how the displacement of a body performing simple harmonic motion varies with time.

Calculate:

a. the angular frequency of oscillations
b. the maximum velocity of the body
c. the velocity after 3.0 s
d. the maximum acceleration.

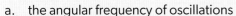

Solutions

a. The period is 5.0 s. $\omega = \dfrac{2\pi}{5.0} = 1.3\,\text{rad s}^{-1}$.

b. The amplitude is 4.0 cm.

$v_{\text{max}} = \omega x_0 = \dfrac{2\pi}{5.0} \times 4.0 = 5.0\,\text{cm s}^{-1}$.

c. The displacement follows a sine function, $x = x_0 \sin \omega t$.

Hence, the velocity after a time t should be modelled with a cosine function, $v = \omega x_0 \cos \omega t$.

At $t = 3.0\,\text{s}$, $v = \dfrac{2\pi}{5.0} \times 4.0 \cos\left(\dfrac{2\pi}{5.0} \times 3.0\right) = -4.1\,\text{cm s}^{-1}$.

d. $a_{\text{max}} = \omega^2 x_0 = \left(\dfrac{2\pi}{5.0}\right)^2 \times 4.0 = 6.3\,\text{cm s}^{-2}$.

Worked example 8

A particle of mass 4.0 g undergoes simple harmonic motion with frequency 25 Hz and amplitude 13 mm. Calculate, when the displacement of the particle is 10 mm:

a. the speed
b. the force acting on the particle.

Solutions

a. The angular frequency is $\omega = \dfrac{2\pi}{T} = 2\pi f = 2\pi \times 25 = 157\,\text{rad s}^{-1}$.

$v = \omega \sqrt{x_0^2 - x^2} = 157\sqrt{13^2 - 10^2} = 1300\,\text{mm s}^{-1} = 1.3\,\text{m s}^{-1}$.

b. $F = ma = -m\omega^2 x = -4.0 \times 10^{-3} \times 157^2 \times 10 \times 10^{-3} = -0.99\,\text{N}$.

Practice questions

9. The graph shows how the displacement x of a particle undergoing simple harmonic motion varies with time t.

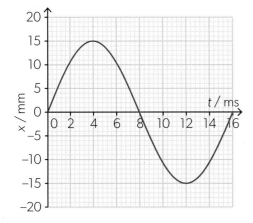

a. Identify the time when the particle has:
   i. the maximum negative velocity
   ii. the maximum positive acceleration.
b. Calculate the velocity of the particle:
   i. at $t = 10\,\text{ms}$
   ii. when $x = 5.0\,\text{mm}$ for the first time.
c. Calculate the maximum acceleration of the particle.

10. The velocity–time graph for an object undergoing simple harmonic motion is shown.

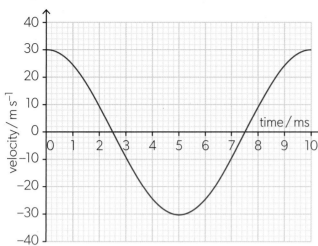

a. Identify the time at which the object has a maximum positive displacement.
b. Calculate the amplitude.
c. Calculate the displacement of the object at 4.0 ms.

11. An object of mass 100 g is suspended from a vertical spring of spring constant 7.8 N m^{-1}. The object is displaced by 12 cm vertically downwards from the equilibrium position and released.

a. Calculate the frequency of the oscillations.
b. Calculate the maximum speed of the object.
c. Calculate the speed of the object when it is 6.0 cm above the equilibrium position.

Modelling simple harmonic motion

- Tool 2: Use computer modelling.

The defining equation for simple harmonic motion is

$$a = -\omega^2 x$$

This can be written in differential form as $\dfrac{d^2x}{dt^2} = -\omega^2 x$ because acceleration is $\dfrac{d^2x}{dt^2}$. This second-order differential equation can be solved by calculus, by spreadsheet modelling or by using modelling software. This is one of many examples in physics of a simple second-order differential equation of the sort that you may meet in IB Diploma programme mathematics.

Simple harmonic motion is used as an example of modelling using a spreadsheet or modelling software in *Tools for physics* (page 363).

Phase angle and phase difference

So far, we have looked at simple harmonic motion that begins at particular positions in the motion, the extreme displacements when $x = x_0$ and at the centre of the motion when $x = 0$. Is it possible to produce an equation that allows for any starting point?

The simple harmonic motion equation is a second-order differential equation, and it can be shown that there are general solutions to this equation. One of these is

$$x = x_0 \sin(\omega t + \phi)$$

This resembles the earlier solutions, but has the addition of the single term ϕ. This quantity is known as the **phase angle** as before.

Look carefully at the displacement–time graphs for two simple harmonic motions in Figure 16. At the beginning of the graph, the blue curve shows a displacement of zero, but the red curve is just about to reach its maximum displacement. It is about one-eighth of a cycle ahead of the displacement. To be precise, the red curve is one radian ahead of the blue curve. As one cycle corresponds to 2π rad, the red curve leads the blue by $\dfrac{1}{2\pi} = \dfrac{1}{6.3} = 0.16$ of a cycle.

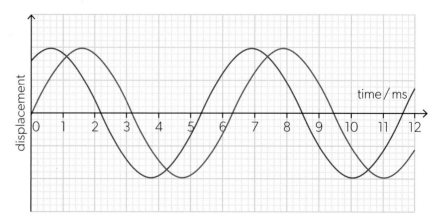

▲ Figure 16 Phase difference in simple harmonic motion. The blue curve lags behind the red curve because it reaches its peak at a later time.

Since the equation for the blue curve is $x = x_0 \sin(\omega t + 0)$, then the equation for the red curve must be $x = x_0 \sin(\omega t + 1.0)$.

The phase difference between the curves in Figure 16 can be modelled using the circular motions for both oscillations, as in Figure 17. Remember that both oscillations have the same ω and therefore travel around the circle at the same angular speed. The **phase difference** is the angle between the radial lines that are tracing out the simple harmonic motion as the blue tracing point chases the red point around the circle.

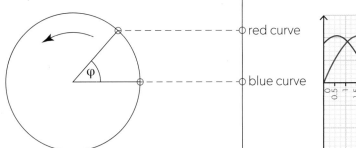 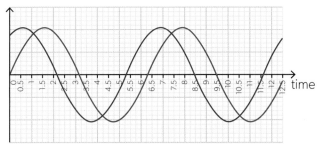

▲ Figure 17 A circular motion projected onto a line gives simple harmonic motion. The red point leads the blue point by one radian.

The equations for displacement, velocity and acceleration in full become:

- Displacement: $\quad x = x_0 \sin(\omega t + \phi)$

- Velocity: $\quad v = \omega x_0 \cos(\omega t + \phi)$

- Acceleration: $\quad a = -\omega^2 x = -\omega^2 x_0 \sin(\omega t + \phi)$

Worked example 9

The graph shows how the displacement x varies with time t for an object undergoing simple harmonic motion.

a. The displacement can be modelled with an equation $x = x_0 \sin(\omega t + \phi)$.
   i. State the value of x_0.
   ii. Calculate the value of ω.
   iii. Determine the phase angle ϕ.
b. Calculate the velocity of the object at $t = 3.0\,\text{s}$.

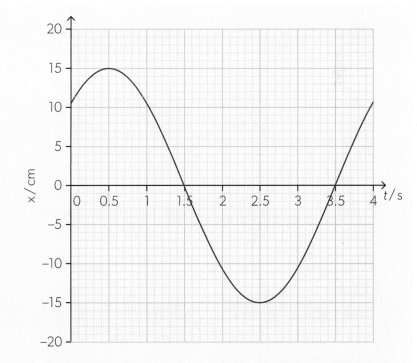

Solutions

a. i. $x_0 = 0.15\,\text{m}$.

   ii. The period of motion is $4.0\,\text{s}$.
   $$\omega = \frac{2\pi}{4.0} = 1.6\,\text{rad s}^{-1}.$$

   iii. The object is at the equilibrium position after $1.5\,\text{s}$. Had the oscillation started at $x = 0$, the object would have returned to the equilibrium position after $2.0\,\text{s}$, which is $\frac{1}{8}$ of the period later than it actually did. The phase angle is therefore $\phi = \frac{2\pi}{8} = \frac{\pi}{4} \approx 0.79\,\text{rad}$.

b. $v = \omega x_0 \cos(\omega t + \phi) = \left(\frac{2\pi}{4.0}\right)(0.15) \cos\left(\frac{2\pi}{4.0} \times 3.0 + \frac{\pi}{4}\right) = 0.17\,\text{m s}^{-1}$.

Worked example 10

The displacement x, in metres, of a particle undergoing simple harmonic motion is given by the equation $x = 7.5 \times 10^{-3} \sin(12t + 2.0)$, where t is the time in seconds.

a. Calculate the period of motion.

b. Calculate the velocity of the particle after $0.30\,\text{s}$.

Solutions

a. The angular frequency is $\omega = 12\,\text{rad s}^{-1}$. $T = \dfrac{2\pi}{\omega} = \dfrac{2\pi}{12} = 0.52\,\text{s}$.

b. $v = \omega x_0 \cos(\omega t + \phi) = 12 \times 7.5 \times 10^{-3} \cos(12 \times 0.3 + 2.0) = 7.0 \times 10^{-2}\,\text{m s}^{-1}$.

Practice questions

12. The graph shows the variation with time t of the displacement x of a particle undergoing simple harmonic motion.

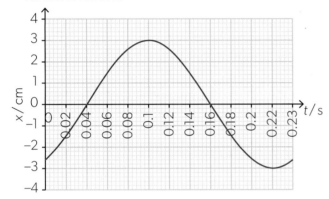

The oscillation can be modelled with an equation $x = x_0 \sin(\omega t + \phi)$.

a. Determine the values of x_0, ω and ϕ.

b. Calculate the maximum velocity of the particle.

c. Calculate the velocity and the acceleration of the particle after $0.08\,\text{s}$.

13. The displacement x, in cm, of a particle undergoing simple harmonic motion is given by $x = 12.0 \sin(0.500t + 1.00)$, where t is time in s.

a. Calculate the period of the oscillations.

b. Calculate the velocity at $t = 0$.

Energy transfer equations

The energy transfers between kinetic E_k and potential E_p that drive harmonic oscillators were described earlier. The simple harmonic motion equations can be used to derive a set of energy equations. Phase differences are ignored, but ϕ can easily be re-introduced into the equations when you need to.

The kinetic equation is related to $\dfrac{1}{2}mv^2$ as usual. The speed $v = \pm\omega\sqrt{x_0^2 - x^2}$ and therefore $v^2 = \omega^2(x_0^2 - x^2)$, so that

$$E_k = \frac{1}{2}m\omega^2(x_0^2 - x^2)$$

where m is the mass of the object undergoing simple harmonic motion.

Immediately, we can see that the total energy (which occurs when the object is moving at its fastest when $x = 0$) is

$$E_{tot} = \frac{1}{2}m\omega^2 x_0^2$$

Also, $E_{tot} = E_k + E_p$ and therefore $E_p = E_{tot} - E_k = \dfrac{1}{2}m\omega^2 x_0^2 - \dfrac{1}{2}m\omega^2(x_0^2 - x^2)$

which is equal to $\dfrac{1}{2}m\omega^2 x_0^2 - \dfrac{1}{2}m\omega^2 x_0^2 + \dfrac{1}{2}m\omega^2 x^2 = \dfrac{1}{2}m\omega^2 x^2$.

▲ Figure 18 E_k, E_p and E_T for simple harmonic motion.

The graphs of the variations of both E_k and E_p with displacement are parabolas. Figure 18 shows E_{tot}, E_k and E_p all plotted against displacement.

Notice that the displacement at which the kinetic energy and the potential energy are equal ($E_k = E_p$) is not at half the amplitude but closer to x_0 than the equilibrium point.

Worked example 11

A mass of 0.15 kg attached at the end of a weightless spring oscillates with simple harmonic motion. The mass passes through the equilibrium position with a speed of 1.4 m s⁻¹.

a. Calculate the total energy of the oscillating system.
b. The spring constant is 6.4 N m⁻¹. Determine the amplitude of the oscillations.

Solutions

a. At the equilibrium position, the potential energy is zero, so the total energy of the system is kinetic only. $E_T = \dfrac{1}{2}mv^2 = \dfrac{1}{2}(0.15)(1.4)^2 = 0.147$ J.

b. We can find the amplitude x_0 by rearranging the equation $E_T = \dfrac{1}{2}m\omega^2 x_0^2 \Rightarrow x_0 = \sqrt{\dfrac{2E_T}{m\omega^2}}$.

For a mass–spring system, we have $a = -\dfrac{k}{m}x$ and so $\omega^2 = \dfrac{k}{m}$.

We combine the equations to get $x_0 = \sqrt{\dfrac{2E_T}{k}} = \sqrt{\dfrac{2 \times 0.147}{6.4}} = 0.21$ m.

Worked example 12

The graph shows how the potential energy of an object executing simple harmonic motion varies with the displacement of the object. The amplitude of motion is 20 cm.

a. State the total energy of the oscillating system.
b. Estimate, using the graph, the displacement of the object when the kinetic and the potential energies are equal.
c. Sketch a graph showing the variation of the kinetic energy of the object with displacement.
d. The mass of the object is 2.6 kg. Calculate the maximum speed of the object.
e. Determine the period of the oscillations.

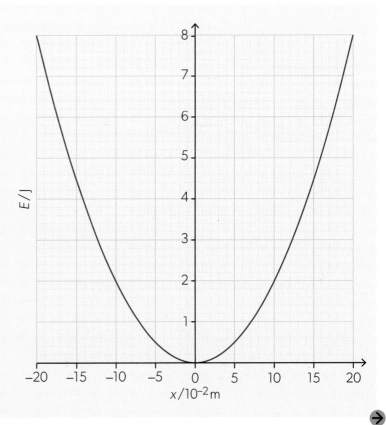

E / J

$x/10^{-2}$ m

Solutions

a.   8.0 J

b.   In this situation both E_k and E_p are equal to 4.0 J. From the graph, this happens when the displacement is approximately ±14 cm.

c.   The graph has a similar parabolic shape but is inverted compared with the potential energy curve.

d.   The maximum kinetic energy is 8.0 J, so the maximum speed can be calculated from

$$v = \sqrt{\frac{2E_k}{m}} = \sqrt{\frac{2 \times 8.0}{2.6}} = 2.5 \, \text{m s}^{-1}$$

e.   It is convenient to first find the angular frequency and then the period T.

$$E_T = \frac{1}{2} m\omega^2 x_0^2 \Rightarrow \omega = \sqrt{\frac{2E_T}{mx_0^2}} = \sqrt{\frac{2 \times 8.0}{2.6 \times 0.20^2}} =$$

12.4 rad s^{-1}.

From here, $T = \dfrac{2\pi}{12.4} = 0.51 \, \text{s}$.

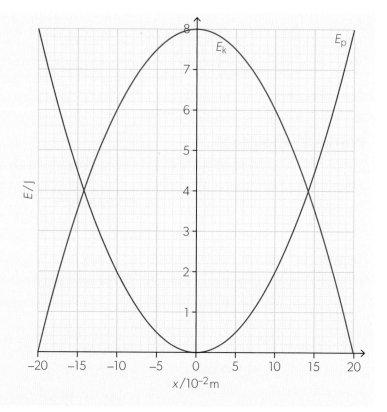

Practice questions

14.   An object of mass 0.060 kg undergoes simple harmonic motion with frequency 4.0 Hz and amplitude 0.25 m. Calculate, when the displacement of the object is 0.10 m:

  a.   the potential energy

  b.   the kinetic energy.

15.   The graph shows how the kinetic energy of an oscillating mass-spring system varies with the displacement of the mass from the equilibrium position. The mass is 0.70 kg.

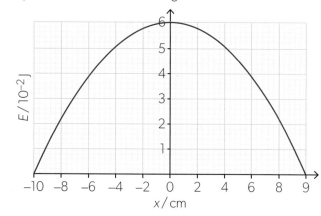

  a.   Calculate the maximum velocity of the mass.

  b.   Determine:

   i.   the period of the oscillations

   ii.   the spring constant.

16.   An object of mass 1.00 kg is attached to a spring with spring constant 4.50×10^2 N m^{-1} and is allowed to undergo simple harmonic motion on a frictionless horizontal surface. The object is initially displaced by 0.200 m and is given an initial velocity of 3.50 m s^{-1}. Determine:

  a.   the total energy of the system

  b.   the amplitude of the oscillations.

17.   An object oscillates simple harmonically with an amplitude x_0. When the displacement of the object is zero, the kinetic energy of the object is E. What is the kinetic energy of the object when the displacement is $\dfrac{x_0}{2}$?

  A. $\dfrac{E}{4}$ B. $\dfrac{E}{2}$ C. $\dfrac{3E}{4}$ D. E

It is useful to repeat the comments about energy variation with time from page 377 algebraically:

A substitution from the simple harmonic motion velocity equation gives

$E_k = \frac{1}{2}m(\omega x_0 \cos \omega t)^2$ which becomes $E_k = \frac{1}{2}m\omega^2 x_0^2 \cos^2 \omega t$.

Using $E_p = E_{tot} - E_k$ leads to $E_p = \frac{1}{2}m\omega^2 x_0^2 - \frac{1}{2}m\omega^2 x_0^2 \cos^2 \omega t$. This simplifies to $\frac{1}{2}m\omega^2 x_0^2 (1 - \cos^2 \omega t)$ and hence, using $\sin^2 \theta + \cos^2 \theta = 1$, to

$$E_p = \frac{1}{2}m\omega^2 x_0^2 \sin^2 \omega t.$$

The energy–time graphs in Figure 13 showed the relationships between E_{tot}, E_p and E_k, and remind you that the frequency of the energy change is double the frequency of the underlying simple harmonic motion.

Total energy E_{tot}	$\frac{1}{2}m\omega^2 x_0^2$	$\frac{1}{2}m\omega^2 x_0^2$
Potential energy E_p	$\frac{1}{2}m\omega^2 x^2$	$\frac{1}{2}m\omega^2 x_0^2 \sin^2 \omega t$
Kinetic energy E_k	$\frac{1}{2}m\omega^2 (x_0^2 - x^2)$	$\frac{1}{2}m\omega^2 x_0^2 \cos^2 \omega t$

▲ Table 2 The energy equations.

 How does damping affect periodic motion?

Strictly speaking, once resistive losses of any sort occur for an oscillating system, then the oscillation is no longer simple harmonic. The graphs for the variation with time of displacement/velocity/acceleration and the energy–time graphs have constant amplitudes, as there are no resistance or energy losses to reduce the amplitude. This is discussed in more detail in Topic C.4, where the effects of damping (friction) are described in detail.

This is an easy question to answer if you use modelling software, as shown in the section on modelling in *Tools for physics* page 362. In the Modellus X model used there, only one change is required to the first equation.

The term $-b \times vx$ must be added to represent a drag force that is proportional to the speed. The drag coefficient is b; vx is the velocity of the oscillating particle.

When b is set to 1.0, then the behaviour of the oscillating system changes to an oscillation that is damped. The amplitude decreases with time, and the motion eventually stops.

You can explore the effects of varying d if you set this model up for yourself. A particularly interesting case occurs with $b = 2.6$. This is **critical damping** and is examined in Topic C.4.

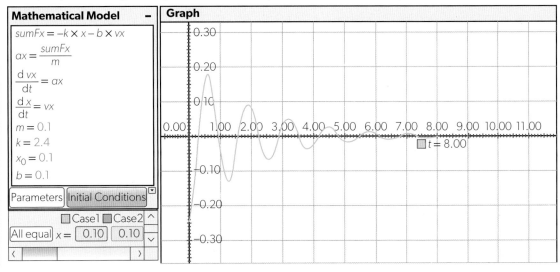

◀ Figure 19 Part of the ModellusX software screen running a model of damped simple harmonic motion and the outcome when the model is run. The graph is displacement against time.

Wave model

What are the similarities and differences between different types of waves?

How can the wave model describe the transmission of energy as a result of local disturbances in a medium?

What effect does a change in the frequency of oscillation or medium through which the wave is travelling have on the wavelength of a travelling wave?

Waves come in many forms. They are produced when an object undergoing a periodic motion interacts with a medium that can transmit a wave. This might be an oscillating electron forming an electromagnetic wave in a vacuum or the string and body of a violin generating sound waves in the air. A stone falling into a still lake produces an initial disturbance that leads to ripples travelling along the water surface. Gravitational waves, generated by the motion of an accelerated mass, have been detected. An understanding of wave theory is vital to a physicist.

Physics focuses on the similarities between different types of wave and wave motions to form general conclusions about wave theory. For example, all travelling waves transfer energy as they move through their medium. Another common feature is that the medium returns to its original state once the wave and its associated energy have passed through.

Finally, the need for a medium (or not, in the case of electromagnetic waves) provokes the question: how is the behaviour of the wave itself influenced by the properties of the medium? Can changes in density or constitution modify the properties of the travelling wave? These are all questions discussed in this chapter and the answers to them link all types of waves.

▲ Figure 1 Buzz Aldrin deploying a seismometer to measure seismic waves travelling through the Moon. These waves were caused by meteorite impacts and moonquakes. Such measurements enable scientists to investigate the internal structure of the moon.

In this topic, you will learn about:

- transverse travelling waves
- longitudinal travelling waves
- wavelength, frequency, time period and wave speed when applied to wave motion

- the nature of sound waves
- the nature of electromagnetic waves
- the differences between mechanical waves and electromagnetic waves.

Introduction

Earlier themes described the transfer of energy through mechanical and thermal means. This energy was transferred either through transfer of kinetic or potential energy, or because a temperature gradient leads to the transfer. In this topic, we study the transfer of energy due to wave motion.

This energy can be transferred by the following types of waves.

- **Mechanical waves**, which require a medium such as a fluid or a solid for propagation. The wave on the surface of an ocean or the movement of a slinky are examples of mechanical waves.

- **Electromagnetic waves**, which can travel through a vacuum or a medium. An electromagnetic wave is a pair of electric and magnetic fields that travel through space. Unlike mechanical waves they can only be detected when the changes in electric and magnetic field strengths cause charged particles in the path of the wave to move.

Describing waves

A wave can be defined as a disturbance that transfers energy from point to point in a medium. The disturbance can be an elastic deformation or a variation in any one of several parameters of the medium. For example, the sounds we hear result from pressure variations that travel to our ear as sound waves from the sound source. When a wave has finished travelling, the medium through which it moved is undisturbed. The energy has been transferred from one place to another without any overall change in the transmitting medium.

Figure 2 shows the passage of two different types of wave along a slinky spring. In both types of motion, the wave progresses along the medium. The medium is the spring in this case. A wave that does this is known as a **travelling wave**.

Figure 2(a) shows what happens when the end of the slinky is moved at 90° to its central axis. The hand provides the movement and therefore the energy. The spring takes a curved shape that moves along the spring coils from one end to the other. The motion may also be reflected at the far end. This type of wave is known as a **transverse wave**. This refers to the relationship between the direction in which the energy is transferred (along the slinky away from the hand) and the direction in which the individual particles (coils, in this case) of the spring are moving.

In a transverse wave the direction of energy transfer is at 90° to the direction in which the particles of the medium vibrate.

An individual coil in the slinky moves in two directions between a displacement maximum and minimum. One of these is called the **crest** of the wave; the other is known as the **trough**.

Figure 2(b) shows a different type of wave in which the end coil of the slinky is moved backwards and forwards along the central axis. In this case, the direction of energy transfer and the direction of coil movement are the same. This is a **longitudinal wave**. The coils are moving towards and away from each other causing the coils to be compressed together (called a wave **compression**) or moved apart (a **rarefaction**).

In a longitudinal wave the direction of energy transfer is parallel to the direction in which the particles of the medium vibrate.

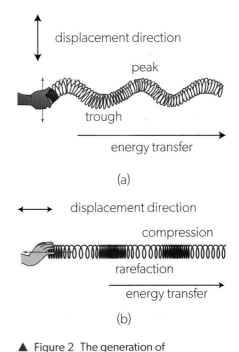

▲ Figure 2 The generation of
(a) a transverse travelling wave and
(b) a longitudinal travelling wave along a slinky spring.

Water waves and slinky waves

You might want to compare the transverse wave on the slinky with the ripples you see on a pond or lake when a stone is thrown into it (Figure 3). In both cases, the waves are obvious.

In fact, although the appearance of the water surface and the slinky wave are superficially similar, there are differences. The individual coil in the slinky is moving in one dimension. In the deep-water wave, the particles are moving in circles to give the characteristic "transverse" shape to the water surface.

Why do physicists generalize models that so that they can be used for related phenomena? In this case, waves that are transverse, longitudinal or some other type.

(a)

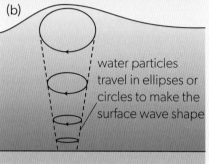

(b)

water particles travel in ellipses or circles to make the surface wave shape

▶ Figure 3 (a) A disturbance in a liquid leads to the propagation of waves on the surface. There are waves in the bulk of the material too, but these are less obvious. (b) Although for deep water the surface waves look like sine waves, they arise because the particles of the liquid are rotating in a vertical plane.

The individual coil in a slinky oscillates about a fixed equilibrium position. So similar vocabulary to that used to describe oscillations in Topic C.1 is used here:

- **Frequency** f is the number of oscillations per second formed by the wave source. This is also the number of crests that pass a point in one second. The unit is the hertz (Hz).

- **Wavelength** λ is the shortest distance between two points on the wave that have the same phase, that is, between two consecutive crests and two consecutive troughs.

- **Time period** T is the time that it takes one wavelength to pass a fixed point. This is the same as the time for one particle in the wave to undertake one cycle of the oscillation.

- **Amplitude** x_0 is the maximum displacement of a particle in the wave from its equilibrium (rest) position.

- **Wave speed** c is the distance a wave moves forward in one second in its direction of propagation.

Graphing wave motion

One way to visualize waves is to use graphs that show the variation of the wave displacement. These come in two forms:

- A **displacement–distance graph** show how displacement of the medium varies with distance along the wave *at one instant in time*. For a mechanical transverse wave, this is essentially a photograph of the **wave profile**.

- A **displacement–time graph** shows the variation with time of the motion *of one specific place in the medium*.

Always take care that you know which type of graph you are dealing with by looking carefully at the axes. You must also be aware of the type of wave (longitudinal or transverse) that is being described.

Displacement terminology

The word "displacement" must be treated with care in wave phenomena, as it may not refer to the physical distance moved by the wave medium. It can, for example, refer to the magnitude of an electric or magnetic field for electromagnetic waves.

Displacement–distance graphs

These graphs have displacement as the y-axis and distance along the wave on the x-axis. These axes are drawn as perpendicular to each other but, for a longitudinal wave, displacement of the medium and distance along the wave are parallel. Of course, for transverse waves, the displacement and distance along the actual wave will be at 90°.

Figure 4 shows the displacement–distance graph for a transverse wave together with the wave profile at the top of the diagram. The red and green dots show the positions of zero displacement at the instant in time for which the graph is drawn.

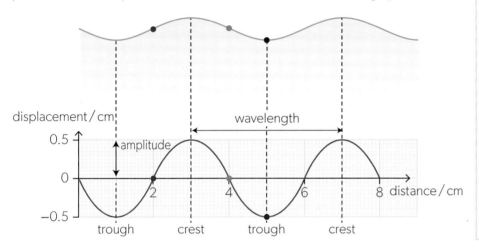

▲ Figure 4 A transverse wave profile.

You can read the amplitude and wavelength directly from the displacement–time graph. Notice that this graph cannot tell you about the time period or the frequency of the wave directly.

Figure 5 shows the displacement–distance graph for a longitudinal wave. The graph looks similar to that for the transverse wave, but the displacements now refer to movements of particles on the wave in the x-direction and this is shown in the top half of the figure. The top row of points represents the non-displaced (equilibrium) locations of positions in the medium. The lower row of points shows the disturbed positions of these points as a result of the wave moving through. Again, this is for one instant in time.

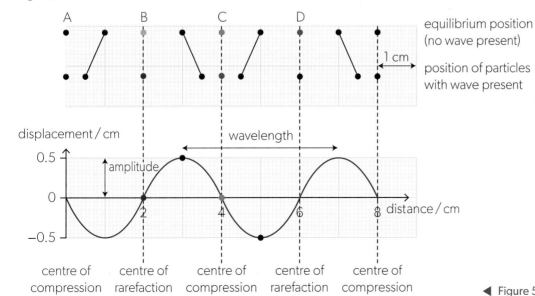

◀ Figure 5 A longitudinal wave profile.

How can the length of a wave be determined using concepts from kinematics?

Topic A.1 was concerned with the extension of space and time to include the concepts of speed and acceleration.

The disturbance that leads to a real wave will have definite starting and stopping positions in both space and time. This means that a non-infinite wave must have an overall length. This is the link between Topics A.1 and C.2.

When the disturbance that creates a wave goes through n cycles (or time periods), then the time for the disturbance is nT. The wavelength of the wave is λ so that the overall length of the wave is $n\lambda$.

To explain this further, look at the graph using the letters at the top of the diagram in Figure 5. Between A and B and between C and D the displacement is negative and, for this diagram, this means a displacement of positions to the left. You can see that the lower row of particles between A and B have indeed moved to the left compared with their undisplaced positions. Similarly, between B and C and to the right of D the particles are displaced to the right according to the graph.

This leads to interesting behaviour by the particles. At A, B, C and D the graph indicates that there is zero displacement, so these particles are at their equilibrium position at the time instant shown. At C, the particles are more bunched up than normal, so C is the centre of a compression. At B, the particles are more spread out than at equilibrium, so B is the centre of a rarefaction.

The compressions and rarefactions do *not* occur where the displacements are at their greatest and least. Both the compression and rarefaction maximum points are where the displacement is zero. At the maximum and minimum displacements, the distance between particles is the equilibrium separation when there is no wave present.

Of course, these compressions and rarefactions are not static places in the travelling wave because it is moving in space. But we cannot infer this from a single displacement–distance graph. To do that we need a series of wave profiles each drawn at a later time than the one before.

Figure 6 shows such a series of graphs for a wave at time intervals of one-quarter of the time period of the wave $\left(\dfrac{T}{4}\right)$. On each wave profile three positions on the wave are marked P, Q and R:

- P and Q are in anti-phase (180° or π rad out of phase)

- Q and R are 90° $\left(\text{or } \dfrac{\pi}{2}\, \text{rad}\right)$ out of phase

- this makes P and R 270° $\left(\text{or } \dfrac{3\pi}{2}\, \text{rad}\right)$ out of phase.

A straight line connects the same zero displacement point on the wave in each diagram. It shows how the wave will appear to move to the right, even though the individual points stay, on average, in the same place. In one time period T, this wave moves forward by 4 cm, so this wave travels with speed $\dfrac{4}{T}\,\text{cm s}^{-1}$.

 Simulations

There are many applets available on the web to help you visualize the motion of waves—both transverse and longitudinal. Search for words like " transverse wave applet" or "longitudinal wave motion applet" to find them.

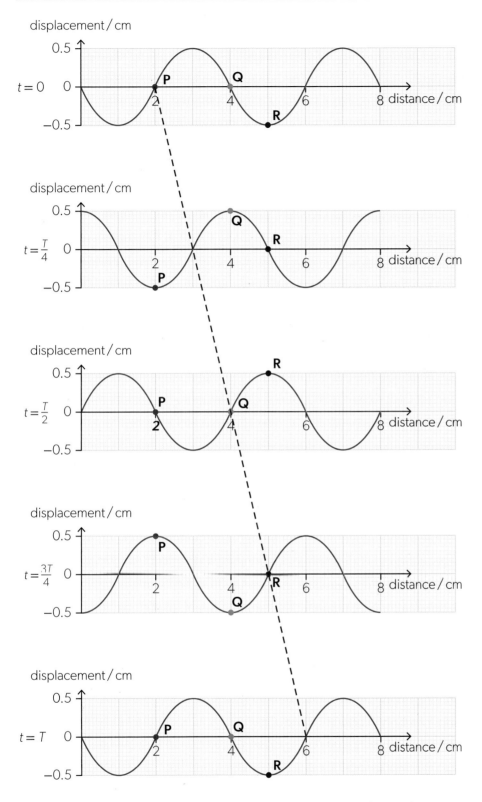

▲ Figure 6 A sequence of graphs of the variation of displacement with distance for a wave at intervals of $\frac{T}{4}$.

Worked example 1

A longitudinal wave travels in a medium from left to right. The diagram shows how the displacement of the particles in the medium at time $t = 0$ varies with the distance x along the wave. Displacements to the right of the equilibrium positions are positive. The point labelled P represents a particle of the medium whose instantaneous displacement is zero.

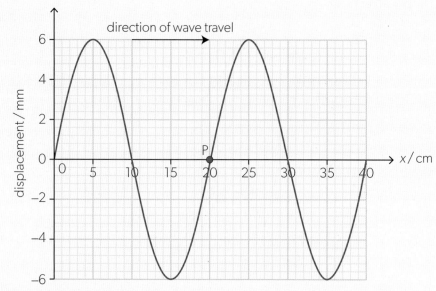

a. State the wavelength and the amplitude of the wave.

b. Explain the direction of motion of P at $t = 0$.

c. Explain why P is at a centre of a rarefaction.

d. Draw, on a copy of the diagram, a graph showing the displacement of the particles at time $t = \dfrac{T}{4}$, where T is the period of the wave.

e. State the displacement and the velocity of particle P at $t = \dfrac{T}{4}$.

Solutions

a. Wavelength = 20 cm, amplitude = 6 mm.

b. The crest of the wave immediately to the right of P will have moved further to the right, so the displacement of P is decreasing at $t = 0$. A decreasing displacement means that P is moving to the left.

c. All particles immediately to the left of P are displaced further left from their equilibrium positions, so away from P. All particles immediately to the right of P are displaced further right, also away from P. The density of the particles is therefore lowest at P.

d. Between $t = 0$ and $\dfrac{T}{4}$ the wave moves by one quarter of the wavelength to the right. The dashed line shows the displacement of the wave at $t = \dfrac{T}{4}$.

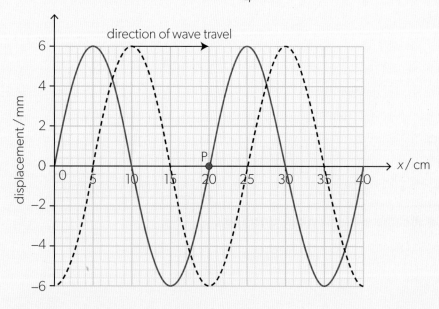

e. The displacement of P is now maximum negative, −6 mm. This is the turning point of the oscillation of P, so its instantaneous velocity is zero.

Displacement–time graphs

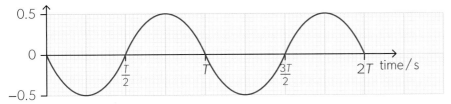

Figure 7 A graph of displacement against time.

A displacement–time graph shows the displacement of *one* particle in the medium as it changes with time. In Figure 7, this is over two complete cycles of the motion.

This graph gives us temporal as opposed to spatial information and gives the time period of the wave and the amplitude of the motion. These must be the same for all the particles (ask yourself why).

Worked example 2

A wave travels in a medium from left to right. The graph shows how the displacement of a particle P in the medium varies with time t.

a. Calculate the frequency of the wave.

b. Another particle Q is directly to the right of P. The equilibrium positions of P and Q are separated by one quarter of the wavelength. Draw a graph to show the variation with time of the displacement of Q.

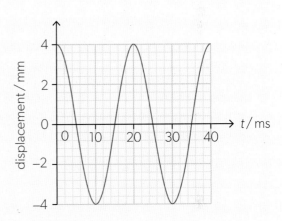

Solutions

a. The period of the wave is $T = 20\,\text{ms}$. The frequency is

$$f = \frac{1}{T} = \frac{1}{20 \times 10^{-3}} = 50\,\text{Hz}.$$

b. P and Q oscillate with the same amplitude and period. The phase difference is $\frac{\pi}{2}$ and the oscillation of P precedes that of Q by one quarter of the period. For any displacement of P, particle Q will be at the same displacement 5 ms later.

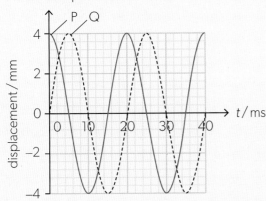

Practice questions

1. The displacement-position graph of a transverse wave travelling in a medium is shown. P, Q and R are three particles in the medium.

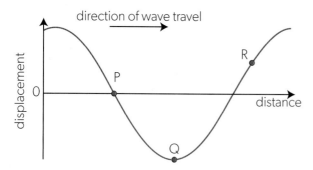

a. State, for each particle, whether the instantaneous velocity is positive, negative or zero.

b. Draw, on a copy of the diagram, a displacement–position graph a time $\frac{T}{2}$ later, where T is the period of the wave.

2. The graph shows how the displacement of the particles of a medium varies with distance. A longitudinal wave is travelling in the medium from left to right. A positive displacement corresponds to displacement to the right of the equilibrium position. P, Q, R and S are four particles of the medium.

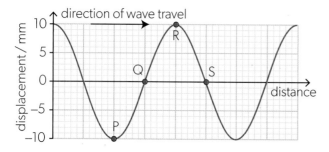

a. State the amplitude of the wave.

b. State and explain which of the particles is at a centre of a compression.

c. A quarter of the period later, the displacement of particle P is zero. State, for the same instant, the displacements of the particles Q, R and S.

3. A displacement-position graph of a travelling wave is shown. X and Y are two particles along the wave.

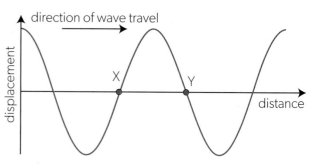

What is the phase difference between the oscillations of X and Y?

A. 0   B. $\frac{\pi}{4}$ C. $\frac{\pi}{2}$ D. π

4. A wave travels along a stretched string. A displacement–time graph is shown for two particles on the string.

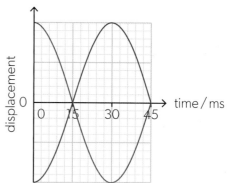

a. Calculate the frequency of the wave.

b. State the phase difference between the oscillations of the particles.

c. The wavelength of the wave is 60 cm. State the smallest possible distance between the particles.

Wave equation

Figure 6 shows that a travelling wave is moving forward to transfer the energy through the medium even though the particles remain, on average, at their equilibrium position. There is a simple relationship between the parameters of the wave and the wave speed v.

During one cycle, which takes time T (the time period), the wave moves forward by one wavelength λ. The wave speed must be given by:

$$v = \frac{\lambda}{T}$$

But $f = \frac{1}{T}$, therefore

$$v = f\lambda$$

This equation seems reasonable because wavelength has the dimensions of distance and frequency has the dimensions of time^{-1}, so the overall dimensions of $f \times \lambda$ are $\frac{\text{distance}}{\text{time}}$, which is the dimension for speed.

Worked example 3

A loudspeaker emits a sound wave of frequency 1200 Hz. The speed of sound in air is 340 m s^{-1}.

Calculate, for this wave:

a. the wavelength

b. the time it takes to travel a distance of 5.0 m from the loudspeaker.

Solutions

a. $v = f\lambda \Rightarrow \lambda = \frac{v}{f} = \frac{340}{1200} = 0.28$ m.

b. time $= \frac{\text{distance}}{\text{speed}} = \frac{5.0}{340} = 15$ ms.

Worked example 4

A transverse wave of wavelength 1.1 m is produced in a stretched string by periodically moving one end of the string up and down with a period of 0.60 s. Calculate the speed of the wave.

Solution

The period of oscillations of the end of the string is the same as the period of the wave.

$$v = \frac{\lambda}{T} = \frac{1.1}{0.60} = 1.8 \text{ m s}^{-1}.$$

Worked example 5

The graph shows how the displacement of a wave on the surface of a lake varies with distance. The red line shows the displacement at $t = 0$ and the blue line shows the displacement at $t = 0.25$ s. Between $t = 0$ and $t = 0.25$ s the wave has moved by less than one wavelength.

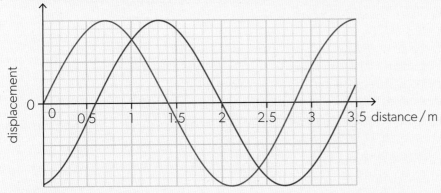

Calculate:

a. the speed of the wave

b. the period.

Solutions

a. The distance travelled by the wave during 0.25 s is 0.60 m. The speed is therefore $v = \frac{0.60}{0.25} = 2.4 \text{ m s}^{-1}$.

b. From the graph, the wavelength is 2.8 m. $v = \frac{\lambda}{T} \Rightarrow T = \frac{\lambda}{v} = \frac{2.8}{2.4} = 1.2$ s.

Practice questions

5. A wave travels along a stretched string at a speed of 30 m s⁻¹. A particular point in the string has a maximum displacement at $t = 0$ and returns to the equilibrium position at $t = 2.0$ ms. Calculate:

    a. the period of the wave

    b. the wavelength.

6. A sound wave of frequency 800 Hz is moving with a speed of 350 m s⁻¹. Calculate:

    a. the wavelength

    b. the minimum distance between a centre of compression and a centre of rarefaction.

7. A wave of frequency 5.0 Hz travels along a metal wire. There are 12 full wavelengths of the wave in a length of 48 m of the wire. Calculate the speed of the wave.

What happens when waves overlap or coincide?

A travelling wave is the result of disturbances travelling through a medium or, for electromagnetic waves, a self-propagating pair of electric and magnetic fields. For both cases, when two or more waves meet, the disturbances add to give the resultant sum of at the position and time of the meeting. This effect is called superposition and is examined in detail in Topic C.3.

The nature of sound waves

Sound waves are longitudinal waves that can travel in gases, liquids and solids. Gases cannot sustain a displacement at right angles to the direction of energy transfer since there is no restoring force; neither can liquids except at their surfaces. Solids can transmit both transverse and longitudinal waves because of the fixed bonds between the atoms and molecules.

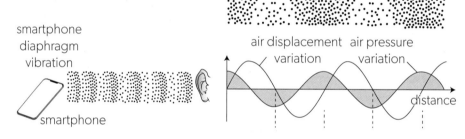

▲ Figure 8 How a longitudinal sound wave moves through the air. The regions of high and low pressure move towards the ear from the smartphone.

We use sound waves to hear. Figure 8 shows the progress of sound from a smartphone to a human ear. Inside the loudspeaker of the smartphone is a flexible material that moves backwards and forwards at the frequency of the sound. This produces high-pressure and low-pressure regions in the gas next to the cone. These pressure regions move away from the loudspeaker as a longitudinal travelling wave. The graph in Figure 8(b) shows the displacement and pressure variations. The amplitude of these waves will usually decrease with distance as the wave "spreads out" and as energy transfers to heat the air slightly. Note the $\frac{\pi}{2}$ rad phase difference between compressions and the corresponding displacement maxima on the graph. Maximum particle displacement corresponds to the average pressure in the gas. Where the particles are not displaced, the pressure has its maximum difference from the atmospheric (normal) pressure of the gas. The pressure varies above and below atmospheric once in one cycle (in one time period).

Eventually, the pressure variations reach the ear and are converted into electrical signals that travel to the brain to be interpreted as sound.

The human ear is a remarkable organ. Atmospheric pressure is roughly 10^5 Pa, but the ear can detect sounds that have pressure variations of only 2×10^{-5} Pa. Sounds become painful to hear at variations of about 20 Pa. The ratio of loudest pressure to quietest is, therefore, $10^6{:}1$.

Measuring the speed of sound

- Tool 1: Understand how to accurately measure time and length to an appropriate level of precision.

- Tool 2: Carry out image analysis and video analysis of motion.

- Inquiry 1: Demonstrate creativity in the designing, implementation or presentation of the investigation.

- Inquiry 3: Compare the outcomes of an investigation to the accepted scientific context.

There are several ways to measure the speed of sound in free air. "Free air" means that the air is not confined in a tube, which would decrease the value measured for the speed.

Using direct timing

To measure the speed of sound directly, you will need a loud sound which has a visual cue. A clapperboard can easily be constructed from two strips of wood and a hinge.

- Observe the clapperboard from a measured distance, ideally about 100 m.

- Measure the time between seeing the clapperboard close and hearing the sound. Videoing this might help you to make these measurements.

- Use the equation $\text{speed} = \dfrac{\text{distance}}{\text{time}}$ to calculate the speed of sound.

You could also use the echo off a large, flat wall to make your measurements.

Using a double-beam oscilloscope

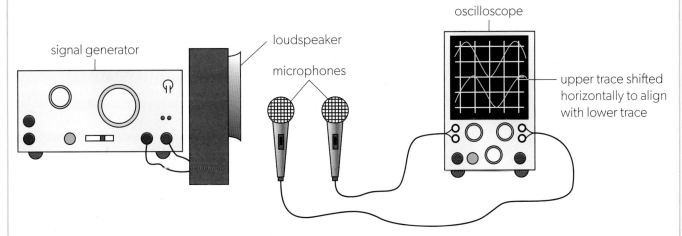

▲ Figure 9 A method for determining the speed of sound in air using a double-beam oscilloscope.

The apparatus (Figure 9) consists of two microphones connected to the inputs of a double-beam oscilloscope, and an audio-frequency signal generator connected to a loudspeaker.

- A frequency f of between 500 Hz and 2 kHz is usually suitable for the experiment.

- Begin with one microphone close to the loudspeaker and put the other microphone about 1.5 m away from it.

- Move the second microphone carefully along the line between it and the first microphone until the two traces on the oscilloscope display are aligned (in phase with each other). Mark the position of the second microphone.

- Now move the second microphone away until the display shows that the two traces are in phase again. Mark the new position.

- The microphone has moved through one wavelength so that the distance between the marks is λ. The speed of sound $v = f \times \lambda$.

- You can improve the experiment by measuring several distances over more than one wavelength and taking an average of the results.

Using smart phones

There are many other ways in which you could measure the speed of sound. Smartphone apps can help you to make measurements. For example, you could use the Doppler shift (see Topic C.5) to calculate the speed of sound by measuring the frequency of a sound when it is stationary and when it is moving at a constant speed.

Some smartphone apps have the facility to measure the speed of sound via the SONAR principle. In this case, the app itself will have instructions for how to do this.

There is another method. If you work with a friend and use two smartphones, then you can use them as timers. You will need an app that starts and stops timing when it is triggered by a sound. Search for "acoustic stopwatch".

- Place each smartphone a measured distance apart – 2 m will work, but larger distances are better (up to 10 m).

- Set each smartphone to start and stop timing on a loud clap. Banging two pieces of wood together, or hitting a hammer on a metal plate, makes a suitable loud and short sound.

- Start the smartphone timers by making the sound exactly in the middle of the two smartphones. This is so that the two timers start at the same time.

- Stop the timers by making the sound on the far side of one of the smartphones. The more distant smartphone should hear the sound a little later and record a slightly longer time.

- Calculate the difference in the two times.

- Repeat the experiment and take an average of the difference between the two times. You should take repeats, making the stopping sound on both sides.

- Calculate the speed of sound from your results.

- You can improve this method by making the starting sound on one side of the smartphones and making the stopping sound on the other. The effective distance over which the sound has travelled is doubled, as is the time difference recorded.

Practice questions

8. In an experiment to determine the speed of sound in air, a sound sensor is placed close to the open end of a tube of length 2.6 m. The other end of the tube is closed. A short sound pulse is emitted at the open end of the tube. The sensor records an echo returning from the closed end of the tube 14 ms after the emission of the pulse. Estimate the speed of sound in air according to this experiment.

9. A student designs an experiment to measure the speed of sound in air by timing a sound pulse travelling between two microphones that are 1.8 m apart. The microphones are connected to a digital timer. The time measurement is triggered by the arrival of a sound pulse at one of the microphones. The student can control the sampling rate of the timer (the number of sound samples it collects per second) in the range from 100 Hz to 5 kHz.

   a. Explain why a sampling rate of 100 Hz is insufficient for the purpose of this experiment. Assume that the speed of sound in air is about $340 \, \mathrm{m \, s^{-1}}$.

   b. The student claims that a percentage uncertainty of the calculated speed of sound will be less than 5% when a sampling rate of 5 kHz is used. Comment on the student's claim.

The nature of electromagnetic waves

While the wave theory developed in this topic applies to all types of waves, a distinction was made earlier between mechanical and electromagnetic waves. Figure 10 shows a glass prism dispersing white light. Our eyes are sensitive to only a small part of the whole electromagnetic spectrum.

Beyond the ends of this spectrum are many regions of both longer and shorter wavelengths. All these regions have the following properties in common:

- They are regarded as transverse waves.

- They can travel through a vacuum.

- They have an identical speed in a vacuum (3.00×10^8 m s^{-1}).

- They consist of a time-varying electric field at right angles to a time-varying magnetic field. The strengths of these fields have their maximum value at the same time (Figure 11).

- They arise from the motion of accelerated electrons (or other charged particles) or occur when charged particles change energy (and photons are emitted).

- Like all waves, electromagnetic radiation has a frequency, and hence a wavelength. However, this wavelength changes when the radiation enters a different medium whereas the frequency remains constant.

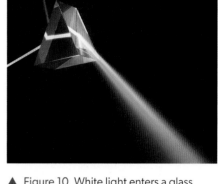

▲ Figure 10 White light enters a glass prism from the left-hand side. The light is dispersed and different wavelengths leave the prism in different directions.

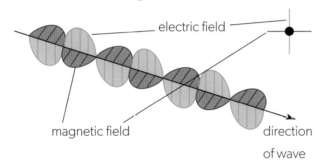

▲ Figure 11 Electromagnetic radiation consists of an electric field and a magnetic field at right angles. The fields are at their maximum field strength at the same instant.

 How can light be modelled as an electromagnetic wave?

The nature of light was hard for scientists to understand before the work of scientists such as Maxwell in the 19th century. Many models had been proposed ranging from the ideas of the Ancient Greeks, who proposed the idea of small bullets fired from the eye, to the corpuscular theory of Newton.

We now take a more ambivalent view of light—on the one hand, we sometimes describe it as a wave (the stance taken throughout Theme C) and, on the other hand, it is sometimes described as a photon (the Theme E view). Perhaps the nature of light is still difficult for scientists today?

All electromagnetic radiation is modelled as pair of electric and magnetic fields. One field gives rise to the other which takes us to the realm of Theme D with its emphasis on the field concept.

Photon or wave?

Throughout Theme C we treat electromagnetic radiation as a wave phenomenon. And this is how it has been regarded by scientists for most of scientific history. However, the 20th- and 21st-century view is different. Electromagnetic radiation can be shown to have particle-like properties as well as wave-like ones. When gamma radiation acts as a particle it is described as a photon. Photon properties are a matter for Theme E, where they are discussed in detail, but it is interesting to discover the history of this subject and to see how important the study of electromagnetic radiation has been in the development of 20th-century science.

To what extent do our perceptions of nature and science depend on past history?

 ## How are electromagnetic waves able to travel through a vacuum?

Topic D.3 shows that a point charge situated in a region of electromagnetic radiation will be accelerated by the electric field. This acceleration leads both to a further electric field (because the charge is being displaced from its starting position sinusoidally) and a magnetic field (because the charge is moving with a speed that varies sinusoidally).

However, it is less clear why an electromagnetic wave transmits through a vacuum. This question was a difficult one for physicists of the late-19th century, who even invoked a medium called the aether to explain how it happens. (The aether was later shown not to exist by Michelson and Morley, see page 166).

One way to avoid the problem is to say that electromagnetic radiation is carried not by a wave but by a photon. This opens the issue of wave–particle duality that is discussed in more detail briefly in this topic and in Topics E.1 and E.2. However, a consideration of electric field theory and the postulates of special relativity, in Topics D.2 and A.5, give a clue to what happens from a wave standpoint.

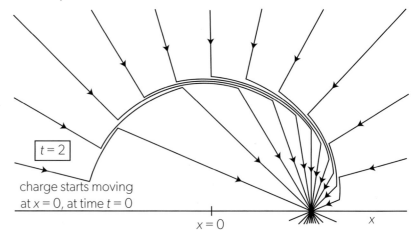

▲ Figure 12 A point charge that is moving away from its initial rest position at $x = 0$.

When a charge moves from an origin, the information about its motion must move away from it at the speed of light. At the points that this information has not yet reached, the field lines must still point to the origin. The distortion between the two regions the information has, and has not yet, reached is moving outwards at the speed of electromagnetic radiation. This gives rise to the electric and magnetic fields of an electromagnetic wave.

The spectrum itself together with the approximate wavelength ranges is provided in the Data Booklet. Note that there are sometimes considerable overlaps between regions (for example, between X-rays and gamma radiation). The distinction between one region and another is usually due to the mechanism by which the radiation is produced. For example, X-rays are produced by the rapid deceleration of electrons and the energy changes in the electron shells close to the nuclei of heavy metal atoms. Gamma radiation is usually considered to originate from nuclear changes, rather than atomic, even though X-rays and gamma radiation can, in principle, have similar wavelengths.

Patterns and trends

It is important to remember that the electromagnetic spectrum is continuous. There are no sharp boundaries between the regions of the spectrum. The range of wavelengths runs from radio waves, with wavelengths more than thousands of kilometres, to gamma rays with wavelengths less than a picometre (10^{-12} m). With such a wide range of wavelengths it is unsurprising that the properties of the waves change throughout the spectrum.

The different properties of the waves can be accounted for in the ways in which they interact with matter. Radio waves have low frequencies and are not absorbed easily, so can travel large distances. Shorter wavelengths such as microwaves and infrared waves can

▲ **Figure 13** Radio waves interact with matter very differently to visible light—as a result, the mirror on this radio telescope does not look like a mirror for visible light.

have frequencies that match molecular vibrations—enabling them to transfer energy to matter and to be absorbed in the atmosphere. Visible light frequencies cannot do this, so our atmosphere is transparent. UV light has sufficient energy to excite and perhaps ionise atoms—UV light is therefore absorbed by gases in our atmosphere. X-rays and gamma rays have wavelengths smaller than an atom which causes them to penetrate matter more effectively; however, they can interact with atomic electrons and nuclei.

Gamma radiation ($\lambda_\gamma \sim <1\,\text{pm}$)

- The shortest wavelength and highest frequency of the spectrum.

- Most gamma radiation produced outside Earth's atmosphere is absorbed by ozone in the atmosphere and does not reach the surface of Earth.

- Has many uses in medicine including imaging inside the body and treating cancer.

- The radiation kills bacteria and is used to sterilize food and medical instruments.

X-rays ($\lambda_X \sim 30\,\text{pm}–3\,\text{nm}$)

- Have many uses for internal investigation, such as in medical diagnoses and airport security.

- Used to irradiate and destroy cancers.

- Highly penetrating and dangerous to living tissues as they can alter DNA by ionization. The medical professionals carrying out X-ray work need to be protected by lead screening.

- Pulsars, supernovae and black-hole accretion discs emit X-rays given their very high temperatures.

Ultraviolet ($\lambda_{UV} \sim 100–400\,\text{nm}$)

- Important to animals because it is a source of vitamins when it illuminates the skin.

- Can also be harmful, causing sunburn and damage to DNA structures.

- Emitted by mercury atoms when excited in electric fields, which is the basis of the fluorescent tube.

- Satellites can be positioned above Earth's atmosphere to make galactic observations in the ultraviolet region.

Visible light ($\lambda_{VL} \sim 400$–700 nm)

- The part of the electromagnetic spectrum that we detect, with a wavelength from about 390 nm (violet) to 700 nm (red).
- The eye is most sensitive to green wavelengths.

Infrared ($\lambda_{ir} \sim 1$–1000 µm)

- Can be sensed by the nerves in the skin.
- Considerable amounts emitted by hot (but not glowing) objects; the basis of night-vision goggles and telescopes.
- Used for cooking and grilling (the red glow from an electric grill element is only a small part of the total power emitted by the grill).

Microwave radiation ($\lambda_{micro} \sim 1$ mm–30 cm)

- Used extensively in radar systems, radio astronomy, satellite communication, mobile phones and cooking, meriting their own named part of the spectrum.
- Microwave ovens heat up food because the molecules of water, sugars and fats in the surface absorb the radiation. The energy then spreads through the food through thermal transfer.
- High-power microwaves can damage living tissues through this heating mechanism.

Radio waves ($\lambda_{radio} \sim 1$ mm–100 km)

- Possess the longest wavelengths of all the radiations.
- Commonly used for communication given their unique properties of reflection from parts of the atmosphere and their ability to diffract around hills and large buildings.
- Radio telescopes are used to observe the emitted signals from objects in the Universe.

▲ Table 1 The regions of the electromagnetic spectrum.

How were X-rays discovered? (NOS)

Never underestimate the importance of serendipity in science.

In November 1895, Wilhelm Röntgen in Würzburg was experimenting with electrical discharges produced in a gas using high voltages. He noticed that a green fluorescence appeared on a nearby platinum–barium screen even when the container of gas was covered in cardboard. Something was leaving the gas that could penetrate the board. Rather than ignore this, he investigated the phenomenon intensively for six weeks until the end of December. By this time, he had already submitted preliminary findings to a scientific journal. By the end of January 1896, he had given a presentation to other scientists of what were to become known as "X-rays".

Within days, X-rays had their first medical uses around the world involving diagnosis of bone fractures. Within the year doctors were treating tumours by irradiating patients with X-rays—a procedure we now call radiotherapy. These techniques are still used extensively today. The basic strategies used to view internal structures, whether living or inert, are unchanged since Röntgen first suggested them, even though the methods have been much refined.

Wilhelm Röntgen was awarded the first Nobel Prize for physics. The prize is given to those who have conferred "the greatest benefit on mankind". It is completely appropriate that Röntgen should have received it for his far-reaching, life-saving and serendipitous discovery.

Can the wave model inform the understanding of quantum mechanics? (NOS)

Quantum physics—the subject of Topic E.2—takes as its starting point the premise that matter has a wave–particle duality. This means that, depending on the nature of the observation, matter can be observed to possess wave-like properties or particle-like properties. Electrons can be treated as though they are particles possessing mass, momentum and energy, or they can be made to diffract through a double-slit in the same way as light.

Erwin Schrödinger developed the concept of wave mechanics by describing the evolution of a particle using a wave equation. His differential equation, named the Schrödinger equation, is a key result in physics that represented a breakthrough in 1935. The equation is the quantum equivalent of Newton's second law. Newton's equation predicts the path of a physical system with time. Schrödinger's full equation predicts the way in which a wave function, which is a description of a physical system in quantum-mechanical terms, evolves with time. It was an example of the application of analytical reasoning to observed patterns and existing hypotheses.

Schrödinger's success in applying a wave model to quantum physics was recognized when he was awarded a Nobel Prize.

How are waves used in technology to improve society? (NOS)

It is impossible to overemphasize the importance to us of mechanical waves and electromagnetic radiation. The whole of life on the planet depends on the energy reaching us from the Sun—and this travels only by electromagnetic radiation through the vacuum of space. Nevertheless, electromagnetic radiation can also travel through solids, liquids and gases, depending on the wavelength of the radiation.

Since their discovery in the 19th century, electromagnetic waves have brought untold benefits. These range widely from their use in medicine for scanning and treatment, through industrial use and weather forecasting, to domestic applications such as the radar systems used in self-driven cars. Our mobile phones rely heavily on microwaves as do our ovens.

Mechanical waves such as sonar (sound reflected from an object in water) allow mapping of the seabed and fishery research. An ultrasound wave with its frequency greater than humans can hear (above about 20 kHz) can be used for flaw detection in materials. Ultrasound scans are used extensively in medical diagnosis and treatment. These and many other applications illustrate the improvements in society that have arisen from the use of wave technology.

Why does the intensity of electromagnetic wave decrease with distance according to the inverse square law?

The intensity of a wave is a measure of the energy transferred by the wave to an area. A point source S of energy (Figure 14) spreads the energy over a complete sphere is space. As the wave moves outwards, at any particular time the sphere has a radius r. This is discussed in greater detail in Topic B.1.

The **intensity** I of the wave is the $\dfrac{\text{the energy transferred per second}}{\text{the area of a sphere of radius } r}$

or $\dfrac{\text{the power of the wave } P}{\text{the area of a sphere of radius } r}$.

This means that $I = \dfrac{P}{4\pi r^2}$ because the area of a sphere is $4\pi r^2$.

Therefore $I \propto \dfrac{1}{r^2}$. The SI unit of intensity is W m⁻², as explained before.

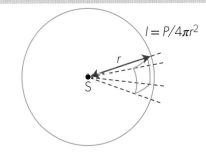

▲ Figure 14 As a wave moves from a point source, it is spread over a larger and larger sphere. The intensity of the wave varies with $\dfrac{1}{r^2}$.

Practice questions

10. A certain mobile network operator uses electromagnetic waves of frequency 1500 MHz to provide wireless services for its mobile phone users. Calculate the corresponding wavelength and identify the part of the electromagnetic spectrum to which these waves belong.

11. A laser pen emits monochromatic light of a wavelength 650 nm. Calculate the frequency of light.

12. The distance to the Moon can be determined by measuring the round-trip time of laser pulses emitted from Earth towards the Moon and reflected by one of the mirrors placed on the Moon's surface during space missions. In one experiment, the round-trip time of a pulse is 2.6 s. Estimate the distance between the surfaces of Earth and the Moon.

How are observations of wave behaviours at a boundary between different media explained?

How is the behaviour of waves passing through apertures represented?

What happens when two waves meet at a point in space?

To explain the behaviour of a wave moving between one medium and another, we need new concepts—the wavefront and the ray—to describe boundary behaviour.

The particles in a single wave close to an obstacle or gap must interact with the obstruction. The fixed edge of a boundary influences the movement of waves in the medium. Particles well away from the boundary are affected less. With so many particles in any wave, you can describe the overall effect by looking at the averages. The phenomenon is known as diffraction and is important in many areas of wave propagation. Figure 1 was taken in 1953 by Rosalind Franklin. Analyzing the diffraction pattern can enable deductions to be made about the structure which caused it and, later in 1953, Franklin's work led to the discovery of the double-helix structure of DNA by Francis Crick and James Watson.

When two or more waves meet at the same point in space, the individual particles of the medium must respond to both waves. Although the response is due to the individual interactions of many particles, it can be treated as a sum of

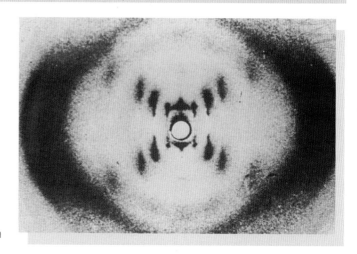

▲ Figure 1 The diffraction pattern created when X-rays pass through a crystalline sample of DNA.

the multiple waves. As in Theme B, we use a macroscopic lens to view what is a microscopic effect. This is the phenomenon known as interference where two waves superpose (add). Again, this has important applications both in science and elsewhere.

In this topic, you will learn about:

- wavefronts and rays
- reflection, refraction, and transmission of waves at a boundary between media
- diffraction around objects and through apertures
- Snell's law, critical angle, and total internal reflection
- superposition of waves and wave pulses
- coherent sources
- double-source interference

- constructive interference and destructive interference
- Young's double-slit interference

- single-slit diffraction
- the modulation of a double slit interference pattern by a single-slit diffraction pattern
- interference at multiple slits and diffraction gratings.

AHL

Introduction

You can move from studying the motion of individual particles to studying the general concepts of waves travelling in two or three dimensions. These are the **wavefront** and the **ray**. For some types of wave, wavefronts are visible and this is where we begin in the modelling of the wave phenomena. Rays help us to visualize the direction in which waves move—and the changes in direction that occur when waves move across boundaries between media or when they are constrained at the edges of a medium.

Wavefronts and rays

A water wave on a still lake spreads out in a series of ripples. They expand without limit until they reach an obstruction. There is implied direction in this motion because the ripples move away from the initial disturbance in the water. Figure 2 illustrates this. The boy is skimming a stone across the water surface. The stone is travelling away from the boy. As it hits the surface it causes waves. These expand outwards from the point where the stone touched the water. The water disturbance closest to him was made first and has expanded the most. The ripples are expanding outwards in circles, so that each point on the wave (ripple) is moving radially away from the centre of the disturbance.

These ideas lead to a way of thinking about wave motion that uses wavefronts and rays (Figure 3):

- **Wavefronts** are surfaces that move with the wave and are perpendicular to the direction of the wave motion. Consecutive wavefronts are imagined to be one wavelength apart.

- **Rays** are lines that show the direction of energy transfer by the wave. They are locally perpendicular to the wavefront.

▲ Figure 2 Skimming a stone on water.

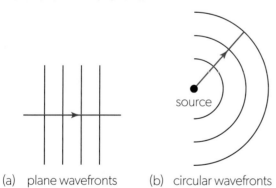

(a) plane wavefronts   (b) circular wavefronts

▲ Figure 3 Plane and circular wavefronts. The red line is a ray; the arrow indicates the ray direction. Sometimes the arrow is not shown on the ray as the direction is implicit from the diagram.

Figure 3(b) shows the relationship between wavefront and ray for circular waves (for example, water ripples) and plane waves. The ray is the straight line with an arrow indicating the direction. Wavefronts have no arrows and show the shape of the wave at successive times in the case of a single disturbance (a stone in a pond). They are essentially a photograph of the wave for the case of a source that keeps generating waves one after another (such as the ripple tank in Figure 4).

The series of wavefronts tells the story of the wave, its origin and subsequent history. For example, the larger the radius of curvature of a circular wavefront, the older the wavefront compared with other drawn wavefronts with smaller radii. This property is used later in the topic.

Ripple tank

- Tool 3: Determine the effect of changes to variables on other variables in a relationship.

- Inquiry 3: Compare the outcomes of an investigation to the accepted scientific context.

A ripple tank (Figure 4(a)) is a convenient way to show the properties of water waves as they move. The transparent tank carries a shallow layer of water illuminated by a single point source of light. The wave shapes on the water surface focus the light to produce a pattern of light and dark areas on a screen below. These areas correspond to the crests and troughs of the water surface. A variety of wave sources (single dippers, multiple dippers and plane dippers) can be used to simulate the behaviour of light for many effects discussed in this topic.

The ripple tank can also show effects of speed changes. The speed of the water waves in the tank depends on the depth of the water. Waves travel more slowly in a region of reduced depth (Figure 4(b)).

- As a wave travels to a shallower part of the tank, the friction between the water and the bed increases, so the wave slows down in shallower water.

- The frequency f cannot change in the shallower region (why not?) and $v = f\lambda$ tells us that the wavelength λ decreases when the speed v decreases. The wave peaks move closer together.

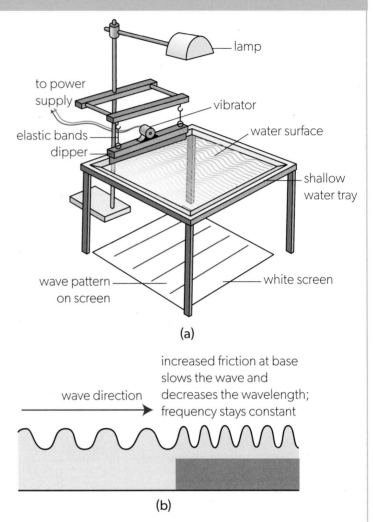

(a)

increased friction at base slows the wave and decreases the wavelength; frequency stays constant

wave direction

(b)

▲ Figure 4 (a) A ripple tank; (b) a wave moving into a region of reduced water depth.

Reflection, refraction and transmission

When a wave travelling in a medium meets a boundary, two outcomes are possible:

- All or part of the wave can be reflected back into the original medium with an unchanged wave speed.

- There can be **transmission** of all or part of the wave into the new medium beyond the boundary, usually at a changed wave speed.

It is the relative wave speeds that determine whether transmission or reflection occurs in the new medium.

When a wave is transmitted from one medium to another, the wave speed usually changes. The effect of this is shown for a water wave in Figure 5. Wave speed and wavelength are linked because the frequency of the wave as it moves across a boundary is unchanged. When speed decreases, wavelength decreases too.

Figure 5 shows ripple-tank images for plane waves that are:

- moving up to a plane metal strip (a barrier that simulates a mirror)

- moving from one water depth to a shallower water depth.

Ripple-tank pictures can be difficult to interpret and line drawings are included to give the important features of the images.

In Figure 5(a), there is no change in wavelength. The wavefronts have the same spacing both before and after they strike the barrier. Further, the **angle of incidence** i is the same as the **angle of reflection** r. Both i and r are measured from the line at 90° to the mirror—known as the **normal**—where the incident ray meets the mirror. The rule that $i = r$ is always true for reflection at a plane mirror.

Visualization

Figure 5 contains images created using a ripple tank. The tanks are an important practical way to model the behaviour of light waves when their wavefronts and rays are not visible to us.

What are the limitations in using these types of visualization in science?

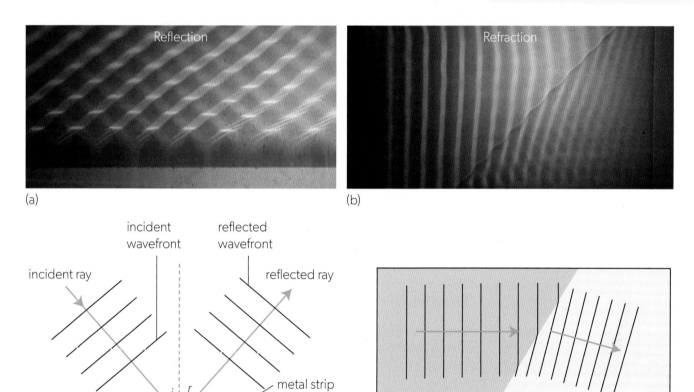

▲ Figure 5 (a) Reflection and (b) refraction shown in a ripple tank.

Figure 5(b) shows the wave travelling from the left and moving into a region where the water is shallower. The increased friction at the base of tank slows the wave down and two things happen:

- The wavelength becomes shorter in the shallower water.

- The wave bends at the interface between the two wave speeds.

This effect is known as **refraction**.

Huygens' principle

Can the future motion of a wavefront be predicted? In 1678, the Dutch scientist Christiaan Huygens thought so.

Huygens suggested that the prediction could be made by assuming that each point on a wavefront acts as a single new point source of circular waves. Each of these circular wavelets (little waves) expands independently. The new wavefront is the tangent of these circular wavelets and the wave itself moves forward. Figure 6 shows the principle operating for both plane wavefronts and spherical wavefronts. Only a small number of wavelets are shown here for clarity, but Huygens imagined many small wavelets for each wavefront.

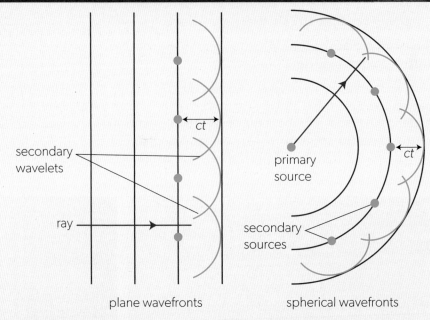

▲ Figure 6 Huygens' principle applied to plane and curved waves.

In fact, Huygens' model of wave propagation is limited. His model suggests that there must also be another part of the spherical wavefront that expands in the reverse direction, whereas light propagates in a rectilinear way (in a straight line). Also, Huygens' principle cannot be true because it requires a medium for electromagnetic radiation such as light. This requirement was refuted by Michelson and Morley in a famous experiment in 1887 (see page 166).

Does this make Huygens' model any less useful as a model? Can flawed reasoning ever lead to useful conclusions?

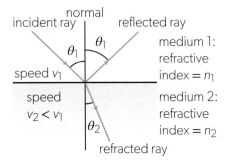

▲ Figure 7 Reflection and refraction of rays when they meet a boundary.

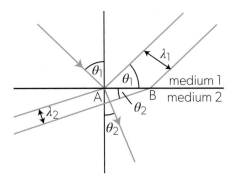

▲ Figure 8 Refraction as imagined by Huygens.

Snell's law

Snell's law, named after Dutchman Snellius, predicts that, when a wave passes from one medium 1 to medium 2 (Figure 8),

$$\frac{v_1}{v_2} = \frac{\sin \theta_1}{\sin \theta_2}$$

where v_1 and v_2 are the wave speeds in media 1 and 2, and θ_1 and θ_2 are the corresponding angles between the rays and the normal to the interface (see Figure 7).

This can be illustrated with Figure 8 which uses Huygens' principle. Two wavefronts (together with their respective rays) are shown moving from medium 1 to medium 2. The wavefronts are one wavelength apart. The wavefronts are different distances apart in the two media because the wave speed has decreased while the frequency of the wave has remained unchanged. The angles of incidence θ_1 and refraction θ_2 are also shown.

When light moves from a vacuum to a medium, then the speed of the wave in medium 1 (vacuum) is c and Snell's law becomes

$$\frac{\sin \theta_{vacuum}}{\sin \theta_2} = \frac{c}{v_2}$$

The ratio $\frac{c}{v_2}$ is known as the **absolute refractive index** n of the medium into which the light is travelling.

The absolute refractive index is a ratio of speeds, so it has no unit, and it is highly dependent on frequency. For the purposes of IB Diploma Programme physics, n is always greater than 1 for radiation travelling from a vacuum to another medium.

The adjective "absolute" is often dropped. The absolute refractive index of air is very close to 1 (1.00028 at $0\,°C$ and a pressure of $10^5\,Pa$). For all practical purposes, the speed of light in air is the same as the speed of light in a vacuum.

This also allows Snell's law to be written in a slightly different way. Because $\frac{\sin\theta_1}{\sin\theta_2} = \frac{v_1}{v_2}$ for two different media, then $\frac{\sin\theta_1}{\sin\theta_2} = \frac{v_1}{c} \times \frac{c}{v_2}$. But this is simply

$$\frac{\sin\theta_1}{\sin\theta_2} = \frac{n_2}{n_1}$$

which is a useful equation when light is leaving one medium of refractive index n_1 and entering another of index n_2.

The refractive index between two media, say from glass to water, is written as $_{glass}n_{water}$ To calculate this use

$$_{glass}n_{water} = \frac{1}{n_{glass}} \times n_{water}.$$

Models—What happens at the boundary interface?

The physics of refraction described so far is no more than a nod to empirical evidence with a link to wave speed. It tells us little about the processes that happen at the interface between, say, air and glass when light crosses it.

In fact, the energy from the light causes electrons in the material to vibrate. Unless the frequency of the wave matches one of the natural frequencies in the atomic systems of the glass, then the energy will be re-emitted. However, this takes time. The delay in the process of energy absorption and re-emission leads to an apparent reduction in the speed of the wave in the new medium. As a rule of thumb, the more atoms per unit volume, the slower the speed of the electromagnetic wave.

Theories

Figure 9 gives one of Huygens' original diagrams that shows refraction into a slower medium. The change of angle is clear in the diagram.

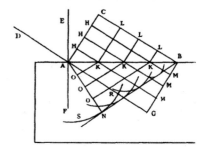

▲ Figure 9 Huygens' original treatise on his principle.

However, even though Huygens had much success in explaining these phenomena, it took over 100 years before his work was accepted over the alternative corpuscular theory of Isaac Newton. The English scientist had such a high reputation among his fellow scientists that his work was regarded as definitive until the later work of Thomas Young and Fresnel showed the merits of Huygens' approach.

ATL Research skills—Acknowledgement of the ideas of others

Willebrord Snellius or Ibn Sahl?

It is sometimes important to look beyond the conventional wisdom of the history of science to find the origins of a theory.

This was not an original discovery by Snell. There is a version of the rule for small angles by Ptolemy who lived in the 2nd century CE and a more comprehensive version by the Persian scientist Ibn Sahl who lived around 1000 CE (Figure 10).

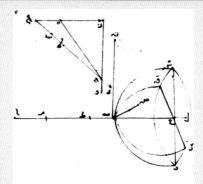

▶ Figure 10 A page from a manuscript by Sahl where he derives "Snell's law" 600 years before Snell.

Measuring the refractive index of a solid

- Tool 1: Understand how to measure angles to an appropriate level of precision.

- Tool 3: Carry out calculations involving fractions and trigonometric ratios.

- Tool 3: Record uncertainties in measurements as a range ($\pm$) to an appropriate precision.

- Inquiry 3: Compare the outcomes of an investigation to the accepted scientific context.

A simple approach to this measurement is to use a semicircular block of a transparent material such as glass or Perspex (Figure 11). (It is possible to measure n for a liquid using a hollow container with a semicircular shape—the technique is almost the same.)

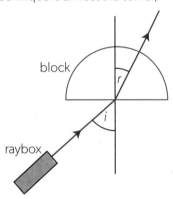

▲ **Figure 11** An experiment to measure the refractive index of a solid.

- Place the block on a piece of paper. Draw around the edges of the block so that you can remove and replace it quickly. Mark the position of the centre of the flat side.

- Use the ray box to send a beam of light into the centre of the flat side so that it leaves through the curved side.

- Mark points on the path of the beam so that you can remove the block and draw the beam positions. At least two points on each of the incident and refracted beams are required.

- Draw in the lines of the beam and use a protractor to measure the incident and refracted angles. It is helpful to construct the normal to the long side at the point where the beams meet (the centre of the side).

- Repeat for several different incident angles.

- You can calculate n from the data either by averaging each value of $\frac{\sin i}{\sin r}$ or by plotting a graph of $\sin i$ against $\sin r$ and calculating the gradient.

- Whichever method you choose, determine the uncertainty in your value of n.

The experiment can be repeated with the beam directed at the centre of the long side but entering through the curved side of the block so that the light is not deviated at its first refraction. This will enable you to establish that:

$$_{block}n_{air} = \frac{1}{_{air}n_{block}}$$

Worked example 1

A ray of light in air is incident on a glass surface at an angle of 65.0° with the normal.
The refractive index of the glass is 1.48.

a. Calculate the angle of refraction.

b. The wavelength of light in air is 532 nm. Calculate the wavelength in the glass.

Solutions

a. The refractive index of air is practically equal to 1. The angle of refraction θ_2 can be calculated

from Snell's law: $\frac{\sin \theta_2}{\sin 65.0°} = \frac{1}{1.48}$.

$\sin \theta_2 = \frac{\sin 65.0°}{1.48} \Rightarrow \theta_2 = 37.8°$.

b. The speed of light in the glass is $\frac{c}{1.48}$. Since the frequency is the same as in air, the wavelength in the glass is reduced by the same factor 1.48 as the speed.

$\lambda_{glass} = \frac{\lambda_{air}}{1.48} = \frac{532}{1.48} = 359$ nm.

Worked example 2

Travelling sea waves pass a boundary between shallow water and deep water. The diagram shows the direction of the waves and some of the wavefronts.

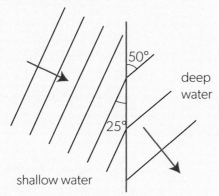

a. Explain what the diagram implies regarding the speed of the waves in shallow and deep water.

b. The speed of the waves in shallow water is $4.0\,\mathrm{m\,s^{-1}}$. Calculate the speed of the waves in deep water.

Solutions

a. The distance between the wavefronts, and therefore the wavelength, is greater in deep water. The period of the wave is unchanged, so the waves must have a greater speed in deep water in order to move by a longer distance during the same time.

b. The ratio of the wave speeds can be determined from Snell's law: $\dfrac{v_{\text{deep}}}{v_{\text{shallow}}} = \dfrac{\sin\theta_2}{\sin\theta_1}$, where θ_1 and θ_2 are the angles of incidence and refraction. The angle that a ray of a wave makes with the normal to a boundary is the same as the angle between a wavefront and the boundary. (You can draw a more detailed diagram to convince yourself why.) Therefore $\theta_1 = 25°$ and $\theta_2 = 50°$.

The wave speed in deep water is $v_{\text{deep}} = 4.0 \times \dfrac{\sin 50°}{\sin 25°} = 7.3\,\mathrm{m\,s^{-1}}$.

Practice questions

1. A glass slab has an absolute refractive index of 1.50. The slab is placed in water with a refractive index of 1.33.

   a. Calculate the relative refractive index of the glass slab with respect to water.

   b. A ray of light enters the glass slab from water. The angle of incidence is 25°. Calculate the angle of refraction.

2. A ray of light goes from air to diamond at an angle of incidence of 30.0°. The ray is refracted at an angle of 11.9°.

   a. Calculate the refractive index of diamond.

   b. The wavelength of light in air is 450 nm. Calculate the wavelength in diamond.

   c. Another light ray goes from diamond to air at an angle of incidence of 24.0°.

      i. Calculate the angle of refraction.

      ii. Draw a wavefront diagram to show how light travels from diamond to air.

Total internal reflection

In a ripple tank where the wave is moving into shallower water, the wave is refracted towards the normal (Figure 8).

The wave speed v_1 in the incident medium is greater than the wave speed v_2 in the refraction medium. As a result, the refracted angle is smaller than the incident angle and there is also a weak reflected ray, sometimes only just visible. As usual, this reflected angle is equal to the incident angle.

However, what about the case where the wave moves into a region where the wave speed is larger? This is the case for light going into air from another medium, as shown in Figure 12.

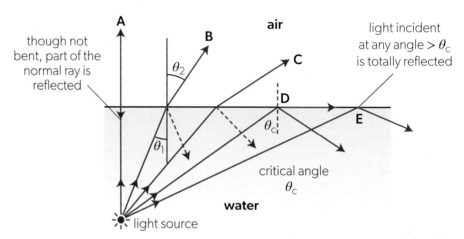

▲ Figure 12 Light passing from a more optically dense medium to a less optically dense medium.

A light source is placed in a medium such as water where the wave speed is slower than in air. The water is said to have a greater **optical density** than the air. Five light rays (**A** to **E**) are shown at various angles from the source through the water. All are incident on the boundary (the **interface**) between the water and the air.

- **Ray A** is incident on the interface with an incident angle of zero. Most of the light travels straight through with a small amount of reflection directly back into the water.

- **Ray B** has a small angle of incidence θ_1 with a larger angle of refraction θ_2 into the air. There is a weak reflection back into the water with the angle of reflection equal to the incident angle.

- **Ray C** is like ray B but now θ_1 is even larger and the refracted ray is deviated by a considerable amount from its original direction in the water.

- For **ray D**, the angle of incidence is so large that the angle of refraction is now 90° and it grazes the edge of the water as it leaves the interface. The angle of incidence for this condition is known as the **critical angle** θ_c. The reflection back into the water is now stronger than before.

- For all angles of incidence greater than θ_c, there is no refraction at all because the speed change is too great to allow propagation of the light (the wave) out into the air. Now there is only a strong reflected **ray E** that obeys the usual rules of reflection. This is known as **total internal reflection** because no light can emerge into the air. The word "total" is important because there is always some internal reflection at the boundary — even though it is usually weak. For angles of incidence greater than θ_c, a strong reflected ray is the only outcome.

 Light is reversible

- Tool 3: Select and manipulate equations.

- Tool 3: Derive relationships algebraically.

- Tool 3: Carry out calculations involving fractions and trigonometric ratios.

Imagine that the direction of ray B in Figure 12 is reversed. The light then enters the water from the air and traces out the original path but in the opposite direction.

So $\dfrac{\sin \theta_{air}}{\sin \theta_{water}} = {_{air}}n_{water}$. However, we know that for the original direction $\dfrac{\sin \theta_{water}}{\sin \theta_{air}} = {_{water}}n_{air}$.

Thus, ${_{water}}n_{air} = \dfrac{1}{{_{air}}n_{water}}$.

For any refraction,

$$\frac{\sin \theta_1}{\sin \theta_2} = \frac{\text{speed of wave in incident medium}}{\text{speed of wave in refraction medium}}$$

At the critical angle, $\sin \theta_1$ becomes $\sin \theta_c$ and $\sin \theta_2 = \sin 90° = 1$. This leads to

$$\sin \theta_c = \frac{\text{speed of wave in incident medium}}{\text{speed of wave in refraction medium}}$$

Worked example 3

Calculate the critical angle for light rays entering air from glass of refractive index 1.60.

Solution

We substitute $\theta_1 = \theta_c$ and $\theta_2 = 90°$ into Snell's law: $\dfrac{\sin 90°}{\sin \theta_c} = \dfrac{v_{air}}{v_{glass}} = \dfrac{n_{glass}}{n_{air}}$.

We know that $\sin 90° = 1$ and $n_{air} = 1$; hence $\sin \theta_c = \dfrac{1}{n_{glass}} = \dfrac{1}{1.60}$.

$\theta_c = \sin^{-1}\left(\dfrac{1}{1.60}\right) = 38.7°$.

Worked example 4

The diagram shows the direction of a light ray travelling from glass of refractive index 1.60 to water of refractive index 1.33.

a. Calculate the critical angle from glass to water.

b. Explain whether light emerges from the glass.

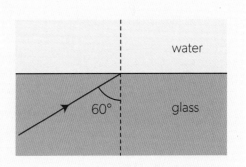

Solutions

a. The critical angle θ_c can be calculated from $\sin \theta_c = \dfrac{n_{water}}{n_{glass}} = \dfrac{1.33}{1.60}$.

$\theta_c = \sin^{-1}\left(\dfrac{1.33}{1.60}\right) = 56.2°$.

b. The angle of incidence is greater than the critical angle ($60° > 56.2°$). Hence the ray is totally internally reflected and no light emerges from the glass.

Practice questions

3. Water ice has a refractive index of 1.3.

    a. Calculate the critical angle for total internal reflection of light travelling from ice to air.

    The diagram shows the direction of a light ray entering an ice cube from air. The light ray makes an angle θ with the normal to side AD of the cube.

    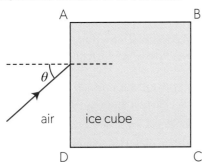

    b. Calculate, for $\theta = 40°$, the angle of refraction of the light ray from side AD.

    c. Explain why the light ray will be totally internally reflected from side AB.

    The angle θ is now increased.

    d. Determine, to the nearest degree, the minimum value of θ required for the light ray to emerge into air from side AB.

4. The diagram shows the direction of a sound wave travelling from air to water. The speed of sound in air is $340\,\mathrm{m\,s^{-1}}$ and the speed of sound in water is $1500\,\mathrm{m\,s^{-1}}$.

    Calculate the minimum value of the angle of incidence θ so that no sound enters the water.

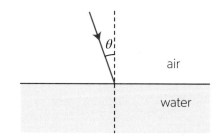

Principle of superposition

When waves of the same type meet at a point in a medium, their individual displacements add. The total displacement of the waves is the vector sum of the individual displacement of each wave separately. This is the **principle of superposition**.

To see this, it is best to use only two waves and to begin with pulses rather than continuous waves. Figure 13 shows what can happen.

Two waves A and B are moving towards each other on a rope. Time is increasing as you move down the diagrams. In Figure 13(a), the displacements of A and B are in the same direction. In Figure 13(b), B is opposite to A. As they arrive at the same point, the individual displacements add vectorially to give a double-height pulse in Figure 13(a). In Figure 13(b), there is a zero displacement for an instant as the waves meet. After superposition, the waves move apart. Try this with a lab partner and yourself at each end of the rope or slinky spring. It is quite hard to get the two pulses with exactly the same height but displaced in opposite directions, but when you succeed the effect is impressive.

When A and B are displaced in the same direction and superpose to give a double-height pulse, the system is showing **constructive interference**. When A and B cancel out because they are in opposite displacement (effectively they are π radians out of phase), the effect is called **destructive interference**.

When continuous travelling waves are involved, the principle still holds, as shown in Figure 14.

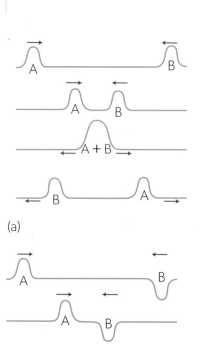

(a)

(b)

▲ Figure 13 Two identical pulses on a rope are initially approaching each other. In (a) the rope is displaced in the same direction for each pulse. In (b) the displacements are in opposite directions, so that at one instant in time there is no disturbance in the rope.

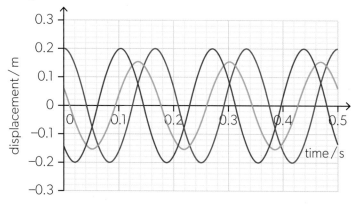

▲ Figure 14 The green wave is the superposition of the red and blue waves. The amplitude of the green wave is less than the two waves from which it is formed.

The red and blue waves are out of phase and superpose. The green wave indicates the variation of the sum of the displacements. This has an amplitude smaller than either of the original waves.

This example uses a displacement–time graph to make the point, but the principle also works with displacement–distance graphs.

If you have not met radian (rad) measure before, you can find a description of it in Tools for physics (page 335).

Practice questions

5. Two wave pulses are moving towards each other on a rope.

Which diagram shows a possible displacement of the rope when the pulses overlap?

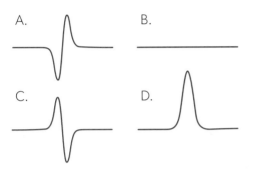

A.

B.

C.

D.

6. Two sound waves travel in a medium. The graph shows how the displacement due to each wave varies with the position x along the wave.

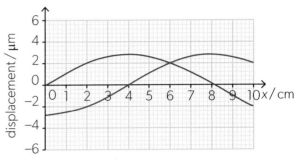

a. State the resultant displacement at $x = 0$.

b. Identify the position(s) along the wave at which the resultant displacement is

   i. zero   ii. maximum.

c. State the amplitude of the resultant wave.

d. Sketch, on a copy of the diagram, a displacement–position graph of the resultant wave.

Diffraction

Diffraction effects can be demonstrated using a ripple tank. Figure 15(a) shows a ripple-tank image of a wave passing the edge of a body, with a ray diagram to help you to make sense of the image. The wave can be seen to spread into the region beyond the edge as a "quarter circle" of circular waves with one of the new ray directions drawn in.

Figure 15(b) shows what happens when an obstacle (sometimes called a body) blocks the path of the wave. The wave continues on each side of the obstacle, but it also spreads into the space behind the obstacle as this is a double version of Figure 15(a).

Observations— Grimaldi and diffraction

The first detailed description of a phenomenon called diffraction was probably made by an Italian priest called Francesco Grimaldi whose work was published in 1665 after his death. (However, there is some evidence that the effect was observed by Leonardo da Vinci who had died roughly 150 years earlier.)

Grimaldi describes what happens when light is obstructed by a narrow rod. He noted that the shadow of the rod did not appear as expected from purely geometrical considerations. The light spreads out in the shadow region, and he also saw bright and dark fringes at the edges of the shadows. These were remarkable results for the time given the instruments that were available to him.

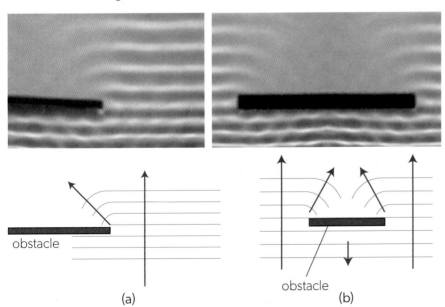

obstacle

obstacle

(a)

(b)

▲ Figure 15 (a) Diffraction at an edge. (b) Diffraction around a body.

Diffraction versus refraction

Be careful not to confuse *diffraction* with *refraction*. There is a change in direction in both cases, but refraction is always associated with a change in wave speed, leading to a change in wavelength. For refraction, only the wave frequency stays the same. For diffraction, it is the amplitude that varies along the wave.

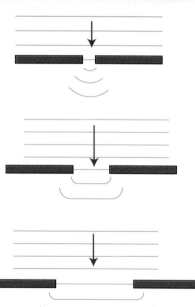

▲ Figure 16 Diffraction through an aperture.

Theories

Throughout the parts of this topic that deal with interference and diffraction, the assumption is that the wave is incident on the slits or the diffraction grating at normal incidence. In other words, the wave and the slit are parallel when they meet.

The theories that are developed here can be extended to cases where there is a non-zero angle between a slit or grating and the wave.

This gradual removal of assumptions as a theory becomes more and more developed is an essential part of the development of a scientific theory.

Figure 16 shows diffraction occurring when a wave is diffracted through an aperture (a gap), in the real case of sea waves entering a narrow inlet. The diagrams show the effect of changing the gap width.

Study of diffraction using ripple tanks and your observations of waves in real situations should convince you that:

- The wave speed, wave frequency and wavelength do not change when diffraction occurs (for example, sound waves can diffract around corners, but the frequency of the sound does not change when this happens).

- Diffraction effects are most obvious when the size of the aperture or obstacle is roughly the same as the wavelength of the diffracted wave. With obstacles much larger than the wavelength, diffraction is observed at the edges, but this is only a small part of the whole wavefront.

- Diffraction is always associated with a change in direction of at least part of the wave.

- The amplitude of the diffracted wave is less than the original wave because the energy is distributed over a larger wavefront.

Huygens' principle can be used to model diffraction. Figure 17 shows a plane wave moving towards an aperture from the top of the diagram. The original wave reaches the aperture. Here it is redrawn as six point sources shown as orange dots. Each source generates its own set of circular waves on this 2D surface. In 3D, these would be expanding hemispheres. The wavelets from each point source combine in the straight-on direction to give a straight wave. At the edges, the Huygens' construction suggests that we should only see the curved parts of the sources at the edge of the aperture. The resulting construction looks remarkably like the diffraction shapes that occur in practice.

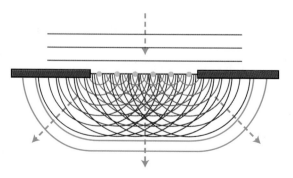

▲ Figure 17 Diffraction as imagined using the Huygens' construction.

Global impact of science—Diffraction by a mountain

Diffraction is of importance beyond the confines of the physics lab. It has real-world implications. One of these is in radio and television reception.

The waves used in some radio transmissions have long wavelengths, of the order of kilometres. These waves will be diffracted by objects of about the same size as the wavelength, in other words, something of the size a hill. It is common for a radio signal to be detected in a valley on the other side of a mountain, even though there is no line of sight to the transmitting antenna from the radio (Figure 18). Shorten the wavelength and there is no reception because the diffraction at this frequency does not allow the waves to reach the valley floor.

The effect can also be observed with sound. Every time you hear sound "around a corner" the waves must have diffracted at the edge of the building or feature to reach your ear.

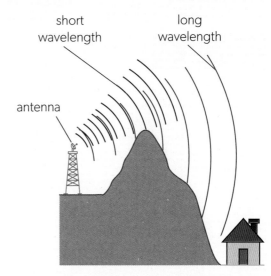
▲ **Figure 18** Diffraction by a natural feature. In this case a mountain top.

Diffraction of light
Single slit diffraction of light

Diffraction patterns (the variation of wave intensity across the wave) can be observed when visible monochromatic light passes through a very small aperture or slit. The experimental arrangement to view these patterns is shown in Figure 19.

Observing diffraction

- Tool 1: Recognize and address relevant safety, ethical or environmental issues in an investigation.

You can use laser pen light and a single slit to produce a diffraction pattern on a suitable screen, as in Figure 19.

Take care not to shine the laser beam or a reflection of the beam into your eye or anyone else's eye. It is usual to keep high light levels in the laboratory to keep the iris of the eye partially closed.

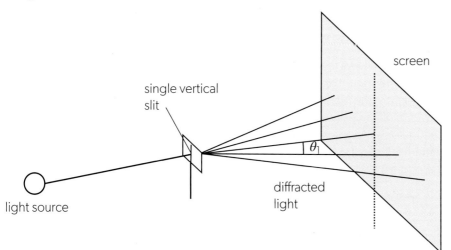
▲ **Figure 19** Producing a diffraction pattern.

Monochromatic light contains only one wavelength and can be produced using a laser. Alternatively, a white light source with a filter will give a narrow range of frequencies that approximate to one colour. The light is incident on a single vertical slit. The width of the slit will need to be small. Remember that diffraction is best seen when the gap and the wavelength match. The light is diffracted by the slit and the diffraction pattern can then be observed using a screen (or more permanently using a camera).

▲ Figure 20 Two diffraction patterns produced by the same slit but with different wavelengths. Blue light has a shorter wavelength than red and the diffraction pattern formed with blue light spreads out from the centre less than the red pattern.

When studied in detail, the diffraction pattern due to a single slit shows an intense (bright) central maximum. On each side of this are less intense areas of colour with dark areas of zero intensity (called minima) between them. Figure 20 shows the pattern produced by the same slit with two colours: red (long wavelength light) and blue (short wavelength light). The images show that:

- the red diffraction pattern is spread out more than the blue pattern
- the red central maximum is broader than the blue central maximum
- the minima are in different places for the two colours.

Effect of slit width on the pattern

The width of the diffracting aperture or obstacle is crucial to the appearance of the diffraction pattern. As the gap or obstacle become narrower, the pattern spreads out. Figure 21 shows two graphs to illustrate what happens.

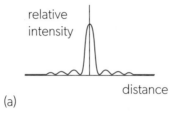

(a)

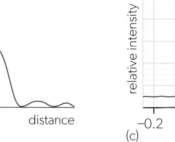

(b)

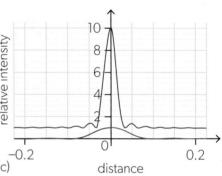

(c)

▲ Figure 21 Diagrams of the diffraction patterns due to (a) a wide slit and (b) a slit that is one-third of its width. The graphs have been normalized. (c) shows the two patterns to scale with the lines displaced vertically for clarity.

Figure 21(a) shows the diffraction pattern due to a wide slit. When the slit becomes one-third of the original width (Figure 21(b)), with the same wavelength, the diffraction pattern expands so that it is three times wider. Fewer minima can now be seen as they are more spread out. There are other changes too. As the slit is one-third of the original width the amplitude of the pattern will also be one-third. This means that the intensity is only one-ninth of the original height.

The reason why this factor of $\frac{1}{9}$ arises is because, as $E_T \propto x_0^2$ and intensity is the energy of a wave per unit area, intensity must be proportional to amplitude^2.

Diffraction patterns in more detail

Figure 22 links the intensity variation of the diffraction pattern for a single slit to the screen observations that are made.

Drawing diffraction patterns

- Tool 3: Sketch graphs, with labelled but unscaled axes, to qualitatively describe trends.

- Inquiry 2: Interpret diagrams, graphs and charts.

It is important to be able to sketch the intensity–position graphs of diffraction patterns quickly and accurately.

- The central maximum is twice as wide as the secondary maxima—which have the same angular width as each other.

- There is a large difference between the intensity of the central maximum and the intensities of the secondary maxima. The first maximum is only about 5% of the height of the central maximum (though you may need to reduce this difference to get all the detail into your sketch). The second and third maxima are even smaller at 2% and 1% of the central maximum. This is another reason why diffraction patterns are not usually obvious in everyday life.

The x-axis shows the angle rather than distance along the screen. This means that the graph does not depend on the slit–screen distance. The angle θ_1 is marked on Figure 22 as well, so that you can relate the diagrams to each other.

The three minimum positions on each side of the central maximum make angles θ_1, θ_2 and θ_3 with the direction of the ray coming from the light source.

Modelling single-slit diffraction

The positions of the diffraction minima in the single-slit pattern can be modelled using the ideas of destructive interference and superposition developed earlier in this topic. The model is shown in Figure 23(a).

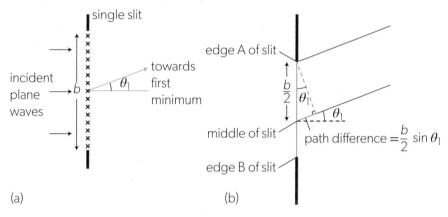

(a)   (b)

Incident plane waves are incident on a single slit of width b. The waves arrive parallel to the plane of the slit (the rays are at 90° to this)—known as **normal incidence** as in reflection and refraction. (You will not be asked to consider incident waves from any other direction in questions about diffraction.)

Using Huygens' principle, the wave at the slit is divided into many individual wavelets which expand as circular waves (the point sources are shown as × in the diagram). Figure 23(b) shows the geometry of the arrangement.

We will model the first diffraction minimum. Consider the first Huygens point source at A—the edge of the slit. There is another point source exactly halfway between A and B. Suppose that the point source at A and the point source halfway down AB interfere destructively in the direction θ_1, which is the angle between the centre of the pattern and the first minimum. Every point source in the top half of the slit has a counterpart in the bottom half of the slit. Every source pair cancels out to give zero intensity at the first diffraction minimum.

This can only happen for the first diffraction minimum at position θ_1. Each pair of wavelets arrives exactly π out of phase with each other. For this to occur, the light from the bottom source of each pair must travel half a wavelength $\frac{\lambda}{2}$ further than the light from the upper source. This is the **path difference** between the two waves. This path difference is shown for the first pair in Figure 23(b). It is the small extra distance that the light from the bottom source must travel to "catch up with" the light from the top source.

The geometry of the arrangement suggests that this distance is also equal to $\frac{b}{2}\sin\theta_1$, where b is the slit width. Equating these algebraically gives $\frac{b}{2}\sin\theta_1 = \frac{\lambda}{2}$ and therefore $\sin\theta_1 = \frac{\lambda}{b}$. The angles in diffraction experiments are usually small so the approximation $\sin\theta_1 \approx \theta_1$ holds.

The angular position of the first diffraction minimum is given by $\theta_1 = \frac{\lambda}{b}$.

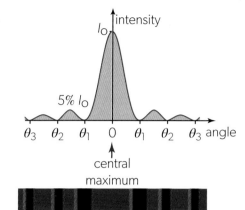

▲ Figure 22 Variation of intensity with angle for the diffraction pattern of a single slit.

◀ Figure 23 A simple model for the position of the minima in the diffraction pattern due to a single slit.

Diffraction at a single slit

- Tool 1: Recognize and address relevant safety, ethical or environmental issues in an investigation.

- Tool 1: Understand how to accurately measure length to an appropriate level of precision.

- Inquiry 2: Carry out relevant and accurate data processing.

- Inquiry 3: Interpret processed data and analysis to draw and justify conclusions.

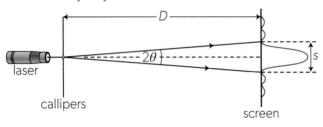

laser

callipers

screen

▲ Figure 24 An experimental arrangement for a diffraction experiment.

A pair of vernier callipers is a convenient instrument to form a single slit. Alternatively, you can rule a single slit with a pin in an opaque film of colloidal graphite that has been painted onto a microscope slide.

Throughout this experiment you should take precautions to avoid looking at the laser light.

- Shine a beam of light from a laser pointer onto the gap or slit. The light should then illuminate a screen (Figure 24). The distance from slit to screen should be of the order of metres.

- Adjust the slit to get a clear diffraction image and then use a metre ruler to measure the separation of the first minima on the screen. This is s.

- Now measure the distance from the slit to the screen. This is D.

- You can calculate 2θ from s and D and hence use $\theta_1 = \frac{\lambda}{b}$, where b is the width of the slit.

- If you used a slit that you drew yourself, you will need a microscope and a scale to measure the slit width.

This experiment can be extended to measure the wavelengths of various laser colours or to measure θ for various slit widths to obtain λ from the gradient of a graph of θ against $\frac{1}{b}$.

Worked example 5

A student investigates the diffraction pattern of monochromatic light incident on a single slit. The graph shows how the intensity of the light on the screen varies with the distance x along the screen.

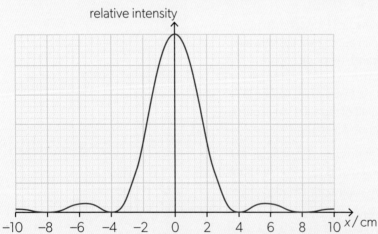

a. Explain, with reference to the principle of superposition, how the intensity minima are formed.

The screen is at a distance of 2.5 m from the slit. The wavelength of the light used in the experiment is 532 nm.

b. Determine the width of the slit.

c. Outline how the width of the central intensity maximum changes when:

  i. the distance between the screen and the slit is decreased

  ii. the light source is replaced by one of longer wavelength

  iii. the slit is replaced by one of greater width.

→
Solutions

a. The slit can be thought of as a series of point sources of light, and the diffraction pattern is modelled as a result of superposition of individual waves emitted by these sources. At an intensity minimum, there is a phase difference of π between the waves emitted by pairs of sources in the opposite halves of the slit, which results in fully destructive interference.

b. The distance s from the centre of the pattern to the first diffraction minimum is about 4.0 cm. This is much less than the distance D from the slit to the screen. Hence the angular position of the first minimum is approximately given by

$$\theta \approx \frac{s}{D} = \frac{4.0}{250} = 0.016 \text{ rad. The slit width is therefore } b = \frac{\lambda}{\theta} = \frac{532 \times 10^{-9}}{0.016} = 3.3 \times 10^{-5} \text{ m} = 0.033 \text{ mm}$$

c. i.   The angular positions of the minima are unchanged. Since the linear width of the central maximum is proportional to the distance from the slit, the central maximum becomes narrower when the screen is moved closer to the slit.

   ii. From $\theta = \frac{\lambda}{b}$, a longer wavelength λ for a given b results in a larger angle of the minimum. The width of the central maximum increases.

   iii. With a greater slit width b and unchanged wavelength, the minima occur at smaller angles and the central maximum becomes narrower.

Practice questions

7. Monochromatic light of wavelength 640 nm is incident normally on a thin slit of width 0.08 mm. A diffraction pattern forms on a screen at a distance 1.2 m from the slit. Calculate:

   a. the angle of the first diffraction minimum

   b. the width, in mm, of the central maximum of the diffraction pattern.

8. The diagram shows the diffraction pattern for monochromatic light of wavelength λ passing through a single slit.

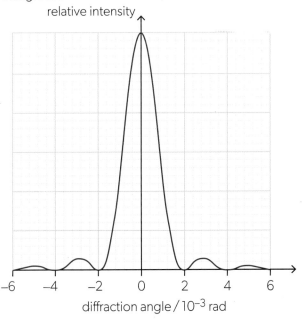

relative intensity

diffraction angle / 10^{-3} rad

What is the width of the slit?

   A. 125λ B. 250λ C. 500λ D. 1000λ

9. Monochromatic light passes through a narrow slit and the resulting diffraction pattern is observed on a screen. Which of these changes will increase the angular separation between the diffraction minima?

   A. Increasing the distance between the slit and the screen.

   B. Using a narrower slit.

   C. Replacing the light source by one of shorter wavelength.

   D. Increasing the intensity of the light.

Modelling diffraction patterns

The full equation for the intensity I_θ of a single-slit diffraction pattern at an angle θ to the centre of the pattern is:

$$I_\theta = I_0 \left(\frac{\sin \beta}{\beta} \right)^2$$

where $\beta = \dfrac{\pi b \sin \theta}{\lambda}$ (with the usual notation).

The quantity β is half the phase difference between the top and bottom of the slit.

The diffraction intensity graphs in Figure 21 were modelled using this equation. You can easily set up this equation using graphing or modelling software such as *Mathematica*, *Desmos* or *GeoGebra* to model the changes that occur when the parameters of the diffraction pattern equation are modified.

Double-slit interference

You met the principle of superposition on page 416. Another example of the principle of superposition in action is the interference of two or more waves.

When two waves of the same type meet at a point, they will always interfere. However, for the effect to be observable, the two waves need to have a constant phase relationship over a long enough time for the observation to be made. This property is called **coherence**.

Interference between two sound waves is easy to demonstrate. Figure 25 shows two loudspeakers driven by the same signal generator so that they emit sound of the same frequency in phase. As a microphone moves along a straight line parallel to the line joining the loudspeakers, the sound detected alternates between loud (L) and soft (S) regions. These L–S distances are equally spaced.

At points L, the waves from both loudspeakers arrive in phase with each other.

At points S, the wave from one loudspeaker arrives exactly 180° (π rad) out of phase with the wave from the other loudspeaker.

Identical, in-phase signals are emitted by both loudspeakers. This is an important feature of the experiment. The two sets of waves from the loudspeakers are said to be **coherent**.

- The phase between the emitted signals must not change for the interference to be observed consistently in one place. This implies that the emitted signals must also have the same frequency.

- Coherence also requires that the waves must exist at the same position in space simultaneously. Interference cannot occur when the identical signals arrive from both loudspeakers, but one sound has ended before the other begins.

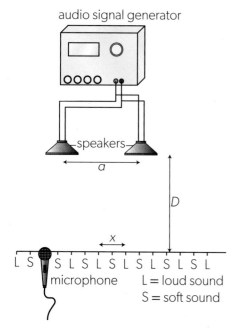

▲ Figure 25 An example of interference with sound waves.

Double-slit interference with light

Like sound, light waves and microwaves also interfere. One way to observe interference is to use a single lamp or laser to illuminate two thin vertical slits. The arrangement is shown in Figure 26. The experiment was first reported by the English physicist and physician Thomas Young in 1801. The slit pair is known as a double slit and the resulting effect is **double-slit interference**. In some books you will also find it called the **Young's slit experiment**.

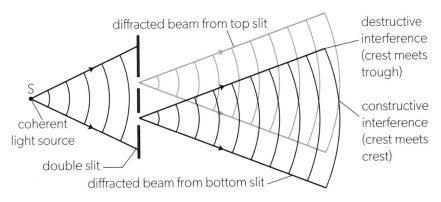

▲ Figure 26 Interference between coherent light from a double slit.

Light from S falls on both slits. If a laser is used, the light is coherent. Therefore, the light at each slit has the same phase since it has travelled the same distance from S. If a lamp is used, a single slit is used to diffract a small portion of the wavefront across both slits—this has the effect of making the light coherent. (This is not shown in the diagram.) Each slit then emits a cone of diffracted light bounded by the central maximum of the diffraction pattern for this cone. When the slits are very narrow, the cone will be wide.

The region where the two diffracted beams intersect is shown on Figure 26. Interference can only occur here as two beams are required. A screen placed parallel to the plane of the slits in the interference region shows a pattern of fringes parallel to the orientation of the double slits (Figure 27).

Compare this with the alternating loud and soft regions of the sound interference pattern. The light fringes show an alternating bright–dark arrangement with the fringes spaced equally on the screen. There is also a region on Figure 27 where some fringes appear much weaker than their neighbours. This effect is discussed later in the additional higher level section of this topic.

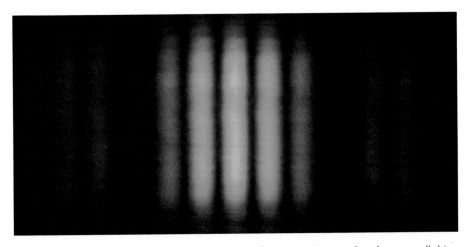

▲ Figure 27 The appearance of a double-slit interference pattern made using green light.

Global impact of science — The laser

Lasers are important because the light they emit is monochromatic (single wavelength) and coherent. Although the theory behind lasers was published by Einstein in 1917, the first working laser was not made until 1960. It was made by Theodore Maiman. At the time, Maiman described the laser as "a solution without a problem".

Since the first laser was demonstrated, the technology has become cheaper, easier to use and smaller. The coherence of the light enables lasers to be used for applications where interference is important. This includes barcode scanners and DVD and CD players.

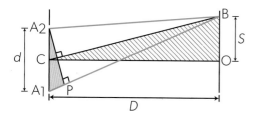

▲ Figure 28 Modelling double-slit interference.

🧪 Similar triangles

Treating the two triangles as similar only works when the wavelength of the light and the slit separation are *much* smaller than the distance from slit to screen. Figure 28 is unrealistic in this respect and not to scale. For a typical demonstration of Young's slits using light, d will be order of a millimetre, while D will be of the order of metres. For the slits shown on Figure 28, the slit–screen distance should be drawn at scale to be about 1.5 km!

You will not be tested on the derivation of the double-slit interference equation, but you do need to know how to use it in the contexts of any type of wave, not just electromagnetic radiation.

The double-slit equation

A simple equation models the distance between adjacent fringes in the interference. This distance is called the **fringe spacing**. The following notation is used in developing the model:

- d is the slit spacing (the distance between the two slits)

- λ is the wavelength of the wave

- D is the distance between the slits and the screen

- s is the fringe spacing (the distance between two neighbouring bright fringes).

In Figure 28, O is the centre of the fringe pattern on the screen, in other words, the point that is the same distance from both slits. This point will be bright because the paths from the slits A_1 and A_2 are the same length. The light from the slits arrives at O in phase and interferes constructively.

B is the position on the screen of the neighbouring bright fringe to the O bright fringe. The light that reaches B takes the routes A_1B and A_2B. These lengths are not the same. A_1B is longer than A_2B by the distance A_1P. However, because B is the neighbouring bright fringe to the one at O, the distance A_1P must be exactly one wavelength in length. This will enable the light from both slits to arrive in phase (or, more correctly, 2π rad out of phase). $A_1P = \lambda$.

The remainder of the derivation is pure trigonometry. There are two similar triangles on the diagram: the one in solid colour and the one hatched. They both have right angles (the larger triangle has a right angle where CO meets OB).

Because they are similar, $\dfrac{BO}{CO} = \dfrac{A_1P}{A_2P}$ which, using the symbols introduced earlier, is $\dfrac{s}{D} = \dfrac{\lambda}{d}$. This rearranges to

$$s \approx \frac{\lambda D}{d}$$

using the "$\approx$" symbol to remind us that this is an approximate expression (albeit a very good approximation when $D \gg d$).

Worked example 6

Two loudspeakers A and B, driven in phase, emit sound of the same amplitude and frequency of 1400 Hz. A microphone is placed at point M, 16.0 m from A and 17.5 m from B.

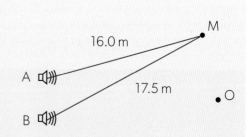

a. The speed of sound is 350 m s⁻¹. Determine the phase difference between the waves arriving at M from the two loudspeakers.

b. Hence, state the nature of the interference occurring at M.

c. The microphone is now moved from M to point O, an equal distance from A and B. State and explain how many times during this motion the microphone will record a minimum of sound.

Solutions

a. The wavelength is $\lambda = \dfrac{v}{f} = \dfrac{350}{1400} = 0.25$ m. The path difference for the waves arriving at M from A and

B is $17.5 - 16.0 = 1.5$ m $= 6\lambda$. This is an integer multiple of the wavelength, so the phase difference between the waves is zero.

b. The waves arrive in phase, so there is a constructive interference at M.

c. At O, the path difference is zero. Between M and O there must be points where the path difference
is $\dfrac{11\lambda}{2}, \dfrac{9\lambda}{2}, \dfrac{7\lambda}{2}, \dfrac{5\lambda}{2}, \dfrac{3\lambda}{2}$ and $\dfrac{\lambda}{2}$. At each of these points, the sound waves from A and B arrive in
antiphase (phase difference $= \pi$). This gives a total of 6 points of destructive interference, and
hence minimum sound intensity.

Worked example 7

Monochromatic light of wavelength 627 nm is incident on a double slit.
The interference pattern is observed on a screen, a distance 1.8 m from the
slits. It is found that 12 bright fringes are present in a length of 5.0 cm of the
interference pattern.
Calculate the slit spacing.

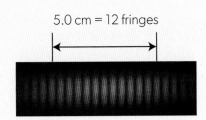

Solution

The distance between two consecutive bright fringes is $s = \dfrac{5.0 \times 10^{-2}}{12} = 4.17 \times 10^{-3}\,\text{m}$.

The slit spacing can be found by rearranging the double slit equation $s = \lambda\dfrac{D}{d} \Rightarrow d = \lambda\dfrac{D}{s}$.

$d = 627 \times 10^{-9} \times \dfrac{1.8}{4.17 \times 10^{-3}} = 2.7 \times 10^{-4}\,\text{m} = 0.27\,\text{mm}$.

Practice questions

10. A ray of coherent monochromatic light is incident
    normally on a double slit of slit separation 0.500 mm.
    A series of bright fringes appears on a screen, a
    distance 2.40 m from the slits. The fringe separation on
    the screen is 2.84 mm.

    a. Calculate the wavelength of light.

    b. The source of light is replaced by one of wavelength
        468 nm. Calculate the new distance between the
        neighbouring fringes on the screen.

11. A source of coherent monochromatic light is used in
    a double-slit experiment. A series of bright and dark
    fringes appears on a distant screen.

    a. Outline how a dark fringe is formed.

    b. The separation between two consecutive bright
        fringes on the screen is 1.0 mm. The distance
        between the slits and the screen is halved and the
        slit separation is doubled. Calculate the separation
        between the fringes after this change.

12. A coherent monochromatic beam of microwaves
    is incident on two identical slits A and B. Two
    consecutive intensity maxima are observed at points
    M_1 and M_2. Point M_1 is at an equal distance from both
    slits. A detector placed at point D records an intensity
    minimum. Point D is 2.84 m from slit A and 2.90 m
    from slit B.

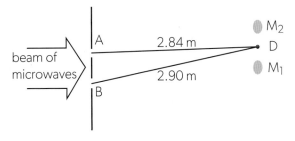

What is the wavelength of the microwaves?

A. 3.0 cm   B. 6.0 cm   C. 12 cm   D. 24 cm

🧪 Measuring wavelength using double slits

- Tool 1: Recognize and address relevant safety, ethical or environmental issues in an investigation.

- Tool 3: Select and manipulate equations.

- Inquiry 2: Collect and record sufficient relevant quantitative data.

- Inquiry 2: Carry out relevant and accurate data processing.

(a) With laser light

This is an up-to-date version of the experiment devised by Thomas Young.

- Observe the usual precautions with laser light.

- The light from the laser is shone through a double slit, so that the interference pattern is formed on a white screen. Figure 29(a) is not to scale. The slit separation will be typically a fraction of a millimetre; the screen will be several metres from the slits.

- You can make a homemade double slit by painting a glass microscope slide with colloidal graphite and scratching the slits with a pin that you run along a metal ruler.

- The slit separation d is measured using a travelling microscope or a microscope with a piece of graph paper and the pin on the microscope stage.

- Mark the positions of 10 fringe separations (10s) on the screen (this will be from fringe number 1 to fringe number 11). Then turn off the laser and use a metre ruler to measure this distance. Hence, find the value of a single fringe spacing s.

- Use a tape measure to measure the distance d from the slits to the screen.

- Use the double-slit interference equation to calculate the wavelength of the laser light.

(b) With microwaves

- Point the transmitter and the receiver towards each other about half a metre apart. Ensure that they are correctly oriented with respect to each other. If there is no reading on the milliammeter, then rotate the transmitter through 90° so that the polarizations of the transmitter and receiver match.

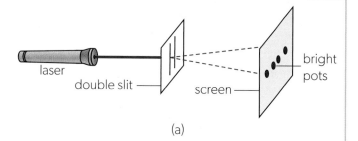

(a)

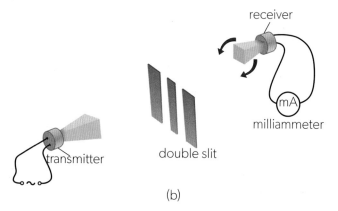

(b)

▲ Figure 29 Measuring the wavelength of (a) laser light, and (b) microwaves.

- Use small sheets of aluminium, as in Figure 29(b), to construct a pair of slits with a separation of about 3 cm. This maximizes the diffraction from each slit. This is the approximate wavelength of the microwaves.

- Move the receiver horizontally, parallel to the slits. You should see the reading on the meter vary between a small value and a maximum. These are the "fringes".

- Measure the values for D, d and s. Read the previous experiment for details of how to achieve a value for s.

- Calculate the microwave wavelength.

- This technique will work for other types of wave, sound waves for example.

Intensity variations in the double-slit interference pattern

Figure 27 showed that the light fringes in the double slit experiment are not of equal brightness. Figure 30 also shows a wider view of fringes in red light with dark regions where the fringes are suppressed.

This inequality in intensity is due to diffraction, an effect ignored in the simple analysis earlier. The assumption there was that only the central diffraction maximum from each slit is involved in the interference because the individual slit widths are infinitesimally small. In practice, this is not the case. Fringe patterns observed in the laboratory contain both diffraction and interference effects at the same time. If the slits were very thin, then, in practice, not enough light energy would be transferred to the screen for the interference fringes to be seen.

Each slit in the double-slit pair produces its own diffracted beam. With one slit open and the other one blocked off, we would observe only a normal diffraction pattern. With identical slits, both diffraction patterns are identical, but there is interference between them.

Figure 31 shows how this happens. Figure 31(a) gives the predicted variation of intensity with angle from pattern centre for the fringe pattern due to two identical slits without diffraction. The pattern has a constant amplitude and an equal fringe spacing, as predicted by the double-slit equation. Figure 31(b) shows the variation of intensity with angle for one of the slits alone. This is the usual single-slit diffraction pattern.

Figure 31(c) shows the combination of the diffraction and the interference. The single-slit intensity **modulates** the intensity of the fringe pattern. The fringes are suppressed where the diffraction equation predicts that there should be a minimum. This applies whether there is an interference maximum at this position or not.

To model what happens around the first minimum of the diffraction pattern, you will need both the fringe spacing equation and the equation for predicting the position of the first diffraction minimum .

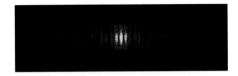

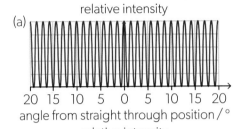

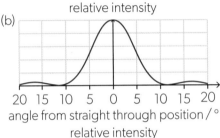

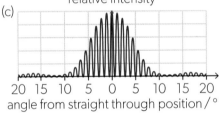

▲ Figure 30 Double-slit interference pattern produced from the light of a helium–neon laser.

▲ Figure 31 The overall effects of diffraction and interference for a double slit (a) Interference alone. (b) Diffraction alone. (c) The combination of both diffraction and interference.

Worked example 8

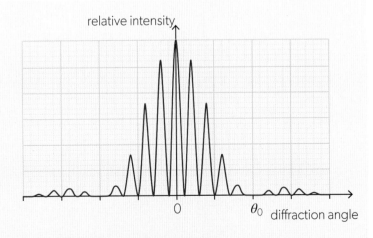

The diagram shows how the intensity of light observed in an experiment with two parallel identical slits varies with the diffraction angle.

The bright double-slit interference fringe occurring at the diffraction angle θ_0 is eliminated.

a. Explain why a minimum intensity of light is observed at the angle θ_0.

b. The width of each slit is 0.20 mm. Determine the distance between the slits.

Solutions

a. The double-slit pattern of bright fringes is modulated by a wider single-slit diffraction pattern due to a finite width of each slit. The single-slit pattern has a first minimum at θ_0, resulting in a minimum overall intensity.

AHL

b. From the single-slit diffraction formula, the angle θ_0 is related to the slit width b: $\theta_0 = \frac{\lambda}{b}$. On the other hand, the same angle θ_0 corresponds to five fringes observed in the double-slit interference pattern: $\theta_0 = 5 \times \frac{s}{D} = 5 \times \frac{\lambda}{d}$, where d is the distance between the slits. By equating these two expressions and eliminating the wavelength of light, we get $5 \times \frac{\lambda}{d} = \frac{\lambda}{b} \Rightarrow d = 5b$. The distance between the slits is five times the slit width. $d = 5 \times 0.20 = 1.0\,\text{mm}$.

Practice questions

13. The wavelength of the light used in a double-slit experiment is 520 nm. The slits are parallel and have equal width. The graph shows how the intensity of light observed on a screen varies with the diffraction angle.

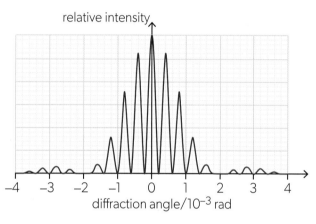

Calculate:

a. the width of each slit

b. the distance between the slits.

14. The diagram shows the intensity pattern observed in a double-slit interference experiment with monochromatic light.

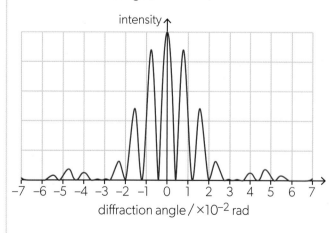

The separation of the slits is increased with no other changes being made. Which graph on the right shows the intensity pattern after the change?

A.
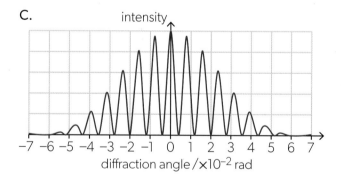

B.
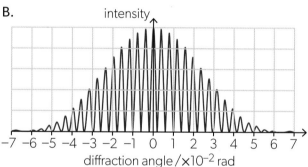

C.
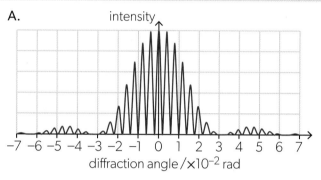

D.
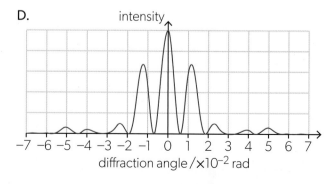

What can an understanding of the results of Young's double-slit experiment reveal about the nature of light?

What evidence is there that particles possess wave-like properties such as wavelength? (NOS)

The work in this topic is devoted to waves, much of it in the context of the transmission of electromagnetic radiation. The models that are developed to explain diffraction and interference do so on the basis that electromagnetic radiation acts as a wave. But does this link between the model and the empirical observations prove that light has wave properties?

In Theme E, you will see strong and equally convincing evidence that electromagnetic radiation has particle-like properties. For example, the concept of the photon as the carrier of the electromagnetic interaction, essentially regarded as an interacting particle.

What is the scientific truth here? Is light a wave or a particle? If it is a particle, how can interference and diffraction occur? If it is a wave, why are photoelectric effects observed?

In a remarkable series of tests in the 20th century, scientists carried out a series of double-slit experiments. The light intensities used were extremely small, so that only one photon could be in the region between the source and the screen at any one time. Even so, interference effects were still observed. What were these single photons interfering with?

Neither wave nor particle view is correct. When we look for wave-like properties, we find them. When we look at light in a particle context, we see those properties too. The description of electromagnetic wave behaviour lies in probability considerations. There are many popular books and articles on this subject if you wish to explore this more.

Choosing the appropriate model for a particular situation is all part of the Nature of Science.

Multiple-slit interference

When the number of illuminated slits is increased in an interference experiment, with a constant spacing between any two adjacent slits, two effects occur:

- The bright fringes stay in the same place but become narrower (or sharper).

- The intensity of the bright fringes increases (and is proportional to N^2, where N is the number of slits).

These effects are shown in Figure 32, although it is hard to reproduce the intensity changes on the printed page. The single-slit pattern is shown at the top for reference (marked 1 slit). Remember that interference fringes are *always* suppressed at the diffraction minimum.

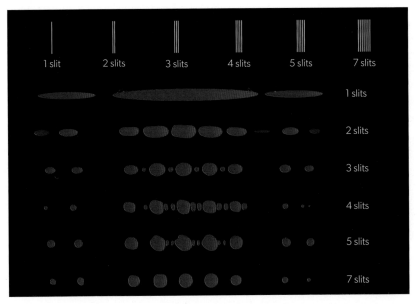

◀ Figure 32 Increasing the number of slits in an interference experiment makes the fringes sharper.

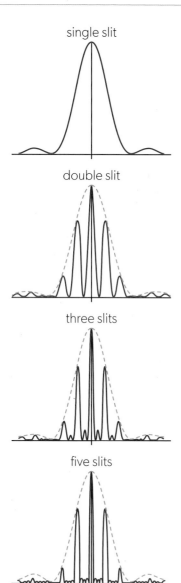

single slit

double slit

three slits

five slits

▲ Figure 33 The effect of increasing the number of interfering slits on the intensity graph.

Graphs of relative intensity against angle from the straight-through position are also helpful (Figure 33).

The usual double-slit pattern is shown as well as the patterns for three slits and five slits. These are relative-intensity plots and the heights of the three-slit and five-slit patterns have been normalized to allow for easy comparison. The three-slit pattern should be 9 times higher to scale than the double-slit pattern, and the five-slit version should be 25 times higher.

As usual the fringe patterns are modulated by the single-slit diffraction envelope. Also notice that between the major fringes from three slits onwards, other subsidiary fringes appear. When N is the number of slits, there are $N - 2$ subsidiary fringes for values of $N \geq 3$.

The analysis of multiple-slit interference is difficult, but it is possible to give a qualitative argument for these effects:

- Remember that the assumption is that the slits have the same slit separation, so that a 100-slit arrangement will be about 30 times wider than a 3-slit arrangement.

- The slits are illuminated uniformly so that the amplitude of the wave at a bright fringe will be about N times more that with a single slit. Because intensity is proportional to the square of the amplitude, the intensity of the bright fringes is $\propto N^2$.

- The fringes are sharper because when N is very large (compared with $N = 2$), only a small angular movement away from the centre of a bright fringe will be needed before there are a pair of slits that have a phase difference of π rad. When a pair are related like this (say the 1st slit and the 22nd) then the 2nd slit and the 23rd will also have the same π rad relationship as will the 3rd and the 24th—and so on across the whole array of slits.

Diffraction grating

The suggestion of multiple-slit arrangements with hundreds of slits begs the question: what happens when N is very large, approaching thousands of very thin, diffracting slits. Such an arrangement is known as a **diffraction grating**.

A diffraction grating, often used to produce optical spectra (Figures 34 and 35), consists of many parallel, equally-spaced slits—usually called "lines" in this context—formed on a suitable transparent medium.

▲ Figure 34 When white light goes through a diffraction grating, the central maximum remains white and the first-order pattern is a spectrum. Note that red light is deviated the most; this is opposite to the effects of dispersion when white light is refracted.

white light source

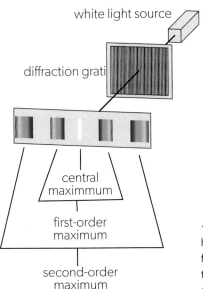

diffraction grati

central maximmum

first-order maximum

second-order maximum

◀ Figure 35 Diffraction gratings are said to have orders of spectra. This diagram shows five orders: the zero-order (central maximum), two first-order spectra and two second-order spectra.

Grating spacing

Diffraction gratings are often specified by the number of grating lines per mm. d is the reciprocal of this. You may want to convert it to metres at the same time to match the rest of a problem:

$$d \text{ (in mm)} = \frac{1}{\text{number of lines per mm}}$$

The lines are often etched or inscribed onto glass or plastic. There are usually hundreds of these lines for each millimetre of the grating surface. For a standard laboratory diffraction grating with 600 lines per millimetre, this means that each line (slit) is $\frac{1}{600} = 0.0017\,\text{mm}$ from its neighbour. Human hairs have an average diameter of 75 μm, so one hair will cover about 50 lines on the grating.

In line with the conclusions about increasing the number of slits above we expect:

- the maxima to be very sharp (because N is large)

- the maximum to be very intense (because N is large and each line diffracts energy)

- the fringe pattern to be well spread out (because the slit separation is very small).

The maxima become isolated because of their sharpness and angular separation.

When white light is incident on the grating, a series of spectra are formed on a screen some distance from the grating. Each spectrum is called an "order" and these orders are labelled from the centre outwards as the first order, second order, third order and so on. The central maximum (white for a spectrum because all the colours have the same zero phase difference) is the zero-order maximum.

The diffraction-grating equation is easily derived (you will not need to recall it in an examination) and uses the symbols (Figure 36):

- n, the order of the maximum

- λ, the wavelength

- d, the grating spacing (distance between adjacent lines)

- θ, the angle between the central maximum and the maximum.

Figure 36 shows diffracted waves from three adjacent slits going to the first-order maximum. The slits are very narrow and the distance to the screen is very large, so that the angles between the straight-on directions and the first-order maximum are almost equal and are all labelled θ to indicate this.

For the direction θ to be a maximum, the path difference between successive slits must be λ. This path difference is also equal to $d \sin \theta$. For the first order, $\lambda = d \sin \theta$.

For the second-order maximum (θ will be much greater now), the path difference must be 2λ, where $2\lambda = d \sin \theta_2$. The general case for the nth order is

$$n\lambda = d \sin \theta_n$$

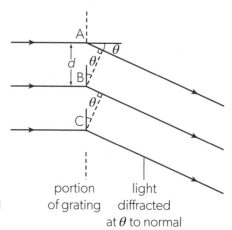

▶ Figure 36 Modelling the first-order maximum for a diffraction grating illuminated by monochromatic light.

portion of grating

light diffracted at θ to normal

Worked example 9

A diffraction grating has 600 lines per millimetre. A beam of monochromatic light is incident at right angles to the grating. A second-order maximum is observed at an angle of 31° to the direction of the incident beam.

a. Calculate the wavelength of the light.

b. Determine the total number of the diffraction maxima present in the transmitted light.

Solutions

a. The grating spacing is $d = \dfrac{1.0 \times 10^{-3}}{600} = 1.667 \times 10^{-6}\,\text{m}$. $n = 2$ and so
$\lambda = \dfrac{d \sin \theta}{n} = \dfrac{1.667 \times 10^{-6}\,\sin 31°}{2} = 429\,\text{nm}$.

b. The value of $\sin \theta$ in the diffraction grating equation cannot be greater than 1. Hence $\dfrac{n\lambda}{d} < 1$. This condition can be used to find the largest possible order of diffraction. $n < \dfrac{d}{\lambda} = \dfrac{1.667 \times 10^{-6}}{429 \times 10^{-9}} \approx 3.9$. The largest order is $n = 3$. Hence a total of 7 maxima will be present in the diffracted light—the central maximum and two secondary maxima for every order up to $n = 3$.

Practice questions

15. Monochromatic light of wavelength 573 nm is normally incident on a diffraction grating. The angle between the central maximum and a first-order maximum is 9.90°.

    a. Calculate the number of lines per millimetre of the diffraction grating.

    b. Determine the number of maxima present in the diffracted light.

    c. The light source is replaced by one of wavelength 640 nm. Calculate the angular separation between the two second-order maxima.

16. A diffraction grating has 600 lines per mm.

    a. Calculate the angle of the third-order diffraction maximum, when light of wavelength 480 nm is incident normally on the grating.

    b. Determine the longest wavelength of light that will produce a third-order maximum under normal incidence.

Data-based questions

A laser pen was shone at normal incidence through a diffraction grating with 300 lines per millimetre. The diffraction pattern was observed on a screen  2.73 m behind the grating. The distance from the central order on the screen to the nth order was measured and the results are shown in the table.

n	Distance from zero order / m ± 0.01 m
0	0
1	0.44
2	0.91
3	1.47
4	2.27
5	3.62

- Using the distance to the screen, tabulate values of $\tan \theta$.
- Considering the uncertainties in x and D, calculate uncertainties in your values of $\tan \theta$.
- Calculate values of $\sin \theta$ and add them to a column of your table.
- Plot a graph of $\sin \theta$ against n and find the gradient.
- By considering the maximum and minimum possible values of $\tan \theta$, find the equivalent maximum and minimum values of $\sin \theta$. Hence add an uncertainty in each value of $\sin \theta$ to your table and to your graph.
- By considering the equation $n\lambda = d \sin \theta_n$, use the value for the gradient of your graph to find the wavelength of the light.
- Using your values for the maximum and minimum gradients, find an uncertainty in your value for the wavelength and express this as a percentage.

 ## Global impact of science—The light interference in a soap bubble

A soap bubble viewed in white light shows colour bands and fringes on its surface that change as the bubble ages (Figure 37). How do these arise?

There is a π rad phase change at the outer surface because the light is reflected at a medium

▲ Figure 37 The bands of colour that can be seen on a soap bubble illuminated by white light. Each band corresponds to a certain thickness of soap film and the colours change as the liquid in the film gradually drains to the bottom of the bubble. This reduces the local thickness and determines the single wavelength that is removed from the light reflected from the bubble.

that is denser than the air. However, at the inner surface there is no phase change because light is reflected at a less dense medium. This is similar to the cases that were described on page 440 where waves move from one string to another for different relative string densities.

The two beams of light are reflected out of the bubble and interfere after leaving the film. One ray has gone through the soap film twice. This introduces an extra travel distance and changes its phase relative to the ray reflected at the outer surface. When the thickness of the film is equal to one-half of a wavelength of the light in the soap film then

one ray travels exactly one wavelength λ more than the other. Superposition occurs and leads to destructive interference because of the π rad phase change of one ray relative to the other.

Destructive interference removes the colour that corresponds to λ from the white light. All the colours of the spectrum are reflected from the bubble except one; this gives colour to the fringe. Variations in the soap film thickness occur as liquid drains to the bottom of the bubble. Different wavelengths become suppressed in different parts of the bubble leading to colour fringes.

Similar colour fringes can be seen on sheets of oil lying on water patches on roads and pavements. There are many applications for this effect: Thin-film interference is used in anti-reflection coatings on lenses. Similar coatings have evolved behind the retinas of some vertebrate's eyes to reflect light forward to the sensitive retinal cells. The beautiful colours of the European Peacock butterfly (Figure 38) are also due to thin-film interference.

◄ Figure 38 The European Peacock butterfly.

What distinguishes standing waves from travelling waves?

How does the form of standing waves depend on the boundary conditions?

How can the application of force result in resonance within a system?

This topic takes the ideas of Topics C.2 and C.3 further to look at the interaction of two or more travelling waves moving in different directions. This can lead to the formation of a standing wave that has a shape which appears to be fixed in space (but not in time). This is why standing waves are sometimes called stationary waves.

These standing waves can form because of reflections at a boundary. The boundary determines the nature of the reflection and the exact shape of the standing wave. The boundary therefore imposes a set of conditions on the system that leads to the formation of the standing wave.

In this topic, you will also examine the creation of resonance in a system. A system that can oscillate can usually be driven by another driver system. The driver exerts a time-varying force on the driven oscillator. This stimulates the driven system into oscillations of its own.

▲ Figure 1 Most musical instruments rely on standing waves to make pleasing sounding notes. Resonance and damping are also important considerations in the design of these instruments.

In this topic, you will learn about:

- the nature and formation of standing waves
- nodes and antinodes
- relative amplitude and phase difference of points along a standing wave

- standing waves patterns in strings and pipes
- resonance, natural frequency and driver frequency
- the effect of light, critical and heavy damping on an oscillating system.

Introduction

Pluck an open guitar string and you generate waves going in opposite directions along the string. The waves reflect at the bridge and at the nut to reverse their direction. The waves superpose and a standing wave forms on the string. The generation of a standing waves is the basis for the pitched notes in many musical instruments.

In this topic, we examine the physics of standing waves in many more contexts than stringed instruments. We also look at the important effects that occur when one oscillating system running at one natural frequency is driven by a system running at another.

Formation of standing waves

Standing waves originate when two or more travelling waves interact. The simplest case is that of two identical **travelling waves** moving in opposite directions in the same medium. The waves have the same wavelength and, because they are identical, the same amplitude. Figure 2 shows what happens.

- The red wave in Figure 2 is moving to the right. The blue wave is moving to the left. The black wave shows the red and blue waves **superposed** (see Topic C.3). The diagram gives a series of positions in time as the two waves move through each other. In Figure 2(a), the blue and red waves are almost in phase. They add to give a wave almost double the amplitude of each wave separately. The black wave is the **standing wave** formed by the superposition of the red and black waves.

- Identify the points where the black wave is zero. These are where the red and blue waves cancel each other out.

- As time increases, the red wave (moving to the right) and the blue wave (moving to the left) drift out of phase and, as a result, the amplitude of the black wave is decreasing. However, the points at which the black wave is zero never change. Here the red and blue waves are always equal in magnitude but opposite in displacement.

- The wave medium—gas particles, or string, or the electric field—always has a zero displacement at these points. They are known as **nodes**. These positions are marked N in Figure 3.

- Eventually (Figure 2(d)), the red and blue waves are 180° out of phase and cancel completely, so that there is zero displacement for the whole of the black wave.

- As time continues to increase, the red and blue waves again give a non-zero sum but this time the part of the black wave that was positive-going earlier is now negative-going (and vice-versa). The black wave oscillates with the same time period as the red and blue waves—but it is not moving to the left or right. On this diagram it is stationary in space (but not in time).

- Finally (Figure 2(g)), the red and blue waves are exactly in phase and the black wave now has its largest displacement, equal to twice the amplitude of the original waves because the red and blue amplitudes are the same.

- The positions where the displacements are as large as possible are known as **antinodes** and are marked A.

Figure 3 summarizes these observations for the black standing wave alone when the red and blue waves are ignored.

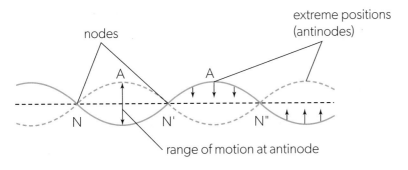

▲ Figure 3 Nodes, N, and antinodes, A, on a standing wave.

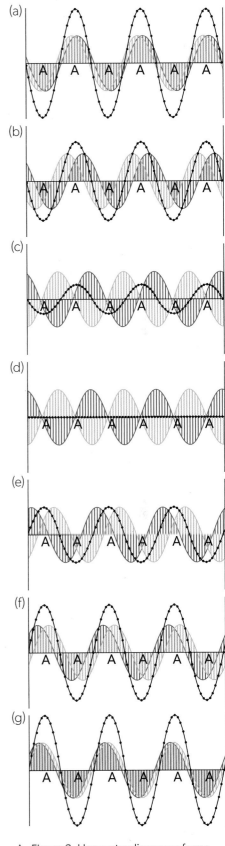

▲ Figure 2 How a standing wave forms from two travelling waves moving in opposite directions.

Seeing standing waves

Many websites contain applets showing animations of the formation of a standing wave. Try searching the web using "applet for standing wave formation". Seeing a standing wave form in slow motion is a good way to understand how they work.

Figure 3 may seem odd when you first meet it, but it shows the standing wave when it is at its maximum displacement at two instants half a cycle apart. One of these positions is shown as a solid line, the other as a dashed line. This representation is frequently used to show the extent and position of a standing wave and is particularly important with an "invisible" wave. An example of this is the standing wave in air formed inside the tube of a musical instrument, such as a flute. The diagram also reminds you that the wave is fixed in space and is not travelling.

It is important to be able to relate the shape of the standing wave to the motion of the particles of the medium.

- Between two adjacent nodes (N to N' in Figure 3) all the particles move **in phase** with each other.

- In adjacent nodal regions (NN' and N'N"), the particles are 180° (π rad) **out of phase** (when particles in NN' are moving downwards, then in N'N" the particles will be moving upwards).

- There is, therefore, only one value for the **phase difference between nodal points** along a standing wave. Figure 3 shows the π (180°) phase difference between region NN' and region N'N".

- The distance between two adjacent nodes (or antinodes) is equal to half a wavelength of the original waves.

- The standing wave oscillates with the same frequency as the travelling waves.

- Within NN', the **relative amplitude** between two points along the standing wave varies. No two points have the same amplitude, which can vary between 0 (at the node) and the maximum (at the antinode).

These statements about phase difference and relative amplitude for the standing wave are in direct contrast to similar statements for a travelling wave. For the travelling case, there is a phase difference between each point on the wave and its neighbour and the amplitude for each point on the wave is the same (assuming no energy loss from the system).

Worked example 1

Two transverse waves with a wavelength of 80 cm each and with equal amplitudes travel in opposite directions on a stretched rope. At time $t = 0$, all the particles of the rope have a displacement of zero. This is represented by the horizontal line in the following displacement–position graph, where x is the position along the rope.

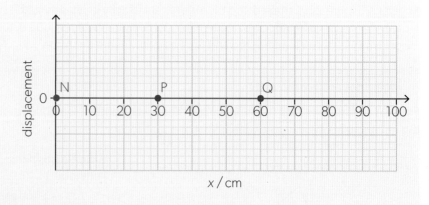

N is the position of one of the nodes of the resulting standing wave. P and Q are two particles in the rope.

a. Label, on a copy of the diagram, the positions of any remaining nodes of the standing wave between $x = 0$ and $x = 100$ cm.

b. Compare the amplitude and phase of oscillations of particles P and Q.

c. At $t = 0$, particle P is moving in the direction of increasing displacement. Draw a graph to show the displacement of the wave at $t = \frac{T}{4}$, where T is the period of the wave.

Solutions

a. Adjacent nodes are separated by one-half of the wavelength of the original travelling waves, so by 40 cm. There will be two more nodes, at $x = 40$ cm and $x = 80$ cm.

b. P and Q are separated by a node at $x = 40$ cm, and hence oscillate 180° out of phase. Q is halfway between two adjacent nodes, at an antinodal position. Because of this, Q has a greater amplitude than P.

c. At $t = \frac{T}{4}$ all the particles of the rope will be at their maximum displacement, positive in the nodal region containing P and negative in the region containing Q.

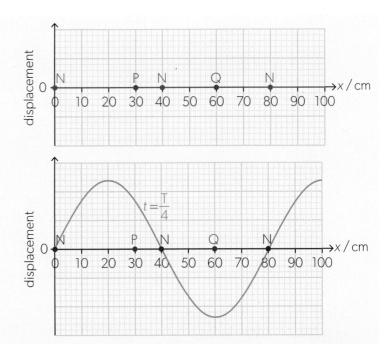

Making standing waves visible

- Inquiry 2: Identify and record relevant qualitative observations.

- Inquiry 3: Compare the outcomes of an investigation to the accepted scientific context.

- Inquiry 3: Identify and discuss sources and impacts of random and systematic errors.

The German physicist Franz Melde invented a way to produce standing waves easily and visibly. Two variants of his method (which originally used a string under tension and a tuning fork) are commonly used.

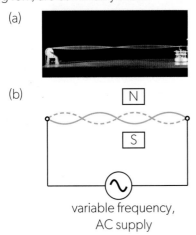

(a)

(b)

variable frequency, AC supply

▲ **Figure 4** (a) Using a vibration generator to view a standing wave. (b) Using the electric motor effect to view a standing wave.

Using a vibration generator

Figure 4(a) shows one variant in which a string is attached to a vibration generator. One end of the string is driven vertically up and down when an alternating current from a signal generator is supplied to the vibration generator. The frequency of vibration can be set by the observer. The other end of the string is fixed.

Adjust the frequency until the standing wave pattern becomes obvious. Sketch the shape of the string. Increase the frequency until another pattern is obvious. Again, sketch the shape. What is the relationship between the frequencies that you have noted? What is the change in successive shapes?

Use a stroboscope to illuminate the string with a frequency close to the signal generator frequency to "freeze" the motion of the string.

When you look carefully at the nodes, you may see that they are not truly zero throughout the cycle. This is because energy is always absorbed by the fixed support for the string and the reflected wave has a reduced amplitude compared with the incident wave. This leads to an incomplete cancellation of the two waves at the node, but it does not affect its position.

Using the electric motor effect

An alternative is to replace the string with a metal wire, to carry electric current from the signal generator, and to add a pair of magnets straddling the wire at its centre, to provide a magnetic field at 90° to the wire (Figure 4(b)).

When the tension is matches the frequency of the alternating current, the wire will be driven up and down by the magnetic interaction and will display the standing wave with the same frequency as the AC.

How does the standing wave form?

In Melde's experiment, when the wave travels along the string from the vibration generator and arrives at the fixed end, it is reflected. The reflected wave is an inversion of the incident wave (this is explained later) and so is 180° out of phase with the incident wave. The incident and reflected waves superpose. For most frequencies, the superposition leads to a disorganized behaviour of the string. However, at certain frequencies a standing wave appears.

As the frequency of the signal generator is increased, at one frequency, the string oscillates with a large amplitude in the middle of the string, with something close to zero at the ends. When the signal generator frequency is increased further, the large oscillation initially dies away, but a different standing wave is observed at a higher frequency. This occurs when the frequency of the first standing wave has doubled. Two loops are observed. (This is the state shown in Figure 4(a).)

These standing waves are known as the **first harmonic** and the **second harmonic**, respectively. Further standing waves with more loops are observed at higher multiples of the first standing wave frequency. These further standing waves are described as third harmonics, fourth harmonics and so on.

Boundary conditions for a wave

In the experiment you just looked at, when the wave comes to the fixed end of the string it reflects, so that it is π rad out of phase with the incident wave.

Fixed end

The fixed end cannot move and so its displacement must always be zero. This is the **boundary condition**. For this condition to apply when the incident wave moves towards the fixed end, the reflected wave must:

*   move in the opposite direction away from the fixed end

*   have an opposite displacement to the incident wave.

This means that the reflected wave is inverted (π rad out of phase) but it has the same wave shape as it travels away from the fixed end.

Free end

A free boundary condition is different. In this case, the reflected wave is generated as the particles in the medium return to their equilibrium positions. This means that the reflected wave moving away from the end of the string will have the same shape as the incident wave and will not be inverted on reflection.

Models—Neither fixed nor free

The cases where a wave travels between media of high and low density are shown in Figure 5. This is an extension of the models for transmission to infinitely large and zero density.

In a guitar, a small-amplitude, incident wave (on the string) meets a fixed end (at the bridge). The energy transfer to the bridge is usually ignored in discussing the formation of the guitar standing wave. However, a small amount of energy is transferred from the standing wave of a guitar to the bridge during each cycle. The body of the instrument vibrates, and a sound wave travels through the air to allow us to hear the music.

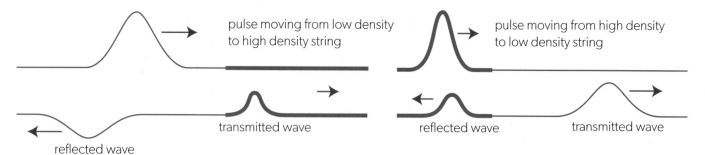

▲ Figure 5 The reflection and transmission of a transverse wave on a string when it meets a boundary between materials of different densities.

Worked example 2

A transverse wave of frequency 2.5 Hz travels towards a fixed end of a stretched rope with a speed of 0.60 m s^{-1}.

a. Explain why a standing wave will be formed in the rope.

b. Calculate the distance from the fixed end of the rope of the first two antinodes of the standing wave.

Solutions

a. The wave will be reflected from the fixed end. The reflected wave has the same frequency and approximately the same amplitude as the incident wave. The waves interfere with each other, leading to the formation of a standing wave pattern, with a node at the fixed end.

b. The wavelength of the original wave is $\lambda = \frac{v}{f} = \frac{0.60}{2.5} = 0.24$ m $= 24$ cm. The nearest antinode is formed one-quarter of the wavelength from the fixed end, at a distance of 6 cm. The antinodes are one-half of the wavelength from each other, so the next antinode is $6 + 12 = 18$ cm from the fixed end.

Practice question

1. A transverse standing wave is formed on a string when a travelling wave of amplitude 4.0 cm and frequency 5.0 Hz is reflected from a loose end of the string. The speed of the wave is 3.0 m s^{-1}. Calculate:

    a. the amplitude of oscillations of the loose end of the string

    b. the distance between adjacent nodes of the string

    c. the time interval between instants when all the particles in the string have zero displacement.

Standing wave patterns in strings

(a) Two fixed ends

The pitch of a note played by a stringed instrument depends on the length of and tension in the string. Varying the tension is the principal way used to "tune" instruments so that the frequencies of the strings are both internally consistent and set to the same base frequency as other instruments. String players then produce the individual notes by varying the lengths of the strings.

The relationship between the length L of a string and its frequency f is a simple one that relates to the harmonic that is being played.

Figure 6 shows the first three harmonics for a string fixed at both ends. The distance between the fixed supports is L.

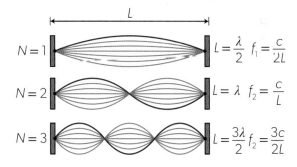

▲ Figure 6 The first three harmonics of a string fixed at both ends.

For the first harmonic ($N = 1$) the standing wave is half a wavelength long and therefore $L = \frac{\lambda}{2}$

The usual wave equation applies: $c = f\lambda$ leading to $c = f_1 \times 2L$, $f_1 = \frac{c}{2L}$

For the second harmonic ($N = 2$): $L = \lambda$ so $f_2 = \frac{c}{L}$ meaning that $f_2 = 2f_1$

For the third harmonic ($N = 3$): $L = \frac{3\lambda}{2}$ so $f_3 = \frac{3c}{2L}$ meaning that $f_3 = 3f_1$.

For the nth, $L = \frac{n\lambda}{2}$ so $f_n = \frac{nc}{2L}$ meaning that $f_n = nf_1$

Although the discussions so far have been in the context of waves on strings, the theory is a general one that applies to all wave motions.

Global impact of science—Music meets physics I

Players in all cultures use harmonics to colour the sound of string instruments. Touch the centre of a sounding string to stop it moving and then every harmonic with an antinode at that point will be suppressed. That means that the first, third, fifth and so on harmonics will disappear from the sound, leaving the second, fourth, sixth and so on. This change in the balance of the harmonics changes the character of the sound.

▲ **Figure 7** These xylophone bars are also designed to promote certain harmonics. The bars are supported at approximately one-quarter and three-quarters of their length. This forms a node at these positions. The shape of the bar is also thinner in the middle. This reduces damping and allows the xylophone bar to be tuned.

Worked example 3

The diagram shows a second harmonic standing wave on a string of length 0.90 m fixed at both ends. P and Q are two particles in the string. The speed of the waves in the string is $270\,\mathrm{m\,s^{-1}}$.

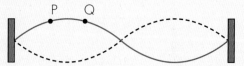

a. State the phase difference between the oscillations of P and Q.

b. Calculate the frequency of the second harmonic.

c. Particle Q is at a distance of 0.30 cm from the left end of the string. The string is now clamped at Q and made to vibrate with the lowest frequency possible in this situation. For this standing wave:

   i.   draw the displacement graph

   ii.   calculate the frequency.

Solutions

a. P and Q are in the same nodal region, so the phase difference is zero.

b. The wavelength of the second harmonic is equal to the length of the string, $\lambda = 0.90$ m. The frequency is therefore $f = \dfrac{v}{\lambda} = \dfrac{270}{0.90} = 300\,\mathrm{Hz}$.

c. i. Q is one-third of the length of the string from the left end. If the string is clamped here, Q becomes a node of the standing wave, and the lowest frequency mode will be the third harmonic.

   ii.   The wavelength is now $\dfrac{2}{3} \times 0.90 = 0.60$ m and the frequency is $f = \dfrac{270}{0.60} = 450\,\mathrm{Hz}$.

Worked example 4

A string of length 0.80 m is fixed at both ends. Two successive harmonic frequencies of this string are 360 Hz and 480 Hz.

a. Determine the frequency of the first harmonic mode.

b. Calculate the speed of the waves in the string.

Solutions

a. The harmonic frequencies of a string fixed at both ends are consecutive integer multiples of the lowest frequency f_1 of the first harmonic mode. It means that f_1 is the difference between any two successive harmonic frequencies.
$f_1 = 480 - 360 = 120\,\mathrm{Hz}$.

b. For the first harmonic, $\lambda = 2 \times 0.80 = 1.60$ m and $v = f\lambda = 120 \times 1.60 = 190\,\mathrm{m\,s^{-1}}$.

(b) One free end, one fixed end

When a string has one free end and one fixed end, the set of harmonics changes. You may wonder how a string can have a free end as there is nothing then to support the string. The trick is to hang the string vertically and to suspend the string using the vibration generator at the top. The mathematics developed here will not work exactly because the tension in the string decreases with distance from the support (you might like to think why). The assumption made here is that the string has a uniform tension.

A practical way to observe this effect is to use a horizontal wire, one that is thin enough to have observable oscillations but strong enough to support its own weight. A vibration generator is used at the fixed end.

Figure 8 shows what happens this time. The fixed end is always a node (as with the case of two fixed ends). However, at the free end, the reflected wave is in phase with the incident wave. This leads to the free end always acting as an antinode.

The change of the boundary conditions changes the allowed patterns of standing waves.

The first harmonic is one-quarter of the full wavelength and, when the length of the string is L, the wavelength must be $4L$ and $c = f\lambda$ becomes $c = f_1 \times 4L$ so that

$$f_1 = \frac{c}{4L}$$

The next harmonic is related to the first harmonic by a factor of three because the wavelength of the new standing wave is $\frac{4L}{3}$, given the boundary conditions, which leads to

$$f_3 = \frac{3c}{4L}$$

This is called the third harmonic *not* the second.

The important notation rule for harmonics is:

The ratio $\dfrac{\text{frequency of the harmonic}}{\text{frequency of the first harmonic}}$ is the number assigned to the harmonic.

As an example, the pattern with $2\frac{1}{2}$ loops has a length that is $\frac{5\lambda}{4}$ with a frequency $f_5 = \frac{5c}{4L}$ and therefore

$$\frac{f_5}{f_1} = \frac{\left(\dfrac{5c}{4L}\right)}{\left(\dfrac{5c}{4L}\right)} = 5$$

This is called the fifth harmonic even though it is only the third in the series as the frequency increases. For this arrangement, there are no even-numbered harmonics.

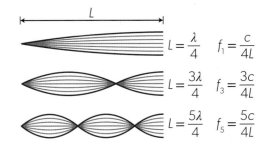

$$L = \frac{\lambda}{4} \qquad f_1 = \frac{c}{4L}$$

$$L = \frac{3\lambda}{4} \qquad f_3 = \frac{3c}{4L}$$

$$L = \frac{5\lambda}{4} \qquad f_5 = \frac{5c}{4L}$$

▲ Figure 8 The first three harmonics of a string fixed at one end and free at the other.

Worked example 5

A stiff wire of length 1.5 m is clamped at one end and has the other end free. The wire is made to vibrate at a frequency of 420 Hz in a standing wave pattern, as shown in the diagram.

a. Calculate the wave speed.

b. The frequency is changed to 140 Hz. Draw the standing wave that will be formed on the wire.

Solutions

a. The standing wave formed here is the third harmonic. The length of the wire is $\frac{3}{4}$ of the wavelength: $\frac{3}{4}\lambda = 1.5\,\text{m}$. The

wavelength is therefore $\lambda = \frac{4}{3} \times 1.5 = 2.0\,\text{m}$, and the wave speed $v = f\lambda = 420 \times 2.0 = 840\,\text{m s}^{-1}$.

b. The speed of the wave is still $840\,\text{m s}^{-1}$, but the wavelength is changed to $\frac{v}{f} = \frac{840}{140} = 6.0\,\text{m}$. This is exactly four times longer than the length of the wire. Hence, the first harmonic wave will be formed, with the node at the clamped end and the antinode at the free end.

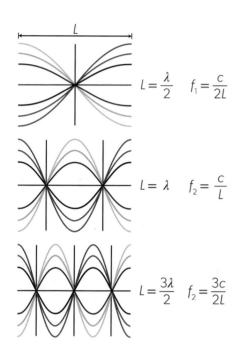

$L = \frac{\lambda}{2} \quad f_1 = \frac{c}{2L}$

$L = \lambda \quad f_2 = \frac{c}{L}$

$L = \frac{3\lambda}{2} \quad f_2 = \frac{3c}{2L}$

▲ **Figure 9** The first three harmonics of a stiff wire free at both ends.

(c) Two free ends

Although it is hard to imagine a string being free at both ends, it is possible to generate a standing wave on a flexible wire. However, it would be necessary to support the wire at a nodal position.

The standing wave will have an antinode at each end. This leads to the sequence of harmonics shown in Figure 9. (The second harmonic will be supressed if the wire is clamped in the centre as this forces the centre to be a node.)

The sequence of harmonics (first harmonic, second harmonic and so on, in this case) is identical to those when there are two fixed ends. The difference is in the appearance of the harmonics as the nodes and antinodes appear at different places on the standing wave.

Worked example 6

A metal rod of length 0.80 m is clamped in the middle and made to vibrate in the first harmonic mode, as shown. The speed of the transverse waves in the rod is $3200\,\text{m s}^{-1}$.

a. Calculate the frequency of the first harmonic mode.

b. Determine whether the rod will have any further harmonics of frequencies below 5 kHz.

Solutions

a. The length of the rod is one half of the wavelength. Hence $\lambda = 1.6\,\text{m}$.
$f = \frac{3200}{1.6} = 2\,\text{kHz}$.

b. The second harmonic mode, of wavelength 0.80 m and frequency $2 \times 2000 = 4\,\text{kHz}$, is suppressed because it requires the mid-point of the rod to be an antinode. The next possible mode is therefore the third harmonic, of wavelength $\frac{2 \times 0.80}{3} = 0.53\,\text{m}$ and frequency $3 \times 2000 = 6\,\text{kHz}$. We can see that, except for the first harmonic, no other standing wave mode has a frequency less than 5 kHz.

Practice questions

2. A guitar string of length 66 cm vibrates in its first harmonic mode of frequency 380 Hz.

   a. Calculate the speed of the waves on the string.

   b. The vibrating length of the string is reduced by pressing one point of the string against the fretboard of the guitar. The string now produces a note of frequency 480 Hz. Calculate the vibrating length of the string.

3. A standing wave is set up on a string of length 60 cm fixed at both ends, as shown. The speed of the wave on the string is 380 m s^{-1}.

   a. Calculate the frequency of the standing wave.

   b. Two points on the string are separated by 30 cm. Neither of the points is a node of the wave. State the phase difference between the oscillations of the two points.

4. A third harmonic standing wave is set up on a string fixed at one end. What is the phase difference between the midpoint of the string and the free end?

   A. 0   B. $\frac{\pi}{4}$ C. $\frac{\pi}{2}$ D. π

5. A stiff wire of length 1.0 m is fixed at one end and has the other end free. The wave speed in the wire is 400 m s^{-1}. What are the three lowest frequencies of standing waves that can be set up in the wire?

   A. 100 Hz, 200 Hz, 300 Hz
   B. 100 Hz, 300 Hz, 500 Hz
   C. 200 Hz, 400 Hz, 600 Hz
   D. 200 Hz, 600 Hz, 1 kHz

6. A metal rod of length 60 cm is clamped at the midpoint and vibrates in the first harmonic mode of frequency 2.3 kHz. Calculate the speed of the wave on the rod.

Standing waves in pipes

Some musical instruments make sounds using pipes rather than strings.

With a string, the vibrating string takes up the shape of the standing wave as it moves and the curved lines on the diagram represent the successive maximum displacements of the string over a half cycle.

However, for sound in a gas, the wave is longitudinal. The standing wave diagrams now relate to the displacement of the air molecules backwards and forwards along the central axis of the pipe. The end (boundary) of a pipe can be open or closed and the type of boundary determines the nature of the reflection of the sound wave.

> The nature of sound is discussed in Topic C.2 (page 398). You should be familiar with the relationship between displacement of the gas particles and the pressure variations in the gas.

Reflection at a closed end

The wave is longitudinal, so the individual particles in the wave are moving parallel to the central axis of the pipe (Figure 10). This leads to a longitudinal wave that travels up to the closed end, moving to the right in the figure. This closed end does not permit these particles to move along the axis there. The reflected wave must move away from the end, exactly cancelling the incident wave at the boundary.

🧪 Assumptions

Our usual assumption, unless stated otherwise, is that the gas in the pipe is air. When a different gas is used, the speed of sound will change for particular temperature and pressure conditions, so that the constant of proportionality between frequency and wavelength will be different.

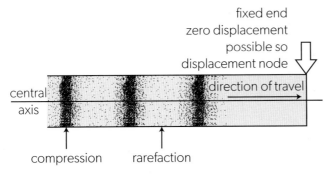

fixed end
zero displacement
possible so
displacement node

central axis

direction of travel

compression   rarefaction

▲ Figure 10 A longitudinal wave travels along a pipe towards a fixed end. The molecules cannot be displaced at the end and so the boundary condition here is a displacement node.

A closed end is a (displacement) node for a sound wave.

Reflection at an open end

It is easier to understand what is happening when a sound wave is reflected at the open end of a pipe by considering the variation in pressure. Because the pipe is open, the pressure at the open end must always be atmospheric.

Begin by visualizing the incident wave as a compression of molecules moving towards the open end of the pipe. The movement of the compression is of one compressed group of molecules pushing against other molecules that are less compressed. This is how the compression wave moves along. Remember that, on average, the individual molecules do not move over time as the wave goes through them.

When the compression wave reaches the open end, it is no longer restrained by the pipe walls. This compression wave corresponds to gas above atmospheric pressure and the wave must begin to spread out from the end of the pipe. It accelerates until its pressure has fallen to atmospheric pressure, at which moment the molecules have their highest speeds away from the pipe and the momentum is at its greatest. The molecules continue to move. This leads to a region of below-average pressure between this high-pressure region and the open end—a rarefaction. This low-pressure region propagates down the pipe away from the open end as the air begins to flood in from outside the pipe. This results in a pulse of high pressure travelling towards the open end which is reflected as a pulse of low pressure travelling away from the open end. There is a π phase change in the pressure wave.

To sum up:

- At an open end in a pipe, the standing wave has a (displacement) antinode because the molecules are free to move.

- At a closed end in a pipe, the standing wave has a (displacement) node because the wall prevents the molecules from moving along the pipe.

Standing wave patterns in pipes

Your knowledge of the boundary conditions for a pipe and your earlier work on the standing waves on a string now enable you to make a prediction about the standing waves allowed in pipes.

The harmonic series for the three cases are shown in Figure 12.

The standing waves for pipes **open at both ends** and **closed at both ends** are very similar. Only the boundary conditions shift, leading to the same sequence of harmonics. All harmonics are present (first, second and so on). The number of half-wavelengths in the pipe increases by one with each harmonic.

The standing waves for pipes **closed at one end** are different. Every even-numbered harmonic is suppressed so that only the first, third, fifth… harmonics appear.

As usual, it is straightforward to determine the sounding frequencies of a harmonic for any of the pipes. Take the third harmonic for the pipe closed at one end:

$$L = \frac{3}{4}\lambda \text{ so } \lambda = \frac{4L}{3} \text{ and } f_3 = \frac{3c}{4L}$$

whereas, for the third harmonic of a pipe open at both ends,

$$L = \frac{3}{2}\lambda \text{ so } \lambda = \frac{2L}{3} \text{ and } f_3 = \frac{3c}{2L}$$

 Figure 11 Organ pipes of various sizes. The circular pipes are open at both ends, whereas the square pipes are closed at one end. The metal collars on the round pipes can be moved to enable a fine adjustment in the tuning. The square pipes have a plunger for this.

Wave diagrams

Remember that the waves in the pipe are longitudinal and the positions of the nodes are due to the movement of molecules backwards and forwards along the axis of the pipe. The lines on the diagrams showing the wave do not represent transverse standing waves as you might at first think. They are graphical representations of the amplitude of the gas molecules at each point in the tube. They show the largest displacements half a cycle apart. Where the lines intersect, there is a node. Where they are at a maximum separation, there is an antinode.

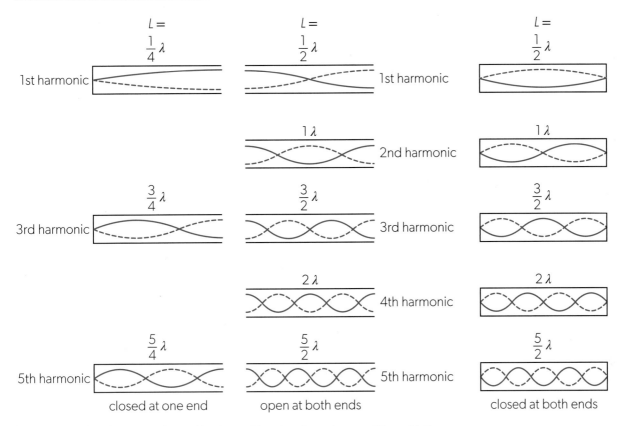

▲ Figure 12 Standing waves formed in pipes with various boundary conditions. L is the length of the pipe.

Notice an important difference between the first harmonics. A pipe closed at one end sounds its first harmonic at half the frequency of the same length of a pipe open at both ends.

Data-based questions

The table below shows the length, L, of some organ pipes measured to the nearest 5 mm. The frequencies, f, of the notes that they produce is also given. The uncertainty in these frequencies is 3%.

L/cm ± 0.5	f/Hz ± 3%
27.3	523
34.9	415
56.2	262
71.9	208
101.9	145
165.2	87
212.7	69

- Tabulate values of f^{-1} and calculate uncertainties in these values.

- Plot a graph of f^{-1} against L and add a line of best fit.

- Add error bars to your graph and find the gradient of your line.

- By considering maximum and minimum uncertainties, find the uncertainty in your value for the gradient.

- The organ pipes were open at both ends. Use your value for the gradient to find the speed of sound.

Global impact of science—Music meets physics II

When a flautist or recorder player blows too hard, the sound suddenly shoots up by an octave in pitch. But when a clarinettist does the same thing, the sound goes up an octave and a half (a twelfth in musical language, or three times the first harmonic). The reason for this difference is straightforward. The flute and the recorder are pipes that are open at both ends. The clarinet is closed at one end by the reed through which the player blows. This means that there is no second harmonic for the clarinet unlike the other two wind instruments. When the instrument is playing the first harmonic and suddenly there is more energy transfer into the pipe (a harder blow by the player), then the oscillation in the pipe jumps suddenly to the third harmonic.

ATL Social skills—Appreciating the diverse talents of others

Many physicists in history have also been good musicians. Albert Einstein is reported to have said "life without playing music is inconceivable for me. I live my daydreams in music. I see my life in terms of music. I get most joy in life out of music."

Both Newton and Young wrote on acoustics and the tuning of musical instruments and this led to further work on waves.

Galileo was born into a musical family (his father was a composer, a lute player and one of the first to explore the relationship between the tension in a string and the note it produces). Galileo himself was an excellent keyboard and lute player. Galileo's experimental methods—combining predictive mathematical theories with experimental observation—provided the basis for modern scientific methods.

▲ Figure 13 Einstein playing the violin.

Worked example 7

The first harmonic frequency of the standing wave in a pipe that is closed at one end and open at the other is 440 Hz. Calculate the first harmonic frequency in a pipe of the same length that is closed at both ends.

Solution

The wavelength of the first harmonic wave in a pipe that is closed at both ends is halved compared with the wavelength in the half-open pipe (because the length of the pipe now fits one-half of the wavelength, compared with only one-quarter of the wavelength for the half-open pipe). The wavelength is halved, so the frequency will double: $f = 880$ Hz.

Worked example 8

A loudspeaker emits a note of a single frequency towards the open end of a pipe. The other end of the pipe is closed with a moveable piston. The position of the piston is adjusted until a loud sound is heard from the pipe. The diagram shows the standing wave pattern formed in the pipe for a particular position of maximum loudness.

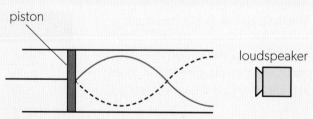

The piston is now moved to the right by 0.24 m and another loud sound is heard from the pipe. The speed of sound in the pipe is 340 m s^{-1}.

a. Calculate the frequency of the note emitted by the loudspeaker.

b. Determine the sounding length of the pipe for the situation shown in the diagram.

Solutions

a. A loud sound is heard again when the piston has moved by one-half of the wavelength, to the position of the next node of the original standing wave.
$\frac{\lambda}{2} = 0.24$ m $\Rightarrow \lambda = 0.48$ m. The frequency is $f = \frac{340}{0.48} = 710$ Hz.

b. The sounding length of the pipe is reduced from $\frac{3\lambda}{4}$ to $\frac{\lambda}{4}$. The original sounding length is $\frac{3\lambda}{4} = \frac{3 \times 0.48}{4} = 0.36$ m.

Practice questions

7. A student blows across the top of a half-open pipe of length 20.0 cm. A first harmonic standing wave of frequency 440 Hz is set up in the pipe.
   Calculate the speed of sound in the pipe.

8. Calculate the first two harmonic frequencies that can be produced in a tube of length 0.50 m that is:
   a. open at both ends
   b. open at one end and closed at the other.
   Take the speed of sound to be 340 m s^{-1}.

9. The speed of sound in air increases with temperature. A student measures the frequency of the first harmonic standing wave in a pipe. The air temperature is increased. Outline the change, if any, of:
   a. the wavelength
   b. the frequency of the first harmonic in the pipe.

Resonance

Natural frequency

Many mechanical systems oscillate. This often occurs when there is some mass (or inertial equivalent) in the system that is coupled to a spring-like component. A simple example of this is the suspension of a car (Figure 14). The car acts as the mass and a substantial spring supports the body of the vehicle on the subframe. This combination will oscillate when free to do so. Going over a bump in the road will provoke the oscillation.

When a mass–spring system is displaced from equilibrium and released, it will oscillate. When there is little or no friction in the system, then the oscillations can continue for a long time. These are known as **free vibrations**. The mass and the spring determine this oscillation frequency which is known as the **natural frequency** f_0. You already know that, for a mass m attached to a spring of constant k,

$$f_0 = \frac{1}{2\pi}\sqrt{\frac{k}{m}}$$

Generally, however, this simple equation for a mass oscillating on a spring does not describe the complex behaviour of a mass–spring system where there is friction and damping.

▲ Figure 14 One type of suspension unit used in motor vehicles. The blue cylinder inside the spring is the damper (shock absorber) and usually consists of a piston filled with oil.

Damping

A completely frictionless oscillation is rare. Usually, friction in the system eventually stops the oscillation. The energy is transferred away from the oscillation through turbulence, friction at moving surfaces, air resistance and so on. These are **damping forces** (or just **damping**).

Damping acts in the opposite direction to the resultant restoring force provided by the system and generally increases with speed. The damping, therefore, has its maximum effect when the system is close to or at its equilibrium point (maximum speed) and is least (often zero) at the maximum displacement (zero speed). The system must do work to overcome the damping and this energy is transferred from the kinetic energy of the oscillator. Instead of maintaining the constant total energy of simple harmonic motion, the energy decreases with time. The damping increases the time period (and decreases the frequency) of the oscillation, although this is only noticeable when the damping is large.

For small damping the variation of displacement with time will be similar to that shown in Figure 15. This oscillator is said to be **lightly (or under-) damped**. The time period (and therefore the frequency) of the oscillator is largely constant throughout the motion.

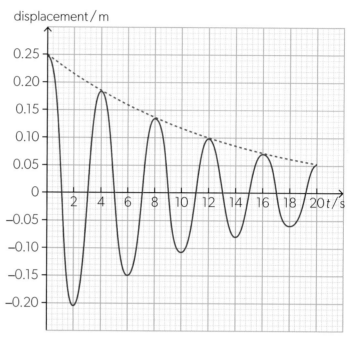

▲ Figure 15 Lightly damped (or under-damped) oscillations. The dashed line shows the exponential nature of the decay.

The dashed line in Figure 15 touches the maximum value of the displacement for one displacement direction. It shows the variation with time of the amplitude of oscillation. Damping halves the amplitude of the oscillation with a constant "half-life". For the damped oscillation of Figure 15, this "time to halve" is always around nine seconds. The greater the degree of light damping, the shorter the time to halve the maximum amplitude.

Not all damping is light, however. As the amount of damping increases, the shape of the displacement–time curve also changes dramatically. Figure 16 shows the original lightly-damped curve along with two other cases. In the over-damped case, the system takes a very long time to return to the equilibrium position after release. There is no oscillatory behaviour at all. This is **heavy damping.**

A transition between light and heavy damping occurs with a case where the oscillator takes the minimum possible time to come to rest. There is no true oscillation, and the system simply goes back to a rest position at equilibrium as quickly as possible. The energy is transferred from the oscillator at the maximum possible rate. This is known as **critical damping**. The minimum return time makes it important in many applications, including fire-door closure mechanisms and in car suspension systems.

When you have studied Theme E, you will recognize the shape of this dashed line, which has the usual properties of exponential change as in radioactive decay.

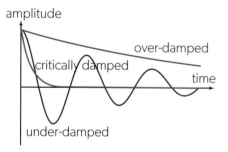

▲ Figure 16 A comparison of the three degrees of damping. The critical-damping case allows the system to dissipate its energy at the fastest possible rate.

Data-based questions

An orchestral tam-tam, sometimes known as a gong (Figure 17), produces a sound that contains many frequencies. Gongs are tuned versions of the instrument that have distinct pitches.

▲ Figure 17 An orchestral gong.

A gong was struck and the sound intensity was measured in the time afterwards. Sound pressure in decibels is a logarithmic quantity. An exponential decay in the amplitude results in a linear decrease in the sound pressure (in dB). The absolute error in each measurement of sound pressure level is ±2 dB.

The results are shown in the table.

• Plot a graph of the data. Include vertical error bars.

• Add a straight line of best fit and find the equation of your line. By considering the error bars, find the uncertainty in your value of the gradient.

• A drop in the sound pressure level of 3 dB means that the intensity of the sound wave has halved. Use your graph to find the time for the sound intensity to halve. Give an uncertainty with your value.

• The background sound pressure level was 30 dB. When the sound pressure level from the gong dropped below this, the sound was longer audible. Using your equation for the line of best fit, calculate the amount of time for which the gong was audible. Give an uncertainty for this time.

Time / s	Sound pressure level / dB (± 2)
0	88
5	83
10	75
15	70
20	63
25	59
30	55

 What is the relationship between resonance and simple harmonic motion?

The frequency equations for the mass–spring system and the simple pendulum are

$$f_0 = \frac{1}{2\pi}\sqrt{\frac{k}{m}} \quad \text{and} \quad f_0 = \frac{1}{2\pi}\sqrt{\frac{g}{l}}$$

It is possible to generalize these expressions to

$$f_0 = \frac{1}{2\pi}\sqrt{\frac{\text{spring term}}{\text{inertial term}}}$$

so that for more "elasticity" in the system the frequency increases and with more "mass" in the system the frequency decreases. Provided that these two contributions can be identified, the frequency equation can be derived for a particular system.

The presence of friction or other damping modifies the simple harmonic motion equation with the addition of extra terms that will depend on the instantaneous speed of the mass (or speed raised to a power). This will change the natural frequency, although, depending on the amount of damping, this frequency shift may be small. To observe a constant amplitude of oscillation, the driver must be transferring the same energy into the system per cycle as damping is transferring energy away.

Thus, resonance will occur at a frequency not dictated by the simple harmonic motion alone but by the interaction between the oscillation of the system and its energy losses.

In fact, once there is damping, then the motion can no longer be regarded as simple harmonic. As described in Topic C.1, simple harmonic motion leads to oscillations that never die out because there is no energy transfer from the system.

▲ Figure 18 A mass–spring oscillator as an example of forced vibrations.

Forced vibrations

Whereas **free vibrations** occur when an oscillating system oscillates at its own natural frequency, **forced vibrations** occur when an oscillating system is driven by another oscillator. A good example of this is a mass–spring oscillator being driven by an electric motor whose rotational speed can be varied (Figure 18). The diagram shows a disc attached to the drive shaft of a motor (not shown) with the top of the spring fixed to the disc off-centre. The spring will be moved up and down at the rate determined by the rotation speed of the motor. This transfers energy into the kinetic and gravitational potential energies of the mass. The angular frequency of the disc can be varied and the amplitude of motion of the mass measured.

The rotational frequency of the motor corresponds to the **driving frequency** of the system. Remember that the mass–spring system has its own natural frequency which does not need to be equal to the driving frequency.

When the driving frequency is very different from that of the mass–spring system, the amplitude of the mass will be small. When the two frequencies are close, then the amplitude will be much larger. When the two are very close or identical, then the amplitude of motion of the driven system will be extremely large, possibly even enough to break the spring.

Different degrees of damping can be provided by submerging the mass in water, or in oil or another viscous liquid.

This behaviour is summed up in Figure 19. The family of graphs is sometimes known as **resonance curves.**

When there is no (zero) damping, then the mass–spring system will be driven to large amplitudes at the natural frequency. This is **resonance**. The **resonant frequency** is defined here as the frequency at which the amplitude is a maximum.

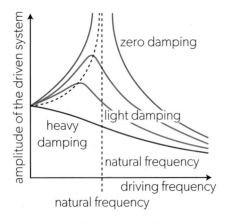

▲ Figure 19 The amplitude–frequency graph for different degrees of damping.

When there is light damping, the system will show a maximum amplitude at a frequency slightly less than the natural frequency. As the degree of damping increases, this frequency for the maximum amplitude drifts to smaller frequencies. You can see that the peak of the damping curves moves to the left as the amount of damping increases.

When there is heavy damping, an obvious maximum may not appear, the maximum amplitude will be at the lowest frequencies and the amplitude then decreases with increasing driving frequency.

 How does the amplitude of vibration at resonance depend on the dissipation of energy in the driven system?

Resonance is caused when the driving and natural frequencies are close (identical if there is no damping). The driver matches the driven system and can transfer energy to it at exactly the correct rate. One issue, however, is the phase at which the energy is supplied. Think about pushing a child on a swing. The push *rate* must be correct, but the push also must be supplied at the right moment in the cycle. The maximum push has to be given when the child is moving at the fastest point in the cycle, at the equilibrium position. In other words, the push for the swing (the driver) has to be 90° ahead of the swing (driven) itself.

 How can resonance be explained in terms of conservation of energy?

At both small and large driving frequencies, well away from the natural frequency, the energy transfer from the driver cannot occur efficiently. This is because the difference in frequencies means that the driven system will receive the "push" at different points of successive cycles. This leads to small amplitudes in the driven system well away from resonance.

Resonance curves

You may also see resonance curves where the resonance and resonant frequency are defined in terms of the *maximum energy* in the system (rather than the maximum amplitude). When this is done, the resonance curves sit vertically above each other (Figure 20).

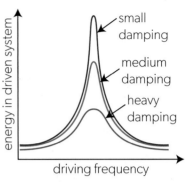

▶ Figure 20 The energy–driving frequency graph when resonance is described in terms of the maximum energy in the system rather than the largest amplitude of the driven system.

Using Barton's pendulum to demonstrate the relationship between a driver and its driven system.

- Inquiry 1: Demonstrate creativity in the designing, implementation or presentation of the investigation.

- Inquiry 2: Identify and record relevant qualitative observations.

- Inquiry 2: Interpret diagrams, graphs and charts.

In the Barton's pendulum apparatus, driven pendulums of small mass and varying lengths are driven by a heavy pendulum of length midway between the longest and shortest driven ones (Figure 21).

Set this array up yourself or find a video on the web by searching for "Barton pendulum".

The paper cones (1–5) are light and provide damping; the brass bob (0) is the driver. Pendulums 0 and 3 are the same length. Cone 3 will oscillate with the greatest amplitude and is 90° behind the driver. Cones 1 and 2 have small amplitudes and are 180° behind the driver. Cones 4 and 5 have small amplitudes and are in phase with the driver.

The phase relationship in Barton's pendulum mentioned above is summed up in Figure 22.

When a driver frequency is smaller than the natural frequency, the driver and driven system are in phase. The driver is dragging the mass and the spring along with it, so they stay in phase.

When the driver frequency is greater than the natural frequency, then the driven system tries to respond more slowly than is being allowed by the driver and so there is a phase shift of 180° between the two.

The amount of damping determines the sharpness of the transition from in phase to out of phase as the system moves through the resonant frequency, as shown in Figure 22.

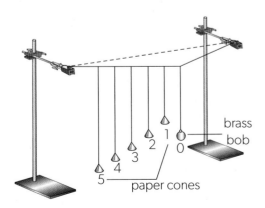

▲ Figure 21 Barton's pendulums.

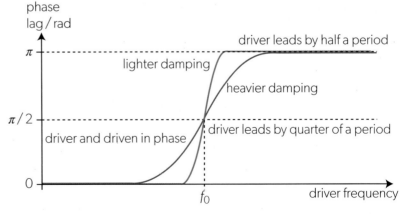

▲ Figure 22 The phase relationships between driver and driven systems as the frequency changes during a resonance experiment.

Investigating resonance

- Inquiry 2: Collect and record sufficient relevant quantitative data.

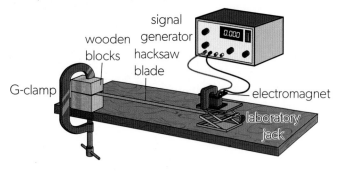

▲ Figure 23 Investigating resonance using a hacksaw blade driven by an alternating magnetic field.

An oscillating hacksaw blade can be used to demonstrate and investigate resonance effects.

- The blade is clamped firmly at one end. The natural frequency of the blade oscillation depends on the projecting length L of the blade.

- The other end of the blade lies close to an electromagnet powered by a variable-frequency signal generator. A vertical ruler measures the amplitude of the blade oscillation.

- Use the apparatus to measure the variation of the amplitude of the end of the blade with:

  ○ signal generator frequency f supplied by the signal generator

  ○ blade length L.

- Plot graphs to display the effects of resonance in this system.

- Vary the damping of the blade by adding strips of card to the blade to increase air resistance. How does damping affect the system?

 Extending the model of forced vibrations

The description of forced vibration and the effects of resonance given here assumes that the systems—driven and driver—have reached a steady state where both have unchanging amplitudes of oscillation. It ignores how the system arrives at this state.

When the system is set in motion for the first time, the driven system receives an initial impulse from the driver that makes it oscillate at its own natural frequency. But because the driver exerts a cyclic force on the driven part of the system, the driving frequency will eventually dominate the motion of the driven oscillator.

When you watch a set of Barton's pendulums oscillating, you should be able to see this transient and steady-state behaviour clearly as the natural frequency of each pendulum is gradually taken over by the frequency of the heavy driver. It can be enhanced by adding small paper cones to the pendulums to increase the damping of each cone.

 How can the idea of resonance of gas molecules be used to model the greenhouse effect? (NOS)

The greenhouse effect is mentioned a number of times during the course, in terms of thermal effects, resonance and in mechanical and atomic terms. All these separate disciplines within physics and chemistry link to give an account of the important effects in our atmosphere. It is in the Nature of Science for scientists from separate disciplines to pool their expertise to solve a global question.

Resonance in practice

Resonance effects offer both advantages and disadvantages in many areas of science and engineering. Examples are listed in Table 1.

Advantages of resonance	Disadvantages of resonance
• Microwave ovens use resonance effects both to produce the radiation and to enable it to excite the water molecules in food. • Ozone in the atmosphere absorbs ultraviolet radiation through a resonance effect in the molecules. This prevents most of the radiation reaching Earth's surface and damaging living tissue. • Nuclear magnetic resonance (NMR) is a technique used extensively in many branches of science, in particular in diagnostic medicine where it is known as magnetic resonance imaging (MRI). • Laser light is produced using resonance effects to set up standing waves at light frequencies in optical cavities.	• Vibrations in bridges can lead to undesirable effects, instability and damage. Example of this are: the collapse of the Tacoma Narrows Bridge in 1940 and the Millennium Footbridge in London. The latter swayed from side to side when it opened, as the footsteps of passengers fell into synchronism with the bridge movement. The fault was cured by fitting dampers. • Damaging or annoying vibrations are common in motor-driven systems. Examples are rear-view mirrors in lorries driven by low-frequency pulses when the engine is at low speed and washing machine vibrations at certain drum speeds during spin drying.

▲ Table 1 Some advantages and disadvantages of resonance.

Worked example 9

A simple pendulum undergoes damped oscillations.

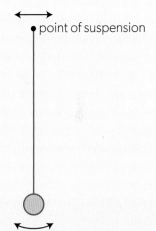

a. Outline **two** ways in which oscillations of a damped pendulum differ from an undamped one.

A time-varying driving force is applied to the pendulum so that its point of suspension vibrates horizontally.

b. Explain why the pendulum is oscillating with a constant amplitude.

The frequency of the driving force is adjusted so that the amplitude of the pendulum is a maximum.

c. Compare the frequency of the driving force to the natural frequency f_0 of the pendulum.

The frequency of the driving force is now increased so that it becomes much greater than f_0.

d. Describe, for the pendulum bob after the change:

   i. the amplitude   ii. the phase of oscillation.

Solutions

a. Undamped oscillations have a constant amplitude, while the amplitude of a damped pendulum decreases with time because the energy is transferred away from the pendulum.
The frequency of a damped pendulum is less than the natural frequency of undamped oscillations, and the difference depends on the degree of damping.

b. A steady state is reached in which the driver transfers energy to the pendulum at the same rate as it is transferred away due to damping.

c. In the presence of light or moderate damping, the amplitude of oscillations is a maximum when the driver frequency is less than the natural frequency f_0 of an undamped pendulum.

d. i.   The system is now far from resonance and the amplitude of the pendulum has decreased.

   ii. The pendulum bob oscillates out of phase (phase difference π) with the point of suspension.

C.5 Doppler effect

What are some practical applications of the Doppler effect?

How can the Doppler effect be explained both qualitatively and quantitatively?

Why are there differences when applying the Doppler effect to different types of waves?

The Doppler effect is a frequency shift that is detected when a source of waves is moving relative to the observer of the waves. It was originally a curiosity when it was first identified in 1842, even though the hypothesis was tested and found to be correct within a few years. Nowadays, techniques using the Doppler effect are used extensively in medicine, astronomy, flow measurements and in many other applications.

The Doppler effect is described in terms of the movement of the wavefronts through a medium as perceived by the observer. These qualitative descriptions are used as a basis for quantitative equations that can predict the frequency shift.

Topic A.5 showed that, when the speed of an object approaches c, we need to apply the theory of special relativity rather than Newtonian mechanics to the situation. This applies to the Doppler effect too, whatever the type of wave under consideration. The Doppler effect for light is always relativistic because c is invariant.

The speed of sound, whether in fluid or solid, is never close to c. The speed of a sound wave through diamond, at $12\,\mathrm{km\,s^{-1}}$, is one of the largest values known and this is still only 0.004% of the speed of light in a vacuum. The

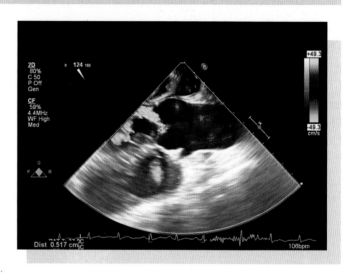

▲ Figure 1 An ultrasound image of the heart. The Doppler effect enables the speed of blood flow to be measured. The different colours represent the different flow speed.

relativistic effects are of the order of 10^{-6}%. For this reason, relativity can be ignored when evaluating how the Doppler affects sound and other mechanical waves. However, when considering Doppler shifts for electromagnetic radiation, we must be more careful to establish the approximations being used.

In this topic, you will learn about:

- the Doppler effect for sound waves and electromagnetic waves

- visualizing the Doppler effect when either the source or the observer is moving

- the relative change in wavelength for a light wave due to the Doppler effect

- shifts in spectral lines that provide information about the motion of astronomical objects

- quantitative Doppler shifts in sound and mechanical waves when either the source or the observer is moving.

AHL

Introduction

An ambulance travels along the road at speed, sounding its siren. You notice a change in the frequency of the siren—high to low—as the vehicle passes you. This is the Doppler effect. It was first suggested by the Austrian scientist Christian Doppler, who showed mathematically that an observed wave frequency depends on the relative speed between the sound source and an observer. In France, the effect is ascribed to Doppler–Fizeau because the Frenchman Amand Fizeau extended the work to the spectral shifts in light from stars shortly before Doppler's death.

The nature of the Doppler effect

As you saw in Topic C.2, wavefronts from a point source spread out in a sphere. Figure 3 is a wavefront diagram showing the effect in two dimensions, but this is easily extended to 3D in your imagination. S is the point source and the observer of the waves is O. In this and subsequent wavefront diagrams, you are acting as a second observer, at rest relative to the medium, and you can see the wavefronts together with the movement of both source and observer.

This description and the later derivation of the Doppler equations on page 459 ignore relativistic effects. In Topic A.5 you met the theory of special relativity. This theory has practical importance when speeds are greater than about $\frac{c}{5}$ (c here is the speed of electromagnetic radiation in a vacuum). When we are dealing with speeds of sound or with observer/source speeds close to c, then we must allow for time dilation and length contraction effects. Wave speeds are invariably very much less than c, so this approximation is a good one.

▲ Figure 2 An observer will hear a change in the frequency of an ambulance's siren as it passes.

The changes to the Doppler equations under relativistic conditions are described on page 464.

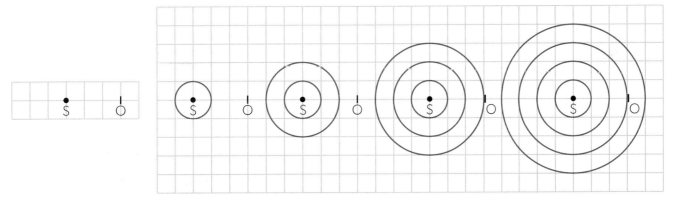

▲ Figure 3 A spherical wave expands. Both source (S) and observer (O) are stationary. O observes the same wavelength and frequency as emitted by S.

Figure 3 shows what happens when neither the source S nor observer O move. In this diagram the source is emitting a wavefront at a regular rate (the frequency of the wave). The red circles show the position of these wavefronts at equal time intervals. Time increases from left to right in the figure. The circular wavefront increases in radius by one square between diagrams. The first wavefront reaches the observer three time intervals after it was generated. The second wavefront crosses the observer one time interval later. The source and the observer agree on the rate at which the wavefronts are being generated and detected.

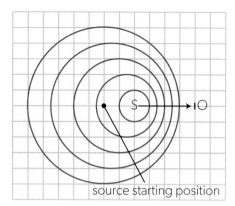

▲ Figure 4 O is stationary and S is moving towards O at half a square per time interval. The wavefronts in the medium are compressed into a smaller volume because of the movement of S.

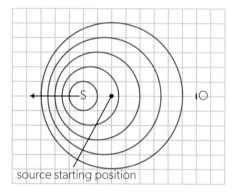

▲ Figure 5 When S moves away from O, the wavelengths observed by O are further apart.

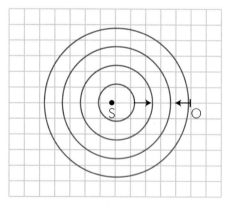

▲ Figure 6 O is now moving towards S. O crosses the wavefronts more often than they are emitted, so observes a higher frequency.

What happens when the observer or source move relative to the wave as it spreads through the medium? There are two cases: source moving and observer stationary, and source stationary with observer moving. These are analysed separately. For both cases, the source and observer move along the straight line that joins them.

The Doppler effect analysed

(a) Moving source — Stationary observer

When the Doppler shift is caused by source movement, there is a change in the wavelength of the waves relative to the stationary medium. This is observed as a shifted frequency because the wavefronts sweep across the observer at a different rate from that at which they were emitted.

In Figure 4, the source S is moving to the right. It moves half a square during the time that a wavefront advances by a full square. Mathematically, the source speed is half that of the wave speed.

Five time intervals elapse over the course of the diagram. There are five wavefronts shown and the source has moved 2.5 squares to the right. The wavefront emitted earliest is about to cross the observer. Because the wavefront moves at the wave speed in the medium, it will cross the observer twice as quickly as it would without the source movement. (Look again at Figure 4, remembering that the radius of each wavefront is constantly expanding by one square every time interval.) The observed frequency increases for the stationary observer (in fact, doubling in this case).

When the source is moving away from the observer, the frequency change is in the opposite direction. The observed frequency is now less than the emitted frequency (Figure 5).

The scale distances between wavefronts as they move across the observer are larger than for the original static situation (Figure 3), showing that the wavefronts pass over the observer less frequently than before.

(b) Stationary source — Moving observer

When the observer moves but the source is stationary, the wavefronts move symmetrically through the medium, expanding as concentric spheres. However, because the observer moves towards or away from them, they are detected more often or less frequently than they were emitted.

The waves that are created in the medium have the usual spherical pattern (circular in 2D).

When O is moving towards S, the wavefronts are crossed more quickly than they were emitted. An increased frequency is observed with an unchanged wavelength (because that wavelength is determined by the medium, not by the observer). Relative to the observer, the speed of the wave has changed.

When O is moving away from S, then the wavefronts take longer to catch the observer up, so that the time period is longer and the apparent frequency (according to the observer) is lower.

Worked example 1

A source of sound S is moving at a constant speed along the line joining two stationary observers A and B. The diagram shows the wavefronts emitted by S in equal time intervals.

a. Explain, with reference to wavelength and wave speed, why observer B will detect sound of a higher frequency than observer A.

b. Determine the ratio $\dfrac{\text{frequency observed by B}}{\text{frequency observed by A}}$.

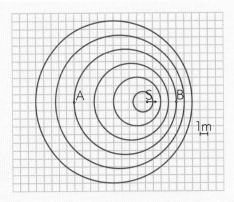

Solutions

a. B observes a shorter wavelength of sound than A, because at the position of B the wavefronts are closer to each other than at the position of A. The speed of the waves is the same according to both observers because they are both stationary relative to the medium. Since $f = \dfrac{c}{\lambda}$, the shorter wavelength observed by B results in a higher frequency.

b. The neighbouring wavefronts are 2.5 m apart at A and 1 m apart at B. Hence $\dfrac{\lambda_A}{\lambda_B} = 2.5$.

$$\frac{\text{frequency observed by B}}{\text{frequency observed by A}} = \frac{\frac{c}{\lambda_B}}{\frac{c}{\lambda_A}} = \frac{\lambda_A}{\lambda_B} = 2.5.$$

Calculating the observed frequency due to the Doppler effect

(a) Moving source — Stationary observer

The change in frequency can be related to the speed of the wave v, the frequency emitted by the source f and the source speed u_s.

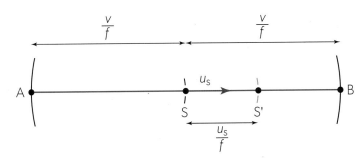

▲ Figure 7 Modelling Doppler shift when the source is moving and the observer is stationary.

Figure 7 shows the source position S as the point source emits its first wave. One time period T later, the source has moved to position S'. T is $\dfrac{1}{f}$. The distance SS' is the source speed × one wave period, which is $\dfrac{u_s}{f}$. The radius of the wavefront after time period T is $\dfrac{v}{f}$. During this first time period, the wave emitted when the source was at S has expanded to become a sphere of diameter AB.

Focus on the wavefront at B which is the position of the observer. At the point S', the source is just emitting another wavefront, so that from the observer's point of view the apparent wavelength of the wave is the distance from S' to B. This is $\left(\dfrac{v}{f} - \dfrac{u_s}{f}\right)$, which can be written as $\dfrac{(v - u_s)}{f}$.

The observed (apparent) frequency f' as detected by the observer is

$$\frac{\text{wave speed in medium}}{\text{apparent wavelength}} = \frac{v}{\dfrac{(v - u_s)}{f}}$$

This can be rearranged to

$$f' = f\left(\frac{v}{v - u_s}\right)$$

This equation predicts that, when the source is moving towards an observer at B, the observed frequency will increase (because the denominator in the fraction is smaller than the numerator).

An observer at A on the other side of the source will see the opposite effect. S is moving away and the apparent wavelength is now increased to $\dfrac{v}{f} + \dfrac{u_s}{f}$. When the equation is rearranged as before, the observed frequency becomes

$$f' = f\left(\frac{v}{v + u_s}\right)$$

and f' is smaller than f.

Worked example 2

An ambulance moves at a constant speed of $30.0\,\mathrm{m\,s^{-1}}$ towards a stationary observer. The siren of the ambulance emits a sound of frequency $1600\,\mathrm{Hz}$. The speed of sound in air is $340\,\mathrm{m\,s^{-1}}$. Calculate:

a.   the frequency of the sound heard by the observer

b.   the observed wavelength.

Solutions

a.   We have $v = 340\,\mathrm{m\,s^{-1}}$ and $u_s = 30.0\,\mathrm{m\,s^{-1}}$. The observed frequency is higher than the emitted frequency, so we must use the minus sign in the Doppler effect equation:

$$f' = f\frac{v}{v - u_s} = 1600 \times \frac{340}{340 - 30} = 1750\,\mathrm{Hz}.$$

b.   The observed speed of sound is $340\,\mathrm{m\,s^{-1}}$, so the wavelength becomes

$$\lambda' = \frac{v}{f'} = \frac{340}{1750} = 0.194\,\mathrm{m}.$$

Worked example 3

A train moving at a speed u along a straight track sounds a whistle of frequency $1300\,\mathrm{Hz}$. The frequency heard by an observer standing next to the track is $1250\,\mathrm{Hz}$.

a.   State the direction of motion of the train relative to the observer.

b.   Calculate the ratio $\dfrac{u}{v}$ of the speed of the train to the speed of sound.

Solutions

a.   The frequency heard is less than that emitted. Hence the train is moving away from the observer.

b.   The observer is stationary, and we can find the speed of the train u by solving the Doppler effect

equation: $f' = f\dfrac{v}{v + u} \Rightarrow u = \left(\dfrac{f}{f'} - 1\right)v$. From here, $\dfrac{u}{v} = \dfrac{f}{f'} - 1 = \dfrac{1300}{1250} - 1 = 0.040$. The train is

moving at 4% of the speed of sound.

Practice questions

In questions 1 and 2, assume that the speed of sound in air is 340 m s^{-1}.

1. A loudspeaker emitting sound with a frequency of 850 Hz moves in a circle at a constant linear speed of 12 m s^{-1}. An observer, at rest relative to the centre of the circle, is some distance away and in the plane of the circle. Calculate the maximum and the minimum frequency of the sound heard by the observer.

2. A train approaching a station sounds a horn of frequency 900 Hz. A stationary observer on the station platform measures a frequency of 970 Hz. Calculate the speed of the train.

3. A police car moves along the line joining two stationary observers P and Q. The speed of the car is 10% of the speed of sound in still air. The siren of the car emits a sound detected by both observers.

The frequency of the sound measured by observer P is f. What is the frequency measured by observer Q?

A. $\frac{11}{10}f$ B. $\frac{10}{9}f$ C. $\frac{12}{10}f$ D. $\frac{11}{9}f$

(b) Stationary source — Moving observer

Figure 8 shows the arrangement. The observer moves towards the source with a speed u_o. The speed of the wave in the medium is v.

▲ Figure 8 Modelling Doppler shift when the observer is moving towards the stationary source.

The observer at O begins exactly one wavelength (in the medium) away from the source at S. At this instant, another wavefront is emitted by the source just as the observer crosses the previous wavefront.

After a time T', the observer is at point P on Figure 8 and observes the second wavefront which has travelled from S in the time T'. This means that the true wavelength λ in the medium is

$$\lambda = u_o \times T' + v \times T'$$

The wavelength λ is equal to the product of the speed v in the medium and the time period T as emitted by the source:

$$\lambda = v \times T$$

Therefore $v \times T = u_o \times T' + v \times T'$ and, because $f' = \dfrac{1}{T'}$ and $f = \dfrac{1}{T}$, this leads to

$$\frac{v}{f} = \frac{u_o}{f'} + \frac{v}{f'}$$

Simplifying this equation gives

$$f' = f\left(\frac{v + u_o}{v}\right)$$

When the observer moves away from the source the sign is negative:

$$f' = f\left(\frac{v - u_o}{v}\right)$$

Worked example 4

Waves of frequency 0.50 Hz travel across the surface of a lake at a constant speed of 2.5 m s⁻¹. A boat is moving at right angles to the wavefronts. A passenger in the boat observes that the boat crosses the wavefronts with a frequency of 0.70 Hz. Calculate the speed of the boat relative to the lake.

Solution

We need to solve the Doppler equation for the unknown speed u_o of the boat. $f' = f\dfrac{v + u_o}{v} \Rightarrow u_o = \left(\dfrac{f'}{f} - 1\right)v$

We substitute $v = 2.5\,\text{m s}^{-1}$, $f = 0.50\,\text{Hz}$ and $f' = 0.70\,\text{Hz}$. $u_o = \left(\dfrac{0.70}{0.50} - 1\right) \times 2.5 = 1.0\,\text{m s}^{-1}$.

Worked example 5

A bat is flying in a cave at a constant velocity of 6.00 m s⁻¹ towards a flat vertical wall. The bat emits an ultrasound pulse of frequency 45.0 kHz towards the wall. The ultrasound is reflected off the wall and returns to the bat. Calculate the frequency of the ultrasound that the bat will hear, assuming that the speed of sound in air is 340 m s⁻¹.

Solution

The bat is a moving source of waves travelling towards the stationary wall, and the frequency of the

ultrasound reaching the wall is $45.0 \times \dfrac{340}{340 - 6.00} = 45.8\,\text{kHz}$. The wall reflects the ultrasound with no

change in frequency, and the bat becomes a moving observer of the approaching wave. The frequency

heard by the bat is therefore $45.8 \times \dfrac{340 + 6.00}{340} = 46.6\,\text{kHz}$.

Practice questions

In questions 4 and 5 assume that the speed of sound in air is 340 m s⁻¹.

4. A stationary siren emits sound of frequency 800 Hz.

    a. A cyclist is moving towards the siren with a constant speed of 9.0 m s⁻¹. Calculate the frequency heard by the cyclist.

    b. Passengers of a car moving away from the siren hear a frequency of 750 Hz. Calculate the speed of the car.

5. A stationary motion sensor emits a sound pulse of frequency 80.0 kHz towards a cart approaching the sensor at a speed of 18.0 m s⁻¹. The sound is reflected off the front of the cart and returns to the sensor. Determine the frequency of the returning sound.

    Hint: consider the cart as (1) a moving observer of the emitted pulse and (2) a moving source of the reflected pulse.

 Predictions—When both source and observer move

The equations can be combined for the general case where the source has speed u_s and the observer has speed u_o:

$$f' = f\left(\frac{v \pm u_o}{v \mp u_s}\right).$$

The upper sign is used when the source and observer are approaching; the lower sign when they are moving apart. You will not be asked questions involving the simultaneous movement of source and observer in the IB Diploma Programme physics examinations.

Doppler effect and light

The nature of electromagnetic radiation means that the Doppler analysis above cannot be used for light or any other part of the electromagnetic spectrum. This is because, as you saw in Topics A.5 and C.2:

- Electromagnetic radiation does not require a medium through which to travel.

- A postulate of special relativity states that the velocity of light waves is constant for all inertial frames (page 164). This is not the case for waves that require a medium such as sound. For example, when a sound source moves, the wavefronts in the medium are closer than they should be.

- The motion of a source and an observer cannot be distinguished and the concepts of source speed and observer speed have no meaning in special relativity. They must be replaced by a relative velocity.

However, when the relative speed v between source and observer is very much less than the speed of light c ($v \ll c$), the equations derived earlier are approximately correct. However, modifications to the equations are required:

- The wave speed symbol v changes to c.

- The source speed or observer speed becomes v. (There is now no need to distinguish between them.)

- The moving source–stationary observer equation $f' = f\left(\dfrac{v}{v + u_s}\right)$ changes to $f' = f\left(\dfrac{c}{c + v}\right)$. The equation $f' = f\left(\dfrac{c}{c + v}\right)$ can be re-written as $f' = f\left(\dfrac{1}{\left(\dfrac{c+v}{c}\right)}\right) = f\left(1 + \dfrac{v}{c}\right)$. The binomial theorem allows this to be expanded as $f' = f\left(1 - \left(\dfrac{v}{c}\right) + \left(\dfrac{v}{c}\right)^2 + \left(\dfrac{v}{c}\right)^3 + \cdots\right)$. As $v \ll c$, only the first two terms are significant so that $f' \approx f\left(1 - \dfrac{v}{c}\right)$. (You are only required to know this result, not its derivation.) A further simplification is possible: the change in observed frequency Δf is $(f' - f)$, so that $\Delta f \approx \dfrac{fv}{c}$, which leads to $\dfrac{\Delta f}{f} \approx \dfrac{v}{c}$ providing an equation for the fractional change in frequency.

It is straightforward to extend the equation to the fractional change in wavelength, giving

$$\frac{\Delta f}{f} = \frac{\Delta \lambda}{\lambda} \approx \frac{v}{c}$$

Remember that this set of equations is only true when the relative speed v between source and observer is much less than that of electromagnetic radiation c.

What are the similarities and differences between light and sound waves?

Light waves and sound waves have many similarities. Both types of wave demonstrate the wave properties described in earlier in Theme C. Reflections of light may be re-named as echoes in sound, but the basic description of the phenomenon is the same. If you fill a toy balloon with carbon dioxide gas (with its high "optical" density), it acts as a very effective lens for sound waves. Both sets of waves can undergo diffraction and interference with diffracting apertures and source separations of an appropriate size.

However, the mechanisms that lead to these effects in light and sound are very different. Light reflection involves absorption and re-emission of photons, whereas sound reflection involves compressions and rarefactions at a solid boundary. Also, electromagnetic radiation in a vacuum has a universal speed (which underpins the theory of special relativity in Topic A.5) which leads to differences between the Doppler behaviour of light and sound waves.

You will only need to use the result for the fractional changes in frequency and wavelength. The derivation given here is an explanation and is not required as part of the course.

What happens if the speed of light is not much larger than the relative speed between the source and the observer?

The Doppler effect for sound can seem quite different from that for electromagnetic waves both algebraically and in its origin. A relativistic treatment should really be used for both. However, as mentioned earlier in this topic, the speed of sound waves is so much less than c, so $\frac{\Delta f}{f} = \frac{\Delta \lambda}{\lambda} \approx \frac{v}{c}$ is a good approximation.

When v and c are closer, then a full relativistic treatment is required. The ideas of Topic A.5 lead to a full relativistic expression for the Doppler effect which applies to all waves:

$$\frac{f_R}{f_S} = \frac{\left(1 - \frac{v_R}{c_S}\right)}{\left(1 + \frac{v_S}{c_S}\right)} \times \frac{\sqrt{1 - \left(\frac{v_S}{c}\right)^2}}{\sqrt{1 - \left(\frac{v_R}{c}\right)^2}}$$

Here v_R and v_S are the speed of receiver and source, respectively. It is no longer appropriate to talk in terms of source and observer. The speed of sound in the medium is c_S. When v_R, v_S, and c_S are much smaller than c, then the value of the term within the square root is close to 1 and

$$\frac{f_R}{f_S} \approx \frac{\left(1 - \frac{v_R}{c_S}\right)}{\left(1 + \frac{v_S}{c_S}\right)}$$

which is the same as the approximate expression earlier.

Another special case is $c_S = c$, which gives the relativistic Doppler equation for light.

Worked example 6

A sodium lamp emits light of wavelength 588.995 nm. Calculate the wavelength according to an observer moving towards the lamp at a speed of 25 km s^{-1}.

Solution

The change in the wavelength can be calculated from the Doppler equation:

$\Delta \lambda \approx \lambda \frac{v}{c} = 588.995 \times \dfrac{25 \times 10^3}{3.00 \times 10^8} = 4.9 \times 10^{-2}$ nm. The distance between the source and the observer decreases, so the observed wavelength is shorter than the emitted wavelength.
$\lambda_{obs} = 588.995 - 0.049 = 588.946$ nm.

Worked example 7

An ultraviolet wavelength of 85.0 nm is observed in the spectrum of a distant galaxy. A corresponding stationary source on Earth emits a wavelength of 78.0 nm.
a. Explain the direction of motion of the galaxy relative to the Earth.
b. Calculate the relative speed of the galaxy.

Solutions

a. The observed wavelength (85.0 nm) is longer than the emitted wavelength (78.0 nm). Hence the galaxy is moving away from Earth.

b. $v \approx \dfrac{\Delta \lambda}{\lambda} c = \dfrac{85.0 - 78.0}{78.0} c = 8.97 \times 10^{-2} c$. The galaxy is receding at approximately 9% of the speed of light.

Practice questions

6. The frequency of light reaching Earth from a distant galaxy is 1.2% lower than the frequency of a corresponding stationary source.

   a. Calculate, in terms of c, the relative speed of the galaxy. State whether the galaxy is approaching or moving away from Earth.

   b. A stationary source emits a spectral line of wavelength 527.0 nm. Calculate the wavelength of the same line observed in the light from the galaxy.

7. A spaceship approaches a space station at a relative speed of 0.050 c. The space station emits a navigation signal of frequency 2.80 MHz towards the spaceship. Calculate the frequency of the signal received by the spaceship.

Applications of the Doppler effect
Medical

Ultrasound consists of sound waves that have a frequency greater than 20 kHz and that are inaudible to most people. It has several uses in medicine. One of these is the non-invasive measurement of the speed of blood flow in blood vessels. An assessment of the speed and quality of flow is important in the diagnosis of some heart and tissue diseases.

Figure 9 shows the arrangement that can be used.

A transducer that transmits and receives high-frequency sound produces a beam that is directed into a blood vessel. The beam is reflected from moving blood cells so that there is a double Doppler effect.

- First, the transducer source is stationary and the "observer" (the blood cell) is moving.

- Then, after reflection, the blood cell, as the source, is moving and the observer (the transducer now in its receive mode) is stationary.

The speed of sound in liquids is roughly 1.5 km s^{-1} and the flow speed of the blood cell is around 1 m s^{-1}. This means that $u \ll v$ (using the notation developed earlier) and the approximation $\dfrac{\Delta f}{f} \approx \dfrac{u}{v}$ is valid. Because of the double Doppler shift the equation becomes

$$\frac{\Delta f}{f} \approx \frac{2u}{v}$$

There is, however, an angle θ (as shown in Figure 9) between the beam direction and the flow direction. The system will measure the component of blood speed in the beam direction so that the fractional change in frequency will be

$$\frac{\Delta f}{f} \approx \frac{2u \cos\theta}{v}$$

The advantage of using ultrasound in this way is that the transducer does not have to be inserted into the blood vessel. It is a non-invasive technique. Lack of corrosion to the instrument or issues surrounding possible infection for the patient are other advantages.

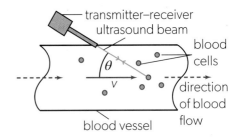

▲ Figure 9 Ultrasound measurements of blood flow.

RADAR

The acronym "radar" stands for radio detection and ranging. It is a technique that has been in use for almost a century and allows the detection and measurement of many moving and stationary objects.

The wavelengths used for radar are usually in the microwave region or slightly longer with some merging into the radio part of the spectrum. Many of the applications involving radar involve static determinations of distance, including collision avoidance in sea and air travel, and density anomaly detection below ground. However, applications that combine the Doppler effect and radar include:

- flow measurements in many context—medical, rain cloud speed measurements, weather forecasting

- vehicle speed determinations (police speed traps)

- remote sensing of ocean currents

- measurement of turbulence in river and ocean flow.

What gives rise to emission spectra and how can they be used to determine astronomical distances?

Doppler originally hypothesized his effect to explain the colour of binary stars. Modern astronomers still use Doppler shift to estimate galactic and stellar speeds.

Topic E.1 explains how emission and absorption spectra from atoms and ions arise. The spectra are characteristic of the chemical element that emits them. When spectra from astronomical objects are observed, the spectral lines are found to have different wavelengths compared with the wavelengths measured with laboratory sources on Earth. The reason for the wavelength change (or "shift") is that the laboratory sources are at rest relative to the observer, whereas the astronomical objects are not. Figure 10 shows typical observations of the absorption line spectra from stars.

The unshifted spectrum is at the top of Figure 10 and this is what would be seen with no relative motion between the source and the observer. Some spectra from distant galaxies are shifted to the red end of the spectrum. This is called "redshift". It shows that the wavelength is longer than expected, so that the frequency is too small. To cause this effect, the source of the radiation must be moving away from us. Similarly, a blueshift – wavelength too short – indicates that the source is approaching Earth.

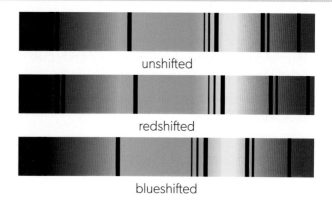

▲ Figure 10 The effects of blueshift and redshift on the observed absorption line spectra from stars.

However, the redshift is not always due to relative motion between source and observer in astronomical observation. There are two other reasons why it can occur:

- **Gravitational redshift** occurs because electromagnetic waves are moving through a strong gravitational field.

- **Cosmological redshift** occurs because spacetime has continued to expand after the Big Bang.

Data-based questions

In 1929, Edwin Hubble published his observations of some distant galaxies. By measuring the redshift of the galactic spectra, he deduced their recessional velocity—the speed at which they are moving away from us. He observed that the more distant a galaxy, the faster it was moving away—an observation which is known today as Hubble's law.

Some modern measurements of some galactic clusters are shown in the table below.

Cluster	Abell reference	Distance / 10^6 light years	Recessional velocity / $km\,s^{-1}$
Centaurus Cluster	Abell 3526	164 ± 12	3121
Hydra Cluster	Abell 1060	208 ± 16	3988
Norma Cluster	Abell 3627	230 ± 16	4707
Perseus Cluster	Abell 426	252 ± 18	5396
Leo Cluster	Abell 1367	327 ± 23	6465
Hercules Cluster	Abell 2151	538 ± 38	11106
Corona Borealis Cluster	Abell 2065	1073 ± 75	22213

- Plot a graph of recessional velocity against distance. Include error bars on your graph to show the uncertainties in the distance. (The uncertainties in the velocity are much smaller and would be hard to show on your graph.)

- Find the gradient of your graph and, using the error bars on your graph, determine the uncertainty in the gradient.

- Quasar MRK 1014 (a very distant and very bright galaxy) is observed to have a spectral line with a wavelength of 763.3 nm. This spectral line is expected to have a wavelength of 656.3 nm in the rest frame.

  ○ Calculate the recessional speed of Quasar MRK 1014.

  ○ Assuming that this quasar also fits the Hubble law trend from your graph, calculate the distance to Quasar MRK 1014. Give an uncertainty with your answer.

ATL Research skills

Astrophysics is a subdiscipline of physics which is concerned with the study of astronomical objects such as stars, galaxies and superclusters of galaxies. Astrophysics follows the same scientific methods as the rest of physics in that theoretical models are compared with experimental evidence. However, the experimental evidence is normally in the form of astronomical observations.

The best quality observations are often made by large telescopes which can generate a large amount of data due to the large numbers of stars and galaxies in the observable universe. The data are often shared in the form of a database so that other astrophysicists can test their models against the observations. A collection of observations from one source is often called a catalogue.

An example of one such catalogue is the Abell catalogue of galactic clusters. In the data-based question, each galactic cluster has a reference (e.g. Abell 3526) so that astrophysicists can refer to galactic clusters unambiguously.

Often many catalogues and sets of data are published in online databases. For example, the NASA Extragalactic Database (NED) contains a database of observations of galaxies and galactic clusters. You should be able to find this database online. Try to find some of the galactic clusters in the data-based question.

How can the use of Doppler effect for light be used to calculate speed? (NOS)

You can use the Doppler effect to determine the speed of a moving object, and it does not necessarily rely on sound waves. For example, you can use a radar speed gun to determine the speed of an automobile. A similar technique can be used to measure the speed of blood flowing in a patients artery, except that microwaves are used. The frequency received by the moving object is doppler shifted relative to the transmitted frequency. The receiver then reflects this shifted frequency back to the stationary source. When the observer receives the signal a double shift has occurred, and the frequency difference yields a value for the relative speed.

This technique can be extended to astronomical measurements too. The small changes in the speed of the Moon along the orbital radius that connects it to Earth can be measured by reflecting radio waves from the Moon's surface and using the shift in frequency.

How can the Doppler effect be utilized to measure the rotational speed of extended bodies?

The Doppler effect is frequently used to measure the tangential speed of nearby stars or planets.

Figure 11 shows three points on the rotating body from which electromagnetic radiation is reaching an observer. The body rotates anticlockwise as viewed in the diagram. Point C is moving towards the observer while point A is moving away. These lead to Doppler shifts to shorter wavelengths from C and to longer wavelengths from A. The wavelength from point B is unshifted because the velocity of the surface of the body is at right angles to the line joining B to the observer.

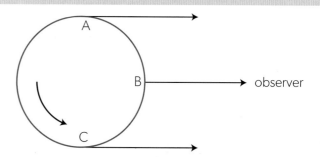

▲ Figure 11 The three points on the rotating object all have different speeds relative to the (stationary) observer.

Measurements of the three frequencies allow the tangential speed of the emitting parts of the surface to be estimated using the Doppler equations. This leads to the rotational period of the body.

Worked example 8

Absorption by mercury atoms is responsible for a spectral line of wavelength 546 nm present in the spectrum of the Sun. When the line is observed in the light coming from the opposite limbs of the solar disk, it is found that the wavelength is shifted by +0.0037 nm and −0.0037 nm, compared with the wavelength observed at the centre of the disk.

a. Describe how this observation provides evidence for the rotation of the Sun.
b. Calculate the speed of the edge of the Sun relative to Earth.
c. The radius of the Sun is 7.0×10^8 m. Determine, to the nearest day, the rotational period of the Sun.

Solutions

a. The wavelength shift indicates that one of the ends of the solar disk is moving towards the Earth and the opposite end away from Earth, as if the Sun is rotating about the axis roughly perpendicular to the line of sight.

b. $v = \dfrac{\Delta\lambda}{\lambda}c = \dfrac{0.0037}{546} \times 3.00 \times 10^8 = 2.0 \times 10^3\,\mathrm{m\,s^{-1}}$.

c. The rotational period is $T = \dfrac{2\pi R}{v}$, where $2\pi R$ is the circumference of the Sun and v is the tangential speed found in part b. $T = \dfrac{2\pi \times 7.0 \times 10^8}{2.0 \times 10^3} = 2.2 \times 10^6\,\mathrm{s} = 25$ days.

Worked example 9

Microwaves of frequency 24 GHz are emitted by a police radar towards an approaching car. The reflected microwaves are recorded by the radar and it is found that their frequency is shifted by 5.6 kHz.

a. Explain why the frequency shift Δf satisfies the equation $\Delta f = \frac{2v}{c}f$, where f is the emitted frequency and v is the speed of the car.

a. Determine the speed of the car.

b. Calculate the wavelength of the microwaves emitted by the radar.

c. Calculate the wavelength shift of the returning microwaves.

Solutions

a. According to the Doppler equation, the frequency of the microwaves incident on the approaching car is shifted by $\frac{v}{c}f$. The car reflects the microwaves and thereby becomes the moving source of the returning wave. Relative to the radar, the reflected microwaves experience another frequency shift of nearly the same magnitude, $\frac{v}{c}f$. The combined frequency shift is therefore $\Delta f = \frac{2v}{c}f$.

b. $v = \frac{\Delta f}{2f}c = \frac{5.6 \times 10^3}{2 \times 24 \times 10^9} \times 3 \times 10^8 = 35\,m\,s^{-1}$.

c. $\lambda = \frac{c}{f} = \frac{3 \times 10^8}{24 \times 10^9} = 1.25 \times 10^{-2}\,m \approx 1.3\,cm$

d. The relative shift in the wavelength is the same as the frequency shift:

$\frac{\Delta\lambda}{\lambda} = \frac{\Delta f}{f}$. Therefore, $\Delta\lambda = \lambda\frac{\Delta f}{f} = 1.25 \times 10^{-2} \times \frac{5.6 \times 10^3}{24 \times 10^9} = 2.9 \times 10^{-9}\,m$.

Practice questions

8. A hydrogen line of wavelength 656 nm is observed in light from a nearby star. The star rotates around an axis perpendicular to the line of sight and the linear speed of the points on the equator of the star is $9.5 \times 10^4\,m\,s^{-1}$. Estimate the width of the hydrogen line. Assume that its broadening is only caused by the Doppler effect of light emitted from different points on the star's disk.

9. Microwaves of wavelength 2.5 cm are emitted towards an airplane that is moving with a speed of $220\,m\,s^{-1}$ away from the source of the microwaves. Calculate:

   a. the frequency of the emitted microwaves

   b. the frequency shift of the microwaves reflected from the airplane.

10. Ultrasound beam of frequency $f = 40\,kHz$ is used in a medical examination to determine the speed of blood flow in an artery of a patient. The frequency of the returning ultrasound is reduced by $\Delta f = 5.3\,Hz$. The speed of ultrasound in body tissue is $v = 1500\,m\,s^{-1}$.

   a. Outline why the relative frequency shift is approximately given by $\frac{\Delta f}{f} = \frac{2u}{v}$, where u is the speed of blood cells in the artery.

   b. Calculate the speed of blood cells.

Theme C — End-of-theme questions

1. Two loudspeakers, A and B, are driven in phase and with the same amplitude at a frequency of 850 Hz. Point P is located 22.5 m from A and 24.3 m from B. The speed of sound is 340 m s⁻¹.

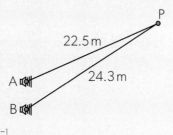

   a. Deduce that a minimum intensity of sound is heard at P.

   b. A microphone moves along the line from P to Q. PQ is normal to the line midway between the loudspeakers.

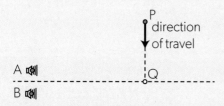

   The intensity of sound is detected by the microphone. Predict the variation of detected intensity as the microphone moves from P to Q.

   c. In another experiment, loudspeaker A is stationary and emits sound with a frequency of 850 Hz. The microphone is moving directly away from the loudspeaker with a constant speed v. The frequency of sound recorded by the microphone is 845 Hz.

   i. Explain why the frequency recorded by the microphone is lower than the frequency emitted by the loudspeaker.

   ii. Calculate v.

2. The red line in the graph shows the variation with distance x of the displacement y of a travelling wave at $t = 0$. The blue line shows the wave 0.20 ms later. The period of the wave is longer than 0.20 ms.

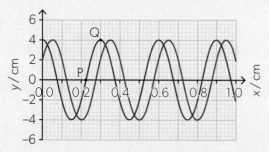

   a. i. Calculate, in m s⁻¹, the speed for this wave.

   ii. Calculate, in Hz, the frequency for this wave.

   b. The graph also shows the displacement of two particles, P and Q, in the medium at $t = 0$. State and explain which particle has the larger magnitude of acceleration at $t = 0$.

   c. One end of a string is attached to an oscillator and the other is fixed to a wall. When the frequency of the oscillator is 360 Hz the standing wave shown is formed on the string.

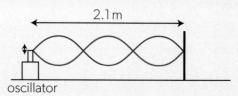

   The frequency of the oscillator is reduced to 120 Hz. Draw the standing wave that will be formed on the string.

3. A vertical solid cylinder of uniform cross-sectional area A floats in water. The cylinder is partially submerged. When the cylinder floats at rest, a mark is aligned with the water surface. The cylinder is pushed vertically downwards so that the mark is a distance x below the water surface.

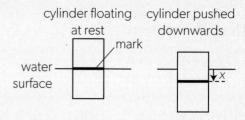

   At time $t = 0$ the cylinder is released. The resultant vertical force F on the cylinder is related to the displacement x of the mark by $F = -\rho Agx$, where ρ is the density of water.

   a. Outline why the cylinder performs simple harmonic motion when released.

   b. The mass of the cylinder is 118 kg and the cross-sectional area of the cylinder is $2.29 \times 10^{-1}\,\text{m}^2$. The density of water is $1.03 \times 10^3\,\text{kg m}^{-3}$. Show that the angular frequency of oscillation of the cylinder is about 4.4 rad s⁻¹.

   c. i. The maximum kinetic energy of the cylinder is $E_{k\text{max}}$. Draw the graph to show how the kinetic energy of the cylinder varies with time during **one** period of oscillation T.

   ii. The cylinder was initially pushed down a distance $x = 0.250\,\text{m}$. Determine $E_{k\text{max}}$.

4. The diagram shows the direction of a sound wave travelling in a metal sheet.

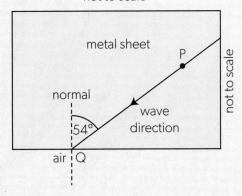

not to scale

a. Particle P in the metal sheet performs simple harmonic oscillations. When the displacement of P is 3.2 μm the magnitude of its acceleration is 7.9 m s⁻². Calculate the magnitude of the acceleration of P when its displacement is 2.3 μm.

b. The wave is incident at point Q on the metal–air boundary. The wave makes an angle of 54° with the normal at Q. The speed of sound in the metal is 6010 m s⁻¹ and the speed of sound in air is 340 m s⁻¹. Calculate the angle between the normal at Q and the direction of the wave in air.

c. The frequency of the sound wave in the metal is 250 Hz. Determine the wavelength of the wave in air.

d. Sound of frequency $f = 2500$ Hz is emitted from an aircraft that moves with speed $v = 280$ m s⁻¹ away from a stationary observer. The speed of sound in still air is $c = 340$ m s⁻¹.

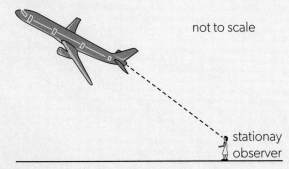

not to scale

Calculate:

i. the frequency heard by the observer

ii. the wavelength measured by the observer.

5. Monochromatic coherent light is incident on two parallel slits of negligible width a distance **d** apart. A screen is placed a distance **D** from the slits. Point M is directly opposite the midpoint of the slits.

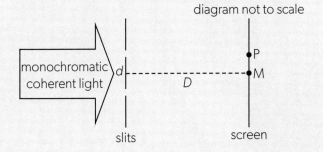

diagram not to scale

a. P is the first maximum of intensity on **one** side of M. The following data are available:

$$d = 0.12\,\text{mm}$$
$$D = 1.5\,\text{m}$$
$$\text{Distance MP} = 7.0\,\text{mm}$$

Calculate, in nm, the wavelength λ of the light.

b. The width of each slit is increased to 0.030 mm. D, d and λ remain the same.

i. Suggest why, after this change, the intensity at P will be less than that at M.

ii. Show that, due to single slit diffraction, the intensity at a point on the screen a distance of 28 mm from M is zero.

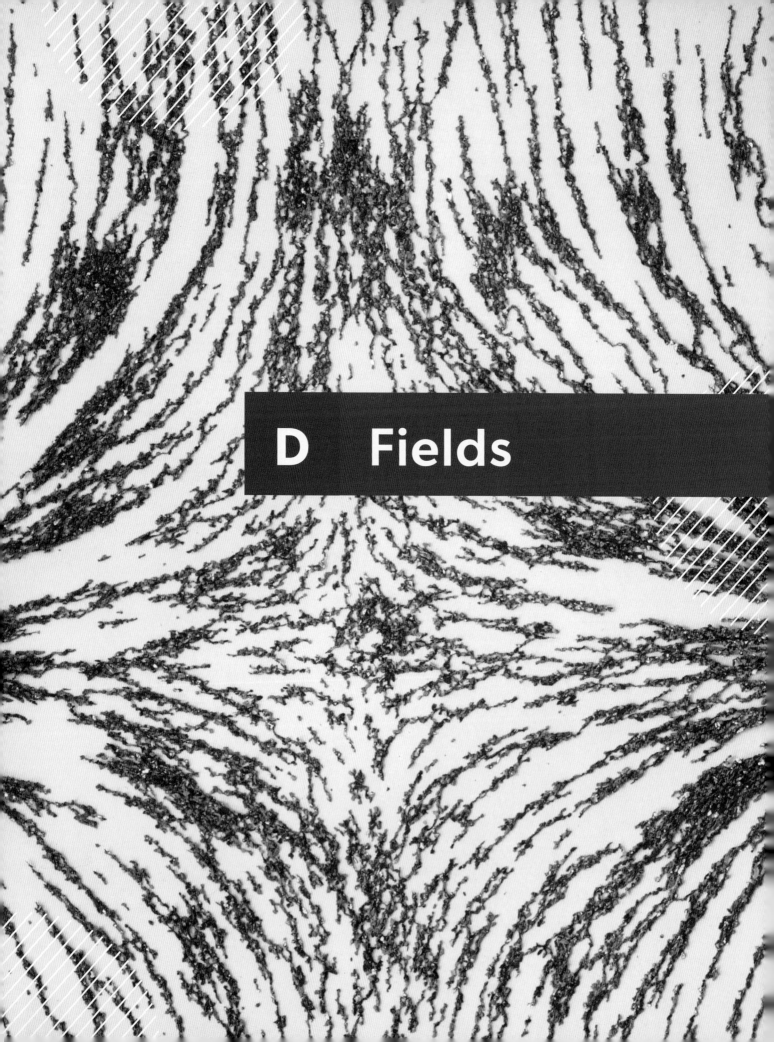

D   Fields

Introduction

The concept of **field** is the central idea of Theme D. Fields are the way we describe and quantify "action at a distance". This is where an object exerts a force on another object even though they are not in contact. Mass gives rise to a gravitational field and another object with mass in this field will experience a force of attraction.

Theme D deals with other fields besides gravity: electrostatic (electric) and magnetic. Charge is the quantity – whether stationary (electric) or moving (magnetic) – that gives rise to the fields in this case.

Patterns and trends abound in this theme. The forces of gravitation and electrostatics are controlled by separate laws that both depend on $\frac{1}{distance^2}$. These laws are known as inverse-square and lead to patterns that are common to both fields. Use the similar patterns that arise to link your learning of these new concepts. They will also help you to link your knowledge of energy transfer from Themes A and B to Theme D.

As always, the over-riding concepts of physics: **energy**, **particles**, and **forces** are intimately bound together within field theory. As a charged particle moves in an electric field, energy is transferred to or from the field–particle system. When a force acts on the particle and the particle moves with a displacement component in the direction of the force then work is done. This leads to new ideas of potential, a property of the field, and potential energy – a measure of the energy stored within the field due to the presence of an object. These are the quantities we use to describe energy transfer within field theory. Strong links exist between the strength of a field and the variation of potential with distance; these are explored in detail in this theme.

This theoretical study of fields has practical consequences: The first use of magnetic fields for recording sound dates to 1888 and magnetic tape was developed in 1928. By storing the 1s and 0s of binary data as a pattern of magnetised areas on a magnetic tape or disc, large quantities of data can be stored and this method of storing data is still used today. A 1 TB hard disk, now considered a small storage amount, can hold roughly 19 million documents each one four pages long or perhaps 40 days of video. Bigger and better forms of computer storage are being developed all the time.

The moon's surface is covered with craters, each of which are due to the transfer of gravitational potential energy into kinetic energy of an asteroid. The size of the crater is related to the work done on the moon's surface by the asteroid's collision which in turn, is related to the gravitational potential energy of the asteroid. Since all these asteroids were in the same gravitational field – that of the moon, an asteroid with more mass will have a greater initial energy and thus cause a larger crater. The pattern of craters on the moon tells the history of these impacts.

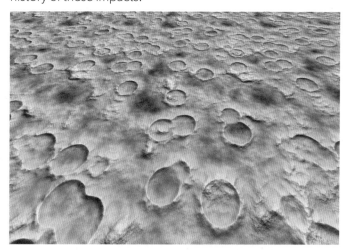

One final but important caveat: What field theories cannot tell us is how the fields arise – we do not yet know what causes one mass to attract another. Nevertheless, the inverse-square laws of Newton and Coulomb, and the field theories to which they lead, enable physicists to measure the properties of the smallest particles in our Universe and to plan journeys deep into interstellar space.

D.1 Gravitational fields

How are the properties of a gravitational field quantified?

How does an understanding of gravitational fields allow for humans to explore the Solar System?

Sometimes the origin of the force acting between two objects is obvious. One example is the friction pad in a bicycle brake rubbing on the rim of the wheel to slow the bicycle down. In other cases there is no physical contact between two objects even though a force exists between them. Examples of this include the magnetic force between the north-seeking poles of two magnets and the electrostatic force between a charged plastic comb and some small pieces of paper. Such forces are said to "act at a distance".

We say that in the case of the comb picking up the paper, the paper is in the **electric field** due to the comb. The concept of a **field** is a powerful one in physics, not least because there are many ideas and pieces of mathematics common to all fields.

We need to quantify what we mean by a field. In this topic, this is done in the context of gravitation. Topic D.2 extends the field concept to electrostatics and magnetism. Similar mathematical rules are imposed on all three fields. Learn the underlying ideas and concepts of one type and you have learned them for the others too.

An over-arching connection between electrostatics and gravitation is the inverse-square law that you have already met in the context of light intensity: double the distance from a point source, and the intensity decreases to one-quarter of its original value. The implication here carries

through to the inverse-square law fields—the influence of the electric or the gravitational field stretches out to infinity. It took the genius of Newton to recognize the truth of this for gravity.

It is this consequence—that gravitational force extends to infinity—that allows us to predict the effects of gravity at large distances from our planet and our Sun. We can calculate to a high degree of precision the future path of a space vehicle as it travels away from Earth. All from a simple law that states that

$$\text{force} \propto \frac{1}{\text{distance}^2}$$

▲ Figure 1 This observatory in Arizona, USA, monitors and searches for near-Earth asteroids. Tracking these asteroids and calculating their future paths can predict whether they are likely to collide with Earth.

In this topic, you will learn about:

- Kepler's three laws of orbital motion

- Newton's universal law of gravitation

- the conditions under which extended bodies can be treated as point masses

- the definition of gravitational field strength

- gravitational field lines

- gravitational potential and gravitational potential energy AHL

- gravitational field strength g as gravitational potential gradient

- the work done in moving a mass m in a gravitational field AHL

- equipotential surfaces for gravitational fields

- the relationship between equipotential surfaces and gravitational field lines

- escape speed and orbital speed

- the effect of the atmosphere on the height and speed of an orbiting body.

Introduction

Isaac Newton developed a theory of gravitation as well as his laws of motion. His law of gravitation is mathematical in nature but cannot be proved. He makes no attempt to explain why two masses are attracted by the force of gravity; he simply accepts that they are. Topic D.1 examines Newton's law of gravitation in detail. As well as setting out the law, it describes the mathematics of field theory and its consequences.

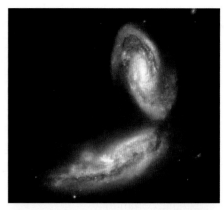

▶ Figure 2 The force of gravity is significant on very large scales. Here, two spiral galaxies are seen interacting and merging 450 million light years from Earth.

ATL Research skills

The realization that Earth orbits the Sun rather than the other way round was one of the great developments in scientific understanding. Galileo Galilei and others in the 16th century overcame cultural, philosophical and religious prejudices to establish this. Some even suffered persecution for the scientific truths they discovered.

However, Galileo was not the first scientist to propose such a model. Copernicus, a Polish mathematician and astronomer, had published a model of the universe with the Sun at its centre shortly before his death in 1543 — almost 90 years before Galileo published his ideas.

An earlier suggestion of heliocentrism (the idea of the Sun being at the centre) was proposed by Aristarchus, an Ancient Greek astronomer (c.310–c.230 BCE). Many Islamic astronomers of the 12th and 13th centuries questioned astronomical models in which the Sun orbited Earth. Copernicus cited the work of these astronomers, naming al-Battānī (c. 858–929), Thābit ibn Qurra (c. 836–901), al-Zarqālī (1029–1087), ibn Rushd (1126–1198) and al-Bitruji (died c. 1204). However, it appears that he removed the section which cited the work of Aristarchus.

Research the lives and discoveries of some of these astronomers.

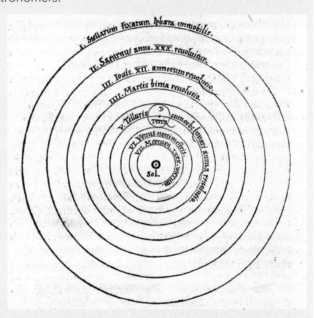

▲ Figure 3 A diagram from *De revolutionibus orbium coelestium* in which Copernicus shows Earth orbiting around the Sun.

Fields

Pick up a stone and you feel the pull of Earth acting on the stone through your hand. Let go and it accelerates towards the ground. There is a gravitational force acting on the stone that is direct and observable.

This gravitational force acts at a distance and has a force field associated with it. Imagine two masses in deep space with no other masses close enough to influence them. One mass (call it A) is in the gravitational force field due to the second mass (B). A force acts on A due to B. B is in the gravitational field of A and it also experiences a force. These two forces have an equal magnitude (even though the masses may be different) but act in opposite directions.

What are the benefits of using consistent terminology to describe different fields? (NOS)

The benefits of both a consistent terminology and a consistent set of concepts to describe field theory are clear and are explained throughout this theme. This consistency goes beyond its convenience when learning the subject and underpins the work of the whole scientific community.

Scientists have to communicate and collaborate with each other. They adopt similar conventions to make this communication unambiguous.

What other concepts in this course involve a consistent terminology that links scientific work in different areas of the subject?

The term **field** is used in physics for cases where two separated objects such as A and B exert forces on each other. A field exists when one object can exert a force on another object at a distance. Every mass has a gravitational field associated with it. Any other mass placed in this field experiences the gravitational force.

The idea that an object must be in the relevant field, but not necessarily in contact with the source of the field, is known as "action at a distance". It is normally assumed that there are no time delays whatever the distance between the two masses. However, a consideration of time delays becomes essential when relativistic effects need to be included, as in Topic A.5.

The concept of a field requires the idea of action at a distance where one object influences another without the need for physical contact. The acceptance, in the history of science, of action at a distance was difficult and represented a significant paradigm shift.

The same idea is applied to electric charge in Topic D.2 where one charged object can attract or repel another.

Theories

Modern theories suggest that fields, in general, do not have a negligible time delay. According to Einstein's theory of general relativity, rapidly accelerating masses should create a change in the gravitational field that ripples outwards as a gravitational wave. These waves were first detected in 2015 by the LIGO detectors. A year later, gravitational waves were detected from two colliding neutron stars. These waves coincided with the detection of gamma rays from the same event. Even though the event occurred about 130 million light years away, the two signals were detected within two seconds of each other. This demonstrated that gravitational waves propagate at the speed of light.

Modern physics explains interactions using the concept of exchange particles that have a finite (and calculable) lifetime. The field is our way of observing the operation of these particles at a macroscopic level. This model suggests that there must be an exchange particle for the gravitational interaction. This hypothetical particle is named a graviton and, like the photon which propagates electromagnetic interactions, should have no mass. The graviton has not been proven to exist and the models and theories which explain it are incomplete.

▲ Figure 4 One of the LIGO detectors which were responsible for detecting gravitational waves in 2015.

Defining gravitational field strength

When both masses are small (the size of a human being), the force of gravity between them is extremely small. Even two large rocks with a mass of 1000 tonnes each only have a gravitational attraction of about one newton when they are touching. The force becomes noticeable to us on an everyday basis when one mass is the size of a planet. Gravitational force is the weakest of the fundamental forces and we need a way to measure the strength of a field that arises from even a small mass.

The strength of a gravitational field is defined using the idea of a small test mass. This test mass must be small so that it does not disturb the field being measured.

If the test object were large, then it would a exert a large enough force on the object that originates the measured field to accelerate it and change the system that is being measured.

The gravitational force that acts on the small test mass in Figure 5 has both magnitude and direction. These are shown in the diagram. The test mass will accelerate in this direction if it is free to move.

Gravitational field strength g is defined as

$$g = \frac{F}{m}$$

where F is the gravitational force on the test mass of size m. The units of gravitational field strength are $N\,kg^{-1}$.

Gravitational field strength is a vector quantity as it has a magnitude and a direction—the direction of the force acting on the test mass.

A formal definition in words is as follows.

Gravitational field strength g at a point is the force per unit mass experienced by a small point mass placed at that point.

This definition requires that the mass of the test object is not only small but is also an infinitesimally small point in space. The presence of any test object distorts and alters the field in which it is placed since the test object carries its own field. This raises the question of how the situation changes when there is more than one mass, excluding the test object itself.

Field strength is independent of the magnitude of the point test mass (because $g = \frac{F}{m}$). The vector field strengths can therefore be added together (Figure 6).

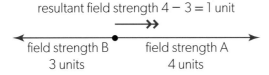

resultant field strength $4 - 3 = 1$ unit

field strength B   field strength A
3 units   4 units

▲ **Figure 6** Two gravitational field strengths act at the same point. The resultant of these two vectors is the total gravitational field strength there. The field strengths are added vectorially.

Gravitational field strength vs acceleration due to gravity

At Earth's surface (using Newton's second law: $F = ma$),

$$\frac{\text{acceleration due to gravity at the surface in m s}^{-2}}{} = \frac{\text{force on a mass at the surface due to gravity in N}}{\text{size of the mass in kg}}$$

The acceleration $= \frac{F}{m}$ but $\frac{F}{m}$ is also the definition of gravitational field strength for a test mass, so

acceleration due to gravity $\equiv$ gravitational field strength $= g$

(The symbol $\equiv$ means "is equivalent to".)

The magnitude of the gravitational field strength (measured in $N\,kg^{-1}$) is equal to the value of the acceleration due to gravity (measured in $m\,s^{-2}$). You should be able to show that $N\,kg^{-1} \equiv m\,s^{-2}$.

The gravitational, electric and magnetic fields described in Theme D are vector fields because the fields and the forces they represent have magnitude and direction. Other types of fields have only magnitude and are known as scalar fields. The scalar potential field is used to describe gravitational and electric fields (Topic D.2). See page 339 for more on vectors and scalars.

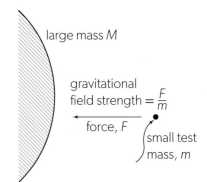

large mass M

gravitational field strength $= \frac{F}{m}$

force, F

small test mass, m

▲ **Figure 5** The definition of gravitational field strength.

Measurements— Defining field strength in general

The definition of field strength as a quantity involves the idea of a **test object**. This object has the property that it is affected by the field. For a gravitational field we need a test mass. For an electric field we need a test charge.

The field strength is defined as

$$\frac{\text{force acting on the test object}}{\text{size of the test object}}.$$

What is meant by "size" here depends on the field. For gravity, it will be mass of the test object; for electric fields, it will be the amount of charge on the object.

The definition is changed slightly for magnetic fields because of the way these fields arise. However, the general shape of

$$\frac{\text{force}}{\text{size of the quantity}}$$ can still be

seen in the definition of magnetic field strength in Topic D.3.

Equivalence principle

That the acceleration due to gravity is equivalent to the gravitational field strength relies on two definitions of mass being equivalent. The first is the concept of **inertial mass** — the property of an object that defines how it accelerates when an unbalanced force is applied. This is the term that appears in Newton's laws ($F = ma$) and in the definition of momentum, $p = mv$. The second is the concept of **gravitational mass** — the term that appears in equations such as Newton's law of gravitation or the equation for weight, $W = mg$. The idea that the inertial mass and the gravitational mass of an object are the same is called the **equivalence principle**.

The equivalence principle was an important consideration in Einstein's development of general relativity. One of Einstein's premises was that we cannot distinguish between the effects of a gravitational field and the effects of acceleration. The equivalence principle has been tested experimentally and shown to be true to one part in 10^{15}.

Should we distinguish between these two types of mass?

Newton's universal law of gravitation

Newton realized that the gravitational force F between two objects with masses m_1 and m_2 whose centres are separated by distance r is:

- proportional to $\dfrac{1}{r^2}$
- proportional to the masses m_1 and m_2
- always attractive.

This leads to

$$F = G\frac{m_1 m_2}{r^2}$$

where G is known as the **universal gravitational constant** and has an accepted value of $6.67 \times 10^{-11}\,\mathrm{N\,m^2\,kg^{-2}}$.

When we combine Newton's law of gravitation with the field strength definition, it allows us to write the gravitational field strength due to a point mass in a more useful way.

A simple case is the field experienced by a test mass of mass m placed a distance r from a large point object of mass M. The magnitude of the force F between point object and test mass is $F = G\dfrac{Mm}{r^2}$. Using the field strength definition $g = \dfrac{F}{m}$ gives

$$g = \frac{F}{m} = \frac{G\dfrac{Mm}{r^2}}{m} = \frac{GM}{r^2}$$

Remember that the distance is measured outwards from the large point object but the force on the test object is inwards along the radius towards the position of the point object.

Inverse-square laws

Laws that depend on $\dfrac{1}{r^2}$ are known as **inverse square**. For Newton's law of gravitation, when the distance between the two masses is *doubled* without changing mass, then the force between the masses goes down to *one-quarter* of its original value.

 Physics utilizes a number of constants such as G. What is the purpose of these constants and how are they determined? (NOS)

A page of the Physics Data Booklet is devoted to fundamental constants. Some of these, such as G, c, h and e are truly fundamental in that they are believed to be constant throughout the universe and unchanging in time. Since 2019, SI units have been defined in terms of seven of them.

The fundamental constants are discussed in *Tools for physics* (page 334) .

Working from these constants, a consistent set of units (A, m, s and so on) can be developed and used to calibrate instruments.

The determination of the constants is, however, ultimately a question of definition. The process used existing measurements to give the best estimate of each of the definitions. These measured values then formed the basis of the new 2019 SI definitions.

The need for a common set of units for mass, length and time affects all of us. The agreements between scientists and engineers that led to these standard units is one of the best examples of the global impact of science.

This is a good example of where mathematics can help you to learn and conceptualize ideas about physics. When you have learned about one situation (in this case, gravitation) then you will be able to apply the same rules to new situations (for example, electric fields, in Topic D.2). Both fields obey the inverse-square law in which the force depends on $\dfrac{1}{\text{distance}^2}$. An inverse-square rule also arises in the context of radiation from the geometry of space (page 229).

The exact value of n in $\dfrac{1}{r^n}$ has been tested many times since the inverse-square behaviour of electric and gravitational fields was suggested. For electric fields, n is known to be within 10^{-16} of $n = 2$.

One consequence of a force law being inverse square is that the force becomes weaker as the distance increases, but never becomes zero at any finite distance. We say that the "force is zero at infinity". There is no actual infinity point in space, but there can be one in our imagination. We use infinity as a useful concept for our energy ideas later and in Topic D.2. In one sense, infinity can be pushed further away as the instruments that measure field strength become more precise!

In mathematical terms, as r tends to infinity, $\dfrac{1}{r^2}$ tends to zero.

Measurements— Direction and sign

Gravity is always attractive, so when the distance is measured *from* the centre of mass M to mass m, then the force on m due to M is *towards* M. In other words, the force is in the opposite direction to the direction in which the distance is measured. Sometimes a negative sign is used here to help predict this direction. Otherwise, keep a careful track of the force direction.

(ATL) Communication skills

There is a story that Newton's insight into the force of gravity arose when he saw an apple falling from a tree. Although there are accounts of Newton telling his friends about this, they were not published until much later. The story was first published after Newton's death by the French writer Voltaire. It is sometimes suggested that the apple hit Newton on the head while he was asleep but there is no evidence to support this.

Newton's insight was to realize that the force of gravity went on beyond the top of the apple tree. It stretched up into the sky, to the Moon and beyond. He realized that the Moon was falling continuously towards Earth under the influence of gravity and, because it was also moving "horizontally" at 90° to the vector direction of the gravitational force, it was in orbit.

The story of the apple has become symbolic of Newton's genius and insight. It has even been noted that, since an apple has a mass of approximately 100 g, the weight of an apple is about 1 newton.

▲ Figure 7 The apple tree near Newton's home in the UK where he is believed to have watched an apple fall. The tree is believed to be over 400 years old.

Worked example 1

Calculate the force of attraction between an apple of mass 100 g and Earth.

Mass of Earth $= 5.97 \times 10^{24}$ kg, radius of Earth $= 6.37 \times 10^6$ m.

Solution

$$F = \frac{GMm}{r^2} = \frac{6.67 \times 10^{-11} \times 0.100 \times 5.97 \times 10^{24}}{(6.37 \times 10^6)^2} = 0.981\,\text{N}$$

Worked example 2

Calculate the force of attraction between a proton of mass 1.7×10^{-27} kg and an electron of mass 9.1×10^{-31} kg when they are at a distance of 1.5×10^{-10} m apart.

Solution

$$F = \frac{GMm}{r^2} = \frac{6.67 \times 10^{-11} \times 1.7 \times 10^{-27} \times 9.1 \times 10^{-31}}{(1.5 \times 10^{-10})^2} = 4.6 \times 10^{-48}\,\text{N}$$

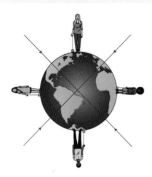

▲ Figure 8 Each person around the sphere thinks that the field direction is towards the centre. The gravitational field lines are radial and in three dimensions.

🧪 Interpreting and sketching field lines

Remember, when interpreting field lines (sometimes called lines of force), that:

- their direction gives the force of attraction on a small mass placed at the position of the field line

- the relative strength of the field is shown by the density of the field lines.

For the radial gravitational field due to a point mass in Figure 8, the lines become closer (more densely packed) as the distance to the point mass decreases and the arrows on the lines point towards the mass.

Conversely, when you are sketching field lines:

- mark the direction of the field as the direction in which the gravitational force is acting

- use the density of the lines to show the variations in field strength.

Practice questions

1. Two masses, 5.4 kg and 1.2 kg, are initially at a distance of 0.25 m, centre to centre. Calculate:

   a. the mutual force of attraction between the masses

   b. the distance between the masses when the force is 1.0×10^{-9} N.

2. A body of mass 7.4 kg is at a distance of 4.0×10^{5} m from the centre of a spherical asteroid. The force of attraction between the body and the asteroid is 5.0×10^{-7} N. Calculate:

   a. the mass of the asteroid

   b. the gravitational field strength due to the asteroid at the position of the body.

Gravitational field lines

Field lines help us to visualize the shapes of fields and to understand important connections between field strength and other field properties.

A gravitational **field line** gives the direction in which gravitational force acts on a mass placed on the field line. The closer field lines are together, the stronger the force.

There is no easy experiment to make gravitational field lines visible (unlike the magnetic and electric field lines of Topic D.2), but the field direction can be determined simply by hanging a weight on a piece of string! The direction of the string gives the local direction of the field pointing towards the centre of the planet (Figure 8).

The arrangement of gravitational field lines is **radial** with the field lines directed towards the mass—because gravitational force is always attractive and directed to the centre of a spherical and uniform Earth.

It turns out that the gravitational field strength outside a sphere of mass M is the same as g for a point mass: $g = \dfrac{GM}{r^2}$. For this to be true, the sphere must be uniform or its distribution of mass must be spherically symmetrical. The gravitational field lines for both the point mass and the sphere are identical outside the sphere. The field lines point towards the centre of mass in both cases.

Looking from outside the sphere, all the mass of the sphere acts as though it is a point mass of size M positioned at the centre of the sphere (the centre of mass). We only need one equation for both a point mass and a sphere.

In Topic D.2, field lines will help you to study magnetic and electric fields.

Worked example 3

Calculate the gravitational field strength at the surface of a. Earth and b. the Moon.

Mass of Earth = 5.97×10^{24} kg, radius of Earth = 6.37×10^{6} m, mass of the Moon 7.35×10^{22} kg, radius of the Moon 1.74×10^{6} m.

Solutions

a. $g = \dfrac{GM}{r^2} = \dfrac{6.67 \times 10^{-11} \times 5.97 \times 10^{24}}{(6.37 \times 10^{6})^2} = 9.81 \text{ m s}^{-2}$. This is the same as the acceleration of free fall near Earth's surface!

b. $g = \dfrac{GM}{r^2} = \dfrac{6.67 \times 10^{-11} \times 7.35 \times 10^{22}}{(1.74 \times 10^{6})^2} = 1.62 \text{ m s}^{-2}$. This is about one-sixth of that on Earth.

Worked example 4

Calculate the gravitational field strength of the Sun at the position of Earth.

Mass of Sun = 2.0×10^{30} kg, Earth–Sun distance = 1.5×10^{11} m.

Solution

$$g = \dfrac{GM}{r^2} = \dfrac{6.67 \times 10^{-11} \times 2.0 \times 10^{30}}{(1.5 \times 10^{11})^2} = 5.9 \text{ mN kg}^{-1}$$

Practice questions

3. An aircraft is flying at an altitude of 10.0 km above sea level. The average radius of Earth is 6370 km. Calculate the ratio

   $$\dfrac{\text{gravitational field strength at the location of the aircraft}}{\text{gravitational field strength at sea level}}.$$

4. Determine the height above sea level at which the gravitational field strength is reduced to 99% of its value at sea level.

5. Two asteroids X and Y can be modelled as uniform spheres of the same density. The radius of asteroid X is twice that of asteroid Y. What is

   $$\dfrac{\text{gravitational field strength at the surface of X}}{\text{gravitational field strength at the surface of Y}}?$$

   A. 1 B. 2 C. 4 D. 8

Extended bodies

Real objects that we deal with in gravitational theory have size and shape. They are known as **extended bodies**. Analysis shows that, when two extended bodies interact gravitationally, we can treat them approximately as point masses. To do this we must treat the whole of the mass of one object as being placed at its centre of gravity. This statement assumes that the objects are in a uniform gravitational field or that they are well away from any other mass.

However, when an object is in a gravitational field that is not uniform (perhaps because the field varies significantly across the object), then the bodies in the system cannot be treated as point masses. More complex mathematics is needed to deal with the forces acting. An example is the tidal force that acts on Earth's oceans. The gravitational field due to the Moon varies significantly across Earth's diameter. The centripetal and gravitational forces balance at the centre of gravity, but, at Earth's surface, forces on the water lead to tidal bulges under which Earth rotates every day. This leads to a tide in most parts of the world every 12.5 hours.

Observations—Centre of mass or centre of gravity?

Centre of mass is the point about which the distribution of mass in an object is the same in all directions. It depends on the geometry of the object and any variations of density inside it.

Centre of gravity is the point at which all the weight of an object appears to act: in other words, the point of balance.

These terms are often used interchangeably and they are the same providing that the gravity field is uniform when the centre of gravity is determined. Centre of mass does not depend in any way on the gravitational field. It is, like moment of inertia, determined by the mass distribution within the object.

Imagine a *very* tall uniform box-shaped object in Earth's gravitational field (Figure 9).

The centre of mass of the object will be in the centre of the box. In a uniform gravitational field, the centre of

gravity of the object will be in the centre too. However, because the gravitational field strength varies from top to bottom of the object, the centre of gravity now lies in the lower half of the object, whereas the centre of mass remains in the centre.

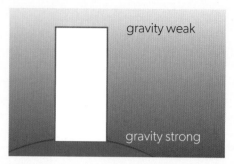

▲ Figure 9 Because the gravitational force varies with height above the surface, the centre of gravity and the centre of mass are not in the same position in the object.

Kepler's laws of orbital motion

Knowledge of Kepler's work was a stimulus for Newton. There had been a major paradigm shift from the Ptolemaic system (which had Earth stationary and everything in the sky moving around it) to the post-Copernican system (where Earth moves around the Sun).

Theories—Paradigm shift

This change to our understanding of the Solar System is one of the great paradigm shifts in Western science. Before it there was complete agreement that Earth was at the centre of the universe. Afterwards the Sun was taken to be the centre of the Solar System with other stars beyond it.

The shift was in no sense immediate nor universal. There are still proponents of the flat-earth theory that originated with the Ancient Greeks. In July 2020, *Physics World*—a publication of the UK Institute of Physics—covered attempts made by professional physicists to convince flat-earthers that their ideas are not true.

When Copernicus's book *De Revolutionibus* was published, someone added a letter at the beginning. This was most likely to have been the editor, Andreas Osiander, worried that the conclusions that were presented in the book would be controversial. He wrote:

> "For it is not necessary for the hypotheses to be true, nor even probable; it is sufficient if the calculations agree with the observations."

Osiander hoped that the book would not be immediately banned. He was suggesting that the heliocentric model

was a mathematical model to help astronomers calculate the positions of planets and not a suggestion that the Solar System actually had the Sun at its centre.

Can there ever be a difference between reality and an accurate model of reality?

▲ Figure 10 A diagram showing the heliocentric model of the Solar System.

Kepler analysed the data collected by the Danish astronomer Tycho Brahe. This led him to three **laws of orbital motion** which were able to account for the motion of Mars as observed by Brahe. The resolution of Brahe's instruments was good enough to show that the Ptolemaic system required major adjustments for it to predict planetary orbits accurately. Essentially, the Ptolemaic description of the universe had been broken.

The first breakthrough by Kepler was the following suggestion:

Planets move in elliptical orbits with the Sun at one focus — Kepler's first law of planetary motion

The ellipse (Figure 11) is one of the conic sections mentioned on page 35. A circle is a special case of the ellipse in which the two axes have the same length so that a circle has only one radius. The ellipse has two foci (the plural of "focus"). Kepler suggested that the Sun — in the case of the Solar System — is positioned at one of these foci. This means that the planets and other astronomical objects, as they travel around their orbits, have distances from the Sun that are constantly changing.

Kepler also recognized that the variation of orbital distance from the Sun implied a variation in orbital speed throughout the orbit. He wrestled with this problem for a long time, but finally realized that the answer was remarkably simple:

The line connecting the planet to the Sun sweeps out equal areas in equal times — Kepler's second law of planetary motion

Figure 11 shows this. When close to the Sun, the planet (or comet) moves fast so that the relatively short line from Sun to planet draws out an area quickly. At the other end of the orbit, the line is long and the planet has a slow orbital speed, so that the areas for the two cases are drawn by the radius line at the same rate.

These two laws were published by Kepler in 1609. He had to analyse his data for nine more years before he could publish his final empirical law. This indicates the following relationship between the orbital time period of a planet and its distance from the Sun.

The square of the periodic orbital time T of a planet is directly proportional to the cube of the semimajor axis of its orbit

For circular or near-circular orbits, this becomes

The square of the periodic orbital time T of a planet is directly proportional to the cube of the orbital radius r: $T^2 \propto r^3$ — Kepler's third law of planetary motion

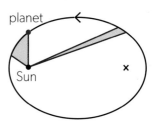

▲ Figure 11 Two equal-time intervals of the elliptical orbit of a planet are shown. Kepler's first law tells us that the Sun is at one of the two foci of the ellipse. Kepler's second law says that the speed of the planet is such that it sweeps equal areas of the ellipse in equal times.

 How can the motion of electrons in the atom be modelled on planetary motion and in what ways does this model fail? (NOS)

The work of Niels Bohr and others described in Topic E.1 was influenced by their knowledge of the orbital theory set out in this topic. The hypothesis was that an electron could orbit the proton in a hydrogen atom in a similar way to the orbit of a satellite around a planetary object. Bohr's theory succeeded because the force laws in both cases are inverse-square. It was unsuccessful because, under a classical theory, the electron must emit energy as a result of its circular motion.

Bohr's theory ultimately failed but — as we see many times in this course — the adoption of an idea that is subsequently rejected is all part of the progress of science. Scientists uncover more and more scientific truth through repeated cycles of hypothesis and falsification.

How is uniform circular motion like — and unlike — real-life orbits?

The work in this topic takes as its basis the assumption that the orbits of the satellites are circular. This is rare in real life. The deviation from circularity is small in the case of many of the planets in the Solar System, but nevertheless cannot be ignored when it comes to the calculation of spacecraft trajectories. Kepler's laws come to our aid and allow the computation of a trajectory when the orbit is elliptical.

For a particular circular path, the kinetic energy and gravitational potential energy of the satellite are both constant. When the path is elliptical, only the sum of these is constant and the planet–satellite system continually transfers energy between kinetic and gravitational potential forms as the orbital distance and orbital speed vary.

Another difference between our assumed model and real life arises when a satellite and planet have similar masses (the Moon, for example, has a mass roughly 1% of that of Earth). It is possible for a binary star system to have similar masses for both stars. In this case, both stars orbit the common centre of mass of the system. A very different pair of orbits from the theory envisaged here.

Patterns and trends — Eccentricity in practice

The eccentricity of an ellipse tells us how "squashed" it is compared with a circle. The value of the eccentricity varies between 0 (a circle) and 1 (a parabola — which can be regarded as an open ellipse). The small value for the eccentricity of Earth's orbit tells us that our orbit around the Sun is almost circular, as are many of the Sun's planets. Earth's nearly circular orbit gives a stable surface temperature. This has had a significant impact on the evolution of life on Earth.

Object	Eccentricity
Earth	0.0167
Moon	0.055
Venus	0.0068
Jupiter	0.048
Pluto	0.25
Halley's comet	0.97

The orbital eccentricities of some objects in the Solar System are shown in the table.

Worked example 5

The orbital period of Earth around the Sun is 365 days. The semimajor axis of Earth's orbit is 1.50×10^8 km and that of the orbit of Venus is 1.08×10^8 km. Calculate, in days, the orbital period of Venus.

Solution

$$\left(\frac{T_V}{T_E}\right)^2 = \left(\frac{r_V}{r_E}\right)^3$$

$$T_V = T_E \left(\frac{r_V}{r_E}\right)^{\frac{3}{2}} = 365 \times \left(\frac{1.08}{1.50}\right)^{\frac{3}{2}} = 223 \text{ days.}$$

Worked example 6

Mars revolves around the Sun with a period of about 1.9 years. Calculate the ratio $\dfrac{\text{orbital radius of Mars}}{\text{orbital radius of Earth}}$.

Solution

$$\frac{r_M}{r_E} = \left(\frac{T_M}{T_E}\right)^{\frac{2}{3}} = 1.9^{\frac{2}{3}} = 1.5.$$

Practice questions

6. A satellite orbits Earth in an elliptical orbit in the direction represented by the arrow.

    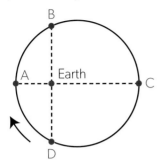

    At which position is the kinetic energy of the satellite increasing?

7. Satellites X and Y move in circular orbits around a planet. The orbital period of satellite X is 16 days. The orbital radius of satellite Y is $\frac{1}{4}$ of the orbital radius of satellite X. Calculate the orbital period of satellite Y.

8. An asteroid moves around the Sun in a circular orbit. The following data are given:

    Orbital period of the asteroid = 3.6 years

    Sun–Earth distance = 1.5×10^{11} m

    Calculate, in m, the radius of the asteroid's orbit.

Hypotheses — Kepler's third law revisited

Despite its apparent success, Kepler's third law was an approximate hypothesis. He had extended his results well beyond the limits of his data. When the masses M and m of both objects in the orbit are considered, then Kepler's third law becomes

$$T^2 \propto \frac{r^3}{(M+m)}$$

The Sun has so much more mass than any of its satellites (planets) that the $T^2 \propto r^3$ form is more than good enough for the Solar System.

Linking orbits and gravity

Newton's genius was his recognition that Kepler's third law implied the inverse-square gravitational law that bears Newton's name. He began with the insight that the gravitational force of a planet provides the centripetal force to keep a satellite in orbit.

Newton used the example of a cannon on a high mountain (Figure 12). The cannon fires its cannonball horizontally and the ball accelerates vertically downwards. On a flat Earth, it will eventually hit the ground. Newton knew that Earth is a sphere. Therefore, the curvature of Earth allows the ball to travel further before hitting the ground (Figure 12(a)).

He then imagined the ball being fired at larger and larger initial speeds (Figure 12(b)). Eventually the cannonball will travel "horizontally" at such a high speed that the curvature of Earth and the curve of the trajectory will be the same. When this happens, the distance between shell and surface is constant and the shell is in orbit around Earth (Figure 12(c)).

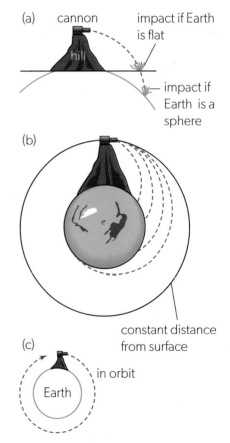

▲ Figure 12 Newton's cannonball thought experiment.

What do you expect to happen to the trajectory of the cannonball when it is fired at even greater speeds? To check your conclusions, find an applet on the Internet that will allow you to vary the firing speed of Newton's cannon. A good starting point for the search is "applet Newton cannon".

This motion of a satellite around Earth can be analysed by combining the ideas of centripetal force and gravitational attraction. The gravitational attraction F_G provides the centripetal force F_C, and (ignoring direction signs)

$$F_C = F_G = m\omega^2 r = \frac{GM_E m}{r^2}$$

where M_E and m are the mass of Earth and the satellite, respectively, and ω is the orbital angular speed. The symbol m cancels from the equation which becomes:

$$\omega^2 r = \frac{GM_E}{r^2}.$$

We also know from Topic A.2 that $\omega = \frac{2\pi}{T}$, where T is the orbital time period.

Therefore $\left(\frac{2\pi}{T}\right)^2 r = \frac{GM_E}{r^2}$ and $\frac{4\pi^2}{T^2} = \frac{GM_E}{r^3}$, which leads to

$$T^2 = \frac{4\pi^2}{GM_E} \times r^3$$

This is an expression of Kepler's third law.

Data-based questions

The Kepler-90 system was first observed by NASA's Kepler mission in 2013. It has eight exoplanets (a planet which orbits a star other than our Sun). The eighth planet — third from the central star — was discovered in 2017.

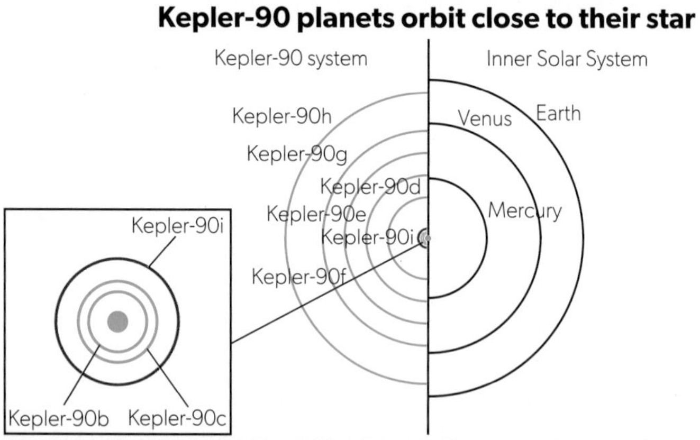

Kepler-90 planets orbit close to their star

▶ **Figure 13** A diagram showing the eight planets of the Kepler-90 system compared with our Solar System.

Orbital radius / AU	Uncertainty in orbital radius / AU	Orbital period / days	Uncertainty in orbital period / days
0.074	0.016	7.008151	0.000019
0.089	0.012	8.719375	0.000027
0.107	0.03	14.44912	0.0002
0.32	0.05	59.73667	0.00038
0.42	0.06	91.93913	0.00073
0.48	0.09	124.9144	0.0019
0.71	0.08	210.60697	0.00043
1.01	0.11	331.60059	0.00037

- Calculate the percentage uncertainty in the orbital radius for the third planet in the table.

- Show that the percentage uncertainties in the orbital period are more than a million times smaller than the percentage uncertainties in the orbital radius.

- $1\,AU = 1.50 \times 10^{11}\,m$. Make a table of orbital radius (and its uncertainty) in metres and orbital period in seconds.

- Tabulate values of (orbital radius)^3 and (orbital period)^2. Calculate the uncertainties in (orbital radius)^3.

- Plot a graph of (orbital period)^2 on the y-axis and (orbital radius)^3 on the x-axis. Add the horizontal error bars to your graph.

- Add a line of best fit to your graph and find the gradient.

- By considering the maximum and minimum gradients, find the uncertainty in your graph.

- By considering the equation $T^2 = \dfrac{4\pi^2}{GM} \times r^3$, find the mass of the star at the centre of the Kepler-90 system.

- Masses in Astronomy are often quoted in solar masses where one solar mass is 1 M$\odot$ = 1.99×10^{30} kg. Express the mass of the Kepler-90 star in solar masses and give an uncertainty in your value.

Worked example 7

Calculate the orbital period of Jupiter about the Sun.

Mass of Sun = 2.0×10^{30} kg, radius of Jupiter's orbit = 7.8×10^{11} m.

Solution

$$T^2 = \frac{4\pi^2 r^3}{GM_s}$$

$T = 3.7 \times 10^8$ s (about 12 years)

Worked example 8

The Moon orbits Earth with a period of about 27 days in an approximately circular orbit or radius 3.8×10^8 m. Calculate the mass of Earth using this information.

Solution

$$T^2 = \frac{4\pi^2 r^3}{GM_E} \Rightarrow M_E = \frac{4\pi^2(3.8 \times 10^8)^3}{6.67 \times 10^{-11} \, (27 \times 24 \times 3600)^2}$$

$$= 6.0 \times 10^{24} \text{ kg.}$$

Practice questions

9. Phobos and Deimos are two moons of Mars that move in approximately circular orbits. The radius of Phobos' orbit is 9.4×10^6 m and its orbital period is 7.7 hours.

   a. Calculate the mass of Mars.

   The orbital period of Deimos is 30 hours.

   b. Calculate the radius of Deimos' orbit.

10. A satellite is placed in a circular orbit 100 km above the surface of the Moon. Calculate the orbital period of the satellite.

    Mass of the Moon = 7.35×10^{22} kg

    Radius of the Moon = 1.74×10^6 m

Gravitational potential energy

When a mass m is moved through a vertical height change of Δh in a gravitational field of strength g,

$$\Delta E_g = m \times g \times \Delta h$$

This equation applies when g is effectively constant over the height change being considered. Remember that g has to be the local value of the field strength. Near Earth's surface, 9.8 N kg^{-1} is the value to use, but further from the surface a smaller value would be required.

The idea that g is constant arises from what Earth's radial gravitational field looks like on a human scale (Figure 14).

The curvature of Earth is almost imperceptible to us, so that the fact that the field is radial and in different directions is equally imperceptible. The field lines close to the surface are effectively parallel to each other and equally spaced so that the field is uniform (Figure 14). The $mg\Delta h$ formulation for gravitational potential energy works most of the time.

In Topic A.3, gravitational potential energy was used for the energy transfer ΔE_g.

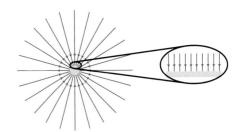

▲ Figure 14 Although the gravitational field of a uniform sphere is a radial field, close to the surface the field is uniform and perpendicular.

AHL

Integrating the "work done" expression

The work done over the whole journey from infinity to the final position is found by integrating the expression for work done from infinity to the distance R from the centre of the object:

$$W = \int_{\infty}^{R} G\frac{Mm}{r^2}\, dr$$

$$= GMm \int_{\infty}^{R} \frac{1}{r^2}\, dr$$

$$= GMm \left[-\frac{1}{r} \right]_{\infty}^{R}$$

$$= GMm \left[-\frac{1}{R} + 0 \right]$$

$$= -\frac{GMm}{R}$$

You may meet this integration in the IB mathematics course.

▲ Figure 16 Systems of gravitationally bound objects are not always restricted to two bodies. This picture shows the globular cluster Omega Centauri—a cluster of about 10 million stars which lies about 17 000 light years away in our galaxy. All of these stars will exert an attractive force on each other, and the system as a whole has gravitational potential energy which can be defined as the work done to assemble the system starting with infinite separation of each component.

On a larger scale this equation will not work for two reasons:

- The gravitational field strength will vary with the distance between the masses.

- The gravitational potential energy arises because there are two masses involved. In the $mg\Delta h$ equation the mass of Earth is included in g.

To see the entire gravitational potential energy stored in a two-mass system, we need to look at the work done (energy transferred) in forming the system from an initial state in which the two masses do not interact. This non-interaction place is called "infinity".

The gravitational potential energy of two objects is the work done (energy transferred) in bringing the two objects from infinity to their present position.

Infinity is not a real place—it is in our imagination—but that does not prevent it from having some interesting and useful properties. These properties arise from the inverse-square law and the implication that, when two objects are an infinite distance apart, there is no gravitational attraction between them. As $r \to \infty$, $F \to 0$ in $F = G\frac{Mm}{r^2}$.

The problem for us is that to use *work done = force × distance*, we must integrate mathematically. This is because the gravitational force varies as the distance between the objects changes. This is different from the approximation of a uniform field (and therefore force) close to Earth's surface. The situation is shown in Figure 15.

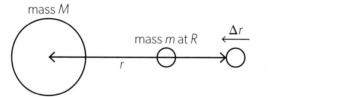

▲ Figure 15 A test mass is moving from infinity to a distance R from a mass M. At one point in the journey, when it is r from the mass M, it moves through a small distance Δr.

The small object of mass m is on a journey from infinity to a final position that is a distance R from the centre of the large object. Figure 15 shows the small object when it is at an intermediate position between infinity and its final position. At this point the gravitational force acting on the small object is $F = G\frac{Mm}{r^2}$ and is directed to the left. The small object now moves a small distance Δr also to the left. We can assume that Δr is so small that the gravitational force is constant over the distance concerned. The equation *work done = force × distance* becomes

$$W = G\frac{Mm}{r^2}\Delta r$$

Over the whole distance from infinity to R, $W = -\frac{GMm}{R}$.

What is the significance of the negative sign in this equation? Remember that at any distance other than infinity, there is an attractive force between the objects. This means that, if the masses are free to move, they will both accelerate towards each other and collide. For this not to happen, the agent that is moving the object from infinity must be transferring energy away from the two-object gravitational system.

When the objects are an infinite distance apart, zero gravitational potential energy is stored in the system. The equation implies that having moved to a finite separation R, then the gravitational potential energy stored in the system is negative. When we want to move the two objects to an infinite separation again, we will need to transfer a positive quantity of energy *into* the system to return to the initial amount of stored energy which was zero at infinity.

The gravitational potential energy E_p in a system consisting of two masses m_1 and m_2 with centres separated by r is

$$E_p = -G\frac{m_1 m_2}{r}$$

For gravity:

- Separating two objects with mass at a constant speed means that you need to *add* energy to the system.

- Allowing two objects to move closer together at a constant speed means that you need to *remove* energy from the system.

> When there are more than two masses, the gravitational potential energy E_p of the system of masses is the work done to assemble the system from infinite separation.

Worked example 9

A satellite of mass 740 kg is to be launched into a circular orbit 320 km above the surface of Earth. The radius of Earth is 6.37×10^6 m and the mass of Earth is 5.97×10^{24} kg. Calculate:

a. the gravitational potential energy of the satellite at the surface of Earth

b. the change in the gravitational potential energy between the surface and the orbit.

Solutions

a. The distance between the satellite and the centre of Earth is equal to Earth's radius.

$$E_p = -\frac{6.67 \times 10^{-11} \times 5.97 \times 10^{24} \times 740}{6.37 \times 10^6} = -4.63 \times 10^{10}\,\text{J}.$$

b. $\Delta E_p = 6.67 \times 10^{-11} \times 5.97 \times 10^{24} \times 740 \times \left(\frac{1}{6.37 \times 10^6} - \frac{1}{6.37 \times 10^6 + 3.2 \times 10^5}\right) = 2.21 \times 10^9\,\text{J}.$

Gravitational potential at a point

Potential is another concept that is strongly connected to the ideas of field and field strength. Potential and potential energy are not the same thing. Potential energy is transferred when a particular object moves—as you see in the gravitational potential energy equation which contains the mass of the object.

Potential at a point, on the other hand, links to the gravitational field, not to a particular object. Potential is an important concept in all field theories.

One way to learn about **gravitational potential at a point** is to think of it as the work done to bring a unit mass (in other words, one kilogramme in our unit system) from infinity to the point concerned. Here the unit mass acts as a test object in the same way as the test objects we used to develop the definition of gravitational field strength.

The gravitational potential V_g at a point a distance r from the centre of a uniform sphere (or point object) of mass M is

$$V_g = -G\frac{M}{r}$$

This equation is obtained by taking the earlier expression for E_p and dividing by m. The smaller mass that was moved from infinity in the earlier section is

AHL

now the test mass with a mass of 1 kg. The negative sign reflects the fact that gravitational potential can never be positive because the gravitational force is always attractive.

Potential is work done per unit mass: in other words, $\frac{energy}{mass}$. The SI unit is J kg⁻¹ (or m² s⁻² in the fundamental base unit form).

Again, for consistency, gravitational potential at a point is defined to be zero at infinity where the test mass and the mass of size M have no interaction. This definition allows a visual interpretation of potential too.

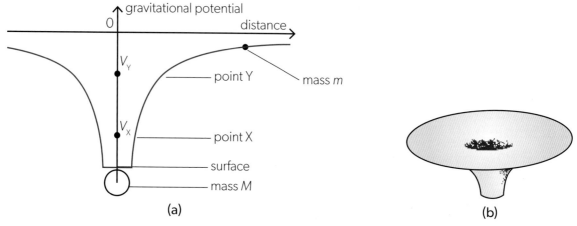

(a)   (b)

▲ Figure 17 (a) The surface of a planet represents a potential well. When a mass is released at infinity with zero kinetic energy it is attracted to the planet. The mass gains kinetic energy at the expense of the gravitational potential energy. As the initial energy was zero and energy is conserved, as the kinetic energy increases (and is positive), the gravitational potential energy decreases below zero. (b) In three dimensions the shape of the potential is that of a well.

Figure 17(a) shows that, at large distances from the planet, the gravitational potential is close to, and just below, zero. Near to the object the gravitational potential is negative and its value is below zero. The gravitational potential resembles a water well (Figure 17(b)).

A test mass on the surface of the planet is deep inside the well and energy must be transferred to the system to move it vertically above the surface.

One advantage of using gravitational potential is that it is a scalar quantity that arises from the vector field and is therefore easier to handle as no direction is involved.

Why zero at infinity?

What makes a good zero reference for potential? One obvious reference point could be the surface of the oceans. When travel is based only on Earth, this is a reasonable choice of reference point. We refer to heights as being above or below sea level so that Mount Everest is 8848 m above sea level. Despite being an appropriate local zero of potential for Earth-based activities, sea level is not good enough once we leave Earth. It would make no sense to refer the surface of Mars to Earth's sea level. This is why infinity makes most sense—although it might seem implausible when you first meet it.

The use of infinity as the reference point for the definitions of zero gravitational potential and zero gravitational potential energy is to ensure that the force between objects is zero at this point. Even though this is unattainable in practice, it can be imagined and used mathematically in calculations.

When a definition of a quantity relies on unattainable measurements, is it a useful definition at all?

Gravitational potential difference

Look again at Figure 17(a). A small object of mass m sits in the gravitational field due to a large object. Two positions for the small object of mass m in the field of the large object are shown at X and Y. The gravitational potentials of these positions are different. At position X, which is a long way down in the potential well, the gravitational potential is V_X. At Y, the gravitational potential is less negative (it is higher up the curve and closer to zero) and is V_Y. Gravitational potential at a point is defined as the work done in moving a unit mass from infinity to the point concerned. Therefore:

- The work done in taking the mass m from infinity to point X is $m \times V_X$.

- The work done in taking the mass m from infinity to point Y is $m \times V_Y$.

- We start at X and finish at Y.

- So the work done in taking the mass from point X to point Y is $m \times (V_Y - V_X)$.

This is

(*difference in potential between X and Y*) × (*mass moved between X and Y*)

and can be written algebraically as $m \times \Delta V_g$, where ΔV_g stands for the difference in potential between the two points.

The work done W in moving a mass m between two points in a gravitational field is

$$W = m \times \Delta V_g$$

where ΔV_g is the gravitational potential difference between the two points.

To use this equation, you have to assume that only gravitational potential energy changes. However, if a mass changes position it must accelerate, which involves a transfer of kinetic energy. Therefore, you have to assume that there is no change of speed.

Equipotential surfaces

Our knowledge of the gravitational potential means that we can assign some values of gravitational potential to Earth.

The average radius of Earth is 6370 km and the best estimate of the mass of Earth is 5.972×10^{24} kg. With these data and the value for G of $6.674 \times 10^{-11}\,\text{N}\,\text{m}^2\,\text{kg}^{-2}$, we can calculate the gravitational potential V_{sl} at sea level:

$$V_{sl} = -\frac{GM}{r} = -\frac{6.674 \times 10^{-11} \times 5.972 \times 10^{24}}{6.370 \times 10^6} = -6.257 \times 10^7\,\text{J}\,\text{kg}^{-1}$$

One metre above sea level the gravitational potential will be $V_{sl} + 9.8\,\text{J}\,\text{kg}^{-1}$ (because near the surface raising 1.0 kg requires a transfer of energy of 9.8 J). Two metres above sea level it will be $V_{sl} + 19.6\,\text{J}\,\text{kg}^{-1}$ and so on. A scale can be imagined drawn vertically up a wall to show the increase in gravitational potential with height. Similarly, a diagram of Earth in space can have gravitational values drawn on it. For a spherical planet these values are distributed symmetrically so that, for example, all the $-5 \times 10^7\,\text{J}\,\text{kg}^{-1}$ points lie on a sphere centred on Earth's centre. This sphere is a surface of equal potential — known as an **equipotential**. When a mass moves around on an equipotential surface, then no energy is transferred either to or from the gravitational field.

The gravitational field lines cut the equipotential surfaces at right angles. Figure 18 shows the relationship between field lines and equipotential.

Rationalizing signs

Care is needed with the signs in using the expression for the gravitational potential difference between the two points X and Y in Figure 17.

$$GM\left(-\frac{1}{r_Y} - \left(-\frac{1}{r_X}\right)\right) = GM\left(\frac{1}{r_X} - \frac{1}{r_Y}\right)$$

where r_X and r_Y are the distances from the the centre of mass M to the points.

Here the magnitude of $\frac{1}{r_X}$ is greater than that of $\frac{1}{r_Y}$, so the overall expression is positive, meaning that energy must be transferred into the system to move the small object from X to Y — as expected.

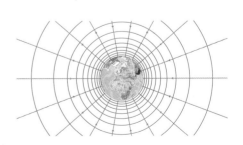

▲ Figure 18 The gravitational field lines (grey) and equipotential surfaces (green) around Earth.

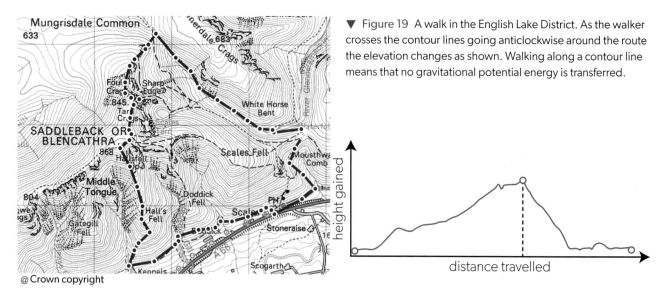

@ Crown copyright

▼ Figure 19 A walk in the English Lake District. As the walker crosses the contour lines going anticlockwise around the route the elevation changes as shown. Walking along a contour line means that no gravitational potential energy is transferred.

If you use a map when walking in the countryside, then you will be very familiar with the idea of an equipotential. The contour lines on maps are lines of equal potential as well as being lines of constant height, so if you walk along one you do not transfer gravitational potential energy.

Worked example 10

A satellite of mass 850 kg is moved from a point where the gravitational potential due to Earth is $-6.0 \times 10^7 \,\text{J kg}^{-1}$ to a point where the gravitational potential is $-4.0 \times 10^7 \,\text{J kg}^{-1}$. The mass of Earth $= 5.97 \times 10^{24}$ kg. Calculate:

a.   the change in the gravitational potential energy of the satellite

b.   the final distance between the satellite and the centre of Earth.

Solutions

a. $\Delta E_p = m\Delta V_g = 850(-4.0 \times 10^7 - (-6.0 \times 10^7)) = 1.7 \times 10^{10}\,\text{J}.$

b. $V_g = -G\dfrac{M}{r} \Rightarrow r = 6.67 \times 10^{-11} \times \dfrac{5.97 \times 10^{24}}{4.0 \times 10^7} = 1.0 \times 10^7\,\text{m}.$

Worked example 11

The graph shows how the gravitational potential due to Earth varies with the distance r from the centre of Earth. The curve begins at the surface of Earth.

A satellite of mass 1200 kg is initially at the surface of Earth. Determine the minimum work that must be done on the satellite in order to place it in a circular orbit of radius 20 000 km.

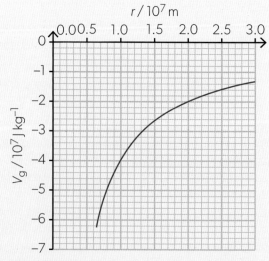

Solution

The gravitational potential difference between the orbit and the surface of Earth is $\Delta V_g = (6.3 - 2.0) \times 10^7 = 4.3 \times 10^7 \,\text{J kg}^{-1}$. The minimum work is equal to the change in the gravitational potential energy, $W = m\Delta V_g = 1200 \times 4.3 \times 10^7 = 5.2 \times 10^{10}\,\text{J}.$

The actual work will be greater because the satellite in orbit will have kinetic energy in addition to gravitational potential energy. Later in this topic you will learn how to include kinetic energy to describe fully the orbital motion of planets and satellites.

Practice questions

11. The diagram shows gravitational field lines (grey) and equipotential surfaces (green) around Earth. The potential difference between any two adjacent equipotential surfaces is $5.0 \times 10^6 \, \text{J kg}^{-1}$.

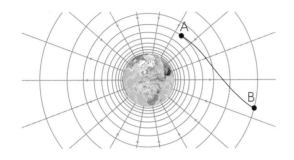

a. An object of mass 80 kg is moved from point A in the field to point B. Calculate the work done against the gravitational force.

b. The gravitational potential at B is one half of the gravitational potential at A. The gravitational field strength at A is $2.3 \, \text{m s}^{-2}$. Determine the gravitational field strength at B.

12. A projectile of mass 1.0 kg is fired vertically up from the surface of Earth with a speed of $7.7 \times 10^3 \, \text{m s}^{-1}$.

a. Calculate the initial kinetic energy of the projectile.

b. Use the graph in Worked example 11 to estimate the maximum height from the surface of Earth that the projectile can reach.

Field strength and potential

There is one more fundamental link between the field quantities discussed so far.

The gravitational potential at a point V_g at a distance r from a point object of mass M is given by $V_g = -\dfrac{GM}{r}$. When this equation is differentiated with respect to r it becomes $\dfrac{dV_g}{dr} = \dfrac{GM}{r^2}$. The right-hand side of this expression is the value for g due to a point object of mass M at a distance of r from the object: $g = -\dfrac{GM}{r^2}$ (including, in this case, the sign). We can write $\dfrac{dV_g}{dr} = -g$ or, in a non-differential form,

$$g = -\frac{\Delta V_g}{\Delta r}$$

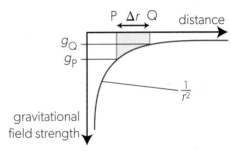

▲ Figure 20 A graph of gravitational field strength against distance.

Two graphical links between gravitational field strength and gravitation potential follow from these equations.

The first link appears in a graph showing the variation of gravitational field strength with distance for a point mass (Figure 20).

The change in gravitational potential in moving from P to Q on the graph is equal to $-g \times \Delta r$. This is the area between the $\dfrac{1}{r^2}$ - shaped curve of gravitational field strength and the distance axis (shaded in blue). On the graph this is a negative area and the negative sign in $-g \times \Delta r$ cancels with it to give an overall positive value for the potential change. This is what we expect, because going from P to Q means moving towards infinity and energy will need to be transferred into the gravitational system.

The second link is in the graph of variation of gravitational potential with distance (Figure 21). This time the gravitational field strength at a point is equal to $-\dfrac{\Delta V_g}{\Delta r}$ which is the gradient of the graph at the point concerned. The sign shows that the direction of the gravitational field strength is opposite to the direction in which gravitational potential increases.

▲ Figure 21 (a) When the distance between two positions is small, then the gravitational field strength is the change in potential divided by the change in distance. (b) The gravitational field strength at a point is the negative of the gradient of the line.

What measurements of a binary star system need to be made in order to determine the nature of the two stars?

The two stars in a binary star system are bound together by gravity. They obey the rules of gravitational fields outlined in this topic. A full investigation of the nature of the stars requires physics from other themes in this course. Visual binary systems can be seen to orbit each other using an Earth-bound or satellite-based telescope with all the optical limits that such an instrument

involves (Theme C). Astronomers can determine the nature of spectroscopic binaries using the emission and absorption spectral lines that are described and explained in Theme E. When the orientation of the binary stars relative to Earth is suitable, then changes in the overall luminosity of the pair (Theme B) will give an astronomer important clues about the nature of the stars.

How is the amount of fuel required to launch rockets into space determined by considering energy?

The graph of potential against distance can help when a rocket is being designed to reach another object in the Solar System.

The graph in Figure 22 shows the variation in gravitational potential between Earth and the Moon. Before take-off, a spacecraft sits in a potential well on Earth's surface. When the rockets are fired, the spacecraft gains speed and moves away from Earth. It only needs enough energy to reach the maximum of the potential at point L (this is known as a Lagrange point and it is also the point where the gravitational field strengths of Earth and Moon are equal and opposite). Once at L, the spacecraft can fall down the other side of the potential hill to arrive at the Moon.

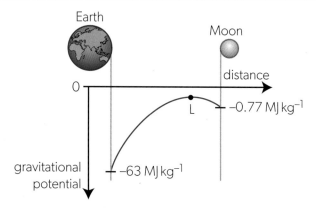

▲ Figure 22 The maximum of gravitational potential between Earth and the Moon is known as Lagrange point L. There are five Lagrange points that represent equilibrium points where the gravitational forces are balanced. L_1 is an unstable equilibrium point which allows a spaceship from Earth at L_1 to fall down the potential hill to the Moon.

Worked example 12

Two point objects of mass M and $25\,M$ are at a distance R from each other. A particle of mass m is moved along the line joining the masses.

a. Determine the distance from M at which the gravitational potential has a maximum value.

b. Calculate the maximum potential energy of the particle.

Solutions

a. Since $g = -\dfrac{\Delta V_g}{\Delta r}$, the maximum of V_g corresponds to zero value of the gravitational field strength g. This can only occur on the line joining the masses, at a point where the field strength from M is equal but opposite to the field strength from $25\,M$.

Let P be a point on the line joining the masses, at a distance x from M and $(R - x)$ from $25\,M$.

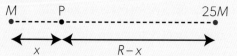

The field strength at P is zero if $G\dfrac{M}{x^2} = G\dfrac{25\,M}{(R-x)^2}$. This can be written as $\dfrac{R-x}{x} = \sqrt{25} = 5$, which implies that $x = \dfrac{R}{6}$. The potential has a maximum value at one-sixth of the distance between the masses.

b. The particle is at a distance of $\dfrac{R}{6}$ from M and $\dfrac{5R}{6}$ from $25\,M$.

$$E_p = -G\frac{Mm}{R/6} - G\frac{25\,Mm}{5R/6} = -36\frac{GMm}{R}.$$

Worked example 13

The graph shows how the gravitational potential V_g between a planet and its moon varies with the distance r from the centre of the planet.

The centre of the moon is 8.0×10^7 m from the centre of the planet. Point A is at the surface of the planet and point C is at the surface of the moon. The gravitational potential is a maximum at point B.

a. Explain why the gravitational field strength at B is zero.

b. Estimate the ratio $\dfrac{\text{mass of the planet}}{\text{mass of the moon}}$.

A space probe of mass 2.4×10^3 kg is launched from the surface of the planet.

c. Calculate the gravitational potential energy of the space probe when it is at a distance of 1.0×10^7 m from the centre of the planet.

d. Determine the minimum work that has to be done on the space probe so that it can reach the moon.

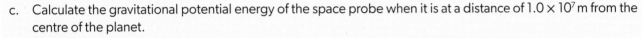

Solutions

a. The gradient of the gravitational potential is zero at B. The field strength is equal to the (negative) gradient of the potential, so it must also be zero at B.

b. For the combined field strength to be zero at B, the field strength due to the planet must be equal but opposite to the field strength due to the moon.

$$G\frac{M_p}{(6.0 \times 10^7)^2} = G\frac{M_m}{((8.0-6.0) \times 10^7)^2}. \text{ Therefore, } \frac{M_p}{M_m} = \left(\frac{6.0 \times 10^7}{2.0 \times 10^7}\right)^2 = 9.$$

c. $E_p = mV_g = 2.4 \times 10^3 \times (-1.8 \times 10^7) = -4.3 \times 10^{10}$ J.

d. The space probe must first reach B. The minimum work to move the probe from A to B is
$W = m\Delta V_g = 2.4 \times 10^3 \times (-0.4 \times 10^7 - (-4.4 \times 10^7)) = 9.6 \times 10^{10}$ J.

Patterns and trends—G and V_g inside Earth

Outside Earth, the gravitational field strength varies as $\dfrac{1}{r^2}$. But what happens to the field strength inside? Is it zero? Is it a constant? Does it become larger and larger, reaching infinity, as we get closer to the centre? You might, at first sight, expect this from Newton's law of gravitation.

In *Journey to the Centre of the Earth*, the novelist Jules Verne imagined going through a volcanic tunnel to the centre of the planet. Visualize his travellers when they have travelled halfway down the tunnel. They stand on the surface of a "smaller" Earth defined by their present distance from the centre (Figure 23).

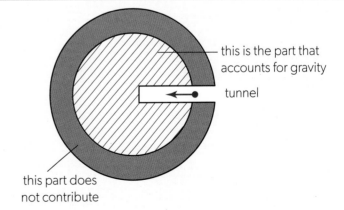

▲ Figure 23 When Jules Verne's travellers have reached a point inside Earth, they have a reduced gravity because the concentric shell with solid shading does not contribute to the gravitational pull.

The shell of Earth "above" the travellers makes no contribution to the gravitational field. All the parts of the outer shell cancel out. Only the mass inside the "small" Earth contributes to the gravitational pull. The mass of this "small" Earth is equal to its density $\rho \times$ its volume V: in other words, $\rho \times \frac{4}{3}\pi r^3$. The effective mass therefore varies with r^3 inside Earth, assuming a constant density for the solid Earth (not a good assumption, in fact). Because the gravitational force varies with $\frac{1}{r^2}$, these two variations mean that, overall, g inside Earth varies with r. The whole graph for the variation of g for a planet (inside and out) is given in Figure 24.

The inner solid sphere, a miniature Earth with a smaller radius, behaves as a normal Earth but with a different gravitational field strength g' given by:

$$g' = G \times \frac{\text{mass of the "small" Earth}}{\text{radius of the "small" Earth}^2}$$

The inverse-square law gives rise to a linear relationship because the $\frac{1}{r^2}$ behaviour of the force cancels with the r^3 variation of the mass of the "small" Earth.

Another surprising result appears when the planet is a spherical shell, like an empty eggshell. In this case, when Verne's travellers break through into the interior, there will be no gravitational force arising from the shell at all.

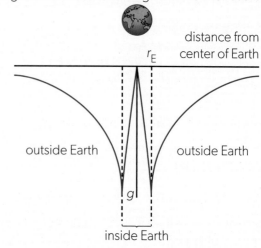

▲ Figure 24 The graph of gravitational field strength g versus distance from the centre of a uniform spherical Earth. Inside the planet g is proportional to the distance from the centre.

Modelling assumptions

The orbits of planets, moons and satellites are often not circular. For example, the orbit of the dwarf planet Pluto around the Sun is a narrow ellipse. However, for these calculations you can assume that planetary orbits are circular.

What is *r*?

Remember that r in the equation for orbital speed is the radius of the orbit—counting from the centre of the planet. It is *not* the distance above the planet surface h. When you are given the orbiting height h of a satellite, you must add the radius of Earth R_E or the planet to this height:

$$r = R_E + h.$$

Orbiting a planet

When a satellite orbits a planet, the centripetal force that keeps the satellite in orbit is provided by the gravitational attraction to the planet.

The derivation of Kepler's third law of planetary motion equated the centripetal force F_C and the gravitational attraction F_G. This time, we use the linear orbital speed $v_{orbital}$ rather than the angular speed ω:

$$F_C = F_G = \frac{m v_{orbital}^2}{r} = \frac{GM_E m}{r^2}$$

where r is the radius of the orbit.

A simple rearrangement of the equation leads to

$$v_{orbital}^2 = GM_E \times \frac{r}{r^2}$$

and hence

$$\text{orbital speed, } v_{orbital} = \sqrt{\frac{GM_E}{r}}$$

The angular orbital speed is $\omega = \frac{v}{r}$, leading to $\omega = \sqrt{\frac{GM_E}{r^3}}$.

The time T that a satellite takes to orbit a planet once is

$$T = \frac{2\pi}{\omega} = 2\pi \sqrt{\frac{r^3}{GM_E}} = \frac{2\pi}{\sqrt{GM_E}} r^{\frac{3}{2}}$$

which for a satellite in low-Earth orbit (so that the orbital radius is effectively Earth's radius) is about 85 minutes. Low-Earth orbit does not mean that the satellite is grazing the rooftops! Such satellites orbit at about 200 km above the surface, but this is a radius difference of only about 3%.

Worked example 14

A satellite orbits Earth in a circular orbit 150 km above the surface.

a. Calculate the orbital speed of the satellite.

b. Show that the mechanical energy of the satellite is given by $E = -\dfrac{GMm}{2r}$, where M is the mass of Earth and m is the mass of the satellite.

c. Determine the energy needed to increase the orbital radius of the satellite to 250 km. The mass of the satellite is 450 kg.

Mass of Earth $= 5.97 \times 10^{24}$ kg, radius of Earth $= 6.37 \times 10^6$ m

Solutions

a. The radius of the orbit is $r = 150 \times 10^3 + 6.37 \times 10^6 = 6.52 \times 10^6$ m.

$$v = \sqrt{\frac{GM}{r}} = \sqrt{\frac{6.67 \times 10^{-11} \times 5.97 \times 10^{24}}{6.52 \times 10^6}} = 7.81 \times 10^3 \, \text{ms}^{-1}$$

b. The kinetic energy of the satellite is $E_k = \dfrac{1}{2}mv^2 = \dfrac{GMm}{2r}$, using the formula for the orbital speed. The potential energy is $E_p = -\dfrac{GMm}{r}$ and the total mechanical energy is therefore $E = E_k + E_p = \dfrac{GMm}{2r} - \dfrac{GMm}{r} = -\dfrac{GMm}{2r}$.

c. The energy needed to increase the orbital radius is equal to the change in the total orbital energy between the initial and the final orbit.

$$\Delta E = -\frac{GMm}{2r_{\text{new}}} - \left(-\frac{GMm}{2r_{\text{old}}}\right) = \frac{GMm}{2}\left(\frac{1}{r_{\text{old}}} - \frac{1}{r_{\text{new}}}\right)$$

$$\Delta E = \frac{6.67 \times 10^{-11} \times 5.97 \times 10^{24} \times 450}{2}\left(\frac{1}{6.52 \times 10^6} - \frac{1}{6.62 \times 10^6}\right)$$

$$= 2.1 \times 10^8 \, \text{J}$$

Worked example 15

A satellite of mass 790 kg is at rest on the surface of Earth. Determine the energy needed to put the satellite in a circular orbit 400 km above the surface.

Solution

Before launch, the satellite only has the gravitational potential energy $E_i = -\dfrac{GMm}{r_E}$, where r_E is the radius of Earth (we ignore the kinetic energy due to the rotation of Earth because it is much smaller than the kinetic energy in the final orbit).

In orbit, the total energy of the satellite will be $E_f = -\dfrac{GMm}{2(r_E + h)}$, using the equation derived in Worked example 14. The energy needed to reach the final orbit is the difference $\Delta E = E_f - E_i = GMm\left(\dfrac{1}{r_E} - \dfrac{1}{2(r_E + h)}\right)$.

$$\Delta E = 6.67 \times 10^{-11} \times 5.97 \times 10^{24} \times 790 \times \left(\frac{1}{6.37 \times 10^6} - \frac{1}{2 \times (6.37 \times 10^6 + 400 \times 10^3)}\right) = 2.6 \times 10^{10} \, \text{J}.$$

Practice questions

13. A satellite orbits a planet of radius R. The satellite is initially in circular orbit A, at a height R above the surface of the planet. The satellite is moved to a new orbit B at a height $2R$.

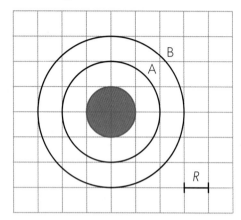

a. Calculate the ratio $\dfrac{\text{speed in orbit A}}{\text{speed in orbit B}}$.

b. The total energy of the satellite in orbit A is E. Calculate, in terms of E, the work done on the satellite to move it from orbit A to orbit B.

14. A space probe has a mass of 1300 kg.
    Calculate the energy needed to:
    a. put the space probe in a circular orbit 300 km above the surface of Earth
    b. increase the orbital radius by a further 10 km.

15. Ida is an asteroid (minor planet) in the Solar System. Ida has a moon, Dactyl, orbiting it with a period of about 20 hours. The exact orbit of Dactyl is not known, but we can assume that it is a circle with radius of about 90 km.
    a. Calculate the orbital speed of Dactyl.
    b. Determine the mass of Ida.

 Models — Delivering the mail!

Earlier we saw that inside a tunnel drilled through Earth the gravitational field strength g' is proportional to the distance from Earth's centre: $g' \propto r$. The force towards the centre in the tunnel is mg' and this linear dependence on r is the condition for simple harmonic motion (see Topic C.1).

A vehicle inside a tunnel drilled right through Earth will therefore perform simple harmonic motion when friction is neglected. The vehicle released at one end of the tunnel will stop momentarily when arriving at the opposite end, 42 minutes after release. This is the half the period of a satellite that is in low-Earth orbit. Get the timing right and mail bags could be exchanged at the surface!

Chordal tunnels dug between two cities (*not* passing through the centre of Earth) will also allow the mail to be delivered with the same time period. Use some of the ideas and equations from the course to assess the engineering issues with putting this plan into operation!

Types of orbit

Although satellites can be put into orbits of any radius (provided that the radius is not so great that a nearby astronomical body can disturb it), there are two types of orbit that are of particular importance.

The **polar orbit** is used for satellites close to Earth's surface. The satellites orbit over the poles in one plane (intersecting the centre of Earth) with Earth rotating beneath it (Figure 25(a)). Over the course of a 24-hour period, the satellite can view every point on Earth.

(a)

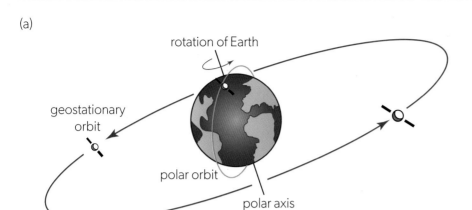

(b)

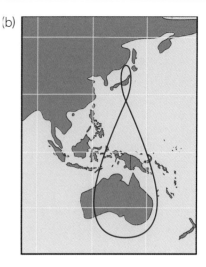

▲ Figure 25 (a) A polar orbit and a geostationary orbit around Earth. (b) The track of a geosynchronous orbit in the sky.

Geosynchronous satellites orbit at much greater distances from Earth and have orbital times equal to one sidereal day, which is roughly 24 hours. This means that the geosynchronous satellite stays in the same area of sky and typically follows a figure-of-eight orbit above a region of the planet. A typical track in the southern hemisphere is shown in Figure 25(b).

A **geostationary orbit** (Figure 25(a)) is a special case of the geosynchronous orbit. In this case, the satellite is placed in orbit above the plane of the Equator and will not appear to move when viewed from the surface.

Worked example 16

Calculate the orbital period for a satellite in polar orbit.

Solution

The orbital period can be found by rearranging the Kepler's third law equation, $T = \sqrt{\dfrac{4\pi^2 r^3}{GM}}$.
The radius of the orbit is the orbital height (100 km) plus the radius of Earth.

$$T = \sqrt{\frac{4\pi^2 \times (100 \times 10^3 + 6.37 \times 10^6)^3}{6.67 \times 10^{-11} \times 5.97 \times 10^{24}}} = 5180\,\text{s (about 86 min)}$$

Worked example 17

A geostationary satellite has an orbital period of 24 hours. Calculate:

a.   the distance of the orbit from the surface of Earth

b.   the gravitational field strength at the orbital radius of a geostationary satellite.

Solutions

a.   24 hours is 86 400 s.

$$T^2 = \frac{4\pi^2 r^3}{GM} \Rightarrow r = \sqrt[3]{\frac{T^2 GM}{4\pi^2}} \quad \Rightarrow \quad r = \sqrt[3]{\frac{86400^2 \times 6.67 \times 10^{-11} \times 5.97 \times 10^{24}}{4\pi^2}} = 4.22 \times 10^7\,\text{m}$$

The distance from the surface is $r - r_E = 3.59 \times 10^7\,\text{m} = 35\,900\,\text{km}$.

b. $g = \dfrac{GM}{r^2} = \dfrac{6.67 \times 10^{-11} \times 5.97 \times 10^{24}}{(4.22 \times 10^7)^2} = 0.223\,\text{N kg}^{-1}$

Escaping Earth

A space launch from Earth may put a satellite into Earth orbit or send a craft to explore a region of space far away from Earth. In the second case, the craft must escape Earth's gravity and eventually come under the gravitational influence of other planets and stars.

What is the minimum energy needed to allow the craft to escape Earth?

The total (mechanical) energy of a satellite is made up of the gravitational potential energy and the kinetic energy (ignoring any energy transferred to the internal energy of the atmosphere by the satellite). To escape from the surface of Earth, work must be done on the satellite to take it to infinity. For an unpowered projectile this is equal to the kinetic energy—which means that the total of the (negative) gravitational potential energy and the (positive) kinetic energy must add up to zero. Thus, to reach infinity,

gravitational potential energy + kinetic energy $= 0$

and therefore

$$-\frac{GMm_s}{r} + \frac{1}{2}m_s v_{esc}^2 = 0$$

where r is the distance of the satellite from the centre of Earth, v_{esc} is the escape speed, and M and m_s are the masses of Earth and the satellite, respectively. The gravitational potential energy is negative because this is a bound system. Kinetic energy is always positive. (It is important to keep track of the signs in this proof.)

To escape Earth's gravitational field completely, the total energy of the satellite must be (at least) zero. For the case where it is exactly zero (for the satellite to just reach infinity), $v_{esc}^2 = \frac{2}{m_s} \times \frac{GMm_s}{r}$.

Thus, escape speed, $v_{esc} = \sqrt{\dfrac{2GM}{r}}$

Notice that:

- the escape speed is independent of the satellite mass, depending only on the properties of the planet

- $v_{esc} = v_{orb} \times \sqrt{2}$

- $v_{esc} = \sqrt{2gr}$, where g is the gravitational field strength *at the surface*

- for Earth, from its surface, v_{esc} is about $11\,200\,\mathrm{m\,s^{-1}}$ ($40\,000\,\mathrm{km\,h^{-1}}$).

The true meaning of escape speed is the speed at which an *unpowered* object, something like a bullet, would have to be travelling to leave Earth from the surface. In theory, a rocket with enough fuel can leave Earth at any speed. All that is required is to supply the $63\,\mathrm{MJ}$ for each kilogram of the mass of the rocket (the gravitational potential of Earth's field at the surface is $-62.5\,\mathrm{MJ\,kg^{-1}}$). However, in practice, it is best to reach the escape speed as soon as possible.

Similarly, when a spacecraft begins its journey from a parking orbit, then less fuel will be required from there because part of the energy has already been supplied to reach the orbit.

Worked example 18

Calculate the escape speed from the surface of the Moon.

Mass of the Moon $= 7.35 \times 10^{22}$ kg, radius of the Moon $= 1.74 \times 10^6$ m.

Solution

$$v_{esc} = \sqrt{\frac{2 \times 6.67 \times 10^{-11} \times 7.35 \times 10^{22}}{1.74 \times 10^6}} = 2.37 \times 10^3 \, m\,s^{-1}$$

Worked example 19

A space probe of mass 2500 kg is in a circular parking orbit 350 km above the surface of Earth. The main engine of the space probe is fired for a short time in the direction of motion so that the space probe reaches the speed needed to escape Earth's gravitational field.

Determine the work done on the space probe by the engine.

Solution

To escape the gravitational field of Earth, the space probe must have the kinetic energy corresponding to the escape speed from the radius of the parking orbit, $E_k = \frac{1}{2}mv_{esc}^2 = \frac{1}{2}m\frac{2GM}{r} = \frac{GMm}{r}$.

The space probe already has the kinetic energy due to its orbital motion, $E_{k,\,orbit} = \frac{1}{2}mv_{orbit}^2 = \frac{GMm}{2r}$.

The work done by the engines is $W = E_k - E_{k,\,orbit} = \frac{GMm}{r} - \frac{GMm}{2r} = \frac{GMm}{2r}$.

The orbital radius is $r = 6.37 \times 10^6 + 350 \times 10^3 = 6.72 \times 10^6$ km.

$$W = \frac{6.67 \times 10^{-11} \times 5.97 \times 10^{24} \times 2500}{2 \times 6.72 \times 10^6} = 7.4 \times 10^{10} \, J$$

Practice questions

16. Calculate the speed needed to escape the gravitational influence of the Sun from the orbital radius of Earth. Mass of the Sun $= 2.0 \times 10^{30}$ kg; Sun–Earth distance $= 1.5 \times 10^{11}$ m.

17. A space probe of mass 640 kg is to be launched from the surface of Earth. Calculate:

a. the energy needed to move the probe from the surface to a parking orbit at a height of 500 km

b. the escape speed from the parking orbit

c. the energy needed to escape the gravitational field of Earth from the parking orbit.

Orbits and the atmosphere

In Theme A, the effects of air resistance on the motion of a projectile were described. Drag force acts to reduce the range of a projectile. It also alters the trajectory to a steeper angle to the horizontal compared with the situation where drag is negligible.

The atmosphere does not suddenly end at one fixed altitude. The gas density decreases with height until eventually there are so few gas atoms that a satellite is effectively travelling in a vacuum. However, at the orbital radii of low-orbit satellites, there is still a significant effect over time.

The collisions between the satellite and the gas atoms transfer energy from the satellite to the atmosphere. The atmosphere heats up at the expense of the total energy of the satellite. As a result, the radius of the satellite's orbit decreases. Paradoxically, this means that the speed of the satellite in its orbit must increase because

$$v_{orbital} = \sqrt{\frac{GM_E}{r}}$$

As the radius decreases, $v_{orbital}$ increases.

This is because the kinetic energy of the satellite is a positive quantity which increases with decreasing radius, while the gravitational potential energy of the satellite decreases (that is, becomes more negative) as the radius decreases.

The total energy E of a satellite is the sum of the kinetic energy E_k and the gravitational potential energy W: $E = W + E_K$. Using the energy quantities from earlier,

$$E = -\frac{GMm_s}{r} + \frac{1}{2}m_s v_{orbital}^2$$

We already know that $v_{orbital} = \sqrt{\frac{GM}{r}}$ and therefore $\frac{1}{2}m_s v_{orbital}^2 = \frac{1}{2}m_s \times \frac{GM}{r}$. This is $\frac{1}{2} \times \frac{GMm_s}{r}$, which is $\frac{W}{2}$. The magnitude of the gravitational potential energy is twice that of the kinetic energy—although remember that the gravitational potential energy is negative, whereas the kinetic energy is positive.

A loss in orbital energy $-\Delta E$ leads to a decrease in the orbital radius of Δr that has two contributions:

- a loss of gravitational potential energy of $-2\Delta E$

- a gain in kinetic energy of $+\Delta E$.

 How can air resistance be used to alter the motion of a satellite orbiting Earth?

The gradual decrease in orbital radius and increase in orbital speed mean that low-orbit satellites that are intended for long-term use need to have their orbits boosted periodically, perhaps three or four times every year. The consequence of not doing so is that the orbit continues to decay and the speed continues to increase within an increasingly dense atmosphere as the orbital radius decreases. The collisions with atoms become more frequent and a runaway process begins, which eventually leads to a burn-up of the satellite in the atmosphere.

Worked example 20

A satellite of mass 700 kg is in a circular orbit of height 380 km above the surface of Earth.

a. Calculate the total energy of the satellite.

Due to residual atmospheric drag, the satellite loses mechanical energy at an average rate of 0.10 W.

b. Explain why the kinetic energy of the satellite is increasing.

c. Calculate the change in the orbital radius of the satellite:

    i.   during one orbital revolution

    ii.   during one year.

Solutions

a. $E = -\dfrac{GMm}{2(r_E + h)} = -\dfrac{6.67 \times 10^{-11} \times 5.97 \times 10^{24} \times 700}{2(6.37 \times 10^6 + 380 \times 10^3)} = -2.1 \times 10^{10}\,\text{J}$

b. The total mechanical energy decreases and hence the orbital radius of the satellite decreases.

The orbital speed is $\sqrt{\dfrac{GM}{r}}$, so it increases as the orbital radius r decreases.

c. One orbital revolution takes $\sqrt{\dfrac{4\pi^2 r^3}{GM}} = \sqrt{\dfrac{4\pi^2 (6.75 \times 10^6)^3}{6.67 \times 10^{-11} \times 5.97 \times 10^{24}}} = 5.52 \times 10^3\,\text{s}.$

    i.   Energy loss of the satellite during one orbit is $\Delta E = 0.10 \times 5.52 \times 10^3 = 552\,\text{J}$. The change Δr of the orbital radius can be calculated from the equation

$$\Delta E = -\frac{GMm}{2r} - \left(-\frac{GMm}{2(r - \Delta r)}\right), \text{ where } r = 6.75 \times 10^6\,\text{m is the original radius.}$$

$$552 = \frac{6.67 \times 10^{-11} \times 5.97 \times 10^{24} \times 700}{2} \left(\frac{1}{6.75 \times 10^6 - \Delta r} - \frac{1}{6.75 \times 10^6}\right).$$

Solving this equation for Δr gives $\Delta r = 0.18\,\text{m}$. The orbital radius decreases by 18 cm per revolution.

    ii.   In one year, the satellite makes about $\dfrac{365 \times 24 \times 3600}{5.52 \times 10^3} = 5700$ orbits. The orbital radius will decrease by $5700 \times 0.18 = 1000\,\text{m}$. This is an approximate answer because the orbital period is gradually decreasing and the magnitude of the drag force is increasing as the satellite is getting closer to Earth.

There is a long experimental story that led to the discovery of the electron. Many scientists have done sustained experimental work based on the theory of electric and magnetic fields. You will read more about this in Theme E. There is a detailed description of the experiment where charged beams are deflected by both electric and magnetic fields in Topic D.3.

Visualization aids our understanding. The field line is one of the best examples of this type of thinking. Scientists began by imagining the shape and strength of a magnetic field through the position and direction of fictitious lines of force (or field lines) that surrounded a magnetic dipole—a bar magnet. The image proved to be so powerful that the visualization was later used for electric and gravitational field lines too. However, such a technique can only go so far and an algebraic representation of field is needed as well.

This topic uses the ideas of field theory that you met in Topic D.1 and applies them to both electrostatics and magnetism. Field theory does not attempt to explain the origins of the electric and magnetic forces. It is a way of describing these fields in an algebraic and quantitative way—in the case of electrostatics using an inverse-square law.

Much of our modern world is driven by the interactions between electric and magnetic fields. The historical and physical consequences of these are described elsewhere in this course. There are forces acting between two wires when both carry electric current. Understanding this is one endpoint in your study of electric and magnetic fields.

▲ Figure 1 Many devices such as laptops and mobile phones have touchscreens. These work by detecting the change in the electric field when your finger (a conductor) is nearby.

In this topic, you will learn about:

- the two types of electric charge and the direction of the electric forces between them
- Coulomb's law
- conservation of electric charge
- Millikan's experiment
- the transfer of electric charge using friction, electrostatic induction and by contact

- electric field lines
- electric field line density and field strength
- magnetic field lines
- electric potential energy and electric potential
- the electric field strength as electric potential gradient
- the work done in moving a charge in an electric field
- equipotential surfaces.

AHL

Introduction

Electromagnetism is an effect observed when charge moves in a circuit. The electric current leads to the appearance of a magnetic field. But it was not the observation of a magnetism arising from current electricity that began the ancient study of magnetism. Early navigators

knew that some rocks are magnetic and that the rocks could move to indicate the direction of magnetic north. However, the true origins of magnetic effects remained obscure for many centuries. Only comparatively recently has our knowledge of the microscopic aspects of materials allowed us to understand how magnetism arises.

Explaining electrostatics

In Topic B.5 the origin of electric current was explained in terms of free charges (electrons) that move through conductors. In the context of electrostatics, static electrons are added to, or removed from, an object.

Experiments show that a positively charged object is attracted to objects with a negative charge but repelled by a positive charge. Simple electrostatic effects are due to the movement of negatively charged electrons. Positive charge arises from the presence of ions in a substance. An object with no *observed* charge has an exact balance between the electrons and the positively charged protons; it is said to be **neutral**.

Figures 2(a) and 2(b) show the interactions between opposite charges and like (same sign) charges. There can be an attraction between charged and uncharged objects due to **charge separation** in the uncharged object. In Figure 2(c), the free electrons in the uncharged sphere A are repelled by the negative sphere B. They move away to the other side of sphere A. Sphere A is now polarized with a surplus of negative charge on its left and a surplus of positive on the right. These positives are closer to sphere B than the negative charges. The electric force increases with decreasing distance and so the overall force on sphere A is towards sphere B.

When explaining the effects described here, always describe negative charge in terms of a surplus of electrons and describe positive charge in terms of a lack (or deficit) of electrons.

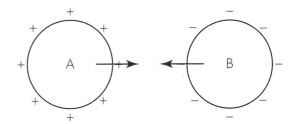

(a) Opposite charges attract.

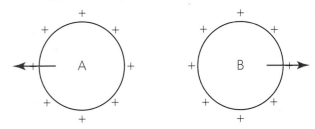

(b) Like charges repel.

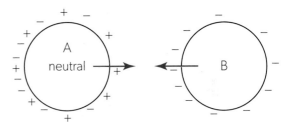

(c) Charge separation means that attraction occurs when one body is charged.

▲ Figure 2 The three interactions between (a) unlike, (b) like and (c) charged bodies and uncharged bodies.

Conservation of charge

Charge carriers in a single conductor must move in and out of the conductor at equal rates. The same applies at a junction. Figure 3 shows a junction with three incoming currents and two outgoing ones. Current splits at a junction but the number of electrons moving into and out of it is the same in each second. So $I_1 + I_2 + I_3 = I_4 + I_5$, which means that the sum of the currents flowing into the junction equals the sum of the currents flowing out. This is a consequence of **conservation of charge**.

This important rule was first quoted by Gustav Kirchhoff in 1845. He devised another electrical conservation rule: **conservation of energy**. In a closed electrical circuit, the energy being converted into electrical energy (by emf sources) must equal the energy being transferred from electrical to internal (by pd sinks).

▲ Figure 3 The sum of the currents into a junction is equal to the sum of the currents away from the junction. This is conservation of charge by another name.

Transferring charge

There are three methods through which charge can be transferred between objects.

(a) Charge transfer by friction

Rub a Perspex rod with a cloth and both objects become charged by friction. (This is known as the triboelectric effect.)

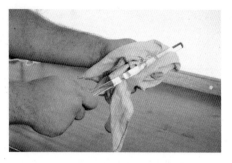

▲ Figure 4   Transferring charge by friction. The Perspex rod and the cloth are initially uncharged. After rubbing the cloth on the rod, the Perspex is positively charged and the cloth is negative.

The Perspex loses electrons and gains an overall positive charge (Figure 4). The electrons that are transferred to the silk give it an overall negative charge. Notice that electrons are not lost in these transfers. If 1000 electrons are removed from the rod, the cloth will be left with 1000 extra electrons at the end of the process. **Charge is conserved** in a closed system.

Two materials charged by friction end up with opposite charges.

In charging by friction, as the silk and the glass rod are rubbed together, the atoms on the surfaces of both materials are in close contact. Initially, the electron clouds surrounding the atoms on the surfaces do not interact. However, because an external force is being applied, the energy barriers between the electron clouds are modified and electrons transfer from one material to the other. When the force is removed, the energy barriers return to the original states and the electrons remain in their new location.

Table 1 shows some materials in the triboelectric series which shows the relative strength of materials with respect to this charging. When one material is rubbed against a material lower on the list, the material closer to the top is likely to become positive while the lower material becomes negative.

The success of plastic wrap (cling film) is due to several factors, including its elasticity and the fact that it becomes charged by friction as it is pulled off the roll. This means that it will cling effectively to a neutral object that has some free electrons.

most positive
glass
leather
lead
silk
amber
polystyrene
rubber
plastic wrap
polypropylene
ebonite
most negative

▲ Table 1   A list of some materials in the triboelectric series.

(b) Charge transfer by contact

A simple method to charge an object is to put one charged object in contact with an uncharged object. Some of the electrons are transferred from one to the other, leaving the initially charged object with a reduced overall charge. The uncharged object is no longer neutral but gains charge with the same sign as the original charged object.

The two materials charged by contact end up with the same sign of charge.

(c) Charge transfer by electrostatic induction

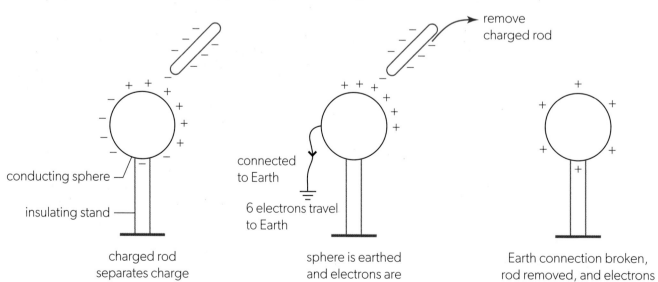

conducting sphere

insulating stand

charged rod separates charge

connected to Earth

6 electrons travel to Earth

sphere is earthed and electrons are repelled to Earth

remove charged rod

Earth connection broken, rod removed, and electrons re-distributed to leave sphere positive after rearrangement

▲ Figure 5   Transferring charge by electrostatic induction.

The process is shown in Figure 5. A charged rod is brought up to a neutral object, in this case a conducting metal sphere on an insulating stand. The rod is negatively charged and this repels some of the free electrons to the bottom of the sphere. There is charge separation and the sphere is still neutral because the charges remain balanced overall. (Its electric potential has been changed, however. The concept of electric potential and how it applies to this charging is explained later.)

The sphere is now earthed (grounded) by connecting a conductor between the ground and the sphere. (The sphere now has zero potential.) Some of the free electrons are repelled by the rod, move through this connection and flow down into Earth.

The loss of electrons leaves the sphere with an overall positive charge. When the rod and Earth connection is removed, the free electrons in the sphere will rearrange themselves to give an even distribution of positive charge on the sphere.

Notice that when the charged rod is removed, there is now a force of attraction between it and the sphere because they now have opposite signs. Work must be done to remove the rod. (The electric potential of the sphere changes again.)

The rod and the sphere have opposite signs of charge at the end of the process.

Worked example 1

A positively charged rod is brought up to an initially neutral conducting sphere. The sphere is grounded and the rod does not make contact with the sphere.

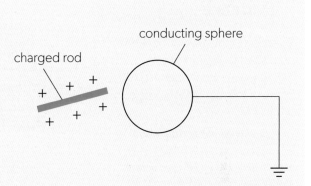

a. As the rod is moved towards the sphere, there is a short pulse of electric current between the sphere and the ground. Explain the direction of this conventional current.

b. The ground connection is now removed. State the sign of the charge induced on the sphere.

Solutions

a. The rod attracts free electrons in the sphere, causing a flow of electrons from the ground towards the sphere. The conventional current represents the flow of positive charge, so it is directed from the sphere to the ground.

b. The excess electrons remain on the sphere, and it is now negatively charged.

Forces between charged objects

In 1785, Charles-Augustin de Coulomb reported his series of investigations into the effects of forces from charged objects. He found, experimentally, that the force between two point charges a distance r apart is proportional to $\frac{1}{r^2}$. This confirmed earlier theories by Daniel Bernoulli, Alessandro Volta, Joseph Priestley and others. Like Newton's law of gravitation, this is an inverse-square law.

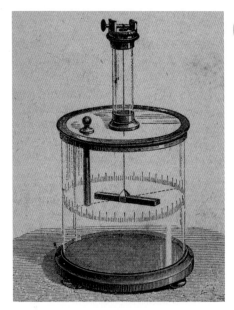

▲ Figure 6 The apparatus used by Charles-Augustin de Coulomb to show that the force between two charged objects obeys an inverse square law.

Evidence

Scientists at the time of Coulomb published their work in very different ways from scientists today. Some of the text of Coulomb's original *Mémoire* is:

Loi fondamentale de l'Électricité

La force répulfive de deux petits globes électrifés de la viéme nature d'électricité, eft en raifon inverfe du carré de la diftance du centre des deux globes.

Coulomb's work will not have been subject to peer review unlike the published research and review papers of scientists working in the 21st century.

 ## Replicating Coulomb's experiment

- Tool 1: Recognize and address relevant safety, ethical or environmental issues in an investigation.

- Inquiry 1: Demonstrate creativity in the designing, implementation or presentation of the investigation.

- Inquiry 2: Identify and record relevant qualitative observations.

- Inquiry 2: Interpret qualitative and quantitative data.

These are sensitive experiments that need care and a dry atmosphere to achieve a result.

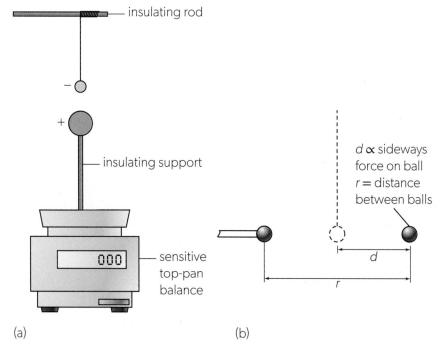

(a)   (b)

▲ Figure 7 Modern versions of Coulomb's experiment.

- Take two small polystyrene spheres and paint them with a metal paint or colloidal graphite or cover them with aluminium foil. Suspend one from an insulating rod using an insulating thread (perhaps nylon fishing line). Mount the other on top of a sensitive top-pan balance, again using an insulating rod.

- Charge both spheres by induction when they are apart from each other, or, alternatively, touch the terminal of a laboratory high-voltage power supply to a sphere. Your teacher will give you instructions about this.

- Bring the spheres together as shown in Figure 7(a) and observe changes in the reading on the balance.

Another method is to bring both charged spheres together as shown in Figure 7(b). The distance d moved by the sphere depends on the force between the charged spheres. The distance r is the distance between the centres of the spheres.

- Vary d and r, making careful measurements of them both.

- Plot a graph of d against $\frac{1}{r^2}$. An experiment performed with care can give a straight-line graph.

Later experiments confirmed that the force is proportional to the product of the size of the point charges q_1 and q_2. Combining Coulomb's results together with the later work gives

$$F \propto \frac{q_1 q_2}{r^2}$$

The magnitude of the force F between two point charges of charge q_1 and q_2 separated by distance r in a vacuum is given by

$$F = k \frac{q_1 q_2}{r^2}$$

where k is the constant of proportionality known as **Coulomb's constant**. In SI units, the value of k is $8.99 \times 10^9 \, N\,m^2\,C^{-2}$.

This equation does not appear to give the direction of the force between the charged objects. Forces are vectors, but the quantities *charge* and *distance²* are scalars. There are mathematical ways to cope with this, but, for point charges, the equation gives an excellent clue when the signs of the charges are included.

Take the positive direction to be from charge A to charge B. In Figure 8 this is from left to right (the positive *x*-direction). Begin with both charge A and charge B positive. When two positive charges are multiplied together in $k\frac{q_1 \times q_2}{r^2}$, the resulting sign of the force acting on charge B due to charge A is also positive. This means that the direction of the force on B will be assigned the positive direction (from charge A to charge B): in other words, left to right. This agrees because charge B is repelled. When both charges are negative, then the answer is the same because multiplying two negatives gives a positive value. The charges are repelled and the force is again to the right.

When one of the charges is positive and the other is negative, then the product of the charges is negative and the force direction will be opposite to the left-to-right positive direction. The force on charge B due to charge A is now to the left. Again, this agrees with what we expect, that is, that the charges attract because they have opposite signs.

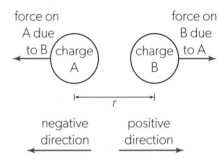

▲ Figure 8 A sign convention for force and charge. Force to the right is considered positive. Force to the left is considered negative.

Measurement—Another form of the force equation

Another way to quote the constant is

$$k = \frac{1}{4\pi\varepsilon_0}$$

The constant ε_0 is called the **permittivity of free space** ("free space" is a historical term for a vacuum). The 4π is added to rationalize electric and magnetic equations—in other words, to give them a similar shape and to retain an important relationship between them (see Topic D.3, page 550).

The equation becomes

$$F = \frac{q_1 q_2}{4\pi\varepsilon_0 r^2}$$

This means that ε_0 takes a value of $8.854 \times 10^{-12}\,C^2\,N^{-1}\,m^{-2}$ or, expressed in base (fundamental) units, $m^{-3}\,kg^{-1}\,s^4\,A^2$.

The equation as it stands applies only to charges that are in a vacuum. When the charges are immersed in a different medium (say, air or water), then the value of the permittivity is different. It is usual to amend the equation slightly. k becomes

$$k = \frac{1}{4\pi\varepsilon}$$

as the "0" subscript in ε_0 is only used for the case of a vacuum. For example, the permittivity of water is $7.8 \times 10^{-10}\,C^2\,N^{-1}\,m^{-2}$ and the permittivity of air is $8.8549 \times 10^{-12}\,C^2\,N^{-1}\,m^{-2}$. The value for air is so close to the free-space value that we normally use $8.85 \times 10^{-12}\,C^2\,N^{-1}\,m^{-2}$ for both air and a vacuum. The table gives permittivity values for some different materials.

Material	Permittivity / $10^{-12}\,C^2\,N^{-1}\,m^{-2}$
paper	34
rubber	62
water	779
graphite	106
diamond	71

Worked example 2

Two point charges of +10 nC and –10 nC in air are separated by a distance of 15 mm.

a.   Calculate the force acting between the two charges.

b.   Comment on whether this force can lift a small piece of paper about 2 mm × 2 mm in area.

Solutions

a.   It is important to take great care with the prefixes and the powers of ten in electrostatic calculations.

The charges are: $+10 \times 10^{-9}$ C and -10×10^{-9} C. The separation distance is 1.5×10^{-2} m. (Notice how the distance is converted right at the outset into consistent units.)

So $F = 8.99 \times 10^9 \dfrac{(+1.0 \times 10^{-8}) \times (-1.0 \times 10^{-8})}{(1.5 \times 10^{-2})^2}$

$= -4.0 \times 10^{-3}$ N

The charges are attracted along the line joining them. (Do not forget that force is a vector and needs both magnitude and direction for a complete answer.)

b.   A sheet of thin A4 paper of dimensions 210 mm by 297 mm has a mass of about 2 g. So the small area of paper has a mass of about 1.3×10^{-7} kg and therefore a weight of 1.3×10^{-6} N. The electrostatic force could lift this paper easily.

Worked example 3

Two point charges of magnitude +5 µC and +3 µC are 1.5 m apart in a liquid that has a permittivity of 2.3×10^{-11} $C^2 N^{-1} m^{-2}$. Calculate the force between the point charges.

Solution

$F = \dfrac{(+5 \times 10^{-6}) \times (+3 \times 10^{-6})}{4\pi \times 2.3 \times 10^{-11} \times (1.5)^2} = 23$ mN; a repulsive force acting along the line joining the charges.

Practice questions

1.   Two point charges of equal magnitude are separated by an air gap of 5.0 cm. The force between the charges is 0.80 mN.

a.   Determine the magnitude of each charge.

b.   Calculate the force between the charges when:

i.   their separation increases to 10 cm

ii.   the space between the charges is filled with a material of permittivity 1.8×10^{-11} $C^2 N^{-1} m^{-2}$.

2.   Two point charges of $+5.6 \times 10^{-7}$ C and -1.2×10^{-6} C in a vacuum are separated by a distance of 45 mm.

a.   Calculate the magnitude of the electrostatic force between the charges.

b.   Calculate the distance between the charges at which the force is 1.0 N.

Electric fields

In Topic D.1, the concept of (gravitational) field was developed and you were introduced to new concepts such as potential and field strength. In field theory, all force laws of the same type (in our case, the inverse-square force) can be described using similar concepts. Topic D.1 gives a good understanding of gravitational fields, but all the concepts are re-introduced here as it gives you a different perspective on field theory.

How are electric and magnetic fields like gravitational fields?

There are many similarities and differences between fields. For example, gravitational fields are always attractive, whereas electric and magnetic fields can be attractive or repulsive; gravitational potentials are always negative, whereas both positive and negative potentials are possible in electric and magnetic fields; and magnetic monopoles are thought not to occur, so true point sources are only possible in electric and gravitational fields.

But there are also other comparisons to make. It is important to remember that Topics D.1 and D.2 give a description of fields in terms of a common set of concepts. There is no attempt to explain how the gravitational or electrostatic forces arise. To show this distinction, gravitational force can be explained in Einstein's general theory as a distortion of spacetime by the mass of the attracting object. This leads to an inverse-square law—and it is at this point that our theory began.

The assumption of inverse-square behaviour for electric and gravitational fields is a useful algebraic description but by no means the whole story.

Electric field strength

The definition of field strength arises from a "thought experiment" involving the measurement of the force acting on a test object. There is a problem in carrying out the measurement of field strength practically. The presence of a test object will distort and alter the field in which it is placed since the test object carries its own field.

The definition of field strength in general is $\dfrac{\text{force acting on the test object}}{\text{size of the test object}}$ (as on page 476). This leads directly, for electric charge, to

$$E = \frac{\text{force acting on a positive test charge}}{\text{magnitude of test charge}} = \frac{F}{q}$$

where q is the size of the test charge.

The unit of electric field strength is $N\,C^{-1}$.

The definition for electric field strength at a point is that it is the force per unit charge experienced by a small positive point charge placed at that point.

The direction of the field is the same as the direction of the force acting on a positive charge. Electric field strength is a vector. It has the same direction as the force F (because the charge is a scalar). Extra attention to direction is required with electric fields because of the two signs of charge.

Electric field strength due to a point charge

Imagine an isolated charge of size q sitting in space. What is the strength of the electric field at a point P, a distance r away from the isolated charge? We use a positive test charge of size q_t at P and measure the force F that acts on the test charge due to the isolated charge. Then the magnitude of the electric force F acting on the test charge is $k \times \dfrac{q \times q_t}{r^2}$.

The electric field strength is given by

$$E = \frac{\text{force acting on a positive test charge}}{\text{magnitude of test charge}} = \frac{F}{q_t}$$

Therefore, $E = \dfrac{\left(k\dfrac{q \times q_t}{r^2}\right)}{q_t} = k\dfrac{q}{r^2}$

This leads to the following definition.

The electric field strength E at a distance r from an isolated point charge q is

$$E = k\frac{q}{r^2} = \frac{q}{4\pi\varepsilon_0 r^2}$$

You will see many similarities between the definition of gravitational fields, on page 476, and electric fields. One important difference is that the definition of an electric field specifies that the small point charge must be positively charged.

Do test charges and small current elements affect the original fields?

Just as a thermometer can alter the temperature of the object it measures, so test objects affect the field that they are being used to measure. They may accelerate the original current element or disturb the pattern of field lines. In practice, test objects are really imaginary constructs that we use to help our understanding just as we cannot travel to points at infinity.

This is food for thought from a theory of knowledge perspective: these are measurements that we can think about but not, in practical terms, carry out. The German language has a word for it: *gedankenexperiment* (thought experiment).

How can a practical subject such as science have a *gedankenexperiment*?

Worked example 4

An oxygen nucleus has a charge of $+8e$. Calculate the electric field strength at a distance of 0.68 nm from the nucleus.

Solution

Recall from Topic B.5 that the elementary charge (the charge of one proton, or the magnitude of the charge of one electron) is $e = 1.6 \times 10^{-19}$ C. The charge on the oxygen nucleus is thus $8 \times 1.6 \times 10^{-19}$ C. The distance is 6.8×10^{-10} m.

$$E = \frac{1.3 \times 10^{-18}}{4\pi\varepsilon_0 \times (6.8 \times 10^{-10})^2}$$

$= +2.5 \times 10^{10}$ N C^{-1} away from the nucleus.

Worked example 5

Calculate the electric field strengths in a vacuum:

a.   1.5 cm from a $+10\,\mu$C charge

b.   2.5 m from a -0.85 mC charge.

Solutions

a.   Begin by putting the quantities into consistent units: $r = 1.5 \times 10^{-2}$ m and $q = 1.0 \times 10^{-5}$ C.

Then $E = \dfrac{1.0 \times 10^{-5}}{4\pi\varepsilon_0 (1.5 \times 10^{-2})^2} = +4.0 \times 10^{8}$ N C^{-1}

The field direction is away from the positive charge.

b. $E = \dfrac{-8.5 \times 10^{-4}}{4\pi\varepsilon_0 (2.5)^2} = -1.2 \times 10^{6}$ N C^{-1}

The field direction is towards the negative charge.

Practice questions

3.   A nucleus of gold has a charge of $+79e$ and a radius of about 7 fm.

    a.   Calculate the electric field strength at a distance of 1.5 nuclear radii from the centre of the gold nucleus.

    b.   A free proton is at a distance of 1.5 nuclear radii from the centre of a gold nucleus. Calculate the magnitude of the electric force on the proton.

4.   The magnitude of the electric field at a distance of 0.25 m from a point charge is 470 N C^{-1}. The field is directed towards the point charge. Calculate the value of the point charge, including its sign.

5.   Calculate the distance from a point charge $+2.4 \times 10^{-8}$ C at which the electric field strength is 1.0×10^{2} N C^{-1}.

Adding electric fields

When there is more than one charge, the strength of an electric field due to two or more components must be found by vector addition. This means using either a calculation or a scale diagram. Vector addition is discussed on page 340.

Figure 9(a) shows a positive test charge that lies on the line between two charges of size $+Q$ and $-q$. The force on the + test charge due to $+Q$ is to the right. The electric field also produces a force on the test charge to the right due to the $-q$ charge. These two forces are drawn as vectors that act in the same direction (they are shown apart for clarity in the figure). They add to give a resultant force and therefore a resultant electric field as shown.

Figure 9(b) shows an arrangement where the situation is no longer collinear (the charges and the test charge do not lie in a straight line). There are three charges. Charges 1 and 2 have charge $+Q$ while the third has charge $+2Q$ but is much closer to the position of the test charge. Again, either a scale diagram or a calculation works. The length representing the force acting on the test charge due to $+2Q$ is much greater than the other two, which leads to a final summed field that is downwards on the diagram.

(a)

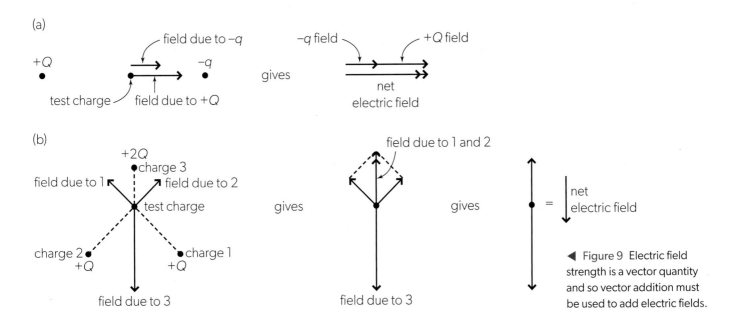

(b)

Worked example 6

Two point charges, a $+25\,nC$ charge X and a $+15\,nC$ charge Y are separated by a distance of $0.5\,m$.

a. Calculate the resultant electric field strength at the midpoint between the charges.

b. Calculate the distance from X at which the electric field strength is zero.

c. Calculate the magnitude of the electric field strength at the point P on the diagram.

Solutions

a. $E_X = \dfrac{2.5 \times 10^{-8}}{4\pi\varepsilon_0 \times 0.25^2} = 3600\,N\,C^{-1}$ $\qquad E_Y = \dfrac{1.5 \times 10^{-8}}{4\pi\varepsilon_0 \times 0.25^2} = 2200\,N\,C^{-1}$

The field strengths act in opposite directions, so the net electric field is $(3600 - 2200) = 1400\,N\,C^{-1}$. This is directed away from X towards Y.

b. For E to be zero, $E_X = -E_Y$ and so

$$\frac{2.5 \times 10^{-8}}{4\pi\varepsilon_0 \times d^2} = \frac{1.5 \times 10^{-8}}{4\pi\varepsilon_0 \times (0.5 - d)^2}$$

Thus

$$\frac{d^2}{(0.5 - d)^2} = \frac{2.5}{1.5} \qquad \text{or} \qquad \frac{d}{(0.5 - d)} = \sqrt{\frac{2.5}{1.5}} = 1.3$$

$$d = 0.65 - 1.3d$$

$$2.3d = 0.65$$

$$d = 0.28\,m$$

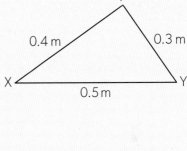

c. $PX = 0.4\,m$, so E_x at P is $\dfrac{2.5 \times 10^{-8}}{4\pi\varepsilon_0 \times 0.4^2} = 1400\,N\,C^{-1}$ along XP in the direction away from X.

$PY = 0.3\,m$, so E_Y at P is $\dfrac{1.5 \times 10^{-8}}{4\pi\varepsilon_0 \times 0.3^2} = 1500\,N\,C^{-1}$ along PY in the direction away from Y.

PX is perpendicular to PY so the magnitude of the resultant electric field strength is $\sqrt{1400^2 + 1500^2} = 2100\,N\,C^{-1}$.

(The calculation of the angles was not required in the question and is left for the reader.)

◀ Figure 9 Electric field strength is a vector quantity and so vector addition must be used to add electric fields.

Practice questions

6. The diagram shows two point charges $-80\,\mu C$ and $+80\,\mu C$. Point P is at an equal distance of $0.25\,m$ from the two charges.

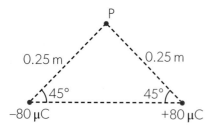

a. Determine the magnitude and direction of the resultant electric field at P.

b. A test charge of $-5.0\,nC$ is placed at P. Calculate the magnitude of the electric force on the test charge and state the direction of the force.

7. Two positive point charges Q_1 and Q_2 are $1.0\,m$ apart. Point P is on the line joining Q_1 and Q_2, at a distance of $0.20\,m$ from Q_1.

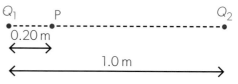

The electric field strength at P is zero. What is $\dfrac{Q_1}{Q_2}$?

A. $\dfrac{1}{16}$ B. $\dfrac{1}{4}$ C. 4   D. 16

8. Two charges $+Q$ and $-q$ are shown in the diagram. The magnitude of $+Q$ is greater than the magnitude of $-q$.

$$A \quad +Q \quad B \quad C \quad -q \quad D$$

At which of the points A, B, C or D can the electric field strength due to the two charges be zero?

Electric field lines

Electric field lines show the direction of the force acting on a test charge and help us to visualize the shapes of electric fields that arise from static charges. The field lines are imaginary, but they give us a clear impression of the variation of the field strength in the space between charged parallel plates, around point charges and charged spheres, and other arrangements.

The concept of the field line was first introduced by Michael Faraday. His original idea was of a set of elastic tubes that repelled other tubes, rather than a set of lines.

There are some conventions for drawing these electric field patterns:

- The lines start and end on charges of opposite sign.

- An arrow is essential to show the direction in which a positive charge would move (that is, away from positive charge and towards negative charge).

- Where the field is strong, the lines are close together. The lines act to repel each other.

- The lines never cross.

- The lines meet a conducting surface at 90°.

🧪 Plotting electric fields

- Tool 1: Recognize and address relevant safety, ethical or environmental issues in an investigation.

- Inquiry 1: Demonstrate creativity in the designing, implementation or presentation of the investigation.

- Inquiry 2: Identify and record relevant qualitative observations.

- Inquiry 2: Interpret qualitative and quantitative data.

Laboratory experiments can be used to make electric fields visible. Patterns of electric field lines can be observed using small particles floating on a liquid. The particles line up in the field that is produced between the wires. The patterns observed resemble those in Figure 11.

- Put some castor oil in a Petri dish and sprinkle some grains of semolina (also known as grits) onto the oil. Alternatives for the semolina include grass seed and hairs cut about 1 mm long from an artist's paint brush.

- Take two copper wires and bend one of them to form a circle just a little smaller than the internal diameter of the Petri dish. Place the end of the other wire in the centre of the Petri dish.

- Connect a 5 kV power supply to the wires. Take care with the power supply!

- Observe the grains slowly lining up in the electric field.

- Sketch the pattern of the grains that is produced.

- Repeat with other wire shapes such as the four examples shown in Figure 10.

- The patterns observed in such experiments will resemble those in Figure 11.

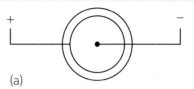
(a)

(b)

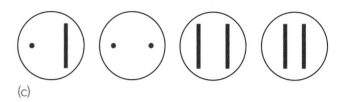
(c)

▲ Figure 10 The Petri dish contains a copper point with a copper collar around the edge. A potential difference between the two pieces of copper produces an electric field in the liquid. Small grains show the shape of this field. Other electrode shapes are shown too.

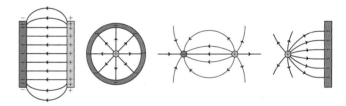

▲ Figure 11 Electric field patterns due to four charge configurations used in the semolina experiment.

The radial field due to a point charge

The electric field lines radiate outwards from a positive point charge and inwards towards a negative point. As with the similar pattern in a gravitational field, this is a **radial field**.

The field between two point charges

It is not hard to imagine how the two separate fields of Figure 12 combine to give the case where the point charges are close (Figures 13 and 14). There are two cases possible: two like charges and two unlike charges. Although a 2-D view is given, the fields are, of course, in 3-D.

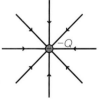

▲ Figure 12 Radial fields for an isolated positive charge and an isolated negative charge.

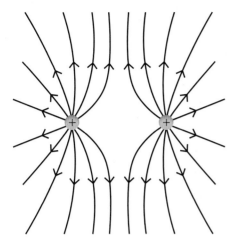

▲ Figure 13 The field around two positive point charges.

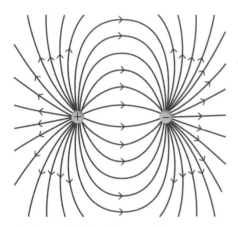

▲ Figure 14 The field around a positive point charge and a negative point charge.

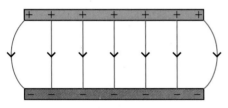

▲ Figure 15 View of the field between two parallel plates including the edge effects.

(a) Like charges

There will be a point where the field is zero somewhere between the charges (where the electric forces on a test charge are equal and opposite). When both charges have the same magnitude, this position of zero field strength will be midway between them.

(b) Unlike charges

The electric field lines now link the two unlike charges in the direction that is from positive to negative charge. There is some resemblance to the field of a bar magnet that you meet later in this topic.

The field between two parallel charged plates

An electric field exists in the space between two identical parallel metal plates that are connected to a dc power supply. A positive test charge in the electric field is attracted to the negative plate and repelled by the positive plate. The field pattern is shown in Figure 15.

This electric field:

* is uniform in the region well within the plates

* becomes weaker at the edges. These are known as **edge effects**. This is where the field changes from the strength inside the plates to the value outside the plates (often zero). You should be able to use the properties of the field lines to explain that there cannot be an abrupt change in field strength. At the plate edges, the field lines curve outwards as the field gradually weakens from the large value between the plates to the much weaker field well away from them. For the purposes of this course, you should assume that this curving begins at the edges of the plates (in practice, it begins a little way in from the edge).

Try to predict the way in which the shape of the field lines might change when a small conducting sphere is introduced in the middle of the space between the two plates. (Use your knowledge of charging by induction.)

Sometimes field lines are called **lines of force**. These lines represent the force acting on a positive test charge (by definition) with direction indicated by the arrow on the field line. When the line is curved, the tangent at a point on a line of force gives the direction of the electric force acting on a positive test charge. The relative density of the lines (how close they are) indicates the strength of the force.

ATL   Thinking skills

Imagine that you are a small test charge sitting in the middle of two horizontal uniformly and oppositely charged parallel plates. You can see the plates stretching out to the distant horizon, much as if you were standing in a huge flat field, except that there is also a plate overhead. The view would be the same in all horizontal directions and so the electric field must be vertical.

Imagine moving to the side a short distance. Since the plates are uniformly charged, your view would hardly change. As a result, the electric field due to the plates will be the same at this new position—the field must be uniform.

Now imagine that you move to the edge of the plates. If you are centrally between the plates, then any horizontal contribution to the field from the positive plate will be cancelled out by the opposite contribution to the negative plate—this would not be true if you were not positioned at an equal distance from each. Now, though, the plates will only occupy half of your field of view. The electric field strength is half the strength it was at the centre.

Field between parallel plates

- Tool 1: Recognize and address relevant safety, ethical or environmental issues in an investigation.
- Inquiry 1: Identify and justify the choice of dependent, independent and control variables.
- Inquiry 2: Identify and record relevant qualitative observations.
- Inquiry 3: Relate the outcomes of an investigation to the stated research question or hypothesis.

A small piece of charged foil can be used to detect the presence of an electric field.

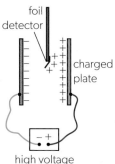

▶ Figure 16 Demonstrating the electric field between two charged parallel plates.

- The detector is made from a rod of insulator. A plastic ruler or a strip of polythene is ideal. Attached to the rod is a small strip of foil: thin aluminium or gold foil or "Dutch" metal are suitable. The dimensions of the foil need to be about 4 cm × 1 cm. The foil can be attached to the rod using adhesive tape or glue.
- Set up two vertical parallel metal plates connected to the terminals of a power supply that can supply

a potential difference (pd) of about 1 kV to the plates. Take care when carrying out this experiment. (Use the protective resistor in series with the supply if necessary.)

- Begin with the plates separated by a distance that is roughly one-third of the length of their smaller sides.
- Touch the foil briefly to one of the plates. This charges the foil. You should now see the foil bend away from the plate it touched.
- The angle of bend in the foil indicates the strength of the electric field. Explore the space between the plates and outside them too. Notice where the field starts to become weaker as the detector moves outside the plate region. Does the force indicated by the detector vary inside the plate region or is it constant?
- Turn off the supply and change the spacing between the plates. Does having a larger separation produce a larger or a smaller field?
- Change the pd between the plates. Does this affect the strength of the field?
- An additional experiment is to place a candle flame midway between the plates. What do you notice about the shape of the flame when the power supply is turned on? Can you explain your observation in terms of the charged ions in the flame?

The foil detector itself can also be used to explore the field around a charged metal sphere such as the dome of a van der Graaf generator.

Potential difference between parallel plates

The uniform electric field between two charged plates provides us with another way to think about electric field strength.

In Figure 17, a positive charge with size q is in a field between two charged plates separated by a distance d. This field has a strength E. The force acting on the charge is (from the definition of electric field strength) $F = E \times q$. Work must be done on the charge to move it towards the positive plate at constant speed. This is *force × distance moved* in the usual way. The field lines are parallel to the displacement, so there is no component of force to worry about. The work done = $F \times x = Eq \times x$, where x is the distance moved. When the charge is moved from the negative plate to the positive, the work done on the charge is Eqd.

From Topic B.5 (page 301), the potential difference V between the plates is

$$V = \frac{\text{work done in moving a charge}}{\text{magnitude of a charge}}$$

from Topic B.5 (page 303). Therefore

$$V = \frac{\text{work done}}{q} = \frac{Eqd}{q} \quad \text{or} \quad E = \frac{V}{d}$$

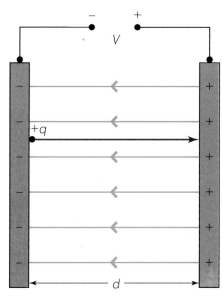

▲ Figure 17 The symbols used to show that $E = \dfrac{V}{d}$ between two parallel plates.

Calculating electric field strength

Electric field strength can be written and calculated in two ways:

- $\dfrac{\text{force acting on a charge}}{\text{magnitude of charge}}$;
  this implies that the unit for electric field strength is $N\,C^{-1}$.

- $\dfrac{\text{potential difference}}{\text{distance moved}}$;
  this implies that the unit for electric field strength is $V\,m^{-1}$.

These give two alternative units for electric field strength as well as two ways to calculate it.

To see this, write

$$\frac{N}{C} \equiv \frac{kg\,m\,s^{-2}}{C} \equiv \frac{kg\,m^2\,s^{-2}}{C\,m} \equiv \frac{kg\,m^2\,s^{-2}}{C} \times \frac{1}{m} \equiv \frac{V}{m}$$

Writing the equations in full for the uniform field between parallel plates gives

$$E = \frac{F}{q} = \frac{V}{d}$$

In practice, it is easier to measure a potential difference with a voltmeter in the laboratory than to measure a force so an electric field strength is commonly expressed in $V\,m^{-1}$.

Worked example 7

A pair of parallel plates with a potential difference between them of 5.0 kV are separated by 120 mm. Calculate:

a.   the electric field strength between the plates

b.   the electric force acting on a doubly ionized oxygen ion between the plates.

Solutions

a. $E = \dfrac{V}{d} = \dfrac{5000}{0.12} = 4.2 \times 10^4\,V\,m^{-1}$

b.   The charge of the ion is $2e = +3.2 \times 10^{-19}\,C$.

$F = qE = 3.2 \times 10^{-19} \times 4.2 \times 10^4 = 1.3 \times 10^{-14}\,N$

The force is directed towards the negative plate.

Worked example 8

A pair of parallel plates are separated by 80 mm. A droplet with a charge of $11.2 \times 10^{-19}\,C$ is in the field.

a.   Calculate the potential difference required to produce a force of $3.6 \times 10^{-14}\,N$ on the droplet.

b.   The plates are now moved closer to each other with no change to the potential difference. The force on the droplet changes to $1.4 \times 10^{-13}\,N$. Calculate the new separation of the plates.

Solutions

a. $E = \dfrac{F}{q} = \dfrac{3.6 \times 10^{-14}}{11.2 \times 10^{-19}} = 3.2 \times 10^4\,N\,C^{-1}$

$V = Ed = 3.2 \times 10^4 \times 0.080 = 2600\,V$

b.   The force changes by a factor of $\dfrac{1.4 \times 10^{-13}}{3.6 \times 10^{-14}} = 3.9$

The separation decreases by this factor, to $\dfrac{80}{3.9} = 21\,mm$.

Practice questions

9. A charged particle of mass 2.5 g is placed in an electric field between two charged parallel plates. The potential difference between the plates is 8.0 kV and the plates are separated by 12 cm.

    a. Calculate the electric field strength between the plates.

    b. The particle accelerates towards the positively charged plate with an acceleration of $1.6\,\mathrm{m\,s^{-2}}$. No other forces than the electric force act on the particle. Determine the magnitude and the sign of the charge on the particle.

10. A particle of charge +15 nC is suspended between two parallel plates separated by 2.0 cm. The electric force acting on the particle is 1.2 mN. Calculate:

    a. the electric field strength at the position of the particle

    b. the potential difference between the plates.

The electronvolt

The energy possessed by an individual electron is very small. When a single electron is moved through a potential difference of 15 V, then, as $W = qV$, the energy gained by this electron is $15 \times 1.6 \times 10^{-19}\,\mathrm{J} = 2.4 \times 10^{-18}\,\mathrm{J}$. This small quantity involves large negative powers of ten. It is convenient to define a new unit for energy and work done.

The electronvolt (symbol eV) is defined as the energy gained by one electron when it moves through a potential difference of one volt.

An energy of 1 eV is equivalent to $1.6 \times 10^{-19}\,\mathrm{J}$. The electronvolt is used extensively in the nuclear and particle physics of Theme E.

Worked example 9

An electron, initially at rest, is accelerated through a potential difference of 180 V. Calculate, for the electron:

a. the gain in kinetic energy

b. the final speed.

Solutions

a. The electron gains 180 eV of energy during its acceleration.

$1\,\mathrm{eV} \equiv 1.6 \times 10^{-19}\,\mathrm{J}$, so $180\,\mathrm{eV} \equiv 2.9 \times 10^{-17}\,\mathrm{J}$

b. The kinetic energy of the electron $= \frac{1}{2}m_e v^2$ and the mass of the electron is $9.1 \times 10^{-31}\,\mathrm{kg}$.

So $v = \sqrt{\dfrac{2E_k}{m_e}} = \sqrt{\dfrac{2 \times 2.9 \times 10^{-17}}{9.1 \times 10^{-31}}} = 8.0 \times 10^{6}\,\mathrm{m\,s^{-1}}$

Worked example 10

In a nuclear accelerator a proton is accelerated from rest and gains an energy of 250 MeV. Estimate the final speed of the particle and comment on the result.

Solution

The energy gained by the proton, in joules, is 4.0×10^{-11} J.

As before, $v = \sqrt{\dfrac{2E_k}{m_p}} = \sqrt{\dfrac{2 \times 4.0 \times 10^{-11}}{1.7 \times 10^{-27}}}$, but using a value for the mass of the proton this time.

The numerical answer for $v = 2.2 \times 10^8$ m s^{-1}.

This is a large speed, 70% of the speed of light. In fact, the speed will be less than this as some of the energy goes into increasing the mass of the proton through relativistic effects rather than into increasing the speed of the proton.

Charged parallel plates

- Tool 1: Recognize and address relevant safety, ethical or environmental issues in an investigation.

- Tool 3: Construct and interpret tables and graphs for raw and processed data including scatter graphs and line and curve graphs.

- Inquiry 1: Justify the range and quantity of measurements.

- Inquiry 2: Collect and record sufficient relevant quantitative data.

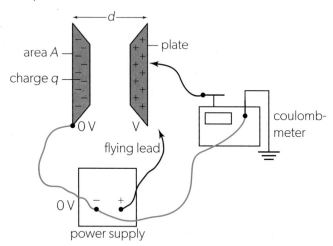

▲ **Figure 18** Measuring the variation of charge on parallel plates.

This experiment shows how electric field strength between parallel plates varies with applied voltage, the area of the plates and the distance between them. **Great care is needed as this experiment uses high voltages**.

- Set up a pair of parallel plates, a 5 kV power supply, a *well-insulated* lead attached to the positive terminal of the power supply and a coulombmeter (a meter that

will measure charge directly in coulombs) as shown in the circuit. Later you will need to replace the parallel plates with ones that have different areas. You will also need to record the distance between the plates with the supply turned off.

- Set and record a suitable voltage V on the power supply. Your teacher will suggest values to use.

- Measure and record the distance d between the parallel plates.

- Turn on the power supply and touch the flying lead to the right-hand plate briefly. Then remove it from the plate. This charges the plates.

- Zero the coulombmeter and then immediately touch its probe to the right-hand plate that you just charged. Record the charge q shown on the meter. This is the charge on the plate. You may wish to repeat this charging and measurement procedure as a check.

- Change the distance between the plates without changing the setting on the power supply.

- Repeat the measurements of d and q. Don't forget to switch off the power supply before you measure the separation of the plates.

- Repeat the experiment with constant power supply voltage V and plate separation d, but use pairs of plates with different areas. Record each plate area A.

- Carry out another experiment in which V is changed but A and d are constant.

- Plot your results as q versus V, q versus A, and q versus d.

Field close to a conductor

The charge q stored on one of the parallel plates in the two-plate arrangement depends on the potential difference V between the plates, the area A of the plates and their separation d:

- $q \propto V$

- $q \propto A$

- $q \propto \dfrac{1}{d}$

These relationships can be combined to give $q \propto \dfrac{VA}{d}$ where the constant of proportionality turns out to be $\dfrac{1}{4\pi k}$ and therefore:

$$q = \frac{VA}{4\pi kd}$$

This expression can give a value for the density of charge on the surface of charged parallel plate. It rearranges to:

$$\frac{q}{A} = \frac{V}{4\pi kd} = \frac{1}{4\pi k} \times \frac{V}{d}$$

Here $\dfrac{q}{A}$ is the charge per unit area, the surface charge density σ. The units of σ are $C\,m^{-2}$. Therefore, between two parallel plates:

$$E = 4\pi k\sigma$$

Each plate contributes half of the field, so that the electric field very close to the surface of any single conductor is $E = 2\pi k\sigma$. "Very close" means close enough for the surface to be considered locally flat.

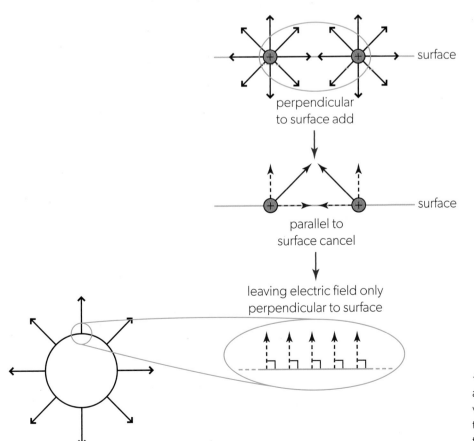

perpendicular
to surface add

parallel to
surface cancel

leaving electric field only
perpendicular to surface

◀ Figure 19 Although from a distance away, the surface of a sphere is curved, very close to the surface it appears flat and the electric field lines are locally normal to the surface.

The reasoning why there is no electric field inside a closed sphere should remind you of the similar suggestion for the gravitational field inside a spherical shell on page 495.

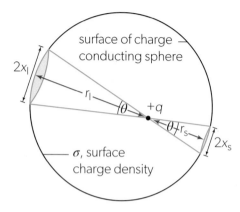

▲ Figure 20 No resultant force acts on the charge $+q$ from the two areas of charge indicated. This means that, overall, there can be no force on the charge. There is no electric field inside the sphere.

▲ Figure 21 In the same way that there is no electric field inside a conducting sphere, any other shape made of a conducting material will have no electric field inside it. This is often called a Faraday cage, although the effect was first noted by Benjamin Franklin. In this picture, a man is wearing a conducting suit of chain mail, which protects him from the large electric fields he is demonstrating. The same effect can be used to shield sensitive electrical equipment from external electric fields.

Imagine going very close to the surface of a conductor. Figure 19 shows what you might see in the case of a conducting sphere. When we are sufficiently close, the surface appears flat (just as we are not aware of Earth's curvature until we can see it from Space). We would also see that the free electrons are equally spaced. This is because any electron has forces acting on it from all the other electrons. The electron will accelerate until all these forces balance out and it is in equilibrium. For this to happen they must be equally spaced.

The electric field vectors radiate out from each individual electron (acting as a point source). Parallel to the surface, all these vectors cancel so there is no electric field along the surface. No electron accelerates along the surface because the component of field strength is zero. Perpendicular to the surface, however, things are different. The field vectors add up and, because there is no field component parallel to the surface, the local field must act at 90° to the surface. Close to any conducting surface, the electric field is at 90° to the surface and has strength $E = 2\pi k\sigma$.

Field due to a conducting spherical body

(a) Outside the sphere

Knowledge of the electric field very close to a conductor can be taken a step further for a conducting sphere—whether hollow or solid. Again, the free electrons on the surface are equally spaced and all the field lines at the surface of the sphere are at 90° to it. The consequence is that the field must be radial, just like the field of an isolated point charge (Figure 19).

To a test charge outside the sphere, the field of the sphere appears the same as that of a point charge. Mathematical analysis confirms that outside a sphere the field indeed behaves as though it came from a point charge placed at the centre of the sphere with a charge equal to the total charge spread over the sphere.

(b) Inside the sphere

Inside the sphere is a different matter. There is no electric field inside a conducting sphere, hollow or solid, a result that was experimentally determined by Benjamin Franklin. It can be shown using Coulomb's law and the ideas of surface charge density.

Because the sphere is a conductor, all the surplus charge must reside on the outside of the sphere. This follows because the charges will move until:

• they are as far apart as possible

• they are all in equilibrium (which, as earlier, means that they must be equidistant on the surface).

To find the field strength inside the conductor, we need to compute the force that acts on a positive test charge q placed at a random position inside the conductor (Figure 20).

What is the force acting on this test charge? Focus on two cones that meet at the test charge. The test charge is close to the conductor surface on one side, so that one of these cones is small and the other is large. The surface charge density on the cone is σ over the whole sphere.

Analysis shows that the forces acting on the test charge due to the two surfaces are equal $\dfrac{kq\sigma\pi x_s^2}{r_s^2} = \dfrac{kq\sigma\pi x_l^2}{r_l^2}$ and opposite in direction. The two forces cancel out, leaving no net force due to these two areas of charge.

This proof applies for *any* pair of cones and *any* test point inside the sphere. There is no net force on the test charge anywhere inside the sphere. Consequently, there can be no electric field either.

(ATL) Self-management skills

Benjamin Franklin was a scientist who conducted research into electricity and electrical fields. He was the first to use some of the language and vocabulary still used to describe electricity. Examples such as "charging and discharging" and the distinction between conductors and insulators came from Franklin. He famously flew a kite during a storm to demonstrate that lightning was an electrical effect and he invented the lightning rod.

Franklin was also a writer, printer and publisher. As a politician he was one of the founding fathers of the Unites States and one of the drafters of the Declaration of Independence.

In his autobiography, Franklin devoted a significant portion to self-management. He identified thirteen virtues (temperance, silence, order, resolution, frugality, industry, sincerity, justice, moderation, cleanliness, tranquillity, chastity and humility) and reflected at the end of each day upon whether he had fallen short in any one of these virtues. He concentrated on one virtue each week to practise and strengthen that virtue. He also constructed an "order" so that "every part of my business should have its allotted time".

Franklin acknowledged that other people might have other routines and that, when his business involved other people, that compromise would be needed.

What will your timetable look like when revising for your exams so that each subject "should have its allotted time"?

Millikan's experiment to determine the charge on an electron

In 1909, the American physicists Robert A Millikan and Harvey Fletcher carried out an elegant experiment to estimate the charge on an electron. Nowadays the experiment is attributed to Millikan alone and is known as **Millikan's method for the determination of *e***. Millikan won the Nobel prize in 1923 for the combination of his work on this and the photoelectric effect (Topic E.2).

The method has two steps:

- **Step 1.** A cloud of small oil drops is sprayed into a chamber (Figure 22(a)). The drops are ionized using X-rays or a source of beta particles, or they charge by friction as they fall through the aperture into the chamber. One drop is selected and allowed to fall through air in the absence of an electric field. The drag force together with the buoyancy force and the gravitational force lead to zero resultant force on the drop during this fall (Figure 22(b)). Measurement of the terminal speed of the drop allows the weight of the drop to be calculated.

- **Step 2.** The same drop is charged while in the apparatus and is held stationary using an upward electric force that is equal and opposite to the weight (Figure 22(c)). This allows the electric force on the drop to be determined (see Worked example 11).

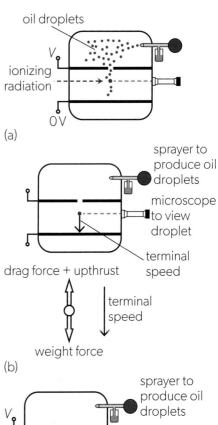

(a)

(b)

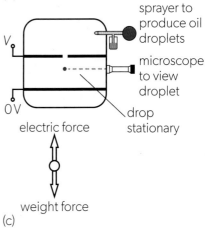

(c)

▲ Figure 22 The stages of Millikan's experiment. A drop is selected and charged, its terminal speed is measured and the charge it carries is estimated.

Charge is quantized. Which other physical quantities are quantized? (NOS)

When a quantity is quantized, it can only take a set of unique values. Theme E shows you that a number of physical quantities besides charge have this formal property. They include the discrete energy levels for an isolated hydrogen atom (and other simple atoms too). This is related to the quantized angular momentum of the system. The spectrum of hydrogen consists of distinct spectral lines for this reason.

Other quantities that are quantized that are not met in this course include the weak hypercharge, colour charge, baryon number, lepton number and spin.

The two sets of measurements allow the excess charge on the drop to be estimated. Millikan and his co-worker found that the excess charge was always an integer multiple of a basic charge. He ascribed this basic charge to the charge of one electron.

When many drops have been measured each with many different charges, it is possible to estimate the lowest common denominator of all the charges. To see how this works, imagine that you have several opaque bags containing identical objects. You know that the mass of each bag is negligible. Can you estimate the mass of one object?

Suppose the masses of four bags are: 420 g, 840 g, 560 g and 1260 g. The highest common factor of these measurements is 140 g, indicating that this is a good estimate for the mass of one object (there will be 3, 6, 4 and 9 objects, respectively, in the bags). However, if the masses had been 420 g, 840 g, 630 g and 1260 g, then the highest common factor is now 70 g and you will need to revise the object mass down.

This was the basis of Millikan's method. After many trials he estimated the elementary charge to be $(1.592 \pm 0.003) \times 10^{-19}$ C. Later experiments have failed to find a lower value for the charge, even though the value for e has increased slightly over the 100 years since the original experiment. Nowadays the electronic charge is a matter of definition and the value assigned to e is $1.602\,176\,634 \times 10^{-19}$ C exactly.

Experiments

Millikan's work shows the importance of error estimates in scientific experiments. He claimed an error of about 0.5% at the time. However, later examinations of the original notebooks showed that of the 175 drops that he measured, only 58 were used for the final result for e.

There are comments in the notebooks about the 100 or so drops that were ignored. These range from *"This is almost exactly right…"* through *"Error high not used"* to *"too high by 1.5%"*.

Experiments that collect data are sampling random points on a normal distribution (which gives a bell-shaped curve). Millikan may have been unconsciously rejecting outliers well away from the curve centre (the true result) and distorting the shape of the distribution, moving the true result away from where it should have been.

Millikan was highly respected and his result carried weight. Richard Feynman noted that the value of e in later experiments only gradually crept up to its modern accepted value as scientists became more and more confident about the correct result and were less influenced by the fame of Millikan.

Worked example 11

In Millikan's experiment, a negatively charged oil drop is introduced into the space between two parallel horizontal plates (see Figure 22).

a. When the electric field between the plates is switched off, the drop falls vertically reaching a constant speed of $0.23\,\text{mm}\,\text{s}^{-1}$.

   i.   Draw a free-body diagram showing the forces acting on the drop.

   ii. Determine the radius of the drop.

   Density of the oil $= 850\,\text{kg}\,\text{m}^{-3}$

   Density of air $= 1.2\,\text{kg}\,\text{m}^{-3}$

   Viscosity of air $= 1.8 \times 10^{-5}\,\text{Pa}\,\text{s}$

b. When the electric potential difference between the plates is adjusted to 1750 V, the drop is brought to rest. The distance between the plates is 1.2 cm.

   Determine the charge on the drop. State the answer in terms of the elementary charge, e.

Solutions

a. i. The weight, W of the drop is balanced by two forces acting upwards: the buoyancy force, B and the viscous drag force, F_d.

   ii. The net force on the drop is zero; hence $F_d = W - B$. Substituting expressions for the individual forces in terms of r and v, we obtain

$$6\pi\eta rv = \frac{4}{3}\pi r^3 (\rho_{\text{oil}} - \rho_{\text{air}})g \Rightarrow r = \sqrt{\frac{9\eta v}{2(\rho_{\text{oil}} - \rho_{\text{air}})g}}.$$

$$r = \sqrt{\frac{9 \times 1.8 \times 10^{-5} \times 0.23 \times 10^{-3}}{2(850 - 1.2)(9.8)}} = 1.5 \times 10^{-6}\,\text{m}.$$

b. The net force on the drop is again zero, but the drag force is replaced by the electric force $F_e = qE = \dfrac{qV}{d}$.

   The equilibrium of the forces gives

$$\frac{qV}{d} = \frac{4}{3}\pi r^3 (\rho_{\text{oil}} - \rho_{\text{air}})g \Rightarrow q = \frac{4\pi r^3 (\rho_{\text{oil}} - \rho_{\text{air}})gd}{3V}.$$

$$q = \frac{4\pi (1.5 \times 10^{-6})^3 (850 - 1.2)(9.8)(0.012)}{3 \times 1750}$$

$$= 8.0 \times 10^{-19}\,\text{C}.$$

   This is equal to $5e$, so there is an excess of five electrons on the oil drop.

Practice questions

11. An oil drop of mass $3.3 \times 10^{-14}\,\text{kg}$ has an excess of ten electrons. The drop is suspended at rest in a uniform electric field in vacuum. Calculate the magnitude of the electric field strength and state its direction.

12. In Millikan's experiment, a negatively charged oil drop of radius 1.8 μm and mass $2.1 \times 10^{-14}\,\text{kg}$ is at rest in the space between two parallel horizontal plates. The distance between the plates is 9.0 mm and the potential difference is 1.9 kV.

   a. Calculate the charge on the oil drop, giving the answer in terms of e. Ignore the buoyancy force.

   b. The charge on the drop is changed and the drop starts to move upwards, reaching a constant speed of $0.11\,\text{mm}\,\text{s}^{-1}$. The electric field strength remains unchanged. Determine the new number of excess electrons on the drop. The viscosity of air is $1.8 \times 10^{-5}\,\text{Pa}\,\text{s}$.

x

AHL

Measurement— Absolute electric potential

Sometimes you may see the term **absolute electric potential** used for this definition when ascribing a zero potential to infinity.

It is possible to choose another location for zero potential. One example of this is Earth. We are usually happy to assume that Earth has a zero potential and we call this value the "earth" or "ground". You can see an example of this near the start of this topic in charging by induction. In fact, the magnitude of the electric field near Earth's surface is about $150\,\mathrm{N\,C^{-1}}$ and it has a downwards direction. This means that Earth has a non-zero potential relative to infinity.

An analogous situation occurs in gravitation (see page 488) where we describe the absolute zero of gravitational potential as being zero at infinity, but for practical purposes we could set our zero on Earth to sea level and count mountains as having positive potential and seabeds as having negative potentials.

Electric potential

In Topic B.5 the quantity **electric potential difference** was introduced. There it was called either "potential difference" or, more colloquially, "voltage" and "pd". Potential differences are associated with the charge movement that we call electric current. Electric potential difference V is the measure of the energy transfer W when a charge q moves through the potential difference:

$$V = \frac{W}{q}$$

This led to the unit of one volt (1 V) as one joule per coulomb ($1\,\mathrm{J\,C^{-1}}$). Electric potential difference is the work done per unit charge.

If you have already studied the gravitational potential ideas in Topic D.1, you will recognize the similarity between this definition and the expression for gravitational potential difference $\Delta V_g = \dfrac{W}{m}$.

Once we have defined a zero of electric potential, then we can extend this idea of electric potential difference to that of an absolute electric potential relative to the zero. As in gravitation, we define the zero of electric potential to be at infinity so that there is no influence of one charge on another when they are separated by an infinite distance.

Our formal definition of **electric potential difference** between two points then becomes $\Delta V_e = \dfrac{W}{q}$ with ΔV_e as the electric potential difference and W as the work done in moving a test charge $+q$ between the two points.

This definition resolves an additional issue: the sign of the test charge. This is not a problem in gravitation because mass is only positive. In electrostatics, the presence or absence of excess electrons leads to positive or negative overall charges. For all definitions in electric field theory, it is conventional to choose a positive test charge.

This leads directly to the following definition of electric potential.

Electric potential at a point is the work done in bringing a unit positive test charge from infinity to the point.

Worked example 12

A particle of charge $+20\,\mathrm{nC}$ is initially at rest. The particle is moved through a potential difference of $1.5\,\mathrm{kV}$ in an electric field. Calculate the work done in moving the particle.

Solution

$W = q\Delta V_e = 20 \times 10^{-9} \times 1.5 \times 10^3 = 3.0 \times 10^{-5}\,\mathrm{J}$.

Worked example 13

The electric potential at the surface of a conducting body is $-300\,\mathrm{V}$. A particle of charge $-15\,\mathrm{nC}$, initially at rest at the surface of the body, accelerates away from it due to electrostatic repulsion. Calculate the kinetic energy of the particle when it has moved to a region of zero potential, a long distance from the conducting body.

Solution

The work done on the particle by the electric force is $W = -q\Delta V_e = -(-15 \times 10^{-9})(0 - (-300)) = 4.5 \times 10^{-6}\,\mathrm{J}$.
This work is equal to the increase in the kinetic energy of the particle. Since the initial kinetic energy of the particle was zero, we have $E_k = W = 4.5 \times 10^{-6}\,\mathrm{J}$.

Equipotential

We know that the field for an isolated point positive charge has the radial pattern shown in Figure 23. Also on the figure are the paths of two identical test charges, both coming from infinity. Although their paths end on different field lines, after their journey from infinity they are the same distance from the positive charge and will therefore be subject to the same final field strength ($k\dfrac{q}{r^2}$, where r is the distance from X or Y to the $+q$ charge). The same work must have been done to take X and Y to their final positions near the point charge. The electric potential at both positions is the same.

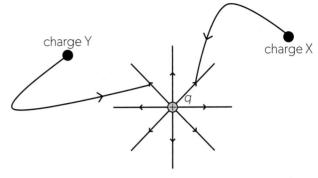

▲ **Figure 23** The electric potential due to a point charge (or a charged sphere) depends on distance from the charge (or the centre of the sphere). The field is conservative (see Topic A.3, page 117).

A spherical surface connects all possible points that have the same electric potential for a point charge. This surface is called the **equipotential surface**. Figure 24(a) shows what a series of equipotentials look like in three dimensions for a positive point charge. Figure 24(b) shows a two-dimensional view that includes the field lines as well. Notice that, as for gravitational field lines and equipotentials, field lines and equipotentials meet at 90°. This is always true.

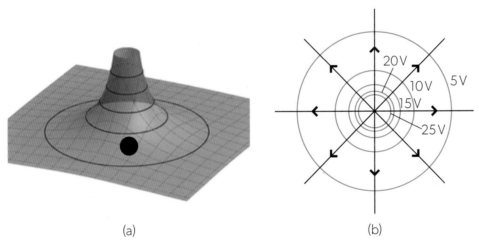

(a) (b)

▲ **Figure 24** (a) The shape of the equipotentials for a point charge in three dimensions.
(b) There is a 5 V difference between each equipotential. They are not separated by equal distances because $V \propto \dfrac{1}{r}$.

No work is done when a charge moves on the surface of an equipotential.

The field pattern due to two parallel charged plates is uniform with electric field lines equally spaced and at 90° to the plates (except close to the edges where the edge effect occurs). Figure 25 shows the arrangement of electric field lines and equipotentials for two charged plates with a potential difference between them of 12 V. Edge effects are not shown in this example. The (blue) electric field lines run from plate to plate; the black equipotentials are equally spaced (and shown here with 2 V intervals).

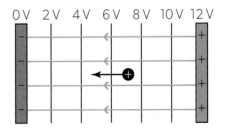

▲ **Figure 25** Work is done by the electric field on the charge when a positive charge moves from a high potential to a low potential. The energy will transfer to the kinetic form because the charge will accelerate when free to do so.

As the positive charge moves to the left in the field from the +12 V plate, it moves from a region of high potential to low potential. Work is done by the electric force (with no other forces acting, the charge will accelerate to the left).

For a negative charge to move to the left in Figure 25 from high to low potential, it would need to have work done by an external force (against the electric force) because the negative plate repels it.

🧪 Measuring equipotentials in two dimensions

- Inquiry 1: Demonstrate independent thinking, initiative, or insight.

- Inquiry 2: Identify and record relevant qualitative observations.

- Inquiry 2: Interpret qualitative and quantitative data.

- Inquiry 3: Compare the outcomes of an investigation to the accepted scientific context.

This experiment shows you how potential varies between two charged parallel plates.

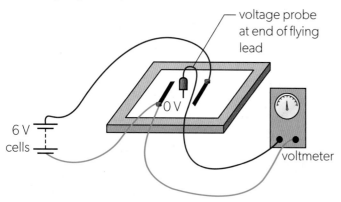

▲ Figure 26 Measuring equipotentials in two dimensions.

You need a sheet of material that has a uniform graphite coating on one side. One type is called "Teledeltos" paper. In addition, you will need:

- a 6 V power supply (domestic batteries are suitable),

- two strips of copper foil

- two bulldog clips

- a high-resistance voltmeter (a digital meter or an oscilloscope will be suitable)

- connecting leads.

- Connect the circuit shown in Figure 26. Take particular care with the connections between the copper strips and the paper. One way to improve these is to paint the connection between copper and paper using a conducting liquid consisting of a colloidal suspension of graphite in water.

- Press the voltage probe onto the paper. There should be a potential difference between the lead and the 0 V strip.

- Choose a suitable value for the voltage (say 3 V) and explore the region between the copper strips. Mark several points where the voltage is 3 V and draw on the paper to join these points up.

- Repeat for other values of voltage. Is there a consistent pattern for the line that represents a particular voltage?

- Try other configurations of plates. One important arrangement is a point charge, which can be simulated with a single point at 6 V and a circle of copper foil outside it at 0 V. One way to create the point is to use a sewing needle or drawing pin (thumb tack). You will need the colloidal graphite to make a good connection between the point and the paper.

- An alternative method, this time in three dimensions, is to use the same circuit with a tank of copper (II) sulfate solution and two copper plates.

The conducting-paper experiment in Figure 26 is two-dimensional. The foil detector on page 517 allowed you to explore a three-dimensional electric field. Equipotentials between the charged plates in three dimensions are sheets parallel to the plates themselves, always at 90° to the field lines. A consequence of extending equipotential ideas to three dimensions is that it is possible to have equipotential surfaces and even volumes.

An example of an equipotential volume is a solid conductor. When there is a potential difference between any part of the conductor and another, then a charge will flow until the potential difference has become zero. All parts of the conductor must therefore be at the same potential as each other. The whole interior of the conductor is an equipotential volume. Consequently, the field lines emerge from the volume at 90° — whatever the shape of the conductor. This links to the earlier work on the electric field lines close to a charged surface.

In summary, equipotential surfaces or volumes:

- link points having the same potential
- are regions where charges can move without work being done on or by the charge
- are cut by electric field lines at 90°
- do not have direction (potential, like any energy, is a scalar quantity)
- can never cross or meet another equipotential with a different value.

Electric potential and electric field strength

Electric potential can be represented graphically too. Figure 27 shows the electric potential plotted against distance r from the 0 V plate in Figure 25.

The graph is a straight line because the electric field is uniform. The gradient of this graph is $\dfrac{\text{change in } V}{\text{change in } r}$, which is $\dfrac{\text{potential difference between plates}}{\text{distance between plates}}$. We saw earlier that this ratio is equal to the magnitude of the electric field strength $\left(E = \dfrac{F}{q} = \dfrac{V}{d}\right)$. Therefore, the magnitude of the electric field strength $= \dfrac{\text{change in } V}{\text{change in } r}$ or, algebraically, $E = \dfrac{\Delta V_e}{\Delta r}$. (As usual, "$\Delta$" stands for "change in".)

This equation is almost complete, but it requires a negative sign. Look at Figure 25 again. The direction of increase in electric potential is to the right. The direction of the electric field is to the left. (This is the direction in which the positive charge will travel when released.)

Travel in the opposite direction to that of the field means, for a positive charge, moving to a position of higher potential and thus a positive potential gain. In Figure 25, when the positive charge moves fully from right to left, ΔV_e is negative (*final state − initial state* is $0\,V − (+12)\,V = −12\,V$). According to the equation, E will be positive and this tells us that the motion is in the direction of the field. In other words, moving against the field means going to higher potential so gaining potential energy (a positive change in ΔV_e): $E = -\dfrac{\Delta V_e}{\Delta r}$

Does the argument change when the moving charge is an electron? No, it does not: when an electron moves from 0 V towards the 12 V plate (a position of higher potential), its potential energy is reduced because the potential energy change is given by $e\Delta V$ as usual, but here e is negative and ΔV is positive. The electron accelerates towards the +12 V plate when it is free to do so, gaining kinetic energy from the field at the expense of its electrical potential energy.

For Figure 25, when the potential difference is 12 V and the distance between the plates is 6.0 cm, then the electric field strength $E = -\dfrac{\Delta V_e}{\Delta r} = -\dfrac{12}{0.06} = -200\,\text{V m}^{-1}$. This can also be written as $-200\,\text{N C}^{-1}$.

Figure 28(a) shows a graph of electric force against distance for a constant force. The work done when this force moves an object through a distance Δr is equal to the area under the graph. In symbols $F \times \Delta r = W$. When the electric force varies with distance (Figure 28(b)), then the work done is still the area under the graph.

Electric field strength is the electric force *per unit charge*, and so the area under a graph of electric field strength against distance is equal to the work done *per unit charge*—in other words, the change in electric potential. In algebraic terms, $E = -\dfrac{\Delta V_e}{\Delta r}$ can be rewritten as $E \times \Delta r = -\Delta V_e$.

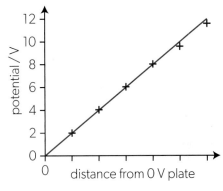

▲ Figure 27 The variation of potential with distance from the 0 V plate.

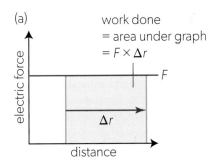

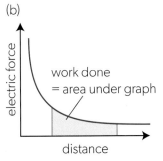

▲ Figure 28 The work done on a charge that moves in an electric field is equal to the area under the graph of force against distance.

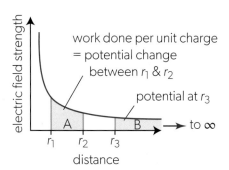

▲ Figure 29 The work done per unit charge on a charge that moves in an electric field is equal to the area under the graph of electric field strength against distance.

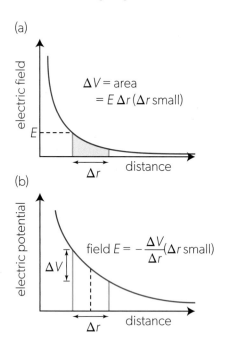

▲ Figure 30 The graphs of electric field strength–distance and electric potential–distance are linked by area and gradient.

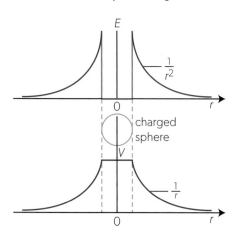

▲ Figure 31 The electric field strength E and electric potential V around and inside a charged conducting sphere at distances r from the sphere centre.

Figure 29 shows the variation of electric field strength against distance for a single point positive charge. As usual, this obeys the inverse-square law. Two areas are shown on this graph:

- Area A shows the *electric potential difference* between r_1 and r_2.
- Area B shows the *electric potential* at r_3. In this case, the area included goes all the way to infinity.

The equation $E = -\dfrac{\Delta V_e}{\Delta r}$ can be written in calculus form as

$$E = -\frac{dV_e}{dr}$$

This calculus version can be applied to the equation for the electric field strength of a single point charge $E = k\dfrac{q}{r^2}$ and it leads to the following definition.

The electric potential at a distance r from a point charge $+q$, is $V_e = +k\dfrac{q}{r}$.

This predicts that the closer we are to a (positive) charge, the greater (more positive) is the potential. Conversely, the closer we are to a negative charge, the more negative is the potential. These predictions agree with the conclusions we reached earlier.

Graphs of field strength and potential against distance for the field due to a single positive point charge are shown in Figure 30.

Potential inside a hollow conducting charged sphere

The point-charge and parallel-plate arrangements have been considered in detail. Another important charge configuration is that of the hollow conducting charged sphere.

Outside a charged conducting sphere, the electric field is indistinguishable from that of a point charge placed at the sphere's centre. The field lines are radial, and an observer outside the sphere cannot distinguish, using the field lines, between a charged sphere and a point charge with the same charge magnitude placed at the sphere's centre. Likewise, outside the sphere the equipotentials are spherical and concentric. The sphere surface itself is an equipotential. (Otherwise free charges would continue to move across it.)

We saw earlier that the electric field inside the sphere is zero. The potential inside the sphere has a constant value. Because the electric field is zero, $\dfrac{\Delta V_e}{\Delta r}$ must be zero too. When V is constant, then no work is required to move a charge at constant speed inside the charged sphere. The potential throughout the interior is equal to the potential at the surface.

We can plot both the electric field strength and the electric potential for a charged conducting sphere, and these graphs are shown in Figure 31.

Charging a sphere by induction — An alternative view

Earlier in this topic, we described how to give a sphere a positive charge using a negatively charged rod (Figure 5, page 506). The charging process can also be discussed in terms of changes in the electric potential V_s of the sphere relative to the earth (0 V) when both rod and sphere are isolated from any other objects.

- Initially V_s is zero when the rod is well away from the sphere (at infinity) because there is no interaction between them and the sphere has no net charge.
- The rod is now brought up to the isolated conducting sphere. The charges on the sphere separate with the mobile free electrons being repelled by the rod

to the opposite side of the sphere. V_s becomes negative. The presence of the negative rod gives the sphere the possibility ("potential") of losing electrons through a connection to a third object.

- The sphere is then earthed (grounded) and some free electrons are repelled to earth. V_s becomes zero as it is connected to the earth (the local zero of potential).
- The connection to earth is broken and the rod is removed to infinity again. The sphere has an overall positive charge after losing some electrons. The negative rod and positive sphere attract each other. An increase in V_s occurs because work is done on the rod–sphere system in separating the negative rod and the positive sphere. V_s is now positive as is expected for an isolated positively charged sphere.

Electric potential energy

Potential is the work done in moving a positive unit charge (from infinity) to a particular point, so the work required to move a charge of size q from infinity to the point will be $V_e \times q$. This is the **electric potential energy** that charge q possesses due to its position in the field that is giving rise to the potential.

When a charge q_1 is in a field that arises from another point charge q_2, then its electric potential energy $E_p = q_1 V_e$, or

$$E_p = k\frac{q_1 q_2}{r}$$

When there are more than two charges, the electric potential energy E_p of the system of charges is the work done to assemble the system from infinite separation.

Worked example 14

The diagram shows electric equipotential lines for an electric field. The values of the equipotential are shown. Explain where the electric field strength has its greatest magnitude.

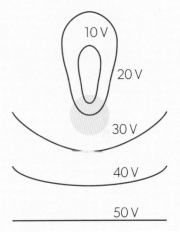

Solution

The work done in moving between equipotentials is the same between each equipotential. The work done is equal to *force* × *distance*. So the force on a test charge is greatest where the distance between lines is least. This is in the region around the base of the 20 V equipotential. Provided that potential change is the same between neighbouring equipotential lines, then the closer the equipotential lines are to each other, the stronger the electric field strength will be.

Worked example 15

A proton is moved through a distance of 15 cm in a uniform electric field. The electric potential energy of the proton decreases by 350 eV.

a. State the potential difference through which the proton is moved.

b. Calculate the component of the electric field strength along the direction of motion of the proton.

Solutions

a. The potential energy of the proton decreases and hence a positive work of 350 eV is done by the electric force and the proton is moved to a position of lower potential. $\Delta V_e = -\dfrac{350\ \text{eV}}{+e} = -350\ \text{V}$.

b. $E = -\dfrac{\Delta V_e}{\Delta r} = -\dfrac{-350}{0.15} = 2.3\ \text{kV m}^{-1}$. The positive sign of E is consistent with the fact that the electric force does positive work on the proton.

AHL

Worked example 16

The electric potential on the surface of a conducting sphere of radius 5.0 cm is 6.0 kV, with the potential defined to be zero at infinity. Determine the charge on the sphere.

Solution

The electric potential outside of the sphere is the same as if the charge of the sphere has been placed at its centre, so it

follows the equation $V_e = \dfrac{kq}{r}$. When $r = 0.050$ m, we have $V_e = 6000$V. Hence $q = \dfrac{V_e r}{k} = \dfrac{6000 \times 0.050}{8.99 \times 10^9} = 3.3 \times 10^{-8}$ C.

Practice questions

13. a. Calculate the electric potential at a distance of 1.5 m from a point charge of −0.48 nC.

   b. Another point charge is placed 1.5 m from the first charge. Calculate the potential energy of the system of the two charges, when the second charge is:

      i. −1.3 nC

      ii. +0.90 nC.

   c. For each of the charges in part b., calculate the work needed to change the separation of the system to 0.50 m.

14. A charged particle of −1.4 μC is placed at a distance of 0.85 m from the centre of the sphere in Worked example 16. Calculate:

   a. the electric potential energy of the system of the sphere and the particle

   b. the work that has to be done in order to move the charge to an infinite separation from the sphere.

15. When a point charge of +56 μC is moved through a distance of 0.50 m along an electric field line, the electric potential energy of the charge increases by 6.7 mJ. Calculate:

   a. the electric potential difference between the initial and the final position of the charge

   b. the average electric field strength.

Patterns and trends — Potential and potential energy for all inverse-square fields.

The connection between field strength and potential gradient is universal and applies to all fields based on an inverse-square law. It tells us about the fundamental relationship between force and the energy that can be transferred. It gives this in a way that is independent of the mass or charge of the test object moving in the field. Field strength is a quantity that represents force but with the mass or charge term of the test object stripped out.

Mass / charge dependent quantities		Mass / charge independent quantities
force	⇔	field strength
potential energy		potential

What are the relative strengths of the four fundamental forces?

Physicists have so far identified four fundamental forces. These are:

- gravitational force

- electromagnetic force

- weak nuclear force (arising from the weak nuclear interaction)

- strong nuclear force (arising from the strong nuclear interaction).

Theme D deals with the ideas of the gravitational and electrostatic forces. The strong nuclear force is described in Topic E.3 as a short-range interaction that acts between nucleons. At very close distances it is repulsive and a greater distances it is attractive. This force is not observed to act outside the nucleus. (The weak interaction is not discussed further in this course. It is responsible for beta decay.)

The table indicates the approximate relative magnitudes and the ranges of these four fundamental forces. These comparative strengths are approximate as there is no standard way to compare one force with another. Compare the relative strengths of the gravitational force and the electrostatic between two protons. How does this ratio change when the two particles are electrons?

Force	Relative strength	Range / m
gravity	10^{-38}	∞
weak nuclear	10^{-13}	10^{-18}
electromagnetic	10^{-2}	∞
strong nuclear	1	10^{-15}

Magnetic fields

The repulsion between the like poles of two bar magnets is a familiar phenomenon. Impressive forces act between magnets of even quite modest strength. Modern materials are used in tiny neodymium magnets (less than 1 cm in diameter and a few millimetres thick) that can easily attract another ferromagnetic material through significant thicknesses of a non-magnetic substance.

Magnetism is another field phenomenon. A **magnetic field** exists at a point when a magnetic force acts on a magnetic pole (in practice, a pair of poles) placed at that point. As with electrostatics and gravitation, magnetic fields are visualized through the construction of field lines.

Magnetic field lines have similar (but not completely identical) properties to electric field lines. The magnetic properties of field line are as follows.

- Magnetic field lines are conventionally drawn from the north-seeking pole to the south-seeking pole or can be closed lines that do not end at a pole. The line directions represent the direction in which a north-seeking pole at that point would move.

- The strength of the field is shown by the density of the field lines. Lines drawn closer together imply a stronger field.

- The field lines never cross.

- The field lines act as though made from an elastic thread. The system acts as if to make the field lines shorter.

These assumed properties suggest ways in which the system can change to reduce the total length of the lines. This can be imagined as a reduction in the total energy stored in the system if there is a fixed "energy per unit length of line".

How can moving charges in magnetic fields help probe the fundamental nature of matter?

Magnetic fields arise when electric charges move. It is not surprising that charges moving through a magnetic field experience a force. They are, after all, interacting with the field due to other moving charges.

This magnetic force, which is examined in much more detail in Topic D.3 allows moving charges to be accelerated to very high speeds in nuclear machines. This means that the charges, which are associated with electrons, protons and other charged nuclear material can collide with each other and nuclei to give us an insight into the structure of matter at the very smallest scale.

The links between Themes A, D and E are clear and important both for science and everyday life.

Observations—Talking about magnetic poles

It can be confusing when you are talking about magnetic poles due to differences in notation. This partly arises from the original observations of magnetism when early navigators noted that a "lodestone" suspended from a thread would align itself north–south. When we write "magnetic north pole" what we really mean is "the magnetic pole that seeks the geographic north pole" (Figure 32(a)). We often talk loosely about a magnetic north pole pointing to the north pole. A misunderstanding can occur here because we also know that like poles repel and unlike poles attract. We could end up with the situation that a magnetic north pole is attracted to the "geographic north pole"—which seems wrong in the context of two poles repelling. In this book we talk about a north-seeking pole meaning "a geographic north-seeking pole" and south-seeking poles meaning "geographic south-seeking". On the diagrams of magnets, N always means geographic-north seeking; S always means geographic-south seeking. Figure 31(a) shows the magnet that must effectively be at the planet's centre to give rise to the observed magnetic field pattern of Earth.

Figure 32(b) shows the patterns for a single bar magnet and Figures 32(c) and 32(d) give two arrangements of a pair of bar magnets of equal strength.

In the magnet pairs, notice the characteristic field pattern when the two opposite poles are close (Figure 32(c)) and when the two north-seeking poles are close (Figure 32(d)). When two north-seeking poles are close (alternatively, two south-seeking poles), there is a position where the field is zero between the magnets (called a null point). When two opposite poles are close, the field lines appear to connect the two magnets, forming a pattern like that of the single bar magnet. In this situation, the magnets will be attracted, so this field pattern implies an attraction between poles.

In terms of the rules for the field lines, poles moving closer together in Figure 32(c) will shorten the field lines between the poles. This implies that magnets will do this if they are free to do so as it is energetically favourable.

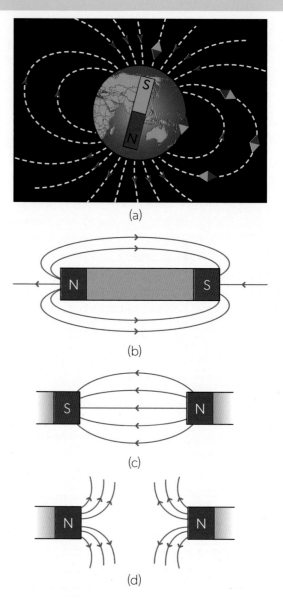

▲ Figure 32 (a) The magnetic field pattern of Earth. (b–d) The magnetic field patterns for a bar magnet and for like and unlike poles.

Data-based questions

While gravitational and electric fields obey an inverse-square law, magnetic fields are not so simple. Along the axis of a bar magnet, theory suggests that the magnetic field will vary with $\frac{1}{r^3}$, where r is the distance from the midpoint of the poles. The data for this question were taken using the sensor on a smartphone and a Magnadur magnet.

You could try this experiment yourself, but don't put strong magnets too close to a smartphone. Your school may have a Hall probe or other means of measuring a magnetic field.

x / cm (± 0.1 cm)	B / µT		
30	43.14	46.92	46.38
25	65.40	66.48	64.84
20	105.64	103.66	104.22
17	155.58	157.46	153.92
14	244.88	241.46	243.66
12	358.80	353.92	355.20
10	536.20	541.98	546.48
9	691.32	677.50	683.38

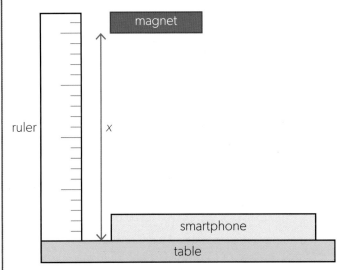

▲ **Figure 33** The experimental arrangement used to investigate the variation of magnetic field strength due to a bar magnet.

- The smartphone was placed flat on a table and a metre ruler was placed vertically next to the smartphone. With the magnet at a large distance (greater than one metre), the vertical background component of the magnetic field was measured. Three readings of 16.92 µT, 17.34 µT and 16.32 µT were recorded.

- The magnet was then held against the ruler vertically above the smartphone, as shown in Figure 33. Readings of distance x between the table and the bottom of the smartphone against the strength of the magnetic field were recorded.

- The exact position of the sensor in the smartphone is unknown. As a result, the distance was measured from the bottom of the magnet to the table. Suggest one other reason why x might be measured like this rather than measuring from the centre of the magnet to the smartphone sensor.

- Using the measurements of the background magnetic field, calculate the average field for each x and give an uncertainty with your answer.

- Tabulate values of the average magnetic field, corrected for the background reading.

- Calculate the uncertainties in the average magnetic field. You should include the uncertainty in the background magnetic field in your calculation.

- Calculate values of $B^{-\frac{1}{3}}$ and add them to your table. Calculate the uncertainty in these values.

- Plot a graph of $B^{-\frac{1}{3}}$ against x.

- The suggestion is that B varies as $\frac{C}{(x-x_0)^3}$, where x_0 is a distance correction (for the position of the sensor) and C is a constant. Use your graph to find the values of x_0 and C.

- By considering the maximum and minimum gradients of your graph, establish uncertainties in your values of C and x_0.

- Using similar apparatus, think about how you could extend or vary this experiment. You might even carry out this experiment.

Observing field patterns of permanent magnets and electric currents

- Inquiry 2: Identify and record relevant qualitative observations.

- Inquiry 3: Compare the outcomes of an investigation to the accepted scientific context.

There are several ways to carry out this experiment. They can involve the scattering of small iron filings, observation of suspensions of magnetized particles in a special liquid or other techniques. This experiment uses iron-filings.

- Take a bar magnet and place a piece of rigid white card on top of it. You may need to support the card along its sides. Choose a non-magnetic material for the support.

- Take some iron filings in a shaker (a pepper pot is ideal) and, from a height above the card of about 20 cm sprinkle filings onto the card. Tap the card gently as the filings fall onto it.

- You should see the field pattern forming as the magnetic filings fall through the air and come under the influence of the magnetic field. Sketch or photograph the arrangement.

- The magnetic field is strong close to the magnet, but becomes weaker further away from it. Make this clear when you draw the field pattern. The lines of force should be drawn at increased spacing as the distance from the magnet increases.

- The iron filings give no indication of field direction. The way to observe this is to use a plotting compass—a small magnetic compass a few

centimetres in diameter. The plotting compass indicates the direction to which a north-seeking pole sets itself. Place one or more of these compasses on the card and note the direction in which the north-seeking pole points.

- Repeat with two magnets in several configurations. Try at least the two in Figure 32.

Electric currents also give rise to a magnetic field. A vertical current-carrying wire can be passed through the hole in a horizontal piece of card to show this.

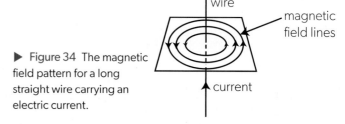

▶ Figure 34   The magnetic field pattern for a long straight wire carrying an electric current.

However, currents small enough to be safe will only give weak fields, not strong enough to affect the filings as they fall. To improve the effect:

- Run a long lead through the hole in the card (the lead will need to be a few metres long).

- Loop the lead in the same direction through the card a number of times (about ten turns if possible). This trick enables one current to contribute many times to the same field pattern.

- You may need at least 25 A in total (2.5 A in the lead) to see an effect.

Magnetic field patterns

(a) Due to a current in a long straight wire

The magnetic field pattern due to a current in a long straight wire is a circular pattern centred on the wire. This seems odd to anyone previously used to the bar-magnet pattern. Observations using plotting compasses show that the direction of the field depends on the direction of the current.

Using the conventional current (i.e. the direction that positive charges drift in the wire, see page 303) the relationship between the current and the magnetic field direction obeys a **right-hand corkscrew rule**.

To remember this, hold your right hand with the fingers curled into the palm and the thumb extended away from the fingers (see Figure 35). The thumb represents the direction of the conventional current and the fingers represent the direction of the field. Another way to think of the current–field relationship is in terms of a screwdriver being used to insert a screw. The screwdriver inserts a right-handed screw by turning clockwise to drive the screw forwards. The direction in which the imaginary screw moves is that of the conventional current, and the direction in which the screwdriver turns is that of the field. You can use any direction rule you prefer but use it consistently. Remember that the rules work for conventional current.

The farther the pattern is from the wire itself, the greater the separation of the field lines. This is telling us that the magnetic field strength decreases as the distance from the wire increases.

(b) Due to the current in a circular coil

The magnetic field due to a coil is shown in Figure 36. A coil is a circular winding of wire with the thickness of the coil much less than its diameter.

To understand how the magnetic field pattern arises, you need to imagine a long straight wire being formed into a one-turn coil shape. As it does so, it carries its circular field with it so that the circular field bends round. Figure 36 shows this.

The field lines run along the centre of the coil and then around the outside. The pattern is identical to the bar magnet pattern (outside the magnet), so that we can assign north-seeking and south-seeking poles to the coil (Figure 36(a)). The easy way to remember is to use N and S to show the north-seeking and south-seeking poles (Figure 36(b)). The arrows on the N and S show the current direction when looking into the coil from outside. When the conventional current is anticlockwise as you look into the coil, then that end is north-seeking. When the current is clockwise looking in, then that is the south-seeking end of the coil.

The strength of the magnetic field in a coil can be increased by:

- increasing the current in the wire

- increasing the number of turns.

(c) Due to the current in an air-core solenoid

A solenoid is a form of coil where the coil length is much greater than the diameter. The magnetic field pattern is similar to that of a coil and it arises in a similar way. Two adjacent turns of wire in the solenoid are shown in Figure 37.

With current in the wire, a circular field is set up in the wire. The familiar circular field adds together with identical fields from neighbouring turns in the solenoid. Figure 35 shows this. Look closely at what happens close to the individual wires. The black lines show the field near the wires. The blue lines show how the fields begin to combine.

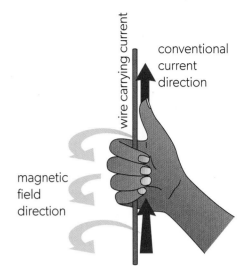

▲ Figure 35 How to apply the right-hand corkscrew rule.

You will meet the equation that describes this variation of magnetic field strength in Topic D.3.

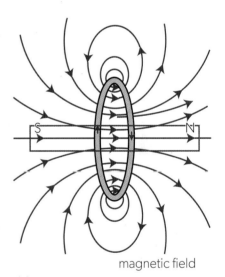

(a)

anticlockwise current gives a north seeking pole

clockwise current gives a south seeking pole

(b)

▲ Figure 36 Predicting the magnetic field direction from a coil or solenoid.

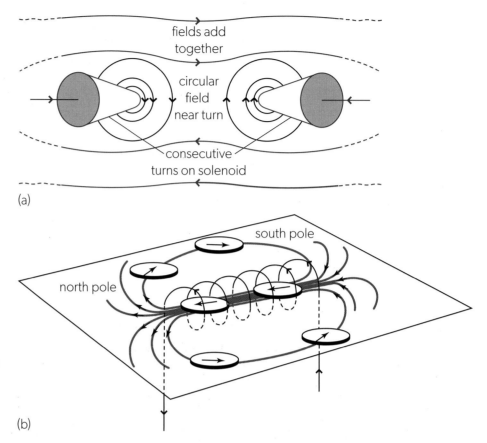

(a)

(b)

▶ Figure 37 (a) The magnetic field pattern of a coil or solenoid is formed from the addition of the magnetic patterns from many parallel straight wires wound together. (b) Outside a solenoid the magnetic field pattern resembles that of a bar magnet but without the poles.

Figure 37(b) gives an overall view of the magnetic field once it is built up.

Remember that the field is only shown in one plane and, in three dimensions, it is rotated through 360° about the axis of the coil. The appearance of the north-seeking and south-seeking poles of the solenoid obey the same rule as for the coil (Figure 37).

The strength of the magnetic field in a solenoid can be increased by:

- increasing the current in the wire
- increasing the number of turns per unit length of the solenoid
- adding an iron core inside the solenoid.

Worked example 17

Four long straight wires are placed perpendicular to the plane of the paper at the edges of a square.

The same current is in each wire in the direction shown in the diagram. Deduce the direction of the magnetic field at point P in the centre of the square.

Solution

The four field directions are shown in the diagram. The sum of these four vectors is another vector directed from point P to the left.

Practice questions

16. Two parallel wires X and Y carry equal currents out of the plane of paper. Point P is at the same distance from X and Y.

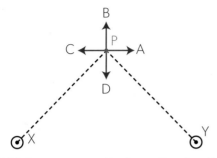

a. Which arrow correctly shows the direction of the resultant magnetic field at P?

b. Current in wire X is reversed and is now into the plane of paper. What is the direction of the magnetic field at P after the change?

17. A circular ring carries a constant current. Point P is at the centre of the ring. The magnetic field at P is directed into the page. What is the direction of the current in the ring and the direction of the magnetic field at point Q?

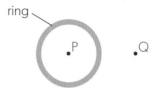

	Current in the ring	Magnetic field at Q
A.	clockwise	out of the page
B.	clockwise	into the page
C.	anticlockwise	out of the page
D.	anticlockwise	into the page

Global impact of science—Why permanent magnets?

The puzzle of permanent magnets or the magnetism of Earth was not explained earlier. Magnetic effects arise when charges move relative to each other. Can this explain the reasons for permanent magnetism (known as ferromagnetism)? Permanent magnetism is rare in the Periodic Table. Only iron, nickel, and cobalt and alloys of these metals demonstrate it.

Permanent magnetism is due to the arrangement of the atomic electrons in these metals. Electrons are known to have the property of *spin*—an internal orbiting motion of the electron. In iron, cobalt and nickel, there is a particular arrangement that involves an unpaired electron. This is the origin of the moving charge needed for a magnetic field to appear.

The second reason why iron, nickel and cobalt are strong permanent magnets is that neighbouring atoms can cooperate and align the spins of their unpaired electrons. Many electrons then spin in the same direction and give rise to a strong magnetic field.

It is thought that a liquid-like metallic core, deep in the centre of Earth, contains free electrons and rotates relative to the rest of the planet. These are conditions that can lead to a magnetic field. In which direction do you predict that the electrons are moving? However, this phenomenon is still the subject of research interest. Why, for example, does the magnetic field of Earth flip every few thousand years? There is much evidence for this, including the magnetic "striping" in the undersea rocks of

the mid-Atlantic ridge and in the anomalous magnetism found in some ancient cooking hearths of the aboriginal peoples of Australia.

▲ **Figure 38** The position of Earth's magnetic north pole has changed over time. The shape of the magnetic field is more complicated than that of a simple bar magnet. The geomagnetic pole is the location of the best-fit model of Earth's magnetic field assuming it acts as a bar magnet.

How do charged particles move in magnetic fields?

What can be deduced about the nature of a charged particle from observations of it moving in electric and magnetic fields?

The beautiful effects of the aurora, cyclotrons used in medical treatment and the colossal 27 km ring that steers and accelerates particles at CERN all rely on the movement of charged particles in magnetic and electrostatic fields.

The kinematics of these motions is straightforward and stems from your knowledge of Theme A. The electric charges and fields in Topic D.2 did not change with time—hence, the older name for that study: electrostatics. Although charges moved through potential differences, they were implicitly stationary at both points of potential. We never considered the effect of kinetic energy on the overall situation. Now we examine the forces that act on moving charges and the link between electrodynamics and magnetism.

This topic shows that there are essential differences when charged particles interact with a magnetic field and an electric field. When the fields are uniform, then these differences result in a fundamental and easily recognizable distinction between the motions of a moving charged object. Observations of the path of a particle give us information about its charge, its speed and its mass. These were some of the techniques used at the beginning of the 20th century to explore the properties of the newly discovered nuclear particles such as the electron.

▲ Figure 1 Solar particles spiral in Earth's magnetic field and ionize atoms in the upper atmosphere to produce the beautiful effects of the aurora borealis over Norway. Similar effects produce the aurora australis in the southern hemisphere.

In this topic, you will learn about:

- the magnitude and direction of the force on a charge moving in a magnetic field

- the magnitude and direction of the force on a current-carrying conductor in a magnetic field

- the force per unit length between two parallel wires

- the motion of a charged particle in a uniform electric field

- the motion of a charged particle in a uniform magnetic field

- the motion of a charged particle in a uniform electric field and a uniform magnetic field at right angles to each other.

Introduction

There is a very close connection between electric and magnetic fields. For example, in Theme C electromagnetic radiation is described in terms of co-oscillating electric and magnetic fields at 90° to each other. The electric field in the radiation cannot exist without its magnetic counterpart. In this topic, you will see that the electric fields associated with a static charge undergo a transformation when the charge accelerates and this leads to a magnetic effect on other nearby charges. A force acts on a moving charge in a magnetic field—remembering that the magnetic field is itself the result of other moving charges.

Forces on moving charges

Force between two current-carrying wires

Magnetism originates in the interactions between conductors when they carry electric current.

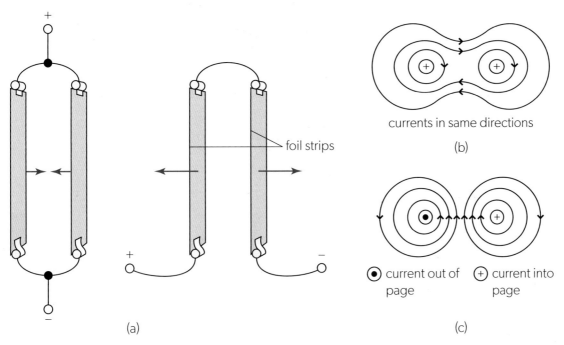

currents in same directions

(b)

⊙ current out of page   ⊕ current into page

(a)   (c)

▲ Figure 2 Two long aluminium foil strips are mounted parallel to each other. There is an electric current in each strip. The magnetic fields due to each current interact so that when the currents are in the same direction the strips attract each other. When the currents are opposed, the strips repel.

Figure 2(a) shows two conducting foil strips hanging vertically. The current directions in the strips can be the same or opposite directions. When the currents are in the same direction, the strips move together due to the force on one foil strip as it sits in the magnetic field of the other strip. When the currents are in opposing directions, the strips move apart. This effect was the basis for the definition of the ampere until 2019 (see page 550).

You can set this experiment up for yourself using two pieces of aluminium foil about 3 cm wide and about 70 cm long for each conductor. The power supply should be capable of providing up to 25 A. So take care! Connections are made to the foils using crocodile (alligator) clips.

The forces on the foil can be explained in terms of the interactions between the fields as shown in Figure 2(b). When the currents are in the same direction, the field lines from the foils combine to give a pattern in which field lines loop around both foils. The notation used to show the direction of conventional current in the foil is explained on the diagram. Look back at Figure 32 on page 534, which shows the field pattern for two bar magnets with the opposite poles close. You know that the bar magnets are attracted to each other in this situation. The field pattern for the foils is similar and leads to attraction too. Think of the field lines as trying to be as short as possible. They become shorter when the foils are able to move closer together.

When the currents are in opposite directions, the field pattern changes (Figure 2(c)). Now the field lines between the foils are close together and in the same direction, and thus represent a strong field.

ATL **Communication skills**

Understanding and interpreting field line diagrams can be difficult. Sometimes it helps to think of the field lines as being elastic in some way (this can work for gravitational and electric fields, not just magnetic fields). Often this leads to an interpretation of the right forces — attraction or repulsion.

However, field lines are not really elastic links between objects. The reason that it works is because field lines can show the direction of a force. When objects are allowed to move, then the work done is the force multiplied by the distance travelled in the direction of the field. Therefore, if the objects move so that the field lines are drawn as being shorter, it is likely that work has been done by the field and that the objects have moved to a state of lower energy.

Sometimes an easy way to communicate a complex idea is appealing, even if the reasoning isn't quite correct.

It seems reasonable that a strong magnetic field (like a strong electrostatic field) represents a large amount of stored energy. This energy can be reduced when the foils move apart, allowing the field lines to separate too. Again, this has similarities to the bar magnet case, but this time with like poles close together.

The magnetic forces can be determined quantitatively too. This is left for later in this topic.

Force between a bar magnet field and a current-carrying wire

One extremely important case is the interaction between a uniform magnetic field and the field due to a current in a wire.

We begin with the field between two bar magnets with unlike poles close together. In the centre of the region between the magnets, the field is uniform because the field lines are parallel and equally spaced.

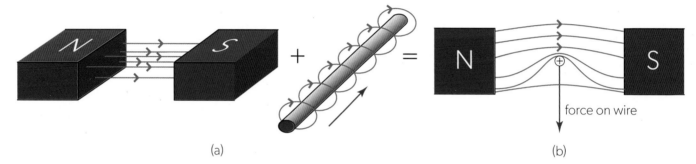

(a)   (b)

▲ Figure 3   Combining the uniform magnetic field between two magnets and the field around a long straight wire gives a distorted field pattern. The new magnetic field leads to a force acting on the wire.

Suppose a long wire carrying a current is in this field. Figure 3(a) shows the arrangement and the directions of the uniform magnetic field and the field due to the current.

The effect can again be explained in terms of the interaction between the two magnetic fields. The circular field due to the wire adds to the uniform field due to the magnets to produce a more complicated field. This is shown in Figure 3(b). Overall, the field is weaker below the wire than above it. Using our ideas of the field lines, the system (the wire and the uniform field) can overcome this difference in field strength by attempting to move either the wire downwards or the magnets that cause the uniform field upwards. This effect is sometimes called the **catapult field** because the field lines above the wire resemble the stretched elastic cord of a catapult just before it fires the object.

This effect is of great importance as it is the basis for the transfer of electrical energy into a kinetic form. It is used in electric motors, loudspeakers and other devices where we need to produce movement from an electrical power source. It is the **motor effect**.

It is possible to predict the direction of motion of the wire by drawing the field lines on each occasion when required, but there are several direction rules that are used to remember the direction of the force easily. One of the best known of these is due to the English physicist Fleming and is known as **Fleming's left-hand rule**.

To use the rule, extend your left hand as shown in Figure 4. Your first (index) finger points in the direction of the uniform magnetic field and your second finger points in the direction of the conventional current in the wire. Then your thumb gives the direction of the force acting on the wire.

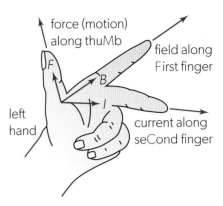

▲ Figure 4 Fleming's left-hand rule.

Explaining the motor effect

The explanation for the motor effect has been given so far in terms of field lines. This is, of course, not the complete story. We should be looking for explanations that involve interactions between individual charges, both those that produce the uniform magnetic field and those that arise from the current in the wire. Electrostatic effects (as their name implies) arise between charges that are not moving. Magnetic effects arise because the sets of charges that produce the fields are moving relative to each other and are in different frames of reference. There have been no electrostatic effects because we are dealing with conductors in which there is an exact balance of positive and negative charges. Magnetism can be thought of as the residual effect that arises when charges are moving with respect to each other.

 How are the properties of electric and magnetic fields represented? (NOS)

As Topics D.1 and D.2 showed, visualizations of electric, magnetic and gravitational fields are commonly used to allow us to imagine the field and its properties. Other representations are those mathematical constructs such as field strength and potential that permit a quantitative description of the fields beginning with an inverse-square dependence on distance. If you study physics beyond school or college, you will meet even more sophisticated treatments that involve vector field theory.

These are all ways to model the abstract notion of a "field". However, these visualizations have nothing to say about the underlying origins of the fields: how they arise and how one object can influence another "at a distance".

What is the value or otherwise of the field concept in terms of explaining the nature of matter and space?

Worked example 1

Wires P and R are equidistant from wire Q.

Each wire carries a current of the same magnitude and the currents are in the directions shown.

Describe the direction of the force acting on wire Q due to wires P and R.

wire P   wire Q   wire R

Solution

Using the right-hand corkscrew rule, the field due to wire P at wire Q is out of the plane of the paper, and the field due to wire R at wire Q is also out of the plane of the paper.

Using Fleming's left-hand rule, the force on wire Q is in the plane of the paper and directed to the left.

Practice questions

1. A straight current-carrying wire passes between two bar magnets. What is the direction of the magnetic force on the wire?

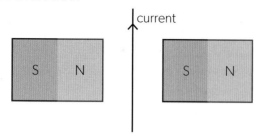
current

A. into the page

B. out of the page

C. to the left

D. to the right

2. A square-shaped conducting loop carries a constant current in the clockwise direction. The loop is placed near a long current-carrying wire.

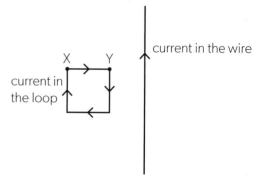
current in the wire

current in the loop

The force acting on the segment XY of the loop is:

A. zero

B. directed upwards

C. directed downwards

D. perpendicular to the plane of the loop

The magnitude of the magnetic force

Experiments show that the force acting on a straight wire in a magnetic field is proportional to:

* the length L of the wire

* the current I in the wire.

The definition of magnetic field strength differs from those of electric field strength and gravitational field strength. We cannot define the magnetic field strength in terms of $\dfrac{\text{force}}{\text{a single quantity}}$ because the force depends on two quantities: current and length. Instead, we define **magnetic field strength** as

$$\frac{\text{force acting on a current element}}{\text{current in the element} \times \text{length of the element}}$$

The "element" is a short section of a wire that carries a current and is at right angles to the magnetic field direction. When a force F acts on the element of length L when the current in it is I, then

magnetic field strength B is defined by $B = \dfrac{F}{I \times L}$

The unit of magnetic field strength is the tesla, abbreviated T. This is equivalent to the base (fundamental) units $kg\,s^{-2}\,A^{-1}$. The tesla can also be thought of as $N\,A^{-1}\,m^{-1}$.

The tesla turns out to be a large unit. The largest magnetic field strength that can be created in a laboratory is a few kT in magnitude and then only for very short times. The magnetic field of the Earth is roughly 10^{-4} T. The largest fields are associated with some neutron stars where the magnetic field strength can be of the order of 100 GT.

The definition leads to the force F acting on a wire of length L carrying current I in a magnetic field of strength B:

$$F = BIL$$

This equation applies when the field lines, the current and the wire are all at 90° to each other (Figure 5(a)), as they are when you use Fleming's left-hand rule. When this is not the case and the wire is at an angle θ to the lines, then we need to use the component of I at 90° to the field (see Figure 5(b)).

When the angle is defined as in Figure 5(b), the equation becomes

$$F = BIL \sin \theta$$

This equation is written in terms of the current in the wire. The current is, as usual, the result of moving charge carriers. The equation can be modified to reflect this. The current is $I = \dfrac{q}{t}$, where q is the charge that flows through the current element in time t.

Substituting gives $F = B\left(\dfrac{q}{t}\right) L \sin \theta = Bq\left(\dfrac{L}{t}\right) \sin \theta$

The term $\left(\dfrac{L}{t}\right)$ is the speed v of the charge carriers through the current element and making this substitution gives the expression

$$F = qvB \sin \theta$$

for the force acting on a charge q moving at speed v at an angle θ to a magnetic field of strength B.

Notice the way that the angle θ is defined in the diagrams. It is the angle between the direction in which the charge is moving (or the current direction—the same thing) and the field lines. Do not get this wrong by using $(90° - \theta)$ in your calculations.

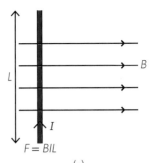

$F = BIL$

(a)

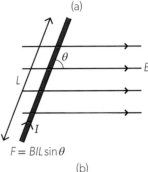

$F = BIL \sin \theta$

(b)

▲ Figure 5 When there is an angle θ between the wire and the field direction then $\sin \theta$ must be included in the magnetic force equation.

🧪 Force on a current-carrying conductor

- Tool 3: Construct and interpret tables and graphs for raw and processed data including scatter graphs and line and curve graphs.

- Inquiry 2: Collect and record sufficient relevant quantitative data.

- Inquiry 2: Assess accuracy, precision, reliability and validity.

- Inquiry 3: Relate the outcomes of an investigation to the stated research question or hypothesis.

▶ Figure 6
The magnetic force acts on balance and wire. The wire is fixed and the balance changes the apparent weight in response to the force.

current carrying lead

magnets on steel yoke

balance

This experiment gives you an estimate of the size of the magnetic force that acts on typical laboratory currents.

It allows you to investigate how the force varies with the length of the conductor and the size of the current.

- You will need some pairs of flat magnets (known as "Magnadur" magnets), a sensitive top-pan balance, a power supply and a suitable long straight lead to carry the current. Arrange the apparatus as shown in Figure 6.

- Zero (tare) the balance so that the weight of the magnets is removed from the balance reading.

- In the first part of the experiment, set the current in the wire as the independent variable and measure the force on the balance. This is the force acting on the magnets. A trick to improve precision is to reverse the current in the wire and take balance readings for both directions. Then add the two together (ignoring the negative sign of one reading) to give double the answer. Draw a graph to display your data of force against current.

- Second, use two pairs of magnets side-by-side to double the length of the field. Take care that the poles match; otherwise the forces cancel out. This may not work as well as the first part of the experiment as you need to assume that the magnet pairs have the same strength. This is unlikely to be true. But, roughly speaking, does doubling the length of wire in the field double the force?

Worked example 2

When a charged particle of mass m and charge q moves at speed v in a uniform magnetic field, then a magnetic force F acts on it. Deduce the force acting on a particle of mass m, charge $2q$ and speed $2v$ travelling in the same direction in the same magnetic field.

Solution

The equation for the force is $F = qvB\sin\theta$

In the equation, $\sin\theta$ and B do not change but every other quantity doubles, so the force is $4F$.

Worked example 3

A straight horizontal wire carries a current of 3.5 A. A length of 4.0 cm of the wire is subject to a horizontal magnetic field between the poles of a magnet. A magnetic force of 14 mN acts on the wire.

a. Determine the magnetic field strength, assuming that it is constant between the poles of the magnet.

b. State the direction of the magnetic force on the wire.

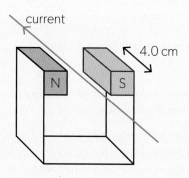

Solutions

a. The field is perpendicular to the current, so $F = BIL$. $B = \dfrac{F}{IL} = \dfrac{0.014}{3.5 \times 0.04} = 0.10\,\text{T}$.

b. The field is directed to the right and the current flows into the page. Using Fleming's left-hand rule, the force on the wire is vertically downwards.

Worked example 4

An alpha particle of mass 6.64×10^{-27} kg moves at a speed of $7.50 \times 10^3\,\text{m s}^{-1}$ through a uniform magnetic field of magnitude 0.240 T. The velocity of the particle makes an angle of 34.0° with the magnetic field. Determine the acceleration of the particle.

Solution

The charge of the alpha particle is $+2e$.

$F = qvB\sin\theta = 2 \times 1.60 \times 10^{-19} \times 7.50 \times 10^3 \times 0.240 \times \sin 34° = 3.22 \times 10^{-16}\,\text{N}$.

$a = \dfrac{F}{m} = \dfrac{3.22 \times 10^{-16}}{6.64 \times 10^{-27}} = 4.85 \times 10^{10}\,\text{m s}^{-2}$. Note that the acceleration is directed at right angles to the velocity; hence the speed of the alpha particle does not change.

Force between two parallel wires

In Topic D.2 the magnetic field pattern due to a long straight wire was described as a series of circular field lines centred on the wire (Figure 34 on page 536).

The magnetic field strength B for this arrangement when the current in the wire is I and the distance from the wire is r is given by:

$$B = \frac{\mu_0 I}{2\pi r}$$

The field strength is a $\frac{1}{r}$ law as opposed to the $\frac{1}{r^2}$ rule that applies to point charges and masses in Coulomb's law and Newton's gravitation law.

We can combine this with the expression for the force acting on a wire to determine the forces that act between two parallel wires, where r is the separation between the two wires.

Wire 1 has a current I_1 that gives rise to a magnetic field of strength B. This field extends to wire 2 which is a distance r away. The field strength at wire 2 due to wire 1 is $B = \frac{\mu_0 I_1}{2\pi r}$. The current I_2 in wire 2 interacts with the field from wire 1 and this leads to a force acting on wire 2 towards wire 1. The magnitude of this force is $F = B \times I_2 \times L$, where both wires have length L. Substituting for B gives

$$F = \frac{\mu_0 I_1}{2\pi r} \times I_2 \times L = \frac{\mu_0 I_1 I_2}{2\pi r} \times L$$

Considering the force per unit length on wire 2 rather than the total force on the whole length means that

$$\frac{F}{L} = \mu_0 \frac{I_1 I_2}{2\pi r}$$

This is the force per unit length on wire 2 which is measured in N m^{-1}. The same force per unit length acts on wire 1 due to wire 2 as a result of Newton's third law of motion.

The constant μ_0 is known as the permeability of free space and you can regard it as the response of a vacuum to the presence of magnetic field. It has the value $4\pi \times 10^{-7} \text{ N A}^{-2}$, which is $1.26 \times 10^{-6} \text{ N A}^{-2}$. The unit of μ_0 can also be written as T m A^{-1}.

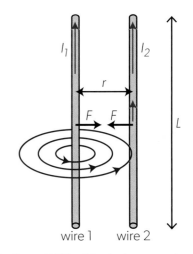

▲ Figure 7 Wire 2 lies in the magnetic field due to wire 1. As both wires carry current, forces act on both of them.

ATL Research skills

The equation $B = \frac{\mu_0 I}{2\pi r}$ is a version of Ampère's law, named after the French physicist André-Marie Ampère (1775–1836). Ampère's father was a wealthy merchant who took a great interest in his son's education but did not believe in formal schooling. As a result, André-Marie Ampère was largely self-taught using his father's library. One of the few areas where Ampère received formal tuition was in Latin, which enabled him to read some of the academic texts in the library.

Ampère went on to make many discoveries in what he called electrodynamics (now referred to as electromagnetism) and the SI unit of current is named after him.

Another scientist who researched the magnetic fields caused by electric currents was British physicist Michael Faraday (1791–1867). Faraday was also largely self-taught. However, his family was not rich. Faraday became an apprentice to a bookbinder, and it was by reading some of these books, that Faraday was able to start learning about electricity.

While Ampère's research was largely theoretical, Faraday was a gifted experimentalist. He was able to confirm Ampère's law and, in doing so, created the first electric motor.

How can independent research augment your own studies?

▲ Figure 8 Michael Faraday.

⊕ Data-based questions

A wire was aligned parallel to the edge of a Magnadur magnet which rested on a sensitive balance. When there was no current in the wire, the balance was zeroed. When the power supply was switched on, there was a current of 14.4 A in the wire. The variation in the balance reading was measured as the distance d between the wire and the edge of the magnet was varied, as shown in Figure 9.

d / mm (± 0.5)	m / g		
2.0	5.51	5.93	5.71
3.0	4.07	3.87	4.19
4.0	3.36	3.29	3.20
5.0	2.80	2.89	2.90
7.0	1.98	1.98	2.01
8.0	1.62	1.77	1.67
9.0	1.43	1.45	1.49
12.0	1.00	0.85	1.10
15.0	0.64	0.63	0.61

- Tabulate values of the average mass recorded on balance. Convert these values into an average force F.

- Calculate uncertainties in your values of the average force.

- By considering the variation in the readings of the mass, calculate the uncertainty in the average force.

Ampère's law suggests that $F \propto \dfrac{1}{d}$. Since the distance was measured to the edge of the magnet, it is expected that there would be an offset to the measured distance. Therefore, we might write $F \propto \dfrac{1}{d + d_0}$ and so a linear relationship between $\dfrac{1}{F}$ and distance is expected.

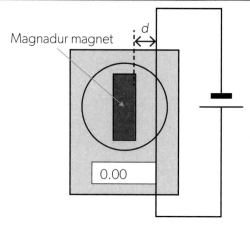

▲ Figure 9 The apparatus to verify Ampère's law.

- Tabulate values of $\dfrac{1}{F}$ and the uncertainty in these values.

- Plot a graph of $\dfrac{1}{F}$ against d. Add error bars to your graph.

- The relationship appears to be linear for values of d up to 10 mm. Find the gradient of the graph for values of d less than 10 mm.

- By considering maximum and minimum gradients, find the uncertainty in the gradient of the graph and express this as a percentage uncertainty.

- It is suggested that a problem with this experiment is that the magnetic field from the wire varies in strength across the width of the magnet. As a result, it is suggested that the relationship is better modelled as $F = e^{-k(d+d_0)}$.

- Plot a graph of $\ln F$ against d to establish whether this trend is a better fit. Use your graph to find values for k and d_0. Try to include uncertainties in these values.

Worked example 5

Electric currents of 8.0 A and 3.0 A are established in two long parallel wires X and Y. The magnetic force on a length of 0.20 m of wire X due to the current in wire Y has a magnitude of 1.5×10^{-5} N.

a. Deduce the magnitude of the magnetic force on a length of 0.20 m of wire Y due to the current in wire X.

b. Calculate the distance between the wires.

Solutions

a. From Newton's third law, the force on Y due to the current in X is equal but opposite to the force on X due to the current in Y. The magnitude of this force is therefore 1.5×10^{-5} N.

b. $\dfrac{F}{L} = \mu_0 \dfrac{I_1 I_2}{2\pi r} \Rightarrow r = \mu_0 \dfrac{I_1 I_2 L}{2\pi F}.$ $r = \dfrac{4\pi \times 10^{-7} \times 8.0 \times 3.0 \times 0.20}{2\pi \times 1.5 \times 10^{-5}} = 6.4$ cm.

Models — Beyond the inverse-square law

The whole of Theme D involves physics based around the inverse-square law. However, Ampère's law features a $\frac{1}{r}$ law. The data-based question on page 535 featured an example of a $\frac{1}{r^3}$ law. Why are these different?

The inverse-square law applies to single point sources. These could be of radiation, charge or mass. It can be thought of as being a consequence of the field lines (or rays of light) spreading out in three dimensions over the surface of a sphere, which increases in area as $A \propto r^2$.

Ampère's law involves a long wire. You can only move away from it in two dimensions, rather than three. The force decreases with a $\frac{1}{r}$ dependence now. The electric field around a long, straight, charged wire would have a similar $\frac{1}{r}$ dependence because the field lines cannot spread out over the third dimension — the length of the wire.

What about an infinite, two-dimensional sheet? You have already seen this situation — this is the same as moving away from the surface of Earth (not so far as to observe Earth's curvature, though). In the case of a two-dimensional object, the field lines are unable to spread out at all — the field is uniform and does not decrease with distance.

The data-based question on page 535 is a different case — a magnet has two poles, north and south, and so we are not dealing with a point source. If the north and south poles were placed one on top of another, they would cancel each other out and there would be no field at all. However, the two poles are displaced slightly, so you might experience a field by being fractionally closer to one pole than the other. This is called a **dipole field**.

Practice questions

3. Three long, parallel wires P, Q and R are separated by a distance of 0.10 m and carry equal currents of 5.0 A in the same direction.

   Calculate the magnetic force per metre of wire length on:

   a. wire P

   b. wire Q.

4. A long straight wire carries a current of 9.0 A. A square loop of side length 15 cm is placed 5.0 cm from the wire. There is a clockwise current of 4.5 A in the loop.

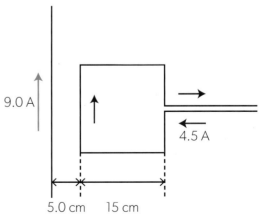

   Determine the magnitude and the direction of the magnetic force acting on the loop due to the current in the wire.

Measurements — The definition of the ampere

Until 2019, the ampere was defined in terms of the force between two current-carrying wires. The old definition was that an ampere is the constant current which, when maintained in two straight parallel conductors of infinite length and negligible cross-section, one metre apart in vacuum, would produce a force between them of $2\pi \times 10^{-7}$ N for every metre of their length.

Fortunately, students no longer need to learn this. The post-2019 definition of the ampere takes as its starting point the charge on the electron — defined to be $1.602176634 \times 10^{-19}$ C. This is also an ampere-second. Because the second is also a defined SI unit, this leads to a direct definition of the ampere as a coulomb of electrons flowing past a point in one second.

This links to the work on electrostatic fields in Topic D.2 and is explained in more detail in the *Tools for physics* section on page 334.

We now have two constants: ε_0, the permittivity of free space and μ_0. ε_0 has the units of $C^2 N^{-1} m^{-2}$, while μ_0 has

$N A^{-2}$. The product $\varepsilon_0 \times \mu_0$ has the units $\dfrac{C^2}{N\,m^2} \times \dfrac{N}{A^2}$. This simplifies to $\dfrac{1}{m^2} \times \dfrac{C^2}{A^2}$. Because $\dfrac{C}{A}$ (charge divided by current) has the unit of a second, then $\varepsilon_0 \times \mu_0$ has the overall unit of $m^{-2} s^2$, which is $\dfrac{1}{\text{speed}^2}$. Multiply the known values of permittivity (8.85×10^{-12}) and permeability (1.26×10^{-6}) for free space and the answer is 1.11×10^{-17}, which is $\dfrac{1}{c^2}$, where c is the speed in a vacuum of electromagnetic radiation. In symbols,

$$c = \frac{1}{\sqrt{\varepsilon_0 \times \mu_0}}$$

This result shows the very strong connection between electric and magnetic effects. It arose from the Maxwell equations that are discussed elsewhere in this book, but is now treated as a definition of ε_0 in the most recent version of the SI.

Patterns and trends — Vectors and their products

It is easy to believe that vectors are simply used for physics scale diagrams. But this would not be true. Vectors come into their own in mathematical descriptions of magnetic force and other parts of the subject.

Scalars can only be multiplied together in one way: run a relay with four stages each of distance 100 m and the total distance travelled by the athletes is 400 m. The product of two scalars is always another scalar.

Vectors, because of the added complication of a direction, can be multiplied together in two ways.

*   The scalar product (sometimes called "dot" product), where, for example, force and displacement are multiplied together to give work done (a scalar) that has no direction. In vector notation this is written as $W = \mathbf{F} \cdot \mathbf{s}$. The multiplication sign is the dot (hence the name). The vectors are written in bold font and the scalars in ordinary font.

*   The vector product (sometimes called "cross" product), where two vectors are multiplied together to give a third vector, this is written as $\mathbf{a} \times \mathbf{b} = \mathbf{c}$.

The multiplication of qv and B to form the vector force F is a vector product. The charge q is a scalar, but everything else in the equation is a vector. A mathematical physicist writes $\mathbf{F} = q(\mathbf{v} \times \mathbf{B})$ to show that the vector velocity and the vector magnetic field strength are multiplied together. The order of $\mathbf{v}$ and $\mathbf{B}$ is important. There is a vector rule for the direction of $\mathbf{F}$ that is consistent with our observations earlier and the $\sin\theta$ appears when the vector multiplication is worked out in terms of the separate components of the vector.

Vector notation is an essential language of advanced physics because it allows a concise notation and because it contains all the direction information within the equations rather than forcing us to use direction rules.

Vector products are not used in the rest of this book. Multiplication signs always relate to the product of two scalars.

Charge moving in a uniform electric field

An electron moves horizontally with constant velocity and enters a region of uniform electric field strength (Figure 10). A simple way to imagine this is for the electron to move between two charged parallel plates. This treatment ignores any edge effects at the plate boundaries. It also ignores the effects of gravity.

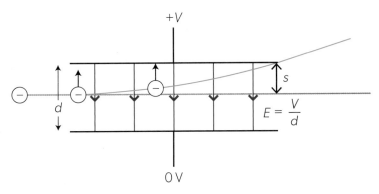

▲ Figure 10 The electron is moving perpendicular to a uniform electric field. An electric force acts on the electron opposite to the field direction. The blue line shows the parabolic trajectory while the electron is between the plates. Once the electron leaves the electric field it continues in a straight path.

The top plate is at a potential $+V$. The bottom plate is at zero potential. The electric field is vertically downwards. The force acting on the electron due to this field is upwards and, therefore, the electron is accelerated vertically upwards. Because the field is uniform, the electric force is constant in both magnitude and direction and the acceleration will also be constant.

This is a familiar situation. The kinematic equations and Newton's second law of motion can be used to analyse the situation. $E = \dfrac{V}{d}$ for the uniform field, and

$$a_{vertical} = \frac{F}{m_e} = \frac{qE}{m_e} = \frac{q}{m_e} \times \frac{V}{d}$$

where m_e is the mass of the electron.

The electron is not accelerated horizontally, so the horizontal component of the velocity does not change. The time t for which the electron is between the plates of length X is $t = \dfrac{X}{v_{horizontal}}$ (the travel time).

The vertical component of the velocity when the electron leaves the plates can be calculated using the kinematic equation $v = u + at$ with an initial vertical speed $u = 0$ as

$$v_{vertical} = a_{vertical} \times t = \left(\frac{q}{m_e} \times \frac{V}{d} \right) \times \left(\frac{X}{v_{horizontal}} \right)$$

The total vertical deflection s of the electron while in the field is proportional to t^2 because $s = \dfrac{1}{2} \times a_{vertical} \times t^2$. This becomes

$$s = \frac{1}{2} \times \left(\frac{q}{m_e} \times \frac{V}{d} \right) \times \left(\frac{X}{v_{horizontal}} \right)^2$$

which is

$$s = \left(\frac{qV}{2dm_e v_{horizontal}^2} \right) \times X^2$$

Because everything in the bracket is constant in this situation, then $s \propto$ *length of the plates*^{2}, so that the trajectory of the electron is a parabola.

This derivation ignores the effects of gravity on the electron. Is this justified? To answer this, compare the gravitational and electric forces acting on an electron when in a reasonable field (perhaps $1\,\text{kV}\,\text{m}^{-1}$) and travelling in Earth's gravitational field. Other data you need are in the IB data booklet. Your answer tells you about the comparative sizes of gravity and electrostatic effects (see also page 533).

Worked example 6

A beam of electrons moving horizontally with an initial velocity $u = 6.7 \times 10^6\,\mathrm{m\,s^{-1}}$ enters a region of a uniform vertical electric field between a pair of parallel plates. The potential difference between the plates is 25 V and the plates are separated by 2.0 cm. The electrons travel a horizontal distance of 5.0 cm and leave the electric field with velocity v.

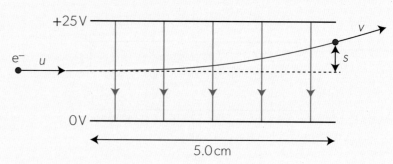

a. Calculate the acceleration of the electrons due to the electric force alone.

b. Hence, explain why gravitational effects on the motion of the electrons can be ignored.

c. Determine:

  i. the final velocity v of the electrons

  ii. the angle that the final velocity makes with the horizontal

  iii. the distance s by which the beam is deflected vertically.

Solutions

a. The electric force on the electron is $F = eE = e\dfrac{V}{d} = 1.6 \times 10^{-19} \times \dfrac{25}{0.020} = 2.0 \times 10^{-16}\,\mathrm{N}$. The acceleration is
$a = \dfrac{F}{m_e} = \dfrac{2.0 \times 10^{-16}}{9.11 \times 10^{-31}} = 2.2 \times 10^{14}\,\mathrm{m\,s^{-2}}$.

b. The acceleration due to gravity is $9.8\,\mathrm{m\,s^{-2}}$, which is negligible compared with the acceleration due to the electric force.

c. i. The horizontal velocity u is constant, so the time of travel is $t = \dfrac{0.050}{6.7 \times 10^6} = 7.5 \times 10^{-9}\,\mathrm{s}$

  The vertical velocity of the electron increases from zero to $v_{vertical} = at = 2.2 \times 10^{14} \times 7.5 \times 10^{-9} = 1.6 \times 10^6\,\mathrm{m\,s^{-1}}$, so the final velocity is $v = \sqrt{u^2 + v_{vertical}^2} = 6.9 \times 10^6\,\mathrm{m\,s^{-1}}$.

  ii. $\dfrac{v_{vertical}}{u} = \tan\theta \Rightarrow \theta = 14°$.

  iii. The vertical deflection can be calculated from kinematics equations, for example,

  $s = \dfrac{1}{2}at^2 = \dfrac{1}{2} \times 2.2 \times 10^{14} \times (7.5 \times 10^{-9})^2 = 6.1\,\mathrm{mm}$.

Practice questions

5. The velocity of the electrons in Worked example 6 is reduced, so that the beam is deflected vertically by $s = 1.0\,\mathrm{cm}$. No other changes are made to the setup. Determine the initial velocity of the electrons after the change.

6. A proton enters a region of uniform electric field with an initial velocity of $5.0 \times 10^3\,\mathrm{m\,s^{-1}}$. The electric field strength is $650\,\mathrm{V\,m^{-1}}$.

Determine the magnitude of the velocity of the proton after a time of $8.5 \times 10^{-8}\,\mathrm{s}$, when the initial velocity is:

a. in the same direction as the electric field strength

b. perpendicular to the direction of the electric field strength.

Charge moving in a uniform magnetic field

For the case of a charge in a magnetic field, our single electron now travels, again horizontally, into a region of a uniform magnetic field (Figure 11). This is the region shaded yellow on the diagram. The field lines are coming out of the page towards you.

Fleming's left-hand rule predicts the effect on the electron. The force is at right angles to both the velocity and the direction of the magnetic field. In using Fleming's rule, remember that it applies to *conventional* current, and that here the electron is moving to the right as it enters the magnetic field—so the conventional current is initially to the left.

The prediction from the left-hand rule is that the magnetic force acts vertically upwards.

Newton's second law of motion tells us that the electron accelerates in response to the magnetic force and that its direction of motion must change. The direction of this change is such that the electron will still travel at right angles to the field and that the magnetic force will continue to be at right angles to the electron's new direction. This is exactly the condition required for the electron to move in a circle. The magnetic force acting on the electron is providing the centripetal force for the electron. As the electron continues in a circle, so the magnetic force direction alters, as shown in Figure 11.

By the time the electron has reached the centre of the region, its velocity is now vertically upwards. The magnetic field direction is still out of the page, so the magnetic force is now horizontal and to the left. The circular motion of the electron continues until it leaves the magnetic field.

The force acting on the electron depends on its charge e, its speed v and the magnetic field strength B. Equating the magnetic force to the centripetal force equation gives $evB = \dfrac{m_e v^2}{r}$. The radius of the circle can be written in several ways:

$$r = \frac{m_e v}{Be} = \frac{p}{Be} = \frac{\sqrt{2 m_e E_k}}{Be}$$

where p is the momentum, E_k is the kinetic energy of the electron and m_e is its mass. The second and third equations for r can be particularly useful and are worth remembering.

When an electron is moving into a magnetic field that is not at 90° to its original direction, then it is necessary to take components of the electron velocity both perpendicular to the field and parallel to it. The perpendicular component leads to a circular motion exactly as before. The only difference will be that $r = \dfrac{m_e v \sin\theta}{Be}$, where θ is the angle between the field direction and the beam direction. The radius of the circle will be smaller than in the perpendicular case.

The component of velocity parallel to the field does not lead to circular motion. The electrons will continue to move at the component speed ($v\cos\theta$) in this direction. The overall result is that as the beam enters the field it moves in a helical path (Figure 12).

ATL **Directions on a diagram**

The "dot in a circle" $\odot$ is a common way to represent a field at 90° to a plane and coming out of it (towards the viewer). For a field going into the page, either $\otimes$, $\oplus$ or just a cross ✗ is used. The symbols are meant to represent the tip of an arrow coming towards you or the flight going away.

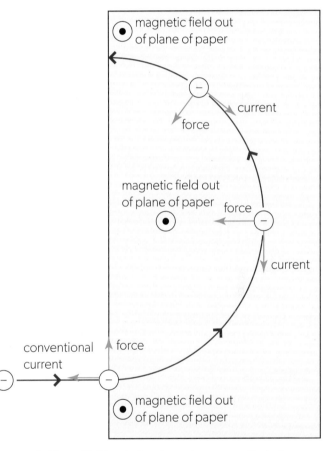

▲ Figure 11 The electron is moving perpendicular to a magnetic field. Unlike the electric field case, the force on the electron is now perpendicular to the field and the electron velocity. This provides a centripetal force for circular motion.

Circular motion is described in more detail on page 94. The problem of electrons moving in circular motion and hence accelerating was also an important consideration in the Bohr model of the hydrogen atom (see page 604).

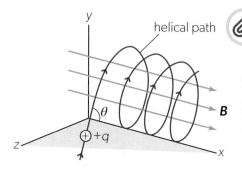

▲ Figure 12 When the velocity of the charged particle is not at right angles to the magnetic field, the path is a helix—a circle with an added linear component. The radius of the orbit is unchanged.

What causes circular motion of charged particles in a field?

How can the orbital radius of a charged particle moving in a field be used to determine the nature of the particle?

Large particle accelerators use magnetic fields to steer charged particles as they are accelerated to higher and higher speeds. The initial acceleration is usually carried out using electric fields.

In Topics A.2 and A.4 you studied the physics of rotation. The magnetic force that acts on a moving charge has the unusual property that it is at 90° to both the direction of movement and the magnetic field direction. The theory in this topic links these together to show that charged particles rotate in a circle or have a helical path.

A simple argument shows that the direction of the particle depends on its charge, while the radius of the path is related to the properties of the particle and the strength of the magnetic field. When we know the kinetic energy of the particle too, then we can estimate its specific charge (charge per unit mass) and therefore gain an insight into its nature.

As so often in physics, many areas of the subject meet to give a practical solution to the real-life challenges that particle physicists meet every day.

Worked example 7

An electron is moving in a uniform magnetic field. The velocity of the electron is perpendicular to the magnetic field. The speed of the electron is $6.5 \times 10^6\,\mathrm{m\,s^{-1}}$ and the magnetic field strength is $1.2\,\mathrm{mT}$.

a. Explain why the kinetic energy of the electron does not change.

b. Calculate the radius of the circular path of the electron.

c. Calculate the period of the motion.

d. The magnetic field is adjusted so that the radius of the path is $4.0\,\mathrm{cm}$. Calculate the magnetic field strength after the adjustment.

Solutions

a. The magnetic force on the electron is always perpendicular to the electron's velocity so the magnetic field does not do any work on the electron. The change in the kinetic energy is equal to the work done. Hence it is also zero.

b. The centripetal acceleration of the electron is provided by the magnetic force. Hence $\dfrac{v^2}{r} = \dfrac{evB}{m_e}$. From here,

$$r = \frac{m_e v}{eB} = \frac{9.11 \times 10^{-31} \times 6.5 \times 10^6}{1.6 \times 10^{-19} \times 1.2 \times 10^{-3}} = 3.1\,\mathrm{cm}.$$

c. $t = \dfrac{2\pi r}{v} = \dfrac{2\pi \times 0.031}{6.5 \times 10^6} = 3.0 \times 10^{-8}\,\mathrm{s}$

d. We use the same equation as in part b., but solved for the magnetic field strength.

$$B = \frac{m_e v}{er} = \frac{9.11 \times 10^{-31} \times 6.5 \times 10^6}{1.6 \times 10^{-19} \times 0.040} = 0.93\,\mathrm{mT}$$

Practice questions

7. A charged particle of mass m moves in a circular path in a region of uniform magnetic field B that is directed into the plane of the paper. The radius of the path is R and the speed of the particle is v.

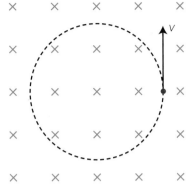

   a. State the sign of the charge of the particle.

   b. Determine, in terms of m, B, R and v, the magnitude of the charge of the particle.

8. The speed of the particle in question 7 and the magnetic field strength are both doubled. What is the radius of the path of the particle after the change?

   A. $\dfrac{R}{2}$ B. R C. $2R$ D. $4R$

9. An alpha particle (charge = $+2e$, mass = 6.64×10^{-27} kg) is moving in a circular path in a uniform magnetic field of magnitude 0.80 T. Calculate:

   a. the radius of the path, when the velocity of the particle is $7.5 \times 10^5\,\mathrm{m\,s^{-1}}$

   b. the velocity of the particle, when the radius of the path is 2.5 cm.

Determining the charge : mass ratio for a charged particle

The ratio $\dfrac{\text{charge}}{\text{mass}}$ for a charged particle can be determined by making measurements of the particle's path in a uniform magnetic field. To show how the method works, take as an example the measurement of the charge per unit mass of an electron (this quantity is also known as the specific charge of the electron).

A beam of electrons is fired through a gas (usually hydrogen or helium) using the equipment shown in Figure 13. The gas is at a very low pressure, so that the electrons do not collide with too many gas atoms and lose too much energy. When a uniform magnetic field is applied at right angles to the beam direction, the electrons move in a circle and their path is shown by the emission of visible light from atoms that have been excited by collisions with electrons along the path. The kinetic energy of an electron (or any charged particle moving in this way) can be regarded as constant in the uniform magnetic field because the magnetic field is perpendicular to the electron velocity and no energy is transferred to the electron.

▲ Figure 13 Electrons moving in a circle as demonstrated in a fine-beam tube. Measurements using this apparatus lead to the specific charge on the electron.

The electrons of mass m_e are accelerated using a potential difference V before entering the field. Their energy is $\frac{1}{2}m_ev^2 = eV$, so that $v = \sqrt{\dfrac{2eV}{m_e}}$.

Using $v = \dfrac{rBe}{m_e}$ from the centripetal acceleration proof earlier gives

$$\frac{e}{m} = \frac{2V}{B^2r^2}$$

This is of historical importance to physics because, even though J J Thomson identified the electron, this measurement of specific charge was the best that physicists could do to measure the properties of the particle until Millikan's determination of the electronic charge (Topic D.2).

For the general case when a particle of mass m with charge q has been accelerated through a pd V into a uniform magnetic field of strength B and moves in a circle of radius r, the equation becomes

$$\frac{q}{m} = \frac{2V}{B^2r^2}$$

Worked example 8

In an experiment to determine the specific charge of the electron, a student uses a beam of electrons accelerated from rest through a potential difference V. The electrons enter a region of uniform magnetic field B and the radius of their circular path is measured.

The student varies V and, for each value of V, adjusts B until the radius of the path is 4.0 cm. The graph shows the values of B^2 plotted against V and the line of best fit.

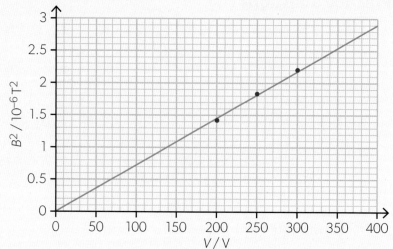

a. Estimate the gradient of the line of best fit.

b. Hence, determine the specific charge $\dfrac{e}{m_e}$ of the electron.

Solutions

a. $\text{gradient} = \dfrac{2.90 \times 10^{-6}}{400} = 7.25 \times 10^{-9}\,\text{T}^2\,\text{V}^{-1}$.

b. Rearranging the specific charge formula derived earlier gives $\dfrac{e}{m_e} = \dfrac{2V}{B^2 r^2} \Rightarrow B^2 = \dfrac{2m_e}{er^2}V$. This equation suggests that the graph of B^2 against V should be a straight line with the gradient equal to $\dfrac{2m_e}{er^2}$. In this experiment $r = 0.040$ m and we have the equation $\dfrac{2m_e}{e(0.040)^2} = 7.25 \times 10^{-9}$.

The specific charge $\dfrac{e}{m_e} = \dfrac{2}{(0.040)^2 \times 7.25 \times 10^{-9}} = 1.72 \times 10^{11}\,\text{C}\,\text{kg}^{-1}$. This is within 2% of the accepted value of $1.76 \times 10^{11}\,\text{C}\,\text{kg}^{-1}$.

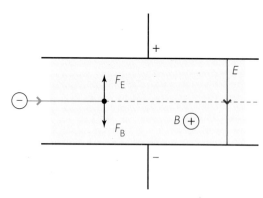

▲ Figure 14 Although electric fields lead to parabolic path and magnetic fields lead to circular paths for charged particles, they can cancel out to give no deflection at all when the electric and magnetic forces are equal and opposite.

Charge moving in perpendicular magnetic and electric fields

Charge movement in a uniform magnetic field leads to a circular orbit. Movement in a uniform electric field leads to parabolic trajectories. However, it is possible to combine the fields in a particular way to do a useful job. The trick is to combine the magnetic and electric fields at right angles to each other. As before, this analysis ignores the negligible effect of gravity on the charged particle.

For the arrangement shown in Figure 14, the electric force F_E is vertically upwards and the magnetic force F_B is downwards. When these forces are equal and opposite, then no net force acts on the charged particle.

For this to be the case $F_E = F_B$ and so $qE = qvB$ using the results from earlier in this topic and from Topic D.2. This leads to

$$v = \frac{E}{B}$$

For a particular ratio of $E : B$, there is a unique speed at which the forces will be balanced. Charged particles travelling more slowly than this speed will have a smaller magnetic force while the electric force is unchanged. The net force will be upwards on the diagram. For faster electrons, the magnetic force will be the larger and the electron will be deflected downwards. This provides a way to filter the speeds of charged particles and the arrangement is known as a velocity selector.

Worked example 9

An electron enters a region of crossed electric and magnetic fields. The fields are perpendicular to each other and to the electron's velocity (see Figure 14). The electron travels undeflected through the fields. The magnetic field strength is 50 mT.

Calculate:

a. the velocity of the electron, when the electric field strength is $7.5 \, kV \, m^{-1}$

b. the electric field strength, when the velocity of the electron is $2.4 \times 10^4 \, m \, s^{-1}$.

Solutions

a. $v = \dfrac{E}{B} = \dfrac{7.5 \times 10^3}{5.0 \times 10^{-2}} = 1.5 \times 10^5 \, m \, s^{-1}$.

b. $E = Bv = 5.0 \times 10^{-2} \times 2.4 \times 10^4 = 1.2 \times 10^3 \, V \, m^{-1}$.

Global impact of science — Bainbridge mass spectrometer

The Bainbridge mass spectrometer is a good example of a number of aspects of electric and magnetic fields being brought together to perform a useful job.

There are two parts to the instrument:

*   a velocity selector with electric and magnetic fields at right angles to each other

*   a deflection chamber with only a magnetic field (in this diagram, into the paper).

It is left as an exercise for you to show that identically charged ions of the same speed but different mass travel in circles of different radii in the deflection chamber. The ions arrive at the photographic plate or electronic sensor at different positions and, by measuring these positions, their mass and relative abundance can be determined.

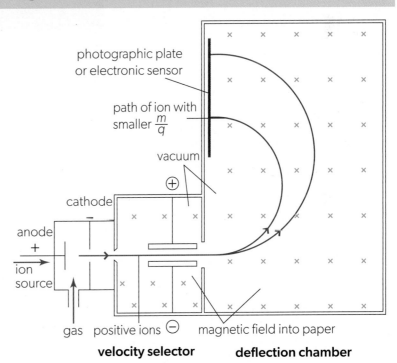

▲ Figure 15 The Bainbridge spectrometer arrangement.

Measurement

As is so often the case, what was originally complex and expensive technology can be improved and refined over time, making it cheaper, more reliable, more compact and more useful.

In the case of the mass spectrometer, as well as being used for analytical chemistry, it can analyse isotopes for dating samples (in carbon dating). Mass spectrometers have been sent to Mars and to Saturn's moon, Titan, to analyse surface samples.

Recent developments now enable a mass spectrometer to provide rapid analyses of body tissue. The onko-knife or iknife can determine whether tissue is cancerous or not. This informs surgeons in real time during an operation, making the procedure quicker and more accurate.

▲ Figure 16 An artist's depiction of NASA's Phoenix Mars Lander.

Practice questions

10. Ions of mass 1.50×10^{-26} kg and charge $+e$ are accelerated from rest through a potential difference of 600 V. The ions enter a region of crossed electric and magnetic fields. The magnetic field strength is 0.250 T.

    a. Calculate the speed of the ions.

    b. Given that the net force on the ions is zero, calculate the electric field strength.

11. Ions of charge $+e$ are accelerated from rest through a potential difference of 300 V and pass undeflected through a region of perpendicular electric and magnetic fields. The electric field strength is 18.1 kV m⁻¹ and the magnetic field strength is 0.200 T.

    a. Calculate the speed of the ions.

    b. Determine the mass of one ion.

12. A positively charged particle moves to the right with a horizontal initial velocity. The particle enters a region of a downward electric field between a pair of parallel horizontal plates. There is also a uniform magnetic field between the plates, directed into the page. The particle is deflected upwards, as shown in the diagram.

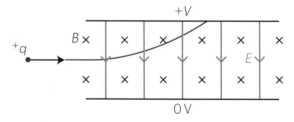

Which single change would allow the particle to pass undeflected between the plates?

A. increasing the magnetic field strength

B. increasing the initial speed of the particle

C. decreasing the potential difference between the plates

D. decreasing the distance between the plates

How can conservation of energy be applied to motion in electromagnetic fields?

Energy is conserved in all energy transfers involving electromagnetic fields. Gravitational and electrostatic fields are conservative (see Topic A.3, page 121). Magnetic fields are not conservative, although in certain circumstances they can appear to be. It all depends on the conditions of the circuit, as will become clear in Topic D.4.

For the case of a point charge orbiting in a magnetic field, energy is conserved. The speed is constant, so that the field is doing no work to transfer energy into the kinetic form. This is a consequence of the fact that the magnetic field is uniform and unchanging.

When point charge is moving at right angles to a uniform electric field it moves in a parabolic path (which links to the work in Topics A.1, A.2 and A.3) and its speed changes with its potential energy. Energy is transferred from the electric field to the kinetic energy of the charge.

 How are the concepts of energy, forces and fields used to determine the size of an atom?

One of the classic experiments of 20th-century physics is described in Topic E.1. Rutherford and his co-workers measured the deflection of alpha particles in a beam that was incident on gold atoms. Some of the alpha particles were moving directly towards a gold nucleus. The initial kinetic energy of the alpha is transferred into the electric potential energy of the system. Eventually the alpha particle stops at a position where its kinetic energy is zero and the electric potential energy is at a maximum. This position can be used as an estimate of the nuclear radius.

The physics involved in these estimates embraces a knowledge of Coulomb's law, and the recognition of energy conservation in the particle–nucleus system. This was the beginning of particle physics.

The simple technique has been improved and extended since Rutherford's work and is now a regular feature of physical investigation. The physics explored throughout this theme is important for these techniques. Electric and magnetic fields are used to originate and accelerate particles to high speeds to collide with other particles. When other, new, particles are created, they move through static magnetic fields maintained in cold liquids. Measurements of the radius of curvature of the particle tracks in the liquid allow estimates of the properties of the new particles.

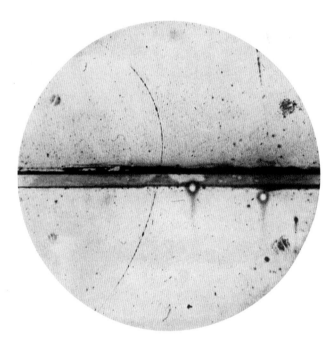

▲ Figure 17 This is a cloud-chamber image of the first positron to be observed. There is a horizontal thick lead plate across the photograph. The positron (an anti-electron) enters bottom left, is slowed by the plate, and leaves top left. Notice the increase in curvature of the track after the positron leaves the plate. This is because it has lost energy crossing the plate.

D.4 Induction

What are the effects of relative motion between a conductor and a magnetic field?

How can the power output of electrical generators be increased?

How did the discovery of electromagnetic induction effect industrialization?

Discoveries of electrical and magnetic effects have changed the conditions of people's lives for the better. The discipline of engineering has used physics to improve lives and society.

Generation of electricity only needs a large rotating coil of wire and a powerful magnet. When there is relative rotational motion between the coil and the magnetic field, then an emf appears across the coil. Under the right conditions, this emf can be translated into a movement of charge and hence a current. Again, using appropriate engineering, this current can be used outside the generator.

The electromagnetic induction occurs because of the relative motion between field and coil. In the case of

the alternating current generator, turn the coil faster and two things happen. The magnitude of the peak emf from the generator increases, and thus increases the potential power output. But the frequency of the supply also increases. This is undesirable for the consumers of the electrical energy and must be avoided. Other changes must be made to increase the power output. This topic looks at these changes.

The origins of the energy that is supplied to turn the coil give rise to debate in modern society. However, the underlying process of generation is clean and relatively efficient. The energy can be transported conveniently through cables and we can provide different sectors of society with its electrical requirements quickly and easily.

In this additional higher level topic, you will learn about:

- magnetic flux
- how a time-changing magnetic flux can induce an emf
- Faraday's law of induction
- a uniform magnetic field inducing an emf in a straight conductor moving at right angles to it

- Lenz's law as a consequence of energy conservation
- a uniform magnetic field inducing a sinusoidal varying emf in a coil rotating in the field and the effect of changing the rotational frequency of the coil on the induced emf.

Introduction

Topic D.2 showed that, when an electric charge moves through a magnetic field, a force acts on the charge. The converse case is when a stationary conductor containing free charges is subject to a changing magnetic field. Here energy can also be transferred in the arrangement and an electric potential difference will be set up across the conductor. This effect is called **electromagnetic induction**.

Induction is the effect used to generate electrical energy and to transform alternating currents from one pd to another. Braking systems in large road vehicles use electromagnetic induction as a way to slow the vehicle down and to reclaim some of the energy from the kinetic form.

The history of induction begins with Michael Faraday, (see page 547) who became the Professor of Chemistry at the Royal Institution in London. He expanded our understanding of electricity and magnetism, paving the way for the later work of Maxwell and others.

▲ Figure 1 Electric vehicles are viable because of electromagnetic induction.

Electromagnetic induction

Faraday's demonstration of electromagnetic induction can be repeated using a bar magnet, a length of wire wound into a solenoid shape (wide enough for the magnet to fit inside) and a sensitive ammeter (sometimes called a galvanometer). The experiment (Figure 2) consists of moving the bar magnet towards and away from the solenoid along the central axis. Different directions of motion and varying magnet speeds allowed Faraday to draw conclusions about the direction of the current that flows in the coil and the magnitude of this current.

Generating emfs and currents

- Inquiry 1: Demonstrate independent thinking, initiative, or insight.
- Inquiry 3: Relate the outcomes of an investigation to the stated research question or hypothesis.

Faraday spent over a decade making the discoveries that you can repeat in a few minutes. This experiment replicates his work and invites you to make the same inferences as this great scientist.

- Begin with a magnet and a coil of wire (Figure 2). Arrange the coil horizontally and connect it to a galvanometer (a form of sensitive ammeter, with the zero in the middle of the scale). You can use a laboratory coil, or you can wind your own from suitable metal wire using a cylindrical former.

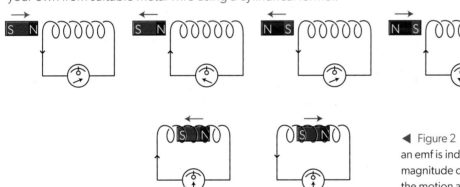

◀ Figure 2 When a magnet is moved near a coil, an emf is induced in the coil. The direction and magnitude of the emf depends on the direction of the motion and the rate at which it takes place.

- Move the bar magnet along the coil axis so that its north-seeking pole approaches one end of the coil and observe the effect on the meter. Record the direction of the current as indicated by the meter and the peak value shown. Relate this to the pole produced by the charge flow at the ends of the coil. Remember the rule from Topic D.2 that enables you to assign a magnetic pole to the end of a coil (Figure 3).

▶ Figure 3 When looking into a coil from the outside, the direction in which conventional charge is flowing gives the polarity of that end of the coil.

- Repeat the movement, moving the bar magnet with its south-seeking pole towards the coil.

- Now move the bar magnet away from the coil with both options of polarity.

- Change the speed at which you move the magnet.

- Compare the current directions for all cases. Also compare the sizes of the currents shown on the meter.

- Now try moving the coil with the magnet at rest. Does this change your observations?

- Now try moving the coil and the magnet at the same speed and in the same direction. What is the size of the current now? If your coil allows it, you might also try making the magnet enter the coil at an angle rather than along its axis. You could also try moving the magnet completely through the coil and out of the other side. Try to interpret this complex situation when you have understood the simpler cases.

- Relate the direction of current flow in the coil to the magnetic poles produced at the ends of the coil using the ideas in Topic D.2.

Several conclusions emerge from these simple experiments:

- Current is registered by the meter when there is relative motion between the coil and the magnet. Moving either the coil or the magnet produces the effect. However, when there is no relative movement between coil and magnet, no current is observed.

  Only movement of the wire in the coil relative to the magnet field gives the effect.

- When the north-seeking pole of the magnet is inserted into the coil, the direction of the charge flow leads to a north-seeking pole at the magnet end of the coil. This induced pole tries to oppose the motion of the magnet by repulsion.

  When a south-seeking pole is pushed into the coil, then a south-seeking pole appears at the magnet end of the coil. It is as though the system is acting to repel the bar magnet and, again, to reduce or prevent its movement.

- When a magnetic pole is moved away from the solenoid, the pole formed by the current in the coil is the opposite of the magnet pole. This attracts the magnet and reduces its speed of motion. Again, the solenoid–magnet system is trying to prevent change.

  The system tries to oppose any change in the magnetic flux.

- The greater the speed at which the magnet moves, the larger the current. Moving the coil at greater speeds relative to the magnet increases the sizes of the currents. The effect is at a maximum when the axis of the magnet between its poles is perpendicular to the area of cross-section of the coil.

  The opposition effect is larger when things happen more quickly.

Explaining electromagnetic induction

Figure 4(a) shows some free electrons in a metal rod. The rod is moving through a uniform, unchanging, magnetic field. The magnetic field direction is into the page and the rod is moving upwards in the plane of the page.

The rod carries the free electrons with it so that each moving electron can be regarded as a current moving up the page, each small current parallel to every other current. A force (according to Topic D.2) acts on each electron (because it is being treated as a current moving in a magnetic field). The direction of this force can be worked out. Remember that a free electron moving up the page is equivalent to a positive charge moving down. The use of Fleming's left-hand rule with conventional current should convince you that the force on the electron is to the right. As you expect from Topic D.2, this direction is perpendicular to both the field and the direction of motion of the rod.

This drift of the electrons to the right leaves the left-hand end of the rod with a deficit of electrons. There is no external connection between the left-hand (L) and right-hand (R) ends of the rod. End R becomes negatively charged and end L becomes positive. There is a potential difference between L and R, with L being at the higher potential.

Charges accumulate at the ends until the electric field along the rod is so large that further charge movement along it stops. At this point the situation in the rod cannot change any more. There is no external circuit outside the rod for charge flow. This means that no work will be required to overcome the resistance of the metal rod, and no transfer of energy occurs.

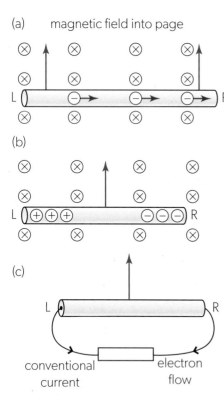

▲ Figure 4 How induced emfs arise in a rod when it is moved in a magnetic field. When there is a return path for the charges, there will be an induced current too.

The keys to understanding these effects lie in ideas you met in Topics B.5 and D.2.

The potential difference between R and L is an electromotive force that can, in principle, drive a current. We use the term **induced emf** to show that the emf arises from an induction effect. (The word "induction" is another term used by Faraday and others in the early days of electromagnetic study.)

The magnitude of the induced emf ε can be determined. When the charge flow has stopped, the forces acting on an electron in the rod are balanced so that the electric force towards L is equal and opposite to the magnetic force towards R. The magnitude of the electric force F_e is Ee, where E is the electric field inside the rod. When the length of the rod is x then $E = \dfrac{\varepsilon}{x}$ and therefore

$$F_e = \frac{\varepsilon}{x} \times e$$

The magnetic force F_m on the electron when the rod (and the electron) is moving at speed v at 90° to a magnetic field of strength B is evB. It follows that

$$F_e = F_m = evB = \frac{\varepsilon}{x} \times e$$

which leads to

$$\varepsilon = vBx$$

When the circuit is closed externally between L and R (Figure 4(c)), a flow of electrons occurs. Inside the rod, the conventional current flows from R to L. The electrons flow out of end R of the rod, and this is a conventional current in the external circuit from L to R. We say that a current has been induced (or generated) in the circuit.

Lenz's law

An important aspect of electromagnetic induction is that the induced emf exists in the conductor *whether the charge flows in a complete circuit or not*. In the case shown in Figure 4(b), the circuit is incomplete and electrons collect at one end until the electric field prevents further flow. It is only when the circuit is completed that charge can flow continuously.

- An induced emf is **always** generated by a magnetic field and conductor moving relative to each other.

- A continuous induced current exists **only** when there is a complete circuit for charge flow.

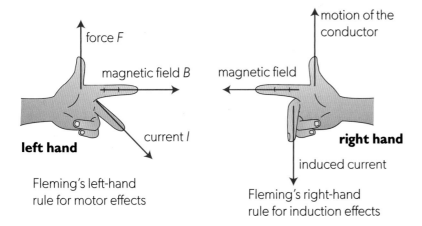

Fleming's left-hand rule for motor effects

Fleming's right-hand rule for induction effects

▲ **Figure 5** A comparison of Fleming's left-hand rule for motor effects and the right-hand rule for induction effects.

Models— Electromotive force

When the circuit is closed, the system moves the electrons through the resistances of the rod and the external circuit. Because this is a transformation of energy *into* an electric form, the term "electromotive force" should be used rather than a potential difference. As usual, we can identify the amount of energy transferred for each coulomb of charge that moves around the circuit.

You can see why the term "electromotive force" was used early in the discovery of electromagnetism. Some physicists object that no force acts when the term emf is applied to electric cells, batteries, solar cells and so on. Therefore, these physicists say that emf is not a good term to use.

It is true that it is difficult to see how the word "force" in its mechanical sense can apply in the case of an electric cell. But in the case of electromagnetic induction, a force is clearly acting on the electrons in the moving rod of Figure 4, and so the term emf continues to be used in physics. The fact that we often use the abbreviation emf rather than the full expression is a reminder to focus on the units of emf (volt or JC^{-1}) rather than the term "force".

This links to the expression derived from an energy standpoint later in this topic on page 565.

An alternative to Fleming

Not everyone prefers Fleming's left-hand and right-hand rules. Another possibility is to use the right hand with its thumb in the direction of the current and fingers in the direction of the magnetic field. Your palm points in the direction of the force on a positive charge or a conventional current. When an electron is the moving charge, point the thumb in the opposite direction to get the force in the opposite direction. This convention can also be used to get the direction of the magnetic field due a conventional current: for example, by curling the fingers around a wire with carrying a current.

How useful are rules such as this in science? Are they a reflection of scientific truth or a concept, or simply devices to help students do well in examinations?

We also need a direction rule for induced current. Fleming's left-hand rule predicts the force on a current of electrons. Because this electron flow is equivalent to a conventional current acting in the opposite direction, you have two choices for a direction rule (Figure 5). Either:

- use Fleming's left-hand rule to work out the force direction from first principles and let this lead you to the conventional current direction (using the argument given above), or

- use another rule (Fleming's right-hand rule as shown in Figure 5), which uses the symmetry between left and right hands to give the relationship between the motion of the conductor, the direction of the field and the direction of the induced *conventional* current.

Faraday's experiments showed that the current in the coil always opposes the change that the moving magnet is trying to impose on it. A rule to describe this was suggested by the German scientist Heinrich Lenz in 1833 and subsequently named after him. This is Lenz's law:

> **The direction of the induced current opposes the change that created the current.**

Check the diagrams in Figure 2 to see whether the results there confirm this rule.

In fact, Lenz's law is no more than a statement of conservation of energy. Suppose that the induced current in the solenoid were to increase the movement of the magnet rather than oppose it. This would imply that when the north-seeking pole approached the solenoid, a south-seeking pole would appear at the solenoid. There would be an attraction instead of a repulsion between magnet and solenoid. The magnet would be pulled into the coil, increasing its speed, and leading to an even greater acceleration. The magnet would move faster and faster, gaining kinetic energy from nowhere. Conservation of energy tells us that this cannot happen.

Another way to look at the consequences of Lenz's law is to realize that you cannot do work without having some opposition. The induced current in the coil produces an induced magnetic field that opposes the motion of the magnet. When the circuit is open, there is no current, no opposition and no electric energy produced. Move a powerful magnet very fast in the presence of a large conducting coil connected to a complete circuit and you will feel the force opposing you!

Electrons had not been discovered in Lenz's time and we can see how his law arises from first principles. In Topic D.3, when a charge flows within a magnetic field, the current produces a magnetic field which distorts the original field pattern. The catapult field (page 542) is an example of this. The result is that a force acts on (and can accelerate) the current-carrying conductor. This effect leads to the basis of an electric motor.

Energy transfers during induction

Electromagnetic induction occurs when there is a current in a conductor because of the conductor's motion through a magnetic field. Figure 6 extends the earlier example of the moving rod.

The rod is now rolling to the right at a constant speed on a pair of metal rails through the magnetic field. The rails conduct and form part of a complete electrical circuit WXYZ. (Rolling means that we do not have to worry about friction between the rod and the rails.) Charges, driven by the induced emf across the rolling rod, flow around the circuit. They give rise to an induced movement of electrons in the direction shown (from X to Y in the rod).

This induced current also interacts with the uniform magnetic field giving rise to the motor-effect force that was discussed in Topic D.3 (page 542). Use Fleming's left-hand rule for this current direction and magnetic field and you will see that a force must act to the left in Figure 6. This is opposite to the direction in which the conductor is moving.

For the rod to move at a constant speed, an external force must be acting on it by Newton's first law of motion. Whatever is pushing the rod to the right is doing work to overcome this motor-effect force to the left. The work done by the external agent to keep the conductor moving at a constant speed is transferred to electrical energy in the conductor. It is only possible for work to be done because of the opposition to the motion provided by the external magnetic field as it produces a magnetic force on the *induced* current.

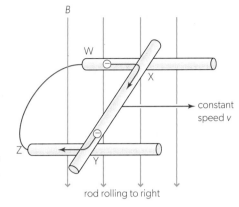

Estimating the induced emf in the rod — An energy approach

We can use the physics from Topics B.5 and D.3 to extend these qualitative ideas. The magnitude of the magnetic force due to the induced current is BIL where B is the magnetic field strength, I is the induced current in the rod and L is the length of the rod (XY in Figure 6). The magnetic force arising from the induced current opposes the original force. In other words, the opposing magnetic force is to the left in Figure 6 when the original applied force is to the right. The net resultant force is zero and the rod moves at constant speed.

From Newton's first law, to keep the rod in Figure 6 moving at a constant velocity, a constant force equal to BIL must act to the right on the rod.

The energy we have to transfer in a time Δt, therefore, is *force × distance moved*, which is $BIL \times \Delta x$, where Δx is the distance moved to the right by the rod in Δt.

The induced emf is equal to the energy per unit charge supplied to the system. In other words,

$$\varepsilon = \frac{\text{energy supplied in } \Delta t}{\text{charge moved in } \Delta t} = \frac{BIL \times \Delta x}{Q} = \frac{BIL \times \Delta x}{I \times \Delta t}$$

Of course, $\dfrac{\Delta x}{\Delta t}$ is the speed of the rod v. Therefore, as with the earlier derivation,

$$\varepsilon = BvL$$

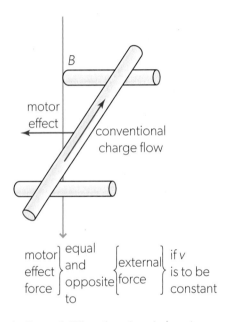

▲ **Figure 6** When there is an induced current in a loop, then there is also a motor effect that opposes the original motion that led to the induced emf.

Worked example 1

A metal rod of length $L = 0.75$ m moves without friction at a constant speed $v = 1.5\,\text{m s}^{-1}$ along a pair of parallel conducting rails. The rails are electrically connected at the left-hand side and form a complete electric circuit with the rod. The system is in a uniform magnetic field of magnitude $B = 0.16$ T directed out of the page.

a. Calculate the magnitude of the electromotive force induced in the rod.

The rod has a resistance of $4.0\,\Omega$. The resistance of the rails is negligible.

b. Calculate the current in the rod and explain its direction.

c. Explain why an external force must act on the rod to maintain the constant speed.

d. Calculate the rate at which the energy is transferred to the system.

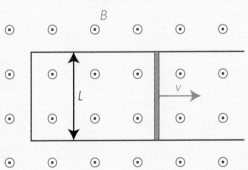

Solutions

a. $\varepsilon = BvL = 0.16 \times 1.5 \times 0.75 = 0.18\,V$

b. $I = \dfrac{\varepsilon}{R} = \dfrac{0.18}{4.0} = 45\,mA$. The electrons in the rod are moving to the right together with the rod. Hence they will experience an upward magnetic force, causing free electrons in the rod to drift upwards. The conventional current has the opposite direction to the flow of electrons, and hence is downwards.

c. Using Fleming's left-hand rule, the magnetic force on the current induced in the rod is to the left, opposite to the velocity of the rod. Constant velocity of the rod means that the net force on it is zero, so there must be an external force acting on the rod to the right that counters the effect of the magnetic force.

Note that the directions of the current and the magnetic force are consistent with Lenz's law: the current in the circuit is induced by the motion of the rod in the magnetic field, and the direction of the induced current is so as to provide a force that opposes this motion.

d. The rate of energy transfer can be calculated from two different viewpoints:

i. As the electrical power dissipated in the circuit, $P = \varepsilon I = 0.18 \times 0.045 = 8.1\,mW$.

ii. As the power developed by the external force F in moving the rod. The magnitude of the force $F = BIL = 0.16 \times 0.045 \times 0.75 = 5.4\,mN$. $P = Fv = 0.0054 \times 1.5 = 8.1\,mW$.

Practice questions

1. The external force acting on the rod in Worked example 1 above is increased to $1.2 \times 10^{-2}\,N$. The speed of the rod quickly increases to a new, constant value.

a. Calculate the current in the rod after the change.

b. Hence, determine:

i. the emf induced between the ends of the rod

ii the speed of the rod

iii the power transferred to the system.

2. An aircraft with a wingspan of 35 m moves at a constant horizontal velocity of $800\,km\,h^{-1}$. At the location of the aircraft, the vertical component of Earth's magnetic field is $2.5 \times 10^{-5}\,T$. Calculate the emf induced between the wingtips of the aircraft.

Patterns and trends — Flux or flow

Flux is an old English word that has the meaning of "flow" and one way to think about flux is to imagine a windsock used to show the direction and speed of wind at an airfield. When the wind is strong, then the flux density is high. The flux is the number of streamlines going through the sock. When the wind has the same speed for two windsocks of different sizes, then the windsock with the larger opening will have a larger flux, even though the flux density is the same for both windsocks.

Magnetic flux and magnetic flux density

The equation $\varepsilon = BvL$ can be developed further. The product $v \times L$ is the rate at which the rod sweeps out the area between the two conducting rails. To see this, $v \times L$ can be written as

$$\frac{\Delta x}{\Delta t} \times L = \frac{L \times \Delta x}{\Delta t}$$

as before. $L \times \Delta x$ is the change of the area swept out by the rod. Thus

$$\varepsilon = B \times \frac{\Delta A}{\Delta t} = B \times rate\ of\ change\ of\ area$$

In words:

induced emf = magnetic flux density × rate of change of area

This introduces you to an alternative term in magnetism for magnetic field strength — magnetic flux density.

Magnetic field strength is numerically equivalent to magnetic flux density.

In Topics D.2 and D.3, the term "magnetic field strength" was used because there we were concerned with basic ideas of field. In both electrostatics and gravity, the term field strength has a meaning of $\dfrac{\text{force}}{\text{charge}}$ or $\dfrac{\text{force}}{\text{mass}}$ depending on the context. We defined magnetic field strength as $\dfrac{\text{force}}{\text{current} \times \text{length}}$. However, this definition does not take account of the old, but helpful, view of a magnetic field being represented by lines directed from a north-seeking pole to a south-seeking pole. The density of the lines is a measure of the strength of the field. It is this visualization of a field in terms of lines that link magnetic field strength to magnetic flux density. You can use either term interchangeably.

Field lines are close together when the magnetic field strength is large. There will be many lines through a given area, so they are densely packed. We say that "the magnetic flux density is large". The total number of lines through one square metre is a measure of the magnetic flux density and therefore the total number of lines in a given area is a measure of the **magnetic flux** in the area.

Field lines

▲ Figure 7 A magnetic liquid forms shapes which seem to show field lines.

Magnetic flux density can be thought of as being the number of field lines (lines of flux) passing through a unit area.

However, field lines are not real. They are a visualization of the magnetic field introduced by Faraday. Strictly, there are an infinite number of field lines for a magnetic field, although we only draw a small number of them to illustrate a situation.

Although there are an infinite number of field lines in a given area for any particular field strength, different flux densities lead to different numbers of field lines over the same area. Infinities can be larger or smaller than each other!

Are field lines a helpful model?

Magnetic flux density B is the number of flux lines per unit area, and therefore magnetic flux Φ must be equivalent to *flux density* × *area A*: in symbols,

$$\Phi = B \times A$$

This equation assumes that B and A are at right angles to each other.

Figure 8 shows what happens when area and magnetic flux density are not perpendicular. The component of the field normal to the plane of the area is now the important quantity. When the normal to the area makes an angle θ with the lines of flux, then $\Phi = B \cos \theta \times A$. For $\theta = 0$ (B normal to area) $\Phi = B \times A$, whereas for $\theta = 90°$ (B parallel to area) the magnetic flux through the area is zero.

- Magnetic flux density B is related to the number of field lines per unit area. It is a vector quantity.

- Magnetic flux Φ is equal to $B \times A$. It is a scalar quantity.

- The equation $\Phi = BA \cos \theta$ is used when the area is not at right angles to the lines.

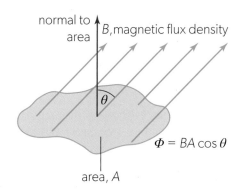

▲ Figure 8 The flux that links an area is the magnetic flux density × the area. When the field direction is not at right angles to the area, then $\cos \theta$ must be used where θ is the angle between the area normal and the field.

AHL

Communication skills

It might seem unnecessary to distinguish between magnetic flux density and magnetic field strength.

The theory behind much of the physics in this topic was condensed into a set of equations by James Clerk Maxwell. Maxwell's equations, as they are known, deal with two different types of magnetic field: B, the magnetic flux density, and H, the magnetic field strength. In the majority of cases, B and H are proportional to each other, and so distinguishing between the two is unnecessary. The distinction becomes important when dealing with magnetic fields interacting with magnetized materials. Since this situation is beyond the scope of this course, magnetic field strength is sometimes used here even though the correct term is magnetic flux density.

Is it unnecessarily pedantic to use precise terminology when such specific language is rarely required?

The unit of flux is the weber (Wb) and is defined in terms of the emf induced when a magnetic field changes. The equation $\varepsilon = B \times \dfrac{\Delta A}{\Delta t}$ can be rewritten as

$$\varepsilon = \frac{\Delta(BA)}{\Delta t}$$

In words, this is

$$\varepsilon = \frac{\text{change in flux}}{\text{time taken for change}}$$

This equation links induced emf and the rate of change of flux, so that a definition of the weber is as follows.

A rate of change of flux of one weber per second induces an emf of one volt across a conductor.

A knowledge of the magnetic flux due to a magnetic field and the rate at which the flux changes, allows a direct calculation of the magnitude of the emf that will be induced in a conductor.

There is a direct link between magnetic flux density and magnetic field strength.

Magnetic flux density $\left(\dfrac{\text{flux}}{\text{area over which flux acts}}\text{ in weber metre}^{-2}\right)$ is

numerically equal to magnetic field strength $\left(\dfrac{\text{force}}{\text{current} \times \text{length}}\text{ in tesla}\right)$.

One tesla (T) $\equiv$ one weber per square metre (Wb m^{-2}).

We therefore also have a link between changes in the magnetic field strength and the induced emf.

Worked example 2

A square loop of side length 15 cm is perpendicular to a uniform magnetic field of magnitude $B = 0.36$ T, directed out of the plane of the paper.

a. Calculate the magnetic flux through the loop.

The direction of the magnetic field is reversed in a time of 0.50 s.

b. Calculate the average emf induced in the loop.

c. Explain the direction of the current induced in the loop.

d. The resistance of the loop is $1.0 \times 10^{-3}\,\Omega$. Determine the total energy transferred to the loop.

Solutions

a. $\Phi = BA \cos 0° = 0.36 \times 0.15^2 = 8.1 \times 10^{-3}$ Wb.

b. After the field has been reversed, the magnetic flux becomes -8.1×10^{-3} Wb (Think of the angle between the normal to the loop and the field becoming 180°, and $\cos 180° = -1$.) The magnitude of the change in the flux is

   therefore $\Delta\Phi = 2 \times 8.1 \times 10^{-3} = 1.62 \times 10^{-2}$ Wb. The average induced emf is $\varepsilon = \dfrac{\Delta\Phi}{\Delta t} = \dfrac{1.62 \times 10^{-2}}{0.50} = 32$ mV.

c. The magnetic field is originally out of the plane of the paper and decreases. From Lenz's law, the induced current must oppose this change. Hence the magnetic field produced by the current in the loop must have its north-seeking pole pointing out of the plane of the paper. From Figure 3 the induced current is anticlockwise.

d. The average electric power developed in the loop is

   $P = \dfrac{\varepsilon^2}{R} = \dfrac{0.032^2}{1.0 \times 10^{-3}} = 1.02$ W. The energy transferred is $E = Pt = 1.02 \times 0.50 = 0.51$ W.

Magnetic flux linkage

There is one more quantity required. The derivation of $\varepsilon = \dfrac{\Delta(BA)}{\Delta t}$ above used a single rod rolling along two rails. Another way to imagine this is as a single rectangular coil of wire that is gradually increasing in area. The single turn gradually includes more and more field lines as the rolling rod moves to the right. As before, the emf across the ends of the coil will be equal to the rate of change of area multiplied by the magnetic flux density. When there are N turns of wire in the coil rather than a single turn, then the induced emf will be N times greater so that

$$\varepsilon = N\frac{\Delta(BA)}{\Delta t} = \frac{\Delta(N\Phi)}{\Delta t}$$

$N\Phi$ is known as the **magnetic flux linkage**.

You may sometimes see the unit of flux linkage written as **weber-turns**. This is equivalent to writing weber because the number of turns is simply a number and you can use weber by itself if you prefer.

The relationships between these interlinked quantities are shown in Figure 9.

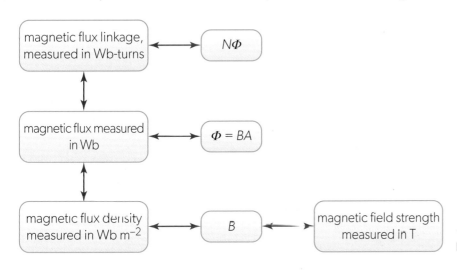

Figure 9 The inter-relationships between flux density and field strength.

Magnetic induction as a whole is summed up in the following law devised by Faraday himself. It is known as **Faraday's law**.

> The induced emf in a circuit is equal to the rate of change of magnetic flux linkage through the circuit.

In our usual notation, this is written algebraically as

$$\varepsilon = -N\frac{\Delta\Phi}{\Delta t}$$

The negative sign is added to include Lenz's law. (In its full mathematical form with the negative sign, the equation is also known as Neumann's equation.)

The equation combines Faraday's and Lenz's ideas. It reminds us that magnetic flux can be changed in three different ways:

- by changing the area of cross-section with time $\left(\dfrac{\Delta A}{\Delta t}\right)$
- by changing θ, the angle between B and A, with time $\left(\dfrac{\Delta \cos\theta}{\Delta t}\right)$
- by changing magnetic flux density with time $\left(\dfrac{\Delta B}{\Delta t}\right)$.

Each of these leads to the generation of an induced emf.

Cutting lines of force

Faraday first introduced the field-line model. However, his interpretation of a magnetic field is not quite the same as the modern view. He considered the lines of force to be at the edges of "tubes of force", like elongated elastic bands. At the time, it was thought that an invisible "aether", having elastic properties, filled space, including a vacuum. Later, Faraday and others took the concept of the field line further by suggesting that it was the action of the conductor "cutting" the tubes of force that led to an induced emf. This is a helpful way to think of the process, although it conceals the link between a charge being moved in a magnetic field and the magnetic force that acts on the charge as a result. But we need to remember that Faraday and the others did not know of the existence of the electron, and that they were very familiar with the ideas of field lines. In the 19th century, Maxwell refined these models by including both electrostatic and magnetic forces into one set of electromagnetic equations.

This illustrates two things about science: the way in which scientists allow a discovery to illuminate prior knowledge in a different way, and the power of the visual image to help us to understand a phenomenon.

Can you identify other powerful visualizations that you have come across in science? How do they help understanding?

Changing fields and moving coils

An emf can be induced in a conductor through magnetic flux changes in several ways that appear, at first sight, different from each other.

*   A wire or coil can move in an unchanging magnetic field (the example of the rolling rod above).

*   The magnetic field can change in strength while the conductor does not move or change its shape.

*   A coil can change its size or orientation in an unchanging magnetic field.

Combinations of these changes can occur.

We will look at these cases in the context of a rectangular coil interacting with a magnetic field that is uniform across the coil (but may change in magnitude).

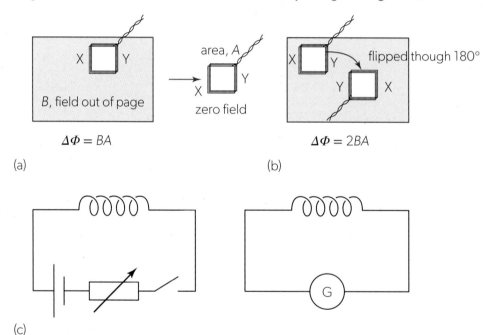

▲ Figure 10   Three ways to generate an induced emf.

Case 1: Straight wire moving in a uniform field

This is the case of the rod rolling on parallel rails above. The change in area per second is $L \times v$, the product of the length L of the rod and the speed v of the rod. The induced emf is therefore $\varepsilon = BvL$ when the wire moves at 90° to the field lines. When the wire motion is not at 90° to the field, then, as usual, the component of field at 90° to the direction of motion should be used.

Case 2: Coil moving

The coil can move as shown in Figure 10(a) from one position in a magnetic field to another position where the field is different. When the coil begins and ends in positions inside the same *uniform* field, then there is no change in the flux linkage and there is no induced emf. Although the coil is cutting lines, the same number of lines are being cut on opposite sides of the coil. Two emfs are induced, but they are in opposite directions and therefore cancel out.

When the coil with N turns moves from a position where the flux is Φ to a position where the flux is zero, the change in flux linkage is $(N\Phi-0)$ and the induced emf ε is

$$\varepsilon = \frac{N\Phi}{\text{time taken for change to occur}}$$

An interesting variant of this occurs where a coil in a field is flipped through 180° (Figure 10(b)). To visualize this, look at the system from the point of view of the coil. The field lines appear to reverse their direction through the coil, and so the change in flux is $\Phi -(-\Phi)$, in other words, 2Φ. The emf induced will be equal to $\frac{2N\Phi}{\text{time taken for change to occur}}$.

When a coil rotates in a field, the emf produced instantaneously depends on the rate of change of the flux linkage, and this, in turn, depends on the angle the coil makes instantaneously with the field. When the coil rotates at a constant angular speed, then the emf output varies in a sinusoidal way. This is the basis of an alternating current generator as you will see later in this topic.

Case 3: Magnetic field changes

Sometimes a stationary coil is immersed in a magnetic field that changes from one value to another. The field gets stronger or weaker. The act of cutting field lines is not so obvious here.

Suppose the field is being turned on from zero. Initially, there are no field lines inside the coil. You can think of the lines as moving from outside into the area bounded by the coil. The flux change stops when the flux density is at its final, unchanging, value. In moving into the coil, the magnetic field lines must have cut through the stationary coil.

Now $\varepsilon = -N\frac{\Delta\Phi}{\Delta t}$ becomes

$$\varepsilon = -NA\frac{\Delta B}{\Delta t}$$

because only B is changing. You need to know the rate at which the field is changing with time (or the total change of magnetic flux density and the total time over which it happens).

Another example is the case of two coils face-to-face as in Figure 10(c). One coil is connected to a galvanometer alone. The other coil is connected to a circuit with a cell, a variable resistor and a switch. In what way will you expect the galvanometer reading to change when the switch is closed and remains closed? Or when the switch is opened? Or when the switch is closed and the resistance in the circuit is varied?

The general case here is that the induced emf is always the rate of change of flux with respect to time. When you have a graph that shows the variation of flux linkage with time, then the induced emf will be the gradient of the graph (Figure 11). When the flux linkage is not changing, then there is a zero induced emf.

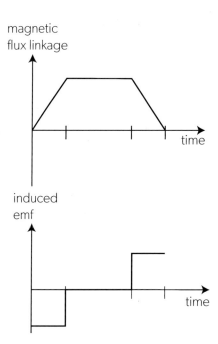

▲ Figure 11 The gradient of the graph of flux linkage versus time is the induced emf in the coil.

Worked example 3

The graph shows the variation of magnetic flux with time through a coil of 500 turns.

Calculate the magnitude of the emf induced in the coil.

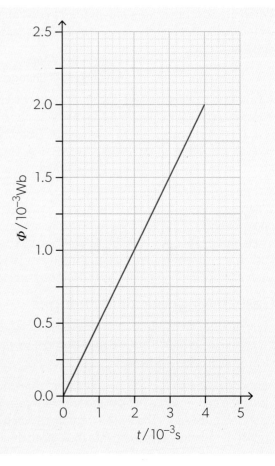

Solution

The change in flux is 2×10^{-3} Wb and this occurs in a time of 4.0 ms. The rate of change of flux is the gradient of this graph (as always). As the flux is proportional to the time, we can use any corresponding values of Φ and t.

$$\frac{\Delta\Phi}{\Delta t} = \frac{2\,\text{mWb}}{4\,\text{ms}} = 0.5\,\text{V}$$

Thus the induced emf = $500 \times 0.5\,\text{V} = 250\,\text{V}$.

Worked example 4

A small cylindrical magnet and an aluminium cylinder (which is non-magnetic) of similar shape and mass are dropped from rest down a vertical copper tube of length 1.5 m.

a. Show that the aluminium cylinder will take about 0.5 s to reach the bottom of the tube.

b. The magnet takes 5 s to reach the bottom of the tube. Explain why the objects take different times to reach the bottom.

Solutions

a. Use a kinematic equation, e.g. $s = ut + \frac{1}{2}at^2$.
    $1.5 = \frac{1}{2} \times 9.8 \times t^2$, which gives $t = 0.55$ s.

b. As the magnet falls, the copper tube experiences a changing magnetic flux, and, as a result, an emf is induced in the walls of the tube. This emf results in a current in the tube. The current leads to another magnetic field that opposes the motion of the magnet by Lenz's law. There is an upward force on the magnet, so that its acceleration is less than the value for free fall. In the case of the aluminium cylinder, no current arises and it falls with the usual acceleration.

Practice questions

3. A conducting ring of radius 3.0 cm is perpendicular to a uniform magnetic field. The magnetic field strength increases from 0.5 T to 2.0 T in a time of 0.25 s.

    a. Calculate the initial magnetic flux through the ring.

    b. Calculate the emf induced in the ring.

    c. The resistance of the ring is 10 mΩ. Determine the total energy transferred to the ring.

4. A square coil of side length 8.0 cm and 250 turns moving at a constant velocity v enters a region of uniform magnetic field B perpendicular to the coil.

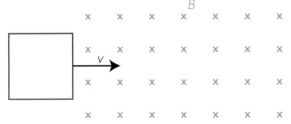

The graph shows how the magnetic flux linkage $N\Phi$ through the loop changes with time.

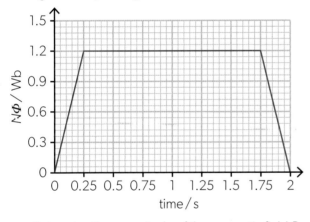

    a. Determine the magnitude of the magnetic field B.

    b. Draw a graph to show the variation of the emf induced in the coil with time.

    c. Calculate the speed v of the coil.

    The coil makes a complete electric circuit of resistance 0.50 Ω.

    d. State the direction of the current induced in the coil:

        i.   between 0 and 0.25 s

        ii. between 1.75 s and 2.0 s.

    e. Determine the total energy transferred to the coil between 0 and 2.0 s.

5. A bar magnet falls through a stationary horizontal conducting ring. The graph shows how the electromotive force (emf) induced in the ring varies with time.

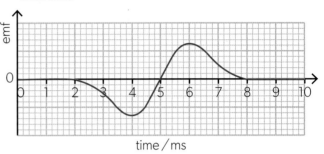

    At what time does the magnetic flux through the ring have a maximum absolute value?

    A   4.0 ms   B   5.0 ms   C   6.0 ms   D   8.0 ms

6. A bar magnet is moved towards a circular coil, with its north-seeking pole facing the coil. Seen from the left, a clockwise current in the coil is positive and an anticlockwise current is negative.

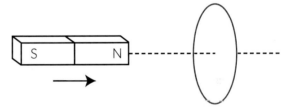

    As the magnet approaches the coil, what is the sign of the induced current and the nature of the magnetic force between the magnet and the coil?

	Current in the coil	Magnetic force
A.	positive	attractive
B.	positive	repulsive
C.	negative	attractive
D.	negative	repulsive

Global impact of science — Applications of electromagnetic induction

There are many applications of electromagnetic induction. They include:

- the generation of electrical energy
- electromagnetic braking, which is used in large commercial road vehicles
- the use of an induction coil to generate the large voltages required to provide the spark that ignites the fuel–air mixture in a car engine
- the generation of the signal in geophones and metal detectors.

In each of these examples, a changing magnetic field leads to the generation of an emf and demonstrates the physics developed in this topic.

(a)

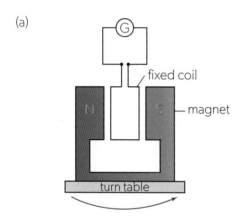

(b)

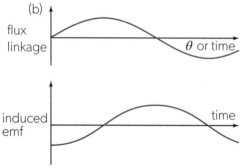

▲ Figure 12 (a) The coil is stationary and the field rotates around it in this simple ac generator. (b) The relationships between flux linkage and time, and induced emf and time for the generator.

Alternating current (ac) generators

From the 1830s onwards, scientists figured out various practical ways of generating electricity, firstly for direct current and then for alternating current. Alternating current (ac) generators are most commonly used. They consist of a coil with many turns; the coil rotates relative to a magnetic field.

For the moment, imagine that there is a fixed coil placed between the poles of a U-shaped magnet that stands on a rotating turntable (Figure 12(a)).

The turntable can turn at different angular speeds and the coil can have different numbers of turns and cross-sectional areas. Only one possibility is shown on the diagram. The static coil is connected to a galvanometer or to a data logger that registers the emf across the coil.

The magnetic field lines are to the right, from N to S, in the diagram. In the position shown there is no flux linked to the coil because the plane of the coil and the field line direction are parallel. When the magnet is turned through 90° (or 270°) from this position, then the field lines and the coil plane are at right angles; now the magnetic flux in the coil is maximum. The emf induced (and therefore current) is minimum (0) and vice-versa.

When the turntable and the magnet are made to turn continuously, then the coil moves from a zero flux to a maximum flux position twice in every cycle. The maximum flux is reversing in direction relative to the coil as the magnet rotates.

Figure 12(b) shows the changes in flux linkage during one cycle when the turntable rotates at a constant angular speed. The flux linkage graph is a sine curve. The emf induced in the coil is equal to the *negative rate of change* of flux linkage. This is proportional to the gradient of the flux linkage–time graph. When the flux linkage varies most rapidly (near the zero positions), the magnitude of the emf is at its greatest. When the flux linkage is at its maximum value (but the gradient of the graph is zero), the emf is zero. The induced emf graph is a negative cosine curve.

Changing the speed of the turntable changes both the frequency and the amplitude of the emf. Increasing the number of turns *or* increasing the area of the coil increases the amplitude of the emf but leaves the frequency unchanged because the turntable speed has not changed.

However, while some ac generators have a rotating magnetic field, others have fixed magnets and a rotating coil (Figure 13). The principle is the same: the direction in which charge flows in the coil varies with the flux linking it and therefore with its instantaneous orientation relative to the magnetic field.

Figure 13 (a) shows a rotating coil with a fixed magnet. One half of the single-turn coil is drawn with a heavier line than the other half to emphasize the coil rotation. The magnetic field direction is always from left to right.

When the left-hand side of the coil is moving upwards as shown in Figure 13(b), the direction of conventional current in this wire is away from you towards the back of the coil. The right-hand wire is moving downwards at the same instant and the current in this wire is towards you. Charge flows clockwise (looking from above) in the coil and out into the external circuit.

One quarter of a cycle later (Figure 13(c)) and the coil is vertical with both sides moving parallel to the field lines for an instant. In this position, there is no emf because there is zero rate of change of flux linkage.

A further quarter cycle later (Figure 13(d)) the sides of the coil have now exchanged positions compared with Figure 13(b). The way the coil is drawn reminds you of this. Conventional current is clockwise as far as we are concerned, but from the point of view of the coil, the current in it is in the opposite direction.

If wires were to be permanently connected between the coil and the meter in Figure 13(a), they would quickly become twisted. Energy needs to be extracted from the generator without this happening. So slip rings are used. The ends of the coil terminate in two rings of metal that rotate with the coil about the same axis. Two stationary brushes, connected to the external part of the circuit, press onto the rotating rings and charge flows out into the circuit through these connections.

The essential requirements for an ac generator are therefore:

*   a rotating coil
*   a magnetic field
*   relative movement between the coil and the magnetic field
*   a suitable connection to the static circuit outside the generator.

ac generator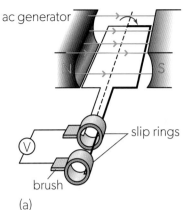

 slip rings

 brush

(a)

(b)

(c)

(d)

▲ Figure 13 This time the coil rotates in the static magnetic field. A slip-ring arrangement is required to allow charge to leave the coil.

 Faraday's law of induction includes a rate of change. Which other areas of physics relate to rates of change? (NOS)

The phrase "rate of change" is one that occurs throughout physics. It is short for "rate of change with respect to time" because the word "rate" has the implication of "per unit time" in its meaning. The phrase has physical meanings in conceptual, graphical and mathematical terms.

The idea of rate of change occurs early—and extensively—in Theme A of this course where the link between displacement, velocity and acceleration is in terms of rates of change. For example, velocity is the rate of change of displacement, or, in mathematical terms, $v = \frac{\Delta x}{\Delta t}$. Similarly, acceleration is "rate of the rate of change" of displacement:

$$a = \frac{\Delta v}{\Delta t} = \frac{\Delta}{\Delta t}\left(\frac{\Delta x}{\Delta t}\right)$$

You may recognize this as either the expression $\frac{d^2x}{dt^2}$ or as the gradient of a velocity–time graph.

Theme E also contains an important rate of change where we will consider the activity of radioactive material. This is the rate of decay of nuclei and turns out also to be proportional to the number of undecayed nuclei. Finally, there is the link between power and energy where power is the rate of energy transfer per unit time.

What other rates of change can you identify in science?

Global impact of science—Generators in real life

Commercial electrical generators are more sophisticated than our simple model and an internet search will allow you to see the variety of different types of ac generator. Figure 14 shows a generator of induced emf for the lamp on a bicycle (a dynamo). This resembles our original rotating-magnet arrangement.

The electrical power taken from a generator must be equal to the mechanical power that is given to it, ignoring friction and electrical resistance losses in the generator. Returning to the simple rod (on page 562) of length L moving at constant speed v through a magnetic field B, we know the induced emf $\varepsilon = vLB$ (the rod acting as a generator) and the force acting on the rod $F = BIL$ (the rod acting as a motor). When the ends of the rod are not connected to a complete circuit, then there will be an induced emf but no induced current. Without this current there is no motor effect, and no work needs to be done to overcome the force acting on the rod.

F cannot be zero; otherwise the generation of energy contravenes conservation of energy.

This is easily demonstrated using a bicycle dynamo like the one in Figure 14. Here a permanent magnet rotates in the gap inside a coil. The magnet is driven by a spindle connected to a wheel that is turned by the bicycle tyre as the two are pressed together. When the lamp is

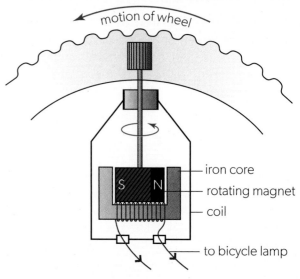

▲ Figure 14 A bicycle dynamo.

switched off so that no induced current is produced, the dynamo is relatively easy to turn. (remember that there will still be an induced emf across the terminals of the coil.) When the dynamo supplies current and lights the lamp, more effort is required to rotate the dynamo at the same speed since an opposing magnetic force will acts on the current in the coil. This is Lenz's law in action.

Modelling an alternating current (ac) generator

Faraday's law can be used to model a simple ac generator. Any current in the generator coils is ignored.

The coil has an average length l and an average width of m with N turns. The average area of the coil A is ml.

When the normal to the coil plane is at an angle θ to the field, the flux linkage through the coil is $N \times BA\cos\theta$. The coil spins at a constant angular speed ω. In time t the angle rotated by the coil is θ ($= \omega t$). The graph of variation of flux linkage with time has a cosine shape with a maximum value of $+NBA$ at $t = 0$ and a minimum value of $-NBA$ when the coil is halfway through one cycle.

The value of the induced emf at any instant is the negative rate of change of the flux linkage and is the negative gradient of the flux linkage–time graph. The induced emf is $\varepsilon = NBA\omega\sin\omega t$, which has maximum and minimum values for the emf of: $\varepsilon = \pm NBA\omega$.

A supply in which the current and voltage vary as a sine wave is an **alternating supply.** For everyday purposes we use the frequency f rather than ω, where $f = \dfrac{\omega}{2\pi}$, the same as in Topic C.1.

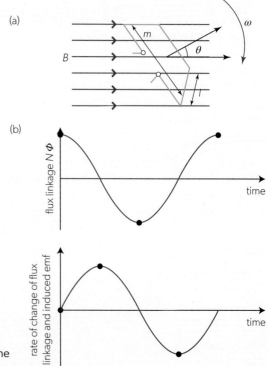

▶ Figure 15 The arrangement for the proof of the ac generator equation.

Worked example 6 shows the effect on the emf of changing the angular speed of the coil (without changing any other feature of the coil or field). When the angular speed of the coil is increased (this is ω in the equations), then:

- the coil will take a shorter time to complete one cycle; there will be more cycles every second and hence **the frequency increases**

- the time between maximum and minimum flux linkage will decrease and therefore (as the flux linkage is constant) the rate of change increases and hence **the peak emf increases**.

Other ways to increase the output of the emf, but without changing angular speed and frequency, include: increasing the magnetic flux density, increasing the number of turns on the rotating coil, or increasing the coil area.

Worked example 5

A rectangular coil of area $6.0 \times 10^{-2}\,m^2$ and 500 turns is placed in a uniform magnetic field of magnitude 53 mT. The coil rotates at a constant angular speed with a period of 0.04 s. At $t = 0$, the coil is perpendicular to the field.

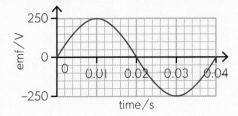

a. Calculate the average emf induced in the coil between $t = 0$ and $t = 0.02\,s$.

The maximum instantaneous emf induced in the coil is 250 V.

b. Draw a graph to show how the induced emf varies with time during the first complete rotation of the coil.

Solutions

a. In a time of 0.02 s the coil makes one half of a complete revolution, and the flux through the coil changes from $+ BA$ to $-BA$. $\Delta\Phi = -BA - (+BA) = -2BA$. Hence the average emf is

$$\varepsilon = -\frac{N\Delta\Phi}{\Delta t} = -\frac{-500 \times 2 \times 53 \times 10^{-3} \times 6.0 \times 10^{-2}}{0.020} = 160\,V.$$

b. At $t = 0$, the flux linkage through the coil is a maximum and hence the induced emf is zero. The emf will be a maximum at $t = 0.01$ s, when the coil is parallel to the magnetic field, and a minimum at $t = 0.03$ s.

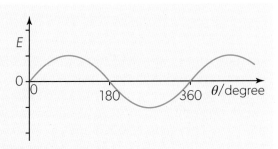

Worked example 6

A coil rotates at a constant rate in a uniform magnetic field. The variation of the emf E with angle θ between the coil and the field direction is shown.

Copy the graph below and, on the same axes, add the emf that will be produced when the rate of rotation of the coil is doubled. Explain your answer.

Solution

The rate of change of the flux linkage will double, so the magnitude of the peak emf will also double.

This is because Faraday's law states that the induced emf is proportional to the rate of change of flux linkage. However, the coil now rotates in half the time, so the time for one cycle will be halved. The new graph is drawn using the same scale.

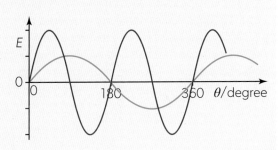

How is the efficiency of electricity generation dependent on the source of energy?

Electricity generation often involves a chain of energy transfers. Wind turbines transfer the kinetic energy of the wind mass to rotational kinetic energy in the turbine and then to electrical energy. Nuclear power stations have a complex chain involving mass–energy conversions leading to thermal energy in steam followed by turbine and generator stages. The final electrical energy often undergoes transformation from one alternating voltage to another.

All these steps involve inefficiency. The balance of input energy to output energy was shown visually using the Sankey diagrams of Topic A.3. The meaning of efficiency is mentioned several times in this course. Thermodynamic efficiency (Topic B.4) also has a part to play in the transformations.

Select a couple of generation methods and research the efficiency of the various stages and the overall efficiency of each method. Present your findings using Sankey diagrams (page 126).

Models—Measuring alternating currents and voltages

The current and voltage of an alternating supply change constantly throughout one cycle. Measuring these quantities is not straightforward because, with positive and negative half-cycles, the average values for current or voltage over one cycle are zero.

One way around this problem is to use the power supplied to a resistance R connected to the generator. The instantaneous power dissipated in the resistance is I^2R, where I is the instantaneous current.

Figure 16 shows both the current–time graph and the power–time graph with the same time axes for the alternating current and power dissipated in a resistor.

Notice the difference between the two:

*   The power–time graph is always positive (which we would expect because the power is I^2R and a number squared is always positive).

*   The power graph cycles at twice the frequency of the current.

To see why the power cycles twice in one cycle of the current, suppose that the time period of the ac generator is very large, taking 10 s to turn once through one cycle. Watch a filament lamp supplied with an ac supply of such a low frequency and you will see the lamp flash on and off *twice* in each cycle. The lamp is on when the emf is near its maximum (positive) and minimum (negative) values. When we look at a filament lamp powered by the ac mains, persistence of vision prevents us seeing its flashing like this because it is switched on and off at 100 or 120 times per second (twice the normal mains frequency of 50 or 60 Hz). Additionally, the time between "flashes" is

so short that the filament does not have sufficient time to cool down and is emitting light for a substantial part of the cycle.

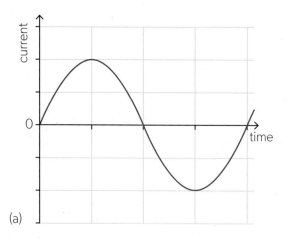

(a)

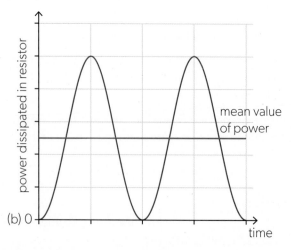

(b)

▲ Figure 16 The relationships between current and power with time for an alternating current.

Alternating values are measured using the equivalent direct current that delivers the same power as the alternating current over one cycle.

A lamp supplied from a dc supply would have the same brightness as the average brightness of our ac lamp flashing on and off twice a cycle. This equivalent dc current gives the average value of the power in the power–time graph. Because the average value of a $\sin^2$ graph is halfway between the peak and zero values and the curve is symmetrical about this line (shown on Figure 16(b)), the areas above and below the average line are the same.

The mean power that an ac circuit supplies is $\frac{1}{2}I_p^2 R$, where I_p is the peak value of

the current. The dc current required to give this power is $\sqrt{\frac{1}{2}I_p^2}$, which is $\frac{\sqrt{2}}{2}I_p$.

This value is known as the root mean square (rms) current: $I_{rms} = \dfrac{I_p}{\sqrt{2}} = \dfrac{I_p}{1.414...}$

In a similar way, $V_{rms} = \dfrac{V_p}{\sqrt{2}}$

The mean power P dissipated in a resistance is $I_{rms} V_{rms} = \dfrac{I_p}{\sqrt{2}} \times \dfrac{V_p}{\sqrt{2}} = \dfrac{I_p V_p}{2}$

with the usual equivalents: $P = I_{rms} V_{rms} = I_{rms}^2 R = \dfrac{V_{rms}^2}{R}$

Many countries use alternating current for their electrical supply to homes and industry. Different countries have made differing decisions about the potential differences and frequencies at which they transmit and use electrical energy. Thus, in some parts of the world, the supply voltage is about 100 V; in others it is roughly 250 V. Likewise, frequencies are usually either 50 Hz or 60 Hz.

> The frequency at which the power cycles being double that of the current is analogous to the comparison between energy and displacement variation in simple harmonic motion. In Topic C.1 you saw that the time for one energy cycle (kinetic or potential) was half that of the shm time period itself.

Worked example 7

The diagram below shows the variation with time t of the emf E generated in a rotating coil.

Calculate:

a. the rms value of the emf

b. the frequency of rotation of the coil.

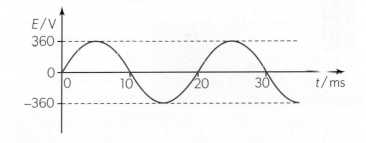

Solutions

a. The peak value of the emf is 360 V, so the rms value is $\dfrac{360}{\sqrt{2}} = 255\,V$.

b. $f = \dfrac{1}{T} = \dfrac{1}{0.02} = 50\,Hz$

Worked example 8

A resistor is connected in series with an alternating current supply of negligible internal resistance. The peak value of the supply voltage is 140 V and the peak value of the current in the resistor is 9.5 A. Calculate the average power dissipation in the resistor.

Solution

The average power $= \dfrac{\text{peak current} \times \text{peak pd}}{2} = \dfrac{1}{2} \times 140 \times 9.5 = 670\,W$

Practice questions

7. A resistor dissipates a power of 3.6 W when the direct current in it is 0.30 A. The resistor is connected in an alternating current (ac) circuit and the root mean square potential difference across it is 6.0 V.

    For the resistor in the ac circuit, determine:

    a. the average power dissipated

    b. the rms current

    c. the peak current.

8. The peak current in a 10 Ω resistor connected to an ac power supply is 2.5 A. Calculate:

    a. the average power dissipated in the resistor

    b. the rms voltage across the resistor.

9. A bicycle dynamo (see Figure 14) drives a load of a constant resistance. The dynamo outputs a power of 3.0 W at a particular bicycle speed. What is the power output of the dynamo when the speed of the bicycle is doubled?

    A   3.0 W   B   6.0 W   C   12 W   D   24 W

▲ Figure 18 A metal detector works by having two sets of coils, a transmitter and a receiver. Nearby metal objects affect the mutual inductance between the coils.

▲ Figure 19 Wireless charging is an application of mutual induction. When objects are charged wirelessly, there is a circuit with a coil in the charging pad (equivalent to circuit A in Figure 17) and a second circuit with a coil in the device that is being charged (circuit B).

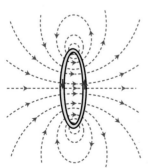

▲ Figure 20 A coil links its own flux and so there is a self-induction effect.

Mutual induction

There is an alternative way to induce an emf – by changing the magnetic flux density near a conductor. This changes the flux linkage.

When the current in A (Figure 17) is switched on or the variable resistance is changed, the magnetic flux density due to solenoid A takes time to reach a steady value. Field lines link the B solenoid, giving a changing flux in B during this time. An emf is induced in solenoid B giving a current and a magnetic field that opposes the field increase in A. This is **mutual induction** and occurs when an emf is induced in one conductor due to variations in flux linkage in another.

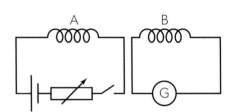

▲ Figure 17 Mutual induction between two coils. The left-hand circuit has a variable resistor, a cell and a switch. The right-hand circuit has only a sensitive ammeter G.

When the current in A is steady, the flux linkage to B is constant and the emf in B is zero.

When the current in A is switched off, the reverse occurs. Circuit B attempts to support the A field. This time the field due to solenoid B must be in the same direction as the (collapsing) field of solenoid A.

Self-induction

Induced emfs can arise in single conductors, including any wire, coil or solenoid which links its own flux; the effect is known as **self-induction**. Again, this is easiest to see in a coil arrangement (Figure 20).

When the current in the coil is changing, then the flux linked inside the coil is also changing. As $\varepsilon = -N\dfrac{\Delta \Phi}{\Delta t}$, an additional induced emf appears in the conductor. The negative sign is important here. The induced emf (by Lenz's law) will attempt to prevent the change of current. Therefore, when the current supplied to the coil is increasing, the induced current will oppose the change by trying to reduce the current. When the current supplied is decreasing, the induced current attempts to increase it. When the supplied current is constant, no induced emf occurs. The induced emf is a **back emf** because it is in the reverse direction to the forward imposed pd across the conductor.

Data-based questions

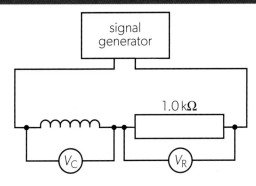

▲ Figure 21 A circuit to investigate self-inductance.

A coil and a resistor were connected in series to a signal generator. An oscilloscope was used to measure the peak-to-peak voltage across the coil and the resistor. The oscilloscope also provided a measurement of the frequency of the signal generator.

The data are shown in the table.

The voltage across the resistor can be used to determine the peak-to-peak current in the circuit:

$$I_p = \frac{V_R}{1000}$$

The "effective resistance" of the coil can then be found using

$$R = \frac{V_C}{I_{pk}} = 1000\frac{V_C}{V_R}$$

This effective resistance should increase with frequency due to the effects of self-inductance. (The technical term for this effective resistance is "reactance".)

- Tabulate values of $\frac{V_C}{V_R}$.
- Calculate values of the uncertainty in your values of $\frac{V_C}{V_R}$.
- Plot a graph of $\frac{V_C}{V_R}$ against f. Include error bars on your graph.
- Find the gradient of your graph.
- Using maximum and minimum gradients, find the uncertainty in your values of the gradient. Express this as a percentage uncertainty.

f / kHz	V_R / V (±0.004 V)	V_c / V (±0.04 V)
12.72	0.112	0.83
23.36	0.086	1.08
31.42	0.080	1.32
43.99	0.076	1.78
52.44	0.076	2.04
62.85	0.072	2.44
73.22	0.068	2.72
83.10	0.068	3.04
91.34	0.066	3.24
103.3	0.064	3.56
112.7	0.062	3.68
128.6	0.060	4.04

Worked example 9

Two stationary coils X and Y are parallel to each other and have a common axis. There is a decreasing current in coil X.

a. Explain the direction of the current induced in Y relative to that in X.

b. Explain the nature of the magnetic force between the coils.

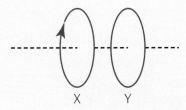

Solutions

a. There is a magnetic flux through coil Y caused by the current in X. This flux is decreasing, so by Lenz's law, the current induced in Y will have a direction so as to oppose this change and produce a magnetic field with the same direction as that of coil X. Therefore, the current in Y has the same direction as the current in X.

b. Magnetic fields from X and Y can be modelled as a pair of parallel bar magnets with opposite poles facing each other. The force is therefore attractive.

AHL

Worked example 10

A coil and a resistor are connected in series. The coil is made of wire of negligible resistance. There is a common current *I* through the coil and the resistor. The potential difference across the resistor is V_R and the potential difference across the entire connection is V_{tot}.

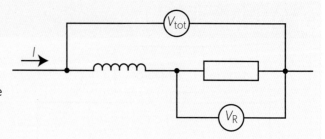

Compare V_{tot} and V_R when the current *I* is:

a. constant

b. increasing

c. decreasing.

Solutions

a. The magnetic field due to the current in the coil is constant and so is the magnetic flux linked by the coil. There is no emf induced in the coil and therefore no potential difference appears across it. $V_{tot} = V_R$.

b. The coil links its own increasing flux, so, by Faraday's law, there will be a self-induced emf in the coil, directed so as to oppose the increase in the current. The coil's emf results in a decrease of the potential as we move across the coil in the direction of the current. The potential decreases once again in the resistor, and

since both changes have the same direction, we have $V_{tot} > V_R$.

c. The situation is reversed compared with b. The direction of the self-induced emf is such that it opposes the *decrease* in the current, so we encounter a positive change (increase) in the potential across the coil and, later on, a decrease in the resistor. The coil's potential difference has the opposite sign to the resistor's potential difference, so $V_{tot} < V_R$.

Global impact of science — The transformer as an example of induction

Devices called transformers are used to change alternating supplies from one pd to another. Some transformers can convert voltages at powers of many megawatts. Others are small devices used to power domestic devices that need a low-voltage supply.

The transformer is an example of induction in action.

A transformer consists of three parts:

*   an input (or primary) coil

*   an output (or secondary) coil

*   an iron core on which both coils are wound.

Figure 22 shows a schematic diagram of a transformer (a) and a real-life transformer (b).

In a transformer:

*   Alternating current is supplied to the primary coil.

*   A magnetic field, produced by the current in the primary coil, links around a core made from a magnetic material, usually soft iron.

*   Because the primary current is alternating, the magnetic field in the core also reverses its direction. It goes first in one direction around the core for half a cycle and then reverses its direction for the remainder of the cycle. The flux in the core is constantly changing.

*   This changing flux also links the secondary coil.

*   Because the secondary coil has a changing field inside it, an induced alternating emf appears at its terminals. When the coil is connected to an external load, charge flows in the secondary circuit.

*   Energy transfers from the primary to the secondary circuit through the magnetic field.

An alternating pd with a peak value of V_p is applied to the primary coil leading to a flux of Φ in the core. The flux linked to the secondary coil of N_s turns is therefore $N_s \times \Phi$ and the induced emf in the secondary coil is

$$V_s = -N_s \times \frac{\Delta \Phi}{\Delta t}$$

The flux produced by the primary coil links itself through self-induction. This gives rise to an emf:

$$\varepsilon_p = -N_p \times \frac{\Delta \Phi}{\Delta t}$$

where ε_p and N_p are the induced emf and the number of turns in the primary coil. A simple theory leads to the equation

$$\frac{V_p}{V_s} = \frac{N_p}{N_s}$$

where V_p is the pd across the primary coil.

This **transformer rule** relates the ratio of the number of coil turns to the ratio of the input and output voltages:

- When $N_s > N_p$, $V_s > V_p$. This is known as a **step-up** transformer.

- When $N_s < N_p$, $V_s < V_p$. This is known as a **step-down** transformer.

- The terms "step-up" and "step-down" refer to changes in the alternating voltages.

This theory, assuming no energy loss in the transformer, suggests that the energy entering the primary is equal to the energy leaving the secondary, so $I_p \times V_p = I_s \times V_s$ where I_p and I_s are the currents in the primary and secondary circuits, respectively.

Many transformers have an efficiency that is close to 100% as energy losses can be reduced by good design. The **efficiency of a transformer** is equal to

$$\frac{\text{energy supplied by the secondary coil}}{\text{energy supplied to the primary coil}} \times 100\%$$

(a)

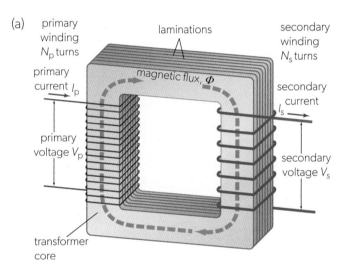

(b)

▲ Figure 22 The ac transformer.

HVDC and international collaboration

High transmission voltages lead to better efficiency because the energy loss in a transmission cable is equal to I^2R and halving the current means a four-fold reduction in resistive losses. This argument is valid for ac and dc. However, historically, the transformation between voltages was easier using ac. As the physics and engineering of electrical transmission improve, it becomes advantageous to use high-voltage direct-current transmission (HVDC). The Baltic Cable, which runs between Germany and Sweden, operates at a voltage of 450 kV. Since this voltage is almost 2000 times higher than the 230 V mains voltage in those countries, the energy loss is reduced by almost 4 million times when the cable is operated at 450 kV compared with 230 V.

Although the cost of the equipment to convert between two dc voltages is greater than the cost of a transformer,

there are other factors in the equation. Countries use different supply frequencies, and this is a major problem when feeding electricity from one country into the grid of another. Using undersea cables over long distances with ac also involves larger currents than might be expected as the cables have self-induction effects. Additional currents are required to move the charge every cycle.

Governments work together to maintain electricity supplies. There are many examples of electrical links between countries so that one nation can supply energy to another during times of shortage. There are short-term fluctuations in the demand for electricity, and energy is fed from one country to another at one time of day and then fed back again later. Examples include electrical links from the Netherlands to the UK and the HVDC link between Italy and Greece.

Theme D — End-of-theme questions

1. A planet is in a circular orbit around a star. The speed of the planet is constant. The following data are given:

   Mass of planet = 8.0×10^{24} kg
   Mass of star = 3.2×10^{30} kg
   Distance from the star to the planet = 4.4×10^{10} m

   a. Explain why a centripetal force is needed for the planet to be in a circular orbit.

   b. Calculate the value of the centripetal force.

   c. A spacecraft is to be launched from the surface of the planet to escape from the star system. The radius of the planet is 9.1×10^3 km.

      i. Show that the gravitational potential due to the planet and the star at the surface of the planet is about -5×10^9 J kg^{-1}.

      ii. Estimate the escape speed of the spacecraft from the planet–star system.

2. a. The moon Phobos moves around the planet Mars in a circular orbit.

      i. Outline the origin of the force that acts on Phobos.

      ii. Outline why this force does no work on Phobos.

   b. The orbital period T of a moon orbiting a planet of mass M is given by $\dfrac{R^3}{T^2} = kM$, where R is the average distance between the centre of the planet and the centre of the moon.

      i. Show that $k = \dfrac{G}{4\pi^2}$.

      ii. The following data for the Mars–Phobos system and the Earth–Moon system are available:

         Mass of Earth = 5.97×10^{24} kg

         The Earth–Moon distance is 41 times the Mars–Phobos distance.

         The orbital period of the Moon is 86 times the orbital period of Phobos.

         Calculate, in kg, the mass of Mars.

   c. The graph shows the variation of the gravitational potential between Earth and the Moon with distance from the centre of Earth. The distance from the Earth is expressed as a fraction of the total distance between the centre of Earth and the centre of the Moon.

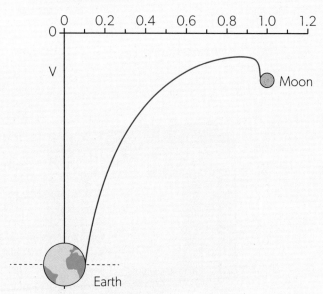

   Determine, using the graph, the mass of the Moon.

3. An electron is placed at a distance of 0.40 m from a fixed point charge of –6.0 mC.

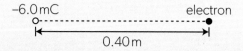

   a. Show that the electric field strength due to the point charge at the position of the electron is 3.4×10^8 N C^{-1}.

   b. i. Calculate the magnitude of the initial acceleration of the electron.

      ii. Describe the subsequent motion of the electron.

4. A proton is moving in a region of uniform magnetic field. The magnetic field is directed into the plane of the paper. The arrow shows the velocity of the proton at one instant and the dotted circle gives the path followed by the proton.

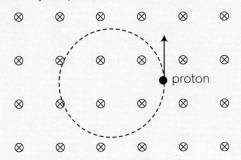

a. Explain why the path of the proton is a circle.

b. The speed of the proton is 2.0×10^6 m s^{-1} and the magnetic field strength B is 0.35 T.

   i. Show that the radius of the path is about 6 cm.

   ii. Calculate the time for **one** complete revolution.

c. Explain why the kinetic energy of the proton is constant.

5. A square loop of side 5.0 cm enters a region of uniform magnetic field at $t = 0$. The loop exits the region of magnetic field at $t = 3.5$ s. The magnetic field strength is 0.94 T and is directed into the plane of the paper. The magnetic field extends over a length 65 cm. The speed of the loop is constant.

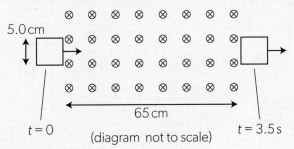

(diagram not to scale)

a. Show that the speed of the loop is 20 cm s^{-1}.

b. Sketch a graph to show the variation with time of:

   i. the magnetic flux linkage Φ in the loop.

   ii. the magnitude of the emf induced in the loop.

c. i. There are 85 turns of wire in the loop. Calculate the maximum induced emf in the loop.

   ii. The resistance of the loop is 2.4 Ω. Calculate the magnitude of the magnetic force on the loop as it enters the region of magnetic field.

d. i. Show that the energy dissipated in the loop from $t = 0$ to $t = 3.5$ s is 0.13 J.

   ii. The mass of the wire is 18 g. The specific heat capacity of copper is 385 J kg^{-1} k^{-1}. Estimate the increase in temperature of the wire.

6. A conducting sphere has radius 48 cm. The electric potential on the surface of the sphere is 3.4×10^5 V.

a. Show that the charge on the surface of the sphere is +18 μC.

b. The sphere is connected by a long conducting wire to a second conducting sphere of radius 24 cm. The second sphere is initially uncharged.

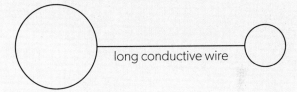

long conductive wire

   i. Describe, in terms of electron flow, how the smaller sphere becomes charged.

   ii. Predict the charge on each sphere.

7. A small magnet is dropped from rest above a stationary horizontal conducting ring. The south (S) pole of the magnet is upwards.

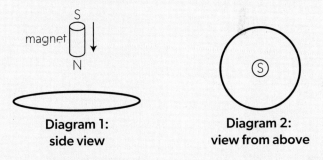

Diagram 1:
side view

Diagram 2:
view from above

While the magnet is moving towards the ring:

a. State why the magnetic flux in the ring is increasing.

b. Sketch, using an arrow on **Diagram 2**, the direction of the induced current in the ring.

c. Deduce the direction of the magnetic force on the magnet.

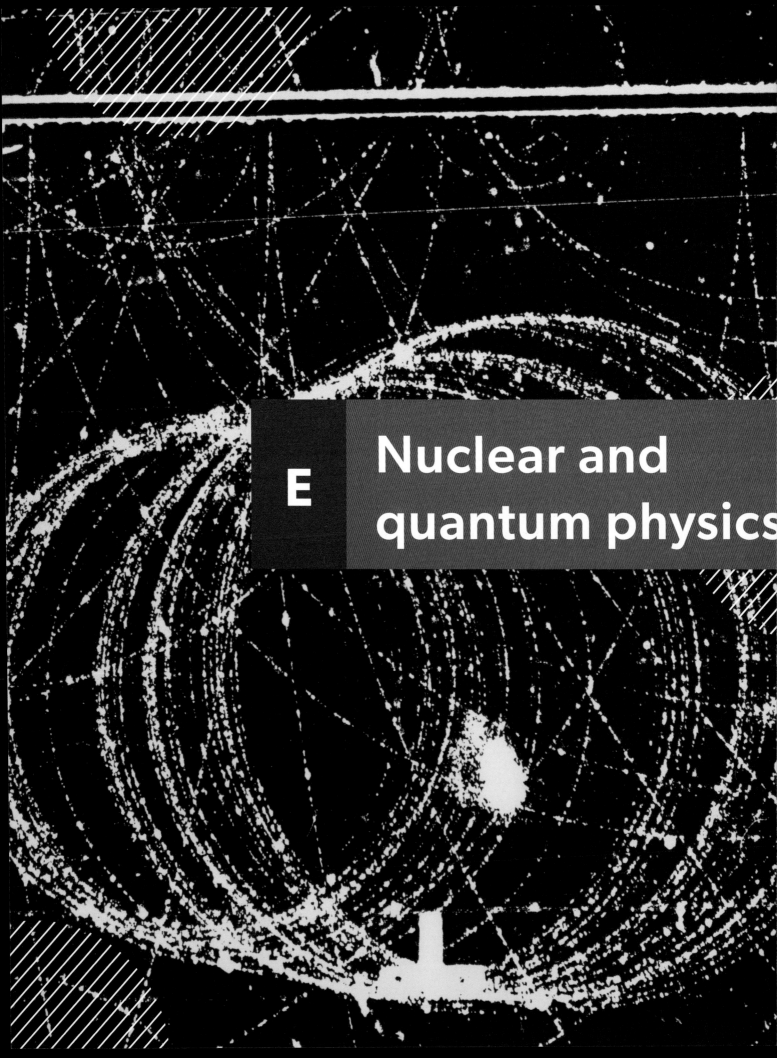

E **Nuclear and quantum physics**

Introduction

The scope of Theme E ranges from the smallest fundamental nuclear particles to the largest structures visible in the night sky. Scientists have been fascinated by these largest and smallest objects in the Universe throughout history.

All three of the overarching concepts of physics merge in Theme E. Nuclear material is in the grip of three interactions: strong, electromagnetic, and weak. These give rise to the **forces** that act between fundamental **particles**. When **energy** is sufficient to overcome the attractive forces then nuclear changes can occur.

Through the centuries, scientists have observed matter on earth and the night sky above with greater and greater magnifications. **Falsification** of theories and subsequent paradigm shifts have been driven by discoveries about atoms and the Universe.

The Greek philosophers Leucippus and Democritus in the fifth century BCE were the first atomists to suggest that atoms were indivisible. This idea remained in scientific thought up until the 19th century. At the turn of the 20th century, arguments about the nature of matter still persisted. Ernst Mach died in 1916 without acknowledging the existence of the modern atom; Boltzmann had already embraced the idea and had argued the point with him in 1897 (Topic B.3). But, despite Mach's reluctance, physics

was about to change forever (Topics A.5 and E.1) with Einstein's application of mechanics to Brownian motion in 1905.

The **global impact** of the new nuclear physics was immense and continues to this day. New sources of energy have been discovered and these can be used for good or for ill. New methods for medical diagnosis and treatments for illness have been developed involving nuclear science. Radioactive tracers give us insight into the world of the invisible. The pursuit of science has ethical consequences. Scientists must assess the risks of their work, weighing these risks against benefits. They should aim to do no harm.

Our starting point for this theme is that atoms are composed of a nucleus and electrons. The nucleus consists of nucleons: protons and neutrons. Protons have positive charge and neutrons, as the name suggests, have no charge. Outside the nucleus are the electrons. The charge on one proton and the charge on one electron have the same magnitude but opposite sign. So, your knowledge of the electric forces described in Theme D will be important. Early theories of the simplest atoms were based firmly on classical physics. Your knowledge of the mechanics of Theme A will be important too. Later theories of the atom involve wave mechanics which relies heavily on concepts from Theme C.

E.1 Structure of the atom

What is the current understanding of the nature of an atom?

In what ways are previous models of the atom still valid despite recent advances in understanding?

What is the role of evidence in the development of models of the atom?

Our understanding of atomic and nuclear structure has come a long way since the work of J J Thomson and his discovery of the electron 130 years ago. We now understand that the atom consists of a nucleus at the centre of a cloud of electrons. Originally one, and then two, types of nucleon were thought to exist. These were the proton and neutron. Physics now has a Standard Model that describes the composition of the proton, neutron and other nuclear particles using more fundamental entities. This model was developed during the second half of the 20th century by many particle physicists building on data from high-energy colliders such as those at CERN and Brookhaven. Work still continues to refine the features of the model.

The results and methods of early Greek thinkers are in strong contrast to those of present-day nuclear scientists. What part does evidence play in modern science? This is an essential question for you throughout this course. In Theme E, you focus on how scientific evidence leads to models of the atom and the nucleus. Atoms are invisible to the naked eye. We need powerful instruments to image the electron distributions that exist outside the nucleus. Accelerators use immense energies to help us infer the properties of the nucleus and the forces that act within it.

Early philosophical world views, such as those of the Ancient Greeks, came in various perspectives. These sometimes focused on the simplest possible explanation. Some Greek thinkers, for example, described matter as being made up of the "elements" earth, air, fire and water.

▲ Figure 1 In 1989, the company IBM demonstrated new technology that enabled them to place individual atoms for the first time. This picture shows 35 xenon atoms spelling the name of the company.

These ideas are a long way from the present list of stable and unstable elements that you will meet in Topic E.3.

We should not dismiss earlier ideas simply because they are old. Scientific method changes with time. The accepted views of both cosmology and nuclear science have altered throughout the past 150 years. Sometimes scientists resisted change for reasons that were not scientific but were based on dogma and bias.

Modern research into the nucleus is carried out by large teams of scientists, usually drawn from all parts of the world. Their work demands large, powerful and expensive machines that collect large amounts of data. The analysis of these data then requires powerful computers and the results of the analysis—usually statistical in nature—allow scientists to infer the presence of fundamental particles. These conclusions are submitted to critical peer review before being accepted into mainstream scientific understanding.

In this topic, you will learn about:

- the Geiger–Marsden–Rutherford experiment and the discovery of the nucleus
- nuclear notation
- emission and absorption spectra and how they provide evidence for discrete atomic energy levels
- photons and how they are released during atomic transitions
- the relationship between a difference in energy levels and photon frequency
- the emission and absorption of photons during atomic transitions

- the relationship between the nuclear radius and nucleon number
- deviations from Rutherford scattering at high energies
- the distance of closest approach in scattering experiments
- the Bohr model for hydrogen
- quantized energy, quantized angular momentum, and orbits in the Bohr model.

AHL

Deductive logic vs experimental evidence

It is easy to dismiss the early ideas of the elements being earth, air, fire and water. These ideas explained why rain falls (so it could reach its natural place in the rivers and seas) and why a stone would fall and sink (because its natural place was with Earth). Fire on the other hand rose upwards. Burning wood released fire and left ash behind—wood must therefore be made of the elements earth and fire.

Today, the purposes of science remain the same—to provide explanations of the world. However, the tests of knowledge have changed. The ancient Greeks valued deductive logic; modern scientists value experimental evidence.

How can we determine the relative merits of these ways of justifying knowledge?

Introduction

In 1858, the German physicist Julius Plücker found that electric discharges through a low-pressure gas caused a fluorescent glow on the walls of the glass container that held the gas. He was able to make the glowing area move using an electromagnet. By 1869, Johann Hittorf, one of Plücker's co-workers had identified **"cathode rays"**. These were produced when electric charge flowed through a low-pressure gas. Eight years later, Joseph J Thomson, a British scientist, found that these cathode rays were deflected by both electric and magnetic fields, and must therefore be charged.

Thomson eventually concluded that the cathode rays were beams of negatively charged particles coming from atoms. Atoms have no overall charge, so this meant that some other part of the atom must be positive.

Figure 2 shows the model that Thomson suggested. It is called the "plum pudding" or "current bun" model of the atom. The electrons were thought to be buried in a diffuse cloud of positive charge.

Thomson's model was the first suggestion that atoms have an internal structure. He had shown that objects smaller than an atom make up the most elementary building blocks of matter. His research prompted work by others. Since then, the attempts to discover these building blocks at ever-decreasing scales have been an important part of physics research.

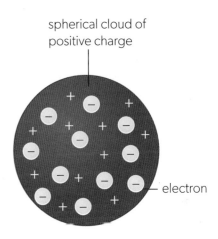

▲ Figure 2 Thomson's plum-pudding model.

Models

Rutherford's and Bohr's models of the atom are described in this Topic. Scientists use models to describe or explain things because a full explanation or description can be too complicated. The topic of atomic structure could easily fill this entire book and, even then, be incomplete and approximate.

It can be tempting to dismiss a model as being wrong or simplistic. The value of a model is whether it provides a useful way of thinking about a subject. The ancient Greeks' model of the atom as an indivisible blob works well when considering particles of gas bouncing off walls (see Topic B.3) and it was sufficient to enable chemists to develop the periodic table.

▲ Figure 3 A flight simulator models flying. Despite its limitations, it is a useful tool for pilots to practise their flying.

Nuclear notation

An atom is composed of a **nucleus** and **electrons**. The nucleus consists of **nucleons**, that is: **protons** and **neutrons**. A proton has a positive charge. A neutron, as its name suggests, has no charge. Outside the nucleus are the electrons. The charge on one proton and the charge on one electron are known to have the same magnitude but with opposite signs of electric charge. So, the numbers of protons and electrons in an uncharged atom must be equal to give an overall neutral charge.

A hydrogen atom—the simplest atom of all—consists of one electron and one proton. The proton is the nucleus of the atom; the electron is outside the nucleus. All atoms have a series of energy states determined by a probability wave that instantaneously describes the electron (and therefore the atom as a whole).

The nucleus of every other atom contains an additional particle: the neutron. Named for its charge-neutral properties, the neutron is a form of nuclear packing that allows protons to co-exist with each other in the confined volume of the nucleus.

The physics of these protons co-existing is explained in Topic E.3.

The numbers of protons, electrons and neutrons in a particular atom are important for the properties of the nucleus. The terms used are:

- **Proton number** Z—this is the number of protons in a nucleus. It is unique to the chemical element that the atom represents. It is also the number of electrons in a neutral, uncharged atom of the element. Historically, the proton number was called the atomic number. The proton number is also the number of a chemical element on the periodic table.

- **Nucleon number** A—this is the total number of protons plus neutrons in the nucleus. Historically, this was called the mass number.

- **Neutron number** N—this is less frequently used than A and Z but its name speaks for itself as the number of neutrons in the nucleus.

Algebraically: $A = Z + N$

There is a shorthand notation to indicate the ingredients of a particular atom:

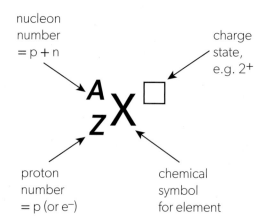

nucleon number = p + n

charge state, e.g. 2+

proton number = p (or e⁻)

chemical symbol for element

This notation is used when writing nuclear equations later.

Examples of this notation are:

helium-4 → $^4_2$He; 2 protons + 2 neutrons — 4 nucleons in total

oxygen-16 → $^{16}_8$O; 8 protons + 8 neutrons — 16 nucleons in total

oxygen-17 → $^{17}_8$O; 8 protons + 9 neutrons — 17 nucleons in total

proton → $^1_1$p
neutron → $^1_0$n
electron → $^0_{-1}$e

See Topic E.3 for why the electron is given a proton number of −1.

Worked example 1

State the number of protons and neutrons in a nucleus of a. $^{235}_{92}$U b. $^{24}_{11}$Na.

Solutions

a. There are 92 protons from the proton number in ^A_ZX and so the number of neutrons must be 235 − 92 = 143.
b. There are 11 protons and 13 neutrons in this nucleus of sodium.

The Geiger–Marsden–Rutherford experiment

In 1909, Ernest Rutherford, from New Zealand, was supervising the research of two students at the University of Manchester in England. One of these was a German researcher, Johannes Geiger; the other was an undergraduate, Ernest Marsden. They were studying the scattering when alpha particles were incident on a very thin gold foil in a vacuum.

Figure 4 shows the basic experiment. When alpha particles collide with a fluorescent screen, light is emitted from the screen. A microscope detects these flashes of light.

Geiger and Marsden expected their alpha particles to be deflected only through small angles as a result of electrostatic deflection by the diffuse atomic charge (assuming that Thomson's plum-pudding model was correct). Rutherford suggested, however, that they should look on the same side of the foil as the alpha source. All three scientists were astonished to find that about one alpha particle in 8000 was reflected (or "back-scattered") by the thin foil back in the direction from which it had come. This could not have been a reflection from the diffuse positive charge of Thomson's plum-pudding as the alpha particles were travelling too fast for this.

Figure 5 shows the paths of alpha particles coming in from the left. Two of these alpha particles have head-on paths and are then deflected through angles close to 180°. Other alpha particles have initial paths that lie further from the nuclei and so are deflected much less.

You will be introduced to the properties of the **alpha particle** in Topic E.3. For now, you need to know that an alpha particle is a positively charged helium nucleus, $^4_2$He^{2+}.

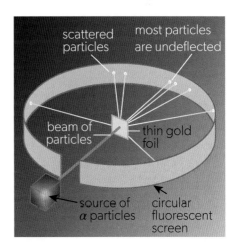

▲ Figure 4 A schematic diagram of the Geiger–Marsden–Rutherford experiment. The apparatus itself was small—a few cm across.

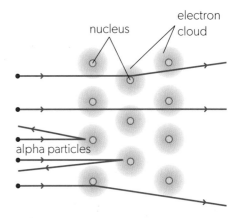

▲ Figure 5 Alpha particles deflected by nuclei in the Rutherford model.

Experiments

On reflecting on the results of the gold foil experiment, Rutherford later wrote that this was …

"… the most incredible event that has ever happened to me in my life. It was almost as incredible as if you fired a 15-inch shell [from a large gun on a warship] at a piece of tissue paper and it came back and hit you. … I realized that this scattering backward must be the result of a single collision, and when I made calculations, I saw that it was impossible to get anything of that order of magnitude unless you took a system in which the greater part of the mass of the atom was concentrated in a minute [he meant "very small"] nucleus. It was then that I had the idea of an atom with a minute massive centre, carrying a charge. …"

▲ Figure 7 Enlarge an atom to the diameter of the meteor crater in Arizona (1200 m across) and the nucleus would be the size of a marble (approximately 1 cm).

(ATL) Managing data

Often, scientific research requires patience and discipline. It is tempting to think that the Geiger–Marsden–Rutherford experiment was based on a single observation — certainly Rutherford's response suggests a reaction to a single event. When Geiger published the data, he stated that over 100 000 scintillations were observed during the course of the measurements.

Geiger or Marsden had to sit in a darkened room and look for a tiny flash of light to record one scintillation. It took about 30 minutes for their eyes to adjust to the dark and be sensitive enough to detect these flashes. Counting the 100 000 flashes of light took weeks of effort.

Rutherford wrote in a letter: "Geiger is a demon at the work of counting scintillations and could count at intervals for a whole night … I […] retired after two minutes."

It is unsurprising that Geiger developed a device for detecting radiation — the Geiger counter.

▲ Figure 6 A Geiger–Müller counter.

The alpha particles have a positive charge with a magnitude twice that of an electron. Rutherford suggested that back-scattering occurs because the atom consists of a small, very dense, positive nucleus with electrons outside the nucleus. His calculations indicated that the diameter of the nucleus was of the order of 10^{-15} m while the diameter of the entire atom was known to be about 10^{-10} m. The atom, taken as a whole, is almost entirely empty space.

Figure 5 is not to scale because the nuclei spacings shown are drawn far too small. The nuclei in the diagram on the page are about 2 mm across. Nuclei at this scale should be separated by 2×10^5 mm, about one-fifth of a kilometre!

The main results of this important experiment are that:

- most alpha particles passed through the gold leaf undeflected or with very small deflections

- a very few alpha particles were deflected through very large angles but, occasionally, alpha particles rebounded in the opposite direction.

These results mean that:

- most of the atom is in a small dense region

- the atom contains small dense regions of electric charge

- this small dense region contains all the atom's positive charge.

An analogue of alpha-particle scattering

- Inquiry 1: Demonstrate creativity in the designing, implementation or presentation of the investigation.

- Inquiry 1: Develop investigations that involve hands-on laboratory experiments, databases, simulations and modelling.

- Inquiry 3: Interpret processed data and analysis to draw and justify conclusions.

- Inquiry 3: Evaluate the implications of methodological weaknesses, limitations and assumptions on conclusions.

Alpha-particle scattering can be simulated using a specially shaped hill and a ramp (Figure 8(a)).

The hill is made so that its height h above the surroundings is inversely proportional to the distance r from the centre of the hill (Figure 8(b)). This means that it simulates, using gravity, the electric potential due to the positively charged nucleus (Topic D.2).

(a)

(b)

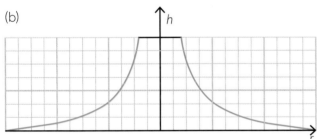

▲ Figure 8 (a) The apparatus itself. (b) The $h \propto \dfrac{1}{r}$ shape of the hill.

- Metal spheres (ball bearings) are rolled down the ramp. The starting height of the sphere determines the sphere's speed as it leaves the ramp—and therefore its kinetic energy.

- Design an experiment that either varies the initial energy of the sphere or the distance between the initial direction of the sphere and the centre of the hill (this is called the "offset"). The meaning of the offset is shown in the plan view of Figure 9.

- You can use the apparatus qualitatively to find out how different initial kinetic energies or different offsets affect the deflection angle.

- Or you can devise a quantitative experiment to check elements of Rutherford's equation.

The equation from the data-based question

$(N \propto \dfrac{1}{\sin^4\left(\dfrac{\phi}{2}\right)})$ is a good place to start.

- You might then investigate how the number scattered at a particular angle varies with the initial kinetic energy.

- Think carefully about the statistics of the experiment and how you will treat the experimental errors.

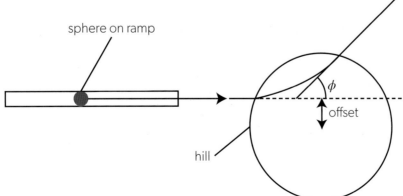

sphere on ramp

ϕ

offset

hill

▲ Figure 9 The apparatus seen from above. The offset is the distance from the centre of the circle to the line along which the sphere was rolling before it met the hill.

Theories

If you have already studied Themes A and D, you may recognize some of the terms in Rutherford's equation. His full equation is

fraction of the alpha particles detected at an angle ϕ

$$= \left(\frac{kq_\alpha Q_{gold}}{R}\right)^2 \times nt \times \frac{1}{\left(\frac{1}{2}mu^2\right)^2} \times \frac{\text{detector area}}{\left(2\sin\left(\frac{\phi}{2}\right)\right)^4}$$

where q_α = charge on the alpha particle, Q_{gold} = charge on the nucleus, n = number of nuclei per unit volume, t = foil thickness, m = mass of an alpha particle, u = initial speed of alpha particle and R = distance between the foil and the detector.

The terms in the equation have been grouped to help you to recognize the factors that determine the deflection of the alpha particles:

- The first term you should recognize as containing elements of the electric-potential equation.
- The second is related to the density of the nuclei.
- The third contains the reciprocal of the initial kinetic energy of the alpha particles.
- The fourth is a term relating to the geometry of the apparatus.

You can see this whole equation in action in a number of simulations available on the internet. Try searching using "alpha particle scattering simulation" or "Rutherford scattering".

Alternatively, you can construct your own model of the scattering system.

Data-based questions

Rutherford's equation predicted that the number N of scattered alpha particles in a given time would vary with angle ϕ according to the relationship $N \propto \dfrac{1}{\sin^4\left(\dfrac{\phi}{2}\right)}$.

Geiger and Marsden's data [H. Geiger and E. Marsden, "The Laws of Deflexion of α Particles Through Large Angles", *Philosophical Magazine*, **25**, 604–623 (1913)] are given in the table.

N, the number of alpha particles observed in a given time, is a discrete quantity.

- Suggest one way in which it is possible to have experimental data which have fractional values for discrete data.
- Make a copy of this table and add columns for $\sin\dfrac{\phi}{2}$, $\log\left(\sin\dfrac{\phi}{2}\right)$ and $\log N$. Calculate these values and record them in your table.
- Plot a graph of $\log N$ against $\log\left(\sin\dfrac{\phi}{2}\right)$.
- Explain which features of this graph support the relationship $N \propto \dfrac{1}{\sin^4\left(\dfrac{\phi}{2}\right)}$.

See page 356 for more on the interpretation of log–log graphs. If you are less confident with the use of logs, try plotting a graph of N against $\dfrac{1}{\sin^4(\phi)}$ and assess whether the resulting graph has a straight line of best fit.

Angle (ϕ)	N
150	33.1
135	43.0
120	51.9
105	69.5
75	211
60	477
45	1435
37.5	3300
30	7800
22.5	27 300
15	132 000

How is the distance of closest approach calculated using conservation of energy?

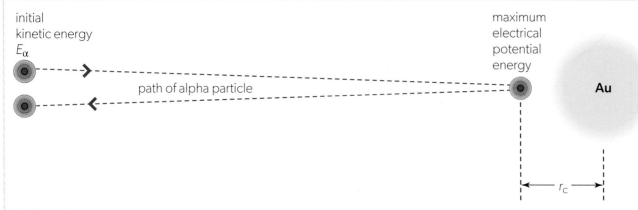

▲ Figure 10 An alpha particle on a head-on collision course with a gold nucleus. The incident and return paths are shown at small angles for clarity.

The data from the Geiger–Marsden–Rutherford experiment can give an estimate for the diameter of a gold nucleus. Figure 10 shows what happens when an alpha particle is scattered and exactly reverses its path. It must originally have been heading straight for the gold nucleus (in other words, on a head-on collision course).

As the alpha particle approaches the nucleus, the initial kinetic energy E_α gradually transfers to electrical potential energy. When the alpha particle is at its closest point to the nucleus, it stops moving just for an instant. Its kinetic energy is zero here. This is the distance of closest approach r_c and we take r_c as the estimate of the upper limit of the nuclear diameter. The positive alpha particle is then repelled by the positive nucleus and reverses its direction.

When the alpha particle has come to rest, all of its initial energy E_α has been transferred to electric potential energy and therefore

$$E_\alpha = \frac{1}{2}m_\alpha v_\alpha^2 = k\frac{q_\alpha Q_{gold}}{r_c}$$

where r_c is the closest approach between alpha particle and nucleus.

When E_α of the alpha particle is equated to the electric potential energy at r_c,

$$E_\alpha = \frac{kZe \times 2e}{r_c}$$

where k is the Coulomb constant, Z is the number of protons in the gold nucleus (so that the positive charge of the nucleus is Ze) and $2e$ is the charge of the alpha particle. You should be able to recognize the origin of this equation from Topic D.2. When the equation is rearranged to make r_c the subject,

$$r_c = \frac{2kZe^2}{E_\alpha}$$

The data for the Geiger–Marsden–Rutherford experiment are $E_\alpha = 7.68\,\text{MeV}$ and $Z = 79$.

So $r_c = \dfrac{2 \times 9.0 \times 10^9 \times 79 \times (1.6 \times 10^{-19})^2}{(7.68 \times 10^6 \times 1.6 \times 10^{-19})}$, which leads to

$r_c = 3.0 \times 10^{-14}\,\text{m}$.

Rutherford's theory makes several approximations, which means that it can only ever be a rough estimate of r_c.

- The gold nucleus is regarded as a point mass that does not move (in fact, it must be repelled by the approaching alpha particle—this is called "recoil").

- We assume that the alpha particle does not penetrate the nucleus. At small values of r_c an alpha particle is influenced by other nuclear forces not just electrostatic repulsion. Research since Rutherford's time shows that an attractive strong nuclear interaction operates at small distance within the nucleus. This complicates the analysis. Because this strong nuclear force is attractive, it reduces the effect of the electrostatic repulsion. There is more detail about the strong nuclear interaction in Topic E.3.

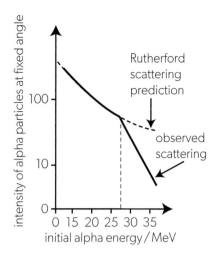

▲ **Figure 11** Rutherford scattering breaks down at initial alpha energies greater than 28 MeV. The alpha particles are now influenced by the strong nuclear interaction.

▲ **Figure 12** The double star system X-7 contains a neutron star thought to have a mass that is 1.4 times the mass of the Sun $(1.4\,M_\odot)$ and a radius of 11 km. This large density is the result of the gravitational collapse of a massive star. An apple made of nuclear matter would have a mass of about 10^{14} kg. This is about 1000 times larger than the total mass of all the humans alive today.

Deviations from Rutherford scattering

At high initial kinetic energies E_α of the alpha particle, the closest approach distance r_c becomes small. A large E_α will lead to an impossibly small nuclear diameter. However, when very energetic alpha particles approach the nucleus, there comes a point where the scattering no longer obeys the assumptions Rutherford made. The observations of this behaviour were made in 1961 by Eisberg and Porter who used a lead-208 target. They showed that, for alpha particles of initial energy greater than about 28 MeV, the predictions of the Rutherford equation were no longer maintained.

You should think of E_α as being the largest energy for which the scattering still obeys the Rutherford model. Then the estimate gives the smallest r_c at which other nuclear forces do not operate. This is the effective size of a nucleus.

Nuclear density

We now know that the total number of nucleons inside the nucleus includes protons and neutrons. Rutherford and his co-workers knew about protons, but it was not until 1932 that the neutron was finally identified. This took many years because the neutron is uncharged and most research techniques early in the 20th century relied on the effects of charge and the ionization it causes to make measurements.

It is possible to arrive at a simple equation that relates the radius R of a nucleus to the total number A of nucleons inside it, the nucleon number. The nuclear volume V must be proportional to the total number of nucleons inside it:

$$V \propto A$$

When we assume that the nucleus is spherical, then $V = \frac{4}{3}\pi R^3$ as usual and $R^3 \propto A$. This is more usually given as $R \propto A^{\frac{1}{3}}$ and it leads to the equation

$$R = R_0 A^{\frac{1}{3}}$$

R_0 is called the **Fermi radius** and it must be measured experimentally. Its value is $R_0 \approx 1.2 \times 10^{-15}$ m.

This analysis can go one step further because the earlier equation for the nuclear volume V can be rewritten as $V = \frac{4}{3}\pi R^3 = \frac{4}{3}\pi A R_0^3$. This leads to an estimate of the density of the nucleus:

$$\rho_n = \frac{\text{mass of the nucleus}}{\text{volume of the nucleus}} = \frac{A \times u}{\frac{4}{3}\pi A R_0^3} = \frac{3u}{4\pi R_0^3}$$

The quantity u in this equation is known as the **unified atomic mass unit** and is roughly equal to the mass of a neutron or a proton. (They are not quite the same.)

1 u is equivalent to 1.661×10^{-27} kg.

These data give an estimate for nuclear density of $\rho_n = \dfrac{3 \times 1.7 \times 10^{-27}}{4\pi \times (1.2 \times 10^{-15})^3}$ which equals 2.4×10^{17} kg m^{-3}. This density is the same for all nuclides since this was the original assumption. The interactions between nucleons are repulsive at short distances so that the nucleons cannot penetrate each other.

Modelling the variation in R with A

- Tool 3: Carry out calculations involving exponents.
- Tool 3: Construct and interpret tables and graphs for raw and processed data including scatter graphs and line and curve graphs.
- Tool 3: Draw and interpret uncertainty bars.
- Inquiry 3: Relate the outcomes of an investigation to the stated research question or hypothesis.

Experiments to estimate nuclear size normally involve firing X-rays or neutrons at a sample. You can use a model to investigate the relationship $R = R_0A^{\frac{1}{3}}$.

You will need some plasticine, a balance with a precision of 0.1 g and some callipers.

- Make about ten small plasticine balls, each of diameter 0.5–1 cm. The balls should be approximately the same mass. Use the balance to help with this. These balls represent the nucleons (protons and neutrons).
- Mould one plasticine ball into a sphere. Measure the diameter of the sphere using the callipers. You should

repeat your measurement so that you have three measurements taken in different orientations since your sphere is unlikely to be perfectly spherical. Take an average of your readings.

- Take a second ball and mould it, together with the first piece, into a new, larger, sphere. Measure the diameter of this sphere as before.
- Continue adding the balls of plasticine and record your results in a table.
- Convert your measured values of the diameter into a radius and find the uncertainty in the radius.
- Add a column to your table for R^3 and the uncertainty in that value.
- Plot a graph of A (number of plasticine balls) against R^3. Include error bars on the graph.
- If your data obey the rule $R = R_0A^{\frac{1}{3}}$, then they should give a straight-line graph. Use your graph to obtain a value of R_0. To what extent is this a good model for a nucleus?

Worked example 2

A nucleus of silver-107 ($^{107}_{47}$Ag) has a charge of +47e.

a. Estimate the nuclear radius of Ag-107.

A beam of alpha particles is directed at an Ag-107 target. The number of alpha particles scattered at very large angles deviates from the Rutherford model at initial energies greater than E_α. The radius of an alpha particle is approximately 1.9 fm.

b. Estimate E_α.

Solutions

a. $R = R_0A^{\frac{1}{3}} = 1.2 \times 10^{-15} \times 107^{\frac{1}{3}} = 5.7$ fm

b. The strong nuclear force affects the results of the scattering when the distance between the particles becomes about equal to the sum of their radii, $5.7 + 1.9 = 7.6$ fm. Assuming head-on approach, the initial kinetic energy of an alpha particle must be equal or greater than the electric potential energy of the particles at a 7.6 fm separation; hence

$$E_\alpha = \frac{8.99 \times 10^9 \times 2 \times 47 \times (1.60 \times 10^{-19})^2}{7.6 \times 10^{-15}} = 2.8 \times 10^{-12} \text{ J. This is equivalent to 18 MeV.}$$

Practice questions

1. The diameter of a nucleus of nucleon number 16 is D. Estimate the nucleon number of a nucleus whose diameter is 2D.

2. Alpha particles are scattered by bismuth-209 ($^{209}_{83}$Bi) nuclei.

a. Calculate the distance of closest approach between an alpha particle of initial kinetic energy of 20 MeV and a bismuth nucleus.

b. Estimate the minimum initial energy of alpha particles so that the distribution of scattered particles deviates from the Rutherford model.

▲ Figure 13 The bottom spectrum is the continuous spectrum of the Sun crossed by three dark absorption lines (an absorption spectrum). The position of these lines corresponds to bright lines in the hydrogen emission spectrum (above).

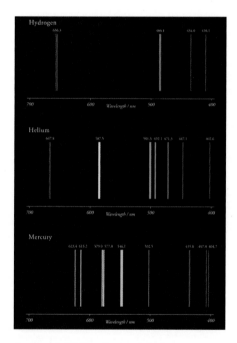

▲ Figure 14 The line spectra for hydrogen (H), helium (He) and mercury (Hg).

Emission and absorption spectra

By the early 19th century, scientists understood that sunlight had a continuous spectrum (Figure 13), but the reasons for it were unknown.

In 1802, William Wollaston observed some strange dark lines in the Sun's spectrum. But it took a further 12 years before Joseph von Fraunhofer invented the diffraction grating and was able to observe these dark lines in detail (see Topic C.3 page 432). We now know that his discovery, **Fraunhofer lines**, is an **absorption spectrum**. We shall see how these lines are formed later.

The technique for observing spectra experimentally is to produce light by passing an electric current through a low-pressure gas or by heating a substance so that it produces a flame. The atoms then emit electromagnetic radiation. When this radiation is incident on a diffraction grating or a prism, it becomes diffracted by the grating (or dispersed by the prism) into its separate wavelengths. These leave the grating or prism at different angles to the original direction of the radiation. They can then be observed using a telescope or by allowing the light to fall on a screen.

The continuous spectrum of the Sun is caused by the interactions of all the atoms in the hot dense gas at the Sun's visible surface. As the pressure of a gas decreases, however, the visual appearance of the spectrum changes dramatically. The individual atoms interact less and less as the pressure falls, until at low pressures (close to a vacuum) the spectrum for a single element becomes a series of separate (discrete) lines.

This low-pressure spectrum is known as an **emission line spectrum** and is characteristic of the element that is producing it. Figure 14 shows the line spectrum for three different elements. The colours and arrangements of the lines are different and unique for each element.

The line spectra extend beyond the ranges of visible light. For hydrogen, an atom that consists of one nuclear proton and a single electron outside the nucleus, there are six series of lines that have been extensively studied, including one in the ultraviolet, one in the visible region and two in the infrared.

We now know that the **emission spectra** are produced by atoms that emit photons of light during energy changes when one excited atom moves to a lower energy state. This electromagnetic radiation arrives as a discrete energy packet (properly called a **quantum** of energy, the plural is quanta). The energy packet itself is now called a **photon**.

The energy E of the photon is linked to its frequency f by

$$E = hf$$

This energy is also equal to the difference in energy level between the two atomic energy states before and after the atomic transition. The exact energy difference in atomic energy states appears as the energy of the photon.

The constant h is the Planck constant which "converts" frequencies and wavelengths into their energy equivalent. Planck had already shown that h has the value 6.63×10^{-34} J s.

The equation can also be written in terms of wavelength (using the equation $c = f\lambda$, where c is the speed of electromagnetic waves in a vacuum) as

$$E = \frac{hc}{\lambda}$$

The atoms in solids, liquids and high-pressure gases are also excited through the energies available at suitable temperatures. But the emitted light observed under such conditions consists of bands of colours rather than lines. In solids, these bands themselves merge to give a continuous spectrum with no single colours at all (Figure 16). This is typical of matter in which the atoms are closely packed. Each atom is modifying the energy states in nearby atoms. For a large group of atoms, the overall energy levels combine to form a series of similar "smeared-out" energy values. This gives an energy band rather than a set of lines.

▲ Figure 15 The 1868 solar eclipse. Astronomers analysed the spectrum of light from the Sun during the eclipse. They discovered a bright spectral line at 587.5 nm which could not be accounted for from the spectral lines of known elements. They proposed that it must be a new element and called this element helium (after *helios*, the ancient Greek word for Sun). Fourteen years later, helium was discovered on Earth.

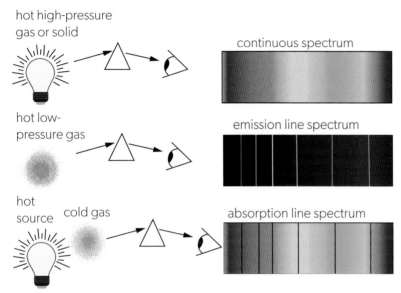

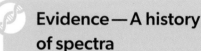

▲ Figure 16 The formation of continuous, emission-line and absorption spectra.

Absorption spectra form in a similar way. A hot object, whether solid, liquid or gas, emits a continuous spectrum. When the hot object is surrounded by a cooler gas, the radiation from the hot object must pass through the cooler gas to be observed. The original continuous spectrum is now seen to be crossed by several dark lines. When a heated tungsten filament is viewed through hydrogen gas, the absorption spectrum shown in the bottom diagram of Figure 16 can be seen. These black lines occur at the precise positions of the lines of the hydrogen emission spectrum.

The absorption occurs when an atom of the cooler gas absorbs a photon of energy identical to the difference between two of its energy levels. This photon promotes an electron from the lower to the higher of the two energy states. The cooler gas therefore removes photons with this energy difference from the continuous range of energies emitted by the light source.

The absorbing atom is now in a higher energy state; this makes it unstable. It reverts to the lower energy level by photon emission. This photon has the original absorbed frequency. However, it is emitted in a random direction, not necessarily in the direction of the original photon. This reduces the intensity of those specific emission-line frequencies in the original direction. The black lines that appear in the spectrum contrast with the higher intensity of non-absorbed photon frequencies.

Evidence — A history of spectra

Observations of atomic spectra began well before the 19th century. A knowledge of optics had existed since Roman times without any break in Middle Eastern countries. However, it was rediscovered in Western Europe around the 16th century. This allowed Europeans of the time to study the light emitted from hot or glowing objects.

Scientists used an optical device, usually a glass prism for these early experiments, to spread white light from the Sun from the shortest wavelengths (violet) to the longest (red). The same order as in a rainbow. Even today, spectra are presented in this linear way. Spectra in this Topic are displayed with the lines coloured correctly and with the colours arranged by wavelength in their correct place.

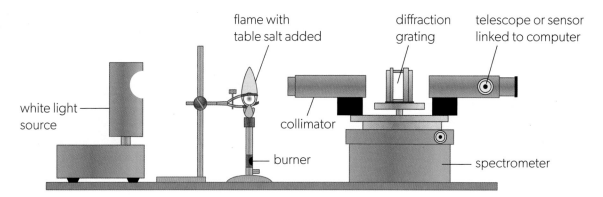

▲ Figure 17 Observing an absorption spectrum.

Absorption spectra for sodium can be demonstrated with the same apparatus as for emission (Figure 17). A white light source emits light which is incident on a diffraction grating on the turntable of a spectrometer. A continuous spectrum can be seen through the telescope or displayed on a computer monitor (using a sensor and software). When a flame is used to heat a vertical wire that has been dipped in sodium chloride, the light from the continuous spectrum is allowed to go through the sodium ions in the flame. The sodium absorption lines are seen in the yellow region of the continuous spectrum.

The observation of both emission and absorption spectra is an effective way to determine the chemical composition of a material. Chemists can quickly identify an anonymous white crystal. By placing the crystal on an inert wire such as platinum, and then putting the crystal and wire in a flame, the colour of the flame gives good indication of the elements in the crystal. For example, brick red means cadmium and green means the presence of copper ions.

A more detailed analysis of the lines in the spectrum also gives a definitive answer to the question of the composition of a material. Figure 14 shows three sets of spectra. You can see that the lines are at completely different wavelengths and in completely different arrangements.

Measurements — Electronvolts

The photon energies and atomic energy level differences are often given in units of electronvolt ($1 \text{ eV} = 1.6 \times 10^{-19}$ J). This is a common unit used in atomic physics (along with the multiplies keV and MeV). It avoids having to use large negative powers of ten. You can find a definition of the electronvolt on page 519.

The IB Physics Data Booklet contains a unit conversion for the product of the Planck constant and the speed of light in a vacuum: $(h \times c) = 1.99 \times 10^{-25}$ J m $= 1.24 \times 10^{-6}$ eV m. You can use this to make the conversion from a change in energy level for an atom to the wavelength of the photon emitted or absorbed during the change.

The change in energy ΔE is equal to hf, where f is the frequency of the photon.

This leads to $\Delta E = hf = h\left(\dfrac{c}{\lambda}\right)$; $\Delta E = hc \times \dfrac{1}{\lambda}$.

When you know the difference between two energy levels in an atom you can use the ΔE expression rearranged as $\lambda = hc \times \dfrac{1}{\Delta E}$. For example, when the atom undergoes a transition between energy levels that differ by 1.88 eV, the wavelength of the emitted photon is

$\lambda = 1.24 \times 10^{-6} \times \dfrac{1}{1.88} = 660 \text{ nm}$.

How can emission spectra allow the properties of stars to be deduced?

How can emission spectra be used to calculate the distances and velocities of celestial bodies?

Our present knowledge of the properties of stars would not be possible without observations of stellar spectra. They reveal the size and nature of the stars themselves (Topic B.1). The peak of the continuous spectrum (equivalent to that from a black body to a good approximation) gives the temperature of the star. This, together with the luminosity of the star (which can be deduced from its apparent brightness and distance), leads to an estimate of the radius of the star (Topic E.5).

Redshifts in the spectra lead to the speeds of stars and galaxies relative to us (Topic C.5) and also rely on precise measurements of spectral wavelengths. The fractional shift in the wavelength is proportional to the relative velocity of a galaxy along the line joining it to Earth.

Worked example 3

a. Outline, with reference to atomic energy levels, how a discrete emission spectrum of a gas is formed.

b. The emission spectrum of hydrogen contains an infrared line at 1282 nm.

   Calculate, in eV, the energy of a photon of wavelength 1282 nm.

Solutions

a. Gas atoms in an excited state move to a lower energy state. Because energy levels of atoms are discrete, photons emitted by the gas can only have discrete energies, equal to differences between atomic energy levels. The photon wavelength λ is related to energy by the formula $\lambda = \dfrac{hc}{E}$. Hence wavelengths in the emission spectrum are also discrete.

b. $E = \dfrac{hc}{\lambda} = \dfrac{1.24 \times 10^{-6}}{1282 \times 10^{-9}} = 0.967\,\text{eV}$

Practice questions

3. The diagram shows some of the energy levels of an atom.

   −1.50 eV ──────────────

   −2.30 eV ──────────────

   −3.90 eV ──────────────

   −6.30 eV ──────────────

   a. Calculate:

      i. the longest wavelength

      ii. the shortest wavelength

      of a photon whose absorption by the atom can result in a transition between two of these energy levels.

   b. Identify the atomic transition that results in an emission of a photon of energy 4.00 eV by this atom.

4. The emission spectrum of sodium contains two lines of similar wavelengths: 588.995 nm and 589.592 nm. The lines are emitted in a transition from one of two narrowly separated levels E_a and E_b to another energy level E_c, as shown in the diagram.

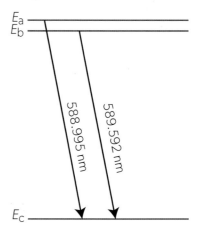

   a. Calculate the energy of each of the photons emitted by a sodium atom.

   b. Hence, calculate the energy difference between levels E_a and E_b.

Remember that an accelerating charge is equivalent to a changing current (Topic B.5), so an accelerating charge must radiate. See Topic D.4 for how a changing current interacts with its own magnetic field through self-induction and why Lenz's law states that the charge must lose energy (in this case, by radiating photons).

▲ Figure 18 Particle accelerators such as this picture of the ALBA synchrotron in Spain cause charged particles to move in circles. They must be accelerated and therefore emit electromagnetic radiation called synchrotron radiation.

Taking Rutherford's model further

The Rutherford model of the atom left physics with a problem: Hydrogen has one proton and one electron outside the nucleus. However, 19th-century classical physics predicted that this situation was unstable.

The argument goes like this. The electron is attracted to the proton and orbits around it. The electron must therefore have a centripetal acceleration (Topic A.2). Charges that are accelerating emit energy in the form of electromagnetic radiation, but a constant radiation from hydrogen is not observed. Further, as the electron transfers energy, it must spiral into the nucleus (according to classical physics) taking a fraction of a second to do so. This means that any hydrogen atom can only exist for a fraction of a second. If hydrogen were unstable in this way, then the very existence of the universe would be in doubt.

However, six years after the work of Rutherford, Danish physicist Niels Bohr realized that the idea of a single electron moving in an unrestricted way around the nucleus was not correct. His breakthrough came when he blended the experimental evidence from atomic spectra with Rutherford's theoretical model. Bohr's thinking about an atomic model began with his knowledge of some 19th-century observations made by Johan Jakob Balmer and others of the spectra emitted by hot gases.

How have observations led to developments in the model of the atom? (NOS)

When a scientist works with patterns of numbers rather than scientific observations, the process is sometimes called numerology. However, this does not necessarily mean that the process is unscientific.

Balmer was a numerologist who believed that numbers alone hold the key to understanding science. The spectral sequence is named in his honour, not because he identified the lines or the reason for them, but because he provided a numerical base on which others could build. Nevertheless, it was the careful observations of the wavelengths of the spectral lines that was the beginning of the search for a model of the atom.

In a similar way, it was observation of the scattered alpha particles by Geiger and Marsden that led Rutherford to suggest a replacement model for the Thomson atom.

▲ Figure 19 This image is from *Harmonices Mundi* (Harmony of the Worlds, 1619) written by Kepler from about 1599 onwards. He makes parallels in the book between planetary orbital speeds and music. This is numerology. The third part of the book contains Kepler's third law of orbital motion (Topic D.1).

Balmer knew of the existence of four lines in the visible region of hydrogen (the ones shown at the top of Figure 14). The wavelengths of these lines had been measured accurately by Ångström. Ångström's four wavelength values are given in Table 1. They are labelled with the chemical symbol (H) and a Greek letter subscript which is used to order the lines in decreasing wavelength.

Balmer line	Colour	λ / nm
H_α	red	656.3
H_β	blue	486.1
H_γ	violet	434.0
H_δ	violet	410.2

▲ Table 1 The wavelengths λ and colours of different lines in the Balmer series.

Balmer eventually found a numerical pattern that connected the four wavelengths together. It was

$$\text{wavelength of the line} / m = 3.6456 \times 10^{-7} \times \frac{m^2}{m^2 - 2^2}$$

where m is an integer between 3 and 6. It is important to remember that this formula is **empirical**. It does not arise from a scientific hypothesis. Balmer did not attempt to explain the science that underpins his formula.

By 1888, the Swedish physicist Johannes Rydberg had realised that the Balmer formula was a special case of an equation he was investigating. Rydberg's formula was also empirical. He had arrived at it through work on the spectra of alkali metals such as sodium and potassium. When Rydberg's formula is applied to hydrogen, it takes the form

$$\frac{1}{\lambda} = R_H \left(\frac{1}{n_1^2} - \frac{1}{n_2^2} \right)$$

where λ is the wavelength of the spectrum line, n_1 and n_2 are integers and R_H is known as the Rydberg constant, which, for hydrogen, takes the empirical value $1.097 \times 10^7 \text{ m}^{-1}$. The Balmer and Rydberg formulas give the same answer.

- For the Balmer formula, when m is 3,

$$\text{wavelength of the line} = 3.6456 \times 10^{-7} \times \frac{3^2}{3^2 - 2^2}$$

$$= 3.6456 \times 10^{-7} \times \frac{9}{9 - 4} = 656 \, nm.$$

- For the Rydberg formula, when $n_1 = 2$ and $n_2 = 3$,

$$\frac{1}{\lambda} = 1.097 \times 10^7 \times \left(\frac{1}{2^2} - \frac{1}{3^2} \right) = 1.097 \times 10^7 \times \left(\frac{1}{4} - \frac{1}{9} \right) = 1.52 \times 10^6 \, m^{-1}.$$

This means that λ is $\dfrac{1}{1.52 \times 10^6} = 656 \, nm$.

However, this gets us no closer to a physical understanding of what the numbers m, n_1 and n_2 in the formulas mean — nor to the physics that is represented by them. This had to wait until the ground-breaking work of Niels Bohr in 1913. His insight allowed him to provide a new interpretation of the physics of the atom.

By 1908, more series of line spectra for hydrogen had been discovered, first in the infrared (the Paschen series), and then in the ultraviolet (the Lyman series). These also obeyed the Rydberg formula but in a way that depends on n_2 in the formula.

For the Lyman series, $\quad \dfrac{1}{\lambda} = R_H \left(\dfrac{1}{1^2} - \dfrac{1}{n_2^2} \right)$ and n_2 must be 2 or greater

For the Balmer series, $\quad \dfrac{1}{\lambda} = R_H \left(\dfrac{1}{2^2} - \dfrac{1}{n_2^2} \right)$ and n_2 must be 3 or greater

For the Paschen series, $\quad \dfrac{1}{\lambda} = R_H \left(\dfrac{1}{3^2} - \dfrac{1}{n_2^2} \right)$ and n_2 must be 4 or greater.

Niels Bohr took these empirical formulas and combined them with earlier work by Planck and Einstein that you meet in more detail in Topic E.2.

The Bohr model

Bohr realized that the Rydberg formula tells us about energies, not just wavelengths. He also realized that the similarities between the formulas for different sets of lines imply an underlying pattern for the single electron in the hydrogen atom. To go further, he had to make four assumptions.

• Electrons can only be in certain discrete, circular orbits. He called these stationary states. ("Stationary" here is related to the idea that the electron is fixed in one energy level, not that it is at rest. We shall see in Topic E.2 that it is wrong to think of an electron at an exact position or moving with a precise speed. The word "stationary" was used by Bohr in the sense of a stationary wave rather than a stationary object.)

• Atoms cannot transfer radiation while in a stationary state.

• Atoms gain or lose energy when they transfer between one stationary state and another.

• The angular momentum of an electron in a stationary state is quantized in integer values of $\frac{h}{2\pi}$.

We now know that Bohr's idea of electrons orbiting a nucleus is not correct (see Topic E.2), but it is a helpful picture for someone beginning a study of the hydrogen atom.

Bohr worked backwards from the results of the line spectra wavelengths (energy changes) and discovered that all the Rydberg formulas can be explained using a single energy-level diagram for hydrogen (Figure 20). The hydrogen atom can exist in any one of these energy states (or levels) and can move between the states. It cannot, however, have any energy other than these values. In other words, the electron–proton system can have an energy of −13.58 eV or −3.39 eV, but it cannot have an energy of −4.52 eV and still be a hydrogen atom.

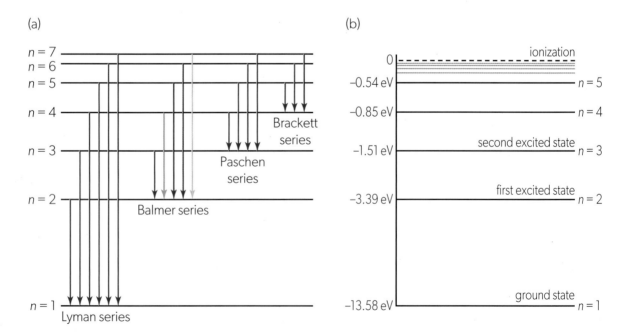

▲ Figure 20 (a) The energy transitions that lead to four of the spectral series in hydrogen. (b) The energy values that correspond to the energy states in hydrogen.

The lowest energy level (−13.58 eV) is known as the ground state and is the lowest energy possible for the hydrogen atom. In this state, the electron is as tightly bound to the atom as it is possible for it to be. This is a **bound state** where proton and electron are locked together.

When the atom has an energy greater than the ground-state value, it is said to be **excited**. The level marked $n = 2$ is the first excited state, $n = 3$ is the second excited state and so on.

The energy levels are **quantized**, which means that they can only take discrete and finite values. For hydrogen, the levels can only have the values given in Table 2 and Figure 20(b).

Within the atom, electrons can only move between levels when an exact amount of energy is transferred to them or away from them. Each line in the spectrum indicates one of the possible energy changes for the atom. Figure 20(a) shows the possible electron transitions between levels and how these link to the various series of spectral lines. (This diagram also includes a further series of lines discovered by Brackett in the far infrared.)

The symbol n that labels the energy levels is known as the **principal quantum number**.

The level that is marked with zero energy represents the state when the electron has left the atom so that the atom is **ionized**. Here the electron and proton are unbound (meaning "not connected"). As the **ionization level** corresponds to an energy value of 0, it follows that every bound state must take a *negative* energy value because atoms in a bound state require energy to be transferred *to* them to become unbound.

Energy level (n)	Energy / eV
1	−13.58
2	−3.39
3	−1.51
4	−0.85
5	−0.54

▲ Table 2 The energy of each energy level.

Data-based questions

The wavelengths of the hydrogen spectrum obey the formula $\frac{1}{\lambda} = R_H \left(\frac{1}{k^2} - \frac{1}{n^2} \right)$, where n is the quantum number of the

upper level, k is the quantum number of the lower level and $R_H = 1.097 \times 10^7 \, \text{m}^{-1}$.

The table gives the wavelengths of some of the lines in the hydrogen spectrum together with n.

Upper level n	Wavelength / nm
3	656.5
4	486.3
5	434.2
6	410.3
7	397.1
2	121.6
3	102.6
4	97.3
5	95.0
6	93.8
7	93.1

- Add two columns to a copy of the table to include values of $\frac{1}{n^2}$ and $\frac{1}{\lambda}$.

- Use these to plot a graph of $\frac{1}{\lambda}$ (y-axis) against $\frac{1}{n^2}$ (x-axis).

- You should see that the data form two linear trends with the same gradient. Find this gradient.

- What is the significance of the x-intercept of each linear trend?

- There is also a spectral line which arises from an upper quantum level $n = 7$ and has a wavelength of 1005.2 nm. Add a point on your graph to represent this spectral line. By assuming that it forms part of a linear trend with the same gradient as the other two trends, deduce the value of the lower quantum number (k) for this transition.

Why are bound states so significant in science?

The language used to describe the atom is important. The energy level relates to the whole atom: the combination of proton and electron in the case of hydrogen. It is better to write about the energy changes and transitions of the *atom* rather than energy changes of the *electron*.

You meet a number of bound states in the IB Diploma Programme physics course. A satellite in orbit around a planet in Topic D.1 is in a bound state. Its gravitational potential energy and its mechanical energy are negative like the energy state of the hydrogen atoms discussed here. Binary stars are bound together in the same way. In Topic D.2, the zero of electric potential is at infinity so that

a system consisting of a positive and negative charge has a negative electric potential.

Although the term is rarely used in this context, the energy state of a nucleus (Topics E.4 and E.5) also represents a bound state because energy must be added to the nucleus to separate the individual nucleons (to infinity). The terms used for this are mass defect and binding energy.

It is important to have a clear understanding of the scalar nature of energy and what it means for the energy of a system to be negative.

Quantifying the Bohr model

The connection between the Bohr model and the historical empirical work on emission spectra is clear. The Lyman series is caused by energy transitions in which the atom moves to its ground state. The Balmer and Paschen series are caused by transitions to lower states $n = 2$ and $n = 3$, respectively. The Brackett series are transitions to $n = 4$ from higher n. Look closely at the Rydberg and the Balmer formulas and you will see how they link to the Bohr model.

The Bohr approach produces an equation that agrees with the Rydberg equation. By considering the total kinetic and electric potential energy of the hydrogen atom, Bohr was able to show that the total energy E (measured in electronvolt) for an electron in the nth energy level is given by

$$E = -\frac{13.6}{n^2}$$

This equation is derived from electric field theory and quantum mechanics, but it agrees very closely with the Rydberg empirical results.

The fourth assumption made in the Bohr model was that:

- the angular momentum of an electron in a stationary state is quantized in integer values of $\frac{h}{2\pi}$.

You met the concept of angular momentum in Topic A.4. Angular momentum is the (vector) product of the moment of inertia I of a particle and the angular speed ω of its orbit. For a particle of mass m in a circular orbit without any external torque acting, the magnitude of the angular momentum will be constant and equal to $mr^2\omega$, which is also equal to mvr (because $v = r\omega$).

Work by de Broglie (explained in more detail in Topic E.2) led to the recognition that particles have wave-like properties and have a de Broglie wavelength λ associated with them. The de Broglie equation links λ to the momentum p of the particle by $\lambda = \frac{h}{p}$. Once again, h is the Planck constant.

This means that for a particle of mass m moving with a speed v, the de Broglie wavelength is

$$\lambda = \frac{h}{mv} \quad \text{(because } p = mv\text{)}$$

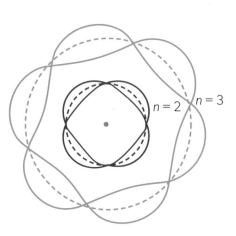

▲ Figure 21 The "standing waves" of the electron orbitals.

The implication of the fourth Bohr assumption is that the electron orbit can be matched to an integer number of de Broglie wavelengths. Another way to put this is that the shape of the de Broglie wave must "join up" when it has gone once around the orbital. This is shown in Figure 21 for the $n = 2$ and $n = 3$ orbitals. Count the wave peaks carefully and you will see that for $n = 2$ there are two complete wavelengths and for $n = 3$ there are three complete wavelengths.

For $n = 2$ the wavelength of each wave for the complete orbital must be $\dfrac{\text{circumference of the orbital}}{2}$

and for $n = 3$ the wavelength is $\dfrac{\text{circumference of the orbital}}{3}$.

Given that the circumference of the orbital is $2\pi r$, where r is the orbital radius, then for the general (nth) orbital $\lambda = \dfrac{2\pi r}{n}$. Combining the two expressions for λ

gives $\lambda = \dfrac{h}{mv} = \dfrac{2\pi r}{n}$.

Rearranging the equation leads to

$$mvr = \frac{nh}{2\pi}$$

where mvr is the angular momentum of the electron.

The Bohr assumption that angular momentum must be quantized:

- leads to the suggestion that the orbitals must have the nature of standing waves, which led scientists to the next development in our understanding of matter at the smallest scales

- confirms the quantization of energy for this model (and for those that subsequently arose from it).

As a detailed and quantitative example of how this all fits together, look at the energy transition from the hydrogen ground state ($n = 1$) to the first excited state ($n = 2$).

Table 2 showed the difference in energy between $n = 1$ and $n = 2$. It is

(13.58 eV − 3.39 eV), which is 10.19 eV. This energy change (using $\lambda = \dfrac{hc}{E}$

$= \dfrac{6.63 \times 10^{-34} \times 3.00 \times 10^{8}}{10.19 \times 1.6 \times 10^{-19}} = 1.2 \times 10^{-7}$ m) is equivalent to a wavelength of

120 nm. This is in the ultraviolet spectrum and was one of the lines measured by Lyman.

To sum up, when a photon of wavelength equal to 120 nm (or of frequency

$f = \dfrac{c}{\lambda} = \dfrac{3.0 \times 10^{8}}{1.2 \times 10^{-7}} = 2.5 \times 10^{15}$ Hz) interacts with a hydrogen atom, the energy

state of the atom increases by exactly the right amount to promote the electron from the ground state to the $n = 2$ level. This makes the atom unstable so that it quickly returns (within a time of about 1 ns) to the $n = 1$ (ground) state. To do this, the atom must transfer exactly the same amount of energy as it absorbed. It does so by emitting a photon. This photon must again have an exact wavelength of 120 nm, which means that the photon frequency is 2.5 PHz.

▲ Figure 22 Some objects fluoresce. When blue or ultraviolet light is shone on them, they glow with a different colour. This picture shows fluorescent fish and coral in an aquarium. Rocks and even some frogs can fluoresce. Photons in the blue or ultraviolet regions have large energies compared with red and yellow light. When they are absorbed by a fluorescent material, the blue photons excite orbital electrons up more than one level. The atoms fall back to their ground states in two or more stages emitting a photon for each energy-level change. For every blue or ultraviolet absorbed photon, two or more longer wavelength photons are emitted.

Theories $-\dfrac{13.6}{n^2}$ —the link between Bohr and Rydberg

Bohr's result that $E = -\dfrac{13.6}{n^2}$, where E is in electronvolts, was not empirical. He had proved it theoretically by

combining classical mechanics with his four assumptions. His *theoretical* expression matched the *empirical* results of Balmer and Rydberg.

When you have studied Themes A and D, you will understand the physics behind Bohr's result. He began by

assuming that the electron in the hydrogen atom has an

energy E that is the sum of its kinetic $\left(\dfrac{1}{2}m_e v^2\right)$ and electric

potential $\left(-\dfrac{ke^2}{r}\right)$ energies:

$$E = \dfrac{1}{2}m_e v^2 - \dfrac{ke^2}{r}$$

where the orbital radius of the electron is r and it has a mass m_e and a speed v. Remember that the proton and electron have charges of magnitude e and opposite signs leading to the negative sign in the equation.

Bohr then assumed—as for any orbiting object—that the

centripetal force $\dfrac{m_e v^2}{r}$ is supplied by the force of attraction

between proton and electron, in this case an electric

force $\dfrac{ke^2}{r^2}$:

$$\dfrac{m_e v^2}{r} = \dfrac{ke^2}{r^2}$$

This produces

$$r = \dfrac{(m_e vr)^2}{m_e ke^2}$$

and substitution into the energy equation gives

$$E = \dfrac{1}{2}\dfrac{ke^2}{r} - \dfrac{ke^2}{r} = -\dfrac{1}{2}\dfrac{ke^2}{r}$$

Using Bohr's suggestion that the angular momentum of

the electron is quantized, $m_e vr = \dfrac{nh}{2\pi}$, means that

$r = \dfrac{(m_e vr)^2}{m_e ke^2}$ becomes $r = \dfrac{n^2 h^2}{4\pi^2 m_e ke^2}$.

This equation shows that $r \propto n^2$. The constant of

proportionality is $\dfrac{h^2}{4\pi^2 m_e ke^2}$, which has the value

5.29×10^{-11} m. This constant has the units of distance and is given the symbol a_0. It is **the Bohr radius of the electron orbit in the ground state of the hydrogen atom**. For the radius of the nth orbital,

$$r = a_0 \times n^2$$

This can be used for an estimate of the hydrogen atom radius. This means that

$$E = -\dfrac{1}{2} \times k \times e^2 \times \dfrac{4\pi^2 m_e ke^2}{n^2 h^2} = -\dfrac{2\pi^2 m_e k^2 e^4}{n^2 h^2}$$

Replacing k with $\dfrac{1}{4\pi\varepsilon_0}$ leads to

$$E = -\dfrac{2\pi^2 m_e e^4}{16\pi^2 \varepsilon_0^2 n^2 h^2} = -\dfrac{m_e e^4}{8\varepsilon_0^2 n^2 h^2}$$

Bohr hypothesized that energy emission or absorption was only possible when the atom moved from one energy state to another. These states correspond to different values of n. The general case for the energy transferred when the state changes from n_1 to n_2 is

$$E = \dfrac{2\pi^2 m_e k^2 e^4}{h^2}\left(\dfrac{1}{n_1^2} - \dfrac{1}{n_2^2}\right)$$

which you will recognize as a form of the Rydberg equation.

The factor outside the brackets is

$$\dfrac{2 \times \pi^2 \times (9 \times 10^9)^2 \times 9.1 \times 10^{-31} \times (1.6 \times 10^{-19})^4}{(6.6 \times 10^{-34})^2}$$

$$= 2.19 \times 10^{-18}\,\text{J}.$$

Converting this to eV gives $\dfrac{2.19 \times 10^{-18}}{1.6 \times 10^{-19}} = 13.6\,\text{eV}-$

as predicted from theory by Bohr and observed from an experiment 20 years earlier by Balmer.

Under what circumstances does the Bohr model fail? (NOS)

Bohr went on to modify his equation so that it could accommodate other one-electron systems such as ionized helium that has lost one electron (and so is left with one electron and two protons) and a lithium ion that has lost two of its electrons. The model could not be extended to other, more complicated, atomic systems. Neither could it explain why certain allowed transitions were found to be more likely to occur than others.

The failure of the Bohr model is an example of falsification in science. It gave apparently good results for the simplest case of the hydrogen atom. More precise observations of the spectra carried out later showed that what were thought to be single lines were in fact two very closely spaced lines. This is caused by an interaction between electron spin (an effect not known to Bohr at the time) and the orbital angular momentum of the electron. The spins can be in the same direction or opposed. The two cases have slightly different energies and therefore correspond to different wavelengths in the emission spectrum.

When there are two or more electrons, they interact with each other. From a classical-physics standpoint, the presence of an additional electron in the helium atom gives this interaction. The electrons cannot occupy the same energy state and so there must be separate energies and separate spectral lines.

Despite these flaws, the Bohr model led to a more fundamental approach to the atom, now called **quantum mechanics** (mentioned in more detail in Topic E.2).

Worked example 4

a. A beam of electrons passes through a gas sample that contains hydrogen atoms in the ground state. Calculate the smallest energy that the electrons in the beam must have in order to excite hydrogen atoms to their energy level $n = 4$.

b. Determine the wavelength of the photon emitted by an atomic transition from energy level $n = 4$ to energy level $n = 2$.

c. The radius of the electron orbit in the Bohr model is a rough estimate of the radius of the hydrogen atom. Compare the volume of the hydrogen atom in the $n = 2$ state with the volume of its atomic nucleus—a single proton. The radius of a proton is approximately 0.84×10^{-15} m.

Solutions

a. When an electron in the beam interacts with a hydrogen atom, the energy loss of the electron is equal to the energy gain of the atom. The electron's energy must therefore at least be equal to the difference between the energies of the excited state and the ground state of hydrogen. This can be calculated using Bohr's formula for energy levels: $-\frac{13.6}{4^2} - \left(-\frac{13.6}{1^2}\right) = 12.8$ eV. Alternatively, the data in Table 2 allow you to find the energy difference correct to two decimal places: $-0.85 - (-13.58) = 12.73$ eV.

b. The energy of the photon is $-\frac{13.6}{4^2} - \left(-\frac{13.6}{2^2}\right) = 2.55$ eV. Rearranging the equation

$E = \frac{hc}{\lambda}$ gives $\lambda = \frac{hc}{E} = \frac{1.24 \times 10^{-6}}{2.55} = 486$ nm. This is the blue H_β line listed in Table 1, and its wavelength can be also calculated using the empirical formulas found by Balmer and Rydberg!

c. The radius of the electron orbit in the $n = 2$ state of hydrogen is $r = a_0(2)^2 = 2.12 \times 10^{-10}$ m and so the volume of the atom is $\frac{4}{3}\pi(2.12 \times 10^{-10})^3 = 4.0 \times 10^{-29}$ m³.

The atomic nucleus (a proton) occupies a volume of $\frac{4}{3}\pi(0.84 \times 10^{-15})^3 = 2.5 \times 10^{-45}$ m³. This is about 10^{16} times less than the volume of the atom, in agreement with the Rutherford model of the atom.

Practice questions

5. The absorption spectrum of a gas is investigated by viewing light emitted by a heated tungsten filament through a sample of the gas. Outline how the absorption spectrum:

   a. provides evidence for discrete energy levels of the atoms of the gas

   b. can be used to deduce the chemical composition of the gas.

6. The diagram shows some of the energy levels of an ionized mercury atom.

   energy / eV
   -2.74 ————————————————

   -3.76 ————————————

   -5.01 ————————————

   a. State how many absorption lines can result from atomic transitions between these energy levels.

   b. On a copy of the diagram, draw an arrow to indicate the atomic transition that gives rise to the absorption line of the **shortest** wavelength.

   c. Show that the line in part b. has the wavelength of about 550 nm.

7. A helium–neon (He–Ne) laser emits monochromatic light of wavelength 633 nm.

   a. Calculate, in J, the difference between energy levels of neon that result in the emission at 633 nm.

   b. The output power of the laser is 2.5 mW. Estimate the number of photons that the laser emits in one second.

8. In the Bohr model of the hydrogen atom, the orbital radius of the electron in the $n = 3$ state is 9 times greater than the orbital radius of the electron in the ground state.

   a. What is $\dfrac{\text{speed of the electron in the } n = 3 \text{ state}}{\text{speed of the electron in the ground state}}$?

   A $\dfrac{1}{9}$ B $\dfrac{1}{3}$ C 3   D 9

   b. For a hydrogen atom in its second excited state:

   i. state the value of the principal quantum number, n

   ii. calculate the energy required to ionize the atom

   iii. show that the frequency of the photon emitted when the atom returns to the ground state is about 3×10^{15} Hz.

   c. The hydrogen atom in the second excited state can also absorb a photon of frequency 2.34×10^{14} Hz. Determine the principal quantum number of the energy state to which the atom is raised.

AHL

E.2 Quantum physics

How can light be used to create an electric current?

What is meant by wave–particle duality?

In 1887, Heinrich Hertz provided the experimental verification that Maxwell's prediction of electromagnetic waves (Topics D.4 and C.2) was correct.

During his experiments, Hertz was using a coil to receive what we now recognize as radio waves. When the waves were detected, an emf was induced in the coil and a spark jumped across an air gap in the circuit. The spark was hard to see and so Hertz put the apparatus into a box to exclude light. He found to his surprise that the maximum length of the spark was less when the apparatus was in the box. The spark length was related to the amount of ultraviolet radiation falling on the air gap. This was the first observation of the **photoelectric effect**.

Hertz had created a current of electrons that were emitted in his apparatus. The energy in the light, carried by photons, was being transferred to free electrons in the metal of his apparatus. However, Hertz could not explain how this was happening. Science had to wait 18 years for an explanation. Einstein received his Nobel Prize in 1921 for a 1905 paper that explained the photoelectric effect. His citation reads "for his services to Theoretical Physics, and especially for his discovery of the law of the photoelectric effect". Einstein never won a Nobel Prize for his work on the theories of relativity and the explanation of Brownian motion, even though the two papers dealing with special relativity and a third on Brownian motion had all been published in the same year.

In Theme C, light was regarded as a wave. In Theme E, electromagnetic energy is transferred by photons. What is the reality? The answer is both. In the 20th century, physics struggled with a way to meld the two opposed concepts and only found a rationale for a description of light after experiments and philosophical debate. As this story unfolds in this Theme, keep this dilemma in your mind: is light a wave or is it a particle? And what are the implications for matter itself?

In this additional higher level topic, you will learn about:

- the photoelectric effect and the evidence it provides
- the photoelectric threshold frequency
- Einstein's explanation of the photoelectric effect

- particle diffraction and wave–particle duality
- the de Broglie wavelength
- Compton scattering and the shift in photon wavelength following the scattering.

(ATL) Social skills — Resolving conflicts

The Nobel Prizes were established in 1895 and first awarded in 1901. They are often regarded as the highest accolade in their field. Einstein was nominated for the prize in every year from 1912 to 1922 with the exception of 1915, often for his work on relativity. However, the Nobel awarding committee was reluctant to award him the prize. It has been suggested that this was because Einstein's theories were too difficult to prove at the time, even though Arthur Eddington's observations of an eclipse in 1919 supported Einstein's theory of general relativity. It has also been suggested that anti-Semitism may have been a factor.

In 1921, the Nobel committee debated whether to award the physics prize to Einstein. Max Planck (the winner of the 1918 award) and others nominated him for his work on relativity, but a report prepared for the committee was highly critical of relativity. Carl Wilhelm Oseen proposed Einstein for his work on the photoelectric effect, but it was felt that this would be too similar to Planck's 1918 Nobel Prize for quantum theory. Eventually, the committee decided not to award the prize that year.

In 1922, Einstein was yet again proposed for the prize. Oseen repeated his nomination on the basis of Einstein's work on the photoelectric effect and Max Planck proposed that the delayed 1921 prize could be awarded to Einstein while the 1922 prize should go to Niels Bohr. The committee finally agreed.

Many of the physicists who appear in this topic also won the Nobel Prize for their work in quantum physics.

▲ Figure 1 Classical theory suggests that this cup of coffee should be emitting visible light and even more harmful UV and X rays. Planck's theory of black-body radiation suggests that the peak wavelength of emission is in the infrared (around 8 μm).

Falsification

The ultraviolet catastrophe is an example of falsification. Classical physics was well established, but improvements in observational techniques revealed a mismatch between theory and experiment at the short wavelengths of continuous spectra. The classical theory was flawed and new hypotheses were required.

Physicists rose to this challenge. The data that they collected and the hypotheses they developed led to a profound shift in our understanding of electromagnetic radiation and its effects.

This constant is the same as the h used in Topic E.1 for the quantized orbital energy of the electrons in the Bohr theory (page 606).

Introduction

By the end of the 19th century, physicists were struggling to explain practical evidence that was not in agreement with their theories. In particular, the intensity of black-body radiation was then modelled using the Rayleigh–Jeans law following the principles of classical physics. The prediction of this law is that

$$\text{intensity} \propto \text{frequency}^2$$

This prediction is accurate for observed visible and longer wavelengths in the electromagnetic spectrum, but breaks down when the ultraviolet region and shorter wavelengths are measured. The problem was so great that the physicists of the day called this "the ultraviolet catastrophe".

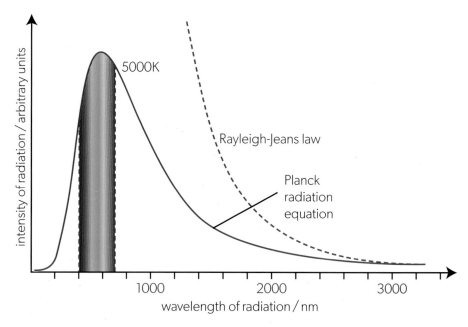

▲ Figure 2 A comparison of the Rayleigh–Jeans model and the Planck theory for electromagnetic radiation.

Figure 2 shows the catastrophe. The Rayleigh–Jeans equation fits the observations at long wavelengths, but as the wavelength decreases the intensity rises well above the intensities that are observed. Max Planck tried to modify the theory to correct the Rayleigh–Jeans law and, in 1900, after a number of partially successful attempts, he made a crucial breakthrough and showed that the catastrophe disappeared when an important assumption was made.

The electromagnetic radiation is produced by oscillating electrons in classical physics (you met this idea in Topic C.2). Planck's assumption was that these electrons must have an energy that is quantized in integer values of hf, where f is the frequency of the electrons (and the radiation they emit) and h is a constant now known as the Planck constant.

Planck could not explain why this quantization was necessary nor what it represented. Einstein used the same assumption of quantization to explain photoelectricity with overwhelming success.

Demonstrating the photoelectric effect

- Inquiry 2: Identify and record relevant qualitative observations.

- Inquiry 3: Compare the outcomes of an investigation to the accepted scientific context.

In the traditional experiment (Figure 3) a piece of zinc plate, freshly cleaned to remove the surface layer of zinc oxide, is attached to a gold-leaf electroscope. A digital coulombmeter can be used nowadays. The electroscope and the zinc are charged using either a

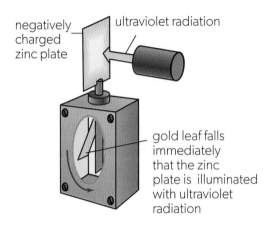

▲ Figure 3 The traditional experiment to demonstrate photoelectricity. When ultraviolet radiation is incident on the zinc, the divergence of the gold leaf decreases.

lead attached to a high-voltage supply or by using a charging-by-induction process (Topic D.2). Then, electromagnetic radiation is shone on the plate. The results in terms of the divergence of the gold leaf (the angle it makes with the vertical) are as follow.

- When the plate is initially negatively charged, the leaf on the electroscope immediately begins to collapse (a coulombmeter will show that the charge on the plate starts to decrease straight away).

- When the intensity of the ultraviolet radiation is reduced even to very low levels, the discharge still begins immediately.

- When visible light or infrared radiation is used for the radiation source, no matter how intense, the effect does not occur, and charge is not lost from the plate.

- Placing a sheet of glass between the source of ultraviolet and the plate prevents the loss of charge (glass absorbs ultraviolet radiation even though it is transparent to visible wavelengths).

- When the plate is positively charged, there is no change to the charge on the plate whatever the wavelength of the radiation.

These observations were found to apply to many metals, but the wavelength below which the radiation can cause the effect varied from metal to metal.

Explaining the photoelectric effect

A breakthrough occurred when Einstein explained the observations in 1905. He adopted Planck's earlier idea of quantization and he added the following predictions about the interaction between electrons and radiation.

- Electromagnetic radiation consists of photons. Each photon has an energy equal to hf (Planck's suggestion).

- Each photon interacts with one, and only one, electron. The photon disappears as a result of the interaction and its energy is entirely transferred to the electron.

- There is a **threshold frequency** f_0, which corresponds to the minimum energy required to release an electron from the metal surface.

- The minimum energy to release the electron is called the **work function** Φ of the metal. This is the energy required to overcome the forces that attract the electron back into the metal. (The release of the electron makes the surface slightly positive.)

- When the photon supplies more energy than the work function, the surplus is transferred to the initial kinetic energy of the electron after it has left the surface.

- When the intensity of the radiation is increased, the number of photons incident on the surface in one second increases.

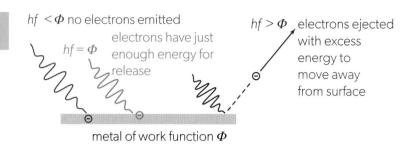

Hypotheses

The term "photon" is used here in the explanation of photoelectricity. However, this was not a term that Einstein used in his 1905 scientific paper. He was familiar with the term "quantum" (plural, "quanta") that Max Planck had coined.

The photon was only named around 1926 after the conclusive work of Compton that is described later in this topic. This is a good example of how a working hypothesis becomes absorbed into both the language and the theories of science by exchanges between scientists with a gradual acquisition of knowledge.

Worked example 1

A copper plate is illuminated by a beam of monochromatic light of wavelength 580 nm and power 0.50 mW. The work function of copper is 4.7 eV.

a. Calculate the number of photons incident on the copper plate in one second.

b. Show that no electrons will be emitted from the plate.

Solutions

a. The energy of a single photon in the beam is
$$hf = \frac{hc}{\lambda} = \frac{1.99 \times 10^{-25}}{580 \times 10^{-9}}$$
= 3.4 × 10⁻¹⁹ J. In one second, the radiation transfers an energy of 0.50 mJ, so the number of photons incident per second is
$$\frac{0.50 \times 10^{-3}}{3.4 \times 10^{-19}} = 1.5 \times 10^{15}.$$

b. In eV, the photon energy is
$$\frac{3.4 \times 10^{-19}}{1.6 \times 10^{-19}} = 2.1 \text{ eV}.$$ This is less than the work function of copper, so no photoelectrons will be emitted.

▲ Figure 4 Observations made in the photoelectric effect for different values of hf.

A metal has a work function Φ. The photon must have an energy greater than this to release an electron from the metal. Algebraically, according to Einstein, $hf > \Phi$, where hf is the energy of the incident photon.

*   When the radiation has a long wavelength then, because only one electron can interact with one photon, the energy of the photon will not be sufficient to release an electron from the metal. This is the "red" photon in Figure 4.

*   When the photon energy is hf_0, then the energy is exactly equal to the work function. This is the greatest energy the electron can gain and still remain in the metal. However, no energy remains to transfer to the electron's kinetic energy. We will not observe its emission from the metal. This is the "green" photon in Figure 4.

*   When the radiation has short wavelength and high frequency, the energy of the photon exceeds the work function. The excess energy is transferred to the initial kinetic energy of the electron. The electron has enough energy to leave the plate (and is then repelled by the already negative plate). This is the "blue" photon in Figure 4.

*   When the plate is positive, even when electrons are released with high-energy photons, they are attracted back to the plate. They do not have sufficient energy to escape the electric field of the plate.

*   When the intensity of monochromatic radiation is increased, the energy of an individual photon is not changed, there are simply more of them arriving every second. The overall effect on the release of an individual electron is not changed (provided the threshold frequency is exceeded). All that happens is that the number of electrons released every second increases.

*   Glass absorbs the ultraviolet radiation and any visible-light photons do not have sufficient energy to exceed the work function.

The Einstein photoelectric equation

Einstein recognized that

maximum kinetic energy of an emitted electron =
energy available from incident photon – energy required to release
electron from surface

Algebraically, this is

$$E_{max} = hf - \Phi$$

Einstein's photoelectric equation

an alternative form is $E_{max} = hf - hf_0$, where f_0 is the threshold frequency.

Millikan's photoelectric experiment

Eleven years after Einstein's explanation, Robert Millikan devised an experiment to test the photoelectric equation.

A cell containing two metal plates in a vacuum is connected to a low-voltage power supply and a sensitive ammeter (shown as a picoammeter (pA) in Figure 5).

The cathode consists of a plate which is the equivalent of the zinc in the gold-leaf electroscope experiment. The other plate (anode) can be given a positive or negative potential relative to the cathode to collect or repel photoelectrons. The potential divider to enable this is not shown in the circuit.

Light of various colours can be shone onto the cathode using filters to select the wavelength range required. The emitted photoelectrons that reach the anode form an electric current through the circuit.

A wavelength of light is selected and shone on the cathode. When this frequency exceeds the threshold, electrons will be emitted with a maximum kinetic energy. When the anode is positive, all electrons will reach it. But if the potential of the anode is now reduced, eventually becoming negative relative to the cathode, there will come a point at which even the most energetic electrons will not be able to reach the anode. The current on the ammeter falls to zero. The potential difference between cathode and anode at this point is known as the **stopping potential** V_s. It is a direct measurement of the maximum kinetic energy E_{max} of a photoelectron.

The photoelectric equation, in these terms, is

$$eV_s = hf - hf_0$$

In wavelength terms (using ideas from Topic E.1) this is

$$eV_s = \frac{hc}{\lambda} - \frac{hc}{\lambda_0}$$

Rearranging the expression yields $V_s = \frac{hc}{\lambda e} - \frac{hc}{\lambda_0 e}$ or $V_s = \frac{hc}{e}\left(\frac{1}{\lambda} - \frac{1}{\lambda_0}\right)$.

A graph of V_s against $\frac{1}{\lambda}$ is a straight line with:

* gradient $\frac{hc}{e}$

* intercept of $\frac{1}{\lambda_0}$ on the $\frac{1}{\lambda}$-axis

* intercept of $V_s = -\frac{hc}{e\lambda_0} = -\frac{\Phi}{e}$ on the V_s-axis.

Notice that:

* The gradient of the graph depends on three fundamental constants, h, c, and e. It does not depend on the metal used for the surface.

* The value of the V_s intercept depends on the gradient and the work function (which affects the threshold wavelength).

* Different metal surfaces give parallel lines with different gradients.

  ## Using the electronvolt

This is one area of physics where the use of the electronvolt (eV) unit is very common. The Millikan experiment measures the stopping potential in volts because this is the calibration of the voltmeter. It therefore seems natural to use the electronvolt rather than the cumbersome negative powers of ten that arise in the energy of an individual electron. You may need to use both the electronvolt and the joule.

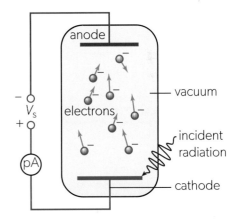

▲ Figure 5 A photoelectric cell. Electrons are emitted from the cathode as photons arrive. When the electrons have sufficient energy, they reach the anode, which is at a negative potential relative to the cathode. The potential of the anode is made more negative until the electrons cannot quite reach it and the picoammeter reading becomes zero.

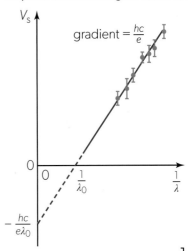

▲ Figure 6 The variation of V_s with $\frac{1}{\lambda}$ for the Millikan photoelectric experiment.

Table 1 gives some typical values for work functions. The work function is affected by the nature of the sample, for example, its crystalline structure.

Element	Work function / eV
caesium	2.1
sodium	2.3
potassium	2.3
calcium	2.9
magnesium	3.7
cadmium	4.1
aluminium	4.1
lead	4.1
silver	4.3
zinc	4.3
iron	4.5
mercury	4.5
copper	4.7
carbon	4.8
cobalt	5.0
nickel	5.0
gold	5.1
platinum	6.4

▲ Table 1 Work functions of elements.

Data-based questions

The table shows some data from Millikan's 1916 paper (Millikan, R. A. 1916. A direct photoelectric determination of Planck's "h". *Physical Review*, Vol. 7, No. 3. pp. 355–388).

(The stopping potentials have been corrected from Millikan's original data to account for the work function of the copper electrode used to detect the current).

Wavelength / nm	Stopping potential / V
546.1	0.36
433.9	0.91
404.7	1.11
365.0	1.49
312.6	2.03
253.5	2.92

- Calculate the frequencies of the six wavelengths of light that Millikan used.

- Plot a graph of stopping potential versus frequency.

- Find the gradient of your graph.

- Millikan suggested that the uncertainties in the stopping potentials were ± 0.01 V. Use this uncertainty to estimate an uncertainty in your value for the gradient.

- Millikan determined the value of h using the equation $h = G \times e$, where G is the gradient of the graph and e is the charge on the electron. Millikan's value for e was $(1.5924 \pm 0.0017) \times 10^{-19}$ C. Determine the value of h and determine its uncertainty.

- Today, Planck's constant is defined as $6.626\,070\,15 \times 10^{-34}$ J s. Is this value consistent with Millikan's data? Calculate the difference between the value from Millikan's data and today's defined value and express this as a percentage difference.

- The work function of the element sodium is often given as 2.3 eV, although a very pure sample would have a work function of 2.75 eV. Millikan's sample is likely to have contained impurities which would have affected his measurement of the work function. (In fact, he did not make use of the work function in his paper.) Use your graph to find the work function that is suggested by Millikan's data.

Worked example 2

The graph shows how the maximum kinetic energy E_{max} of photoelectrons emitted from a metal surface varies with the frequency f of the incident electromagnetic radiation.

a. Explain why, for radiation of wavelength 500 nm, no electrons are emitted from the surface.

b. Estimate, using the graph:

  i. the maximum kinetic energy of the electrons emitted when radiation of wavelength 250 nm is incident on the surface

  ii. the work function of the metal

  iii. Planck's constant h, giving the answer in J s.

c. Another metal surface has a work function of 1.8 eV. Draw a graph showing how E_{max} varies with f for this surface.

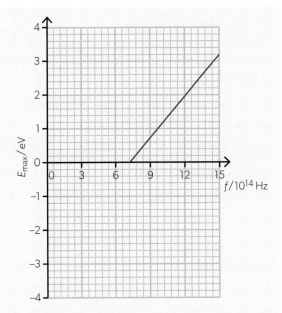

Solutions

a. The frequency of radiation is $f = \dfrac{c}{\lambda} = \dfrac{3.00 \times 10^8}{500 \times 10^{-9}} =$
    6.0×10^{14} Hz and is below the threshold frequency of about
    7.2×10^{14} Hz. The energy of the photons is therefore less than the work function and insufficient to remove electrons from the surface.

b. i. For this wavelength, the frequency
        is $\dfrac{3.00 \times 10^8}{250 \times 10^{-9}} = 1.2 \times 10^{15}$ Hz. The maximum
        energy of photoelectrons can be directly read off the
        graph, $E_{max} = 2.0$ eV.

  ii. The work function is equal to the negative intercept with the energy axis of the extrapolated graph.
       Intercept $= -3.0$ eV $\Rightarrow$ work function $= 3.0$ eV.

  iii. The equation of the line is $E_{max} = hf - \Phi$. Hence the
        Planck's constant is equal to the slope of the graph.
        $h \simeq \dfrac{3.2 - (-3.0)}{15 \times 10^{14}} = 4.1 \times 10^{-15}$ eV s.

        Conversion to J s is straightforward:

        $h = 4.1 \times 10^{-15} \times 1.6 \times 10^{-19} = 6.6 \times 10^{-34}$ J s.

c. Both lines have the same slope h, hence must be parallel to each other. The new graph has an intercept of -1.8 eV with the energy axis.

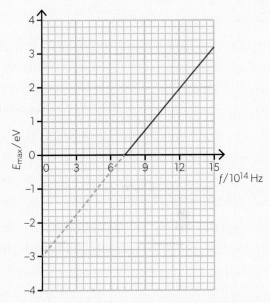

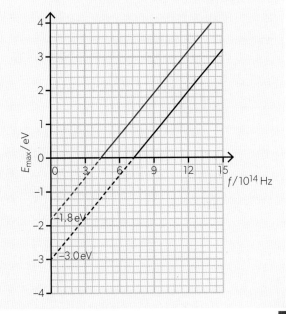

Worked example 3

In an experiment to investigate the photoelectric effect, monochromatic light of wavelength 430 nm is incident on a metal surface.

a. Calculate, in eV, the energy of the photons incident on the photosensitive surface.

A variable voltage V is applied across the photoelectric cell. The current in the ammeter is zero when V is 0.60 V or more.

b. Determine:

   i.   the work function of the metal

   ii. the photoelectric threshold frequency for the metal.

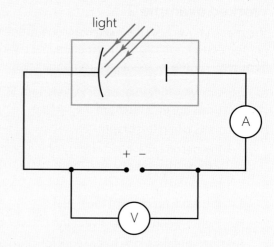

Solutions

a. $E = \dfrac{hc}{\lambda} = \dfrac{1.24 \times 10^{-6}}{430 \times 10^{-9}} = 2.9\,\text{eV}$

b. i. The maximum kinetic energy of the photoelectrons is $E_{max} = 0.60\,\text{eV}$.
(This is the work done by the electric field in stopping the most energetic electrons.)
The work function can be found by rearranging Einstein's photoelectric equation,
$\Phi = hf - E_{max} = 2.9 - 0.6 = 2.3\,\text{eV}$.

  ii. Photons of threshold frequency f_0 have energy equal to the work function: $hf_0 = \Phi$.

Hence $f_0 = \dfrac{\Phi}{h} = \dfrac{2.3 \times 1.60 \times 10^{-19}}{6.63 \times 10^{-34}} = 5.6 \times 10^{14}\,\text{Hz}$. Note that the work function is

expressed in J, because the calculation involves the value of the Planck's constant in J s.

Practice questions

1. The work function for silver is 4.3 eV.

   a. Calculate the photoelectric threshold frequency for silver.

Electromagnetic radiation of wavelength 240 nm is incident on the silver surface.

   b. Calculate, for the electrons emitted from the sliver surface, their maximum:

     i.   kinetic energy

     ii. speed.

   c. The intensity of the radiation is increased. State the effect of this change on:

     i.   the maximum speed of the photoelectrons

     ii   the rate at which the electrons are emitted from the surface.

2. Monochromatic electromagnetic radiation is incident on a photoelectric cell with a sodium cathode, causing the emission of photoelectrons. A potential difference of at least 1.5 V applied across the cell prevents the photoelectric current from flowing in the cell. Work function for sodium = 2.3 eV.

   a. Calculate the maximum initial speed of the electrons emitted from the sodium surface.

   b. Determine the wavelength of the incident radiation.

3. The photoelectric threshold wavelength for a particular metal surface is 480 nm.

   a. Calculate, in eV, the work function of the metal.

   b. Radiation of frequency 9.0×10^{14} Hz is incident on the surface. Calculate the stopping voltage for this frequency.

What are the defining features and behaviours of waves?

How did the explanation of the photoelectric effect lead to the falsification that light was purely a wave? (NOS)

Photoelectrons are emitted instantly when radiation is incident on the metal surface, no matter how small the intensity of the radiation. The wave theory cannot explain this. The properties and defining features of waves were described in Topics C.2 and C.3. In particular, the energy transfer properties of waves were outlined there.

A progressive wave transfers energy at a continuous rate and the intensity of the wave is proportional to the *wave amplitude*[2] (Topic C.2). Wave theory predicts that, given a long enough wait, even a low-power source will deliver enough energy to release a photoelectron eventually *whatever the radiation frequency*. This is not what is observed. No photoelectric emission occurs below the threshold frequency, and instantaneous emission occurs above the threshold even with weak incident radiation.

This falsified the theory that electromagnetic radiation in general, and visible light in particular, possessed only wave behaviour. It drove physics into new areas of study and creativity by creating a dilemma.

The dilemma was that a wave theory is needed to explain properties such as interference and diffraction, but a particle-like photon is required to explain photoelectricity. Light appears to have the properties of either waves or particles depending on the type of experiment or phenomena that we are observing. Put another way, we can invoke the wave properties or the particle properties depending on what we want to explain. This flexibility in approach is known as the **wave–particle duality**. The light has properties that can be either wave-like or particle-like.

The Compton effect

By the early 1920s, physicists were still discussing the implications of Einstein's explanation of photoelectricity. In 1923, Arthur H Compton, a scientist from the USA, was investigating the behaviour of high-energy, short-wavelength X-radiation. The X-rays were scattered by elements with small proton number. Compton found that the wavelengths of the X-rays increased after scattering.

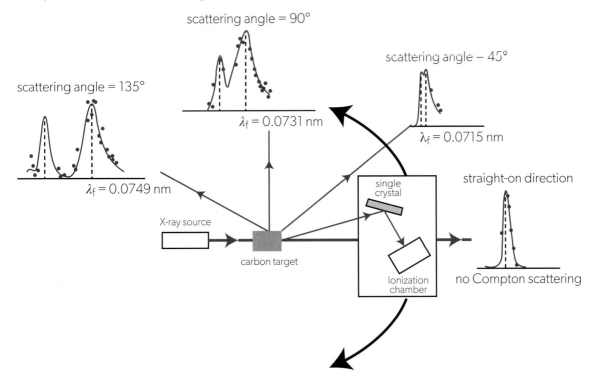

▲ Figure 7 The arrangement and results from Compton's scattering experiment.
The wavelength of the X-ray source is $\lambda = 0.0709$ nm.

X-radiation photons of very large energy are incident on a carbon target. Compton produced the radiation by accelerating electrons through a potential difference of 17 kV to strike a molybdenum target. The deceleration of the electrons in the target produces an intense beam of X-rays that Compton focused onto the carbon block. Figure 7 shows Compton's experimental arrangement and his results at different scattering angles.

The scattered X-ray photons are observed using a detector consisting of a single diffracting crystal and an ionization chamber (in which an electric current is proportional to the number of air atoms ionized in the chamber). A spectrum with two peaks is seen at each scattering angle except for the straight-on direction. The smaller peak is due to atom scattering; the larger peak is due to free-electron scattering. As the scattering angle increases, the wavelength shift increases too.

The spectrum plot of scattered X-ray intensity against the wavelength of the X-rays allows the energy of the scattered X-ray photon to be estimated because

$$E = \frac{hc}{\lambda}$$

The X-rays are scattered by free electrons in the carbon. Only a small fraction of the incident X-ray energy is needed to ionize carbon atoms in the target (the ionization energy of carbon is about 1 eV). The remaining energy of the X-ray photon is involved in its interaction with a free electron. Both an X-ray photon and the electron are scattered away from the initial photon direction (Figure 8). The electron recoils. The photon scattered from the interaction has an increased wavelength and therefore a smaller energy than the incident photon.

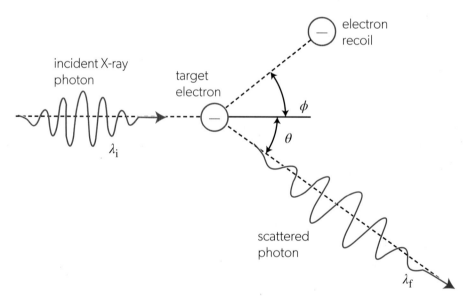

▲ Figure 8 In the Compton effect, an incident X-ray photon interacts with a free target electron. The electron gains energy and moves away while an X-ray photon of a different frequency is scattered in a different direction.

The situation is very like the situation analysed in Topic A.2 where two bodies undergo a two-dimensional collision. Compton proved mathematically that the difference between the scattered and incident wavelengths, λ_f and λ_i, of the X-radiation was

$$\lambda_f - \lambda_i = \Delta\lambda = \frac{h}{m_e c}(1 - \cos\theta)$$

where θ is the scattering angle and m_e is the mass of the electron. It is also possible to predict the angle through which the electron recoils, although this was not measured in Compton's experiment. When the electron recoils at an angle ϕ, then

$$\frac{1}{\tan\phi} = \left(1 + \frac{hf}{m_e c^2}\right)\tan\left(\frac{\theta}{2}\right)$$

Figure 7 shows the results for a series of scattering angles ranging from straight-on (0°) to back-scattered through 135°. For all angles except the straight-on direction, there are two peaks for the photon wavelengths. The larger peak (at λ_f) is due to Compton scattering. This relies on the prior ejection of the electron from the atom by the incident X-ray photon, as explained above.

However, another possibility is that the photon is scattered by the whole of the carbon atom. (It is likely that this scattering occurs at the inner, tightly bound electrons of the atom.) In the whole-atom case, the mass m_e in the Compton equation changes to the mass of the whole carbon atom, some 20 000 times greater than the electron mass. The wavelength shift in this case is correspondingly much less and effectively zero. This is the reason for the second smaller peak in the intensity graphs at the same wavelength as the incident photon.

Compton's measurements confirmed his prediction for the relationship between $\Delta\lambda$ and θ. The prediction had been made on the basis that photons have:

*   energy given by $E = \dfrac{h}{f} = \dfrac{hc}{\lambda}$

*   momentum given by $p = \dfrac{hf}{c} = \dfrac{h}{\lambda}$.

Compton's result cannot be explained on the basis of a wave model. The classical explanation of scattering of light by a charged particle does not predict wavelength shifts when the radiation has a small intensity. Compton's incident X-rays had a small intensity and so the classical theory does not explain the effect. If the radiation intensity were to be very large, then other effects, different from the Compton effect, would be observed under a classical model.

Ⓛ Data-based questions

Use Figure 7 to complete the following tasks.

*   Use the three points of data in the diagram to make a table. Tabulate values of $1 - \cos(\theta)$.

*   Plot a graph of λ_f on the y-axis against $1 - \cos(\theta)$ on the x-axis.

*   Add a line of best fit.

*   By referring to the equation for Compton scattering, deduce what the gradient and the y-intercept of your graph represent. Hence find the value of λ_i.

How is photon scattering off an electron similar to and how is it different from the collision of two solid balls?

Compton proved the result for his effect by assuming that:

- Einstein's light quanta possessed momentum even though this is the attribute of a particle

- there was only one interaction between a quantum and a free electron.

In the collision of two solid balls in mechanics, there are many atoms involved and we use the centre of mass of the balls as an averaging process.

Additionally, Compton could not assume that the final speed of the electron was slow enough for the interaction to be treated as Newtonian. He had to use the full relativistic expression for the invariant energy of the electron.

Finally, two solid balls have similarities in property even though their masses may be different. Likewise, the interaction between them is, for any practical case, inelastic. The derivation of the Compton equation assumes that no energy (in its broad sense, including mass energy) is lost in the interaction. Momentum is also conserved in the solid ball collision and in the Compton effect.

The energy change of the incident solid ball also depends on the "scattering" angle, although through a different equation. In both cases, the energy change of the scattered objects can be predicted using conservation laws, which links to the discussion in Topic A.2.

Can the Bohr model help explain the photoelectric effect? (NOS)

Bohr's model of the atom (Topic E.1), like many early models of a phenomenon was an over-simplification. It is taught to students today largely because of its historical importance and because it represents a superb synthesis of observation and physical hypothesis.

Quantum mechanics dictates that we can no longer model the hydrogen atom as a system of an orbital electron and a stationary proton. Nor can we use the Bohr model to explain photoelectricity. Bohr himself knew that

his model was limited to the simplest atoms with one electron. In the photoelectric effect, the incoming quanta are interacting with free electrons near the surface of a metal. The bonding between electron and metal cannot be modelled using Bohr's ideas.

All physical models have limitations. It is the role of the scientist to appreciate these limitations and either work within them or strive to improve the scope of the model.

Why is Compton scattering more convincing evidence for the particle nature of light than that from the photoelectric effect? (NOS)

In the photoelectric effect, the incoming photon disappears without trace. Its energy is transferred to the metal undergoing the effect and what remains appears as kinetic energy of the emerging photoelectron. Einstein needed to infer the properties of what he called the "quantum" from the link between its disappearance and the resulting electron emission.

The Compton effect is more direct. The energies of both the incident X-ray photon and the emerging photon

can be measured. The change in them can be directly related to the final state of the scattered electron. Further, coherent scattering—when the incident photon interacts with the whole atom—can also be observed and agrees with theory. There is a correlation between the absorption of a photon and the release of a photoelectron that can be explained using Einstein's theory. However, the link between the wavelength shift of the X-ray photon and the gain in energy of the electron in the Compton effect is both measurable and causal.

Worked example 4

A photon of energy 15 keV interacts with a free electron that is initially at rest. The electron recoils and the photon is scattered at an angle of 60° relative to the original direction.

a. Explain why the energy of the photon decreases as a result of the interaction with the electron.

b. Calculate the wavelength of:

   i. the incident photon

   ii. the scattered photon.

c. Determine the energy of the recoil electron.

Solutions

a. During the photon–electron collision, some of the energy and momentum of the photon are transferred to the electron. Since the total energy is conserved, the energy gain of the electron is equal to the energy loss of the photon.

b. i. The wavelength of the incident photon is $\lambda_i = \dfrac{hc}{E} = \dfrac{1.24 \times 10^{-6}}{1.50 \times 10^4} = 8.27 \times 10^{-11}$ m.

   ii. The change in the wavelength can be calculated from the Compton formula, $\Delta\lambda = \dfrac{hc}{m_e c^2}(1 - \cos\theta)$

$$= \frac{1.24 \times 10^{-6}}{0.511 \times 10^6}(1 - \cos 60°) = 1.21 \times 10^{-12}\,\text{m.}$$ (We have expressed the electron mass in MeV c^{-2} in order to simplify the calculation of the ratio $\dfrac{h}{m_e}$. MeV c^{-2} as a unit of mass is described on page 641 in Topic E.3.) The

wavelength of the scattered photon is therefore $\lambda_f = \lambda_i + \Delta\lambda = 8.27 \times 10^{-11} + 0.12 \times 10^{-11} = 8.39 \times 10^{-11}$ m.

c. The energy of the recoil electron is equal to the energy loss of the photon, $E_i - E_f = 1.50 \times 10^4 - \dfrac{1.24 \times 10^{-6}}{8.39 \times 10^{-11}} = 220$ eV.

Worked example 5

In a Compton scattering experiment, an X-ray photon of wavelength 1.20×10^{-11} m collides with a free, stationary electron. The outgoing photon is detected at an angle of 90° to the original direction.

a. Calculate the wavelength of the outgoing photon.

b. Determine the change in momentum (magnitude and direction) of the photon.

c. Hence, state the angle through which the electron recoils.

Solutions

a. $\lambda_f = \lambda_i + \Delta\lambda = 1.20 \times 10^{-11} + \dfrac{1.24 \times 10^{-6}}{0.511 \times 10^6}(1 - \cos 90°) = 1.44 \times 10^{-11}$ m.

b. The initial momentum of the photon is $p_i = \dfrac{h}{\lambda_i} = \dfrac{6.63 \times 10^{-34}}{1.20 \times 10^{-11}} = 5.53 \times 10^{-23}$ N s and the final momentum is

$p_f = \dfrac{h}{\lambda_f} = \dfrac{6.63 \times 10^{-34}}{1.44 \times 10^{-11}} = 4.60 \times 10^{-23}$ N s. The vectors representing p_i and p_f are at right angles to each other, so the magnitude of their difference is $\Delta p = \sqrt{p_i^2 + p_f^2} = 7.19 \times 10^{-23}$ N s. The angle that Δp makes with the

original direction is $180° - \tan^{-1}\left(\dfrac{p_f}{p_i}\right) = 140°$. This is illustrated in the following diagram, in which the incident photon propagates horizontally to the right and the scattered photon vertically upwards.

c. The momentum of the photon–electron system is conserved so the momentum gain of the electron is equal but opposite to the momentum change Δp of the photon. This means that the electron recoils diagonally to the right, at an angle of 40° below the horizontal. You should compare this problem with examples of two-dimensional collisions in Topic A.2—the conservation of momentum applies to both situations!

Practice questions

4. An X-ray photon of energy 25.0 keV collides with a stationary electron. The scattered photon has energy 23.3 keV.

   a. Calculate the energy of the recoil electron.

   b. Calculate the wavelength change of the photon.

   c. Determine the angle through which the photon is scattered.

5. A photon of energy 20 keV is scattered at an angle of 130° off a stationary electron.

   a. Calculate the wavelength of the incident and the scattered photon.

   b. Calculate the energy of the recoil electron.

6. In a Compton scattering experiment, a beam of X-rays of wavelength 2.50×10^{-11} m is incident at a carbon target. The graph shows the intensity spectrum of scattered X-rays observed at a particular angle.

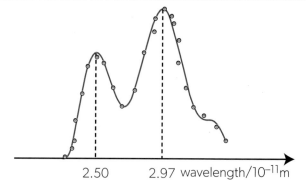

a. Outline how the presence of a maximum at 2.97×10^{-11} m supports the particle model of electromagnetic radiation.

b. Calculate the angle at which scattered X-ray photons are observed.

c. Determine the magnitude of the momentum of recoil electrons.

Other ways in which matter and photons interact

The Compton effect and photoelectricity are just two of the mechanisms by which photons are known to interact with matter.

- The photoelectric effect occurs at small photon energies from a few eV to a few keV. It leads to the ejection of loosely bound electrons from the surface of a metal. These energies run from visible light through to the least energetic X-rays. The photon is completely absorbed by the material.

- The Compton effect occurs for energies up to about 1 MeV. The photon is scattered with a frequency shift.

- Photons with energies of 1.022 MeV and greater can interact with a nearby nucleus to undergo pair production. In this mechanism, the photon disappears and is replaced by two particles, one the antiparticle of the other. Typically, at the lower energy limit the pair will be an electron (e−) and an anti-electron called a positron (e+). The energy of the incident photon must provide at least both rest mass energies of the particles. Any remaining energy goes into the kinetic energies of the particles. Momentum must also be conserved in the interaction.

- At even higher energies, more than 2 MeV, photodisintegration can occur in which a photon is absorbed by a nucleus which enters a higher energy state and later decays with the ejection of a nuclear particle, such as a neutron, proton or alpha particle.

Matter waves and the de Broglie hypothesis

The French aristocrat Louis de Broglie (pronounced "de Broi") took Einstein's revolutionary explanation of photoelectricity and used its ideas to derive the Wien displacement law (Topic B.1) from first principles in quantum terms. The following year, 1923, he presented a series of short articles to the Paris Academy of Sciences in which he extended the wave–particle duality ideas to include matter as well as waves. He hypothesized that electrons and other particles could exhibit wave-like properties under the appropriate experimental condition. This is known as the **de Broglie hypothesis**.

You met this idea in Topic E.1 in the discussion of the Bohr model where the electrons are assumed to have a standing-wave arrangement for the different energy states of the hydrogen atom. This aspect of the Bohr model arises directly from the work of de Broglie.

In his special theory of relativity, Einstein recognized that the mass m of a particle and its energy E were equivalent. This equivalence is described in his famous equation, $E = mc^2$. This is discussed in more detail in Topic E.3. For a moving particle the total energy becomes

$$E = \sqrt{p^2 c^2 + m_0^2 c^4}$$

where p is the momentum of the particle (m_0 is sometimes known as the rest mass of the particle).

The photon is a particle of zero rest mass ($m_0 = 0$) and therefore $E = pc$ or $p = \dfrac{E}{c}$.

According to Einstein, for a photon, $E = hf = \dfrac{hc}{\lambda}$ and substituting for E,

$$p = \frac{hc}{c\lambda} = \frac{h}{\lambda}$$

De Broglie assumed that the equation $p = \dfrac{h}{\lambda}$ applied equally to particles as well as photons and so, for particles, one final arrangement gives

$$\text{the de Broglie wavelength } \lambda = \frac{\text{Planck constant}}{\text{momentum of particle}} = \frac{h}{p}$$

A baseball pitcher can throw a fast ball in excess of $160\,\mathrm{km\,h^{-1}}$. The baseball has a mass of $0.15\,\mathrm{kg}$. A simple estimate shows that the de Broglie wavelength for the ball treated as a single particle is $\dfrac{6.6 \times 10^{-34}}{44 \times 0.15} \approx 10^{-34}\,\mathrm{m}$. This is an impossibly small wavelength. To observe diffraction of the baseball would require an aperture on the same scale—far smaller than a proton in a nucleus. However, choose something with a much smaller mass and the de Broglie wavelength will increase in inverse proportion. This was the basis of an experiment to verify de Broglie's hypothesis.

 Is energy conserved under special relativity?

In Topic A.5 you met some invariant quantities that do not change between reference frames: proper time, proper length and the invariant interval. Another is the invariant energy, the total relativistic energy of a system as measured by an observer at rest relative to the system.

The invariant energy can contain many types of energy, including the rest mass, field potential energies, internal energies and energies due to interactions with radiations. Many of these energy types have appeared earlier in this course in Themes B, D and E.

For the equation in the text, $E = \sqrt{p^2 c^2 + m_0^2 c^4}$, notice that this is also

$E = m_0 c^2 \sqrt{1 + \left(\dfrac{p}{m_0 c}\right)^2}$, which can be expanded.

When $v \ll c$, this becomes

$$E \approx m_0 c^2 \left(1 + \frac{1}{2}\left(\frac{m_0 v}{m_0 c}\right)^2\right) \text{ which is}$$

$$E \approx m_0 c^2 + \frac{1}{2} \times m_0 c^2 \times \left(\frac{m_0 v}{m_0 c}\right)^2 \text{ or } E \approx m_0 c^2 + \frac{1}{2} m_0 v^2$$

The second term in the expansion is effectively the classical Newtonian result for the kinetic energy of a moving body travelling slowly in relativistic terms.

The Davisson–Germer experiment

Two years after de Broglie's work was announced, Clinton Davisson and Lester Germer in the USA verified his hypothesis experimentally. They fired a beam of electrons at a nickel target and observed diffracted electrons being emitted by the nickel surface (Figure 9). This confirmation together with a slightly later one by George Thomson at the University of Aberdeen convinced the Nobel committee to award the 1929 Nobel Prize to de Broglie.

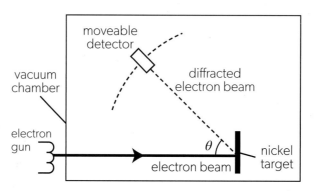
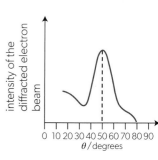

▲ Figure 9 Davisson and Germer fired electrons at a nickel target. The nickel atoms acted like a diffraction grating and the electrons were diffracted back to a detector. The intensity–scattering angle graph had a pronounced maximum rather like the optical diffraction patterns you met in Topic C.3. This allowed the two scientists to estimate the "grating spacing"—the separation of the nickel atoms which agreed with estimates made in a different way. The electrons had a wave-like behaviour when they interacted with the nickel atoms.

Electron diffraction can now be easily demonstrated in a school laboratory using a beam of electrons incident on a thin film of carbon (graphite), as shown in Figure 10.

Electrons are emitted by a heated cathode via the process of **thermionic emission.** In this, the electrons gain sufficient energy to leave the surface—rather like the process in which electrons gain energy from incident photons in photoelectricity. The electrons are then accelerated through a potential difference of about 3 kV in a cylinder at a positive potential relative to the cathode. The final speed of the electrons is about 10% of the speed of light.

If the electrons behaved as particles when interacting with the carbon film, then they would form a bright region in the centre of the fluorescent screen at the far end of the tube. However, they do not do this. What is seen on the screen is a series of concentric bright and dark lines, or fringes, similar to those seen in diffraction gratings due to the interference of light.

You may wonder why there are rings rather than points of light which is what a diffraction grating would produce. This is because the carbon film has many small crystals of graphite with the planes of atoms all at different angles to each other. The individual beams—one from each small crystal—are diffracted at the same angle but rotated about the line of the beam so that the characteristic ring shape is produced.

If you get the chance to use this apparatus, take the opportunity to change the accelerating potential difference of the electrons. This changes their final speed and, according to de Broglie, their wavelength. A smaller pd means a smaller electron speed and therefore a larger wavelength. Reducing the pd should make the rings spread out more and this is what is observed.

(a)
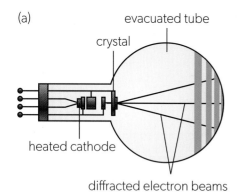

(b)
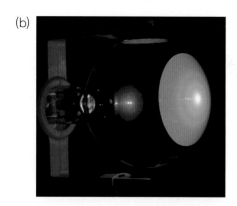

▲ Figure 10 Electrons are emitted from a heated cathode and are incident on a crystalline sample of graphite. They are diffracted by the sample and the resulting diffraction pattern as a series of bright and dark rings is observed on the face of the fluorescent screen. The power supply and its connections are not shown in diagram (a).

To see this in more detail: The kinetic energy gained by one electron as it is accelerated through a pd V is eV, which is equal to $\frac{1}{2} m_e v^2$. This kinetic energy can be expressed in terms of momentum p because

$$\frac{1}{2} m_e v^2 = \frac{1}{2} \frac{(m_e v)^2}{m_e} = \frac{p^2}{2m_e}$$

Therefore, $\frac{p^2}{2m_e} = eV$ and $p = \sqrt{2m_e eV}$. Invoking the de Broglie expression $\lambda = \frac{h}{p}$ gives $\lambda = \frac{h}{\sqrt{2m_e eV}}$.

Decreasing V should increase λ. A comparison with the diffraction grating equation $n\lambda = d \sin \theta$ reminds you that, as λ increases, so does θ.

Worked example 6

Calculate the de Broglie wavelength for:

a. a neutron moving with a speed of $2.5 \times 10^3 \, \text{m s}^{-1}$

b. an electron accelerated from rest through a potential difference of $500 \, \text{V}$.

Solutions

a. The neutron has a momentum $p = m_n v$, so the de Broglie wavelength is $\lambda = \frac{h}{m_n v} = \frac{6.63 \times 10^{-34}}{1.675 \times 10^{-27} \times 2.5 \times 10^3}$
    $= 1.6 \times 10^{-10} \, \text{m}$. This is about the size of an atom.

b. The kinetic energy of the electron is $500 \, \text{eV}$. Since $p = \sqrt{2m_e E}$,
    $\lambda = \frac{h}{\sqrt{2m_e E}} = \frac{6.63 \times 10^{-34}}{\sqrt{2 \times 9.11 \times 10^{-31} \times 500 \times 1.6 \times 10^{-19}}} = 5.5 \times 10^{-11} \, \text{m}$.

Worked example 7

The atomic structure of a crystal can be investigated using a beam of electrons that diffract from parallel planes of atoms in the crystal, in a similar way to light diffracting at parallel slits.

Distances between atomic planes of a particular crystal are of the order of $1.5 \times 10^{-10} \, \text{m}$.

Estimate the energy of electrons in a beam suitable for a diffraction experiment with this crystal.

Solution

Electrons will form a diffraction pattern when their de Broglie wavelength is comparable to or shorter than a typical interatomic distance. Since $\lambda = \frac{h}{p}$, this gives a condition for the momentum of electrons: $p > \frac{h}{d}$, where $d = 1.5 \times 10^{-10} \, \text{m}$ is an estimate of the interatomic distance. This means that the energy of the electrons must be greater than $\frac{p^2}{2m_e} = \frac{h^2}{2d^2 m_e} = \frac{(6.63 \times 10^{-34})^2}{2(1.5 \times 10^{-10})^2 \times 9.11 \times 10^{-31}} = 1.1 \times 10^{-17} \, \text{J} \simeq 70 \, \text{eV}$.

Practice questions

7. Calculate the de Broglie wavelength for:
    a. an alpha particle of kinetic energy $200 \, \text{MeV}$
    b. an electron accelerated from rest through a potential difference of $1.0 \, \text{kV}$.

8. Estimate the minimum energy of neutrons that will form a diffraction pattern when scattered from molecules whose interatomic distance is of the order of $10^{-10} \, \text{m}$.

Models—Wave–particle duality

Is a beam of light a series of photons or is matter a wave? Is light a wave or does matter consist of particles?

Neither of these interpretations is completely correct. We need to use wave models at some times and particle models at others. The visible-light photons interfering in a two-slit experiment are interacting with the slits on a probabilistic basis, so that only chance determines the outcome. All we can say is that there is more chance of a photon arriving at a region where photon densities are high. There is less chance of photon arrival where there is a predicted minimum in the interference pattern.

Figure 11 shows the build-up of an interference pattern as individual photons arrive.

When interference experiments are carried out with small photon incidence rates at the slits, an interference pattern is still observed, even though the photons cannot possibly be interacting (interfering) with each other. The interference wave theory predicts a probability pattern and the photons obey this.

Exactly the same results are obtained when electrons diffract rather than photons. The build-up of the patterns is identical whether there is one electron in transit in the diffracting apparatus or many.

This strange behaviour takes us into the realms of quantum mechanics.

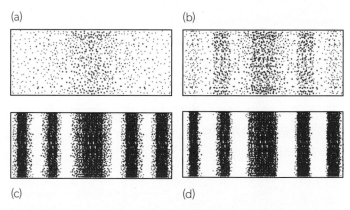

▲ **Figure 11** Each dot represents the impact of an individual photon with the screen. The exposure time increases from (a) through to (d). In (a) it is impossible to see the diffraction pattern clearly, but as time goes on the pattern of maxima and minima becomes more obvious.

What evidence indicates the diffraction of a wave? How can particles diffract?

There are strong similarities between the wave diffraction patterns described in Topic C.3 and the observed behaviour of electrons when they interact with crystalline lattices. This provides exceptionally strong evidence to allow us to describe the electrons as undergoing diffraction.

Wave theory, when applied to waves of visible-light wavelengths and below, is very successful in accounting for the diffraction patterns that are observed. Experiments and advanced wave theory are in good agreement about the origins of the patterns. These observations began as early as the lifetime of Grimaldi in the 1660s. The full theory reached complete acceptance in 1820 with the work of Fresnel. However, it was only when experimentation became more sophisticated in the 20th century that scientists could observe the passage of individual photons through a double-slit apparatus. This had implications that forced a paradigm shift in our understanding of the nature of light and matter.

The work described in Topics C.3, E.1 and E.2 merges into a new but coherent understanding of the interaction of matter that is known as **quantum mechanics**. The Bohr model and the ideas of de Broglie stimulated scientists such as Heisenberg, Schrödinger and Dirac to develop quantum mechanical models, and this work continues today.

In quantum mechanics, particles are described using probability waves. These probability waves superpose with each other to produce interference patterns, whether for light or for particles. The work of Schrödinger was crucial here. He postulated a wave function to describe the state of a particle. It can predict the observed behaviour of many of the simpler atomic systems such as hydrogen.

As an example of the power of the wave function, imagine a double-slit interference experiment in which only one photon is in transit at any one time. Quantum mechanics suggests that the photon has all possible states and positions until the instant at which it is observed (that is, when we see it on a screen). At this moment, the wave function reduces to the classical case as photon detection occurs. This is one way to interpret quantum mechanics—it is known as the Copenhagen interpretation as named by Heisenberg—but it is not the only interpretation possible. In summary, it says that nothing can be treated as real until it is observed. When light is observed as a wave, it is a wave. When it is observed as a photon, it is a particle.

(ATL) Communication skills

▲ Figure 12 Erwin Schrödinger (1887–1961).

Erwin Schrödinger was an Austrian physicist awarded the Nobel Prize in 1933 for his mathematical formulation of quantum mechanics. He is famous for his "Schrödinger's Cat" thought experiment. He corresponded with Einstein about the Copenhagen interpretation of quantum mechanics and used this idea to illustrate the problems of applying quantum mechanics to large systems. It was his attempt to comment on criticism made by Einstein,

Podolsky and Rosen who argued that quantum mechanics gave an incomplete picture of reality.

Schrödinger described a thought experiment in which the life of a cat in a box depends on the state of a radioactive nucleus. If the nucleus decays, the emitted particle from the decay triggers, via a mechanism, the release of a poison gas. This kills the cat. However, if the nucleus remains undecayed, the cat lives. The outcome for the cat depends on the state of a quantum particle in the nucleus that is said to be in a superposition of states: decay/non-decay. Until the box is opened and the state of the large-scale cat system is observed, the state of the quantum particle cannot be deduced. The Copenhagen interpretation implies that the cat is both alive and dead until the quantum state has been observed. In quantum-mechanical terms there is a pre-box-opening wave function to describe the system which includes both outcomes. This wave function reduces to one state (cat alive) or the other (cat dead) as the outcome is observed.

The Schrödinger's Cat thought experiment is still used to test quantum mechanical interpretations. Take your knowledge of this area of quantum mechanics further by carrying out web research into the work and interpretations of John von Neumann-Eugene Wigner, David Bohm (pilot-wave interpretation), Hugh Everett (many-worlds interpretation), and others.

Measurement and observation

The implications of quantum mechanics can seem confusing to us. In order to observe interference from a two-slit system, a contribution from each slit is required so that, when the phase difference between them is π rad, destructive interference occurs. When a dim beam of light is used (so that we have only one photon in the system at a time) or a stream of electrons, then we might wonder which slit each particle travels through. The answer must be that any individual photon or electron passes through both slits.

So, what would happen if we attempted to measure which slit a particle travels through? We could imagine a two-slit experiment with a stream of photons travelling through it. Compton scattering would be one suitable method to detect the presence of photons. A beam of electrons

could be fired across the system and the photons of light would interact with them. The deflection of the electrons could be measured to determine information about the photons. However, as we saw earlier in the topic, in Compton scattering, the photon is scattered into a different direction. This scattering of the photons destroys the two-slit interference pattern and we are left with the single-slit diffraction pattern (which is to be expected when photons of light only pass through one slit). It appears that any measurement system which is sensitive enough to determine which slit the photons travel through will disturb the experiment sufficiently to remove the two-source interference pattern.

Are there other situations where measurement affects the observation? Are there any where it does not?

E.3 Radioactive decay

Why are some isotopes more stable than others?

In what ways can a nucleus undergo change?

How do large, unstable nuclei become more stable?

How can the random nature of radioactive decay allow for predictions to be made?

The story of the discovery of radioactivity is well known. On 1 March 1896, Henri Becquerel developed one of his photographic plates. He saw on the developed plate the outline of some uranium salts that had been lying on top of the wrapped photographic package. Realizing that something about the uranium material caused the effect, he looked further and confirmed that the radiation from the uranium was the cause. Becquerel was inadvertently repeating the same discovery made by Abel Niépce de Saint-Victor forty years earlier. Niépce, however, did not pursue the idea, even though he reported it to the French Academy of Science at the time.

The uranium material near Becquerel's plates was decaying to form a more stable substance. What drives these changes to stability? This is a matter of internal balance within the nucleus and is driven by the interactions that occur between particles in the nuclei of all atoms (except hydrogen, the simplest atom of all). As you may expect from work in earlier themes, the changes are driven by energy considerations. When the nucleus can move to a state where the amount of internal binding is greater, then it will do so. The degree of initial instability is crucial to determining the outcome of the radioactive decay process.

This topic describes the principal ways in which radioactive materials decay. You may already know these as alpha decay, beta decay and gamma-photon emission. For each radioactive nucleus, the decay process is dictated by the causes of the nuclear instability. Some nuclei can have too many protons; others have too many neutrons. Radioactive-decay processes correct these imbalances. As a nucleus undergoes a decay, it moves to an increasingly stable state.

The decay chain is the basis of the way in which a large, unstable nucleus moves to stability. This happens not by a single one-off change but by a series of incremental, and simple, shifts in the balance of the nucleons.

Radioactive decay is a random process. However, even random events can be managed using statistics. The large number of atoms in any manageable sample of a material means that averages can be predicted and that the predictions are accurate to a high degree. Randomness means that the mathematics of radioactive change is well understood because it depends on the knowledge that radioactivity is proportional to the number of active nuclei present in the sample. Predictions can be made from this straightforward starting point.

In this topic you will learn about:

- isotopes
- nuclear binding energy and mass defect
- the variation of the binding energy per nucleon with nucleon number
- mass–energy equivalence
- the existence of the strong nuclear force
- the random and spontaneous nature of radioactive decay
- alpha, beta and gamma radioactive decay
- neutrinos and antineutrinos
- the penetration and ionizing ability of alpha and beta particles and gamma rays

- activity, count rate and half-life for radioactive decay processes and their changes for integral values of half-life
- background radiation
- evidence for the strong nuclear force
- the stability of nuclides
- discrete nuclear energy levels and the evidence for them
- the continuous spectrum of beta decay as evidence for the neutrino
- the radioactive decay constant and the radioactive decay law $N = N_0 e^{-\lambda t}$
- the relationship between half-life and the radioactive decay constant.

AHL

Introduction

Radioactivity is nuclear change. An unstable atomic nucleus will, sooner or later, decay into a different element with the emission of one or more nuclear particles. These particles transfer energy from the unstable nucleus through ionization and other processes to other atoms with which they collide. These could be atoms in a living cell. The damage from these collisions drives cellular change and can lead to the evolution of the organism. But it can also lead to cell mutation. Radioactive decay has been an essential process but is also a potentially damaging one. We need a balanced view on its benefits and drawbacks.

(ATL) Self-management skills

In the history of science, some discoveries have been serendipitous—that is, they were unintentional. Becquerel observed radiation from uranium compounds because he left them lying on a photographic plate (Figure 1). Many other phenomena have been observed when scientists were not looking for them. For example, Wilhelm Röntgen had discovered X-rays largely by accident, the year before Becquerel's discovery. Playing with the properties of these new rays soon led to their ability to see the bones inside the human body (Figure 2).

Serendipity has led to the discovery of drugs and antibiotics, the development of new chemicals and the invention of many devices from the microwave to the Post-it Note. In these cases, the scientists were able to use what might have been a failed experiment and realize the potential in their observations.

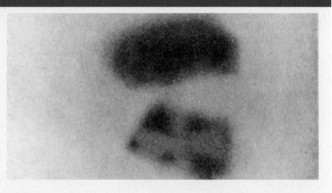

▲ Figure 1 Henri Becquerel placed this photographic plate in a drawer with some uranium compound wrapped in a black cloth. When he later looked at the plate, it appeared to have been exposed to light. The shadow of a metal Maltese Cross, located between the plate and the uranium compound, is also visible. Becquerel realized that the uranium compound was producing invisible radiation.

Isotopes and isotones

The chemistry of an element is determined largely by the configuration of the electrons outside the nucleus. Their numbers are equal to the number of protons in a neutral atom. Nuclei that have identical numbers of protons must therefore have very similar chemical properties and are known as **isotopes**.

> **Isotopes are two or more atoms that have the same proton number but different nucleon numbers.**

Carbon has fifteen known isotopes. The two stable forms of carbon are carbon-12 (also written as C-12; C is the chemical symbol for carbon) and carbon-13. Carbon-14 is the longest-lived unstable isotope of the element with a half-life of about 5700 years.

These three isotopes are written as: $^{12}_{6}C$, $^{13}_{6}C$ and $^{14}_{6}C$. They all have 6 protons in the nucleus. C-12 has 6 neutrons, C-13 has 7 neutrons and C-14 has 8. Each neutral isotope has 6 electrons and all isotopes share the same chemical properties.

The term "isotope" needs a little care. It is often misused. Do not use it as a general name for a stable or unstable nucleus—saying "a $^{1}_{1}H$ isotope" is incorrect unless you are speaking in the context of deuterium ($^{2}_{1}H$) as well. The proper term for a distinct type of nucleus with a characteristic number of protons and neutrons is a **nuclide**.

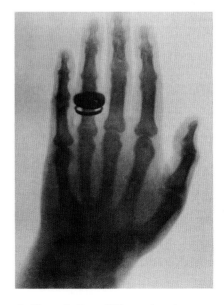

▲ Figure 2 One of Röntgen's early X-ray images.

There is an example of a radioactive series on page 664.

There is an example of a radioactive series on page 664.

There is an example of a radioactive series on page 664.

(ATL) Social skills

Marie Curie is sometimes described as the most famous female scientist ever. Her work on radioactivity led her to discover two new elements, radium and polonium. She shared the 1903 Nobel Prize for Physics with her husband Pierre Curie and Henri Becquerel. She was later awarded the 1911 Nobel Prize for Chemistry and became not only the first woman to win any Nobel Prize, but the first person to win two Nobel Prizes. She remains the only person to have won a Nobel Prize in two different sciences.

During the First World War, she endeavoured to use her scientific discoveries for good. She established mobile X-ray units and acted as a driver to assist wounded soldiers. She also realized the uses of radioactive sources for sterilizing wounds.

While her scientific achievements brought her increasing fame, she rarely accepted awards and donated prize money to scientific institutions. Einstein is reputed to have said that she was probably the only person not corrupted by the fame that she gained.

▲ **Figure 3** Marie and Pierre Curie pictured in c.1903.

Occasionally, it is helpful to have a name for two nuclides that share neutron numbers but have different proton numbers. These are called **isotones**. An example of a set of isotones that have 20 neutrons is: ^{36}S, ^{37}Cl, ^{38}Ar, ^{39}K and ^{40}Ca.

Can you complete the full notation for these isotones? The first is $^{36}_{16}$S.

Radioactive decay

Radioactive decay occurs naturally. The nucleus of an unstable atom changes into a different nuclear arrangement by emitting combinations of particles. The original unstable atom is known as the **parent nuclide**, the element that results from the change is the **daughter nuclide**. The daughter nuclide may or may not be stable. If it is not, then it will go on to decay again, eventually reaching stability after a long series of decays called a **radioactive series** or decay chain.

Radioactivity is:

- **random**; we cannot predict which nucleus in a sample of a radioactive material will decay next

- **spontaneous**; we cannot influence the rate of decay by changing the physical conditions of the sample such as its temperature, pressure, etc.

Nuclear changes during radioactive decay

The principal types of decay possible are:

- alpha (α) decay—a helium nucleus is emitted by the decaying nucleus

- beta (β) decay—an electron or its anti-particle is emitted or captured by the nucleus

- gamma (γ) emission—electromagnetic radiation of high frequency (a gamma photon) is emitted by the nucleus as it transfers energy away when moving to a lower energy state.

Alpha (α) decay

An unstable alpha emitter emits a particle that consists of two protons and two neutrons bound together. This particle is identical to the nucleus of a helium-4 atom ($^4_2$He). The overall change from parent to daughter is represented by a **nuclear equation** such as the decays of uranium-232 and polonium-210:

$$^{232}_{92}\text{U} \rightarrow {}^{228}_{90}\text{Th} + {}^4_2\text{He}$$

$$^{210}_{84}\text{Po} \rightarrow {}^{206}_{82}\text{Pb} + {}^4_2\alpha$$

Notice that in one of these equations the emitted alpha particle is shown as a helium nucleus, in the other with the symbol α. Both are correct.

The equations balance. The left-hand and right-hand totals for both A and Z are the same.

Energy is also released during the decay. This is not usually shown in the equation. In alpha decay, energy is transferred to the kinetic energies of both the daughter nucleus and the alpha particle. This has consequences for the motion of the products.

A fixed quantity of energy from the decay is available to the two decay products. This energy must be conserved as it is shared. But momentum must also be conserved. When the parent nucleus is stationary before the decay, then the total momentum after the decay must also be zero. In an obvious notation, $m_d \times v_d = -m_\alpha \times v_\alpha$. With a fixed amount of energy at the disposal of the system, the speed of the alpha particle away from the site of the decay and the recoil speed of the daughter nucleus are completely determined. Although the momenta are shared evenly, energy is not. The alpha mass is much less than the nuclear mass of the daughter and so the initial alpha energy is likely to be much more than the initial energy of the daughter nucleus.

The emitted alpha speeds are initially the same and the alpha particles lose a roughly fixed amount of energy per collision with each air atom as they travel. This means that alphas with the same energy from a particular nuclear decay travel about the same distance as each other, as shown by the alpha tracks in Figure 4.

▲ Figure 4 Polonium-212 is an alpha emitter. This is a photograph of the paths of the alpha particles made visible as they pass through the supersaturated water vapour in a cloud chamber. Most of the emitted alphas have the same initial energy and travel the same distance through the chamber. A single alpha particle has a higher initial energy and can travel further before stopping. This photograph was taken during the 1920s by Norman Feather.

Patterns and trends

One of the reasons why a large particle with four nucleons is emitted rather than the smaller proton is the energy bound up in the alpha particle. This is large; alpha particles are exceptionally stable. Once outside the nucleus, the alpha particle is a helium-4 nucleus and can exist as a stable entity. It is energetically preferable to allow the alpha particle rather than two protons and two neutrons to leave the nucleus.

A typical alpha particle will have an absolute potential energy of around 6 MeV and this is substantially less than the potential barrier that prevents particles leaving the nucleus. (Imagine the barrier as a fence over which particles must jump to escape. Under a classical view, when there is too little energy, then the particles cannot reach the top of the fence.)

The alpha particles leave the nucleus through quantum-mechanical tunnelling. The probability wave that describes the alpha particle gives a small, but finite, probability that the particle can exist outside the nucleus.

Worked example 1

Radon-222 $\left(^{222}_{86}Rn \right)$ decays by alpha-particle emission into polonium (Po).

a. Write down the nuclear equation for this decay.

b. Calculate the ratio $\dfrac{\text{kinetic energy of alpha particle}}{\text{kinetic energy of polonium}}$. Assume that the radon nucleus is at rest immediately before the decay.

Solution

a. $^{222}_{86}Rn \rightarrow\ ^{218}_{84}Po +\ ^{4}_{2}\alpha$

b. The combined momentum of the decay products is zero; hence $p_\alpha = -p_{Po}$.

The ratio of kinetic energies is $\dfrac{E_\alpha}{E_{Po}} = \dfrac{\frac{p_\alpha^2}{2m_\alpha}}{\frac{p_{Po}^2}{2m_{Po}}} = \left(\dfrac{p_\alpha}{p_{Po}}\right)^2 \dfrac{m_{Po}}{m_\alpha} = \dfrac{m_{Po}}{m_\alpha}$. The mass ratio of

the nuclei is, to a good approximation, equal to the ratio of their nucleon numbers;

therefore, $\dfrac{E_\alpha}{E_{Po}} = \dfrac{218}{4} = 54.5$. This means that the alpha particle accounts for about

98% of the kinetic energy of the decay products.

Practice questions

1. Thorium-234 $\left(^{234}_{90}\text{Th}\right)$ is a product of alpha decay of an isotope of uranium (U).

   a. Write down the nuclear equation for this decay.

   b. Calculate the number of neutrons in the parent nucleus of thorium-234.

2. Americium-241 $\left(^{241}_{95}\text{Am}\right)$ decays into an isotope of neptunium (Np) by an emission of an alpha particle.

   a. Write down the nuclear equation for this decay.

   b. The kinetic energy of the alpha particle is 5.5 MeV. Estimate the kinetic energy of the neptunium nucleus. Assume that americium is stationary when it decays.

Beta (β) decay

There are three variants of beta decay:

- the emission of a beta-minus (β⁻) particle (an electron)

- the emission of a beta-plus (β⁺) particle (a positron)

- electron capture.

Negative beta (β⁻) decay

A neutron in the nucleus is converted into a proton and an electron. The decay occurs in nuclides with a ratio of neutron:proton that is large and where, therefore, it is preferable to lose a neutron and gain a proton.

The electron leaves the nucleus because confining the electron in a very small nuclear region would require an enormous energy to be available in the nucleus. This is not the case, so the electron leaves the daughter nucleus as a result of the interaction. A third particle is emitted as well (unlike the alpha emission with two particles). This particle is known as an electron antineutrino (which has the symbol $\bar{\nu}_e$).

In terms of nuclear notation, the beta-minus decay of a thorium (Th) nucleus to protactinium (Pa) is

$$^{231}_{90}\text{Th} \rightarrow \,^{231}_{91}\text{Pa} + \,^{0}_{-1}\beta + \bar{\nu}_e$$

Again, the equation balances because neutrinos are small particles with no effective mass or charge. Neutrinos do not require A or Z values, but if you include them, they must be both zero. What is important in writing its symbol is to remember the "bar" over the Greek lower-case nu. This signifies that this electron antineutrino is an antiparticle.

Because there are three emitted particles in this equation (the daughter, the beta-minus particle and the antineutrino), there is no single solution to the momentum equation. The beta-minus particle can have a range of energies unlike the alpha particle earlier. It was the observation of this beta energy spectrum that led to the discovery of the neutrino. This is discussed on page 638 (AHL).

Positive beta (β⁺) or positron decay

Here a proton is converted to a neutron and a positron. As you might expect, it is observed in nuclei that are proton-rich. The positron is the antiparticle of an electron with one positive electronic charge and an identical mass to that of the electron.

The third particle here is an electron neutrino. This is an antiparticle to the electron antineutrino involved in beta-minus decay.

The decay of magnesium-23 to sodium-23 is typical of this type of decay. The magnesium nuclide does not occur naturally, but the daughter product is a stable form of sodium.

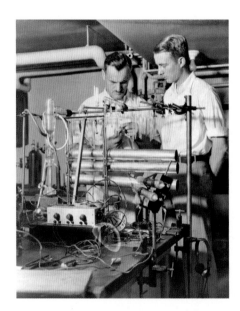

▲ Figure 5 This picture shows Arthur Compton (discoverer of the Compton effect) with Luiz Alvarez (on the right). Alvarez went on to observe electron capture for the first time in 1937 and was awarded the Nobel Prize in 1968 for his development of the bubble chamber. In 1980, in collaboration with his son and two nuclear chemists, he proposed that a large asteroid impact was responsible for the extinction of the dinosaurs—a theory which is now widely accepted.

$$^{23}_{12}\text{Mg} \rightarrow {}^{23}_{11}\text{Na} + {}^{0}_{+1}\beta^+ + \nu_e$$

The equation balances with the ν_e as an allowed shorthand for ${}^{0}_{0}\nu_e$.

Electron capture

An alternative way in which a proton-rich nucleus can increase the neutron:proton ratio is to capture an electron. This electron and a proton interact to produce a neutron and an electron neutrino. For any radioactive decay to take place, the difference in energy between parent and daughter nuclei must be sufficiently large to permit the decay. In nuclides that decay by electron capture, the energy difference is small and the positron cannot be created. Instead, the nucleus captures one of the inner electrons from the atomic shells outside the nucleus.

Subsequently, an outer-shell electron replaces the captured electron with the emission of a photon (of large energy usually in the X-ray region).

The equation for the electron-capture process as argon-37 decays is

$$^{37}_{18}\text{Ar} + {}^{0}_{-1}\beta^- \rightarrow {}^{37}_{17}\text{Cl} + \nu_e$$

where the outcome is a chlorine nucleus and the emitted electron neutrino.

(You might expect a positron-capture process to occur. Although this would be a symmetrical outcome equivalent to electron capture, it is not observed to occur naturally.)

Worked example 2

Write down nuclear equations for:

a. negative beta decay of cadmium-113 $\left({}^{113}_{48}\text{Cd}\right)$ into an isotope of indium (In)

b. positive beta decay of oxygen-15 $\left({}^{15}_{8}\text{O}\right)$ into an isotope of nitrogen (N)

c. electron capture by iron-55 $\left({}^{55}_{26}\text{Fe}\right)$ transmuting into an isotope of manganese (Mn).

Solutions

a. Decay particles are indium nucleus, an electron and an electron antineutrino.
$$^{113}_{48}\text{Cd} \rightarrow {}^{113}_{49}\text{In} + {}^{0}_{-1}\beta + \bar{\nu}_e.$$

There are other ways to represent an electron in nuclear reactions, some of them are ${}^{0}_{-1}e$, e^- or simply e.

b. Decay particles are nitrogen nucleus, a positron and an electron neutrino.
$$^{15}_{8}\text{O} \rightarrow {}^{15}_{7}\text{N} + {}^{0}_{1}\beta^+ + \nu_e$$

Alternative ways to represent a positron include ${}^{0}_{+1}e^+$, e^+ or $\bar{e}$ (the bar as usual denotes an antiparticle).

c. The products of the reaction are a manganese nucleus and a neutrino.
$$^{55}_{26}\text{Fe} + {}^{0}_{-1}\beta \rightarrow {}^{55}_{25}\text{Mn} + \nu_e$$

Practice questions

3. Thorium-234 $\left({}^{234}_{90}\text{Th}\right)$ decays into protactinium-234 $\left({}^{234}_{91}\text{Pa}\right)$. State the two remaining decay products of thorium-234.

4. Oxygen-18 $\left({}^{18}_{8}\text{O}\right)$ is formed when an isotope of fluorine (F) decays with an emission of a positron. Write down the nuclear equation for this decay.

Gamma-photon (γ) emission

A daughter nucleus is often left in an excited state after alpha or beta decay. An excited state is one in which the nucleus has a surplus amount of energy relative to the state of lowest energy. The following decay scheme is part of the path that begins with the neutron bombardment of naturally occurring cobalt-59:

$$_{27}^{59}\text{Co} + {}_{0}^{1}\text{n} \rightarrow {}_{27}^{60}\text{Co}^{m} \qquad\qquad\qquad\text{(A)}$$

$$_{27}^{60}\text{Co}^{m} \rightarrow {}_{27}^{60}\text{Co} + \gamma \qquad\qquad\qquad\text{(B)}$$

$$_{27}^{60}\text{Co} \rightarrow {}_{28}^{60}\text{Ni}^{m} + {}_{-1}^{0}\beta + \bar{\nu}_{e} \qquad\qquad\text{(C)}$$

$$_{28}^{60}\text{Ni}^{m} \rightarrow {}_{28}^{60}\text{Ni} + \gamma \qquad\qquad\qquad\text{(D)}$$

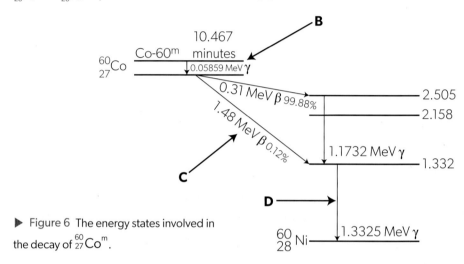

▶ Figure 6 The energy states involved in the decay of $_{27}^{60}\text{Co}^{m}$.

Figure 6 shows part of the energy-level diagram for this scheme.

- In A (not shown on the diagram) a single neutron interacts with naturally occurring cobalt to produce an excited state of cobalt-60. The excited nature of the state is indicated by the superscript "m" to the right of the chemical symbol.

- The original excited state is shown in the diagram (B) together with the energy level to which the nucleus falls after emitting a gamma photon of energy 0.059 MeV.

- The cobalt-60 now undergoes beta-minus decay to form the excited nickel nucleus $_{28}^{60}\text{Ni}^{m}$. This decay (step C) occurs in 99.9% of all cobalt-60 decays by emission of a β^{-} particle of maximum energy 0.31 MeV or, in 0.12% of all cobalt-60 decays, by the emission of a β^{-} particle of maximum energy 1.48 MeV. Whether path B or path C is taken, the nickel is produced in an excited state.

- Finally (step D), the metastable nickel decays by the emission of further photons to arrive at its lowest energy (ground) state.

The gamma photons (or gamma rays) emitted during radioactive decay obey all the usual rules of electromagnetic radiation, although they are likely to have high energies measured in MeV. The energy of the photon is given by $E = hf$ and they possess momentum according to $p = \dfrac{hf}{c}$.

Discrete nuclear energy levels

Figure 7 is an example of the energy changes that occur in gamma emission. There are only three photon energies that are observed in the de-excitation of the cobalt-60.

Earlier the nuclear equation for the decay of uranium-232 to thorium-228 was given. Figure 8 shows the three lowest energy states of the thorium. The figure also shows the alpha and gamma emissions that are observed most often.

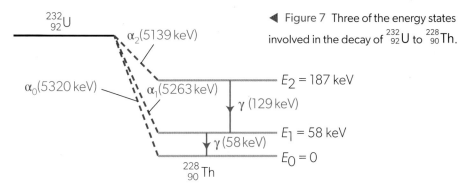

◀ Figure 7 Three of the energy states involved in the decay of $^{232}_{92}U$ to $^{228}_{90}Th$.

The discrete nature of the photon energies emitted in gamma decay and the discrete nature of the alpha energies should remind you of the line spectrum emitted by the hydrogen atom. This is a different phenomenon—the transition of the atom to a lower energy state—but nevertheless, the non-continuous nature of the energies is strongly indicative of a discrete energy structure in the nucleus.

Discrete spectra always provide evidence for discrete energy levels whether in the nucleus or in the atomic electron shells.

 Are there differences between the photons emitted as a result of atomic versus nuclear transitions?

Physics makes no distinctions between photons other than by energy. However, as Topic C.2 shows, we do distinguish these energies by assigning different names to the regions of the electromagnetic spectrum. Often these names reflect the origin of the photons.

A good example of this is the distinction made between photons in the X-ray region and gamma photons. Although both sets of photons can easily have the same energy (and thus the same frequency), they are usually imagined as having different origins:

- X-ray photons arise when high-energy electrons are decelerated to rest when they interact with matter. As the electrons lose the energy, it is radiated as photons in the X-ray region by a process known as bremsstrahlung (German for "braking radiation"). The spectrum of such X-rays is continuous (like black-body radiation).

  X-ray photons can also be produced when high-energy electrons eject the electrons from the tightly bound inner energy states of atoms of heavy metal elements such as tungsten. This radiation has a line spectrum and is known as a characteristic spectrum because it depends on the element from which the inner electrons are ejected.

- Gamma photons arise from interactions, but this time at the level of the nucleus. As described above, changes in nuclear energy states can lead to the emission of photons of high frequency. An example of this is the 58 keV photon emitted when the thorium-228 E_1, energy state changes to E_0 (Figure 8). This gamma photon comes from a nuclear energy change. An X-ray photon of almost exactly the same energy is emitted during the de-excitation of a tungsten atom following its the excitation by high energy electrons in the formation of an X-ray spectrum.

Additionally, nuclear transitions generally have energy gaps of the order of MeV whereas the orbital electron transitions from Topic E.1 have gaps of the order of eV.

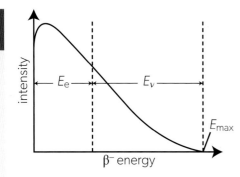

▲ Figure 8 The graph of intensity against beta-minus energy for the beta particles emitted in beta decay.

Continuous spectra of beta decay

Unlike alpha particles that have only a single initial energy (or at most a few discrete energies), the beta particles from a particular decay are observed to take any energy value between zero and the maximum (Figure 8). This maximum value depends on the decay in question and is related to the energy released as a result of the decay.

In alpha decay, the daughter nucleus and the alpha particle share the available energy. With an initially stationary parent nucleus, the momentum of the two particles must sum to zero. Two equations (conservation of energy and momentum) and two unknowns (speeds of the emitted particles) means that there is one unique mathematical solution. The alpha particles can take only one initial speed for each available discrete energy.

In beta decay, energy and momentum must still be conserved, but with three particles and two equations, there is no longer a unique solution. The total energy can be divided between the kinetic energies of daughter, neutrino and electron in an infinite number of ways. In fact, because the mass of the daughter atom is very large compared with that of the electron and neutrino, its mass is by far the largest factor in calculating its momentum, and both its speed and kinetic energy must be small. The energy is therefore effectively split between the neutrino E_v and the beta particle E_e, as shown in Figure 8. The vertical dotted line can be placed anywhere between zero and E_{max} to give the infinite number of possibilities for the energy split.

The detection of a neutrino is exceptionally difficult. The particle interacts only weakly with matter. The first detection experiment involved the use of antineutrinos from a nuclear reactor that then interacted with protons in nuclei to produce neutrons and positrons. The positron almost immediately annihilates with an electron to produce two gamma photons. The neutron is also captured with the emission of a third photon. The coincidence of the photon arrivals at detectors indicates that an antineutrino interaction must have occurred.

There is a constant and very large flux of neutrinos incident on (and through) Earth. For example, the neutrinos emitted as a result of the fusion reactions in the Sun give rise to a flux of 10^{15} neutrinos every second through one square metre of Earth's surface. Experiments carried out in gold mines 3.5 km below Earth's surface and elsewhere have confirmed the properties of the neutrinos as being uncharged, having essentially zero mass and having a very weak interaction with matter. The chance of a single neutrino undergoing one interaction as it passes through Earth along a diameter is less than 1 in 10^{11}.

 How did conservation lead to experimental evidence of the neutrino? (NOS)

Differences between alpha decay and beta emission led Wolfgang Pauli in 1930 to the prediction that a third type of particle was involved in beta decay. The evidence of monoenergetic alpha decay and continuous-spectrum beta decay was an enormous problem for the physicists of the day. Niels Bohr even proposed, at one point, that the principle of conservation of momentum would have to be discarded.

The Solvay conference of 1933 was officially devoted to the discovery of the neutron, but Enrico Fermi used the event to propose that an electron–neutrino pair is the result of beta decay in just the same way that a photon is emitted

by an excited atom. This was the precursor of the modern theory of the weak interaction.

However, it took over 20 years before Frederick Reines and Clyde Cowan found evidence for the existence of the neutrino in 1956. This shows the difficulties that must be overcome to observe interactions involving the neutrino.

This is an example of a successful experiment that provides evidence as a result of a hypothesis. It led to the theory of the weak interaction originally propounded by Murray Gell-Mann one year before the observation of the neutrino that had started it all.

Properties of ionizing radiation

The emitted alpha, beta-minus and beta-plus particles and the gamma photons are classed as **ionizing radiation**. This means that, as they pass through matter, they ionize the material (creating separated electrons and positive ions from previously neutral atoms). This ionization requires a transfer of energy from the energetic particle to separate the electron from the ion.

In the case of alpha and beta particles, ionization results in a reduction in the kinetic energy of the particle. (It slows down.) This is not an option for the gamma photon, which travels at the speed of light in the medium concerned. The loss of energy will either be complete—that is, the photon is completely absorbed—or it will experience a frequency shift. This occurs in Compton scattering (Topic E.2) where a photon and an electron interact, with the electron gaining energy and the photon shifting to a lower frequency, longer wavelength.

Alpha particles are the most massive of the emitted particles. They also have the largest charge. For a given energy, they travel slowly and take the longest time of the particles to pass an atom. The charge and mass of the alphas mean that they have a high chance of interaction with neighbouring atoms and of transferring energy to them. This interaction can be via an inelastic or elastic collision, through ionization, or by some other effect.

Table 1 lists the properties of the radioactive emissions. These are approximate and there are exceptions to the suggested absorption ranges.

It is particularly important to prevent alpha particles coming in contact with animal tissue, whether externally on the skin, or internally through inhaling or swallowing the radioactive emitter.

The strong nuclear force

Earlier themes have discussed the electromagnetic interaction between charges and the gravitational interaction between masses. Another interaction that acts only within the nucleus is the **strong nuclear force**, which is short-range and attractive.

Properties of the strong nuclear force:

- relative strength compared with gravity is approximately 10^{38}

- range is between nucleons ~ 1 fm

- involves nuclear particles such as protons and neutrons—the strong interaction does not affect electrons and neutrinos.

There was an implicit problem with the model of the nucleus in the first half of the 20th century. The discovery of the neutron by Chadwick in 1932 accounted for the nuclear mass, but not for the interactions inside the nucleus. The protons are positively charged and, for every nuclide from helium upwards, there must be electrostatic repulsion between protons in the nucleus. Hydrogen, having only one proton, is exempt.

The nucleus, on this basis, is inherently unstable. It should fly apart under the effects of the electrostatic repulsion. Evidently this does not happen, so there must be other attractive forces, or interactions, that overcome this.

A simple description of the effects of the strong nuclear force and electrostatic repulsion on two protons is given in Figure 9.

- In region X, both strong and electrostatic forces are repulsive and the protons repel each other.

- In region Z, the strong nuclear force is very weak and approaches zero asymptotically on the separation axis. The electrostatic force dominates and the overall force is repulsive.

Alpha particles:

- are very highly ionizing
- penetrate matter very poorly
- are absorbed by a few centimetres of air
- are absorbed by a thin sheet of paper.

Beta-minus particles:

- have a moderate ionizing power compared with alpha particles of the same energy
- have a moderate ability to penetrate matter
- are absorbed by around 25 cm of air
- are absorbed by a few centimetres of animal or plant tissue
- are absorbed by a few millimetres of aluminium.

Gamma photons:

- are weakly ionizing
- eject electrons from metals (via the photoelectric effect) that cause secondary ionization
- are highly penetrating
- are absorbed by a few centimetres of lead
- are absorbed by a few metres of concrete
- are only weakly absorbed by animal tissue.

▲ Table 1 Properties of the radioactive emissions.

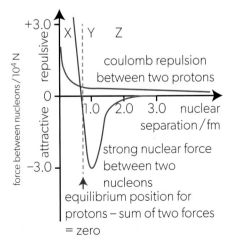

▲ Figure 9 The variation of the forces acting between two nuclear protons with distance between them.

The strong nuclear force is discussed in more detail in the additional higher level section on page 647 later in this topic.

- In region Y, the strong nuclear force is strongly attractive and dominates the repulsive electrostatic force. There is an overall attraction between the two protons and this range of separations is their preferred distance apart.

For a nucleus of helium, $A = 4$ and the radius of the nucleus is about 2 fm. This suggests that the protons will be within region Y of each other. However, for large nuclei—for example uranium where the nuclear radius is about 7.5 fm—there will still be problems of repulsion between protons on opposite sides of the nucleus. This is where the neutrons help. They are subject to the strong nuclear force but *not* electrostatic repulsion because they have no charge. Packing the nucleus with neutrons allows the amount of attraction to increase because now protons will be attracted by nearby neutrons.

 Would a nucleus be able to exist if only gravitational and electric forces were considered?

The existence of equilibria between opposing forces accounts for many areas of physics. The force that balances the electrostatic repulsion between protons is not gravity. The table on page 533 shows the relative sizes of the four known interactions and it is clear that gravity cannot hold the key. The theory behind this table is covered in Topics D.1 and D.2.

The nature of the attractive force—now named the strong interaction—was originally proposed by Hideki Yukawa in 1935. He suggested that the nucleus was stable because nucleons exchanged particles between themselves rather as the massless photon connects one object with another through what we call electromagnetic radiation. Imagine two children, each with a ball, throwing the ball to the other child. While the balls are in the air the children are connected by this ball transfer. Yukawa predicted the existence of these nuclear particles (the balls) that he named mesons. The pion, one type of meson, was discovered in 1947 by Cecil Powell when he flew balloons carrying photographic plates into the high atmosphere.

 How does equilibrium within a star compare with stability within the nucleus of an atom?

The stable nucleus of an atom has a balance between strong interaction and electrostatic repulsion. In a similar way, a star has a dynamic equilibrium. The outwards expansion of the star is due to thermal and radiation pressures arising from the extremely high temperatures of the interior. The compression comes from the gravitational forces that are tending to pull the star's material inwards towards the centre. For most of its life the star is stable, even though these two forces are slowly changing. This is described in Topic E.5.

Equilibria of different sorts signal the end of stellar evolution too. Some stars end as neutron stars in which the equilibrium is a balance between the outwards pressure of neutrons (the end point of electron capture by a proton) and the gravitational attraction, as before. Other, smaller stars end as white dwarfs where the equilibrium is between electron pressure and gravity. Finally, when a neutron star exceeds a certain mass, the neutron pressure outwards cannot resist the inwards attraction of gravity. This time the star collapses completely to form a black hole from which not even photons can escape.

Mass defect and nuclear binding energy

Radioactive change has so far been described simply in terms of instability in the parent nucleus. What are the energy transfers that occur during a decay or a nuclear fusion or fission?

The fact that the nucleus is a stable and bound system means that energy is required to break the nucleus apart and separate the individual neutrons. Figure 10 gives the process for separation and recombination of the simplest nucleus that contains more than one nucleon: deuterium.

Deuterium has one proton and one neutron. To separate these (Figure 10(a)), energy must be transferred to the nucleus to break the strong nuclear interaction between the proton and neutron. (There is no electrostatic repulsion.) This energy is known as the **binding energy**.

When more energy than the binding energy is provided, the excess will appear as kinetic energy of the two nucleons so that they move apart (Figure 10(b)).

Conversely, a proton and a neutron can collide to form a deuterium nucleus and release the binding energy as they form a bound system. This energy is released as a photon.

Energy and mass are related through the Einstein mass–energy equivalence: $E = mc^2$. This is better expressed as

$$\Delta E = c^2 \Delta m$$

where ΔE is the energy transfer and Δm is the mass difference. The release of energy when the nucleus forms from its component nucleons is equivalent to a loss of mass. The total mass of the separate nucleons in any nucleus is greater than the mass of the nucleus, since energy must be provided to separate the nucleons from the bound system. This difference in mass is called the **mass defect**.

It is important to realize that this change of mass is implicit in any energy transfer. When 1.0 kg of water is boiled completely into steam, an energy of 2.26 MJ is required to change the water molecules from the liquid to the gas phase. But this (when divided by c^2) is a mass change of about 10^{-11} kg, so is not noticed on a regular basis.

Nuclear mass

Changes in mass during a nuclear reaction drive the reactions and it will be important to have suitable units for mass measurement in nuclear physics. Measuring in kilograms is not suitable as the small mass of a nucleus leads to inconvenient and large negative powers of ten. Instead, it is usual to use a quantity that is defined by experiment, in this case the **unified atomic mass unit**. The term is usually shortened to atomic mass unit (or even amu) and has a unit symbol of u.

The value of the atomic mass unit is $\frac{1}{12} \times$ (mass of a $^{12}_{6}C$ atom).

The full value of the atomic mass unit is
1 atomic mass unit (u) $\equiv 1.660\,539\,066\,60(50) \times 10^{-27}$ kg $\equiv$ $931.494\,102\,42(28)$ MeV c^{-2}. (The brackets around the final pair of digits is intended to indicate the uncertainty in the measurement. There are further details in the *Tools for Physics* section page 349.)

The final alternative unit that is used is the energy equivalent of mass. This allows masses to be measured in units of MeV c^{-2}. The rearranged equation $\Delta m = \dfrac{\Delta E}{c^2}$ allows Δm to be measured in energy units divided by c^2.

To check that this is correct: an energy of 931.5 MeV is equivalent to $931.5 \times 10^6 \times 1.6 \times 10^{-19} = 1.49 \times 10^{-10}$ J. When this is divided by c^2 (9×10^{16} m^2 s^{-2}), the answer is 1.66×10^{-27} kg, as stated above.

This makes the:

- mass of a proton $= 1.673 \times 10^{-27}$ kg $= 1.007\,276$ u $= 938$ MeV c^{-2}

- mass of a neutron $= 1.675 \times 10^{-27}$ kg $= 1.008\,665$ u $= 940$ MeV c^{-2}.

The neutron is slightly more massive than the proton.

Chemists often use the unit "dalton" (symbol: Da) rather than refer to it as the "unified atomic mass unit". It is certainly shorter! Because 1 Da $\equiv$ 1 u, there is no need to learn anything new, but the Da will not be used in the IB Diploma Programme physics examinations.

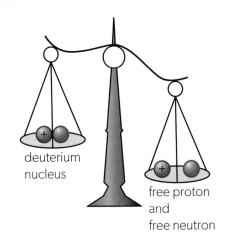

deuterium nucleus

free proton and free neutron

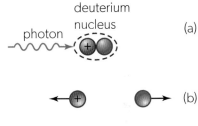

photon → deuterium nucleus (a)

proton → ← neutron (c)

deuterium nucleus → photon (d)

▲ Figure 10 A deuterium nucleus can be separated into a proton and neutron when energy is supplied to it (a & b). When a proton and neutron combine, this energy must be released (c & d). The total mass of the deuterium nucleus is less than that of the free proton and free neutron. The difference is the mass defect, which is equivalent to the binding energy.

Worked example 3

The mass of an atom of copper-63 $\left(^{63}_{29}Cu\right)$ is 62.929 597 u.

Calculate, for a nucleus of copper-63, its: a. mass defect   b. binding energy.

Solutions

a. We need to subtract the mass of the electrons from the atomic mass to get the mass of the nucleus of copper: $m_{nucleus} = m_{atom} - Zm_e = 62.929\,597 - 29 \times 0.000\,549 = 62.913\,676\,u$.
The nucleus of copper has 29 protons and 34 neutrons. The mass defect is $\Delta m = Zm_p + Nm_n - m_{nucleus}$
$= 29 \times 1.007\,276 + 34 \times 1.008\,665 - 62.913\,676 = 0.591\,938\,u$.

b. The binding energy of a nucleus is the energy equivalent of its mass defect. Numerically, the conversion between mass and energy units is easiest when the mass is expressed in MeVc^{-2}: $\Delta m = 0.591\,938 \times 931.5$
$= 551.4\,MeV\,c^{-2}$. From here, the binding energy $= \Delta mc^2 = 551.4\,MeV$.
If needed, this can be easily converted to joules: $551.4 \times 10^6 \times 1.60 \times 10^{-19} = 8.82 \times 10^{-11}\,J$.

Practice questions

5. For each of the following nuclei, calculate its:
i. mass defect   ii. binding energy.

a. oxygen-16 $\left(^{16}_{8}O\right)$, atomic mass $= 15.994\,915\,u$.

b. nickel-58 $\left(^{58}_{28}Ni\right)$, atomic mass $= 57.935\,342\,u$.

c. lead-208 $\left(^{208}_{82}Pb\right)$, atomic mass $= 207.976\,652\,u$.

6. The mass of a typical grain of coarse sand is 12 mg. Calculate the energy equivalent of this mass. Express the answer in (a) joules and (b) MWh.

Data-based questions

The table shows the masses of nuclides which all have $A = 197$.

Nuclide	Z	Nuclear mass / u
^{197}Ir	77	196.969 65
^{197}Pt	78	196.967 34
^{197}Au	79	196.966 57
^{197}Hg	80	196.967 21
^{197}Tl	81	196.969 58
^{197}Pb	82	196.973 43
^{197}Bi	83	196.978 86
^{197}Po	84	196.985 66
^{197}At	85	196.993 19
^{197}Rn	86	197.001 58

- Tabulate values of the mass defect m_D for each nuclide. ($m_p = 1.007\,825\,0\,u$ and $m_n = 1.008\,664\,9\,u$)

- Plot a graph of the mass defect m_D against Z.

- The mass defect for these nuclei can be modelled by the equation $m_D = c_1 - c_2(Z - c_3)^2$. Taking a value of $c_3 = 78.5$, plot a graph of m_D against $(Z - c_3)^2$. Use your graph to find the values of c_1 and c_2.

- The formula is empirical (that is to say, it is based on observation rather than testing a hypothesis). Evaluate how well this equation predicts the mass defect.

- What is the maximum percentage deviation of the mass defect from the predicted value?

Variation of binding energy per nucleon

As the nucleon number increases, the binding energy of the nucleus increases. There are more strong nuclear force "bonds" because each nucleon sees every other nucleon through the strong interaction. Going from four nucleons to five nucleons means going from six nucleon–nucleon bonds to ten. More energy *per nucleon* will be required to break up a large nucleus compared with a small one.

This makes the quantity the **average binding energy per nucleon** a useful one.

To calculate the average binding energy per nucleon, divide the total binding energy of a specific nucleus by the nucleon number (total number of nucleons in the nucleus):

$$\text{average binding energy per nucle on} = \frac{\text{total binding energy of nucleus}}{A}$$

A plot showing average binding energy per nucleon against nucleon number is an important tool in our understanding of nuclear change (Figure 11). Each point on the chart shows one nuclide.

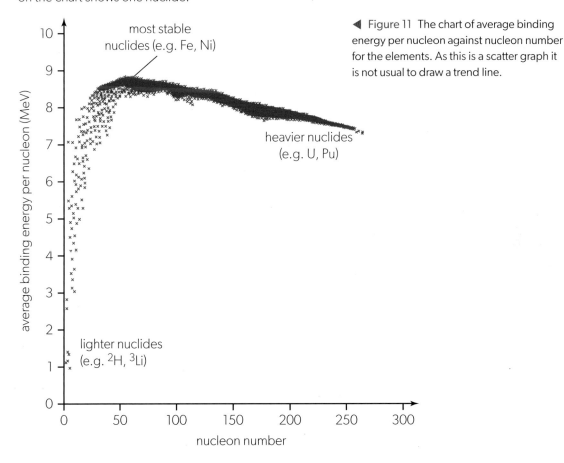

◀ Figure 11 The chart of average binding energy per nucleon against nucleon number for the elements. As this is a scatter graph it is not usual to draw a trend line.

General features of the plot are:

- The larger the binding energy per nucleon (in other words, the higher up the chart) the more stable is the nuclide, up to the point where $A = 60$. The most stable nuclides (for example, iron and nickel) are in this region.

- For larger nucleon numbers beyond this stability region, the general stability decreases, which means that it is energetically favourable for very large nuclides (U, Pu, etc) to split up to form smaller but more stable nuclides. This process is known as **nuclear fission**. It is examined in more detail in Topic E.4.

- For values of nucleon number up to the stability region, it is energetically favourable for some nuclides to join forming larger nuclei. This process is called **nuclear fusion**. It is described in the context of the formation and evolution of stars in Topic E.5.

- The region from $A = 1$ to $A = 20$ is of particular interest (and is shown in more detail in Figure 12). Certain nuclides (^{4}He, ^{12}C, ^{16}O) are markedly above the trend of the plot. This means that they are more stable than other nuclides of similar mass. In particular, ^{4}He is an especially stable nuclide and this accounts for its important role in stellar evolution.

- Hydrogen (^{1}H) cannot have a binding energy with only one proton, so is not generally shown on the chart.

- The most massive nuclides on the far right have roughly 1 MeV per nucleon less binding energy than the most stable. This means that splitting a large nucleus up is likely to release significant amounts of energy during fission.

- A similar argument applies to the formation of ^{4}He from smaller nuclei. Significant amounts of energy will be released (roughly 6 MeV per nucleon when forming from ^{2}H).

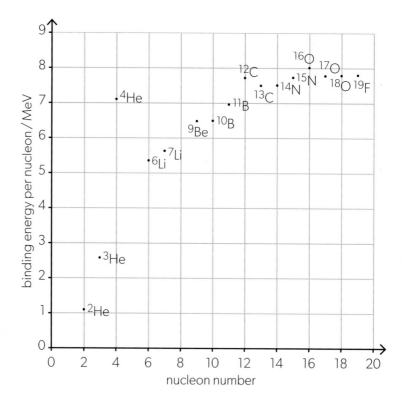

▶ Figure 12 The fusion region of the binding energy per nucleon chart. Nuclides that are formed from integer multiples of helium-4 nuclei are locally more stable than other nearby nuclides.

Worked example 4

Calculate the binding energy per nucleon of copper-63 $\left(^{63}_{29}\text{Cu}\right)$.

Solution

The total binding energy (BE) of copper-63 is 551.4 MeV (see Worked example 3).

$$\frac{\text{BE}}{A} = \frac{551.4}{63} = 8.752 \text{ MeV.}$$

Worked example 5

Radium-223 decays into radon-219 according to the reaction $^{223}_{88}\text{Ra} \rightarrow \, ^{219}_{86}\text{Rn} + \, ^{4}_{2}\alpha$.
The following data are given.

Nuclide	Binding energy per nucleon / MeV
$^{223}_{88}\text{Ra}$	7.6853
$^{219}_{86}\text{Rn}$	7.7238
$^{4}_{2}\alpha$	7.0739

Determine the energy released in this decay.

Solution

The total binding energy of the radium nucleus is $223 \times 7.6853 = 1713.8$ MeV. The total binding energy of the decay products is $219 \times 7.7238 + 4 \times 7.0739 = 1719.8$ MeV. The binding energy has increased as a result of the decay, indicating that the mass of the products is less than the mass of the radium nucleus.

The energy released in the decay is equal to the difference between the binding energies, $1719.8 - 1713.8 = 6.0$ MeV. This energy appears as the kinetic energy of the alpha particle and the radon nucleus.

Worked example 6

Nitrogen-13 $\left(^{13}_{7}\text{N}\right)$ decays into carbon-13 $\left(^{13}_{6}\text{C}\right)$.

a. Identify this type of radioactive decay.

b. The atomic mass of nitrogen-13 is 13.005 739 u and that of carbon-13 is 13.003 355 u. Determine the energy released in the decay.

Solutions

a. The nucleon number is unchanged, but the daughter nuclide has one proton less, so a positively charged particle must have been emitted. This particle is a positron and the decay is an example of positive beta emission.

b. The nuclear equation for the decay is $^{13}_{7}\text{N} \rightarrow \, ^{13}_{6}\text{C} + \, ^{0}_{1}\beta^{+} + \nu_{e}$. Neutral atoms of carbon and nitrogen contain a different number of electrons and we need to subtract the mass of electrons from each of the atomic masses to obtain their nuclear mass.

$m_{\text{nitrogen}} = 13.005\,739 - 7 \times 0.000\,549 = 13.001\,896\,\text{u}$, $m_{\text{carbon}} = 13.003\,355 - 6 \times 0.000\,549 = 13.000\,061\,\text{u}$. The mass difference in the decay is $\Delta m = m_{\text{nitrogen}} - (m_{\text{carbon}} + m_{e}) = 0.001\,286\,\text{u}$. The mass of the positron is the same as that of an electron and we can treat the neutrino as a massless particle. The energy released is $E = \Delta m c^2 = 0.001\,286 \times 931.5 = 1.198$ MeV. This energy is shared between the positron and the neutrino, as the recoil energy of carbon is negligible due to its large relative mass.

Practice questions

7. For each of the nuclides in practice question 5 (page 642), calculate the binding energy per nucleon.

8. Calculate the energy released in the following decays:

    a. $^{238}_{92}\text{U} \rightarrow \, ^{234}_{90}\text{Th} + \, ^{4}_{2}\alpha$.

    Binding energies per nucleon of $^{238}_{92}\text{U}$: 7.5701 MeV, $^{234}_{90}\text{Th}$: 7.5969 MeV, $^{4}_{2}\alpha$: 7.0739 MeV.

    b. $^{18}_{9}\text{F} \rightarrow \, ^{18}_{8}\text{O} + \, ^{0}_{1}\beta^{+} + \nu_{e}$.

    Atomic masses of $^{18}_{9}\text{F}$: 18.000 937 u, $^{18}_{8}\text{O}$: 17.999 160 u.

Models—Viewing binding energy another way

Strictly, binding energy is a negative quantity as it is the energy that must be transferred to the nucleus to split it up into its components. Some physicists draw the chart of binding energy per nucleon "upside down" to emphasize this (Figure 13(a)). The top of the plot is now the energy level at which the nucleons are free and the distance from a nuclide's position to the zero line is the binding energy required per nucleon to separate everything.

Another aspect of this upside-down plot is that it also shows mass per nucleon (after the y-axis has been scaled appropriately). A lower mass per nucleon on this chart means that a larger fraction of the mass of the original nucleons is "hidden" as binding energy (mass defect).

Yet another possible way to show the importance of binding energy per nucleon plots is to draw a 3D plot with axes: proton number Z, neutron number N and average binding energy per nucleon (Figure 13(b)). This plot is sometimes known as the **valley of stability**.

- The plane at the top of the diagram (BE per nucleon = 0) is the region of free protons and neutrons. When these form a nuclide, the position of the nuclide is somewhere on the valley floor.

- For small N and Z (the back left corner of the plot) the fusion hill runs down and eventually meets the iron lake $\left(^{56}_{26}\text{Fe}\right)$ at the most stable part of the diagram. Nuclides on this slope are most likely to increase in size through fusion.

- Around the iron lake is a region where beta-minus and beta-plus decay occurs. Excess neutrons lead to β^- emission. Excess protons lead to β^+.

- As N and Z increase from the iron lake, towards the front right corner, the N increases more rapidly than Z. More neutrons are required in a particular nucleus to counteract the proton–proton repulsion. Nuclides on this gradual slope are likely to undergo radioactive decay via alpha emission, removing 2 protons and 2 neutrons in a single event.

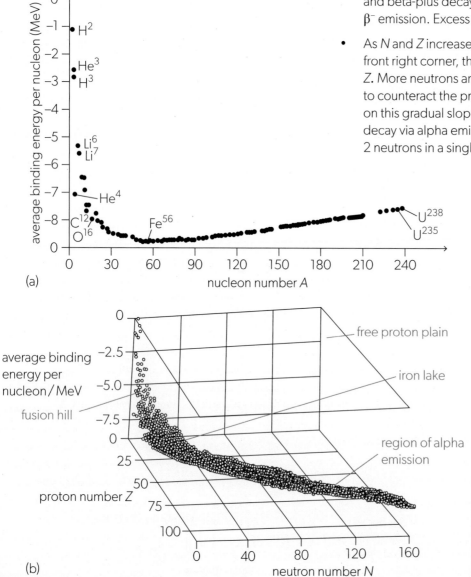

(a)

(b)

◀ Figure 13 (a) The binding energy per nucleon chart replotted to show that the binding energies are negative quantities. (b) This "negative" shape is sometimes replotted in 3D to show the "valley of stability".

Evidence for the strong nuclear force

Topic E.1 contains a description of the Geiger–Marsden–Rutherford experiment in which alpha particles were scattered by the nuclei of gold and other metals. When Rutherford, in 1919, examined the scattering data for aluminium he realized that there was a deviation from the expected result. The equation (page 594) that models the alpha scattering assumes the Coulomb law, predicting that the number N of alpha particles scattering at an angle ϕ to the incident direction is given by

$$N \propto \frac{1}{\sin^4\left(\dfrac{\phi}{2}\right)}$$

You can see this dependency clearly in the Rutherford equation.

This implies that, when the Coulomb law holds, $N \times \sin^4\left(\dfrac{\phi}{2}\right)$ should be constant.

Rutherford found that the equation began to break down at large ϕ. This is because aluminium has a relatively small nuclear positive charge compared with the original gold atoms of the experiment. The alpha particles therefore approach closer to the nucleus in aluminium. They are sufficiently close for the strong force to act on them.

Figure 11 of Topic E.1 shows Eisberg and Porter's published data for a fixed scattering angle of 60° with varying initial alpha energies.

Up to initial energies of 28 MeV, the scattering obeys the Rutherford prediction. However, at energies larger than 28 MeV, the Rutherford prediction is no longer obeyed and the intensity drops rapidly with energy. This is the regime where the strong nuclear force takes over.

Further evidence of the internal structure of the nucleus comes from other types of scattering experiment in the late 1960s. Beams of electrons can be accelerated towards nuclei in particle accelerators. The electrons—with their wave-like properties; see Topic E.2—can be diffracted by the nuclei. Topic C.3 shows that a minimum appears at a diffraction (scattering) angle θ given by $\theta \approx \dfrac{\lambda}{b}$, where b is the diameter of the nucleus and λ is the de Broglie wavelength of the electron $\left(= \dfrac{h}{p}\right)$. This scattering, which is elastic because no energy is lost, can be used to determine the diameter of the nucleus.

However, as the initial energy of the electron increases, the scattering becomes inelastic and the scattering behaviour deviates from what is expected. This is because the electrons—which are not subject to the strong nuclear force—are able to interact with the internal structure of the protons. Experiments of this type have revealed that protons and neutrons contain quarks. These are a more fundamental particle than either a proton or a neutron. It is transitions between quark types that allow a neutron to convert to a proton and an electron. This change in quark configuration is the basic mechanism that drives beta decay.

Further evidence is provided by the approximately constant value of the binding energy per nucleon plot from the most stable nuclides ($Z \approx 60$) up to the highest values of nucleon number. Imagine a particular nucleon in the centre of a thorium-234 nucleus. This nucleon is only influenced by the strong interaction from a relatively few "nearest-neighbour" nucleons around and close to it. The effects of the strong interaction on nucleons at the surface of the nucleus are shielded from the central nucleus. Compare this with a uranium-238 nucleus that

Hypotheses— Superheavy elements

Elements with high proton numbers are unstable. The element with a proton number of 118—Oganesson—has a half-life of less than a millisecond and was first synthesized in 2002.

Some theories of nuclear stability predict that elements with proton numbers around 114 might have some isotopes which are more stable than the currently known isotopes. This is referred to as the island of stability. If found to exist, it may point to the possibility of long-lived isotopes of super-heavy nuclei.

has four more nucleons imagined to be at the surface. These additional nucleons will make little or no difference to the strong forces that act on our central nucleus, so the binding energy per nucleon will not change significantly.

This is a crude argument, because the strength and distance variation of the strong nuclear force vary in subtle ways. However, the change of about 1 MeV per nucleon over the range from $Z \approx 60$ to $Z \approx 240$ is enough to drive nuclear fission but also constant enough to provide us with evidence of the strong nuclear force and its effects.

Nuclear stability

The plot of binding energy per nucleon against A gives valuable insights into the behaviour of nuclides in various parts of the Periodic Table. There are other plots that provide us with information about the nature of decay.

Figure 14 shows the plot of neutron number N versus proton number Z for all the known nuclides whether stable or unstable.

- Stable nuclides are shown in red. When N and Z are small (up to about 15), this is a straight line so that there are equal numbers of neutrons and protons. The region defined by these stable nuclides is known as the **zone of stability**.

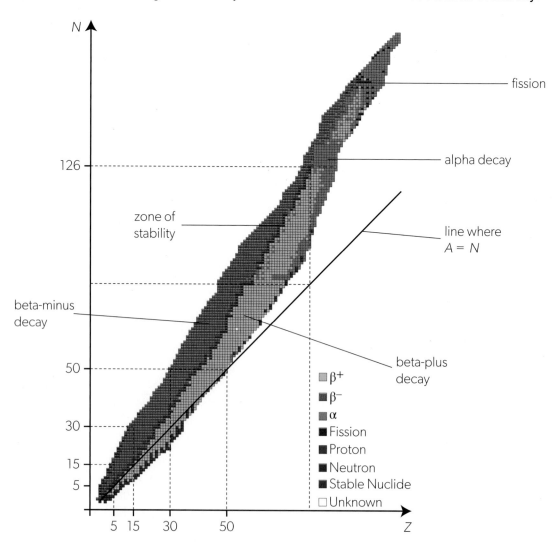

▲ Figure 14 The zone of stability in the plot of the known nuclides (both stable and unstable) of N against Z.

- A line of $N = Z$ is also drawn on the plot. When N and Z are greater than around 15, the zone of stability gradually moves above this line, showing that more neutrons than protons are required for stability. This is the consequence of the increased electromagnetic repulsion at high Z. The extra neutrons supply the attractive strong nuclear force to overcome this.

- Each side of the zone of stability, nuclides are unstable. Generally, the further from the line, the more unstable the nuclide. Greater instability means that the nuclei are likely to decay more quickly (with a larger decay constant, shorter half-life; see later in this topic).

- Above the stability line are the nuclides that decay by beta-minus (β^-) emission. These nuclei are neutron-rich and move diagonally downwards to the right towards the zone of stability, converting a neutron to a proton as they go. Figure 15 shows the change.

- Below the stability line are proton-rich nuclides. These are positron emitters. The change moves the nuclide diagonally upwards to the left.

- At high Z and N are the alpha-particle emitters. Recalling the binding energy per nucleon plot reminds you that the most stable nuclides are around $A = 60$. At higher A, the binding energy per nucleon decreases, so that the higher nuclides are slightly less stable even though their proton:neutron balance is correct. These nuclides decay via alpha emission because this changes both Z and N at once. The nuclide moves two diagonals down and to the left on the plot.

  Well away from the zone of stability at low N and Z lie some very unstable nuclides that emit neutrons (coloured purple on the plot) and protons (coloured red) directly.

- Finally, at large N and Z are some rare nuclides that fission spontaneously (rather than emit alphas). These are coloured green.

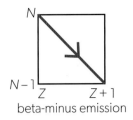

beta-minus emission

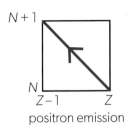

positron emission

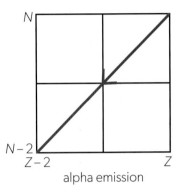

alpha emission

▲ Figure 15 Diagram for decay by beta-minus emission, positron emission and alpha emission.

Worked example 7

Aluminium has only one stable isotope, aluminium-27 $\left(^{27}_{13}\text{Al}\right)$.

Predict the likely decay mode of the radioactive isotopes of aluminium:

a. aluminium-26

b. aluminium-28.

Solutions

a. Aluminium-26 is 'proton-rich' (it has one neutron less than the stable isotope), so it is likely to decay with an emission of a positively charged particle, a positron (positive beta emission).

b. Aluminium-28 has one neutron more than the stable isotope and is therefore 'neutron-rich'. It will likely decay via negative beta emission, converting one of its neutrons into a proton.

Radioactive half-life

Radioactivity was described earlier as a process that is random and spontaneous. We cannot predict which nucleus will decay next, or when it will happen.

Radioactive phenomena obey the laws of random statistics. These tell us that, when large numbers of objects are involved, we can make accurate predictions about behaviour. We will deal from now on, not with the details of an individual nuclear change, but with the behaviour of many nuclei in a sample. This

When the initial number of parent nuclei is N_0, the number of parent nuclei remaining:

- after 1 half-life is $\dfrac{N_0}{2}$

- after 2 half-lives is $\dfrac{N_0}{4}$

- after 3 half-lives is $\dfrac{N_0}{8}$

- and so on…

number is very large indeed: 14 g of radioactive carbon-14—the molar mass is 14 g mol^{-1}—will have about 6×10^{23} atoms in the sample. If your school has access to radioactive materials for experiments, these materials may typically have masses around 1 μg. This still means that the whole sample consists of about 4×10^{16} atoms.

Each of the atoms in a radioactive nuclide has a constant probability of decay, independent of the sample size. This means that the **activity of the sample**—the total amount of decay that happens in one second—depends on the size of the sample. This is the **rate of decay**: the rate (per second) at which the radioactive nuclei of the chemical element are decaying.

The activity of a radioactive sample is the total number of nuclei that decay in the sample in one second.

The SI unit of activity is the becquerel (Bq). 1 Becquerel is an activity of 1 decay per second.

The larger the sample, the greater the activity. As an example, consider a sample that consists of 4×10^{12} atoms and that each atom has a 1% chance of decaying in the next second. That means that 4×10^{10} atoms will decay in the next second. But if there were only 4 million atoms to begin with, then only 40 000 would decay in the next second.

The ratio of $\dfrac{\text{number that decay in the next second}}{\text{total initial number in the sample}}$ is always the same for a particular radioactive decay.

The fact that the probability is constant means that the nuclide has a half-life that depends only on the nuclide in question. This link between decay probability and half-life is developed in more detail in the additional higher level (AHL) section of this topic.

The half-life of a radioactive nuclide can be defined in two ways:

The half-life of a radioactive nuclide is the time taken for half the initial sample to decay.

or

The half-life of a radioactive nuclide is the time taken for the initial activity of the sample to halve.

Half-lives vary over a vast range of time intervals: from the 4.5 billion years of uranium-238 through to the 2.5 ms of fluorine-29 and even shorter times.

Figure 16 shows the progress of a series of decays for a small number of nuclei (100).

Initially, there are 100 parent nuclei (red). In the first half-life, 50% of them decay into the daughter nucleus (assumed stable). After successive half-lives, the amount of parent nuclei that remain falls to 25%, 12.5%, 6.25% and so on.

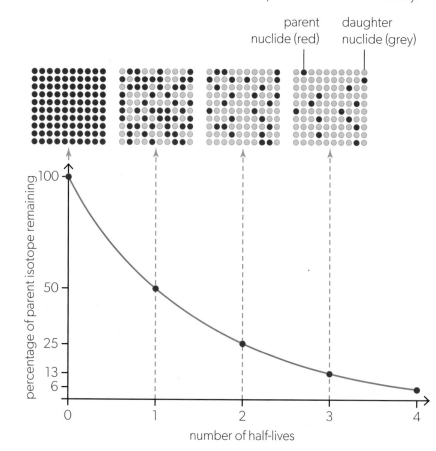

▲ Figure 16 Every half-life the number of the radioactive parent nuclei halves. This is exponential decay.

When the initial number of parent nuclei is N_0, the number of parent nuclei remaining after 1 half-life is $\dfrac{N_0}{2}$.

After n half-lives, the amount remaining is $\dfrac{N_0}{2^n}$.

This behaviour is known as **exponential decay**. It is a consequence of the fact that

$$\text{activity of the sample} \propto \text{number of parent nuclei remaining.}$$

Which areas of physics involve exponential change? (NOS)

Exponential growth and decay are found in many areas of physics and in science as a whole. All that is required is that the rate of change of a quantity N depends on the quantity N itself. In mathematical terms,

$$\frac{dN}{dt} = \pm kN$$

Where the sign is plus, then the equation indicates growth in N; when negative, the change is a decay.

Whenever an effect can be described with this equation, then it can be modelled using the solutions to the differential equation which are of the form

$$N = N_0 e^{\pm kt}$$

Examples from science include:

- the growth or decay of charge on a capacitor

- the change in height of a foam with time

- the change in amplitude with time of a damped harmonic oscillator (Topic C.4).

- a nuclear reaction

- growth of bacteria in a culture

- the change in size of a seashell—as the organism grows its food input is proportional to the size of its mouth.

Exponential change is an example of patterns and trends in science where a solution of one phenomenon leads directly to a solution of another in a different field.

Worked example 8

The graph shows how the activity of a sample of a radioactive nuclide varies with time.

a. State the half-life of the nuclide.

b. Predict the activity of the sample after a time of 10 minutes.

Solutions

a. From the graph, the activity halves after 120 s. Hence $T_{\frac{1}{2}} = 120\,\text{s}$.

b. 10 minutes is 5 half-lives of the nuclide. The initial activity is $8 \times 10^6\,\text{Bq}$.

Activity after 10 minutes

$$= 8 \times 10^6 \times \left(\frac{1}{2}\right)^5 = 2.5 \times 10^5\,\text{Bq}.$$

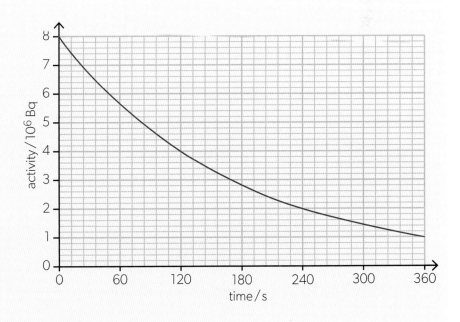

Worked example 9

Nuclide X decays into stable nuclide Y with a half-life of 7 days. A freshly prepared sample contains 240 g of pure nuclide X. Calculate:

a.   the mass of nuclide X remaining in the sample after 14 days

b.   the time after which the sample contains 225 g of nuclide Y.

Solutions

a.   14 days is two half-lives. The remaining mass of X is $240 \times \left(\frac{1}{2}\right)^2 = 60\,g$.

b.   The remaining mass of X is $240 - 225 = 15\,g$. $\frac{15}{240} = \frac{1}{16}$ so the remaining mass of X is $\frac{1}{16}$ of the initial mass.

Because $\frac{1}{16} = \left(\frac{1}{2}\right)^4$, this happens after 4 half-lives, or 28 days.

A more systematic way to find the number of half-lives is by solving the equation $240 \times \left(\frac{1}{2}\right)^n = 15 \Rightarrow n = 4$.

Practice questions

9.   The initial activity of a sample of radioactive nuclide is A_0. The graph shows how the relative activity $\frac{A}{A_0}$ of the sample varies with time.

   a.   State the half-life of the nuclide.

   b.   Predict the activity of the sample after 15 days.

   c.   Determine the time for the activity to decrease to $\frac{A_0}{16}$.

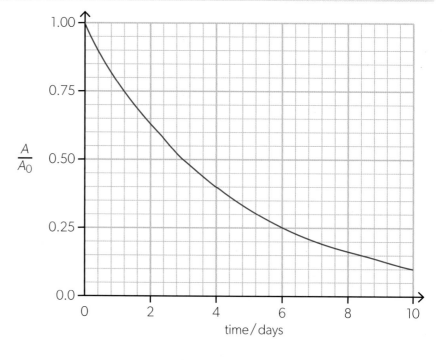

$\frac{A}{A_0}$

time / days

10.   Nuclide X decays into stable nuclide Y with a half-life of 4.0 hours. A sample initially contains 64 g of pure nuclide X.

   a.   Calculate the mass of nuclide Y in the sample after 12 hours.

   b.   Calculate the time after which there will be 1.0 g of nuclide X remaining in the sample.

   c.   Draw a graph to show how the masses of nuclides X and Y in the sample vary with time during the first 24 hours of the experiment.

11.   Iodine-131 decays into stable xenon-131. Activity of iodine-131 in a sample decreases to $\frac{1}{16}$ of its initial value in 32.1 days. Determine the half-life of iodine-131.

12.   A sample initially contains equal masses of radioactive nuclides X and Y. Nuclide X decays with a half-life of 12 hours and nuclide Y decays with a half-life of 6 hours. What is the ratio $\frac{\text{mass of X remaining in the sample}}{\text{mass of Y remaining in the sample}}$ after 24 hours?

   A. 2   B. 4   C. 8   D. 16

Data-based questions

In 1911, Hans Geiger and John Nuttall proposed a rule which related the half-life of alpha decaying isotopes to the range of the alpha particles in air. Since the range of the alpha particles is related to their energy, this rule can be used to relate the energy of the alpha particles to the half-life of decay.

The Geiger–Nuttall rule can be expressed as

$$\log_{10} T_{\frac{1}{2}} = \frac{c_1}{\sqrt{E}} + c_2$$

where c_1 and c_2 are constants.

The table shows some data for isotopes of radon which decay through α-decay.

Isotope	Isotope mass / u	Daughter isotope mass / u	Half-life / s
^{196}Rn	196.002 115	191.991 335	0.0047
^{198}Rn	197.998 679	193.988 186	0.065
^{212}Rn	211.990 704	207.981 246	1430
^{214}Rn	213.995 363	209.982 874	2.70×10^{-7}
^{216}Rn	216.000 274	211.988 868	4.50×10^{-5}
^{218}Rn	218.005 601	213.995 201	0.035
^{220}Rn	220.011 394	216.001 915	55.6
^{222}Rn	222.017 577	218.008 973	3.30×10^5

- The equation for the decay of ^{196}Rn can be written as $^{196}_{86}\text{Rn} \rightarrow {}^{192}_{84}\text{Po} + {}^{4}_{2}\alpha$. Using the masses given in the table and the mass of an alpha particle, $m_\alpha = 4.002\,603$ u, calculate the mass lost in the decay. Using the energy–mass equivalence of $1\,\text{u} \equiv 931.49\,\text{MeV}$, find the energy released in the decay.

- Tabulate the energy released in the decay of the other radon isotopes.

- Tabulate values of $\frac{1}{\sqrt{E}}$ and $\log_{10} T_{\frac{1}{2}}$.

- Plot a graph of $\log_{10} T_{\frac{1}{2}}$ against $\frac{1}{\sqrt{E}}$ and use your graph to determine the values of c_1 and c_2.

- Assume that the uncertainties in your values of $\frac{1}{\sqrt{E}}$ are approximately 3%. Add uncertainties to your graph and use maximum and minimum gradients to estimate the uncertainties in c_1 and c_2.

Radioactive decay measurements

Instruments used to measure radioactivity in some school laboratories include the Geiger–Müller tube with its associated counter and the spark counter.

- The **Geiger–Müller tube** (sometimes abbreviated to Geiger counter or GM tube) is used to detect beta-minus particles and gamma photons. It consists of a sealed metal tube filled with a mixture of gases at low pressure (Figure 17).

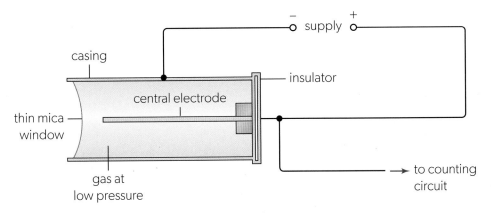

▲ Figure 17 The internal structure of a Geiger–Müller tube.

One end of the tube is made of a thin sheet of a mineral called mica that allows the beta-minus particles to enter the gas. A pd of a few hundred volt is maintained between the inside of the tube and a central electrode. When the beta particle or gamma photon enters the space inside the tube, an atom of the gas is ionized by an interaction with the radiation. The electric field accelerates the ion and the electron in opposite directions (Topic D.4). This leads to more ionization as the ion and electron gain energy and collide with other atoms. The resulting current across the tube is detected by a counting circuit.

There is a mixture of gases in the tube. One of the gases is there to suppress the spark (otherwise it would persist for a significant time) so that the GM tube can detect the arrival of the next particle. This process is called quenching and the minimum time possible between detections is called the "dead time" of the tube.

- The **spark counter** is used to detect alpha particles (Figure 18). A thin metal wire in air lies a few millimetres beneath a metal grid.

A high pd (a few kV) is maintained between the grid and the wire. When alpha particles enter the space between the wire and the grid, they ionize air molecules and a spark passes between wire and grid. This can be seen and counted either visually or electronically.

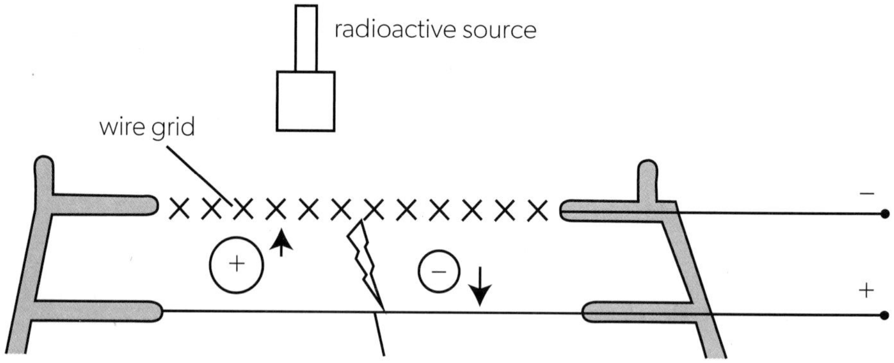

▲ Figure 18 A cross-sectional view of a spark counter. A spark jumps between the grid (or mesh) and the thin wire.

Activity and count rate

There is a distinction between activity and count rate.

- **Activity** is the total number of disintegrations per second in a sample of radioactive material. Activity is measured in becquerel.

- **Count rate** is the number of counts detected by a measuring apparatus in one second. Count rate is measured in counts per second.

A radioactive source sends its emitted particles through a 360° angle. A detector cannot usually sample the whole of this angle. Typically, only a small fraction of the whole activity can be sampled. Some of the particles will be absorbed between source and detector and this will also reduce the count rate. Provided that the geometry of the detector–source arrangement is constant, there should be a proportional relationship between count rate and activity.

Randomness

Throughout this topic, it has been noted that radioactivity is random. While we cannot predict when a particular nucleus may decay, we can assign a probability to the event and use statistics to model outcomes.

Suppose that an experiment has an average background count rate of 5.3 counts in ten seconds. The most likely number of background counts in a ten-second period is 5 (with a probability of 17.4%). However, there is a small probability (0.5%) that there will be no counts in that ten-second period and there is a similar probability of measuring 12 counts. There will be a 73% chance of measuring between 3 and 7 counts in ten seconds. Such considerations enable us to establish an uncertainty in such a measurement. It may appear that we lack knowledge of the system when dealing with random events, but if we can establish an average value and an uncertainty, is this knowledge worse than any other measurement?

Background radiation

When a GM tube is left to count for several minutes in the absence of any radioactive source, it will detect ionizing radiation. This is the background radiation and the count rate will depend on the location of the laboratory and the type of rock from which, and over which, it is built.

It is important to eliminate this **background count** from any counting measurements in radioactivity. When the background has been eliminated from a count, the result is known as the **corrected count rate**:

corrected count rate (in counts per second)

= observed count rate (in counts per second) – background count rate (in counts per second)

There are many contributions to background radiation from both natural and artificial radiation sources. Figure 19 shows a pie chart of some of these contributions.

A major source of radiation is from the Earth's rocks. As the elements in them decay, radon gas is emitted at low levels. This radioactive gas can accumulate in buildings—again, only to low and safe levels.

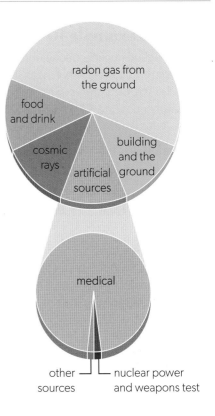

▶ Figure 19 Some of the contributions to the background radiation.

🧪 Evaluating half-life from a graph

- Inquiry 2: Interpret qualitative and quantitative data.

- Inquiry 2: Interpret diagrams, graphs and charts.

- Inquiry 3: Discuss the impact of uncertainties on the conclusions.

- When you have a graph of the variation with time of the corrected count rate from a sample, then the graph can be used directly to determine the half-life.

- The error bars are not shown on this graph, but you will see some scatter about the line when you look closely. This scatter is normal in radioactivity experiments.

- Estimate the time for the corrected count rate to halve. Do this at least three times.

- A common error is to halve the initial corrected count rate, then halve it again, and then halve it for a third time. This is undesirable because this will take you to $\frac{1}{8}$ of the initial rate. In this region of the graph the errors are more significant. It is better to use three separate starting points but to take all of them within the first or second half-life. Then you can take an average of the three results.

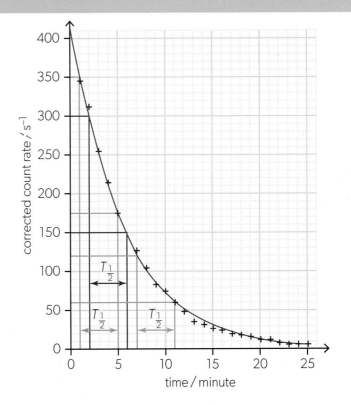

▲ Figure 20 At least three estimates of half-life are required. These are best taken from the region where the corrected count rate is large.

Worked example 10

A radioactive nuclide decays into a stable nuclide with a half-life of 25 days. A Geiger–Müller detector measures the count rate from a sample of the nuclide. The detector is exposed to background radiation of a constant average count rate. The table shows how the count rate in the detector varies with time.

Time / days	Count rate / counts minute^{-1}
0	650
25	330
50	170

a. Determine the average background count rate.

b. Predict, for a time of 100 days:

   i.   the count rate from the sample corrected for background radiation

   ii. total count rate as recorded by the detector.

Solutions

a. The background count rate is constant, so the measured count rate only changes due to the decreasing activity of the radioactive sample. During the first 25 days, the count rate from the radioactive nuclide in the sample decreases to one-half of its initial value. The initial count rate of the sample, corrected for background radiation, must have been $2 \times (650 - 330) = 640$ counts minute^{-1}. The background count rate is therefore $650 - 640 = 10$ counts minute^{-1}.

b. i.   100 days is 4 half-lives. The count rate corrected for background radiation is $640 \times \left(\frac{1}{2}\right)^4 = 40$ counts minute^{-1}.

   ii. The constant background count rate of 10 counts minute^{-1} will be added to the count rate from the sample. The detector will measure $40 + 10 = 50$ counts minute^{-1}.

Measurements—Gray and Sievert

Two measures have been developed to assess the impact of radiation on animal tissue.

The gray (Gy) is the energy absorbed by one kilogramme of tissue. Therefore 1 Gy occurs when 1 J is deposited in tissue.

The sievert (Sv) is a measure of the biological effect that the radiation has, and this depends on the type of radiation and the type of tissue. A dose of 1 Gy of X-rays to the body will lead to a 1 Sv dose equivalence, whereas 1 Gy of the much more damaging alpha particles will lead to a 20 Sv equivalent dose.

These ideas link to the absorption and ionization properties of alpha, beta and gamma radiation described earlier.

The gray and the sievert are not tested in the IB Diploma Programme physics examination.

Practice questions

13. The graph shows how a count rate from a radioactive source varies with time. The background count rate is constant at the location of the detector. The half-life of the source is 4 hours.

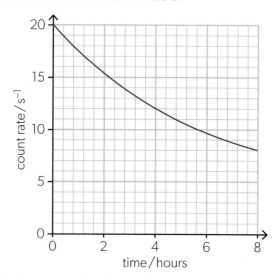

a. Determine the average background count rate.

b. Predict the count rate in the detector after 16 hours.

14. Without a radioactive source present, a Geiger–Müller tube measures an average background activity of 20 counts minute^{-1}. When a radioactive source is placed near the tube, the count rate increases initially to 340 counts minute^{-1}.

   Determine the count rate detected by the tube after three half-lives of the radioactive source.

Radioactive dating

The changes in radioactivity can be used to determine the age of rocks and carbon-based materials using the technique of **radioactive dating** (or radiometric dating).

The general principle is best illustrated using the carbon-dating technique used for materials that are up to about 60 000 years old.

Living plants and animals absorb carbon during their lifetime. At death, the carbon is fixed and no more can be added. Naturally occurring carbon on Earth contains the stable carbon-12 nuclide and the unstable carbon-14 nuclide that decays to nitrogen-14 via beta-minus emission. After the death of the organism, the carbon-14 begins to decay. The ratio of $\frac{\text{carbon-14}}{\text{carbon-12}}$, previously in equilibrium with the environment and therefore constant, begins to decrease. Carbon-14 has a half-life of 5700 years.

The assumption is made that the value of $\frac{\text{carbon-14}}{\text{carbon-12}}$ for living material does not change with time. This means that a present-day measurement of the ratio in the dead material will reveal the length of time since the death of the organism.

Geologists can determine the age of rocks using several possible nuclides. The decays of uranium-238 and uranium-235 (both to lead isotopes) have half-lives of 470 billion years and 700 million years, respectively. This allows them to be used as a cross-check on the dating ages.

Ground water on the surface seeps into underground lakes where it can be trapped for hundreds of thousands of years. Chlorine-36 and krypton-81 are nuclides that decay over long timescales and dating techniques can be applied to assay the age of the water in the lakes. Any uranium or thorium salts in the rocks surrounding the lake will decay releasing alpha particles that acquire electrons to become helium atoms. The detection of the helium also allows an estimate of the lake's age.

Radioactive nuclides and medicine

Radioactivity has many uses in medicine both for diagnosis and therapy.

Diagnostic uses involve the ingestion, injection or inhalation of a suitable radioactive nuclide. The material then collects in the organ under investigation and emits gamma photons during radioactive decay. These photons can be imaged either by photographic or sensor techniques to study the region of interest. Nuclides include:

- Technetium-99m which is used to diagnose a number of problems in the bones, the heart and other organs. The material is usually made on site at the hospital from the decay of radioactive molybdenum-99 as the technetium half-life is only 6 hours and this makes its transport impracticable. (The molybdenum, on the other hand, has a half-life of 2.7 days.) The technetium is often combined with biologically active materials that allow it to be taken to most parts of the body easily and quickly. The low-energy gamma photons from its decay to the stable form can be detected and used for diagnosis.

- Radioactive iodine becomes located in the thyroid gland, which is otherwise difficult to image. Higher doses can then be used to treat the thyroid when a diagnosis is complete.

Models — Carbon dating

The assumption behind carbon dating is that there is a constant ratio in the abundance of $^{14}C : ^{12}C$.

^{14}C is formed in the atmosphere when cosmic rays interact with ^{14}N in the atmosphere. Since the intensity of cosmic rays and the amount of nitrogen in the atmosphere are unlikely to have changed significantly over the past 60 000 years (the maximum age of an object that might be carbon dated), this seems a reasonable assumption.

Studies of tree rings enabled a sequence of measurements to be taken to confirm this. Each tree ring represents a time increment of a year and the ring can be considered to be dead at that point in time. This showed that the amount of ^{14}C has fluctuated in the past. More sophisticated carbon-dating techniques take account of this fluctuation.

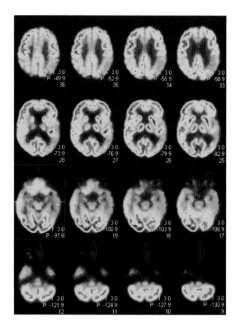

▲ Figure 21 Images of the brain using the technique of positron emission tomography.

- The PET scan (positron emission tomography) is a sophisticated technique in which a nuclide is injected into a region under investigation. The nuclide emits a positron that combines with a nearby electron to emit two gamma photons. These move off in opposite directions due to momentum conservation and are subsequently detected by two gamma cameras which, after computation, gives a very precise position of the point where the positron and electron combined.

Fluorine-18, a β^+ emitter, is commonly used in this way for PET-scan imaging. It is also used to detect infections because the chemical fluorodeoxyglucose has an increased uptake at the site of inflammation. When the fluorine isotope in the chemical is radioactive, then imaging can take place.

Oxygen-15, another β^+ emitter, is used as a tracer for measuring the blood flow in the body. The radioactive oxygen is produced by bombarding nitrogen nuclei with deuterium nuclei at kinetic energies of about 7 MeV. This is often done at hospitals which have small cyclotrons for producing these and other radioactive materials. The reaction is

$$^{14}_{7}N + {}^{2}_{1}D \rightarrow {}^{15}_{8}O^{2-} + n$$

The radioactive oxygen isotope is then reacted with hydrogen to produce water that can be injected and used for a PET scan. A challenge for the medical practitioners is that the half-life of O-15 is 2.04 minutes, so the production and use of the nuclide must be carried out quickly.

Therapeutic uses for radioactive materials include:

- Small low-activity packages of gamma emitters that are placed in the body in order to deliver a localized dose of radiation to that region.

- The gamma knife (teletherapy) focuses the gamma photons from cobalt-60 sources onto a single small region within the brain to destroy tumours there.

Radiation is also used to sterilize medical materials of all types once they have been sealed in their protective package.

Radioactivity and materials testing

Radioactive materials are used increasingly in industry. Thickness measurements can be made rapidly and accurately (the intensity of the radiation emerging from a material varies exponentially). The nuclide used depends critically on the thickness measurement in question. For example, beta-minus particles are commonly used to control the rolling of aluminium into a sheet form. Alpha particles would be unlikely to be useful as either all, or a large percentage, would be absorbed by the metal. Equally, gamma photons would not be absorbed at all by the aluminium.

Flow rates and the presence of rubbish in rivers are often measured using radioactive tracers. Again, the nuclide must be tailored to the application. Half-lives must generally be short so that the material does not damage the environment.

The law of radioactive decay and its consequences

Earlier, the ratio $\dfrac{\text{number of nuclei that decay in the next time interval}}{\text{initial number of nuclei in the sample at the beginning of the time interval}}$ was introduced as a constant for a particular nuclide.

This ratio is the **probability of decay of an individual nucleus** in the given time interval. This time interval must be short, so that the change in the number of nuclei during the time interval is small compared with the initial number. (Earlier, a time interval of one second was used for simplicity.)

When the time interval is sufficiently small (and, therefore, tends mathematically to zero), the probability of decay is known as the **decay constant** λ. This is the constant of proportionality that links the activity A (the number of decays per unit time) and the initial number of nuclei N at the start of the time interval. As an equation this is

$$A = \lambda \times N$$

The activity itself is the rate of change of N which leads to

$$A = -\frac{dN}{dt} = \lambda N$$

The negative sign is important as it models the *decrease* in N as time *increases*. This negative sign is not present when the $A = \lambda N$ equation is quoted in the Physics Data Booklet, but it is important for you to realize that it is always implicit in the definition of activity as a decreasing quantity.

This equation, here in the context of radioactive decay, has a much wider application. The mathematical physics that follows can be applied to any situation where:

rate of change of a quantity $N \propto$ amount of the quantity N remaining to change

The equation $\frac{dN}{dt} = -\lambda N$ rearranges to $\frac{dN}{N} = -\lambda dt$.

This expression can be integrated. The following symbols are used:

- At time $t = 0$ the initial (original) number of nuclei is N_0.

- At a later time t the number of nuclei remaining has fallen to N.

Thus,

$$\int_{N_0}^{N} \frac{dN}{N} = -\int_0^t \lambda dt$$

which integrates to give

$$[\ln N]_{N_0}^{N} = -[\lambda t]_0^t = -(\lambda t - 0).$$

Therefore, $\ln N - \ln N_0 = -\lambda t$ or $\ln\left(\frac{N}{N_0}\right) = -\lambda t$. Taking the natural exponential

function of both sides of the equation,

$$\frac{N}{N_0} = e^{-\lambda t} \text{ or } N = N_0 e^{-\lambda t}$$

Because $A = \lambda N$, this can also be written as

$$A = A_0 e^{-\lambda t} \text{ or } A = \lambda N_0 e^{-\lambda t}$$

where A_0 is the initial activity and $A_0 = \lambda N_0$.

Decay constant or probability of decay?

The link between decay constant and probability of decay is a subtle one. The radioactive decay constant λ is defined, for a nuclide, as the probability P per unit time that a given nucleus of the nuclide will decay. For a time interval Δt this is $\frac{P}{\Delta t} = \lambda$ or $P = \lambda \Delta t$. Put this way, it is clear that Δt must be much smaller than the time scale over which the nuclide decays. Therefore, the statement that the probability of decay is the same as the decay constant is only true when Δt is small compared with the mean lifetime of a nucleus of the nuclide. (This mean lifetime is *not* the same as the half-life.)

Worked example 11

The initial activity of a sample of radioactive polonium-210 (Po-210) is 2.3×10^8 Bq. The decay constant of Po-210 is 5.8×10^{-8} s^{-1}.

a. Determine the initial mass of Po-210 in the sample.
b. Calculate the activity of the sample after 100 days.

Solutions

a. $A_0 = \lambda N_0$, so initially the sample contains $N_0 = \dfrac{A_0}{\lambda} = \dfrac{2.3 \times 10^8}{5.8 \times 10^{-8}} = 4.0 \times 10^{15}$ nuclei.

This amounts to $\dfrac{N_0}{N_A} = \dfrac{4.0 \times 10^{15}}{6.02 \times 10^{23}} = 6.6 \times 10^{-9}$ mol. The molar mass of Po-210 is

approximately 210 g mol^{-1}, so the initial mass of the sample is $6.6 \times 10^{-9} \times 210 = 1.4 \times 10^{-6}$ g $= 1.4$ µg.

b. 100 days $= 100 \times 24 \times 3600 = 8.64 \times 10^6$ s. The activity decreases exponentially and after 100 days it becomes $A_0 e^{-\lambda t} = 2.3 \times 10^8 e^{-5.8 \times 10^{-8} \times 8.64 \times 10^6} = 1.4 \times 10^8$ Bq.

Worked example 12

The count rate, corrected for background radiation, of a sample of radioactive fluorine-18 decreases from 45 count s^{-1} to 28 count s^{-1} during a time of 4500 s. Determine the decay constant of fluorine-18.

Solution

The count rate follows the exponential decay law, $28 = 45 e^{-4500\lambda}$. The equation can be solved for the decay constant graphically or using logarithms.

$\ln 28 - \ln 45 = -4500\lambda$. Therefore $\lambda = \dfrac{\ln 45 - \ln 28}{4500} = 1.1 \times 10^{-4}$ s^{-1}.

Worked example 13

An old piece of cloth is examined for the content of radioactive carbon-14. It is found that the mass ratio of carbon-14 to stable carbon-12 has decreased to 70% of its value in living matter.

The half-life of carbon-14 is 5700 years. Determine the age of the cloth.

Solution

The exponential decay law, written in terms of the ratio of C-14 to C-12, is $0.70 = 1.0 \times \left(\dfrac{1}{2}\right)^n$, where n is the

number of half-lives of carbon-14 since the plants used to make the cloth were harvested. $n = \dfrac{\ln 0.7}{\ln 0.5} = 0.51$.

The cloth is $0.51 \times 5700 = 2900$ years old.

Practice questions

15. The decay constant of caesium-134 (Cs-134) is 1.06×10^{-8} s^{-1}. A laboratory sample contains 6.00 µg of Cs-134.

    a. Determine the initial activity of the sample.
    b. Determine the time for the activity to decrease to 80% of its initial value.
    c. Calculate the mass of Cs-134 remaining in the sample after one year.

16. A sample contains 7.0×10^{12} atoms of radioactive oxygen-15. The initial activity of the sample is 4.0×10^{10} Bq.

    a. Calculate the decay constant of oxygen-15.
    b. Calculate the activity of the sample after 1.0 hour, correct to one significant figure.

Half-life and the decay constant

The earlier treatment of half-life can now be extended.

After one half-life, the activity will have fallen to half of its initial value $\left(so \; \dfrac{A_0}{2}\right)$, and

there will be half of the initial number of nuclei $\left(\dfrac{N_0}{2}\right)$. We can use the symbol $T_{\frac{1}{2}}$ to represent half-life.

Therefore, $\dfrac{N_0}{2} = N_0 e^{-\lambda T_{\frac{1}{2}}}$. Taking logs to base e gives $\ln N_0 - \ln 2 = \ln N_0 - \lambda T_{\frac{1}{2}}$,

which becomes $\ln 2 = \lambda T_{\frac{1}{2}}$ and

$$T_{\frac{1}{2}} = \frac{\ln 2}{\lambda} \approx \frac{0.693}{\lambda}.$$

Worked example 14

The half-life of phosphorus-32 (P-32) is 14.3 days. A sample initially contains 2.0 mg of P-32.

a. Calculate, giving your answer in s^{-1}, the decay constant of P-32.
b. Determine the initial activity of the sample.
c. Calculate the mass of P-32 remaining in the sample after 30 days.

Solutions

a. $\lambda = \dfrac{\ln 2}{T_{\frac{1}{2}}} = \dfrac{\ln 2}{14.3 \times 24 \times 3600} = 5.61 \times 10^{-7}\, s^{-1}$.

b. The molar mass of P-32 is $32\, g\, mol^{-1}$, so the initial number of nuclei in the sample is

$N_0 = \dfrac{2.0 \times 10^{-3}}{32} \times 6.02 \times 10^{23} = 3.8 \times 10^{19}$. The initial activity is $A_0 = 5.61 \times 10^{-7} \times 3.8 \times 10^{19} = 2.1 \times 10^{13}\, Bq$.

c. Expressing time in seconds, mass in mg and using the radioactive decay law, we get mass remaining
$= 2.0 \times e^{-5.61 \times 10^{-7} \times 30 \times 24 \times 3600} = 0.47\, mg$.

Alternatively, the time elapsed can be expressed in terms of the number of half-lives: 30 days $= \dfrac{30}{14.3} = 2.10$ half-lives.
The mass remaining is $2.0 \times \left(\dfrac{1}{2}\right)^{2.10} = 0.47\, mg$, the same as calculated with the first method.

Practice questions

17. The activity of a sample of radioactive nuclide decreases from $24.0 \times 10^7\, Bq$ to $21.0 \times 10^7\, Bq$ over a time period of 10.0 days. Calculate:

a. the decay constant, in s^{-1}, of the nuclide
b. the half-life, in days, of the nuclide
c. the time required for the activity of the sample to decrease to 10% of the initial value.

18. Radium-226 has a half-life of 1600 years.

a. Calculate, in s^{-1}, the decay constant of radium-226.
b. Determine the expected number of nuclei decaying per second in a $1.0\, \mu g$ sample of radium-226.

Determining half-lives
Long half-life

Long half-lives are those where the half-life is much longer than the time interval over which the measurement is made. They are estimated directly using the equation $A = \lambda N$ or with a substitution and rearrangment: $T_{\frac{1}{2}} = \dfrac{0.693}{A} \times N$.

When the sample is pure, a determination of the mass of the sample together with its molar mass will give the total number of atoms N. A careful determination of the activity A of the sample can be made in several ways. These include a knowledge of the total energy transfer per second from the sample and the energy of the emissions. This will yield $T_{\frac{1}{2}}$ directly.

Worked example 15

Thorium-232 (Th-232) is a naturally occurring isotope of thorium. It decays to radium-228 by alpha decay. A sample containing 5.0 g of Th-232 is prepared.

a. Calculate the number of nuclei of Th-232 in the sample.
b. The average activity of the sample is 2.0×10^4 Bq. Determine:
    i.   the decay constant
    ii. the half-life of Th-232.

Solutions

a. As usual, we estimate the molar mass of the nuclide by its nucleon number, 232 g mol^{-1}.

$$N = \frac{5.0}{232} \times 6.02 \times 10^{23} = 1.3 \times 10^{22}.$$

b. i. $\lambda = \dfrac{A}{N} = \dfrac{2.0 \times 10^4}{1.3 \times 10^{22}} = 1.5 \times 10^{-18}\,\text{s}^{-1}.$

    ii. $T_{\frac{1}{2}} = \dfrac{\ln 2}{\lambda} = \dfrac{0.693}{1.5 \times 10^{-18}} = 4.5 \times 10^{17}\,\text{s}.$ This corresponds to 1.4×10^{10} years.

    Thorium-232 is a very long-lived isotope; it occurs on Earth as a primordial nuclide—it has existed since before Earth was formed.

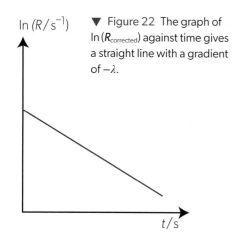

In (R/s^{-1})

▼ Figure 22 The graph of In $(R_{\text{corrected}})$ against time gives a straight line with a gradient of $-\lambda$.

t/s

A point to note is the way the label on the y-axis is written. A logarithm has no unit and so the units for R must be included with R within the brackets.

Short half-life

Short and medium half-lives can be determined using one of two approaches:

*   The graphical approach outlined on page 655. In this method, the analysis is based on the graph of corrected count rate against time. The analysis is based on a curve and depends on the accuracy with which this curve is drawn.

*   Another way to make a graphical analysis, this time with more confidence in the result, is to recognize that the radioactive-decay equation can be written as

$$\ln R = -\lambda t + \ln R_0$$

where R is the corrected count rate and R_0 is the initial corrected count rate. This matches $y = mx + c$ and so a graph of $\ln R$ against t should be a straight line with a gradient $-\lambda$ and an intercept of $\ln R_0$ on the y-axis. The plot is shown in Figure 22. This method has the advantage that the half-life can be found even when R is measured over a period of time that is less than the half-life.

When the straight line has been drawn and the gradient $(-\lambda)$ calculated, then the half-life comes from $T_{\frac{1}{2}} = -\dfrac{0.693}{\text{gradient}}.$

Measuring short half-lives in the laboratory

- Tool 1: Recognize and address relevant safety, ethical or environmental issues in an investigation.

- Tool 3: Construct and interpret graphs using logarithmic scales. Plot linear and non-linear graphs showing the relationship between two variables with appropriate scales and axes.

- Tool 3: Record uncertainties in measurements as a range (±) to an appropriate precision.

- Inquiry 1: Appreciate when and how to take background radiation into account.

Half-life can be determined easily in the laboratory. If you have access to the necessary materials, the procedure is straightforward. You will require a suitable detector of ionizing radiation, such as a Geiger–Müller tube with its counter and a stopwatch. Alternatively, a data logger with a radiation sensor can be used.

A number of possible radioactive substances can be used safely in this experiment. Your teacher will need to explain which one you are to use, and the safety precautions you must take when using it.

- Measure the background count rate in the room with no radioactive sources present. This should be done over several minutes, leading to the rate per second. You should repeat the determination when the main experiment is complete.

- The decaying radioactive material is now brought into the laboratory and placed close to the window of the GM tube or sensor.

- Take readings of the observed count rate at time intervals appropriate to the source until the count rate has fallen to a value close to the background.

- The corrected count rate is (*observed count rate − background count rate*).

- The analysis follows the pattern given earlier: construct a plot of ln (corrected count rate) against time and use the line of best fit to obtain the gradient and then the half-life.

- This is a better method than plotting corrected count rate against time. Why is this?

- You should be able to compare your value with the accepted half-life for the nuclide. This gives an overall view of the accuracy of your experiment.

Errors in radioactive experiments.

When using a data logger, you may need to set the data logger up with a scanning time over which it computes the average count rate. Too long and the radioactive material will have decayed significantly; too short and the total counts measured will be small. A compromise is needed.

Radioactivity is a random process. A rule of thumb for the statistics of radioactive decay is that the error in measuring N radioactive events is $\pm\sqrt{N}$. The error when $N = 100$ is ± 10—that is, 10%—but the error when $N = 10\,000$ is ± 100—that is, 1%. In other words, count for as long as possible provided that the material will not have significantly decayed in the time interval.

Another approach, particularly for use when half-lives are longer, is to time the arrival of the same number of counts which keeps the error per count measurement the same.

Simulating radioactive half-life

- Tool 3: Construct and interpret tables and graphs for raw and processed data including scatter graphs and line and curve graphs.

- Inquiry 1: Pilot methodologies.

- Inquiry 3: Evaluate the implications of methodological weaknesses, limitations and assumptions on conclusions.

It is not always possible to carry out experiments with radioactive nuclides. This experiment is a simulation of a half-life measurement.

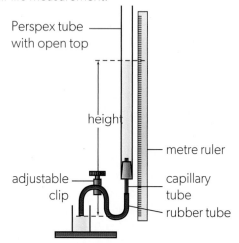

▲ Figure 23 How to measure the change in height of a water column to simulate radioactive decay.

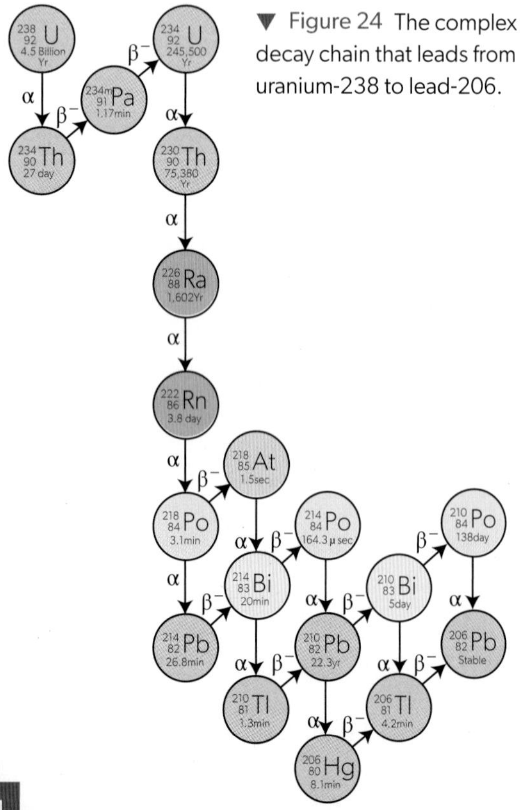

▼ Figure 24 The complex decay chain that leads from uranium-238 to lead-206.

Radioactive decay can be simulated in a number of ways. This practical version has the advantage of simple apparatus. A chemical burette can replace the Perspex tube and clip.

- The Perspex tube is filled with water after closing the adjustable clip at the bottom of the apparatus. There should be a mark on the outside of the tube and the water must be filled above this level.

- The clip is then opened completely so that the height of water in the tube begins to decrease.

- As the water passes the mark on the tube, a clock is started. The water height is measured at regular time intervals. A preliminary experiment will inform you of the ideal interval between time readings for your tube.

- The experiment should be repeated twice more, always beginning the timing when the water passes the mark on the tube. Calculate the average height for each time.

- Plot graphs to show the variation of height with time and ln (height / cm) with time.

- Use the graphs to determine the half-life and the decay constant of this system.

- To what extent does the system model radioactive decay? In what way is it different?

Decay chains—Growth and decay

The section on radioactive dating above (page 657) mentions the decay of uranium isotopes to the element lead. The change in A is from 238 to 206 in this decay and this change cannot occur in one step. This decay is part of a naturally occurring decay chain in which unstable nuclides decay step-by-step via alpha and beta decay. As the chain progresses, the nuclides tend towards greater stability until eventually a completely stable element remains.

The uranium-238 decay is shown in Figure 24.

The circle for each nuclide contains the proton and nucleon numbers and the half-life of the nuclide. The chain links to the next nuclide and gives the decay mode.

During the long series of decays, several separate nuclides form, including three lead isotopes, three polonium isotopes and two isotopes of bismuth.

You can find references on the internet to the other naturally occurring decay chains (such as the thorium and actinium series) and also some that involve synthetic elements.

Models—Radioactive growth

In many ways, radioactive decay is only half the story. As the atoms of a parent nuclide decay, the number of daughter nuclei grows. When the daughter is stable, then the graphs of decay and growth are straightforward (Figure 25). The number of daughter nuclei at any moment is the difference between the initial number of parent nuclei and the present number of parent nuclei.

When the daughter is also unstable, the situation is more complicated. Suppose parent A decays to an unstable daughter B, which then decays to form the stable element C. The variation with time of the numbers of each nuclide is shown in Figure 26. The proportions of the graphs now depend on the relative decay constants of A and B.

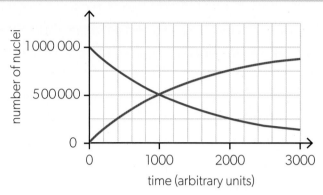

▲ **Figure 25** The simple case where, as the parent nuclide (red curve) decays, the stable daughter (blue curve) grows. The sum of both lines is constant (and the initial number of parent nuclei).

These graphs can be easily modelled:

- The rate of loss of A depends on its decay constant λ_A: $\dfrac{dN_A}{dt} = -\lambda_A N_A$.

- The rate of change of B depends on its growth from A and its own decay constant λ_B: $\dfrac{dN_B}{dt} = +\lambda_A N_A - \lambda_B N_B$.

  Note the signs here and assume that initially $N_B = 0$.

- The rate of change of C (assumed stable) is $\dfrac{dN_C}{dt} = +\lambda_B N_B$.

This model can be constructed and run using suitable software to give the graphs shown here. Details of how to approach this are given in *Tools for physics*. You will need to choose the initial number of nuclei and suitable decay constants for A and B.

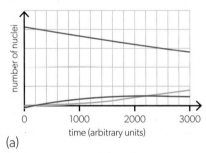

(a)

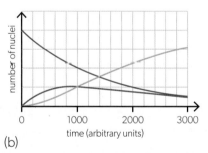

(b)

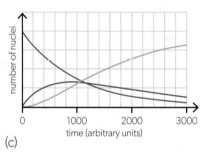

(c)

▲ **Figure 26** The daughter product also decays to form a stable granddaughter (green curve). (a) Here the half-life of the daughter nuclide is less than that of the parent. Only small amounts of the daughter are present in the nuclear mix. (b) When the daughter half-life is equal to that of the parent, the fraction of daughter nuclide in the mix is greater. (c) This is the case when the parent half-life is less than that of the daughter.

In which form is energy stored within the nucleus of the atom?

How can the energy released from the nucleus be harnessed?

Fission leads to the release of energy. This is a consequence of the plot of binding energy per nucleon against nucleon number from Topic E.3.

Figure 1 shows the binding energies per nucleon for elements with high binding energies per nucleon. On the right are the elements with large A. Uranium-235 is indicated together with the energy difference between it and the most stable elements around iron-56. When the uranium fissions—whether by induced or spontaneous fission—it will produce two nuclei that must be around the stable region where $A \approx 120$. The binding-energy difference is released in several energy forms, such as the kinetic energy of the particles after fission and as high-energy photons. Binding energy is the source of nuclear energy.

This topic also considers the use of U-235 in a practical reactor. Problems to overcome include the following.

- Neutrons are easily absorbed by U-238 nuclei to produce plutonium-239. This is unavoidable and undesirable because it removes available neutrons from the reactor. Unfortunately, the naturally occurring U-238 and U-235 occur together in mined uranium ore.

- Each neutron is emitted with an energy of around 2 MeV of kinetic energy. This is a high speed, roughly 10% of the speed of light. However, fission is best initiated by a neutron with a much slower speed with an energy of a factor of 10^8 lower. The neutrons from the fissions must be slowed down without allowing them to interact with other nuclei (principally U-238) in the reactor.

The solutions to these and other engineering problems form the basis of this topic. They involve physics from other areas of the subject too. The uranium fuel must be enriched by using a gaseous form of it in a centrifuge. The heavier U-238 is separated from the lighter U-235 and proportionally less of the heavier isotope is in the final material.

The speed of the emitted neutrons must be moderated down to the equivalent of the kinetic energy of the average particle in a gas at room temperature (Topic B.3). This process involves the mechanics of Topics A.2 and A.3. Repeated elastic collisions redistribute the energy between fast neutrons and the atoms of the moderator material.

This topic looks closely at how physicist and engineers harness nuclear energy in practice, and how they deal with the practical issues of radiation and nuclear waste.

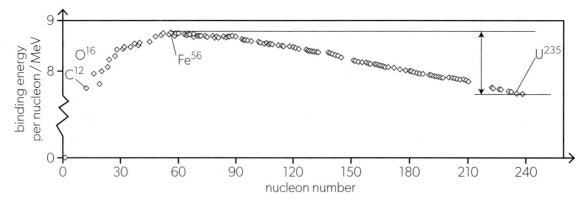

▲ Figure 1 The difference between the binding energy per nucleon of the most stable nuclei and heavy elements is enough to make nuclear fission a sensible method for the transfer of energy.

In this topic, you will learn about:

- spontaneous and neutron-induced fission

- energy released in fission

- chain reactions

- control rods, moderators, heat exchangers and shielding in a nuclear power plant

- management of the waste products of nuclear fission.

Introduction

In her long life, Elise Meitner won many prizes and honours for her scientific research. She received 29 separate nominations for the Nobel Prize in Physics. However, she never received a Nobel Prize, even though her collaborator Otto Hahn won the award for their joint work. In 1939, she used the term "fission" for the first time in a scientific paper that she wrote with her nephew Otto Frisch.

Almost 90 years later, the term nuclear fission is now commonplace, and we use the process to generate electrical energy. But we have also used fission in more sinister ways over the past 90 years, so that the use of nuclear power continues to cause disagreement in society. To have a considered view on nuclear power in society, you need to understand the physics of fission. In other words, the physics of the release of energy because of an induced change in a nucleus and how this release can be safely harnessed for good.

ATL Social skills—Reflecting on the impact of behaviour

Elise Meitner was forced to flee Germany during the Second World War due to the rise of anti-semitism. She fled to Sweden which was neutral in the war and refused to join other scientists in their pursuit of making nuclear weapons.

Her collaborator, Otto Hahn remained in Germany. In later life, he became outspoken against the use of science for military purposes. He later came to regret his conduct during the war saying *"Come the year 1933 I followed a flag that we should have torn down immediately. I did not do so, and now must bear responsibility for it."*

Meitner was critical of many German scientists, including Hahn, for their involvement in the war. She also regretted staying in Germany before the war with the words *"it was not only stupid but very wrong that I did not leave at once".*

Meitner died in 1968 aged 89. The inscription on her grave, written by her nephew, says "Lise Meitner: a physicist who never lost her humanity".

▲ Figure 2 Elise Meitner meeting some young American physicists in 1946.

Nuclear fission—Induced and spontaneous

Nuclear fission occurs when the nucleus of a heavy element (with a nucleon number greater than about 230) splits into two or more nuclei. This nuclear splitting occurs in two ways:

- **Spontaneous fission** This is a rare form of radioactive decay in which the parent nucleus splits into smaller nuclei with the additional release of nuclear particles. These particles are likely to be neutrons, so that the daughter nuclei have proton:neutron ratios that are closer to the zone of stability (Topic E.3, page 648) although they are unlikely to be on it.

  Spontaneous fission is confined to decays observed in naturally occurring thorium-232, uranium-235 and uranium-238. It is also seen more commonly in artificially produced elements in the actinide and transactinide groups of the Periodic Table.

  In the case of the thorium and uranium nuclei, it is much more likely that alpha emission will occur, but geological evidence indicates that they can fission spontaneously and have done so during the Earth's history. The half-life for the alpha emission is around 700 million years, whereas the half-life for spontaneous fission is 11 billion years. This emphasizes the rarity of a spontaneous fission event.

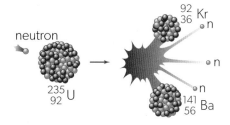

▲ Figure 3 An incoming neutron induces the fission of a U-235 nucleus into two smaller nuclear fragments with the emission of more high-speed neutrons.

Data-based questions

The number of neutrons produced from the fission of U-235 varies. The table below gives the probability P of different numbers of neutrons N being released.

N	P
0	0.033
1	0.174
2	0.335
3	0.303
4	0.123
5	0.028
6	0.003

- Calculate the average number of neutrons emitted (express your answer to 1 decimal place).

- Theory suggests that $\ln P = k(N - N_{av})^2 + A$ where N_{av} is the average number of neutrons emitted and k and A are constants. Plot a graph of $\ln P$ against $(N - N_{av})^2$ to test this theory.

- Use your graph to find values of k and A. Hence, find the probability of a fission reaction releasing 7 neutrons.

Data from N. Ensslin: *The Origin of Neutron Radiation*.

- **Neutron-induced fission** This is the more important type of fission in which a neutron from outside the nucleus interacts with it to produce a new unstable isotope of the original with an increased neutron number (Figure 3).

  This is the type of fission used in nuclear engineering, where it is often known as a "nuclear reaction". It is important to recognize that induced fission is not a radioactive decay of the type outlined in Topic E.3. An initiating neutron is required to induce the fission. The emission of neutrons as some of the final products is essential to maintain the reactions in a sustainable way — to establish a **chain reaction**.

The fission mechanism

Two common nuclear fuels in use today are uranium-235 and plutonium-239. The uranium occurs naturally. Plutonium is only present in trace amounts in the Earth's crust and, when used in nuclear reactors, must be produced synthetically. The trace amounts of plutonium in the rocks are produced when natural uranium-238 captures a neutron as a by-product of cosmic rays or from a natural uranium-235 decay.

Here is the uranium-235 fission process in detail.

- An incoming neutron interacts with the uranium-235 nucleus and is absorbed. This creates the unstable uranium isotope U-236:

$$^{235}_{92}U + {}^{1}_{0}n \rightarrow {}^{236}_{92}U^*$$

The asterisk indicates the unstable nature of the U-236.

- In 82% of the neutron interactions, the U-236 fissions in a very short time to produce two fission fragments of approximately equal size together with several high-speed emitted neutrons.
  In the remaining 18% of interactions, the U-236 emits gamma rays to become a more stable form of the nuclide. In this form, it has a half-life of about 23 million years and is a small but significant problem in nuclear reactors as it builds up in the fuel.

- The main fission process does not have a definite endpoint as, providing the numbers of protons and neutrons balance, many combinations of final nuclides are possible. One typical fission leads to the creation of barium and krypton:

$$^{236}_{92}U^* \rightarrow {}^{144}_{56}Ba + {}^{90}_{36}Kr + 2{}^{1}_{0}n + \Delta E$$

where ΔE is the energy released.

- ΔE can be calculated using $\Delta E = c^2 \Delta m$, where Δm is the overall difference between the total mass before the fission (including the incoming neutron) and the total mass after the fission:

$$\Delta m = m_{U\text{-}235} - m_{Ba\text{-}144} - m_{Kr\text{-}90} - m_n$$

- The barium–krypton endpoint is often quoted in books because it is one of the most common outcomes of U-235 fission. For example, the reaction

$$^{236}_{92}U^* \rightarrow {}^{144}_{56}Ba + {}^{89}_{36}Kr + 3{}^{1}_{0}n + \Delta E$$

also occurs with one more neutron emitted than for Kr-90.
Other possible fissions include

$$^{236}_{92}U^* \rightarrow {}^{140}_{54}Xe + {}^{94}_{38}Sr + 2{}^{1}_{0}n + \Delta E \quad \text{and} \quad {}^{236}_{92}U^* \rightarrow {}^{132}_{50}Sn + {}^{101}_{42}Mo + 3{}^{1}_{0}n + \Delta E$$

- Either two or three neutrons are released in all the possible fissions quoted here. However, other numbers of neutrons are also possible with different products.

Figure 4 gives the % yield of nuclides with peaks around the 36–38 and 54–58 regions. The outcome with equal proton number is rare. (This *y*-axis scale is logarithmic.)

- The two nuclei have greater stability than the U-235 because their binding energies per nucleon are greater than that of the uranium. So, as well as the two nuclei and several neutrons, energy is released as a result of the fission.

- Electrostatic repulsion controls the energy release. The two nuclei suddenly occupy similar regions in space and the large electrostatic repulsion overcomes the strong nuclear force. All the emitted nuclei and particles gain kinetic energy with some photon emission.

  The average binding energy per nucleon in U-235 is in the region of 7.6 MeV per nucleon. For the products it is around 8.5 MeV per nucleon. Thus, the energy released is about 1 MeV per nucleon, or around 200 MeV per U-235 nucleus.

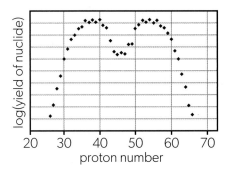

▲ Figure 4 The two peaks in this logarithmic plot indicate that the outcome with two equal proton numbers is rare. Such nuclei would be likely to be neutron-rich and very unstable.

 In which form is energy released as a result of nuclear fission?

The 200 MeV of energy per nucleus that is transferred from the binding energy comes in a number of forms. Some is released immediately (with a total of about 180 MeV), some later. The table gives a broad picture of the types and percentages of the emitted fission energy.

Much of this energy can be recovered as the various particles are brought to rest in a nuclear reactor. This increases the internal energy of the reactor. A coolant is used to transfer the energy away from the reactor vessel.

Some energy (about 20 MeV), however, is released over a much longer time scale as alpha and beta decays from the fission fragments (handling this material is considered later in this topic). It is not practical to recover all the energy because some is emitted as antineutrinos (about 9% of the total energy). Similarly, the neutron kinetic energy is not all immediately available as it can be used to breed plutonium-239 from U-238 interactions.

The various forms that the energy takes are considered elsewhere in the course. For example, treatment of the kinetic energies is in Topic A.3 and the photon energies are discussed in Topic E.1.

	Energy type	Energy released / MeV
immediate release	fission product kinetic energy	165
	neutron kinetic energy	5
	gamma photons (during fission)	7
later release from activity of fission products	beta particles	7
	gamma emission	7
	antineutrinos	9
	net energy released / fission	**200**

How is binding energy used to determine the rate of energy production in a nuclear power plant?

The 200 MeV of energy emitted when one U-235 nucleus fissions is about 3.2×10^{-11} J of energy. On the face of it, to transfer 1 MWh of energy should require 10^{20} fissions.

Of course, not all this binding energy from fission can be transferred to electrical energy. Energy is removed from the reactor vessel using steam at temperatures around 300 °C that drives turbines. The steam is ejected from the turbines at around 50 °C. Topic B.4 reminds us that the second law of thermodynamics will give a maximum theoretical efficiency of

$$\eta_{Carnot} = 1 - \frac{T_c}{T_h} = 43\%$$

The reactor is a heat engine too and is not close to a reversible state. Overall, the maximum efficiencies of the whole reactor plant approach 30%, so that to obtain 1 MWh of electrical energy, 3 MWh of fission energy must be transferred.

▲ Figure 5 A nuclear power station.

Worked example 1

In neutron-induced fission, a nucleus of uranium-235 $\left(^{235}_{92}U\right)$ absorbs a neutron and yields a nucleus of xenon-140 $\left(^{140}_{54}Xe\right)$ and a nucleus of strontium-94 $\left(^{94}_{38}Sr\right)$, according to the reaction

$$^{235}_{92}U + ^{1}_{0}n \rightarrow ^{140}_{54}Xe + ^{94}_{38}Sr + x^{1}_{0}n$$

a. Calculate the number of neutrons released in the reaction.

The atomic masses of the nuclides are given in the table.

b. Calculate, in MeV, the energy released in the reaction.

c. Estimate, in J, the nuclear energy transferred when 1.0 kg of pure uranium-235 undergoes fission.

Nuclide	Atomic mass
$^{235}_{92}U$	235.0439 u
$^{140}_{54}Xe$	139.9216 u
$^{94}_{38}Sr$	93.9154 u

Solutions

a. The number of nucleons on the left-hand side of the reaction must be the same as the number of nucleons on the right-hand side. $235 + 1 = 140 + 94 + x \Rightarrow x = 2$. The fission releases two neutrons.

b. The loss of mass in the reaction is $\Delta m = (m_{U\text{-}235} + m_n) - (m_{Xe\text{-}140} + m_{Sr\text{-}94} + 2m_n)$. The process involves atomic nuclei, but since the proton number does not change, we can work with atomic masses instead and do not need to subtract the mass of the electrons.

Mass difference $\Delta m = 235.0439 - 139.9216 - 93.9154 - 1.0087 = 0.1982$ u, where 1.0087 u is the mass of one net neutron produced in the reaction.

As $u = 931.5$ MeV c^{-2}, $\Delta m = 0.1982 \times 931.5 = 184.6$ MeV c^{-2}. The energy equivalent of this mass is $\Delta E = \Delta mc^2 = 184.6$ MeV.

c. The fraction of mass of uranium converted to energy is $\frac{\Delta m}{m} = \frac{0.1982}{235.0439} = 8.4 \times 10^{-4}$. For every 1 kg of pure uranium-235 undergoing fission, the mass converted to energy is 8.4×10^{-4} kg $= 0.84$ g. This corresponds to $\Delta E = \Delta mc^2 = 8.4 \times 10^{-4} \times (3.0 \times 10^8)^2 = 7.6 \times 10^{13}$ J. This is an approximate value as not every fission reaction yields the same amount of energy. There are many other combinations of fission fragments than Xe-140 and Sr-94, with small differences in energy released from one pair of fragments to another. Moreover, the calculation does not include energy released during subsequent decays of fission products.

Worked example 2

Xe-140 and Sr-94 produced in neutron-induced fission of uranium-235 in Worked example 1 are radioactive and undergo further decays. The stable end products of their respective chains of decays are cerium-140 ($^{140}_{58}$Ce) and zirconium-94 ($^{94}_{40}$Zr).

a. Explain why the combined proton number of Ce-140 and Zr-40 is different from the proton number of U-235.

The binding energies per nucleon are given in the table.

b. Calculate, in MeV, the total energy released as a result of this fission reaction of U-235.

Nuclide	Binding energy / A
$^{235}_{92}$U	7.591 MeV
$^{140}_{58}$Ce	8.376 MeV
$^{94}_{40}$Zr	8.667 MeV

Solutions

a. The combined proton number of Ce and Zr is 98 and that of uranium is 92. Since the nucleon number is unchanged compared with the initial fission products, the decays leading to the production of Ce and Zr must have been beta decays, resulting in an emission of charged particles and a change in the proton number of the resulting nuclei.

b. The total binding energy increases as a result of the initial fission and subsequent decays. Since Ce-140 and Zr-94 are the final stable products, the total energy released can be approximately taken as the binding energy difference between these two nuclei and U-235.

$$\Delta E = 140 \times 8.376 + 94 \times 8.667 - 235 \times 7.591 = 203.5 \text{ MeV}$$

This energy includes 184.5 MeV released in the initial fission (see Worked example 1) and about 19 MeV in the subsequent decays of fission products.

Practice questions

1. Consider the neutron-induced fission reaction

$$^{235}_{92}U + ^{1}_{0}n \rightarrow ^{144}_{56}Ba + ^{89}_{36}Kr + 3^{1}_{0}n$$

The following data are given about the binding energies per nucleon of these nuclides.

Nuclide	Binding energy / A
$^{235}_{92}$U	7.591 MeV
$^{144}_{56}$Ba	8.265 MeV
$^{89}_{36}$Kr	8.615 MeV

a. Calculate, in J, the energy released in the reaction.
b. Estimate the fraction of the mass of uranium-235 converted to energy in this reaction.
c. A nuclear power station outputs 1.3 GW of electrical power. Use your answer in part b. to estimate the mass of uranium-235 that undergoes fission in one day. Assume that the overall efficiency of nuclear to electrical energy transfer in this power station is 0.30.

2. Consider neutron-induced fission of plutonium-239,

$$^{239}_{94}Pu + ^{1}_{0}n \rightarrow ^{134}_{54}Xe + ^{103}_{40}Zr + x^{1}_{0}n$$

a. State the number of neutrons released in this reaction.
b. Calculate, in MeV, the energy released. The following data are available for this question.

Nuclide	Atomic mass
$^{239}_{94}$Pu	239.0522 u
$^{134}_{54}$Xe	133.9054 u
$^{103}_{40}$Zr	102.9272 u

Chain reactions

So far, we have concentrated on the process of fission of a single nucleus. However, a key requirement for the continuous production of energy is the **chain reaction**. Neutrons from a single fission must be able to initiate further nuclear fissions as they interact with other fissionable nuclei.

A working chain reaction in a real-life nuclear reactor involves both careful physics and meticulous engineering. Figure 6 shows the process. The three neutrons from the first fission go on to cause another fission each of which leads to further fissions and so on.

This process was first envisioned by Leo Szilard, a Hungarian physicist, who realized that the emission of more than one neutron from a fission could lead to multiple further fissions. The trick is to maintain the reaction in such a way that it can be controlled: by allowing the number of neutrons to increase to produce more power from a reactor, and by reducing the number to stop the reactor working.

The way these limitations are overcome is described in the next section.

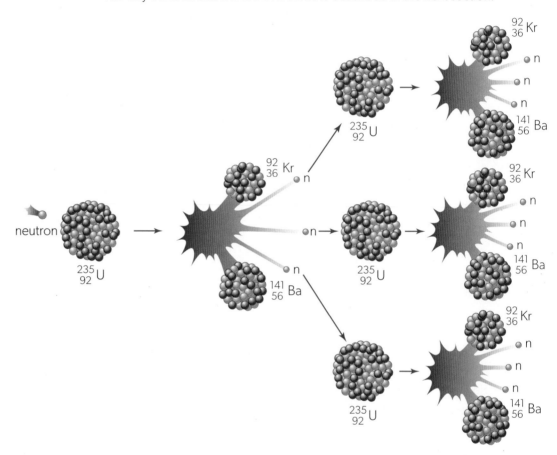

▲ Figure 6 A chain reaction in a nuclear reactor—emitted neutrons go on to cause further fissions.

Practical nuclear reactors

There are many types of reactor in common use around the world. This description focuses on the thermal fission reactors. One example of this type is the pressurized-water reactor (PWR) which uses uranium-235 as fuel. The aim of a nuclear power station is to take energy transferred from the nucleus in the nuclear fission and to use this to create high-pressure steam. The steam then turns turbines connected to an electrical generator using techniques discussed in Topic D.4.

Figure 7 shows a schematic of a PWR with the final output of steam to, and the return pipe from, the turbines.

Uranium is mined as an ore in various parts of the world. About 99.3% of the ore as it comes directly from the ground is made up of uranium-238, with the remainder being U-235; it is the U-235 not U-238 that is required for the fission process. This means that an initial extraction process is required to boost the ratio of U-235 to U-238. The fuel needs to contain about 3% U-235 before it can be used in a reactor. This is because U-238 is a good absorber of neutrons and too much U-238 in the fuel will prevent the fission reaction becoming self-sustaining. The fuel with its boosted proportions of U-235 is said to be **enriched**.

The enriched material is then formed into fuel rods—long cylinders of uranium that are inserted into the core of the reactor.

Immediately after emission, the neutrons are moving at very high speeds of the order of $10^4\,\text{km s}^{-1}$. However, for them to be as effective as possible in causing further fissions to sustain the reaction, they need to be moving with kinetic energies much lower than this, of the order of 0.025 eV (with a speed of about $2\,\text{km s}^{-1}$). This slower speed is typical of the speeds that neutrons have when they are in equilibrium with matter at about room temperatures. Neutrons with these typical speeds are known as **thermal neutrons**. The slowing is achieved by moderation.

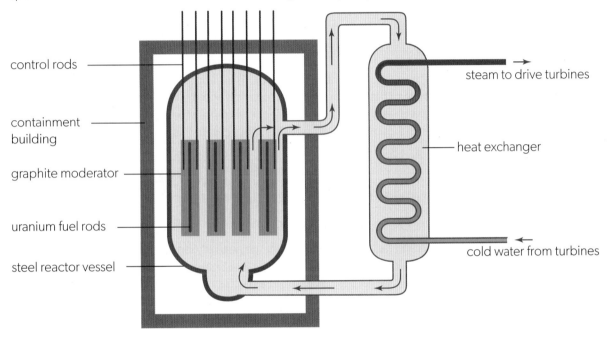

▲ Figure 7 A schematic of a pressurized water reactor (PWR).

labels on figure:
control rods
containment building
graphite moderator
uranium fuel rods
steel reactor vessel
steam to drive turbines
heat exchanger
cold water from turbines

Moderators

The requirement to reduce the kinetic energy of the neutrons is not only so that the neutrons can stimulate further fissions effectively, but so that their energy can be efficiently transferred to the later stages of the power station. The removal of energy is achieved using a **moderator**, so-called because it moderates (slows down) the speeds of the neutrons.

▲ Figure 8 These bundles of fuel rods for a nuclear reactor are about 0.5 m long and 0.1 m in diameter. Each bundle generates about 4 TJ of energy during the time it is in the reactor.

Typical moderators for the PWR type include water, heavy water (deuterium oxide) and carbon in the form of graphite. The transfer of energy is achieved when a fast-moving neutron strikes a moderator atom elastically, transferring energy to the atom and losing energy itself. After a series of such collisions, the neutron will have lost enough kinetic energy for it to be moving at thermal speeds and to have a high probability of causing further fission.

When a neutron strikes a stationary carbon atom, momentum must be conserved. You should be able to show using ideas from Topics A.2 and A.3, that the change in kinetic energy ΔE_k of a neutron of mass m and initial kinetic energy E_k when it collides head-on with a moderator nucleus of mass M is

$$\Delta E_k = \frac{4mM}{(m+M)^2} E_k$$

For one collision with each of the common moderator nuclei:

- Hydrogen: $\dfrac{\Delta E_k}{E_k} = \dfrac{4mM}{(m+M)^2} = \dfrac{4m_n \times m_p}{(m_n + m_p)^2} \approx 1$

  (The neutron stops and the hydrogen gains all the neutron's kinetic energy.)

- Deuterium: $\dfrac{\Delta E_k}{E_k} \approx 0.9$

- Carbon: $\dfrac{\Delta E_k}{E_k} \approx 0.3$

This is a simplified model as few collisions will be head on. But on this basis, with carbon as a moderator, the number of collisions n required to reduce the energy of the neutron by a factor of 10^8 is roughly $(1-0.3)^n \approx 10^{-8}$. This leads to $n \approx 50$. (In practice, the average number required is around 120 because the collisions are not head on.)

A further problem is that U-238 is very effective at absorbing high-speed neutrons. When the slowing down is carried out in the presence of the U-238, then most neutrons will be absorbed. Reactor designers make the moderators close to, but not part of, the fuel rods. The neutrons then slow down in the presence of moderator only. The fuel rods and the moderating material are separated, and neutrons move from one to the other at random. The reactor vessel and its contents are designed to facilitate this.

The criteria for a material to be a good moderator include:

- being a poor absorber of neutrons (absorption would lower the reaction rate and possibly stop the reaction altogether)

- being inert in the extreme conditions of the reactor.

You should be able to predict from the analysis above that the best moderator of all should be a hydrogen atom (a single proton in the nucleus). The maximum energy can be transferred when a neutron strikes a proton. However, hydrogen itself is a very good absorber of neutrons, leading to the formation of deuterium, and it cannot be used as a moderator in this way.

Data-based questions

Modelling the moderation process

The probability of a neutron causing the fission of a uranium-235 nucleus depends on the energy of the neutron. Nuclear physicists often use the concept of a fission cross-section σ to represent a probability. A large cross-section represents a larger target and therefore has a higher probability of the neutron "hitting" that target. The unit is a barn, abbreviated to b, $(1\,b = 10^{-28}\,m^2)$. The table gives σ for uranium-235 at a number of neutron energies E.

- Tabulate values of $\log E$ and $\log \sigma$.

- Plot a graph of $\log \sigma$ against $\log E$.

- Find the gradient of your graph and explain how it demonstrates that $\sigma \propto \dfrac{1}{\sqrt{E}}$.

- Plot a graph of σ against $\dfrac{1}{\sqrt{E}}$ and find the gradient of your graph.

- Assume that the uncertainty in the values of σ are 10%. Add error bars to your graph and hence evaluate the uncertainty in your gradient.

- In a nuclear reactor, the energy of a neutron will be reduced by a factor of 10^8. Estimate the increase in the cross-section from this reduction in energy.

E/eV	σ/b
1.00×10^{-4}	9780
2.00×10^{-4}	6910
5.00×10^{-4}	4370
1.00×10^{-3}	3080
2.00×10^{-3}	2170
5.00×10^{-3}	1360
1.00×10^{-2}	954
2.00×10^{-2}	665
5.00×10^{-2}	399
1.00×10^{-1}	259
2.00×10^{-1}	182

Control rods

The power output from the reactor must be regulated. It is also necessary to shut down the operation when required. These are achieved through the use of **control rods**. These are rods, often made of boron or some other element that absorbs neutrons very well, that can be lowered into the reactor. When the control rods are inserted a long way into the reactor, many neutrons are absorbed in the rods and fewer neutrons will be available for subsequent fissions. The rate of the reaction will drop. By raising and lowering the rods, the reactor operators can keep the energy output of the reactor (and therefore the power station) under control.

Heat exchangers

The last part of the nuclear power station that needs consideration is the mechanism for conveying the internal energy from inside the reactor to the turbines. This is known as the **heat exchanger** and is shown on the right of Figure 7.

The energy exchange cannot be carried out directly as in, say, a fossil-fuel station. There needs to be a closed-system heat exchanger that collects energy from the moderator and other hot regions of the reactor and delivers it to the water. The turbine steam cannot be piped directly through the reactor vessel because there is a chance that radioactive material could be transferred outside the reactor vessel. The use of a closed system prevents this.

The pressurized water reactor is given its name because it transfers the energy from moderator and fuel rods to the boiler using a closed water system under pressure. Water is not the only substance available for this. In the advanced gas-cooled reactors (AGR) used in the UK, carbon dioxide gas is used rather than water, but the principle of transferring energy safely through a closed system is the same.

Shielding a nuclear reactor and its operators

The emissions from the nuclear process have large energies and take many forms: neutrons, gamma photons and so on. Some of these have high penetration through solid material. There needs to be a range of safety measures provided at the site of a nuclear reactor to protect the work force, the community beyond the power station and the environment.

Typically:

▲ Figure 9 The control room of a nuclear power station.

- The reactor vessel is made of thick steel to withstand the high temperatures and pressures present in the reactor. This has the benefit of also absorbing alpha and beta radiations together with some of the gamma photons and stray neutrons.

- The vessel itself is encased in layers of very thick reinforced concrete that also absorb neutrons and gamma rays.

- The whole reactor is within a containment building designed to withstand major events such as earthquakes. In the worst scenario, the entire contents of the reactor would be trapped in this construction.

- There are emergency safety mechanisms that operate quickly to shut the reactor down in the event of an accident.

- The fuel rods are inserted into and removed from the core using robots, so that human operators do not come into contact with the spent fuel rods, which become highly radioactive during their time in the reactor.

Managing nuclear waste

The U-235 and U-238 nuclides in the fuel rods are subject to extensive change in the reactor as the fission occurs.

- Some U-235 nuclei are converted to pairs of elements with smaller A during fission.

- Some U-238 absorbs a single neutron to become plutonium-239.

- Some U-235 absorbs a neutron to become U-236 with a long half-life.

These processes cause significant problems for the operators of a nuclear plant.

- The rods gradually convert from uranium into a cocktail of many chemical elements (krypton, barium, molybdenum, etc). These nuclei are generally neutron-rich compared with their stable isotopes and are therefore likely to decay via beta-minus emission. Typically, such products have half-lives that range from hours to tens of years. This requires safe handling and effective medium-term storage.

- The creation of two nuclei in the place of the original U-235 nucleus will distort the container that contains the fuel rod and it can, in a worst case, jam in its channel in the reactor. Reactor operators only allow a limited amount of fission to occur in any one fuel rod to prevent jamming. The rod is then taken from the reactor vessel, its fission products are chemically removed, and the remaining U-235 is recycled into another fuel rod.

- The formation of plutonium-239 has benefits and disadvantages. The plutonium-239 is fissile and so can be used in a different type of reactor to generate more power. This is the basis of the so-called fast-fission (fast-breeder) reactor. Here, a layer of U-238, placed as part of the shield of a Pu-239 reactor, converts into Pu-239. This type of reactor therefore creates its own fuel. It also does not need a moderator as Pu-239 can fission easily with fast neutrons. However, this type of reactor has not proved to be economically viable.

The medium and short half-life products produced in the fuel rods during the reaction need to be treated carefully. They are highly active materials. Remember that a short half-life means a large decay constant (Topic E.3) and, because $A = \lambda N$, this implies large activity too.

The spent fuel rods are initially transferred to cooled underwater storage in something resembling a large swimming pool (Figure 10). The rods are stored for roughly five to ten years before undergoing treatment to recycle the uranium into fresh fuel rods. This is called reprocessing.

Figure 10 shows a blue light at the bottom of the pond, which is **Cerenkov radiation** emitted as the high-speed alpha and beta particles are slowed down, and finally absorbed, in the water. During this time the rods must be handled remotely given their high radioactivity.

Once the very active elements have been removed from the uranium, the resulting waste can be treated in a number of ways. Remember that some of it can remain active for hundreds of thousands of years (half-life of ^{93}Zr = 1.53 million years; half-life of ^{107}Pd = 6.5 million years, etc). Disposal of this low-activity waste can involve storage:

- in well-sealed drums on the surface

- or in underground vaults such as disused mines.

Every nuclear-reactor plant produces large amounts of low-level waste, for example, gloves, over-shoes and so on used by the workers. Such material is produced by

▲ Figure 10 A cooling pond near a nuclear reactor where the fuel rods are stored before further processing.

other organizations too, such as hospitals and engineering facilities. This is generally disposed of by ground burial under secure conditions.

There may be other problems involving the chemical toxicity of waste material, which mean that it is vital to keep it separate from biological material and thus the food chain. The technology required to achieve this is still developing.

At the end of its life (of the order of 25–50 years at the moment), a reactor plant has to be decommissioned. This involves removing all the fuel rods and other high-activity waste products and enclosing the reactor vessel and its concrete shield in a larger shell of concrete. It is then necessary to leave the structure alone for up to a century to allow the activity of the structure to drop to a level similar to that of the local background. Such long-term treatment is expensive, and it is important to factor this major cost into the price of the electricity as it is being produced during the lifetime of the power station.

To what extent is there a role for fission in addressing climate change? (NOS)

Topic B.2 suggests that global society cannot continue to use fossil fuels in the way that historically it has. Alternative sources of energy include wind, wave and solar forms. However, fission also provides a means—perhaps temporary—for providing us with the energy on which the modern world relies. The alternative sources too have their critics as they require costly and scarce minerals for their construction.

The issue is a balanced one. Fission provides a convenient, controlled source of energy. However, at the end of the reactor's life, the costs of keeping the active parts of the reactor safe are considerable. There have also been accidents at nuclear power plants with the release of radioactive material. Perhaps nuclear power is a short-term solution? Scientists must continue to work together in shared endeavour to create the fusion reactors (Topic E.5) that use mechanisms found in the stars domestically to fuse hydrogen into helium.

At the end of the day, this is a question for society as a whole. The global impact that science has requires scientists to assess risks and to consider all the consequences of their work. The scientists and engineers then have the responsibility to report their research and findings to the general public free of any political filter.

ATL Social skills—Collaboration

Once the principles of nuclear fission had been described by Elise Meitner and Otto Hahn, scientists started to consider how to develop a working nuclear reactor. A large group of scientists led by Enrico Fermi and overseen by Arthur Compton worked towards building such a reactor. Their reactor was built at the University of Chicago and first achieved self-sustaining fission in December 1942.

Such a project required a large team of scientists, technicians and workers. It was, itself, a smaller part of the Manhattan Project which, near the end of the Second World War, employed almost 125 000 people.

The Manhattan Project remains controversial. However, it is widely regarded as an early and successful collaborative scientific project. Today there are many examples of significant international collaborations which embark on large-scale experiments such as CERN. Such collaborations can involve academic institutions, industry and the military.

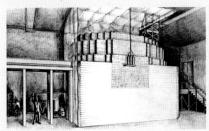

▲ Figure 11 A drawing of the Chicago Pile-1 which was the world's first nuclear reactor in 1942.

Does science need an ethical basis?

Progress in physics and technology is linked. Technological developments enable better measurements and more precise experiments to test new theories. The improved theories can then be used to develop more sophisticated technology. Sometimes technological applications may only be discovered long after a theory is formed.

In the case of nuclear fission, discovered in 1938 although hypothesized earlier, scientists quickly saw that it had the potential for developing weapons. In 1939, Hungarian physicists Leo Szilard and Eugene Wigner wrote a letter, signed by Albert Einstein, to President Roosevelt of the United States. In it they warned of the potential application in the creation of "*extremely powerful bombs of a new type*".

Should scientific research be regulated? Is it possible to pursue scientific research independently of any concerns over the possible applications of discoveries?

E.5 Fusion and stars

What physical processes lead to the evolution of stars?

How are elements created?

Can observations of the present state of the universe predict the future outcome of the universe?

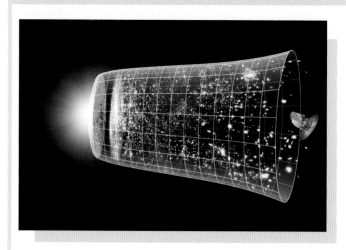

▲ Figure 1 The evolution of the universe with time since the Big Bang.

- Very little can be deduced about the start of the universe and the first fractions of a second of its existence because there was probably quantum fluctuation in the behaviour of spacetime. For the first 10^{-43} s, the universe has a diameter of about 10^{-35} m. All the fundamental forces are likely to have the same strength. As time increases, first gravitation, and then the strong nuclear force, separate from the others. Rudimentary elementary particles begin to form. Cosmic inflation occurs, and the electroweak force (electromagnetic + weak interactions) drives the creation of bosons, and eventually quarks, electrons and neutrinos.

- By the end of the first second, the temperature has dropped to 10^{12} K and protons and neutrons with their antiparticles can begin to form from quarks. These hadrons annihilate frequently, with only the overall charge and energy of the universe remaining constant.

- After three minutes the temperature has fallen to 10^9 K and nucleosynthesis occurs for the next 17 minutes, forming hydrogen, helium and some lithium. At the 20-minute point, the temperature has fallen below that at which nuclear fusion can occur.

- For the next quarter of a million years, the energy of the universe is dominated by the presence of photons which interact with the protons, electrons and any simple nuclei that have formed.

- By 300 000 years from the Big Bang, the temperature has fallen to 3000 K and the ionized atoms capture electrons. This is the point at which the universe becomes transparent to electromagnetic radiation and photons can now travel freely. The universe is a fog of hydrogen and helium gas with traces of lithium.

- The next 150 million years are dominated by darkness. Little changes. The stars have not yet formed, and the universe is controlled by the behaviour of dark matter.

- Now, the first quasars form from gravitational collapse and the intense radiation emitted ionizes the gas so that a plasma re-forms.

- Within the first fractions of a second after the Big Bang, small irregularities in the distribution of energy are thought to have occurred. These are now amplified by gravitational force so that pockets of gas form and begin to collapse under their own gravity. These denser regions of hydrogen gas become hot enough to trigger the nuclear fusions that create short-lived massive stars. These soon burn out and explode in supernovae, the contents of the explosion providing the material for second-generation and third-generation stars. Larger volumes of matter condense to form galaxies in groups, clusters and superclusters. This is the universe we observe today.

At the moment, cosmologists are undecided on the future of the universe. One possible answer is heat death, in terms of the thermodynamics in Topic B.4. They say that the universe may eventually run out of transferable energy in 10^{42} or more years from now. However, this depends on whether protons are truly stable or have a very long radioactive half-life (of the order of at least 10^{35} years).

Astrophysics and cosmology tell a different story. Here physicists make observations in the present that look back through time to make predictions about the future. There is a critical density for the matter and energy in the universe that defines whether it expands forever or contracts or whether it stops expanding after an infinite time. Whatever the endpoint for the universe, whether heat death or expansion/contraction, Earth will not be in existence by then. The Sun will run out of nuclear fuel in a few billion years.

In this topic, you will learn about:

- the stability of stars
- fusion as the source of energy in stars
- conditions leading to fusion
- the effect of stellar mass on the evolution of a star

- the Hertzsprung–Russell (HR) diagram and its main regions and features
- stellar parallax
- how to determine stellar radii.

Introduction

The Sun radiates about 10^{26} J of energy every second. Stellar matter undergoes nuclear fusion at vast rates. What is the source of this enormous power?

Primordial hydrogen was created as a result of the Big Bang. The gravitational collapse of clouds of this hydrogen led to fusion processes. More hydrogen was released in the supernovae of the earliest stars. The evolution of a star leads to the creation of elements heavier than hydrogen. What are the fusion processes that have led to life in the universe?

Evidence — Fusion in the Sun

At the beginning of the 20th century, the source of the Sun's energy was not known. Gravitational collapse was a possible theory, but it predicted an age for the Sun of about 20 million years. There was geological evidence that Earth was at least ten times older.

In the 1920s, Sir Arthur Eddington proposed that the temperature and pressure at the centre of the Sun were sufficient for nuclear fusion to occur. Today, we believe that the Sun will exist in its current state for about ten billion years and that it is nearly halfway through this lifetime.

The evidence for nuclear fusion as the Sun's energy source comes from the detection of solar neutrinos (see Topic E.3 for more on neutrinos). These neutrinos hardly interact with matter and so the majority of them travel out of the Sun unimpeded. The large rate of fusion reactions in the Sun is predicted to release a huge number of neutrinos (almost 100 billion neutrinos pass through 1 cm^2 on Earth every second). The fact that their probability of interaction with matter is so small makes them difficult to detect.

In the 1960s, solar neutrinos were discovered for the first time. This provided evidence for fusion in the Sun. However, the number of neutrinos detected was lower than predicted. It was later found that there are three types of neutrino and that they can change from one type to another. As a result, the experiment, which was looking only for electron neutrinos, found a third of the number of neutrinos that it was expecting.

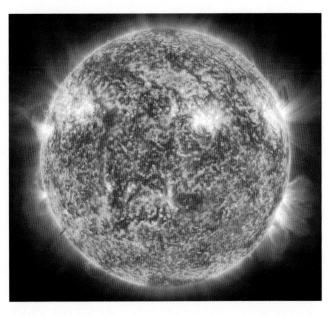

▲ Figure 2 Evidence for the nuclear reactions occurring in the core of the Sun comes from the detection of neutrinos.

In both fission and fusion, nuclei changes lead to the release of binding energy. As shown in both this topic and Topic E.4, this is because the initial mass of all reactants is greater than their final mass. The final products are more tightly bound than the initial materials and the excess energy is emitted.

Nuclear fusion in stars

There are necessary conditions for hydrogen fusion to begin. The hydrogen nuclei collide at high speeds so that the protons are within range of the strong force (see Topic E.3) to interact. Even then, the majority of collisions will not result in a fusion reaction. This means that the gas cloud must have:

- a high temperature (to ensure that collisions are energetic enough for the hydrogen nuclei to get sufficiently close)

- a high density and therefore a high pressure (to ensure a high rate of collisions).

When the conditions are right, then the **proton–proton (p–p) cycle** in main sequence stars begins. This process is sometimes called "hydrogen burning" because four hydrogen nuclei overall fuse to give one helium atom. It occurs in Sun-like stars with masses in the range 0.07–4 solar masses, until no hydrogen remains unconverted. Stars that are more massive than this have additional processes.

The cycle proceeds in three stages. This form of the cycle dominates at core temperatures up to 10 MK.

The overall reaction is

$$4 \, {}_1^1\text{H}^+ + 2 \, {}_{-1}^0\beta^- \rightarrow {}_2^4\text{He}^{2+} + 2\nu_e$$

with an energy release of 26.7 MeV. Some of this energy is lost to the neutrinos.

stage I   stage II   stage III

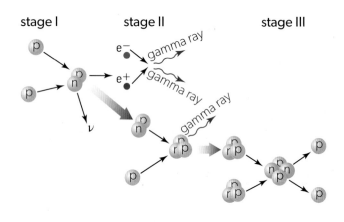

overall reaction

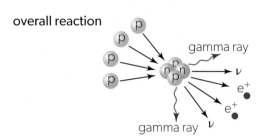

▲ Figure 3 The proton–proton cycle. (a) The individual stages. (b) The overall reaction—this does not include the two positron annihilations but shows them as emitted particles.

- **Stage I** ${}_1^1\text{H} + {}_1^1\text{H} \rightarrow {}_1^2\text{H} + {}_{+1}^0\beta^+ + \nu_e$

  Two protons (hydrogen-1 nuclei) fuse into a hydrogen-2 nucleus (also emitting a positron and a neutrino). The emitted positron is almost immediately annihilated by a nearby electron to give two gamma photons. The net energy release in stage I is 1.4 MeV including the positron annihilation. Stage I relies on the two protons remaining bound together long enough for a β^+ decay to occur. This is very unlikely and most of the time the bound protons fall apart again. Therefore stage I happens at a low rate—the average proton in the Sun will wait a few billion years before successfully completing the reaction in Stage I.

- **Stage II** ${}_1^1\text{H} + {}_1^2\text{H} \rightarrow {}_2^3\text{He} + \gamma$

  A third proton fuses with the hydrogen-2 to form a helium-3 nucleus + a gamma-ray photon. Stage II has an energy release of 5.5 MeV. This happens quickly, within one second of the production of the deuterium (hydrogen-2), on average.

- **Stage III** ${}_2^3\text{He} + {}_2^3\text{He} \rightarrow {}_2^4\text{He} + {}_1^1\text{H} + {}_1^1\text{H}$

  The helium-3 nuclei from two stage II reactions fuse to produce helium-4 and two hydrogen-1 nuclei. Stage III releases 12.9 MeV. The average helium-3 nucleus waits 400 years for the interaction to occur.

The overall reaction is shown in Figure 3.

Models — Other branches to the p–p cycle

The reactions given here are not the only possible pathways for this cycle. There are three more: two observed and one theoretical.

When the core temperature is:

- **between 10 MK and 23 MK** there is sufficient energy to drive together a helium-3 nucleus from stage II and an already-formed helium-4 nucleus. This involves the creation of a beryllium nucleus that itself decays into a lithium nucleus which finally reacts with another proton:

$$^3_2He + ^4_2He \rightarrow ^7_4Be + \gamma$$

$$^7_4Be + ^0_{-1}\beta^- \rightarrow ^7_3Li + \nu_e$$

$$^7_3Li + ^1_1H \rightarrow 2^4_2He$$

The beryllium and lithium do not survive the chain of reactions and are said to be catalytic.

- **greater than 23 MK** the catalytic nuclei are now beryllium (as before) and boron with a subsequent fission of the beryllium into two equal helium fragments:

$$^3_2He + ^4_2He \rightarrow ^7_4Be + \gamma$$

$$^7_4Be + ^1_1H \rightarrow ^8_5B + \gamma$$

$$^8_5B \rightarrow ^8_4Be + ^0_{+1}\beta^+ + \gamma$$

$$^8_4Be \rightarrow 2^4_2He$$

In theory, it ought to be possible for a helium-3 nucleus to interact with a proton to form the helium-4 in one step. However, this has not been observed in the Sun. Theory shows that only 1 helium in 30 million would be produced this way.

83% of the total energy output of the Sun comes from the three-stage process given in the main text.

Worked example 1

Deuterium (^2_1H) and tritium (^3_1H) undergo fusion into helium-4 (^4_2He) according to the reaction

$$^2_1H + ^3_1H \rightarrow ^4_2He + ^1_0n + \text{energy}$$

Binding energies per nucleon of these nuclides are given in the table.

a. Calculate, in J, the energy released in this reaction.

b. Estimate the fraction of the mass of deuterium and tritium converted to energy.

Nuclide	Binding energy / A
^2_1H	1.112 MeV
^3_1H	2.827 MeV
^4_2He	7.074 MeV

Solutions

a. The binding energy of the helium nucleus is greater than the sum of the binding energies of both isotopes of hydrogen, and the difference is transferred to the kinetic energy of the products.
$\Delta E = 4 \times 7.074 - (2 \times 1.112 + 3 \times 2.827) = 17.591$ MeV. The neutron is released as a free, unbound particle. Hence it is not included in the binding energy balance. Energy expressed in J is $\Delta E = 17.591 \times 10^6 \times 1.60 \times 10^{-19} = 2.81 \times 10^{-12}$ J.

b. The mass equivalent of the energy calculated in a. is $\Delta m = \dfrac{\Delta E}{c^2} = 17.591 \text{ MeV } c^{-2} = \dfrac{17.591}{931.5} = 1.89 \times 10^{-2}$ u.
The sum of the masses of deuterium and tritium is roughly 5 u, using their nucleon numbers as an estimate of mass.
The fraction of mass released as energy is therefore $\dfrac{1.89 \times 10^{-2}}{5} \approx 4 \times 10^{-3} = 0.4\%$. This is almost five times more than the fraction of initial mass transferred to energy in neutron-induced fission of uranium (see Worked example 1 in Topic E.4). Fusion of light nuclei releases more energy per unit mass of fuel than fission!

Worked example 2

Consider the proton–proton chain of fusion reactions, whose net result is the fusion of four protons into a nucleus of helium:

$$4\,^1_1\text{H} \rightarrow\ ^4_2\text{He} + 2\,^{0}_{+1}\beta^+ + 2\nu_e + 2\gamma$$

a. Determine, in MeV, the energy released in the reaction. The mass of the helium nucleus is 4.001506 u.

b. Calculate the additional energy released when the two positrons produced in the first stage of the cycle annihilate with electrons in the surrounding matter.

c. Estimate the total energy, in J, transferred when 1 kg of hydrogen undergoes fusion according to the proton–proton cycle.

Solutions

a. The change of mass in the reaction is $\Delta m = 4m_p - m_{He} - 2m_e = 4 \times 1.007\,276 - 4.001\,506 - 2 \times 0.000\,549 = 0.026\,500\,\text{u}$. The energy transferred is $\Delta E = \Delta m c^2 = 0.0265 \times 931.5 = 24.7\,\text{MeV}$.

b. When two positrons annihilate with two electrons, the total mass–energy of the particles is transferred to the energy of the emitted gamma photons. The mass–energy of one electron is 0.511 MeV (this is provided in your Physics Data Booklet), so $\Delta E = 4 \times m_e c^2 = 4 \times 0.511 \approx 2.0\,\text{MeV}$.

c. The total energy from one reaction is $24.7 + 2.0 = 26.7\,\text{MeV}$, which corresponds to a mass of

$\dfrac{26.7}{931.5} = 2.87 \times 10^{-2}\,\text{u}$. The fraction of mass converted to energy is $\dfrac{\Delta m}{4m_p + 2m_e} = \dfrac{2.87 \times 10^{-2}\,\text{u}}{4.03\,\text{u}} = 7.1 \times 10^{-3}$.

The energy available to fusion reactions in 1 kg of hydrogen is $7.1 \times 10^{-3} \times (3 \times 10^8)^2 = 6.4 \times 10^{14}\,\text{J}$. This is 10^7–10^8 times more than chemical energy in 1 kg of fossil fuels such as coal or crude oil.

Practice questions

1. The second stage of the proton–proton cycle is $^1_1\text{H} + ^2_1\text{H} \rightarrow\ ^3_2\text{He} + \gamma$. The binding energies per nucleon are given in the table.

Nuclide	Binding energy / A
^2_1H	1.112 MeV
^3_2He	2.573 MeV

    Show that the energy released in this stage is about 5.5 MeV.

2. Two nuclei of deuterium (^2_1H) undergo fusion into helium-3 (^3_2He) according to the reaction
    $$^2_1\text{H} + ^2_1\text{H} \rightarrow\ ^3_2\text{He} + ^1_0\text{n} + \text{energy}.$$
    The atomic masses are $m_{\text{H-2}} = 2.014\,102\,\text{u}$ and $m_{\text{He-3}} = 3.016\,029\,\text{u}$.

    a. Calculate the energy released in the reaction.

    b. Determine the energy released as a result of fusion of 1 kg of deuterium according to this reaction.

3. Deuterium–tritium fusion in Worked example 1 is a proposed energy source for future fusion power stations. A disadvantage of this reaction is that natural tritium is a very rare isotope. It is suggested that neutrons emitted in the reaction can induce fission of lithium-6 (^6_3Li) yielding tritium for further fusion. The equation for this process is $^1_0\text{n} + ^6_3\text{Li} \rightarrow\ ^3_1\text{H} + ^4_2\text{He}$. Use data provided in Worked example 1 to calculate the energy released in this reaction. The binding energy per nucleon of Li-6 is 5.332 MeV.

4. The Sun radiates $3.8 \times 10^{26}\,\text{J}$ of energy every second.

    a. Estimate, using the result of Worked example 2c. the mass of hydrogen that undergoes fusion in the Sun every second. Assume that 2% of the energy released in fusion is carried away by the neutrinos and does not contribute to the radiated power.

    b. Estimate the decrease of the mass of the Sun every second as a result of fusion reactions.

How is fusion like — and unlike — fission?

Induced fission was discovered in 1932 by Ernest Walton who split lithium into alpha particles using high-speed protons. However, the idea that nuclei can fuse was recognized earlier. In 1920, the UK scientist Francis Ashton found that the mass of four hydrogen atoms is greater than one helium atom. This suggests that there can be a transfer from binding energy by fusing four hydrogen atoms together. Much theoretical work on stellar processes then followed, with Arthur Eddington suggesting the proton–proton cycle as the main mechanism operating in the Sun.

The variation of binding energy per nucleon with nucleon number A shows the possibilities at both ends of the Periodic Table. Whereas the plot for large A is relatively smooth, at small A there are points of stability where a nucleus can be considered to be made of integer numbers of alpha particles: helium, beryllium, carbon and so on. Figure 4 shows this plotted on a logarithmic scale for A (the log plot emphasizes the differences for small nuclei). The larger the magnitude of the binding energy per nucleon, the more stable the nuclide.

Another obvious similarity between fission and fusion is that the processes both drive the nuclei towards the iron–nickel stability region but — an obvious difference — in opposite directions.

A numerical difference is that the changes in binding energy per nucleon are relatively high in fusion and much smaller in fission.

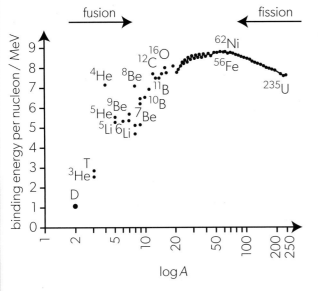

▲ Figure 4 Binding energy per nucleon plotted against log A.

Luminosity and temperature for a star

It is clear from the Stefan–Boltzmann equation that large stars with a high temperature must be bright. But what does the balance of radius R and temperature T tell us about a particular star? The tool that helps us to answer this question was devised in 1900 independently by Ejnar Hertzsprung from Denmark and Henry Norris Russell from the USA: the **Hertzsprung–Russell diagram** (HR diagram). This is a scatter plot showing the variation of star luminosity with temperature.

HR diagrams have been helpful in the classification of stars by finding patterns in their properties. Which other areas of physics use classification to help our understanding? (NOS)

Pattern recognition is a particular ability of the human brain. It allows us to match existing information in the brain with new inputs. It is not surprising that the identification of patterns and trends forms an important part in the development of science. The naming of biological species, due to Linnaeus in 1753, was a crucial step in the history and development of modern biology.

Physics, too, has classification tools. The division of quantities into vector and scalar types, the identification of energy sources and sinks, and the important and developing classification that is the Standard Model are all examples of classification in this subject.

Divergences from a classification system can lead to paradigm shift, as in the case of the known facts about beta-minus decay implying the existence of a new particle.

What other areas of classification can you identify in physics?

In Topics B.1 and B.2, it is shown that the Stefan–Boltzmann law can be written to give a quantity called luminosity L for a star:

$$L = \sigma A T^4$$

where σ is the Stefan–Boltzmann constant, A is the surface area of the star, and T is the temperature of the star. Luminosity is the total energy emitted by the star every second (also, the emitted power). In terms of the radius R of the star this becomes

$$L = 4\pi\sigma R^2 T^4$$

The HR diagram has several points of note.

- The plot is logarithmic on both axes with luminosity on the y-axis.
- The logarithmic temperature axis is plotted with lowest temperature on the right, with temperature increasing to the left.
- The logarithmic luminosity is often plotted relative to that of the Sun, which is therefore taken to have a luminosity of 1.
- Alternative x-axes that you may see are the spectral class of the stars and the BV index of the stars (a measure of the colour).
- The equation $L = 4\pi\sigma R^2 T^4$ can be rewritten as $R^2 = \dfrac{L}{4\pi\sigma T^4}$, so that, for a fixed star radius, $L \propto T^4$ is a straight line on the log–log HR diagram. Some of these **lines of constant radius** are shown on Figure 5(a). The radius sizes are shown relative to that of the Sun $R_\odot$ for the lines of constant radius.

The regions of the HR diagram shown on Figure 5(a) represent large numbers of stars as this is a scattergram. Figure 5(b) shows another representation of the HR diagram with individual stars plotted to make this clear.

The HR diagram shows stars at all ages in their evolution. As a star forms and ages, its position changes because both its temperature and luminosity change. Figure 5(a) shows the location of stars that are hot or cool, dim or bright.

These are the main features of the HR diagram.

- A **main sequence** of stars like the Sun (which is shown in both (a) and (b) versions of the diagram). These stars are fusing hydrogen into helium and constitute about 90% of all stars.

- **Red giants** above and to the right of the main sequence. These are stars cooler than the Sun (hence "red"), but with a large surface area (hence "giant"). Although they emit less energy per square metre than the Sun, their larger area means that they emit much more energy in total.

Red giants have a central core that is hot and that is surrounded by an envelope of a thin tenuous gas.

- **Supergiants** are large and very bright. They can emit 10^5 times the energy of the Sun or greater and therefore have a surface area 10^5 larger too. Their radius, in this case is $\sqrt{10^5} \approx 300$ times that of the Sun. The inner planets of the Solar System and the Sun will easily fit into the volume of a supergiant.

Red giants and supergiants make up only 1% of the total number of stars. This indicates that their lifetime is comparatively short.

- **White dwarfs** are the remains of old stars. They have a small luminosity even though their temperature is high. Their surface area must be small. They make up about 9% of all stars. These stars take billions of years to cool. A typical white dwarf can have a mass roughly that of the Sun but a radius similar to that of Earth. The density of a white dwarf can be millions of times that of Earth.

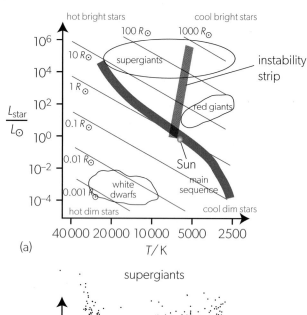

(a)

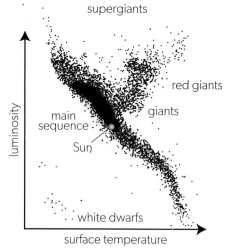

(b)

▲ Figure 5 (a) A Hertzsprung–Russell diagram showing the principle regions. (b) A Hertzsprung–Russell diagram with individual stars.

- The **instability strip**, a narrow region almost vertical on the HR diagram, which contains variable stars. Stars, usually more massive than the Sun, enter this region when they leave the main sequence. They then become unstable and pulsate, changing their size cyclically. As a result, their luminosity varies too. The stars in the instability strip include the Cepheid variables which can be used for astronomical distance estimates.

> A common notation in astrophysics is the symbol $\odot$. This stands for "the Sun" so that the mass of the Sun can be written as $M_\odot$ and its radius as $R_\odot$.

Worked example 3

In the following examples, assume that the surface temperature of the Sun is $T_\odot = 5800\,\text{K}$.

Vega is a bright star in the constellation of Lyra. The luminosity of Vega is $40L_\odot$, where $L_\odot$ is the luminosity of the Sun. The average surface temperature of Vega is $9600\,\text{K}$.

Determine the radius of Vega. State the answer in terms of the solar radius, $R_\odot$.

Solution

As $L = 4\pi\sigma R^2 T^4$, using ratios we get

$$\frac{L_{\text{Vega}}}{L_\odot} = \left(\frac{R_{\text{Vega}}}{R_\odot}\right)^2 \times \left(\frac{T_{\text{Vega}}}{T_\odot}\right)^4 \implies \frac{R_{\text{Vega}}}{R_\odot} = \left(\frac{T_\odot}{T_{\text{Vega}}}\right)^2 \sqrt{\frac{L_{\text{Vega}}}{L_\odot}}.$$

$$R_{\text{Vega}} = \left(\frac{5800}{9600}\right)^2 \left(\sqrt{40}\right) R_\odot = 2.3 R_\odot$$

Worked example 4

Mirach is a star in the constellation of Andromeda. The radius of Mirach is approximately $100 R_\odot$ and its surface temperature is $3800\,\text{K}$.

Estimate, in terms of $L_\odot$, the luminosity of Mirach.

Solution

$$L_{\text{Mirach}} = \left(\frac{R_{\text{Mirach}}}{R_\odot}\right)^2 \times \left(\frac{T_{\text{Mirach}}}{T_\odot}\right)^4 L_\odot = 100^2 \times \left(\frac{3800}{5800}\right)^4 L_\odot = 1800 L_\odot$$

Worked example 5

Mark approximate locations of Vega and Mirach in the HR diagram. Hence, state the stellar type of each star.

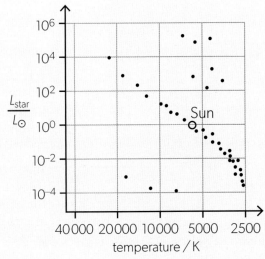

Solution

Vega is a main sequence star hotter and larger than the Sun. Mirach is a red giant.

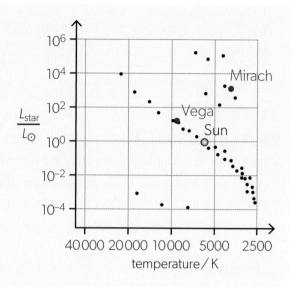

Practice questions

5. Procyon A is a star of surface temperature 6500 K and radius $2.0 R_\odot$.

   a. Calculate $\dfrac{\text{luminosity of Procyon}}{L_\odot}$.

   b. Suggest the likely stellar type of Procyon A.

6. Star Theta Centauri has surface temperature 5000 K and luminosity $60 L_\odot$. Calculate, in terms of $R_\odot$, the radius of Theta Centauri.

7. Stars P and Q have the same luminosity. The surface temperature of star P is 5000 K and the surface temperature of star Q 20 000 K.

   What is $\dfrac{\text{radius of P}}{\text{radius of Q}}$?

   A. $\dfrac{1}{16}$ B. $\dfrac{1}{4}$ C. 4   D. 16

Stellar evolution and stellar mass

The region between the star systems in a galaxy is composed of the **interstellar medium**. This is mainly hydrogen atoms, with some helium and traces of carbon, oxygen and nitrogen. There is also a small quantity of dust together with the inevitable cosmic rays. The density of the interstellar medium is extremely small. It changes depending on the local temperature conditions, but can vary from 10^{-4} ions cm^{-3} where the medium is hot to 10^{6} ions cm^{-3} in cooler regions. A good working average is that there is 1 hydrogen atom in every cubic centimetre. By contrast, the best vacuum that we can achieve on Earth is around 10^{4}–10^{5} molecules per cm^{3}.

The role of the interstellar medium is crucial in the evolution of a star. A region of the medium begins to contract under gravity and as it does so the temperature of the gas cloud begins to increase. This happens because gravitational potential energy of the particles must be released (they are becoming bound together) and this energy is transferred into thermal energy.

The temperature continues to rise as the cloud compresses more and more. It is now a **protostar.** When the temperature reaches about 10^{7} K, nuclear fissions can begin. The star moves into the HR diagram on the extreme right-hand side as a main sequence star. The exact position on the diagram depends on its temperature and luminosity.

During its time on the main sequence, the star is stable and in a **hydrostatic equilibrium**. Two competing sets of forces make up this equilibrium:

- inwards force due to the gravitational attraction between the interior of the star and the outer layers

- outwards forces due to the thermal and radiation pressure that are trying to expand the star.

The greater the mass, the greater the gravitational attraction and, therefore, in equilibrium the greater the thermal and radiation effects.

- The core temperature of the star must be larger in a more massive star to provide the greater outwards pressure.

- The higher the temperature, the more probable are the fusion events inside the star.

- There is a greater rate of nuclear reaction and more energy per unit time is emitted.

- The mass of the star and its luminosity must therefore be related.

Nuclear physics theory indicates that the mass M of a star on the main sequence is related to its luminosity L by $L \propto M^{3.5}$. (You are not required to be able to use this equation.)

When the luminosity is taken to be the mean power radiated by the star over its lifetime T and by assuming that all the mass of the star is converted to energy (it is not), then

$$L = \frac{\text{total energy released by star}}{T} \propto M^{3.5}$$

The total energy and the star's mass are proportional ($E \propto M$ and related by $E = c^2 M$) and therefore

$$\frac{c^2 M}{T} \propto M^{3.5} \text{ and } T \propto M^{-2.5}$$

The larger the mass, the shorter the time the star remains on the main sequence.

The differences in lifetime can be dramatic because of the 2.5 power dependence. The Sun is likely to spend 10 billion years on the main sequence. A star of 10 times the Sun's mass will have a lifetime that is a factor of $10^{2.5}$ smaller, about 30 million years. The larger star has a luminosity 3200 times that of the Sun.

Remember that this analysis applies to stars on the main sequence. Figure 6 gives some examples of stellar lifetimes on a HR diagram together with the positions of some of the more famous stars in the night sky.

As the hydrogen inside the star is converted to helium, it is no longer available for fusion and the radiation and thermal pressures begin to decrease. The star shrinks and heats up again. The hydrogen in the layers around the smaller core can now fuse. This raises the temperature of these layers and they expand to give the star a much greater diameter (but a smaller core). The star has moved off the main sequence to become a red giant. The core continues to shrink and heat up and the helium in the core fuses to form heavier elements such as oxygen and carbon. But the extent to which this can continue depends, again, on the mass of the star.

▶ Figure 6 Some typical stellar lifetimes matched to position on the HR diagram.

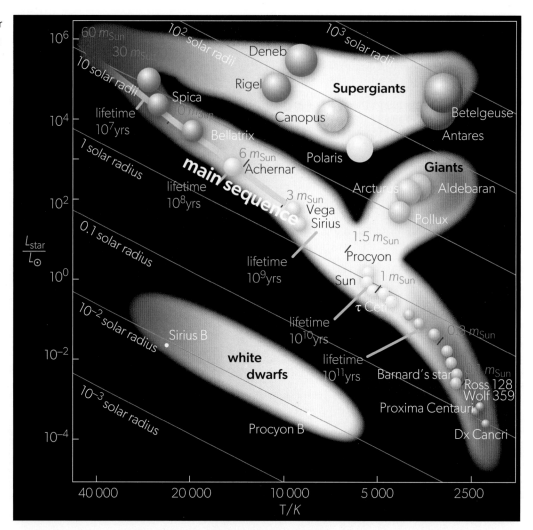

Data-based questions

The mass-luminosity relationship for main sequence stars is $L \propto M^{3.5}$ for stars above $2\,M_\odot$ in mass. For other masses, the relationship can be slightly different.

Measuring the mass of a star is difficult unless another object is interacting with its gravitational field. Binary stars therefore form useful systems to measure the mass of stars. The following data for the mass M and the luminosity L of some stars are taken from a survey of binary systems: *Semi-detached double-lined eclipsing binaries* (Malkov, O.Y., 2020. Monthly Notices of the Royal Astronomical Society, Vol. 491, No. 4, pp. 5489–5497).

- Tabulate values of $\log\left(\dfrac{M}{M_\odot}\right)$ and $\log\left(\dfrac{L}{L_\odot}\right)$.

- Plot a graph of $\log\left(\dfrac{L}{L_\odot}\right)$ against $\log\left(\dfrac{M}{M_\odot}\right)$.

Star	$\dfrac{M}{M_\odot}$	Absolute uncertainty in $\dfrac{M}{M_\odot}$	$\dfrac{L}{L_\odot}$	Absolute uncertainty in $\dfrac{L}{L_\odot}$
Y Cam	2.08	0.09	36	5
V716 Cen	2.39	0.05	66	3
UX Mon	3.38	0.4	309	94
MP Cen	4.40	0.2	1000	480
mu01 Sco	8.30	1.0	4600	1100
V448 Cyg	13.7	0.7	37200	2600
AQ Cas	17.6	0.9	87000	15000
XZ Cep	18.7	1.3	112000	16000

- Show that your graph is consistent with the observation that $L \propto M^{3.5}$.

- Tabulate values of $L^{\frac{1}{3.5}}$ including uncertainties in these values.

- Plot a graph of $L^{\frac{1}{3.5}}$ against M. Include error bars on your graph.

- The mass–luminosity relationship can be written as $\frac{L}{L_\odot} = k\left(\frac{M}{M_\odot}\right)^{3.5}$. Use your graph to determine a value for the constant k and use maximum and minimum gradients to determine an uncertainty for your value.

How can gas laws be used to model stars? (NOS)

Scientists represent physical phenomena as models—artificial representations of reality. The scientists then extend these models into more and more complex systems. One example of this is the work undertaken to model the interior processes in stars. Direct observations of such processes are extremely difficult if not impossible. We can see the visible surface of our own Sun but cannot send spacecraft to measure the properties of stellar matter.

Scientists make reasonable assumptions about the nature of the stellar material in the light of their observations of stars and then hypothesize about the plasma that suns must contain. This knowledge requires not just the theory of gases (Topic B.3) but also knowledge of nucleons and electrons that are the constituents of the stars (Topics E.3 and E.4).

Evolution of stars of moderate mass (<4 $M_\odot$)

The temperature of the core in these stars is not sufficient to allow fusion to form elements beyond carbon. When the helium in the core is used up, the core can only shrink as it continues to emit radiation.

A double shell now forms outside the core: an inner helium-fusing shell and an outer hydrogen-fusing shell. (The terms "helium burning" and "hydrogen burning" are often used by astronomers, but no combustion as a chemical process is implied here.)

This process forces the outer layers of the star away from the centre to form a **planetary nebula** around the core. The core will shrink to an object the size of Earth containing ions together with a free electron gas. (The term "planetary" is misleading—no planets are formed.)

This core cannot shrink further. The electron gas exerts a pressure called the **electron degeneracy pressure**. This arises from an important rule about particles known as the Pauli exclusion principle which forbids two electrons from having the same quantum state. The degeneracy pressure prevents further collapse and the star is now known as a white dwarf. It will cool over billions of years. A simple calculation using the mass of the Sun (about 2×10^{30} kg) and the radius of Earth (about 6400 Km) shows that the final density of the Sun will be of the order of 10^9 kg m^{-3}. A cup of a liquid with this density has a mass of 1000 tonnes.

Modelling versus experiments

The nature of astronomy and astrophysics is such that scientists rely on models to gain insights into how things work. Is knowledge gained through modelling and simulation as valuable as knowledge gained through direct experiment?

The progress of a star like the Sun to the white dwarf stage is shown in an HR diagram (Figure 7(a)) and diagrammatically (Figure 7(b)).

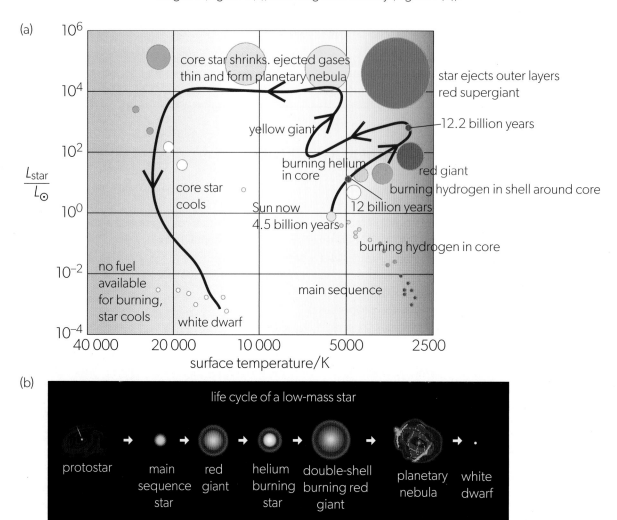

(a)

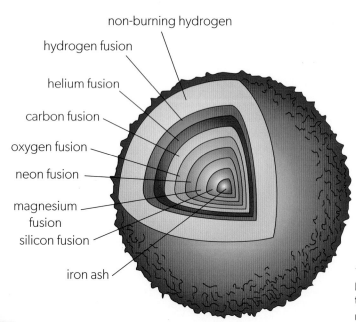

▲ Figure 7 (a) The journey of the Sun through the HR diagram. (b) The separate states of a star like the Sun.

Evolution of stars of large mass (>4 $M_\odot$)

Stars with significantly larger masses than the Sun have a very different evolution.

As main sequence stars (with a higher temperature as they are more massive), the core is large. When the star moves into the red giant phase, the higher temperature in the core leads to the creation of elements heavier than carbon. The giant phase ends when the star has a layered structure, like an onion. Different elements form in these layers with the heaviest (highest proton number) in the core and the proton number decreasing with distance from the core (Figure 8).

◀ Figure 8 The structure of a heavy star at the end of the giant phase. Note that the diagram is not to scale. The core (where fusion takes place) may be 10^4 km across while the outer surface of the star may be over 10^9 km in diameter.

As with stars the mass of the Sun, gravitational contraction occurs and is opposed by electron degeneracy pressure. However, with the increased mass, stability is not possible.

The **Chandrasekhar limit** predicts that stars with a remnant mass greater than 1.4 times that of the Sun ($1.4\,M_\odot$) cannot form a white dwarf. These large stars must go down a different evolutionary path.

As the core continues to contract, the electrons and protons in the core interact to form neutrons as

$$_{-1}^{\ 0}\beta^- + _1^1p \rightarrow _0^1n + \nu_e$$

with the emission of electron neutrinos, as expected. The core collapses still further and eventually the free neutrons are separated by nuclear distances. This collapse is very rapid, taking about one second. The outer layers of the star also contract rapidly, so rapidly that they bounce off the core creating a huge explosion known as a **supernova**. The outer layers containing the formed elements are blown away to leave only the core.

This small but dense core of neutrons obeys a similar exclusion principle to the electrons and a neutron degeneracy pressure is established to prevent further collapse of the core.

Two outcomes are now possible: the core can remain as a dense **neutron star** with the neutron degeneracy balancing the gravitational collapse, or further collapse can occur with the core remnant forming a **black hole**. Work by Robert Oppenheimer and George Volkoff using ideas from Richard Tolman established the limiting case for this transition: it is a mass somewhere between 2 and 3 solar masses.

The **Oppenheimer–Volkoff limit** (OV limit) of 2–3 solar masses is the maximum mass that a neutron star can have without collapsing to form a black hole. For stars smaller than the limit, neutron degeneracy can resist the ultimate collapse to the black hole. Stars greater than the mass limit will collapse further.

The OV limit relates to the mass of the star (the core) after the supernova, not before. When the mass after the supernova stage is 2–3 solar masses, the original mass of the star will have been roughly 15 to 20 solar masses.

The black holes formed by the collapse of these most massive stars are singularities in spacetime: objects so dense with a gravity so strong that nothing can escape them, not even electromagnetic radiation. Supermassive black holes have now been directly observed (Figure 9). This figure shows the supermassive black hole at the centre of galaxy M87 together with the effects of a strong magnetic field associated with the black hole.

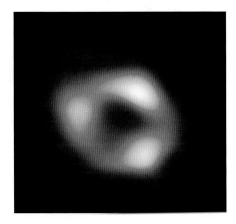

▲ Figure 9 An image taken by the Event Horizon Telescope of a supermassive black hole at the centre of M87. The dark centre is the shadow of the event horizon and outside this is a heated accretion ring that radiates in the radio region. The radius of the accretion ring is about ten times the radius of Neptune's orbit around the Sun.

Other ways to infer the existence of a black hole include the following.

- X-radiation is emitted as mass spirals towards the edge and heats up as it travels. X-ray telescopes mounted on satellites have observed this.

- Jets of matter are emitted from the cores of some galaxies. It is conjectured that only a rotating black hole could produce these.

- Some stars are observed to be influenced by strong gravitational fields that cause the star to spiral for no apparent visual reason. There is the suggestion that this is caused by the presence of a massive black hole nearby.

Figure 10 shows the life cycle for stars of large mass with the two possible outcomes.

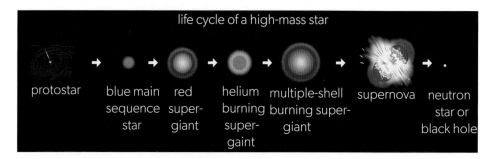

life cycle of a high-mass star

protostar → blue main sequence star → red super-giant → helium burning super-gaint → multiple-shell burning super-giant → supernova → neutron star or black hole

▲ Figure 10 An illustration of the typical history of a large star.

(ATL) Social skills — Resolving conflict

The upper limit for the mass of a white dwarf star ($\approx 1.4 M_\odot$) is called the Chandrasekhar limit after the American–Indian astrophysicist Subrahmanyan Chandrasekhar (1910–95).

Chandrasekhar was only 24 years of age when he presented his work at a meeting of the Royal Astrophysics society in the UK. He described how a star with a mass greater than his limit of $1.4 M_\odot$ would collapse and he predicted the existence of black holes as a result. Sir Arthur Eddington, a highly respected astronomer and physicist, was in the audience but did not believe Chandrasekhar's findings. Eddington publicly refuted these ideas, calling them "*outlandish*". Chandrasekhar felt humiliated saying "*Eddington made a fool of me ... I did not know whether to continue my career.*"

Chandrasekhar did continue, but he moved to the USA and turned his attention to other problems in physics. Later discoveries confirmed that Chandrasekhar was correct.

Chandrasekhar was also renowned for his diligence and love of teaching. In the 1940s, Chandrasekhar was working at the Yerkes observatory. Every weekend he drove 240 km on the return journey to the University of Chicago to teach an astrophysics class of only two students. Those two students (Tsung-Dao Lee and Chen-Ning Yang) went on to be awarded the 1957 Nobel Prize for Physics. Chandrasekhar had to wait until 1983 before he won his Nobel Prize.

▲ Figure 11 Subrahmanyan Chandrasekhar.

Practice questions

8. Mu Columbae is a luminous main sequence star of a mass of approximately $16M_\odot$.

    a. Outline why Mu Columbae is likely to remain a shorter time on the main sequence than the Sun.

    b. Compare and contrast fusion processes that will take place in Mu Columbae and the Sun after the stars leave the main sequence.

    c. Explain, by reference to the Chandrasekhar and Oppenheimer–Volkoff limits, the likely final evolutionary stages of Mu Columbae and the Sun.

9. a. Describe main physical properties of white dwarfs.

    b. Outline how a white dwarf maintains a constant radius.

    c. State the origin of the energy radiated by white dwarfs.

Astronomical distances and their determination

Atomic and nuclear physics has developed units designed to cope with the small masses and energies involved in this area of the subject. In a similar way, astrophysics has non-SI units to cope with the immense distances that arise in astronomy. Although these are not strictly part of the SI, they are common currency amongst astrophysicists and allow relative distances to be compared more easily.

Light year (ly)

The speed of light is (as a result of the 2019 revision of the SI) one of the seven defined constants. The value assigned to it is $299\,792\,458\,\mathrm{m\,s^{-1}}$. The **light year** is the distance travelled by a photon in one Julian year (365.25 days exactly, rather than the Gregorian year of 365.2425 days). This is $9\,460\,730\,472\,580\,800\,\mathrm{m}$ exactly—or $9.46 \times 10^{15}\,\mathrm{m}$ to three significant figures.

Concepts such as the light minute and light second are also used. The light year is of most use when indicating distances to stars in our galaxy and to all astronomical objects at distances greater than this.

Astronomical unit (AU)

Although the distance of the centre of Earth from the centre of the Sun varies throughout the year, the average Earth–Sun distance, known as the **astronomical unit,** is of use when discussing distances within the Solar System. Light takes about 8 minutes to travel from the Sun to Earth's orbit and therefore 1 AU is 8 light minutes which is $1.5 \times 10^{11}\,\mathrm{m}$.

Parsec (pc)

The most commonly used distance unit in astrophysics is the parsec. It links to the measurements of nearby stars using the stellar-parallax method. One parsec is 3.26 light years and therefore $3.1 \times 10^{16}\,\mathrm{m}$. Typical distances to nearby stars are orders of pc whereas distant stars in our galaxy are kiloparsec (kpc) away and galaxies can be Mpc and Gpc distant.

Determination of stellar distance using stellar parallax

Triangulation is a surveying technique in which a baseline is constructed some distance from an object whose position is to be determined. The angle between the object and the baseline is measured at each end of the baseline. The distance of the object from the baseline can then be calculated.

Stellar parallax uses the same principle. The term parallax comes from the movement of one object relative to another when viewed from different positions. When travelling in a car, distant hills are virtually stationary, but nearby objects move quickly relative to the hills and, apparently, in the opposite direction to that of the car.

The baseline for stellar parallax is on a larger scale than Earth-bound surveying — the diameter of Earth's orbit is used. Figure 12 shows the principle of the technique. The yellow stars are very distant and are known as the "fixed" stars. The red star is much closer and its position amongst the fixed stars varies during the year.

The scientific method in astronomy

Experiments are important in science. They underpin the scientific method, and their outcomes determine whether a hypothesis is supported. This may lead to further experiments and testing. A contradiction will lead to the hypothesis being modified or rejected. As Richard Feynman said: *"Experiment is the sole judge of scientific 'truth'"*.

In astronomy and astrophysics, experiments are not possible in the usual sense. A star cannot be constructed in a laboratory. It is impossible to carry out an experiment where the mass of a star is changed in order to investigate the change in its luminosity. Instead, astrophysics relies on observations of a different sort. Astrophysicists make measurements of the millions of observable stars and test their hypotheses against the observed data.

Do astronomy and astrophysics qualify as a science?

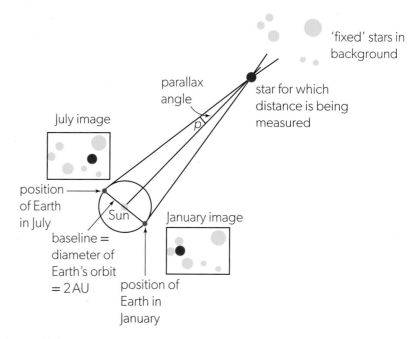

▲ Figure 12 The basis of astronomical distance measurements using stellar parallax.

The baseline distance is known. It is 2 AU. The angles between baseline and the star are measured six months apart (January and July in the figure) and the total angle between the star's position is halved. This new angle is p. Because the angle is small, it is generally measured by astronomers not in degrees but in a subdivision of degrees known as arcseconds: $1° \equiv 60$ arcminutes $\equiv 60 \times 60 = 3600$ arcseconds.

The distance d to the star from the baseline can then be given directly in parsec as

$$d = \frac{1}{p}$$

In which ways has technology helped to collect data from observations of distant stars? (NOS)

Atmospheric turbulence and light absorption in Earth's atmosphere limit stellar parallax to a minimum p of about 0.01 arcsec. From Earth's surface this is a distance of 100 pc. Telescopes mounted on orbiting satellites give better resolution and the Gaia satellite is the second generation of these. The Gaia mission began in 2013 and is predicted to go on until 2025. The resolution of the instruments onboard is about 7 μ arcsec for bright stars giving a maximum distance of up to about 100 000 pc.

The aim of the Gaia mission is to measure the positions of one billion stars in three dimensions, including stars not just in the Milky Way but in the Local Group galaxies too. Other aims include the search for further evidence for Einstein's general theory of relativity and the measurement of the orbital properties of planets outside the Solar System.

Such satellite development and launching are very expensive. They are often the work of national consortia showing the benefits of undertaking science and engineering as a shared endeavour.

Determination of stellar radius

The luminosity L of a star (the power emitted by the star's surface area A) is given by the Stefan–Boltzmann law

$$L = \sigma A T^4$$

as a star is a good approximation to a black body. When the radius of the star is R, this becomes

$$L = \sigma \times 4\pi R^2 T^4$$

assuming a spherical star. The luminosity can be derived from a knowledge of the apparent brightness b, which is the emitted power of the star that arrives on one square metre of Earth's surface:

$$L = 4\pi b d^2$$

where d is the distance of the star from Earth.

- The distance d is obtained from stellar-parallax or using other methods.

- The apparent brightness can be measured and so the luminosity is known.

- The temperature T of the star can be obtained from measurements of the peak wavelength in the emitted black-body spectrum.

- All these quantities can be compared with the same quantities measured for the Sun. This gives the equation

$$\frac{L}{L_\odot} = \frac{R^2 T^4}{R_\odot{}^2 T_\odot{}^4}$$

where the symbol $\odot$ signifies the Sun.

- Therefore

$$\frac{R}{R_\odot} = \frac{T_\odot{}^2}{T^2} \times \sqrt{\frac{L}{L_\odot}}$$

and an estimate of the stellar radius can be made.

Science as a shared endeavour — Sharing data

The Gaia mission is just one example of a project which will collect vast amounts of data on stars. Astronomers share these data among themselves so that they can all construct models and test their theories using the data.

An internet search for "Gaia data" will allow you to access the data yourself. Many other star catalogues and datasets are also available for anyone to access.

Determining the radius of a distant star uses techniques discussed in Topic B.1.

Worked example 6

The following data are given about star Epsilon Tauri: surface temperature = 4900 K and luminosity = 3.7×10^{28} W. Calculate the radius of Epsilon Tauri.

Solution

This is a direct application of the Stefan–Boltzmann law.

$$L = 4\pi\sigma R^2 T^4 \Rightarrow R = \sqrt{\frac{L}{4\pi\sigma T^4}}.$$

$$R = \sqrt{\frac{3.7 \times 10^{28}}{4\pi \times 5.67 \times 10^{-8} \times 4900^4}} = 9.5 \times 10^9 \, \text{m}.$$

Worked example 7

The parallax angle for star Tau Ceti is 0.274 arcsecond.

a. Calculate, in light years, the distance to Tau Ceti.

The apparent brightness of Tau Ceti as observed from Earth is $1.18 \times 10^{-9} \, \text{W m}^{-2}$.

b. Determine the luminosity of Tau Ceti relative to the luminosity of the Sun, $L_\odot = 3.83 \times 10^{26}$ W.

The peak wavelength of the black-body spectrum of Tau Ceti is 540 nm.

c. Estimate the radius of Tau Ceti. Use $T_\odot = 5800$ K.

Solutions

a. The distance in parsec can be calculated directly from the parallax angle in arcseconds.

$$d = \frac{1}{p} = \frac{1}{0.274} = 3.65 \, \text{pc}.$$ As 1 pc = 3.26 ly, the distance in light years is $3.65 \times 3.26 = 11.9$ ly.

b. The apparent brightness b is related to the luminosity L of the star by the equation

$$b = \frac{L}{4\pi d^2};$$ hence $L = 4\pi d^2 b = 4\pi (11.9 \times 9.46 \times 10^{15})^2 \times 1.18 \times 10^{-9} = 1.88 \times 10^{26}$ W.

Note that, for this calculation, the distance to the star must be expressed in metres, 1 ly = 9.46×10^{15} m.

Finally, the luminosity in terms of $L_\odot$ is $L = \left(\frac{1.88 \times 10^{26}}{3.83 \times 10^{26}}\right) L_\odot = 0.49 \, L_\odot$.

c. The peak wavelength provides the information of the surface temperature of the star.

$$T = \frac{2.9 \times 10^{-3}}{540 \times 10^{-9}} = 5370 \, \text{K}.$$ Since $L \propto R^2 T^4$, the radius of Tau Ceti can be estimated from

$$R = \left(\frac{T_\odot}{T}\right)^2 \sqrt{\frac{L}{L_\odot}} \, R_\odot = \left(\frac{5800}{5370}\right)^2 \sqrt{0.49} \, R_\odot = 0.82 \, R_\odot$$

Practice questions

10. a. Discuss limitations of the parallax method for measuring distances to stars.

    The parallax angle for star Arcturus is 8.9×10^{-2} arcsecond and its apparent brightness is $4.3 \times 10^{-8}\,\mathrm{W\,m^{-2}}$.

    b. Calculate:

        i.   the distance to Arcturus, in m

        ii. the luminosity of Arcturus, in W.

    c. The black-body radiation curve of Arcturus has a maximum at 680 nm. Calculate the surface temperature of Arcturus.

    d. Hence, estimate the radius of Arcturus.

11. Sirius, the brightest star in the night sky, is 8.7 ly from the Sun.

    a. Calculate the distance to Sirius in AU.

    b. Calculate the parallax angle for Sirius, in arcsecond.

    c. The luminosity of Sirius is $25\,L_{\odot}$. Calculate the apparent brightness of Sirius, in $\mathrm{W\,m^{-2}}$.

12. The following data are given about the star Denebola.
    Parallax angle $= 9.09 \times 10^{-2}$ arcsecond
    Surface temperature $= 8500\,\mathrm{K}$
    Radius $= 1.2 \times 10^{9}\,\mathrm{m}$

    a. Calculate, in ly, the distance to Denebola.

    b. Calculate the luminosity of Denebola relative to the solar luminosity $L_{\odot} = 3.83 \times 10^{26}\,\mathrm{W}$.

    c. Describe how the knowledge of the luminosity and the temperature of a star helps to predict its future evolution.

How can the understanding of black-body radiation help to determine the properties of stars?

The link between luminosity and stellar radius emphasizes the importance of Topic B.1 to astrophysics. Our terrestrial knowledge of the properties of black-body radiation is extended to cosmological observations. By assuming that stars are black bodies—or close approximations to them—we can determine the properties of local stars and use these as standard candles to allow us to calibrate the universe in terms of stellar and galactic distance. A standard candle is an object in astronomy that has a known energy output. By comparing this output with the brightness of the object as we observe it, we can estimate the distance from Earth to the object.

How do emission spectra provide information about observations of the cosmos?

In a similar way, the work on terrestrial emission spectra in Topic E.1 can help to extend our knowledge of the universe. The emission spectra themselves give invaluable information about the elements in the outer atmospheres of stars and the absorption spectra tell us about any materials between a star and the observer. Wien's law and a determination of the peak wavelength give us a measure of stellar temperature (Topic B.1). Finally, the shifts in the spectra (redshifts and blueshifts) described in Topic C.5 link our knowledge of galactic motion to our hypotheses about the universe and the existence of dark matter.

Theme E — End-of-theme questions

1. The diagram shows the position of the principal lines in the visible spectrum of atomic hydrogen and some of the corresponding energy levels of the hydrogen atom.

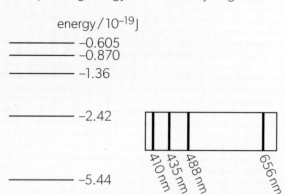

energy / 10^{-19} J

———— −0.605
———— −0.870
———— −1.36

———— −2.42

———— −5.44

410 nm
435 nm
488 nm
656 nm

a. Determine the energy of a photon of blue light (435 nm) emitted in the hydrogen spectrum.

b. Identify, with an arrow on the diagram, the transition in the hydrogen spectrum that gives rise to the photon with the energy in (a).

2. a. In a classical model of the singly-ionized helium atom, a single electron orbits the nucleus in a circular orbit of radius r.

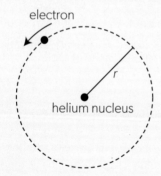

electron

r

helium nucleus

i. Show that the speed v of the electron with mass m, is given by $v = \sqrt{\dfrac{2ke^2}{mr}}$.

ii. Hence, deduce that the total energy of the electron is given by $E_{tot} = -\dfrac{ke^2}{r}$.

iii. In this model the electron loses energy by emitting electromagnetic waves. Describe the predicted effect of this emission on the orbital radius of the electron.

b. The Bohr model for hydrogen can be applied to the singly-ionized helium atom. In this model the radius r, in m, of the orbit of the electron is given by $r = 2.7 \times 10^{-11} \times n^2$, where n is a positive integer.

i. Show that the de Broglie wavelength λ of the electron in the $n = 3$ state is $\lambda = 5.1 \times 10^{-10}$ m.

ii. Estimate for $n = 3$, the ratio
$$\frac{\text{circumference of orbit}}{\text{de Broglie wave length of electron}}.$$

3. A nucleus of fluorine-18 ($^{18}_{9}$F) decays by beta plus (β^+) decay into a nucleus of oxygen-18 ($^{18}_{8}$O).

a. Write down the nuclear reaction for this decay.

b. The atomic mass of $^{18}_{9}$F is 18.000937 u and that of $^{18}_{8}$O it is 17.999160 u.

i. Determine the energy released in this decay.

ii. Explain why every beta particle emitted in this decay can have a different kinetic energy.

c. The graph shows how the count rate from a sample of $^{18}_{9}$F varies with time.

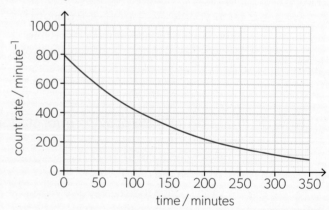

i. Determine the half-life of $^{18}_{9}$F.

ii. Calculate, in s^{-1}, the decay constant of $^{18}_{9}$F.

iii. The sample initially contains 5.0 mg of $^{18}_{9}$F. Determine the mass of $^{18}_{9}$F remaining in the sample after 8 hours.

AHL

4. Radioactive uranium-238 ($^{238}_{92}$U) produces a series of decays ending with a stable nuclide of lead. The nuclides in the series decay by either alpha (α) or beta-minus (β⁻) processes.

   a. Uranium-238 decays into a nuclide of thorium-234 (Th). Write down the complete equation for this radioactive decay.

   b. The graph shows the variation with the nucleon number A of the binding energy per nucleon.

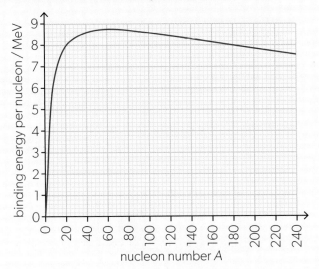

   i. Outline why high temperatures are required for fusion to occur.

   ii. Outline, with reference to the graph, why energy is released both in fusion and in fission.

   iii. Uranium-235 ($^{235}_{92}$U) is used as a nuclear fuel. The fission of uranium-235 can produce krypton-89 and barium-144.

   Determine, in MeV and using the graph, the energy released by this fission.

5. Eta Cassiopeiae A and B is a binary star system located in the constellation Cassiopeia.

   a. The following data are available:

Apparent brightness of Eta Cassiopeiae A	$= 1.1 \times 10^{-9}\,\mathrm{W\,m^{-2}}$
Apparent brightness of Eta Cassiopeiae B	$= 5.4 \times 10^{-11}\,\mathrm{W\,m^{-2}}$
Luminosity of the Sun, $L_\odot$	$= 3.8 \times 10^{26}\,\mathrm{W}$

   i. The peak wavelength of radiation from Eta Cassiopeiae A is 490 nm. Show that the surface temperature of Eta Cassiopeiae A is about 6000 K.

   ii. The surface temperature of Eta Cassiopeiae B is 4100 K. Determine the ratio

   $$\frac{\text{radius of Eta Cassiopeiae A}}{\text{radius of Eta Cassiopeiae B}}.$$

   iii. The distance of the Eta Cassiopeiae system from the Earth is 1.8×10^{17} m. Calculate, in terms of $L_\odot$, the luminosity of Eta Cassiopeiae A.

   b. A Hertzsprung–Russell (HR) diagram is shown.

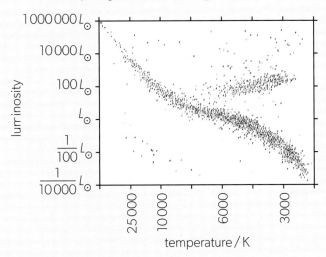

   i. On the HR diagram, draw the present position of Eta Cassiopeiae A.

   ii. State the star type of Eta Cassiopeiae A.

Extended-response questions

Paper 2 in the IB Diploma Programme physics course includes extended-response questions—one at standard level or two at higher level. They make up about 40% of the marks available in Paper 2, so it is important that you give yourself enough time to complete them. The questions will involve more than one theme. You will be required to use a range of skills and techniques, and apply these to unfamiliar situations or contexts. You will be asked to perform specific calculations, reflect on the solutions and generalize findings.

The following are representative of the types of extended question you will meet in Paper 2. Read quickly through the whole question before you start to write your answer.

Question 1

Marian rides an electric bicycle. The combined mass of Marian and the bicycle is 75 kg. The bicycle starts from rest at time $t = 0$. The graph shows how the net force on the bicycle varies with t during the initial acceleration.

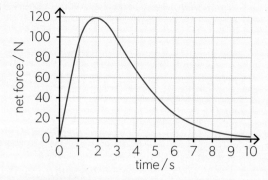

a. Explain why the kinematic equation $v = u + at$ cannot be used to calculate the speed of the bicycle after time t. [2 marks]

b. Estimate, using the graph:
   i.   the maximum acceleration of the bicycle [1 mark]
   ii. the speed of the bicycle at time $t = 10$ s. [3 marks]

c. Energy is transferred to the bicycle's motor from a battery that has an emf of 48 V. When the motor develops the maximum power, the terminal potential difference of the battery is 42 V and the current from the battery is 6.2 A.
   i.   Calculate the electrical power output of the battery when the motor is developing its maximum power. [1 mark]
   ii. Determine the internal resistance of the battery. [2 marks]

   Marian rides up a hill of constant gradient. She starts at sea level and reaches the top of the hill at 420 m above sea level in a time of 40 minutes. The motor develops the maximum power during the entire ride.

iii. Determine the rate at which the gravitational potential energy of Marian and the bicycle increases with time. [2 marks]
iv. Outline why the answers in parts c.i. and c.iii. are different. [2 marks]

d. Show, with an appropriate calculation, that the gravitational field of Earth can be considered uniform across the range of heights above sea level that Marian encounters during her ride. The radius of Earth is 6370 km. [2 marks]

e. Marian's bicycle is equipped with a disc brake system, which produces a frictional force between brake rotors and brake pads. The diagram shows the working principle of the brakes. The total mass of brake rotors and pads is 300 g and they are made of a material of specific heat capacity 850 J kg^{-1} K^{-1}.

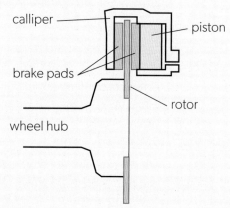

   i.   The bicycle comes to rest from an initial speed of 25 km h^{-1}. Estimate the increase in the temperature of the brake system. [3 marks]
   ii. Discuss any assumptions you made in working out your answer to e.i.. [2 marks]

Question 2

a. The diagram shows the position of some of the lines in the laboratory line spectrum of helium.

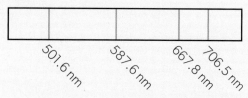

   i.   Explain how the observation of spectral lines provides evidence for atomic energy levels. [2 marks]
   ii. Calculate the energy of the most energetic photon in the line spectrum above. [2 marks]

b. The helium line at 587.6 nm is present in the absorption spectrum of a star. The observed wavelength of the line varies between 586.5 nm and 588.7 nm with a period of several days.

   i. Explain, without calculation, what the variation in wavelength suggests about the nature of the star. [3 marks]

   ii. Calculate, relative to the Earth, the maximum velocity of the star along the line of sight. [2 marks]

c. The spectrum of the star has a maximum intensity at a wavelength of 340 nm.

   i. Calculate the surface temperature of the star. Assume that the star radiates as a black body. [1 mark]

   The parallax angle of the star is 6.5×10^{-3} arc-second.

   ii. Calculate, in light years (ly), the distance to the star. [2 marks]

   The apparent brightness of the star is also measured.

   iii. State what is meant by apparent brightness. [1 mark]

   iv. Explain, without calculation, how the radius of the star can be determined from these data. [3 marks]

d. Kepler deduced his laws of orbital motion by analysing observations of the planets in the Solar System.

   i. Discuss why these same laws can be applied to orbital motion of objects outside the Solar System. [2 marks]

   Astronomers have discovered many planets orbiting stars other than the Sun. The orbital periods have been directly measured for some of these planets.

   ii. Outline how Kepler's laws can be used to determine the orbital radii of these planets. [2 marks]

Question 3

a. State what is meant by an elastic collision. [1 mark]

b. Two stones collide on a horizontal frictionless ice surface. Stone A, of mass 0.10 kg, moving with a velocity of 1.6 m s^{-1} hits stone B of mass 0.20 kg that is initially at rest. After the collision, A moves with a velocity of 1.0 m s^{-1} at an angle of 60° to the original direction of motion.

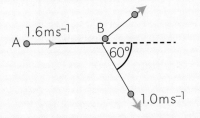

i. Calculate the magnitude of the momentum of B after the collision. [2 marks]

The stones are in contact for a time of 4.0×10^{-4} s.

ii. Estimate the average force between the stones during the collision. [2 marks]

iii. Determine whether the collision is elastic. [2 marks]

The collision of two stones is an analogue of Compton scattering. In a Compton scattering experiment, a monochromatic beam of X-rays is incident on a graphite target. Scattered X-rays are observed at an angle of 60° relative to the direction of the incident beam.

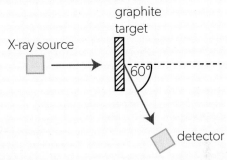

c. Some of the scattered X-rays have a longer wavelength than the incident X-rays. Outline how this observation supports the particle model of electromagnetic radiation. [1 mark]

d. The wavelength of the incident X-rays is 1.0×10^{-10} m.

   i. Calculate the change in the wavelength of the scattered X-rays. [1 mark]

   ii. Outline how conservation laws allow us to predict the initial motion of the recoil electron in this experiment. [2 marks]

   iii. Show that the energy gained by the recoil electron is about 150 eV. [2 marks]

e. A recoil electron leaves the graphite target and enters a region of a uniform magnetic field of strength 2.0 mT perpendicular to the initial velocity of the electron. The electron moves in a circular path.

   i. Determine the radius of the electron's path. [3 marks]

   ii. Explain why relativistic effects on the motion of the electron can be ignored. [1 mark]

f. The energy of the recoil electron calculated in d.iii. is only about 1% of the energy of the incident X-ray photon. The experiment is repeated with X-rays of shorter wavelength. Discuss the effect, if any, of this change on the percentage of the photon's energy transferred to the electron. [2 marks]

g. State one other example of an analogy between two areas in physics. Go on to outline how your example helps us to understand the phenomenon. [1 mark]

The inquiry process

Figure 1 shows the cycle of inquiry used in science. Although the description of the stages is written here in terms of an internal assessment, this cycle applies to all the science you carry out in a practical context. You should regard your experimental work in physics as a preparation for the internal assessment.

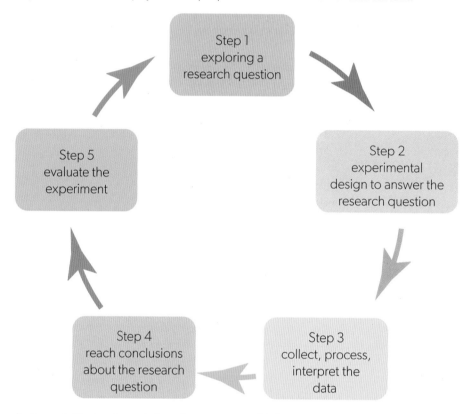

▲ **Figure 1** The inquiry cycle for science expressed in five steps.

The experiment on page 454 in Topic C.4 is a good example of deciding what data to collect to answer a particular research question.

1

Exploring a research question

Scientists formulate a **research question** that usually features a **hypothesis**. Hypotheses lead to predictions that need to be clearly stated and explained in terms of scientific understanding. To support the research questions, a variety of existing relevant **sources of information** must be consulted. There must be sufficient of these sources to justify the research proposal. The sources of knowledge can be in many forms: research papers, advanced books, suggestions from the Internet and so on. It is important to demonstrate independent thought and initiative in generating the research question(s) and you will need insight to recognize the relevance of your proposal.

2

Experimental design to answer the research question

The research question leads to an **experimental design**. Here you will need to be creative at all stages of your investigation: in the initial work, when it is implemented and when it is finally presented. The investigation should consist of real laboratory experiments that you can safely carry out, investigations of databases, or simulations. It could even be a combination of these.

You need to decide on your **independent** and **dependent variables** and propose ways to maintain the **control variables** that you select. You will also need to choose suitable ranges for your measurements and to decide how much data you can reasonably collect in the time available to you.

You need to be able to justify the methodology of your final experiment. All these factors may be informed by early trial (pilot) experiments when you can decide on the nature and quantity of your experimentation.

You need to make decisions about the instruments and sensors that you use for your data collection. Do they, for example, require calibration or zero-setting? Can this be done once or should it be done more frequently? To what extent do you need to consider the environment of the laboratory? Is control required? It is difficult to carry out experiments involving oscillations of a light pendulum when a window is open and a strong wind affects your apparatus. Will friction be a problem? Is there excessive uncontrolled electrical resistance in a circuit?

Practise experimental design using the example in Topic A.4 on page 144.

The range of approaches to data collection in an IA is very great. You can carry out hands-on lab work, use spreadsheets or simulations for modelling and subsequent analysis of the models, and extract data from a scientific database for your own analysis.

See Topic A.3 (page 121) for an example of devising the best method for an experiment.

There is advice about the use of data-logging equipment in *Experimental tools for physics* (page 343).

3

Collect, process and interpret your data

Sufficient data are required (they may be qualitative or, more usually, quantitative) and they must be relevant to your research question. The data may prove to be insufficient for some reason or there may be unforeseen issues in their collection. If this is the case, you will need to address these issues or return to, and modify, your research question.

Your data processing must be accurate and relevant. Always check where possible that the collection itself has been carried out with accuracy and that these data have been manipulated appropriately and correctly. There is advice on these matters in the *Tools for physics* section of the book.

Your data need to be interpreted to decide the extent to which your research question has been answered. Again, there is advice here, but you will need to decide on the presentation of your data. Will it be as a graph, a chart or a table? Will it be quantitative or qualitative? Some of these decisions may have arisen naturally as a result of your chosen research question.

- What relationships can you identify from your data?
- Are there obvious patterns and trends?
- Do any of your data need to be rechecked because they are outliers?
- Are there good reasons for you to disregard outliers or to include them in the final data set?
- Are you satisfied about the accuracy and precision of your measurements?
- Are your measurements reliable and valid?

There is more on processing and presenting data in *Experimental tools* on page 353.

See page 263 (Topic B.3) for an example of constructing and interpreting graphs.

See page 663 in Topic E.3 for an example of dealing with uncertainties carefully in an experiment.

Carry out the experiment on page 422 and compare the outcomes with the accepted scientific context covered in Topic C.3.

There is advice in *Tools for physics* (page 362) about the use of physics simulation software.

4

Reach conclusions about the research question

When you have processed and examined your data critically and thoroughly, you will need to decide on the **conclusions** that they offer. Do the conclusions answer the research question and are they aligned with the present scientific context? At this step or in the previous stage, you will need to consider how measurement uncertainties impact on the conclusion. Are the uncertainties large, so that your conclusion can only be provisional, or do you have more confidence in it because measurement uncertainties are small? You should ask yourself whether there are any simulations that can support your conclusions.

Practise evaluating the outcomes of your experiment with the practical simulation on page 665 in Topic E.3.

There is an example of reviewing and extending your experiment in Topic A.2 on page 80.

5

Evaluate the experiment

The final stage is to **evaluate** your whole experiment.

- To what extent is the strength of your conclusion affected by the impacts of random and systematic errors?

- Can you confirm or reject your initial hypotheses?

- When you reflect on your work as a whole, does it have weaknesses in approach that you would correct in further work?

- What are the limitations in your work and how did these impact the conclusions?

- What further work do you propose to move the research question forward?

- Finally, to what extent did you manage to answer your research question?

It is very important to recognize that, in an internal assessment, it is the *processes* that are important not necessarily the *results*. If you achieve a null result, that can be as important as a new discovery. Remember Michelson and Morley (page 166) and their "failure" to discover the aether.

The experiment on page 614 in Topic E.1 gives a useful example of developing your research further in the light of the conclusions you reach.

6

Refining research

In professional science, this is not the end. Scientists will refine their research questions in the light of their conclusions and evaluation and move on to different or similar experiments depending on the outcomes.

The internal assessment (IA)

Approaching your internal assessment

Internal assessment structure

Your internal assessment is marked in terms of the inquiry process and is assessed using these criteria:

- research design
- data analysis
- conclusion
- evaluation.

Each criterion is marked with equal weight and is out of six marks.

To attain full marks in each criterion, you need to demonstrate the following qualities in your work.

- Your research question must be described within a specific and appropriate context. You must explain your methods for the collection of relevant data with enough data to answer the research question. Your investigation must be described clearly enough for it to be reproduced by someone else.

- The recording and processing of your data must be clear and precise. You must provide evidence that you have shown an appropriate consideration of experimental uncertainties. You must show that the processing of your data is relevant and appropriate to the research question and is completely accurate

- Your conclusion must be justified and relevant to the research question. It should be fully consistent with your analysis. It should also be justified by a relevant comparison with the accepted scientific context such as research papers, your textbooks or other sources. You must cite anyone else's materials fully and correctly.

- Your report must discuss the relative impact of specific methodological weaknesses or limitations that you have identified in your work. You should give realistic and meaningful improvements to your work.

Choosing and planning

You need to discuss your plans at an early stage with your teachers. Try to keep an open mind on your work for as long as possible. You may need to adapt your plans based on what apparatus is available and on your teacher's advice. Listen carefully to what they have to say and modify your plans in the light of their comments.

Perhaps let your project grow out of your personal interests. If you play a musical instrument, you could investigate the way the sound is produced. A string player might look at the variation of note frequency with string tension or some other parameter of the string (Topic C.3). If you play a sport, consider some aspect of the mechanics of the game: the "sweet spot" on the racquet in tennis or badminton, for example (Topic A.4). Devise an investigation of why a baseball bat sometimes jars the wrists or about the power delivered to a pedal cycle and so on. Photographers, artists, stage technicians can all find some aspect of their art or craft to investigate. Just think creatively.

The best topics enable you to collect data and are well defined. *What happens when I heat maple syrup?* may not be as good a suggestion as *How do the flow properties of maple syrup change with temperature?*

When you think you have arrived at a topic, ask yourself these difficult questions:

- Is the physics at IB Diploma Programme physics standard?
- Can a sufficient amount of data be collected in the time available?
- Will preliminary work during the first few hours allow the project to develop to a better and more refined experiment?
- Will you be able to draw conclusions from your experiments?
- Are you going to be as interested in the experiment and its results on the last day of the project as on the first?

If the answers to all these are "yes", then this is a possible project for you.

Choose an investigation in which you can:

- use your understanding of physics to plan out an investigation and choose appropriate experiments or data sets
- demonstrate that any experimental working is safe
- show initiative and work independently or within a group setting
- take and record observations appropriately using instruments correctly or manipulate data sets in a creative and accurate way
- analyse results and evaluate what you have done, checking the reliability of the results
- communicate your results and conclusions well using appropriate techniques, including graphs, charts, tables, diagrams and so on.

If possible, try out some preliminary experiments before the time set for the proper investigation, to see if your idea is going to work. Retain the preliminary data and incorporate them into your report. You should be able to get a feel for the results you are likely to get and the size of the effects that you hope to measure. It will also show if there are any major defects in your apparatus design that will need to be corrected before the proper investigation begins.

There may be some aspects of your work that can be researched through books or the Internet. See what others have done by all means, but if you do use anyone else's work (whether their text or data) you must credit this in your report—and it is sensible to gain your teacher's permission to use the work of other people before you begin.

You will need to have a sense of flow and development. Good investigators at any level are rarely doing exactly the same experiment at the end of a project as at the beginning. Begin with simple ideas and extend these as your experience of the experiment changes.

Recording data

It is vital that you keep an accurate, daily record of what you do. Many professional scientists keep their records in a hard-backed notebook. You could keep your records on loose paper in a file, or in an electronic form in a computer. Make sure that you back-up your computer records and do not lose any papers.

At the end of each day of the project, spend some time reviewing the work you have done. Work out any results, draw the graphs of any data you have taken during the day. Think about the direction of the work and about what you want to achieve next. Then write down your aims and targets in your notebook. Plan your work ahead. Make the most of laboratory access, so do this planning and thinking at home.

Keep asking yourself the questions: Where is the work going? Can you explain the results you have so far? Do they make physical sense? Can you present the results in ways that enhance their meaning and quality?

Writing-up an internal assessment

You are required to produce a formal report of your internal assessment. Here are some ideas to consider.

- Begin by writing a brief summary of the whole project, state your research question at the outset and indicate the extent to which you achieved these aims.

- Discuss early on in the report the underlying science on which your work is based. You need to cite any data that come from someone else. There are particular ways to do this that are agreed by all scientists, but you can use any citation style providing that someone else can find the particular article or book that you are referencing.

- Mention in your report any difficulties that you had or any blind alleys you followed for a while. An internal assessment is as much about your ability to carry work through as it is about producing innovative science. Put down everything in the daily record of your work, including setbacks as well as triumphs. Give details of your preliminary results, including the experiments, and the graphs or tables showing the results themselves.

- Write about the safety arrangements. If there are corrosive chemicals or electrical experiments, your teachers may need a risk assessment. Include this too as part of your evidence.

- Include details of all the apparatus you used. Include pictures of your experimental setup if necessary.

- Graphs and diagrams should be added in the text at the place where they are discussed first, rather than collected together at the end. On the other hand, large tables of data are best placed in an appendix at the conclusion of the report.

- The data analysis should come after your account of the experiment, followed by the conclusions you reached.

- Thoroughly evaluate your work. Comment on the results themselves, drawing attention to any anomalies, although you should have tried to eliminate these. Discuss the extent to which your results support your conclusions and try to account for why this support may be less than you wish.

- Be critical of your work. No one does perfect science. One of the marks of a good scientist is the ability to be self-critical.

Index

Page numbers in *italics* refer to end-of-theme questions.

absolute electric potential 526
absolute refractive index 410–11
absolute temperature 214
absolute zero 202
absorption spectra 242, 598, 599–600, 697
acceleration 17–22, 131
    and air resistance 36–7, 38, 39, 68–9
    angular 131–3, 134–5, 143–4
    average/instantaneous 17
    centripetal 96–8, 100–2, 131
    constant 22–3
    force, mass and acceleration 43–6, 48–9, 51–2
    kinematic equations 22–5, 27, 28–30
    object in a lift 51–2
    projectile motion 28–33
    rockets 48–9, 77–9
    rolling objects 156–8
    simple harmonic motion 372–3, 380–2, 383
acceleration due to gravity 26–30, 117, 477, 478
acceleration-time graphs 36, 380
adiabatic change 280–2
air resistance 35–9, 68–9, 114, 138, 501–3
albedo 237–8, 239–40, 241
alpha decay 632–4, 638, 652
alpha particles 559, 591–6, 639, 647
alternating current (ac) 578–9
    generators 560, 574–7
    transformers 582–3
ammeters 308, 309
ampere 300–1, 550
Ampère's law 547, 549
amplitude
    oscillators and SHM 371, 385
    waves 390, 391, 395
angle of incidence 409, 410
angle of reflection 409
angle of refraction 410, 412, 414
angular acceleration 131–3, 134–5, 143–4
angular displacement 94, 96, 132–5
angular frequency 373
angular impulse 152–3
angular momentum 130, 147–54
    electron in Bohr model 606–7, 608
angular speed 94–5, 96
angular velocity 94–5, 132–5
antinodes 437
approximation 338–9
Archimedes' principle 60–1
Aristotle 41, 42
astronomical distances 236, 466–8, 693–7
astronomical observations 162, 229–31, 466–8, 486–7
astronomical unit (AU) 236, 693
atmosphere 232–3, 240–2, 246–7, 270
    energy balance 236–7, 244–5, 247, 250
    muon decay 179–80
    and orbits 501–3
atomic structure 588, 590–1, 698
    Bohr model 170, 483, 602, 604–10, 622
    energy levels 598–601, 604–6, 607, 608
    Geiger–Marsden–Rutherford experiment 559, 591–5
    plum pudding model 589
    Standard Model 588
atomic vibrations 309
average acceleration 17
average speed/velocity 14–16

Avogadro number 255
Avogadro's law 257

background radiation 655, 656
Bainbridge mass spectrometer 557
Balmer series 602–3, 606
batteries 307, 324–5
Becquerel, Henri 630, 631
beta decay 632, 634–5, 638
beta particles 639
Big Bang 678, 679
binary star systems 494
binding energy 640–6, 666
    nuclear fission 669, 670
    nuclear fusion 683
black holes 130, 691–2
black-body radiation 225–31, 612, 695, 697
Bohr model 170, 483, 602, 604–10, 622
Bohr radius 150
Boltzmann, Ludwig 296
Boltzmann constant 203, 258
bound state 605
boundary conditions 440–4, 445–7
Boyle, Robert 262
Boyle's law 256–7, 258
bremsstrahlung 637
Brownian motion 262–3
buoyancy 59–62, 70

car safety 91
Carnot cycle 283–5
catapult field 542, 564
cathode rays 589
cells 302–5, 307, 324–9
Celsius scale 200, 201
centre of gravity 481–2
centre of mass 143–5, 482
centre of percussion 144
centrifugal force 99, 162
centripetal acceleration 96–8, 100–2, 131
centripetal force 99–102
Cerenkov radiation 676
chain reactions 668, 671–2
Chandrasekhar limit 691, 692
charge 299, 300–2, 505–10, 523–5
charge carriers 300, 302
charge density 302
charge separation 505
charge transfer 505–7, 530–1
charged particles 55, 150, 550–8, 559
Charles's law 257, 259
circuit diagrams 307
circuits *see* electric circuits
circular motion 94–105
    angular displacement 94, 96
    centripetal acceleration 96–8, 100–2
    centripetal force 99–102
    charge in a magnetic field 553–6
    frequency and period 95
    and simple harmonic motion 373, 379–81, 383
    speed and angular velocity 94–5, 96
    in a vertical plane 104–5
Clausius law 290
climate change 238–9, 241, 246–9, 677
climate models 238–9, 243–5, 249
coherence 424, 425
collisions 73–4, 79–81
    elastic 81–3, 86–7
    electrons in a conductor 83, 299–300
    gas particles 86, 253, 264–5
    inelastic 83
    nuclear fusion 680–1

in nuclear reactor moderators 673–4
    photon-electron interaction 620–1, 622, 623
Compton effect 619–24, 629
conduction 214–21, 224, 225, 299–300
conduction electrons 299–300
conductors 83, 215, 216, 299–300
    charge carrier speeds in 302
    electrical resistance 308–12
    good/bad 215, 216, 219–20, 225
    Ohm's law 310–12
    resistivity 313–15
    thermal conductivity 216–18, 219–21
    thermal resistivity 314
conservation of charge 505
conservation of energy 81–2, 109, 117, 118, 274
    alpha particle scattering experiments 595
    determining moment of inertia 159
    electrical energy 505
    electromagnetic induction 564
    motion in electromagnetic fields 559
    radioactive decay 632–3, 638
    resonance 452
    rolling objects 156–7
    special relativity 625
conservation of momentum 79–81
    angular momentum 130, 147–8
    linear momentum 81–5
    nuclear reactor moderators 673–4
    radioactive decay 632–3, 638
    real world examples 89–92
    in two dimensions 85–8
contact forces 51, 55
control rods 675
convection 221–3, 224
conventional current 303, 304
coordinate systems 10, 161–2
Copenhagen interpretation 628, 629
Coriolis force 162, 222
coulombs 300–1
Coulomb's constant 509
Coulomb's law 507–10
couples 139–40, 143–4
critical angle 414–15
critical damping 387, 450
Curie, Marie 632
current-time graph 578
cycle (oscillation) 371

damping 387, 449–50, 452, 455
data handling 348–65, 703–4
    displaying data 352–60
    errors 348–9, 524
    interpolation/extrapolation 360
    modelling and simulation 362–5
    plotting/drawing graphs 355, 357–8
    uncertainties 349–51, 352, 358–9
    using graphs 355–6, 360–2
data loggers 27, 345–6
Davisson–Germer experiment 626–7
de Broglie hypothesis 624–5, 626–7
de Broglie wavelength 606–7, 625, 626, 627
decay chains 632, 664
decay constant 658–9, 660–1
degeneracy pressure 689, 691
density 59–60, 61, 117, 359
deuterium nucleus 640–1
diffraction 406, 417–19, 432–5, 471
    double-slit interference 424–31
    electron diffraction 626–7
    emission spectra 598
    of light 419–23, 425, 427–31

multiple-slit interference 431–2
single-slit 419–24
diffraction gratings 432–5, 598, 600
diffusion 262–3
dimensional analysis 336–7
dipole field 549
direct current (dc) 579
displacement 9–10, 11, 12, 19–21
angular displacement 94, 96
kinematic equations 22–5, 27, 28–30
oscillators and SHM 371, 372–3, 380–4,
385
projectile motion 28–33
wave displacement 390–6, 398, 416–17
displacement-distance graphs 390–4
displacement-time graphs
projectile motion 28
simple harmonic motion 378, 380–1,
382–3
wave motion 390, 395–6, 416
distance 9, 10–12
astronomical 236, 466–8, 693–7
kinematic equations 22–5, 27, 28–30
projectile motion 28–33
distance-time graphs 13–16, 28, 36, 38
Doppler effect 172, 456–9, 471
applications 465–9
calculating observed frequency 459–62
and light 463–5, 468
double-slit interference 424–31, 628, 629
drag see viscous drag force
drift speed 302, 312

Earth 151, 222, 232–51, 330–1
atmosphere 233, 240–4, 501–2
energy balance 236–9, 243–5, 250
gravitational field strength inside 495–6
gravity 117, 479, 480–1
magnetic field 539
orbit 236, 485–6
rotation 151
satellites 485–6, 496–503
tunnels through 495–6, 498
eccentricity 484
edge effects 516
efficiency 207, 284, 285, 288, 578
Einstein, Albert 160, 161, 292, 448, 611
gas behaviour 263, 266
general relativity 181, 476, 478
mass–energy equivalence 625, 641
photoelectric effect 611, 613–14
special relativity 160, 164, 167, 171, 176,
177, 182, 625
elastic constant 57
elastic potential energy 122–3, 377
electric cells 302–5, 307, 324–9
electric circuits 305, 307, 330–1
measuring current and pd 308
parallel connections 308, 315–16, 317–19
potential difference (pd) 302–3, 308
potential divider 321–2
resistance 308–12
resistors in parallel 316, 317–19
resistors in series 316–17, 318–20
series connections 308, 315, 316–17,
318–20
variable resistors 321–2
electric current 150, 300–4, 330–1
chemical effect 304–5, 324
in gases and liquids 300
heating effect 304–5, 323–4
induced 561–4
measuring 308
power, current and pd 305–6
and resistance 308–12
electric fields 299, 510–11, 584
adding 512–14
close to a conductor 521–2

conductor in 299–300
due to conducting sphere 522–3, 530
field lines 514–16
motion of a charged particle 550–2, 556–7
perpendicular to magnetic field 556–7
radial 515
between two parallel plates 516–19, 520–1
between two point charges 515–16
electric field strength 511–12, 517–19, 520,
529–31
electric potential 526–31, 532
electric potential difference 526
electric potential energy 531, 532
electric power 305–6, 323–4, 329
electricity generation 206, 247, 574–80
electricity supply 579, 582–3
electromagnetic fields 540, 585
charge to mass ratio for charged particle
555–6
conservation of energy 559
force between two parallel wires 546–50
force on moving charges 541–6
motion of a charged particle 550–8, 559
motor effect 542–3
electromagnetic force 55, 533
electromagnetic induction 560–83, 585
alternating current (ac) generators 574–80
changing fields/moving coils 570–3
energy transfers 564–6
Faraday's law 569
generating emfs and currents 561–5, 570–3
Lenz's law 563–4
magnetic flux/magnetic flux density 566–9
mutual induction 580
real world applications 560, 574, 576, 578,
582–3
self-induction 580–2
transformers 582–3
electromagnetic radiation 223, 372, 431
absorption by greenhouse gases 240–3,
379
black-body 225–31, 612
Doppler effect 463–4
emission spectra 466, 598
gamma emission 632, 636–7
particle-like properties 402, 619
photoelectric effect 613–14
wave-like properties 401–5, 619
electromagnetic spectrum 401–4
electromagnetic waves 389, 401–5, 463–5,
466–9
electromagnetism 504, 536–9
electromotive force (emf) 305, 326–7, 580
induced 563, 564–6
electron antineutrinos 634
electron capture 635
electron diffraction 626–7, 628
electron neutrinos 634–5
electron orbitals 170
electronic band theory 300
electrons
beta decay 634, 635
Bohr model 604–9
charge to mass ratio 555
charge transfer 505–7
Compton effect 620–1
in a conductor 83, 215, 216, 299–300
in a current 302–3
energy levels 604–6, 607, 608
estimating charge on 523–5
motion in perpendicular magnetic and
electric fields 556–7
motion in uniform electric field 550–2
motion in uniform magnetic field 553–7
photoelectric effect 613–14
spin 539
electronvolt (eV) 519–20, 600, 615
electrostatic force 639–40, 669

electrostatic induction 506–7, 530–1
electrostatics 505
emission spectra 598–9, 600–1, 697
black body 226–7
determining astronomical distances 466–7
hot gases 602–3
emissivity 233–5, 238
energy 106–9, 126–9
Earth's energy balance 236–9, 243–5, 250
mechanical energy of a system 118
and momentum 81–4
of satellite in orbit 497–8, 500
see also conservation of energy; kinetic
energy; potential energy
energy density 128–9
energy levels 604–6, 607, 608, 636–7
energy transfer 106–8, 109
air resistance 35
conduction in metals 299–300
efficiency 124–5
elastic potential energy 122–3
electromagnetic induction 564–6
equations 384–7
internal energy 202–4
kinetic and gravitational potential energy
115–21, 377–8
nuclear fission 668–71, 673–4
nuclear fusion 680–1, 683
nuclear reactor moderators 673–4
power 113–14
radioactive decay 640–6
Sankey diagrams and energy flow 126–8
simple harmonic motion 377–9, 384–7
temperature and phase changes 199–200,
202–13
work done 110–13
see also thermal energy transfer
energy transitions 604–6, 607, 637
entropy 291–7
equilibrium 81, 108, 370, 372–3
translational 52–4, 139
equipotentials 491–3, 527–9
equivalence principle 478
errors 348–9, 357, 524, 663
escape speed 500–1
estimation 338–9
exchange particles 470
experimental skills 343–7, 348–9, 351
inquiry process 702–4, 705
internal assessment (IA) 705–7
exponential decay 651
extended bodies 481
extended response questions 700–1
extrapolation 360

falling objects 26–30, 36–8, 50, 68–72
models and modelling 363–4, 365
falsifiability and falsification 84, 169, 612, 619
Faraday's law 569, 576
Fermi question 339
Fermi radius 596
field lines 480–1, 487
electric fields 514–16
magnetic fields 533–4, 567, 570
fields 299, 301, 474, 475–6
fission see nuclear fission
Fleming's left-hand rule 542–3, 562, 564
Fleming's right-hand rule 564
fluids
buoyancy 59–62
convection 221–3
energy transfer 286–7
laminar/turbulent flow 69, 70
motion in 35–9, 68–72
force 40–1, 55
action–reaction pair 47–9
buoyancy force 59–62, 70
centripetal force 99–102

between charged objects 507–10
conservative/non-conservative 117
contact/non-contact 55
defining unit of force 43
elastic restoring force 56–9
force, mass and acceleration 43–7, 48–9, 51–2
free-body force diagrams 50–2
graphs 75–7, 112–13, 122
and momentum 72–3, 74, 75–6, 80–1, 93
Newton's laws of motion 42–9, 52, 74, 78, 99, 137–8
normal force 51, 54, 101
resultant force 52, 74
simple harmonic motion 372–3, 374
translational equilibrium 52–4
triangle of forces 53, 55
work done 110–13, 114–15
see also friction; torque; viscous drag force
force-distance graphs 112–13
force-extension graphs 122
force-time graphs 75–7
forced vibrations 452–4
fossil fuels 126, 127, 128–9, 246, 247
Fourier analysis 369
Franklin, Benjamin 523
Fraunhofer lines 598
free electrons 216, 299–300, 562–3
free vibrations 449
free-body force diagrams 50–2
frequency 95, 372
    electromagnetic waves 401, 403
    harmonics 441–2, 443, 444, 446–7
    oscillators and SHM 371, 373, 449
    waves 390, 397–8
    see also Doppler effect
friction 62–8, 110
    and centripetal force 100–2
    charge transfer 505–6
    coefficient of 63, 64–6, 100
    dynamic 63, 64, 65–6
    static 63, 64, 65, 66, 67
    theories 68
    see also viscous drag force
fringe spacing 426
fuel rods 673, 674, 676
fundamental constants 333–5, 478
fundamental forces 533
fundamental units 218, 255, 300–1, 334–5
fusion (melting) 204, 210–12

galaxies 231, 464–5, 467
Galilean relativity 160, 162–5, 182
Galilean transformations 163–4, 169, 171–2
Galileo 13, 42, 44, 369–70, 448, 475
gamma emission 632, 636–7
gamma photons 402, 637, 639
gamma radiation 402, 403
gas laws 254, 256–61, 266
gases 199–200, 252
    Brownian motion 262–3
    conduction 215, 300
    diffusion 262–3
    emission/absorption spectra 598–9
    greenhouse gases 240–1, 242–3, 246–7, 379
    internal energy 202–4, 268–9
    kinetic model 264–8
    real gas approximation 269–71
    specific latent heat 210
    see also ideal gases
Gay-Lussac's law 257
Geiger–Müller tube 653–4
geostationary orbits 499
geosynchronous satellites 499
global positioning systems (GPS) 180–1
global warming 238–9, 241, 246–9

graphical analysis 355
graphs 13–23, 353–62
    area under 14, 19–20, 22–3, 75–6, 112, 133, 277, 362
    gradient 13–16, 19, 22–3, 360
    intercepts 361
    interpolation/extrapolation 360
    line of best fit 358–9
    logarithms 356
    plotting/drawing 355, 357–8
    pressure-volume diagrams 276–83
    rotational mechanics 133–5
    wave motion 390–6
    see also specific types
gravimetry 117
gravitational field strength 476–7, 478, 480–1, 493–6
gravitational fields 117, 475–6, 511, 584
    escaping 500–1
    extended bodies 481
    field lines 480–1, 487, 491
    motion of an object in 35
gravitational force 47–8, 50–1, 117, 475–80, 533
gravitational mass 44, 478
gravitational potential 489–95
gravitational potential difference 491
gravitational potential energy (GPE) 117–21, 378, 487–9
graviton 476
gravity 479, 485–7
    acceleration due to 26–30, 117, 477, 478
    centre of 481–2
    greenhouse effect 232–3, 240–5
    Earth's energy balance 240–5, 249–50
    emissivity 233–5, 238
    enhanced 240–1, 247
    global warming/climate change 246–8
    solar radiation 235–40
    temperature balance 238–9, 243–4
greenhouse gases 240–1, 242–3, 246–7, 249, 379
grey bodies 233–5, 238
ground state 605, 606, 607
gun recoil 89
gyroscopes 150, 151

half-life 649–53, 655, 661–4
harmonics 440
heat engines 283–90, 305
heat exchangers 675
heat pumps 286–7
heavy elements 666, 667
Heisenberg uncertainty principle 172
helicopters 91–2
helium nucleus 632
Hertzsprung–Russell diagrams 683–6
Hooke, Robert 56
Hooke's Law 56–8, 372
hot air balloons 254
hot gases 599, 602–3
Huygens' principle 410, 411, 418, 421
hydrogen atom
    emission spectra 598, 602–3
    energy levels 604–5, 607, 609
hydrogen fusion 680, 681
hydrogen spectrum 602–3

Ibn Sahl 411
Ibn Sīna 42–3
ideal gases 86, 256–61, 330–1
    equation of state 257–8, 268, 270
    first law of thermodynamics 277, 279–83
    internal energy 268–9
    kinetic model 264–8
    pressure-volume diagrams 276–83
impulse 74–5, 77–8

incident ray 409
inertia 41–2
inertial frame of reference 99, 162, 167–8
inertial mass 44, 478
infrared 404
instability strip 685
instantaneous acceleration 17
instantaneous speed/velocity 14–16
intensity 226, 405
interference
    constructive/destructive 416–17
    diffraction gratings 432–5
    double-slit 424–31, 628, 629
    light 425, 427–31, 432–5
    microwaves 428
    multiple-slit 431–2
    particle diffraction 628
    sound waves 424, 426–7
interferometers 166
intermolecular potential energy 202–3
internal assessment (IA) 705–7
    inquiry process 702–4, 705
internal energy 202–4, 215, 274–5
    and electricity 310
    ideal gas 268–9, 279–81
internal resistance 326–8
interpolation 360
interstellar medium 686
invariant hyperbola 178, 184–8
invariant quantities 177–8
    invariant energy 625
    invariant hyperbola 178, 184–8
    proper length 174–5
    proper time interval 167–8, 170
    spacetime interval 177–9, 184–6
inverse-square laws 230, 405, 474, 478–9, 530, 549
ionization 605, 639
ionizing radiation 639
ions 215
isobaric change 279
isothermal change 280
isothermals 256
isotones 632
isotopes 631
isovolumetric change 279, 281

Joule, James 274

Kelvin scale 200, 201
Kepler-90 system 486–7
Kepler's laws of orbital motion 150, 482–5, 496
kinematic equations 22–5, 27, 28–33
kinematics 8–39
    describing motion 9–25
    motion in fluids 35–9
    projectile motion 26–34
    suvat equations 22–5, 27, 28–33
kinetic energy (KE) 115–16
    energy transfer with GPE 118–21
    internal energy 202–4
    and momentum 81–4
    rolling objects 156–8
    rotational motion 145–7, 148–9
    simple harmonic motion 377–8, 384–7
kinetic model of an ideal gas 264–8

Lagrange point 494
laminar flow 69
lasers 422, 425
length contraction 174, 179, 187–8, 190
Lenz's law 563–4
light 401, 404, 431
    diffraction gratings 432–5
    Doppler effect 463–5, 468
    double-slit experiment 425–31
    refraction 410–11, 414–15

single slit diffraction 419–23
speed of 165, 166, 171–2, 463–4
total internal reflection 414–15
wave-particle duality 619, 628
light year (ly) 693
light-dependent resistors (LDRs) 320, 322
line spectra 598–9
lines of force 516
liquefaction 269–70
liquids 199–200
conduction 215, 300
pressure 253
specific heat capacity 205, 207–8
specific latent heat 210–13
logarithms 356
longitudinal waves 389, 391–4, 398–400
standing waves 445–9
Lorentz factor 169–70
Lorentz transformations 169, 171–2, 173, 176, 184
luminosity 229–31, 683–5, 688–9, 695–6
Lyman series 603, 606

macrostates 294–5
magnetic field strength 537, 538, 544–5, 546–7, 566–7, 568
magnetic fields 533–9, 584–5
due to current in air-core solenoid 537–8
due to current in circular coil 537
due to current in long straight wire 536–7, 541–2, 544–5
field lines 533–4, 536–8, 567, 570
motion of a charged particle 553–7
perpendicular to electric field 556–7
see also electromagnetic fields; electromagnetic induction
magnetic flux 567–8
magnetic flux density 566–8, 569
magnetic flux linkage 569, 574
magnetic forces 55, 533, 541–6
between two current-carrying wires 541–2, 546–7
on charged particle 553–4, 556
on current-carrying conductor 542–3, 544–6
electromagnetic induction 563, 565–6
per unit length of wire 547
magnetic poles 534
main sequence stars 684, 687
mass 44, 478
centre of mass 143–5, 482
force, mass and acceleration 43–6, 51–2
and momentum 72–3
moving in a vertical circle 104
rotational motion 130, 135–7
on a spring 56–9, 140, 370–1, 372, 374, 376–7, 449, 451–2
of a star 686–92
mass defect 641–2
mass spectrometers 557–8
materials testing 658
maths skills 333–42
Maxwell, James Clerk 161, 166
Maxwell-Boltzmann distribution 270
mean square speed 265
mean-speed theorem 23
measurements 10–11, 343–6, 511
errors 348–9, 524
uncertainties 82, 349–51
mechanical energy 118
mechanical waves 389, 405
medicine 403, 455, 465, 657–8
Melde's experiment 439–40
metals 215, 216, 299–300
microstates 293–5
microwave radiation 404, 425, 428
Millikan's oil drop experiment 523–5
Millikan's photoelectric experiment 615–16

Minkowski diagram see spacetime diagrams
modelling 17–18, 39, 238, 249, 362–5
moderators 673–4
molar mass 255–6
mole (unit) 255
moment of inertia 135–7, 159
momentum 73–93
angular momentum 130, 147–53
change in 73–6, 92–3
collisions 73–4, 79–88, 264–5
impulse 74–5, 77–8
transfer 73–4, 93
see also conservation of momentum
Moon 479, 481
motion 8–13
charged particles 55, 150, 550–8, 559
in fluids 35–9, 68–72
graphing 13–16, 18–22
horizontal 30–3
kinematic equations 22–5, 27, 28–33
laws see Newton's laws of motion
rolling and sliding 155–8
vertical motion 28–32
see also circular motion; falling objects; orbital motion; projectile motion; rotational motion
motor effect 542–3, 565
multiple-slit interference 431–2
muons 179–80
mutual induction 580

natural frequency 449
neutral buoyancy 60–1
neutrinos 634–5, 638, 679
neutron number (N) 590–1, 632
neutron stars 136, 640, 691
neutron-induced fission 668–9
neutrons 590–1, 639–40
nuclear fission 668–9, 671–2, 673–4
radioactive decay 632, 633, 634
Newton, Sir Isaac 48, 164, 479
Newton's cradle 73–4
Newton's law of cooling 223
Newton's law of gravitation 475, 478, 485–6
Newton's laws of motion 25, 41–9, 165
first law 42, 52, 137–8
and momentum 74, 78, 79
rotational motion 137–8, 145, 154
second law 43–5, 52, 74, 78, 99, 137, 154
third law 47–9, 52, 79, 145
Newton's postulates of special relativity 164
Nobel Prizes 611
nodes 437–9
Noether's theorem 109
non-ohmic behaviour 312
normal 409
normal force 51, 101
nuclear density 596–7
nuclear energy levels 636–7
nuclear equation 632
nuclear fission 643, 648, 649, 666–7, 699
chain reactions 668, 671–2
energy released 668–71
moderators and control rods 673–5
reaction process 668–9
reactors 672–7
spontaneous/induced 667–8
weapons 677
nuclear fusion 644, 679–83, 684, 687
nuclear mass 641, 642
nuclear power 91, 672–7
energy production rate 670
fuel 668, 672–3
safety 675–7
nuclear stability 648–9
nuclear waste 676–7
nucleon number (A) 590–1, 596–7, 631
nucleons 590–1, 596–7

average binding energy 643–4, 645, 647–8
nucleus of an atom 590–1
alpha particle scattering experiments 591–6
binding energy and mass defect 640–6, 647–8
density 596–7
electrostatic force 639–40
excited state 636
radioactive decay 630, 631, 632
stability 640
strong nuclear force 639–40
nuclides 631–2, 635
medical uses 657–8
stability 643–4, 648–9

Ohm's law 310–12
Oppenheimer–Volkoff limit 691
optical density 414
orbital motion 35, 482–7, 496–7, 584
orbital speed 496–7
orbits 96, 97, 150, 152
electrons 606–7
and gravity 35, 485–7
heliocentric 475, 482
Kepler-90 system 486–7
Kepler's laws 482–5
low-Earth 497, 501–3
see also satellites
orders of magnitude 338
oscillations 57, 368–70
damping 387, 449–51
forced vibrations 452–4
natural frequency 449
resonance 449–55
see also simple harmonic motion
Otto cycle 288

pair production 624
parabolas 32, 35
paradigm shifts 161, 297, 482–3
parsec (pc) 693
particle diffraction 626–7, 628
particles 181–2, 199
collisions 86, 253, 264–5, 673–4
diffusion and Brownian motion 262–3
in an ideal gas 264–8
motion in a travelling wave 391–6
motion of charged particles 55, 150, 550–8, 559
see also specific types
Paschen series 603, 606
path difference 421
pendulums 373
Barton's pendulum 453, 454
damped oscillations 455
energy transfer 378
simple pendulum 337, 374–5, 378, 455
timing mechanism 369–70
Wilberforce pendulum 140
period
circular motion 95
oscillators 371, 374
waves 390, 395
permanent magnetism 539
permittivity 509
phase angle 380, 382–4
phase changes 203, 204, 210–13
phase difference 382–4
phases of matter 199–200
phase changes 203, 204, 210–13
photoelectric effect 611, 613–19, 622
photoelectric equation 614–15
photoelectrons 615, 617, 619, 622
photons 624–5, 637
in the atmosphere 242
Compton effect 619–24
emission 598–9

photoelectric effect 613–14
worldlines 181–2, 186
Planck constant 598, 600, 606
planetary nebula 689
planets 240, 241–2, 482–5
polar orbits 498–9
Popper, Karl 84
positron decay 634–5
potential difference (pd) 302–4
    measuring 308
    between parallel plates 517–18
    power, current and pd 305–6
    and resistance 308–12
potential divider 321–2
potential energy 377–8, 384–7
    elastic 122–3, 377
    electric 531, 532
    gravitational 117–21, 378, 487–9
    intermolecular 202–3
potentiometer 321–2
power 113–14, 145
    alternating currents 578–9
    electric power 305–6, 323–4, 329
power-time graph 578–9
pressure 59–60, 253–4, 359
    of a gas 256–61, 265–6, 279
    and sound waves 398
pressure (Gay-Lussac's) law 257, 259
pressure-volume diagrams 276–83
principle of moments 137–8
projectile motion 23, 26–34
    with air resistance 35–9
    horizontal/vertical components 30–3
    in two dimensions 28–34, 38–9
proper length 174–5
proper time interval 167–8, 170
proportionality 339
proton number (Z) 590–1, 619, 631–2
proton–proton (p–p) cycle 680–2
protons 590–1, 604–5, 639–40
    proton–proton (p–p) cycle 680–2
    radioactive decay 632, 634
protostars 686

quantities 333–5, 343–4, 351
quantization 605, 606–7, 612
quantum mechanics 404–5, 628–9
quarks 647

radar 466, 469
radiation 402, 403, 632–9, 653–6
    ionizing 639
    measuring 653–5, 656
    nuclear waste 676
    solar 235–8, 240, 241
    thermal 223–9
    uses 657–8
    see also electromagnetic radiation
radiative forcing 249
radio waves 404, 419
radioactive dating 657
radioactive decay 630–1, 632, 698–9
    activity and count rate 654
    alpha decay 632–4, 638, 652
    applications 657–8
    beta decay 632, 634–5, 638
    decay chains 632, 664
    decay constant 658–9, 660–1
    energy transfers 640–6
    gamma emission 632, 636–7
    half-life 649–53, 655, 661–4
    measurements 653–6, 663
    nuclear stability 648–9
    probability of decay 658–9
    randomness 632, 649, 654
    rate of decay 650
    simulating 664

spontaneous fission 667
radioactive growth 665
radioactive series 632
randomness 293–5, 632, 649, 654
range (projectiles) 31
rates of change 113, 575
Rayleigh–Jeans law 612
rays 407, 409
red giants 684, 687, 690
redshift 466–7
reference frames 99, 161–2, 167–8, 176, 182–4
reflected ray 409
reflection 408–9, 410
    standing wave formation 436, 440–1
    total internal reflection 414–16
refraction 408–13
refractive index 410–12
refrigerators 286–7
relativity 160–93
    Einstein's general relativity 181, 476, 478
    Einstein's special relativity 160, 164, 167, 171, 176, 177, 182, 625
    Galilean relativity 160, 162–5, 182
    Galilean transformations 163–4, 169, 171–2
                                              length
    contraction 174, 179, 187–8, 190
    linear motion in relativistic context 168
    Lorentz transformation 169, 171–2, 173, 176, 184
    Newton's postulates 164
    reference frames 161–2
    relativistic effects on GPS 180–1
    simultaneity 164, 182–4, 187, 188–9
    speed of light 165, 166, 171–2
    time dilation 168–9, 172, 179–80, 187–8
    time travel 186
    twin paradox 193
    two postulates of special relativity 164
    velocity addition 164, 176–7
    see also invariant quantities; spacetime
        diagrams
resistance 308–12
resistivity 216, 313–15
resistors
    non-ohmic conductors 312
    ohmic conductors 310–11
    in parallel 315, 316, 317–19
    in series 315, 316–17, 318–20
    variable resistors 320–4
resonance 242–3, 449–55
resonance curves 452–3
restoring force 56–9, 372, 374
resultant force 52
rheostat 321–2
right-hand corkscrew rule 138, 536–7
rigid body mechanics see rotational motion
ripple tanks 408, 409
rockets 48, 77–9, 494
root mean square (rms) 266
rotational acceleration see angular acceleration
rotational motion 130–59
    angular acceleration 131–3, 134–5, 143–4
    angular momentum 147–53
    angular velocity 131, 132–5, 148–9
    centre of mass 143–4
    conservation of energy 159
    equations of motion 132–3, 158
    graphs 133–5
    moment of inertia 135–7, 141, 159
    Newton's laws of motion 137–8, 145, 154
    rolling objects 155–8
    rotational kinetic energy 145–7, 148–9
                                              torque 137–9
Rutherford scattering 559, 591–6, 647
Rydberg formula 603, 604, 608

Sankey diagrams 126–8

satellites 485–6, 496–503
    and atmosphere 501–3
    energy of 497–8, 500, 501–3
    global positioning systems (GPS) 180–1
    launching 500–1
    types of orbit 100, 498–9
scalar quantities 108, 340
scalars 9, 10, 12, 339–40
Schrodinger's Cat 629
scientific collaboration 229, 230, 238, 677
scientific method 694
self-induction 580–2
semiconductors 312
sensors 345–6
shielding 675
SI units 334–6, 478
sign conventions 274–5
significant figures 337–8
simple harmonic motion 57, 368–77, 470–1
    and circular motion 373, 379–81, 383
    damping 451
    energy transfer 377–9, 384–7
    and greenhouse gases 243, 379
    mass-spring systems 140, 370–1, 372–3, 374, 376–7, 451
    modelling 363, 382
    pendulums 373, 374–5, 378
    requirements for 372–3
    and torque 140
simultaneity 164, 182–4, 187, 188–9
single-slit diffraction 419–24
smartphones 345, 359, 398, 400, 535
Snell's law 410–11
solar cells 325
solar constant 235–6
solar radiation 235–8, 240, 241
Solar System 152, 475, 482–5, 487
solids 199–200
    conduction 215
    emission spectra 599
    internal energy 202–3
    pressure 253
    refractive index 411
    specific heat capacity 205, 206
    specific latent heat 210–13
sonar 405
sound waves 398–400, 463
    Doppler effect 456, 457–62, 465
    interference 424
spacetime 161
spacetime diagrams 181–2, 188–92
    simultaneity 183, 188–9
    spacetime interval 184–6
    time dilation and length contraction 187–8, 190
    twin paradox 193
    worldlines 181, 182, 183, 188, 190
spacetime interval 177–9, 184–6
specific charge 555–6
specific heat capacity 204–9, 258
specific latent heat 204, 210–13
spectrometers 226
speed 12–16
    angular speed 94–5, 96
    instantaneous and average 14–16
    kinematic equations 22–5, 27, 28–30
    and kinetic energy 115–16
    linking angular and linear 95, 96
    maximum speed of a car 114
    measuring using Doppler effect 468–9
    projectile motion 28–33
    terminal 36, 68–9, 70
speed of light 165, 166, 171–2, 463–4
speed of sound 271, 399–400
speed-time graphs 22–3
    falling objects 36–7, 69
    motion in a fluid 69, 71

projectile motion 29
spreadsheet models 17–18, 363–4
spring constant 56–7, 58, 123–4
springs 56–9, 122–4, 140
  elastic potential energy 122–3
  natural frequency 449
  resonance 140, 451–2
  simple harmonic motion 140, 370–1,
    372–3, 374, 376–7, 451
standard candles 697
standing waves 436–41, *470*
  boundary conditions 440, 443, 445–6
  harmonics 440, 441–4, 446–9
  in pipes 445–9
  in strings 439, 440–5
stars *699*
  apparent brightness 230–1
  binary star systems 494
  distance using stellar parallax 694–5
  emission spectra 601
  equilibrium 640
  evolution 684–5, 686–92
  HR diagrams 683–6
  luminosity 229–31, 683–5, 688–9, 695–6
  mass 686–92
  neutron stars 136, 640, 691
  nuclear fusion 679–82, 684, 687
  radius 695–6
  temperature 683–7
  variable 685
static charge 299
steady state 217
steam engines 273
Stefan-Boltzmann law 227–8, 229–31, 233,
  684, 695–6
stellar parallax 694–5
Stokes' law 69–70
stopping potential 615, 616
strong nuclear force 533, 639–40, 647–8
Sun 226, 679
  luminosity 230
  solar constant 235–6
  solar radiation 235–8, 240, 241
supergiants 684
superheavy elements 648
supernovae 678, 679, 691
superposition 398, 416–17, 437
surface frictional force 63
suvat equations 22–5, 27, 28–33
symbols 12
systems, defining 211, 272

temperature 199–202, 274
  absolute temperature 214
  change 204–9, 214
  Earth's 240–1, 243–4, 246–8
  of a gas 256–61, 280
temperature scales 200, 201
terminal speed 36, 68–9, 70
test objects 477
theory, developing 229
thermal energy transfer 83, 198–200, 214–25
  black-body radiation 225–9
  conduction 214–21, 224, 225
  conductivity 216–18, 219–21
  convection 221–3, 224
  thermal radiation 223–9
  thermal resistivity 314
  see also thermodynamics
thermal equilibrium 202
thermal neutrons 673
thermal radiation 223–9
thermionic emission 626
thermistors 320, 322
thermocouples 201
thermodynamics 272, *330–1*
  Carnot cycle 283–5
  first law 274–5

heat cycles and engines 283–90
implications for universe and life 296, 297
refrigerators and heat pumps 286–7
second law 290–7
zeroth law 288
Thomson, Joseph J 589
thought experiments 42, 183–4, 511
threshold frequency 613
time dilation 168–9, 172, 179–80, 187–8
time travel 186
torque 130, 137–42
  couples 139–40, 143–4
  Newton's laws of rotational motion 137–8,
    145
  rolling and sliding objects 155–6
total internal reflection 414–16
trajectories 28, 31, 32, 38, 181–2
transformers 582–3
translational equilibrium 52–4, 139
transmission 440
transmittance 244–5
transport phenomena 215
transverse waves 389, 390–1, 397, 398, 401,
  439–45
travelling waves 389–400, 436–8
  see also electromagnetic waves
triangle of forces 53, 55
triboelectric effect 505–6
turbulent flow 69, 70
twin paradox 193

ultrasound 405, 465
ultraviolet 403
uncertainties 82, 349–51, 352, 358–9
unified atomic mass unit 596, 641
uniform circular motion 94
units 11, 43, 59, 333, 334–6
  angular impulse 152
  astronomical distances 693
  energy 108, 274
  fundamental units 218, 255–6, 300–1,
    334–5, 478
  resistivity 313
  rotational mechanics 134
  temperature 201, 205
universal gravitational constant 478
upthrust *see* buoyancy

V-I graphs 310–12
vacuum 401, 402, 410–11
valley of stability 645
Van der Waals equation 270
vaporization 204, 210, 212–13
variable resistors 320–4
variables 333–4, 355–6, 703
vector quantities 9–10, 17, 40, 50, 94, 147, 340
vectors 9–10, 12, 17, 339–42
  addition 340–1, 512–13
  products 550
  resolution of 53
velocity 12–17
  angular velocity 94–5, 132–5
  instantaneous and average 14–16
  kinematic equations 22–5, 27, 28–30
  and momentum 72–3, 74, 81, 83–8
  Newton's laws of motion 42
  particles in an ideal gas 264–6
  projectile motion 28–33
  relativistic velocity addition 176–7
  simple harmonic motion 380–2, 383
velocity addition 164, 176–7
velocity addition equation 164
velocity-time graphs 17–22
  projectile motion 29
  simple harmonic motion 380, 382
vibration generator 439–40
viscosity 69–71
viscous drag force 35, 68–72

air resistance 35–9, 68–9, 114, 138, 501–3
voltmeters 308, 309
volume of a gas 256–61, 279

water
  density 60–1
  specific heat capacity 205, 206, 207, 210
  specific latent heat 210–12
water hoses 89–90
water waves 374, 407, 408, 409
Watt, James 114
wave equation 397–8
wave phenomena 406–35, 470–1
  Huygens' principle 410, 411, 418, 421
  reflection 408–9, 410, 440–1
  refraction 408–13
  Snell's law 410–11
  superposition 416–17
  total internal reflection 414–16
  transmission 408–9
  see also diffraction; interference
wave profiles 390–4
wave speed 390, 397–8
  at boundary between two media 408, 409,
    410–11
wave-particle duality 404–5, 619, 624–5, 628
wavefronts 407, 409, 410, 457–9
wavelength 390, 391, 397–8
  Compton effect 619, 620–1
  Doppler shift 463, 466–9
  electromagnetic radiation 401–4
  harmonics 441, 443
  measuring with double slits 426, 428
  standing waves 437, 438
waves 379, 388–90, *470–1*
  compression/rarefaction 389, 392
  electromagnetic waves 401–5
  energy transfer 109
  graphing wave motion 390–6
  phase difference 438
  sound waves 398–400
  standing waves 436–49
  types of 389–90
  see also wave phenomena
weak nuclear force 533
weber (Wb)/weber-turns 568, 569
white dwarfs 684, 689–90
Wien's displacement law 227, 233
Wilberforce pendulum 140
work done 110, 112–13, 115, 274–5
  on charge between parallel plates 517
  by/on a gas 276–9, 280–3
  by gas particles 268
  heat engines 283–4, 287
  moving a charge in electric field 517,
    526–7, 531
  by objects in gravitational field 488, 489
  rate of doing work 113–14
  against a resistive force 110–11
  stretching a spring 122–3
  by torque 145–6
work function 613–14, 615–16
worldlines 181–2, 186, 188

X-ray photons 637
X-rays 402, 403, 404, 619–21, 631

Young's slit experiment 425–31

zeroth law of thermodynamics 288
zone of stability 648–9